Electrochemistry in Transition

From the 20th to the 21st Century

Electrochemistry in Transition

From the 20th to the 21st Century

Edited by

Oliver J. Murphy
Supramaniam Srinivasan

Texas A&M University
College Station, Texas

and

Brian E. Conway

University of Ottawa
Ottawa, Ontario
Canada

Plenum Press • New York and London

Library of Congress Cataloging in Publication Data

Electrochemistry in transition: from the 20th to the 21st century / edited by Oliver J.
    Murphy and Supramaniam Srinivasan and Brian E. Conway.
        p.   cm.
    Includes bibliographical references and index.
    ISBN 0-306-43946-8
    1. Electrochemistry. I. Murphy, Oliver J. II. Srinivasan, S. (Supramaniam),
1932–   . III. Conway, B. E.
QD555.E42   1992   92-25250
541.3'7—dc20   CIP

ISBN-13: 978-1-4615-9578-6   e-ISBN-13: 978-1-4615-9576-2
DOI: 10.1007/978-1-4615-9576-2
ISBN 0-306-43946-8

© 1992 Plenum Press, New York
Softcover reprint of the hardcover 1st edition 1992

A Division of Plenum Publishing Corporation
233 Spring Street, New York, N.Y. 10013

Contributors

Wesley Aldred • Case Center for Electrochemical Sciences and the Chemistry Department, Case Western Reserve University, Cleveland, Ohio 44106

Terrell N. Andersen • Kerr–McGee Technical Center, Oklahoma City, Oklahoma 73125

Benedict Aurian-Blajeni • ChemLogic, Bellingham, Massachusetts 02019

F. Barbir • Clean Energy Research Institute, University of Miami, Coral Gables, Florida 33124

Donald J. Barclay • IBM United Kingdom, Winchester, Hampshire SO21 2JN, United Kingdom

Bede Bicamumpaka • Laboratoire de Thermodynamique Electrochimique, Faculté des Sciences, Université Libre de Bruxelles, B-1050 Brussels, Belgium

J. O'M. Bockris • Department of Chemistry, Texas A&M University, College Station, Texas 77843

Vlasta Brusic • IBM T. J. Watson Research Center, Yorktown Heights, New York 10598

Jose L. Carbajal • Inland Steel Company, Research Laboratory, East Chicago, Indiana 46312

Per Carlsson • Department of Physical Chemistry, Chalmers University of Technology/ Goteborg University, Goteborg, S-41296, Sweden

B. E. Conway • Chemistry Department, University of Ottawa, Ottawa, Ontario, Canada K1N 6N5

A. Damjanovic • Allied-Signal Incorporated, Corporate Technology, Morristown, New Jersey 07962

Joseph M. Derby • Kerr–McGee Technical Center, Oklahoma City, Oklahoma 73125

Aleksandar R. Despić • Faculty of Technology and Metallurgy, University of Belgrade, 11000 Belgrade, Yugoslavia

Hari P. Dhar • BCS Technology, Incorporated, College Station, Texas 77840

D. M. Dražić • Faculty of Technology and Metallurgy and Institute of Electrochemistry ICTM, University of Belgrade, Belgrade, Yugoslavia

V. J. Dražić • Faculty of Technology and Metallurgy and Institute of Electrochemistry ICTM, University of Belgrade, Belgrade, Yugoslavia

Michio Enyo • Catalysis Research Center, Hokkaido University, Sapporo 060, Japan

N. P. Fitzpatrick • Alcan International Ltd., Kingston, Ontario, Canada K7L 4Z4

Akira Fujishima • Department of Synthetic Chemistry, Faculty of Engineering, The University of Tokyo, Tokyo 113, Japan

Atsushi Fukuoka • Catalysis Research Center, Hokkaido University, Sapporo 060, Japan

Marvin A. Genshaw • Diagnostics Division, Miles Incorporated, Elkhart, Indiana 46515

F. Gutmann • Xsirius Scientific, Incorporated, Fullerton, California 92635, and School of Chemistry, Macquarie University, North Ryde, New South Wales 2109, Australia

Nejat Guzelsu • UMDNJ—School of Osteopathic Medicine, Piscataway, New Jersey 08854, and Biomedical Engineering Program, Rutgers University, New Brunswick, New Jersey 08903

M. Ahsan Habib • Department of Physical Chemistry, General Motors Research Laboratories, Warren, Michigan 48090

W. Halliop • Alcan International Ltd., Kingston, Ontario, Canada K7L 4Z4

R. P. Hamlen • Alupower Incorporated, Warren, New Jersey 07059

Shiro Haruyama • Department of Metallurgical Engineering, Tokyo Institute of Technology, Tokyo, Japan

G. Duncan Hitchens • Lynntech, Incorporated, Bryan, Texas 77803

W. H. Hoge • Alupower Incorporated, Warren, New Jersey 07059

Bertil Holmström • Department of Physical Chemistry, Chalmers University of Technology/Goteborg University, Goteborg, S-41296, Sweden

Jean Horkans • IBM T. J. Watson Research Center, Yorktown Heights, New York 10598

Henri D. Hurwitz • Laboratoire de Thermodynamique Electrochimique, Faculté des Sciences, Université Libre de Bruxelles, B-1050 Brussels, Belgium

G. C. Huth • Xsirius Scientific, Incorporated, Fullerton, California 92635, and Institute of Physics, School of Medicine, University of Southern California, Marina del Rey, California 90291

Masaru Ichikawa • Catalysis Research Center, Hokkaido University, Sapporo 060, Japan

Kiminori Itoh • Institute of Environmental Science and Technology, Yokohama National University, Yokohama 240, Japan

André Jenard • Laboratoire de Thermodynamique Electrochimique, Faculté des Sciences, Université Libre de Bruxelles, B-1050 Brussels, Belgium

V. Jevtić • Faculty of Technology and Metallurgy and Institute of Electrochemistry ICTM, University of Belgrade, Belgrade, Yugoslavia

Vladimir Jovancicevic • W. R. Grace & Co., Washington Research Center, Columbia, Maryland 21044

Lamine Kaba • Center for Electrochemical Systems and Hydrogen Reserch, Texas A&M University, College Station, Texas 77843

Shahed U. M. Khan • Department of Chemistry, Duquesne University, Pittsburgh, Pennsylvania 15282

Hideaki Kita • Department of Chemistry, Faculty of Science, Hokkaido University, Sapporo 060, Japan

W. Kobasz • Alupower Incorporated, Warren, New Jersey 07059

Zlata Kovac • IBM Almaden Research Center, San Jose, California 95120-6099

Duckhwan Lee • Department of Chemistry, Sogang University, Seoul, Korea

Ken-ichi Machida • Catalysis Research Center, Hokkaido University, Sapporo 060, Japan

Yoshiharu Matsuda • Department of Industrial Chemistry, Faculty of Engineering, Yamaguchi University, Ube 755, Japan

James McBreen • Department of Applied Science, Brookhaven National Laboratory, Upton, New York 11973

Klaus Müller • Battelle Europe, Geneva Laboratories, CH-1227 Carouge, Switzerland

Oliver J. Murphy • Lynntech, Incorporated, Bryan, Texas 77803

Hiroshi Nakajima • Department of Chemistry, Faculty of Science, Hokkaido University, Sapporo 060, Japan

Hiroo Numata • Department of Metallurgical Engineering, Tokyo Institute of Technology, Tokyo, Japan

Woon-kie Paik • Department of Chemistry, Sogang University, Seoul, Korea

Milan Paunovic • IBM T. J. Watson Research Center, Yorktown Heights, New York 10598

B. J. Piersma • Houghton College, Houghton, New York 14744

H. J. Plass, Jr. • Clean Energy Research Institute, University of Miami, Coral Gables, Florida 33124

Jai Prakash • Chemical Technology, Argonne National Laboratory, Argonne, Illinois 60439 and Case Center for Electrochemical Sciences and the Chemistry Department, Case Western Reserve University, Cleveland, Ohio 44106

B. M. L. Rao • Alupower Incorporated, Warren, New Jersey 07059

Alvin J. Salkind • UMDNJ—Robert Wood Johnson Medical School, Piscataway, New Jersey 08854, and Biomedical Engineering Program, Rutgers University, New Brunswick, New Jersey 08903

Carlos J. Sambucetti • IBM T. J. Watson Research Center, Yorktown Heights, New York 10598

Benjamin R. Scharifker • Departamento de Química, Universidad Simón Bolívar, Caracas 1080-A, Venezuela

Katsuaki Shimazu • Department of Chemistry, Faculty of Science, Hokkaido University, Sapporo 060, Japan

Supramaniam Srinivasan • Center for Electrochemical Systems and Hydrogen Research, Texas A&M University, College Station, Texas 77843-3577

Marek Szklarczyk • Department of Chemistry, Warsaw University, 02-089 Warsaw, Poland

Haruyuki Takagi • Department of Metallurgical Engineering, Tokyo Institute of Technology, Tokyo, Japan

Donald Tryk • Case Center for Electrochemical Sciences and the Chemistry Department, Case Western Reserve University, Cleveland, Ohio 44106

Kohei Uosaki • Department of Chemistry, Faculty of Science, Hokkaido University, Sapporo 060, Japan

T. N. Veziroglu • Clean Energy Research Institute, University of Miami, Coral Gables, Florida 33124

W. Visscher • Laboratory for Electrochemistry, Faculty of Chemical Technology, Eindhoven University of Technology, 5600 MB Eindhoven, The Netherlands

G. Vitiello • Dipartimento di Fisica, Università di Salerno, 84100, Salerno, Italy

Ralph E. White • Chemical Engineering Department, Texas A&M University, College Station, Texas 77843-3122

R. Woods • CSIRO Division of Mineral Products, Port Melbourne, Victoria 3207, Australia

Halina S. Wroblowa • Ford Motor Company, Research Staff, Dearborn, Michigan 48121. *Present address*: 5924 Dunmore Drive, West Bloomfield, Michigan 48322

Ernest Yeager • Case Center for Electrochemical Sciences and the Chemistry Department, Case Western Reserve University, Cleveland, Ohio 44106

Piotr Zelenay • Department of Chemistry, Warsaw University, 02-089 Warsaw, Poland

Preface

This book originated out of the papers presented at the special symposium, "Electrochemistry in Transition—From the 20th to the 21st Century," scheduled by the Division of Colloid and Surface Science during the American Chemical Society meeting in Toronto. The symposium was in honor of Professor J. O'M. Bockris, who received the ACS award on "The Chemistry of Contemporary Technological Problems" (sponsored by Mobay Corporation) during this meeting and who also reached his 65th birthday in the same year. The symposium was of a multidisciplinary nature and encompassed the fields of theoretical and experimental electrochemistry, surface science, spectroscopy, and electrochemical technology. The symposium also had an international flavor in that the participants represented several countries—Australia, Belgium, Canada, Chile, England, Japan, Korea, the Netherlands, Poland, Switzerland, Venezuela, Yugoslavia, and the United States. The symposium was graciously sponsored by the ACS (Petroleum Research Fund and Division of Colloid and Surface Science), Alcan International, Dow Chemical Company, EG&G, Electrolyzer Corporation, Exxon, General Electric Company, IBM, Institute of Gas Technology, International Association of Hydrogen Energy, Johnson Matthey, Inc., Kerr-McGee Corporation, Medtronics, and Texas A&M University (Center for Electrochemical Systems and Hydrogen Research and the Hampton Robinson Fund).

The "theme" of the papers presented at the symposium covered not only significant contributions made to electrochemistry in the twentieth century, but also "New Horizons in Electrochemistry" for the twenty-first century. Thus, the scientists who presented papers were invited to contribute chapters to this book, having the same titles as the symposium. The response was excellent in that the volume contains 41 of the 55 papers presented at the symposium. The unique feature of this book is that it covers the range from fundamental electrochemistry to electrochemical technology—structure of the double layer, *in situ* examination of electrodes, quantum mechanics of electrode processes, electro-organic reactions, bioelectrochemistry, corrosion, passivation, fuel cells, batteries, and hydrogen technologies. We earnestly hope that it will be of high educational value for graduate students and researchers in electrochemistry.

One of the chapters in the book, which is by Professor J. O'M. Bockris, may be eye-catching, in that it gives an overview of the contributions made by the author and his co-workers in virtually all the research areas in electrochemistry and reminisces about his experiences with his colleagues.

The editors of this volume are most grateful to Mrs. Lily Bockris for the arduous task of assistance before and during the symposium, Mrs. Marilyn Hulkovich for the excellent editorial assistance, and Plenum Press—in particular, Amelia McNamara and Kenneth Schubach—for their patience and cooperation during the production of this volume.

Finally, we would like to recognize the many outstanding original scientific contributions of Professor John O'Mara Bockris over almost 50 years to the various research areas of electrochemistry as exemplified by the chapters in this book, his training and development of numerous graduate students and postdoctoral fellows (a significant number of whom have contributed chapters to this book), his authorship of many textbooks in electrochemistry and related research topics that have promoted these subjects worldwide, and his encouragement and stimulation in organizing the symposium and in making this book a reality.

Oliver J. Murphy
Supramaniam Srinivasan
Brian E. Conway

College Station, Texas and Ottawa, Canada

Contents

Part III: Electrode Kinetics and Electrocatalysis

Part IV: Photoelectrochemistry

Part IX: Electrochemical Energy Conversion

Electrochemistry in Transition

From the 20th to the 21st Century

Double Layer

The Concept of Electrochemical Double Layer in the Case of Weakly Adsorbed Anions

Henri D. Hurwitz, André Jenard, and Bede Bicamumpaka

1. INTRODUCTION

For a large variety of reasons, the understanding of the electrochemical double layer has been a subject of interest for many years. Indeed, the solution of the problem is essential to the correct interpretation of most molecular mechanisms at charged interfaces (electrodes, colloids, etc.). Moreover, it constitutes *per se* a challenging theoretical problem of the statistical mechanics of strong electrolytes. Furthermore, particularly since many of the recent experimental studies of biological cell membranes have been closely associated with concurrent theoretical physicochemical developments, it has become of great importance to biophysicists and electrophysiologists. One of the attributes of the model of the electrochemical double layer is that it provides a simple framework in which to sort out physical from biologically active effects. Thus, unusual ion flows during nerve excitation are actually expected from the influence of the double layer[1]; the selectivity of ion channels across membranes, the electrokinetic potentials during blood flow,[2] etc. can be understood on its basis. A number of apparently unrelated phenomena in biological systems might arise from special properties of charged interphases, such as, for example, spacing between muscle proteins during contraction[3] and the folding of the thylakoid membrane as related to photosynthetic activity.[4]

The main purpose of this work is to shed some light on the properties of the electrochemical double layer and, more precisely, to summarize some recent research aimed at gaining a fundamental understanding of systems of weakly adsorbed anions. Current models of the electrochemical double layer have been revised by several authors with the purpose of achieving a better theoretical consistency. However, the question of the adequacy of the experimental criteria which should be used as guidelines in such treatments has not received sufficient attention. This is particularly evident in the case of the electrocapillary behavior of weakly specifically adsorbed ions. This category of electrolyte should give rise to experimental criteria which are essential for an assessment of the validity of any double-layer

Henri D. Hurwitz, André Jenard, and Bede Bicamumpaka ● Laboratoire de Thermodynamique Electrochimique, Faculté des Sciences, Université Libre de Bruxelles, B-1050 Brussels, Belgium.

Electrochemistry in Transition, edited by Oliver J. Murphy *et al.* Plenum Press, New York, 1992.

theory, even if it is believed to apply to nonspecifically adsorbed systems. The assumption of the absence of specific effects corresponds unavoidably to a limiting state of the ion–surface interaction, and no ions are expected to display an adsorption governed by interactions without short-range effects with the electrode. This is even the case for the F^- ion, which is classically taken as the ideal nonspecifically adsorbed reference system at the Hg electrode. Thus, the theoretician needs to be equipped with accurate experimental data on specifically adsorbed ions. From this viewpoint, and also considering the extensive amount of experimental investigations already performed, the thorough study of strongly specifically adsorbed ions seems less crucial. Because the electrocapillary data for such ions correspond to much higher adsorbed amounts, they are less discriminating between isotherms of one type or another. Actually, in these systems the prevailing effect is the contact with the metal, and most of the stress has to be laid on the so-called partial charge transfer mechanism and on the role of solvent adsorption. It is noteworthy that Bockris and co-workers recognized the importance of systems involving weakly adsorbed anions in 1965, and we will be able to derive some of our conclusions from their results.[5]

Considering the abundance of studies on ion adsorption at various interfaces, we wish to restrict our report to the recent progress made in elucidating the nature of halide ion adsorption at the Hg electrode. However, before going into the specialized aspects of our research, we feel that it is necessary to review briefly some fundamental concepts and their present status in the evolution of the double-layer theory. We will therefore first describe the model as treated under the assumption of the absence of any specific adsorption.

2. THE DOUBLE-LAYER MODEL IN THE ABSENCE OF SPECIFIC ADSORPTION

The aqueous electrolyte ideally polarized Hg electrode is the most widely studied system. Its properties have been investigated primarily through the variation of the interfacial tension γ and differential capacitance C as a function of the applied potential difference E with respect to a reference electrode. Except for some earlier studies,[6] all works published up to this time have been treated in term of the Gouy–Chapman–Stern model suggested by Grahame.[7] The foundations of this model are as follows:

(i) The metal surface is considered as a geometrical plane located at $x = 0$ and bearing a uniform charge density q^M.

(ii) The Stern layer or inner layer, which accounts for the finite ionic size, the dielectric polarization of the solvent, and the phenomenon of specific adsorption, is confined to the region between the electrode and the outer Helmholtz plane (OHP) at $x = x_0$. The thickness of the Stern layer is usually set equal to the radius of the hydrated ion occupying the OHP. Bockris et al.[8] added to it the thickness of one monomolecular layer of the adsorbed solvent.

(iii) The OHP constitutes the boundary toward the electrode of the domain of the diffuse layer for which the Poisson–Boltzmann equation can be expressed as:

$$\frac{d^2\phi(x)}{dx^2} = -\frac{1}{\varepsilon\varepsilon_0}\sum_\gamma ez_\gamma\rho_\gamma(x) \tag{1}$$

and

$$\rho_\gamma(x) = \rho_\gamma^\beta g_\gamma(x) = \rho_\gamma^\beta \exp[-W_\gamma(x)/kT] \qquad x \geq x_0 \tag{2}$$

where $\rho_\gamma(x)$ and ρ_γ^β are, respectively, the molecular densities of the ionic species γ at a distance x from the electrode and in the bulk at $x = \infty$, $\phi(x)$ is the average electrostatic potential at x, $g_\gamma(x)$ is the singlet distribution function, and $W_\gamma(x)$ is the potential of mean force. The electroneutrality condition for the interface requires that

$$e \sum_{\gamma} z_{\gamma} \rho_{\gamma}^{\beta} \int_{x_0}^{\infty} g_{\gamma}(x) \, dx = q_D = -D(x_0) \tag{3}$$

where $D(x_0)$ represents the electric displacement at x_0, and $q_D = -(q^M + q_I) \cong -q^M$, the electrode charge, according to the assumption that the density of charge of specifically adsorbed ions, q_I, is zero. Gouy and Chapman solved the system of Eqs. (1) and (2) in a self-consistent way for point charges embedded in a dielectric continuum, assuming that

$$g_{\gamma}(x) = \begin{cases} 0 & \text{for } x < x_0 \\ \exp[-e \cdot z_{\gamma} \phi(x)/kT] & \text{for } x \geq x_0 \end{cases} \tag{4}$$

Because of the one-dimensional symmetry, the system of eqs. (1)–(4) can be solved rigorously for symmetrical and 1:3 electrolytes.

Many approaches have been taken in attempts to improve this model. We will restrict ourselves to describing briefly some results of the most recent ones.

In one group of methods,[9,10] the Poisson–Boltzmann equation is expressed in a form which formally makes allowance for self-atmosphere interactions between ions adsorbed in the diffuse layer. In this case, the Boltzmann term includes a pair correlation function $g_{\alpha\gamma}(r_{\alpha\gamma}, x)$ and the Poisson equation a pairwise potential $\phi_{\gamma}(r_{\alpha\gamma}, x)$ which is due to the ions α which surround each ion γ at a distance $r_{\alpha\gamma}$. Because of the $r_{\alpha\gamma}$ and x spatial dependence, a numerical solution is usually required. An analytical result can be obtained only within the limit of the development of the Boltzmann factor to the linear term in ϕ_{γ}. This corresponds to a generalization of the Debye–Hückel theory to the interfacial region and is therefore correct at low ionic strength and charge densities.[10] This set of theoretical treatments has evolved over the past few years. Their recent versions are called the modified Poisson–Boltzmann theories (MPB). They include hard-core repulsion of the ions and remain valid only for small values of $\kappa\sigma$, where $1/\kappa$ is the Debye length and σ is the hard-core diameter of the ions.

Among other theories, the prevalent direction today is represented by the works of Henderson et al.,[12] Blum,[13] and Carnie and Chan.[14] Initially, the original impetus was given by Henderson et al., who suggested that one specific ion, at infinite dilution, but exceedingly large in size, can play the role of the electrode. As a result, he developed a procedure starting from the Ornstein–Zernike integral equation for a mixture of ions in a fluid, taking the limit of one ionic radius tending toward infinity. In order to solve this equation, an approximation was made for the expression of $g_{\alpha\gamma}(r_{\alpha\gamma}, x)$. This corresponds to the so-called hypernetted-chain (HNC) theory. An expansion of this expression of $g_{\alpha\gamma}(r_{\alpha\gamma}, x)$ in powers of $\delta g_{\alpha\gamma}(r_{\alpha\gamma}, x)$ and truncation after the linear term yields a second approximation, which is called the mean spherical approximation (MSA). The MSA is to the HNC theory what the linearized Poisson–Boltzmann expression is to the Gouy–Chapman treatment. Because of this linearization, the MSA should in principle be applicable only to small charge densities. The use of the HNC approximation yields, after some mathematical manipulations, an integral equation which was solved numerically in the case of two distinct models.[15] Both models have in common that no distinction is made between the Stern or inner layer and the diffuse layer; instead, they rely on a continuous model in which the inner layer merely appears as the consequence of the choice of the finite diameter σ of the hard-core particles and of the description of the metal as a charged impenetrable wall.

The simplest representation of the metal/electrolyte interface is given by the so-called primitive model, in which the ions are described as hard-core spheres, having the same diameter σ, and the solvent is considered as a continuous fluid with constant dielectric permittivity. In principle, the HNC procedure is more self-consistent and more accurate than the HNC/MSA. Therefore, one should expect the former method to give the most reliable results. In practice, however, the HNC/MSA is often better. This must be the result of cancellation of errors. Comparison with Monte Carlo simulations permits a rigorous test of the reliability of the mathematical analysis that is developed and of the consistency of the results with the initial molecular model. Torrie and Valleau performed such computer simulations for the interface.[16-18] The profile of the singlet distribution function for 1:1 electrolytes as derived from HNC/MSA is nearly in perfect agreement with the Monte Carlo (MC) simulation. The results given by the Gouy-Chapman theory are in fairly good agreement. If, however, a bivalent ion becomes the counterion, an interesting effect can be detected. A layering of charges is seen in both the MC and HNC/MSA treatments. The shape of the density profile indicates that the total charge carried by the counterions in the boundary layer $x_0 \le x \le 2\sigma$ exceeds the electrode charge q^M. Therefore, the initial layers are followed by a layer of co-ions in excess. The Gouy-Chapman-Stern (GCS) theory is unsuccessful in reproducing this type of triple-layer behavior. This layering appears not to be confined to the 1:2 case with a bivalent counterion, but is also apparent in symmetrical electrolytes at the high concentrations obtainable in molten salts. The theoretical charge adsorption isotherms $q_\pm$ obtained in the presence of a monovalent counterion (in a 2:1 electrolyte), as displayed on the right-hand side of Fig. 1, show no significant deviation from the MC simulation results. However, with a bivalent counterion, as displayed on the left-hand side of Fig. 1, the co-ion adsorption isotherm becomes nonmonotonic in all cases except the Gouy-Chapman case. Henderson[15] has pointed out that a good qualitative agreement is to be expected for all theoretical results of the density profiles and for their integration across the double layer in symmetrical electrolytes. This occurs essentially as a consequence of the electroneutrality condition in Eq. (3), which fixes the value of the total area under both counterion and co-ion density profiles, and of the force balance consideration which requires that the following condition be fulfilled:

$$kT \sum_\gamma \rho_\gamma(x_0) = P + \frac{D^2(x_0)}{2\varepsilon\varepsilon_0} \tag{5}$$

where P is the osmotic pressure of the electrolyte, and ε is the dielectric constant of the

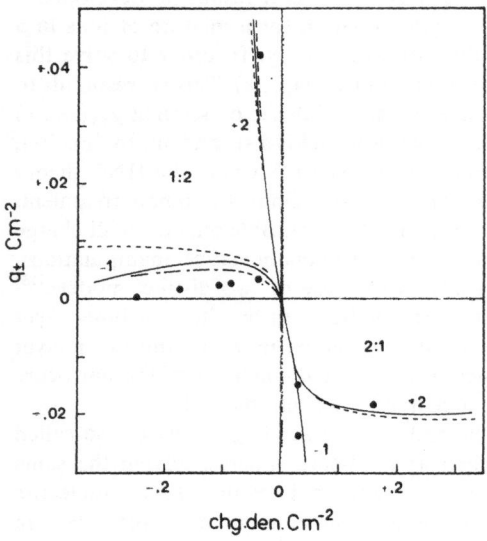

FIGURE 1. Excess charge adsorption isotherms of $0.05M$ 1:2 and 2:1 electrolytes as a function of q^M near an electrode. ●, Monte Carlo simulated values; ——, HNC/MSA results; —·—, MPB results; - - -, GCS results ($\varepsilon = 78.5$; $\sigma = 0.425$ nm, $T = 298$ K).[15]

fluid. The second term on the right-hand side of Eq. (5) becomes rapidly dominant, so that the total ionic density at x_0 comes under the control of q^M and ε.

Both conditions just expressed above do not apply to the potential profile $\phi(x)$ and to the value ϕ_0 of $\phi(x)$ at the OHP ($x = x_0$). Generally speaking, the double-layer thickness is larger in the GCS model than in distributions obtained from HNC/MSA and MC simulation. The GCS values of ϕ_0 also exceed the other theoretical values, and discrepancies become serious in the case of bivalent counterions.

The determination of theoretical values of the differential capacitance C requires a more realistic description of the solvent than is provided by the primitive model. Several approaches were published based on models of multiple orientations of adsorbed dipoles in the inner layer. These theories, which follow the line of thought given in the work of Watts-Tobin,[19] were developed by Bockris et al.[8] and more recently by Fawcett[20] and Guidelli.[21] Let us recall that, in the Grahame model, arguments concerning the inner-layer molecular configuration are essentially based on the experimental determination of the inner-layer differential capacitance C_I. This can be accomplished by using the following relationship predicted by the Grahame model:

$$C^{-1} = C_I^{-1} + C_D^{-1} \tag{6}$$

where C_D is the diffuse-layer differential capacitance. In order to derive the value of C_I, Parsons and Zobel[22] suggested that C^{-1} be plotted as a function of C_D^{-1}, as depicted in Fig. 2 for the $H_2PO_4^-$ system. These data produce straight lines with slopes equal to one at low concentrations, where C_I becomes independent of the ionic strength.

With respect to the continuous ionic distribution used in the HNC theory, it was suggested that the solvent be represented as a fluid of hard spheres of diameter σ_s with embedded dipole moment μ. This constitutes the simplest nonprimitive model where the electrolyte still remains a mixture of ions of equal sizes σ. According to this hard-sphere ion and dipole model, the concept of the inner layer does not emerge as a distinct contribution. The dielectric permittivity of the solvent in this model becomes itself a function of μ and σ_s. One is led to choose a value of $\mu = 2.2$ D and $\sigma_s = 0.3$ nm for water at 20°C. Carnie and Chan[14] evaluated the surface potential $\phi(x = 0)$ by means of the HNC/MSA. They obtained at low electrolyte concentration:

$$\frac{\phi(x = 0)}{q^M} = C^{-1} = \frac{1}{\varepsilon^0 \varepsilon^*} \frac{\sigma}{2} + \frac{1}{\varepsilon \varepsilon^0 \kappa} \tag{7}$$

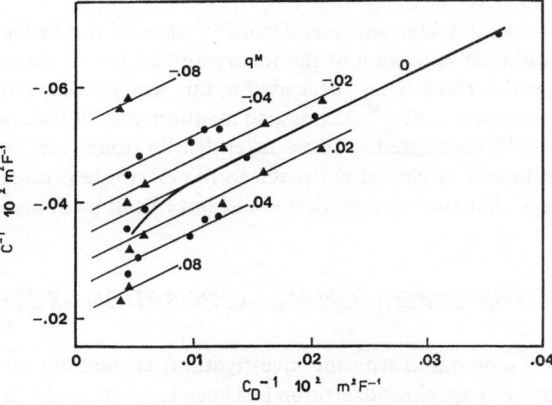

FIGURE 2. The Parsons–Zobel plot for a system of an aqueous solution of NaH_2PO_4 at 25°C at the Hg electrode. The values of $q^M (Cm^{-2})$ are indicated on the figure. The points represent experimental data associated with the GCS model (Ref. 22). The heavy curve gives the full MSA results, and the straight lines the MSA results at low concentration with $\sigma = \sigma_s = 0.276$ nm; $\varepsilon = 78.5$.[15]

Obviously, the second term on the right-hand side of Eq. (7) represents C_D^{-1} in the linearized Gouy–Chapman theory. The first term contains the quantity

$$\varepsilon^* = \frac{\varepsilon}{1 + \dfrac{\varepsilon - 1}{\lambda}\dfrac{\sigma_s}{\sigma}} \tag{8}$$

where λ is a dimensionless constant expressed by the relation $\lambda^2(1 + \lambda)^4 = 16\varepsilon$. For $\varepsilon = 80$ and $\sigma_s = \sigma$, ε^* becomes equal to 2.6. At first sight, it might appear that ε^* is equivalent to a dielectric constant of the inner layer. However, since no ion or dipole is allowed to penetrate within the distance $\sigma/2$ from the electrode, this cannot be the case. In fact, it is the solvent configuration over the entire double layer which has an effect on $\phi(x = 0)$ leading to a virtual dielectric constant ε^*. The function ε^* contributes to a term in Eq. (7) that plays the role of an inner-layer capacitance. Actually, the calculation yields a polarization density or dipole moment per unit volume which changes in an oscillatory way as a function of x. Such a cooperative alignment of dipole molecules parallel to the electrode surface in the diffuse layer induces a depolarizing field. The presence of an enhanced depolarizing field is tantamount to introducing a virtual dielectric constant ε^* in C_I. This aspect of the theory illustrates the difficulty of treating the molecular system by using a macroscopic concept such as the dielectric constant. Rosinberg[24] found that Eq. (7) gives a good description of the solvent contribution to the capacitance of a wide range of solvents. Schmickler and Henderson[25] and also Badiali et al.[26] emphasized the fact that an additional concentration-independent term has to be included in Eq. (7). Thus, for an Hg electrode:

$$C_I^{-1} = \frac{1}{\varepsilon^* \varepsilon_0}\frac{\sigma}{2} - 2.6 \, \text{m}^2/\text{F}$$

This new term can be ascribed to the contribution of the metallic electrode. If this electrode is represented by a jellium model, one would expect the electron profile corresponding to the spillover of the electrons into the fluid to affect the solvent and ion dipole profile. In practice, however, the metal electrons contribute to C_I^{-1} but do not penetrate into the electrolyte enough to disturb the distribution functions derived from the HNC treatment.

It is of interest to note that the application of the HNC/MSA at high electrolyte concentration yields some deviation from the linear dependence of C^{-1} on C_D^{-1} as expressed by Eq. (7). This deviation from the low-concentration linear expansion seems to confirm that the experimental deviations at high concentration from the straight line observed in Fig. 2 are not due to inaccuracies or to the onset of some specific adsorption.[27] The theoretical and experimental data at the pzc recorded in the Parsons–Zobel plot in Fig. 2 agree extremely well.

Schmickler and Henderson[28] applied the MSA to larger values of q^M by means of a nonlinear extension of the theory and an improved model of the electron distribution at the metal surface. They succeeded in this way in computing values of C_I, which displays a peak, as a function of q^M. Let us also mention that Valleau and Torrie[29] and subsequently Bhuiyan et al.[30] evaluated a diffuse-layer distribution function for ions of different sizes. The different distances of closest approach to the electrode produce a shift of the pzc. It is interesting to note that such a deviation could have been attributed to ionic specific adsorption.

3. THE EXPERIMENTAL ADSORPTION ISOTHERMS

Compared with the investigations carried out on strong specific ionic adsorption, work on weak specific adsorption has been less successful in terms of giving a simple and physically reasonable account of the double-layer structure. As yet, there are no broadly accepted models to explain the form of the adsorption isotherms[31] or the differential capacitance.[32] The object of this chapter is to assess the validity of some experimental results in a manner that is free of the ambiguities associated with complicated models of the double layer. In view of this goal, we wish to handle first the case of a solution of a single binary electrolyte and second that of a mixed electrolyte.

3.1. Solution of a Single Binary Electrolyte

In principle, the ionic molar surface excess Γ_γ can be divided into two terms:

$$\Gamma_\gamma = \Gamma_{\gamma I} + \Gamma_{\gamma D} \tag{9}$$

and the electroneutrality condition can be written as

$$F \sum_\gamma z_\gamma \Gamma_\gamma = q_I + q_D = -q^M \tag{10}$$

where the subscripts I and D refer, respectively, to the specific and diffuse-layer adsorption. The quantity Γ_γ is derived from the Gibbs equation. For a $1:1$ electrolyte A^+X^-, this equation takes the form

$$d\xi \equiv d(\gamma + q^M E_\pm) = -\Gamma_{\mp,w} (d\mu_{AX})_{T,P} + E_\pm \, dq^M \tag{11}$$

where we have introduced the Parsons surface state function ξ, the subscripts $+$ and $-$ denoting the cation and the anion, respectively, and $E_\pm$ being the potential measured with respect to an electrode reversible to either A^+ or X^-. The definition of the relative ionic molar surface excess is

$$\Gamma_{\mp,w} \equiv \Gamma_\mp - \Gamma_w W m_{AX} \tag{12}$$

where W and Γ_w are the molecular weight and the adsorbed molar surface excess of water, and m_{AX} is the molality of the electrolyte.

In not too concentrated solutions, unless $\Gamma_\mp$ is itself very small, one might neglect the second term on the right-hand side of Eq. (12). Harrison et al. considered this problem.[33] It should be stressed that the determination of Γ_{-I} from electrocapillary measurements requires that A^+ remains absolutely specifically unadsorbed in the domain of potential and of values of m_{AX} investigated. Thus, we assume that

$$\Gamma_+ = \Gamma_{+D} \tag{13}$$

It is further established by the Gouy–Chapman theory that

$$(\Gamma_- - \Gamma_{-I})^{-1} + (\Gamma_{+D})^{-1} = -F/b\sqrt{J} \tag{14}$$

where $b = (2\varepsilon\varepsilon_0 RT10^3)^{1/2}$, and J, the ionic strength (or salt concentration), is set equal to $|J| \simeq |m_{AX}|$. Upon consideration of Eqs. (9)–(14), Grahame initiated the systematic quantitative study of specific ionic adsorption on Hg. Much of our present knowledge derives from the subsequent series of publications by him and by Parsons. It is manifest from their measurements that for most anions the inequality $F\Gamma_{-I} > q^M$ holds over a large domain of m_{AX} and q^M values, and hence the diffuse-layer charge q_D is most frequently positive although q^M is itself positive. This overequivalent specific adsorption is even present in the case of weakly adsorbed electrolytes such as Cl^-. It could be expected that in this case, by decreasing m_{AX} at constant q^M, the value of $\Gamma_{Cl^-,I}$ would pass through the state of equivalency, where $F\Gamma_{-I} = q^M$, and would tend asymptotically to zero. Usually, this behavior is not observed, as can be seen from several examples shown in Fig. 3. There, we have presented the systems Cl^-,[34] Br^-,[35] N_3^-,[36] and CNS^-.[5] At low values of m_{AX}, the charge $F\Gamma_{-I}$ tends to some

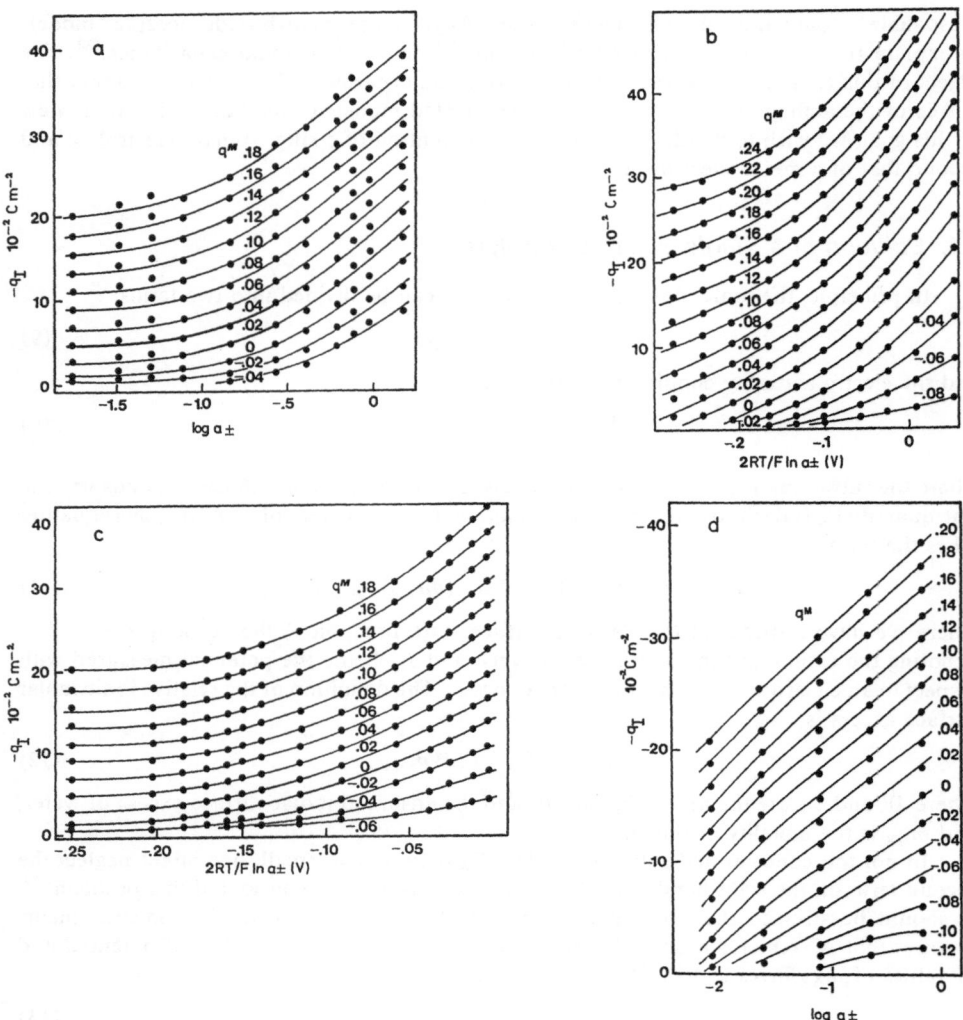

FIGURE 3. Adsorption isotherms of specifically adsorbed anions X^- at the Hg electrode at 25°C: (a) aqueous KCl (Ref. 34); (b) aqueous KBr (Ref. 35); (c) aqueous NaN_3 (Ref. 36); (d) aqueous NaCNS (Ref. 5).

finite value nearly equal to the value of q^M. Consequently, q_D can never reach large negative values for $q^M > 0$. Furthermore, q_I and q^M cannot be varied independently at low concentration.

Many attempts have been made to explain this effect. In fact, at low concentration and at the most positive charge, a very low accuracy in the computation of Γ_{-I} is to be expected. The error in this quantity depends on the value of $(d\Gamma_{-D}/d\Gamma_{+D})_{q^M}$, which is equal to $(q^M/F\Gamma_{+D})^2$ and happens to become quite large at $q^M > 0$ and $J < 10^{-2}M$.

Such limited accuracy was one of the reasons why several authors advocated the use of systems of mixed electrolytes at constant J. They were surprised, however, to discover that effects of charge leveling were not apparent in systems of the type AX-AF. On the contrary, these systems displayed apparently normal behavior in which $\Gamma_{X^-,I}$ drops to zero as a function of decreasing m_{AX}. Damaskin[37] argued that, if provision is made for underequivalent adsorption and if the adsorption obeys a virial-type isotherm, $\Gamma_{X^-,I}$ should become independent of m_{AX} in the case of a single binary electrolyte whereas this should not be the case for a mixed electrolyte. Yet no complete and reasonable explanation was put forward to explain the discrepancies between the systems of single binary and mixed electrolyte, and no concrete answer was given as to why in the single electrolyte q_I becomes quite generally equal to $-q^M$. In the following we shall show that this charge leveling effect can be derived from the Gouy-Chapman theory and from Langmuir and Henri isotherms, which correspond thermodynamically to the simplest and limiting behavior of weak specific adsorption; both isotherms correspond to the experimental condition that $|q_I| < 0.2 \ C/m^2$.

It is assumed that equilibrium is reached between the amount of X^- specifically adsorbed and the amount $\rho_{X^-}(x_0)$ nonspecifically adsorbed at the OHP. This condition gives

$$\frac{a_{X^-}\Gamma_{-I}}{1 - a_{X^-}\Gamma_{-I}} = N_A^{-1}\rho_{X^-}(x_0)\kappa_{X^-} \tag{15}$$

where a_{X^-} is the partial molar surface phase area of X^- in its specifically adsorbed state, N_A is the Avogadro number, and κ_{X^-} is the adsorption constant. As follows from the Gouy-Chapman theory:

$$\phi_0 = \frac{2RT}{F}\sinh^{-1}y = \frac{2RT}{F}\ln(y + \sqrt{y^2 + 1}) \tag{16}$$

where

$$y = (q^M + q_I)/2b\sqrt{J} \tag{17}$$

Thus, the relation given in Eq. (15) may be rewritten as follows:

$$\ln\left(\frac{a_{X^-}\Gamma_{-I}}{1 - a_{X^-}\Gamma_{-I}}\right) = \ln\kappa_{X^-} + \ln N_A^{-1}\rho_{X^-}^\beta + 2\sinh^{-1}y \tag{18}$$

If $(q^M + q_I) > 0$ and $y \gg 1$, the asymptotic expansion of Eq. (16) can be performed in Eq. (18). By considering the condition $a_{X^-}\Gamma_{-I} \ll 1$, Schmickler et al.[38] obtain the equation

$$\frac{a_{X^-}}{z_-F}q_I = \frac{\kappa_{X^-}}{b^2}(q^M + q_I)^2 \tag{19}$$

This equation demonstrates that $q_I = z_-F\Gamma_{-I}$ becomes independent of $\rho_{X^-}^\beta$ hence of m_{AX}. The solution of Eq. (19) is $q_I \cong -q^M$ if z_- and q^M have opposite signs and $|a_{X^-}/z_-F| \ll 2\kappa_{X^-}|q^M|/b^2$. The latter condition is not satisfied if $\kappa_{X^-} \ll 1$, which corresponds to a too weak adsorption, making it impossible to reach the state of equivalent adsorption $q_I = -q^M$. Intuitively speaking, such a result can readily be understood. Let us take a system submitted to overequivalent specific adsorption at large values of m_{AX}. After having reached the point of equivalency through a decrease of m_{AX}, the system attains a negative value of q_D. The onset of a positive potential ϕ_0 facilitates the entrance of the counterion X^- into the OHP and makes the value of $\rho_{X^-}(x_0)$ insensitive to the lowering of m_{AX}. Accordingly, any further change of q_I is prohibited.

In order to have an idea about what Eq. (18) quantitatively represents, Schmickler et al. have solved it numerically[38] for a variety of adsorption energy values $A = \ln\kappa_{X^-}$ with κ_{X^-} expressed in liters per mole. In this calculation, we have taken a_{X^-} equal to $9 \times 10^{-2} \ nm^2$ per ion and have varied q^M in the range $0 < q^M < 20 \times 10^{-2} \ C/m^2$. For $A = 10$, overequivalent adsorption is found at large m_{AX}, for which the Langmuir isotherm is anyhow not valid; at small m_{AX}, $|q_I|$ becomes practically equal to $|q^M|$, as depicted in Fig. 4a. For $A = 0$ in Fig. 4b, no overequivalent adsorption can be detected at large m_{AX}, and the limiting value of $|q_I|$

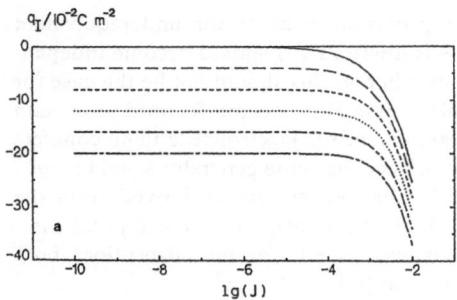

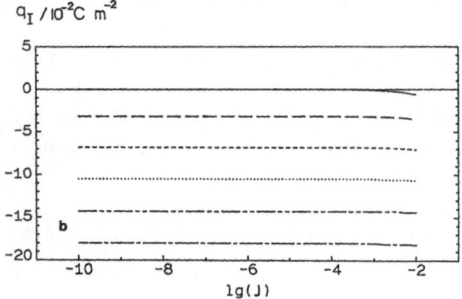

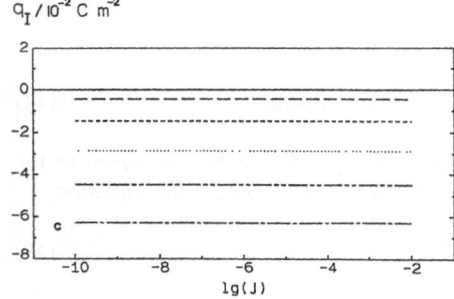

FIGURE 4. Calculated adsorption isotherms at 25°C q^M values of 0 (——), 0.04 (— — —), 0.08 (- - -), 0.12 ($\cdots$), 0.16 (— - - —), and 0.20 (— - — —) (C m^{-2}). (a) $A = 10$; (b) $A = 0$; (c) $A = -5$.

remains somewhat below $|q^M|$. It may be noted that the lower is the value of A, the larger the difference between $|q_I|$ and $|q^M|$ and the more negative the value of q_D. This is particularly evident for $A = -5$ in Fig. 4c, where all isotherms are straight horizontal lines at constant q^M. These predictions fall in line with the experimental observations recorded in Fig. 3. It is easy to prove that in the case of strongly adsorbed ions, such as I$^-$, the bulk concentrations required for reaching the state of charge equivalency are outside the experimental range.

Let us consider some consequences of our results:

(i) It becomes clearly impossible to avoid any specific adsorption in a binary electrolyte at anodic polarization of the electrode when $A \geq -6$, even by infinite dilution. When the substance is present in trace amounts, the adsorbed excess is still constant provided that the mass balance is fulfilled, and hence that $M_w m_{AX}/S \geq \Gamma_{-I}$, where M_w and S are, respectively, the total weight of the solvent and the overall area of the surface.

(ii) The absence of specific adsorption cannot be inferred from the criterion of constancy of the capacitance C_I as a function of m_{AX}. Obviously, a constant q_I will maintain C_I constant at constant q^M in the Parsons–Zobel plot (cf. Fig. 2). It is thus inappropriate to derive the value of C_I in the absence of specific adsorption from this diagram at values of q^M relevant to specific adsorption by means of a linear extrapolation from the region of high dilution.

(iii) Accordingly, one is led to question the choice of F^- as the ion yielding the double-layer properties in the absence of specific adsorption at anodic polarization of Hg. With regard to this matter, one should be reminded of the fact that Grahame[39] estimated $\Gamma_{F^-,I}$ as about $-3 \times 10^{-2} C/m^2$ at $q^M = 10^{-1} C/m^2$. In view of the uncertainties in calculating small values of q_I at very negative q^M, he considered, however, these results as spurious. In very concentrated KF solutions, Melekhova et al.[40] have observed a significant specific adsorption of F^- on Hg. Schiffrin,[23] working at 0 and 15°C in $0.1M$ and $1M$ KF solutions, measured the isotherms reproduced in Fig. 5 with an inaccuracy of about 5×10^{-3} to $10^{-2} C/m^2$. Further evidence of weak adsorption of F^- on Hg was provided by studies of mixed electrolytes AX–AF at constant J. Verkoost et al.[41] applied this method to the system $KF–KHF_2$. Unfortunately, HF_2^- was apparently coadsorbed in an amount sufficient to make the interpretation uncertain. Hills and Reeves[32] investigated the system $KPF_6–KF$ at 25°C. By assuming an *a priori* adsorption isotherm for PF_6^-, they inferred the existence of F^- adsorption and obtained a value of $F\Gamma_{F^-,I} \cong 2 \times 10^{-2} C/m^2$ at $q^M = 12 \times 10^{-2} C/m^2$ and $m_{F^-} = 0.2m$.

It follows from these examples that the adsorption of F^- requires a detailed analysis which might best be carried out for systems of mixed electrolytes. The next section is devoted to this analysis.

3.2. The Mixed Electrolyte System

Many investigations have been performed with mixed electrolytes of the type AX–AF. The most frequent approach consists in keeping J or $m_{AX} = m_{X^-} + m_{F^-}$ constant.[42] The main assumption made in this method is:

$$\frac{\Gamma_{X^-,D}}{\Gamma_{F^-,D}} = \frac{m_{X^-}}{m_{F^-}} = \alpha \tag{20}$$

This condition specifies that, unless specifically adsorbed, all ions of the same sign have the same densities in the electrochemical double layer. As a consequence,

$$\Gamma_{X^-,I} - \alpha\Gamma_{F^-,I} = \Gamma_{X^-,w} - \alpha\Gamma_{F^-,w} \tag{21}$$

A notable feature of the method appears in the fact that $\Gamma_{F^-,I}$ does not affect seriously the left-hand side of Eq. (21) at any α, as long as $\Gamma_{F^-,I}$ obtained from a pure J molar AF solution

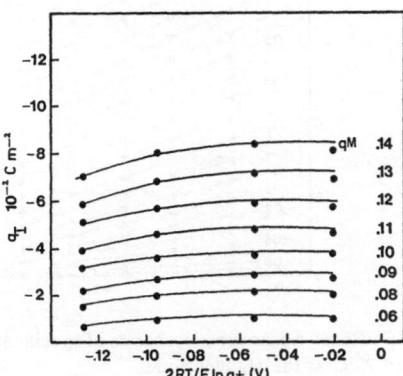

FIGURE 5. Adsorption isotherms of specifically adsorbed F^- in the aqueous KF/Hg electrode system at 15°C.[23]

is small compared to $\Gamma_{X^-,I}$ obtained from a pure J molar AX solution. This is one of the reasons why the specific adsorption of F^- has been generally disregarded in the interpretation of the data, although, as shown later, it may influence the pattern of adsorption. It is worth further noting that $\Gamma_{X^-,I}$ is a true absolute surface molar excess. From these preliminary remarks, we might infer that the values of $\Gamma_{X^-,I}$ in a system containing mixed electrolytes rely less stringently on the Gouy-Chapman hypothesis than in a system containing a single binary electrolyte. The Gibbs equation takes now the form

$$(d\xi)_{T,P,J} = -\Gamma_{X^-,I}(d\mu_{AX})_{T,P,J} + E_+ \, dq^M \tag{22}$$

It is an easy matter of straightforward cross differentiation to derive the following expression:

$$-\frac{1}{RT}\int_{q^{M_0}}^{q^M} \left(\frac{\partial E}{\partial \ln m_{AX}}\right)_{T,P,J,q} dq = (\Gamma_{X^-,I})_{q^M} \tag{23}$$

where the boundary q^{M_0} has been chosen at a potential (-1.5 V versus NCE on Hg) at which $(\Gamma_{X^-,I})_{q^{M_0}} = 0$.

The potential E is determined with respect to a calomel electrode. Corrections which might arise from ionic activity coefficients were evaluated.[42,43] Their contributions are negligibly small in the halide systems which are treated here. Two systems have been studied at 25°C; the KF–KCl system at $J = 1\,M$ and the NaF–NaBr system at $J = 0.1\,M$. The adsorption isotherms derived from the application of Eq. (22) are recorded in Figs. 6 and 7. They both show an asymptotic tendency toward the abscissa at low m_{X^-}.

Any further analysis in terms of specific adsorption of F^- requires that different external conditions be applied to the mixed electrolyte system and that a second equation thus be found in order to solve for the second unknown, q_I. This method, proposed by Lakshmanan and Rangarajan,[44] is based on measurements performed under the simultaneous conditions of variable J and constant α. Thus, the Gibbs equation may be written in the form

$$(d\xi)_{T,P,\alpha} = -\Gamma_{X^-,w}\,(d\mu_{AX})_{T,P,\alpha} - \Gamma_{F^-,w}\,(d\mu_{AF})_{T,P,\alpha} + E_+ dq^M \tag{24}$$

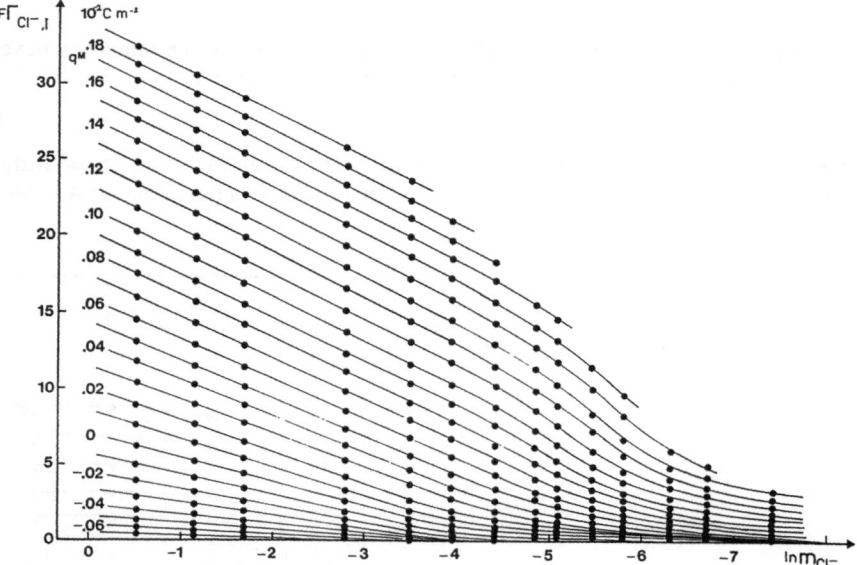

FIGURE 6. Adsorption isotherms of specifically adsorbed Cl^- in a system of aqueous KF–KCl at $J = 1\,M$ and 25°C at the Hg electrode.

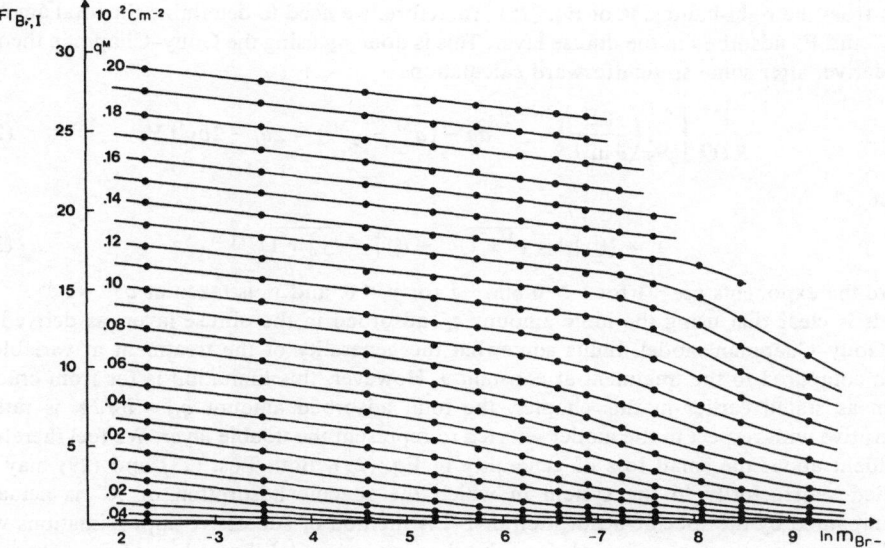

FIGURE 7. Adsorption isotherms of specifically adsorbed Br^- in a system of aqueous NaF–NaBr at $J = 0.1M$ and 25°C at the Hg electrode.

After having carried out the usual cross differentials, one obtains

$$\frac{F}{RTQ} \int_{q^{M_0}}^{q^M} \left(\frac{\partial E}{\partial \ln J} \right)_{T,P,\alpha,q} dq - (q^M - q^{M_0}) = -2F(\Delta\Gamma_{X^-,w} + \Delta\Gamma_{F^-,w}) \qquad (25)$$

where

$$\Delta\Gamma_{X^-,w} \equiv (\Gamma_{X^-,w})_{q^M} - (\Gamma_{X^-,w})_{q^{M_0}} \qquad (26a)$$

$$\Delta\Gamma_{F^-,w} \equiv (\Gamma_{F^-,w})_{q^M} - (\Gamma_{F^-,w})_{q^{M_0}} \qquad (26b)$$

The derivation of Eq. (25) must be supplemented by careful analysis of the influence of ionic activity.[45]

Turning to this problem, we may assume that, as in Eq. (22), the contributions depending on changes of γ_{AX} and γ_{AF}, the activity coefficients of AX and AF, respectively, with respect to α at constant J are negligible. Rocha-Filho et al.,[46] who also studied these effects in mixed electrolytes, came to an identical conclusion about the small variation of γ_{AX} and γ_{AF} with the solution composition.

Nevertheless, there remain non-negligible corrections due to changes of γ_{AX} and γ_{AF} with J. The function Q in Eq. (25) makes provision for these influences. Upon application of Guggenheim's approximation for the chemical potentials of salts in mixed electrolyte solutions,[47] one estimates a value of Q which can be written in the general form[45]

$$Q \equiv 1 - \tfrac{1}{2}(\ln \gamma^{DH}) \left(\frac{1 + 2B^{DH}\sqrt{J}}{1 + B^{DH}\sqrt{J}} \right) + \tfrac{1}{4} \ln \gamma^0_{AX} \gamma^0_{AF} \qquad (27)$$

where

$$\ln \gamma^{DH} = A^{DH}\sqrt{J}/(1 + B^{DH}\sqrt{J}) \qquad (28)$$

is the Debye–Hückel relation, and γ^0_{AX} and γ^0_{AF} are equal to the activity coefficients of AX and AF, respectively, in pure J molal AX and AF solutions. The next step is to get the value of q_I from the right-hand side of Eq. (25). Therefore, we need to determine the total amount of X^- and F^- adsorbed in the diffuse layer. This is done by using the Gouy–Chapman theory. We derive, after some straightforward calculations

$$\frac{F}{RTQ}\int_{q^{M_0}}^{q^M}\left(\frac{\partial E}{\partial \ln J}\right)^{T,P,\alpha,q} dq - (q^M - q^{M_0}) = 2q_I - 2b\sqrt{J}\,Y \qquad (29)$$

where

$$Y \equiv \{(|y| + \sqrt{y^2 + 1})^{\pm 1} - (|y| + \sqrt{y_0^2 + 1})^{-1}\} \qquad (30)$$

where the exponents are $+1$ for $y > 0$ and -1 for $y < 0$, and y_0 is taken at $q^M = q^{M_0}$.

It is clear that using the ionic amount q_D adsorbed in the diffuse layer, as derived in the Gouy–Chapman model, limits somewhat the generality of the treatment at variable J when compared to the treatment at constant J. However, this limitation is far from crucial since, as stated earlier in this chapter, the total adsorbed amount $q_D = b\sqrt{J}\,Y$ is rather insensitive with respect to the model selected to represent the double layer. We feel therefore confident about the small loss of generality in Eq. (29). Both Eqs. (23) and (29) may be applied satisfactorily to the system in which the specific adsorption of X^- is actually accompanied by the specific adsorption of F^-. A method of successive approximations was used in a computer program in order to solve these equations.[45] The values of $\Gamma_{F^-,I}$ obtained in this way are shown in Figs. 8a and b.

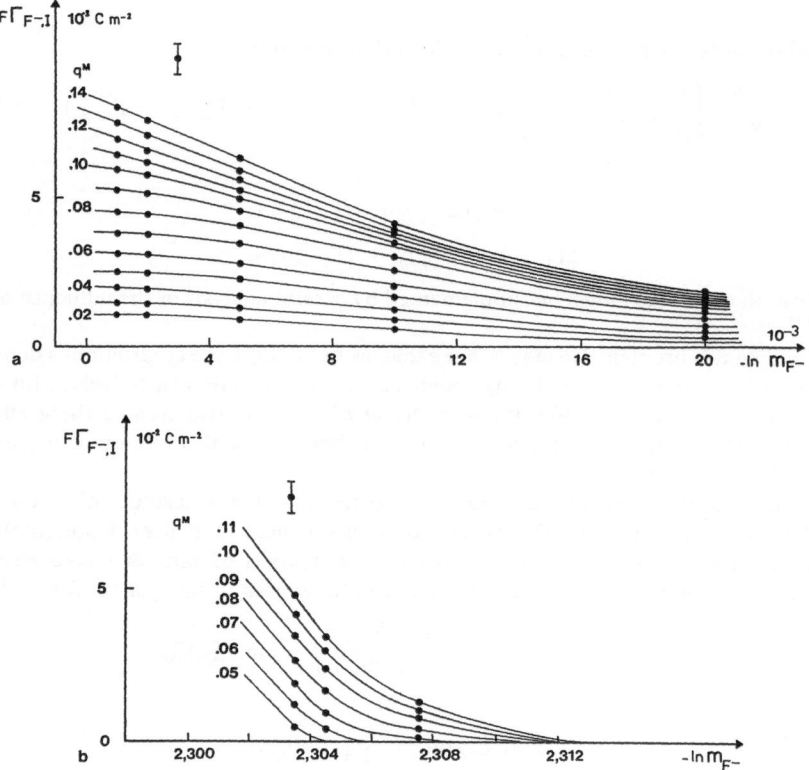

FIGURE 8. Adsorption isotherms of specifically adsorbed F^- at the Hg electrode at 25°C: (a) in the system aqueous KF–KCl at $J = 1M$; (b) in the system aqueous NaF–NaBr at $J = 0.1M$.

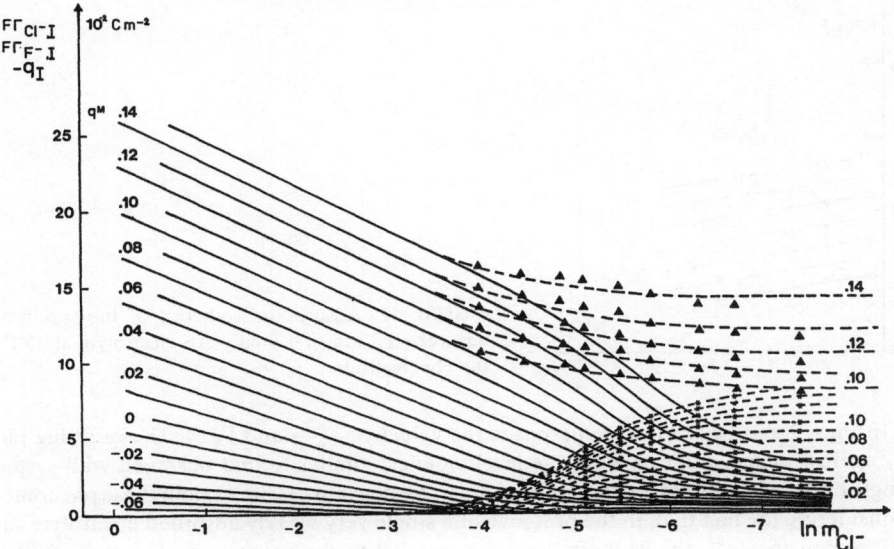

FIGURE 9. The variation of q_I (▲), $F\Gamma_{Cl^-,I}$ (——), and $F\Gamma_{F^-,I}$ (*) as a function of m_{Cl^-} at constant values of q^M. The system is the same as in Fig. 6.

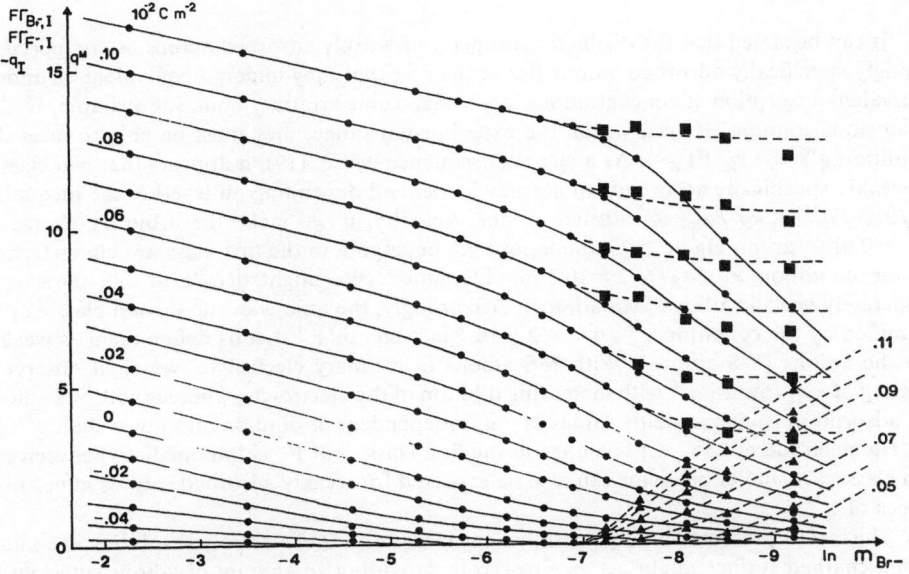

FIGURE 10. The variation of q_I (■), $F\Gamma_{Br^-,I}$ (●), and $F\Gamma_{F^-,I}$ (▲) as a function of m_{Br^-} at constant values of q^M. The system is the same as in Fig. 7.

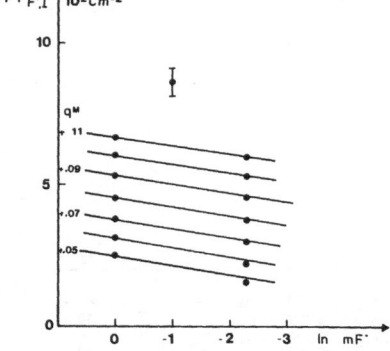

FIGURE 11. Adsorption isotherms of the specifically adsorbed F^- ion in a binary AF electrolyte at 25°C at the Hg electrode.

In Figs. 9 and 10, we have superposed the values of $\Gamma_{X^-,I}$ and $\Gamma_{F^-,I}$. The resulting plots of q_I against $\ln m_{AX}$ indicate a charge leveling effect similar to that observed with systems of single binary electrolytes at low values of m_{AX}. We may now easily explain these phenomena by considering the fact that, in the presence of a single very weakly adsorbed electrolyte such as AF at constant q^M, the isotherm becomes parallel to the axis at a finite value of $F\Gamma_{F^-,I}$ (Fig. 4b, c). In a mixed electrolyte, AX–AF, this same behavior must be attained asymptotically since, at constant q^M, the value of q_I tends to $-F\Gamma_{F^-,I}$ in the limit $|m_{F^-}| \to |J|$. The intercepts after extrapolation of the isotherms in Figs. 8a and b with the vertical axis will allow estimation of these limiting values $F\Gamma_{F^-,I}$ at $1m$ and at $0.1m$. These two points are used to outline the nearly horizontal pattern of adsorption of F^- depicted in Fig. 11.

4. CONCLUDING REMARKS CONCERNING TWO CLASSES OF WEAKLY SPECIFICALLY ADSORBED IONS

It can be stated that the distinctive property of weakly adsorbed anions as compared to strongly specifically adsorbed anions lies in the fact that they undergo equivalent or under-equivalent adsorption at concentrations larger than some arbitrary limit, for example, $10^{-4}M$ in aqueous solution. Hence, within the experimental range, they must be able to meet the condition $q^M \geq -z_X-F\Gamma_{X^-,I}$. As a clear consequence of Eq. (19), it appears that two classes of weakly specifically adsorbed anions may be defined depending on whether the inequality $|b^2/2z_X-Fq^M| < \kappa_{X^-}/a_{X^-}$ is satisfied or not. Actually, if we make the arbitrary choice of $q^M = 0.01$ C/m² on Hg at 25°C, the anions X^- belonging to the first class are characterized by the condition $\kappa_{X^-}/a_{X^-} > 2 \times 10^{-6}$ m^{-2} liter mol^{-1}. We might decide to call these ions moderately specific adions (MS adions). Accordingly, the anions of the second class can be identified by the condition $\kappa_{X^-}/a_{X^-} < 2 \times 10^{-6}$ m^{-2} liter mol^{-1}. Let us define them as weakly specific adions (WS adions). With MS adions in a binary electrolyte, we shall observe a leveling of $-q_I$ toward q^M with increasing dilution of the electrolyte, whereas with WS adions the adsorption isotherm yields values of $-q_I$ independent of dilution and lower than q^M. At the Hg electrode at 25°C, Cl^- belongs to the first class, and F^- is intermediate between the two classes. A similar behavior can also be expected for weakly adsorbed cations at negative values of q^M.

The results presented here suggest that any large volume of electrolytic solution in contact with a charged surface might act as a reservoir, providing an amount of adions sufficient to keep the specific adsorption independent of dilution. Thus, the possibility for determining properties of charged surfaces free of specific adsorption appears very low.

ACKNOWLEDGMENTS

These studies were supported by the Fonds de la Recherche Fondamentale Collective (F.R.F.C.), of Belgium and by the S.P.P.S. (A.R.C. contract no. 86/91).

REFERENCES

1. M. Blank, in: *Electrical Double Layers in Biology* (M. Blank, ed.), p. 119, Plenum Press, New York (1986).
2. R. Schmukler, J. J. Kaufman, P. C. Maccaro, J. T. Ryaby, and A. A. Pilla, in: *Electrical Double Layers in Biology* (M. Blank, ed.), p. 201, Plenum Press, New York (1986).
3. G. F. Elliot, G. R. S. Waylor, and A. E. Woolgar, in: *Ions in Macromolecular and Biological Systems* (D. H. Everett and B. Vincent, eds.), Scientechnica Press, Bristol (1978).
4. J. Barber, J. Mills, and A. Love, *FEBS Lett.* **74**, 174 (1977).
5. H. Wroblowa, Z. Kovac, and J. O'M. Bockris, *Trans. Faraday Soc.* **61**, 1523 (1965).
6. O. A. Ershler and V. Shikov, *Zh. Fiz. Khim.* **17**, 236 (1943).
7. D. C. Grahame, *Chem. Rev.* **41**, 441 (1947).
8. J. O'M. Bockris, M. A. V. Devanathan, and K. Müller, *Proc. Roy. Soc. (London), Ser. A* **247**, 55 (1963).
9. A. L. Loeb, *J. Colloid Sci.* **75**, 75 (1981).
10. W. E. Williams, *Proc. Phys. Soc. A* **66**, 372 (1953).
11. S. Levine, C. W. Outhwaite, and L. B. Bhuiyan, *J. Electroanal. Chem.* **123**, 105 (1981).
12. D. Henderson, F. F. Abraham, and J. A. Barker, *Mol. Phys.* **31**, 1291 (1976).
13. L. Blum, *J. Phys. Chem.* **81**, 136 (1977).
14. S. L. Carnie and D. Y. C. Chan, *J. Chem. Phys.* **73**, 2949 (1980).
15. D. Henderson, in: *Trends in Interfacial Electrochemistry* (A. F. Silva, ed.), p. 473, D. Reidel, Dordrecht (1986).
16. G. M. Torrie and J. P. Valleau, *Chem. Phys. Lett.* **65**, 343 (1979).
17. G. M. Torrie and J. P. Valleau, *J. Chem. Phys.* **73**, 5807 (1980).
18. G. M. Torrie and J. P. Valleau, *J. Phys. Chem.* **86**, 3251 (1982).
19. J. Watts-Tobin, *Phil. Mag.* **6**, 133 (1961).
20. W. R. Fawcett, *Isr. J. Chem.* **18**, 3 (1979).
21. R. Guidelli, in: *Trends in Interfacial Electrochemistry* (A. F. Silva, ed.), p. 387, D. Reidel, Dordrecht (1986).
22. R. Parsons and F. R. G. Zobel, *J. Electroanal. Chem.* **9**, 333 (1965).
23. D. Schiffrin, *Trans. Faraday Soc.* **67**, 3318 (1971).
24. M. L. Rosinberg, Ph.D. thesis, Université M. et P. Curie, Paris (1983), p. 149.
25. W. Schmickler and D. Henderson, *J. Chem. Phys.* **80**, 3381 (1984).
26. J. P. Badiali, M. L. Rosinberg, F. Vericat, and L. Blum, *J. Electroanal. Chem.* **158**, 253 (1983).
27. L. Blum, D. Henderson, and R. Parsons, *J. Electroanal. Chem.* **161**, 389 (1984).
28. W. Schmickler and D. Henderson, *J. Chem. Phys.* **85**, 1650 (1986).
29. J. P. Valleau and G. M. Torrie, *J. Chem. Phys.* **76**, 4623 (1982).
30. L. B. Bhuiyan, L. Blum, and D. Henderson, *J. Chem. Phys.* **78**, 442 (1983).
31. B. B. Damaskin, N. S. Polyanovskaya, U. Palm, and M. Salve, *J. Electroanal. Chem.* **246**, 247 (1988).
32. G. J. Hills and R. M. Reeves, *J. Electroanal. Chem.* **31**, 269 (1971).
33. J. A. Harrison, J. E. B. Randles, and D. J. Schiffrin, *J. Electroanal. Chem.* **25**, 197 (1970).
34. D. C. Grahame and R. Parsons, *J. Am. Chem. Soc.* **83**, 1291 (1961).
35. J. Lawrence, R. Parsons, and R. Payne, *J. Electroanal. Chem.* **16**, 193 (1968).
36. C. V. D'Alkaine, E. R. Gonzalez, and R. Parsons, *J. Electroanal. Chem.* **32**, 57 (1971).
37. B. B. Damaskin, *J. Electroanal. Chem.* **65**, 799 (1975).
38. W. Schmickler, D. Henderson, and H. D. Hurwitz, *Z. Phys. Chem. N.F.*, **160**, 191–198 (1988).
39. D. C. Grahame, Office of Naval Research, Technical Report No. 6, Research Contract NS.GR.66903 (1951).
40. N. I. Melekhova, V. F. Ivanov, and B. B. Damaskin, *Elektrokhimiya* **5**, 370 (1969).

41. A. W. M. Verkoost, M. Sluyters-Rehbach, and J. H. Sluyters, *J. Electroanal. Chem.* **24,** 1 (1970).
42. H. D. Hurwitz, *J. Electroanal. Chem.* **10,** 35 (1965).
43. S. K. Lakshmanan and S. K. Rangarajan, *J. Electroanal. Chem.* **27,** 270 (1970).
44. S. K. Lakshmanan and S. K. Rangarajan, *J. Electroanal. Chem.* **27,** 127 (1970).
45. B. Bicamumpaka, Ph.D. thesis, Université Libre de Bruxelles (1988).
46. R. C. Rocha-Filho, E. R. Gonzalez, and L. A. Avaca, *J. Electroanal. Chem.* **184,** 179 (1985).
47. E. A. Guggenheim, *Phil. Mag.* **19,** 588 (1935).

2

Electroflotation
From the Double Layer to
Troubled Waters†

Klaus Müller

1. INTRODUCTION

Among technologies with an electrochemical basis, electroflotation appears certain to find expanding applications in the 21st century. Electroflotation (EF) is a separation technology for aqueous fluids containing suspended matter, designed to produce water pure enough to discharge and residues concentrated enough for further workup. After additional compatible processes, the water is suitable for recycle, while the residues may, depending on the particular application, constitute a raw material for food, feed, and fuels, allow mineral and metal values to be extracted, and/or simply become disposable waste.

Electroflotation has a long history of successful short-term applications as well as of failures. Most of the problems have now been solved on a technical scale. The reference list of long-term operations is growing steadily, but little notice has been taken in the recent electrochemical literature. It is a different matter, of course, to decide whether EF, at a given time and in a given situation, is an economically feasible technology. For some of the environmentally most troublesome effluent situations, the answer is "yes" now. For more of them, and for a great many recycling and extraction problems in the next century, the answer is "very likely so."

2. PRINCIPLES

2.1. The EF Blackbox

Electroflotation is an electrochemical version of flotation. It differs from flotation mainly in the mechanism of bubble generation. The bubbles are generated by electrolysis, usually of the substrate stream itself or of an auxiliary electrolyte stream. The bubbles produced at electrodes are very fine, are uniform, and rise very slowly, and bubble generation produces

---

† *Troubled waters*: A situation or condition of disorder or confusion (Webster's Third New International Dictionary).

Klaus Müller ● Battelle Europe, Geneva Laboratories, CH-1227 Carouge, Switzerland.

Electrochemistry in Transition, edited by Oliver J. Murphy *et al.* Plenum Press, New York, 1992.

little undesired convection (though probably more than has been thought for a long time, according to recent Soviet work[1]).

Flotation processes separate suspended matter from fluids. In EF, this occurs without a size classification. The gas bubbles become attached to the suspended particles, and these are lifted to the top of the fluid, where they are collected as a sludge; hence, the suspended matter and the fluid can be used or disposed of separately. For a separation of species originally present as true solutes, precipitation or coprecipitation must precede flotation. For the separation of colloidal matter, coagulation/flocculation or adsorption must precede flotation.

EF equipment can be sketched as shown in Fig. 1. Into the EF tank go:

- a stream of the fluid to be treated (e.g., effluent, fruit juice, or a mineral slurry, part or all of which may have been pretreated),
- streams of fluid containing chemicals aiding flotation (the point of entry actually is upstream from the tank),
- dc power to the electrodes.

Out of the EF tank come:

- a purified fluid stream, usually from the bottom,
- concentrated residue (sludge), usually from the top,
- the electrolysis gases (some dissolved in the off-fluid, some contained in the sludge, and some directly vented).

Floor-space requirements for the EF tank depend on throughput; they will be between square meters and tens of square meters. Tank height is on the order of 1 m. Throughputs of operating plants are cubic meters to hundreds of cubic meters per day. This is much below the scale of (nonelectro) flotation equipment found in the minerals industry. The EF units are relatively quiet in their operation.

More technical background will be provided in the last part of the present section and in the sections on applications that follow.

2.2. The Double-Layer Connection

Legion are the examples where fundamental inquiries into the structure and properties of the electrical double layer at interfaces have been justified by the importance of the results, in terms of practical applications or their understanding. Colloid science constitutes the area where the loop between theory and practice has been closed most successfully. A similar claim cannot be made in the case of EF. So where is the double-layer connection?

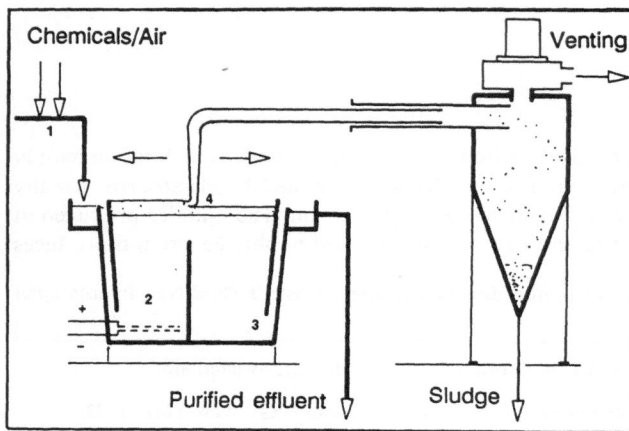

FIGURE 1. Schematic of electroflotation unit. Effluent enters from the left, after appropriate addition of chemicals. Flotation and separation occur in the central unit holding electrodes (+ and −). Sludge collects at the top from where it is removed; it then undergoes degassing (venting) and further dehydration. Purified effluent leaves the EF unit. (Kindly provided by Dr. E. Baer.[36])

It is somewhat indirect. It will not suffice to say that all interfaces are charged, though this is a correct opening statement.

2.2.1. The Suspended Matter

The substrate in EF is a suspension of finely divided matter (liquid and/or solid; e.g., oil emulsions). In such matter, the charge *may* be a highly important factor in stability (before the treatment, at the stage of the "problem") and destabilization (during pretreatment; emulsions and similar, stable colloidal systems must be broken prior to separation by EF). In a technical report on the treatment of oil-sand production recycle water from the University of Alberta,[2] which reads like a dissertation on the subject, the double layer (around the particles) has been discussed as the starting point of an apparently successful practical evaluation of EF.

2.2.2. The Bubbles

The "motor" in any flotation process (not merely in EF) is the gas bubbles. They, too, are charged, and their charge, too, will be more important, the finer they are. The claim is made that, when the bubbles are generated electrolytically, the parent electrode influences their charge; also, electrolytic bubbles are finer (how much so will be a question of solution pH and electrode polarity). Unfortunately, bubbles have been much less popular than mercury drops in double-layer studies. Moreover, though it may be attractive to think of the elegant attraction of positively charged bubbles to negatively charged particles, flotation actually has an optimum at zero zeta potentials of the bubbles and is most efficient when the particles are at their point of coagulation/flocculation.

2.2.3. Electroflotation and the Hamaker Constant

Not many papers can be consulted which provide information concerning the fine details of the kinetics and mechanism of individual EF steps. Panov and Kravchenko[3] discussed the inevitable supersaturation which must exist in the solution (because of the Kelvin equation) when very small bubbles are present in equilibrium. Kul'skii et al.[4] discussed the effect of the degree of contaminant dispersion on coagulant and energy requirements. Fukui and Yuu studied flotation kinetics with model dispersions to see the effect of bubble diameter and bubble charge[5,6]; they also reported that several companies in Japan have succeeded in treating effluents by EF. Fukui and Yuu's treatment considers the effect of Hamaker constants and zeta potentials as the important factors in particle collection by charged bubbles.

Among relevant papers in colloid chemistry, Watanabe's[7] is a fairly recent review of the oil/water interface; adsorption and the interaction between model drops were considered. Elsewhere,[8] particle interactions have been discussed in the context of flotation in terms of surface potential and surface charge of the particles.

Bubble properties have been studied for a long time. Small bubble sizes are found for the cathodic gas in alkaline solutions, but for the anodic gas in acidic solutions[9]; bubble size also is a function of electrode diameter (40-μm bubbles were produced at 0.2-mm-diameter wire,† 130-μm bubbles at 1.5-mm wire, and there is a sharp size distribution maximum),[10] temperature and electrode material have influence on bubble size, and an optimum current density of 20 to 30 mA/cm^2 was reported.[11] Typical bubble sizes in EF are between 20 and 70 μm. Bubble rise velocities decrease with increasing electrolyte concentration and with

---

† In a recent example,[119] optimum EF treatment of meat processing wastewater was achieved with a wire diameter of 0.2–0.5 mm, mesh grid size of 2.5–5.0 mm, and an inclination of the electrodes of 30–45° to the horizontal.

surfactant addition (this was found for relatively large bubbles[12]). Clean bubbles rise faster (there is an analogy to the fall of mercury drops: the clean interfaces are mobile, which is responsible for the higher speed), while bubbles rigidified with surfactant rise more slowly.[13] Bubble charge was found to be positive at pH < 2 and negative at pH > 3; that is, the bubbles have an isoelectric point or point of zero charge at a pH of 2 to 3,[9] which was confirmed quantitatively by bubble electrophoresis.[13] Double-layer structure appears to be governed by negative adsorption of ions (e.g., H^+ or OH^-) at the water/gas interface. In bubble growth, there is an induction time (supersaturation must be reached), a period where growth is sustained by diffusion, and finally a period where growth is sustained faradaically.[9]

In the area of flotation kinetic studies, some more papers are available which take the same direction as Fukui and Yuu's. It was noticed that a finite time is required for bubble and particle to become permanently joined.[14] Charge on the particles and bubbles may be detrimental, and the molecular component of the forces may be preponderant; maximum floatability was found to occur at the isoelectric point.[15] There is a hydrodynamic factor, since liquid streaming around the bubbles is important in the stage preceding attachment; the Reynolds number in surfactant-free systems (but this would appear to be a rather exceptional case!) should be between 1 and 40. Microbubbles are likely to get attached to hydrophobic sites of the floc, and high shear should be avoided.[16] The benefits of using small bubbles seem to be large: flotation rates rise with the inverse third power of bubble radius.[17,18] In model systems, it was found that minute particles first will become attached to, or deposit on, the surface of small bubbles, and such aggregates then are floated by larger gas bubbles.[19] This mechanism is helped by the fact that the microbubbles actually are stabilized by the sheaths of colloidal particles which become attached.[19]

2.2.4. From the Double Layer to Troubled Waters

It must be doubted, despite the theoretical work cited in Section 2.2.3, that double-layer theory has as yet been a quantitative help in putting EF processes to work. Its relevance and value as a guide is evident, and anybody with a background in double-layer structure and/or colloid science should enjoy the introduction to EF afforded by this background. It is gratifying to see that a process in which double-layer properties (but not merely the electrostatic ones) play a basic role can convert some of the most troublesome effluents back to usable water.

2.3. Early Hopes and Failures

To look back is not the purpose of this contribution. May it suffice, therefore, to say that the history and applications of EF have been reviewed, for example, by Kuhn,[20] that many companies and workers have been involved in the past, and that spectacular separation and cleanup operations have been described (from hog farm effluents to diamond fines), yet it must be suspected that many initially successful operations did not continue forever. The reasons are:

  (i) technical complications: many of the early engineers in the field put the steps of flocculant generation, emulsion breaking, and flocculation right into the EF tank's electrode region, which may have brought an untractable situation in the long run;
  (ii) technical difficulties: viz., formation of undesirable deposits on the electrodes and undesirable corrosion of the electrodes; and
  (iii) competition and price: a chicken farm might not be able to support the bill for electric power† and advanced electrode designs, a hog farm's effluent volume and

---

† However, it has been reported that a poultry waste digester could produce biogas to the extent of about 10 W h/day per caged layer.[120]

its contaminant load probably are too high, dissolved-air flotation (or an altogether different method) may have provided a solution at a lower price or, lastly, environmental concerns may not yet have motivated a cleanup at all.

2.4. Breakthrough to a Viable Technology

The following is a list of requirements which must be met for fluid treatment by EF.

- The fluid must be sufficiently conductive for economy of the electrolysis process producing the gas; if it is not, electrolyte must be added (e.g., industrial waste brines or seawater), and it has been suggested that a special electrolyte loop be created only around the electrodes.
- The suspended matter in the fluid must be floatable; this may require prior steps such as emulsion breaking/coagulation, aggregation/flocculation, or the attachment to carrier particles (hydroxide flocs) as well as surface modification of the minute particles by special chemicals. Also, this requirement implies limitations with respect to the specific gravity and number density (concentration) of the suspended particles.
- There may be optimization requirements such as using one or both electrolysis gases, simultaneously producing disinfectant (anodically from chloride ions), or using cathodic pH variation and anodic dissolution to produce hydroxide particles, but all this must occur without interference with the EF lifting act. Also, any chemicals added should not contribute to cleanup problems, and they should preferably be recycled (such as iron and aluminum for floc).

What has persistently caused trouble in long-term operation was incrustation of the electrodes, particularly the cathodes. The phenomenon is not perfectly understood; precipitation by pH variation, electrophoretic deposition, or cathodic (and anodic) electrodeposition may be involved. Mechanical cleaning of the electrodes not only is cumbersome, disruptive, and expensive, but also actually very difficult. A good solution to this problem, polarity change of the electrodes during operation, which in itself will not upset the EF process, was unsuccessful because of excessive corrosion until the quite recent development of stable, nonconsumable electrodes which will operate without corrosion and passivation as anode *and* as cathode, in alternation. This must be regarded as a true breakthrough (no less so than the success of the stable metal-oxide anodes in chloralkali electrolysis).

The structure of these electrodes has been disclosed as being $Ti/TiO_{2-x}, Pt$[21]; they are available from Heraeus Hanau, Germany, and their different applications (in addition to electroflotation) have been described. Important points are their high surface area and highly open design (see Fig. 2). Among other companies supplying platinized titanium electrodes, Engelhard can be mentioned.

2.5. Where to Look

While practical applications recently have multiplied (see below), other strong technologies for separations and water treatment are available or under development, and in some recent symposia about water treatments and separation technologies,[22,23] EF was not a central topic. Literature reviews preceding the "electrode breakthrough" (see above) should, of course, be digested "with a grain of salt." One speculates that investment requirements for the process have become somewhat higher than figures provided occasionally in the more distant past, on account of superior electrode design, though operating costs should be relatively lower now. With this in mind, the interested reader can go back and consult earlier reviews and book chapters.[24-30] Romanov has restated a great many of the salient points

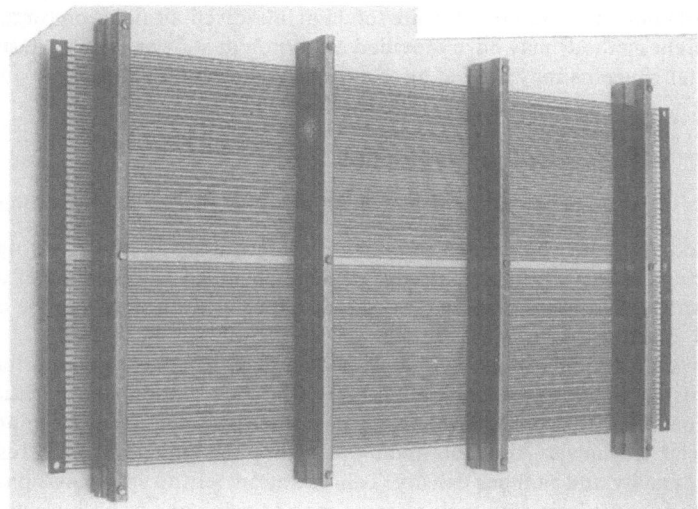

FIGURE 2. Heraeus activated titanium electrode for electroflotation. (Kindly provided by Dr. B. Busse, Heraeus Elektroden GmbH, D-6463 Freigericht/Hanau.)

about EF (speed, selectivity, process parameters) in his 1985 reviews, but the new aspects rightly stressed by him are the environment and the rational, efficient use of natural resources.[31,32] These are the reasons why EF has the potential of being among the key technologies with electrochemical background in the 21st century.

Books are not available on the subject, except one in Russian,[33] and this predates the "breakthrough." There have been few recent reviews on EF as such, but one should watch for information about processes which will favorably complement EF, such as, for instance, electroacoustic dewatering[34] (which appears to have application to EF sludge), all the other electrochemical water treatments,[35] and, of course, all membrane technologies which would be applicable to the EF effluent water.

Electroflotation is one of the easiest subjects to search in data banks by computer. However, as is often the case for technical matter, not all that is published works, and not all that works is published. With this in mind, the reader is invited to look through the section on applications that follows. For plants, one reference is to Dr. Baer Verfahrenstechnik in Frankfurt,[36] who kindly provided photos and processing schemes of operating plants (see Figs. 3 and 4). The Baer technology and its applications to wastewaters from railroads, army vehicles, and steel and photochemical plants has been described in a recent review from Italy.[37] In the United Kingdom, Simon-Hartley had been one of several successful suppliers of EF equipment in the past. A manufacturer might specialize in EF or offer it as an option in a line of flotation equipment.

3. ELECTROFLOTATION APPLICATIONS

3.1. Oil–Water Emulsions

Spent emulsions and effluents in the form of oil-in-water emulsions are among the effluents hardest to treat. They come from metal working (cutting oil; in the United Kingdom,

10^8 gal/yr in 1974[38] and engine cleaning operations (road vehicles, aircraft, ships), and also from rolling mills, petroleum refineries, chemical processing, general manufacturing, tankers, and spillage.[39] The grease, detergents (emulsifiers), and often the metal content make them unfit for discharge and cause problems in ordinary waste treatment installations. A successful treatment could yield recycle water and recycle oil.

Details for the Swissair plant in Zurich Kloten were made available soon after commissioning.[40-44] Here the effluents from workshops, galvanic shops, and aircraft maintenance (especially engine cleaning) are treated in the plant shown in Fig. 3. The system has an 80% yield of technically pure water for recycling and offers a high degree of environmental protection; the load left to the Kloten town clarification plant is very low. The overall process is a combination of EF and reverse osmosis. The inorganics are heavy metals, cyanide, alkalies, acids, and salts; the organics are detergents, various oils, and solvents. Emulsification is strong. The original installation used cathodes of ferritic stainless steel and anodes of platinized titanium. Operation was at 15 to 25 A/m^2 and 6 to 9 V. Sixty to 80 ppm of aluminum (as alum) must be added as the primary flocculant, and a further 1.5 ppm of an anionic polymer. The total investment was 17 million SFr for a capacity of 40 m^3/h in the EF plant. Full operation started in June 1977. The plant is operating; the original cathodes have been replaced by the newly developed electrodes mentioned above. The 1977 operating costs were SFr 0.43 per cubic meter for EF and SFr 1.24 per cubic meter total (including the 20% makeup water), as compared to SFr 0.90 per cubic meter for town water, which means that decontamination and environmental protection costs were SFr 0.34 per cubic meter of the effluent.

Baer plants have been commissioned for similar situations in a number of countries. The Swissair plant is one of the reference setups. Other plants operate for the German Federal Railways, the German Army, Alitalia in Rome Fiumicino (see the scheme in Fig. 4), Opel in Rüsselsheim near Frankfurt, Daimler-Benz, Fiat, Michelin Torino, and the 3M Ferrania plant in the troubled Val Bormida in northern Italy, and also in Czechoslovakia and Hungary. The ability to deal with phenol, cyanide, and heavy metals, the possibility of reducing chromate and treating nitrite, and the fact that the purification effect is very high yet energy requirements are modest (said to be comparable to those in compressed-air flotation, surprisingly)[36] and

FIGURE 3. View of two EF units at the Swissair Kloten (Zurich) installations. The overall scheme includes reverse osmosis. (Kindly provided by Dr. E. Baer.[36])

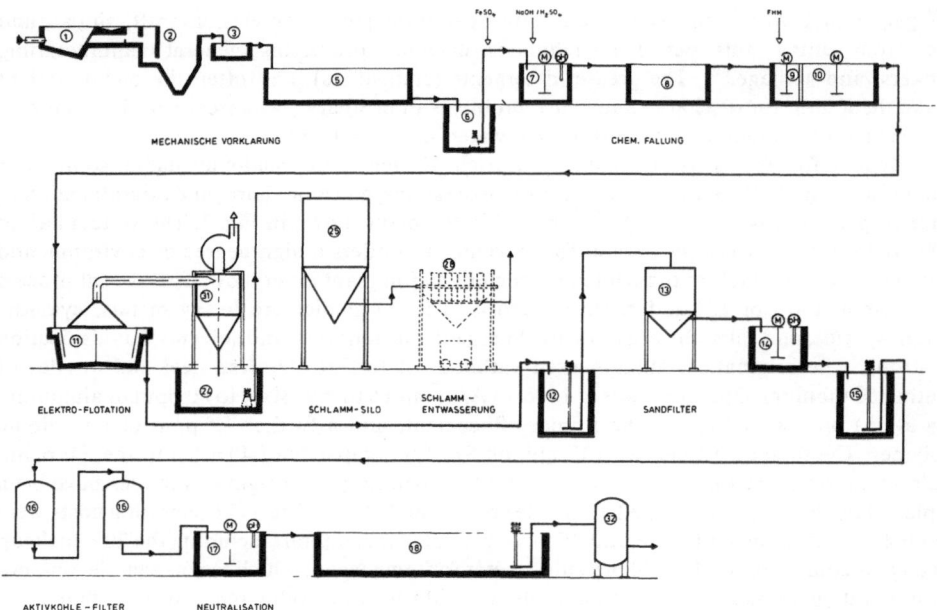

FIGURE 4. Full process scheme of the EF units at the Alitalia Fiumicino (Rome) installations. The consecutive operations include mechanical preclarification, chemical precipitation, electroflotation, sludge dehydration, sand and active carbon filtration, and neutralization. (Kindly provided by Dr. E. Baer.[36])

chemical requirements minimal imply that the installations can be adapted to a number of tough effluent situations. EF plants with a capacity of up to at least 300 m³/h have been built; the process is designed for tough situations and ought to be situated at points close to effluent generation, before any dilution has occurred.

3.2. Metal-Plating Shops, Electrochemical Machining

The effluents from metal-plating shops have a high content of metals which are too valuable and too toxic for direct discharge. The particularly undesirable Cr(VI) was found to be readily removed by an electrocoagulation–electroflotation treatment[45] or by reduction to Cr(III) with metallic iron followed by EF. Zinc also responds well, though at different optimum pH values: >9–10 versus <9 for Cr(VI). Nickel and copper are eliminated most readily at pH 7.[46] Multimetal recovery by EF after ferrocyanide precipitation was described.[47]

Cadmium and cyanide from a finishing process were removed from the effluent by EF according to a laboratory demonstration[48]; 2–6 g of Cd were removed per kilowatt-hour in a system to which seawater was added as a bottom layer, in the space holding the horizontal electrodes; magnesium oxide acted as the floc. Mercury could be removed by electroflotation when precipitated as the sulfide.[49]

Electrochemical machining (ECM) produces particularly high concentrations of suspended and ionic metal dissolution products. Apart from metal recovery (which may be of interest even in the case of iron[50]), sludge removal will extend electrolyte life, and purification schemes have been described long ago.[51] It was found that oxygen should not be used for

EF in ECM, since Fe(II) (green electrolytes) is oxidized by Fe(III) hydroxide (red electrolytes); a diaphragm between the electrodes was said to help.[52]

3.3. Dairy Industry

Dairy effluents are high-strength waste; the BOD5 values are typically ten times those of domestic sewage.[53] The quantities of effluent produced can be very large, and important quantities of fat, proteins, and lactose are lost.[54] Effluent cleanup could become attractive because of the recovered values.[55] To this end, the effluent is adjusted with HCl to pH 4, which is the average isoelectric point for milk proteins (casein, albumin, globulin); a fat–protein complex will precipitate and can be floated by EF.

No more than pilot operations have been described. It was pointed out that the CaO–protein interface structures forming at the bubbles in these applications should be temperature-dependent, and hence temperature control of dairy effluent EF could be the way to optimize the process.[56]

3.4. Food Industries

Meat factories have provided early, drastic examples of the efficiency of EF relative to more conventional separation techniques (Swift, Chicago[57,58]). Grease recovery for nonfood applications was found to be feasible. A recent example of successful laboratory evaluation comes from Bratislava,[59] where 99.8% of the initial fat was removed and the fat–protein concentrate obtained was sufficiently pure to use in feeds. Many other food industries present effluent problems, and EF was tried for sugar plants,[60] in the manufacture of starch from corn,[61] in the isolation of nutrient yeast,[62] and in the purification of grape[33] and recently apple juice.[63] In the treatment of palm oil mill effluents,[64] EF was combined with anodic oxidative destruction of soluble constituents.

3.5. Livestock Farming

From many countries, EF applications to farm effluents have been reported. The technical feasibility appears to have been demonstrated insofar as the effluent is concerned. However, energy requirements are high. A value of 300 A h/m^3 was reported for a hog farm,[65] where the effluent carried 6 kg of suspended solids per cubic meter. Probably because of electrode passivation (fouling, scale formation), instead of the expected 1.9 V,[66] voltages between 3 and 530 V (!) have been reported, and hence excess energy consumption between 0.5 and 80 kW h/m^3. Apparently the problems have not been resolved, though the savings potential of EF relative to conventional processes has been calculated to be enormous[67-69]—provided higher reliability could be achieved.

A fairly positive report on the EF treatment of egg-washing wastewaters and subsequent land disposal of the solid waste, including a description of the facility, was given at a 1984 conference.[70]

3.6. Cellulose and Paper Industry

Early applications used the EF process with integrated electrocoagulation (soluble anodes in the EF tank producing floc for flotation, e.g., Mg anodes for sulfate cellulose effluents).[71] For cardboard manufacturing effluents, insoluble anodes were used and disinfectant anodically produced.[72,73] A 1980 Soviet patent describes EF of lignin from wood-processing wastewaters (using Ti anodes with an active MnO_2 layer).[74]

3.7. Fiber, Textile, and Leather Industry

In rayon production, where a high pollutional load of carbon disulfide, hydrogen sulfide, sulfur, coagulated viscose, hemicellulose, surfactants, etc. arises, EF was examined and found to be an efficient method. Over 99% of the zinc along with many of the other pollutants could be extracted from the effluent; a pilot plant was operated.[75,76]

For the decontamination of textile production effluents containing organic dyes and surfactants, a scheme using dimensionally stable anodes (titanium anodes with an active surface layer of isomorphic metal oxides) was proposed,[77] since active chlorine is required. Owing to the oxidation step, energy requirements here are very high.

For the recovery of concentrated mercerizing liquor, formation of a peroxide adduct and its separation by EF were reported.[78] Apparently successful tests of EF were reported for tannery wastewaters.[79]

3.8. Chemical Industry

Electrocoagulation and EF was proposed for the treatment of effluents from the production of acetylene by oxidative methane pyrolysis.[80] Here the efficiency of purification was found to be very good, at the price of about 1 kW h of electric power and 10 g of aluminum per cubic meter of the effluent. Work concerning the recovery of process water in a heavy-oil extraction facility has been mentioned earlier[2]; speed and price of recycle water production were found to be acceptable.

3.9. Paint and Print Shops

A relatively pessimistic outlook has been presented[81] for the treatment of paint effluents, considering the large volume to be treated and the requirements for electrolyte addition. Yet settling efficiencies demonstrated for EF were two orders of magnitude better than in nonelectrolytic treatment. There should be room for reevaluation. Some results were reported from this trade in Refs. 27 and 82.

3.10. Shipboard Applications

A detailed feasibility study of the use of electroflotation for bilge water purification was performed for the U.S. Coast Guard.[83] The only preliminaries required are the addition of NaOH to pH 10 and of 10 to 15 ppm of anionic polyelectrolyte. Simulated bilge and ballast water were purified to <10-ppm residual oil when influent streams contained as much as 3000 to 4000 ppm of emulsified oil. Pilot plant data were used to estimate system costs, which at the time appeared very modest. Energy consumption was estimated to be 5.6 kW h/1000 gallons, of which only 1.2 kW h was for EF itself, with the rest for pumps, stirrers, and control equipment. Cathode scale formation was a problem then. For seawater duty, Pt–Ir mesh spotwelded to a niobium substrate was proposed as the anodes. A more recent report deals with diesel fuel emulsions in seawater.[84] A Japanese patent describes a three-reactor scheme [electrolysis to produce $Al(OH)_3$, flocculation, electroflotation with graphite electrodes] for the treatment of bilge wastewater.[85]

3.11. Urban Effluents

Kuhn[24] described the treatment of urban effluents by EF under conditions where seawater is available. Then flocculant is produced from the magnesium ions at the cathode owing to alkalinity induced by hydrogen evolution. At the time, a large plant was operating on Guernsey. More recently, Electricité de France in collaboration with the Société Lyonnaise des Eaux has looked into possibilities for simultaneous EF and disinfection (anodic chlorine) of urban sewage.[86,87] Ferric chloride and an anionic polymer had to be added, and 200 A/m^2 were employed. Coliform counts decreased by six powers of ten, yet the activated sludge was found to remain viable, and hence available for recycling, an important consideration since EF would not be a stand-alone operation here. In this study, it was concluded that electrocoagulation [coagulation with the aid of anodically dissolved iron and anodically generated alkalinity (added lime)] is too expensive, while disinfection was regarded as economically competitive even in large plants. Scale formation on and corrosion of the electrodes were cited as problems—this would no longer be true in view of recent developments.

In the Soviet Union, EF was proposed as the preferred alternative to settling tanks in the case of arctic settlements.[88] For ordinary conditions, EF was suggested as a feasible technology for the thickening of excess sludge produced in regular effluent purification schemes[89]; for technical details, see Ref. 90.

3.12. Mining

Hogan et al.[91,92] have carefully reviewed and analyzed the situation of EF in mining applications. Against a theoretical minimum energy requirement of 1.6, true requirements are expected not to fall below 30 kW h/ton of ore floated, and, apart from that, settling tendencies are too high in most instances for trouble-free operation, except in the case of extreme fines posing an environmental threat.

Optimistic papers have come from the Soviet Union for a long time, for example, about EF of chromite.[93] A single American document found EF to be a viable possibility for the beneficiation of strategically important, low-grade U.S. chrome ores (oxygen bubbles and the acidic environment developing in the anolyte compartment are favorable for chromite flotation).[94] Many more instances have been reported, such as EF of polymetallic tin ores[95] or pyrite.[96-99] An early paper worth reading is that about diamond EF,[100] where details were given and EF was reported to compare favorably to the conventional process.

Mining slurries, which are a problem of considerable magnitude resulting from the working of increasingly poorer ore deposits, are somewhat more likely candidate substrates for a technical EF operation. Many examples have been analyzed; an instance of large improvement over ordinary flotation was found in the case of manganese ore processing.[101,102] Anglesite fines EF was studied quite recently.[103]

Operating experience had been reported early in the treatment of mining wastewaters, for example, for molybdenum and uranium recovery[104] in the United States or nickel recovery[105] in the Soviet Union. In the latter case, the effluent contained 14 g Ni/m^3 and also Cu, Co, Fe, Zn, and Pb. Lime (0.2 kg/m^3) was used for precipitation, then poly(vinyl alcohol) was added (1 g/m^3), and EF performed with 0.2 kW h/m^3, 70–80 A/m^2, a flow rate of 2 m^3/(m$^2 \cdot$h), and an effluent with <0.1 g Ni/m^3.

3.13. Silver Recovery

A technological scheme for a 99% recovery of silver from the effluents of a photochemical plant was described by Polish workers.[106] Silver levels were reduced from 70–80 down to less than 0.5 g/m^3. Ferrous sulfate and sodium hydroxide were added together with a flocculating agent. A 1.3 m^3 vessel was reported to handle 9 m^3/h and yield a concentrate containing 10–20% silver.

3.14. Magnesium from Seawater

Electroflotation was described as an efficient method to collect magnesium hydroxide from seawater. It precipitates because of cathodic alkalinity, without addition of any reagents.[107] A current efficiency of 85–98% was reported for its collection in the froth layer. It can be further used for magnesium production, for on-site effluent treatment,[108] and even for oil-spill cleanups.

3.15. Radioactive and Toxic Metal Effluents

Shvedov and Yakushev,[109-111] after reviewing several interesting methods for the removal of radioactive material from the process or waste stream, have examined the possibilities for the removal of ^{90}Sr and ^{137}Cs from wastes by electrocoagulation and electroflotation. A hydroxide collector (soluble titanium electrodes were reported to be most effective) or ferrocyanide collector and the correct pH value (close to six) should be used. Further improvements were seen when a precipitate of nickel ferrocyanide was present (the nickel ions come from the anodic dissolution of stainless steel) and soap is added.

Arguments have been presented for the removal of traces of toxic metal from effluents by flotation with adsorbing colloids,[112] which will only be successful when the very small bubbles of EF are used.

3.16. Biotechnology

In biotechnology, microbiological processes are used to make or destroy products. In both cases, liquids and solids (the latter including the microorganisms) must subsequently be separated.[113] Dissolved-air flotation and EF are available for this task; the former can damage cells due to decompression, and the latter can do so because of active oxygen or chlorine set free at the anode. This may be desired, for example, in the treatment of excess activated sludge; if not, it may be circumvented by the use of membranes to separate the anode compartment from the active area.

Mycelial waste from an antibiotics production was thickened by EF, according to a recent Soviet paper[114]; not only flocculant [poly(acrylamide)] but also foam depressant must be added,[115] since these wastes are high in surfactants and proteins.

4. ADVANTAGES AND DISADVANTAGES

From examples of applications and some insight into the working of electroflotation, one can prepare a summary of advantages and disadvantages. From a technical viewpoint, advantages are:

- Transport efficiency is high.
- Current density is available as a parameter to control bubble number and size.
- Charged bubbles (of like sign) do not coalesce.
- No agitation implies high floc stability.
- High temperatures are acceptable.
- Electrochemistry can be added on.
- Fewer chemicals are used.
- Disinfectant action from anodic chlorine.

Disadvantages are:

- DC power is required.

- Substrate liquors should be sufficiently conductive.
- Tank heights are less than 1 m.
- Electrode corrosion occurs (but this can be avoided now).
- Electrode passivation occurs (but this can be avoided now).
- Poorer performance is obtained with high-specific-gravity materials.

Electroflotation is an expensive technology; it is preferred when

- Emulsions are hard to break.
- Effluents are active.
- Effluents are warm.
- Recycle incentives are high.
- Considerable values can be recovered.
- Demand fluctuates.
- Discharge standards are rigid.
- Cr(VI), phenol, CN⁻, or heavy metals are added problems.
- A very thorough extraction is desired.

With the realization that the coagulation step should occur separately, that coagulant should be recycled, and that only electrodes able to withstand polarity reversal (which are now available) should be used, EF is now a state-of-the-art technology. With the added, if not compulsory, incentive of protection of the environment and the conservation of resources, EF has become a necessity. Romanov summarized in 1985, when reviewing EF as an ecologically expedient technology, that "we are now at the threshold of broad application of electroflotation."[116]

It has been said so often that EF is more expensive than dissolved-air flotation (DAF); however, it should be noticed that fewer chemicals are needed in EF and that compressed air for DAF also has a price tag. The electric bill in EF is highly dependent on the voltage requirements; therefore, EF is at a disadvantage in most mining applications and in the case of drinking water preparation, and electrodes should not passivate. Existing electrode problems in EF probably have their counterpart in jet problems in DAF. Total energy requirements are said to be about the same for EF and DAF. The cleanup efficiency sometimes is 99.9 versus 90% (this ratio may actually be the point to closely examine, since if it proves to be in this neighborhood, this may mean a decision in favor of EF); examples where river discharge is possible are known for EF.

As in any other process, the advantages and disadvantages, the price and the effect, the possibility to meet effluent standards defined by regulations, and the convenience or the productivity and installation time must be weighed when EF is considered and compared with alternative technologies.

5. RESEARCH AND INFORMATION NEEDS

We present here a list of actions needed to enhance the application of EF technology; of course, it cannot be complete.

(a) Separate fact from fiction. The process must work in the laboratory, on the balance sheet, in the pilot tests, and in the full-scale installation—long-term. It must be compatible with other process steps.
(b) Combine theory and practice. Further studies on model systems, interface studies, and mechanical and physicochemical studies on the bubble and particle scale are needed (consider, for instance, the ingenious equipment proposed for studying the

induction time during attachment and the force during detachment of bubbles at particle surfaces that was recently described[117]). Explanations are needed for cases of success and failure.

(c) Find on-site energy sources. Would biogas or solar energy drive installations in some of the locations where they are needed?

(d) Reclaim values from the sludge: calories, fat, fibers, metals—the values are there, but perhaps the right add-on technology is not.

(e) Develop productive uses. Some have been mentioned; the technology is ready to develop beyond the image of a last resort in pollution. Biotechnologies and ceramic materials look promising.

(f) Find optimum integration conditions with membrane and electrochemical technologies—more often than not EF will provide part, and perhaps a key part, of the solution.

6. THE 21ST CENTURY

From the many examples and comments presented here, it can certainly be concluded that electroflotation will be among the key electrochemically based technologies in the next century. To what extent it will be employed will depend on the problems which must be solved in the management of resources and in the management of the wastes produced by human activities. It also will depend on mankind's resolve to deploy the full set of technologies required to keep this planet a pleasant place to inhabit.

It is a tribute to Professor Bockris to conclude that this chapter on electroflotation is but a close-up view of one of the many aspects of a very strong future role of electrochemistry.[118]

ACKNOWLEDGMENTS

Facilities at the Brown Boveri Baden (now ABB Switzerland) and Battelle Europe Geneva Research Laboratories were used in the literature study. Dr. Baer, Frankfurt, Messrs Roth and Förtsch, Swissair Zurich, and Professor Kuhn, Old Stevenage, supplied information, which is gratefully acknowledged.

REFERENCES

1. I. L. Markhasin, V. D. Nazarov, A. G. Tikhomirov, N. K. Tikhomirova, V. I. Shchur, and Yu. A. Lukanin, *Khim. Tekhnol. Vody* **8**(4), 54 (1986).
2. J. Nagendran and S. E. Hrudey, Electrolytic Flotation of In-Situ Oil Sand Production Recycle Water, Environmental Engineering Technical Report No. 80-1, Department of Civil Engineering, University of Alberta, 101 pp.
3. V. A. Panov and Zh. A. Kravchenko, *Elektrokhimiya* **10**, 1427 (1974).
4. L. A. Kul'skii, O. P. Smirnov, and E. M. Balandin, *Khim. Tekhnol. Vody*, **2**, 359 (1980).
5. Y. Fukui and S. Yuu, *Chem. Eng. Sci.* **35**, 1097 (1980).
6. Y. Fukui and S. Yuu, *AIChE J.* **31**, 201 (1985).
7. A. Watanabe, *Surf. Colloid Sci.* **13**, 1 (1984).
8. D. Chan, T. W. Healy, and L. R. White, *J. Chem. Soc., Faraday Trans. 1* **72**, 2844 (1976).
9. N. P. Brandon and G. H. Kelsall, *J. Appl. Electrochem.* **15**, 475 (1985).

10. B. M. Matov and B. R. Lazarenko, *Elektron. Obrab. Mater.* **1969**(3), 44.
11. V. A. Glembotskii, A. A. Mamakov, and V. N. Sorokina, *Elektron. Obrab. Mater.* **1973**(5), 66.
12. Yu. S. Gorodetskii, T. D. Kubritskaya, and V. K. Rotar', *Elektron. Obrab. Mater.* **1979**(4), 50.
13. N. P. Brandon, G. H. Kelsall, S. Levine, and A. L. Smith, *J. Appl. Electrochem.* **15**, 485 (1985).
14. A. Jowett, in: *Fine Particle Flotation* (P. Somasundaram, ed.), Vol. 1, pp. 720-754, AIME, New York (1980).
15. B. V. Deryagin, S. S. Dukhin, and N. N. Rulev, *Surf. Colloid Sci.* **13**, 71 (1984).
16. J. A. Kitchener and R. J. Gochin, *Water Res.* **15**, 585 (1981).
17. G. L. Collins and G. J. Jameson, *Chem. Eng. Sci.* **32**, 239 (1977).
18. D. Reay and G. A. Ratcliff, *Can. J. Chem. Eng.* **51**, 178 (1973).
19. T. Z. Sotskova, A. A. Vinnichenko, O. S. Chechik, and L. A. Kul'skii, *Dopov. Akad. Nauk Ukr RSR Ser. B* **1981**(12), 52.
20. A. T. Kuhn, *Chem. Ind. (London)* **1971**, 946.
21. B. Busse and S. Kotowski, *DECHEMA-Monogr.* **98**, 357-366 (1985).
22. Advances in Separations, A Battelle Technical Inputs to Planning Conference, Colombus, Ohio April 11-12, 1989.
23. S. Stucki (ed.), *Process Technologies for Water Treatment*, Plenum Press, New York (1988).
24. A. T. Kuhn, in: *Electrochemistry of Cleaner Environments* (J. O.'M. Bockris, ed.), pp. 98-130, Plenum Press, New York (1972).
25. A. T. Kuhn, in: *Electrochemistry, the Past Thirty and the Next Thirty Years* (H. Bloom and F. Gutmann, eds.) pp. 355-371, Plenum Press, New York (1977).
26. B. Cooke, *Process Engineering* **1974** (October), 93.
27. W. R. T. Cottrell, *Effluent Water Treat. J.* **16**, 563-567 (1976).
28. R. H. Marks and R. J. Thurston, *Environ. Pollut. Manage.* **7**(4), 94 (1977).
29. C. Camilleri, *Tech. Mod.* **67**(9), 29 (1975).
30. C. Camilleri, *Trib. CEBEDEAU* **30**, 302-308 (1977).
31. A. M. Romanov, *Elektron. Obrab. Mater.* **1985**(4), 29.
32. A. M. Romanov, *Elektron. Obrab. Mater.* **1985**(5), 59.
33. B. M. Matov, *Effluent Purification by Electroflotation* (in Russian), Kartya Moldovenyaske, Kishinev (1982), 172 pp.
34. B. C. Kim and H. W. Johnson, in: Advances in Separations, A Battelle Technical Input to Planning Conference, Columbus, Ohio, April 11-12, 1989.
35. G. Kreysa, in: *Process Technologies for Water Treatment* (S. Stucki, ed.), pp. 65-85, Plenum Press, New York (1988).
36. E. Baer Verfahrenstechnik GmbH, Frankfurt, Technical Leaflets.
37. C. Cristoforetti and G. Pratolungo, *Inquinamento* **28**(7-8), 88 (1986).
38. B. Cooke, *Process Eng.* **1974** (April), 64.
39. K. L. Wang, J. Y. Yang, and D. B. Dahm, *Chem. Ind. (London)* **1975**, 562.
40. Anon., *Abwassertechnik* **28**(6), 11 (1977).
41. E. H. Baer, *Muenchner Beitr. Abwasser Fisch. Flussbiol.* **28**, 265 (1977).
42. E. H. Baer, ACHEMA Frankfurt 85-6-17 (exhibition materials).
43. H. P. Roth, *Wasser Energiewirtsch. Baden* **66**, 185 (1974).
44. H. P. Roth and P. V. Ferguson, *Desalination* **23**, 49 (1977).
45. V. G. Revenko, A. I. Kushnir, and A. A. Mamakov, *Elektron. Obrab. Mater.* **1976**(6), 47.
46. A. A. Mamakov, A. I. Kushnir, R. V. Drondina, and L. F. Ignatova, *Elektron. Obrab. Mater.* **1977**(4), 67.
47. V. I. Zelentsov and K. A. Kiselev, *Elektron. Obrab. Mater.* **1986**(4), 50-54.
48. C. P. C. Poon and K. P. Soscia, *Ind. Water Eng.* **17**(2), 28 (1980).
49. E. I. Sorkin, E. I. Kucheryavykh, and V. V. Vykhovanets, *Elektron. Obrab. Mater.* **1983**(3), 68.
50. A. T. Kuhn, in: *Modern Aspects of Electrochemistry, No. 8* (J. O'M. Bockris and B. E. Conway, eds.), pp. 273-340, Plenum Press, New York (1972).
51. N. G. Kharlan, A. A. Mamakov, A. N. Yagubets, and V. V. Karyakin, *Elektron. Obrab. Mater.* **1969**(5), 47.
52. B. M. Matov, P. M. Stepanov, B. L. Prisyazhnyuk, and B. A. Grabois, *Kolloidn. Zh.* **32**, 91 (1970).
53. M. A. Bull, R. M. Sterritt, and J. N. Lester, *J. Chem. Technol. Biotechnol.* **31**, 579 (1981).

54. D. C. Lewin and C. F. Forster, *Effluent Water Treat. J.* **14**, 142 (1974).
55. D. Lewin, *Water Waste Treat.* **18**(11), 42 (1975).
56. L. Kh. Utyasheva, I. L. Markhasin, V. N. Izmailova, and V. D. Nazarov, *Khim. Tekhnol. Vody* **7**(3), 12 (1985).
57. E. C. Beck, A. P. Giannini, and E. R. Ramirez, *Food Technol.* **28**(2), 18 (1974).
58. E. R. Ramirez, D. L. Johnson, and O. A. Clemens, *Proc. Ind. Waste Conf.* **31**, 563 (1977).
59. J. Ladicky and M. Piatrik, *Vodni Hospod. B* **37**(7), 176 (1987).
60. N. A. Arkhipovich and V. A. Lagoda, *C. R. Assem. Gen. Comm. Int. Tech. Sucr.* **16**, 179 (1979).
61. B. M. Matov and A. I. Chernyi, *Elektron. Obrab. Mater.* **1973**(3), 61.
62. B. L. Prisyazhnyuk, *Gidroliz. Lesokhim. Prom-st.* **1978**(8), 6.
63. A. M. Romanov, T. D. Kubritskaya, V. N. Sorokina, I. S. Panashesku, S. I. Gnilyuk, and G. N. Olaru, *Elektron. Obrab. Mater.* **1984**(2), 85.
64. C. C. Ho and C. Y. Chan, *Water Res.* **20**, 1523 (1986).
65. A. L. Sergeev, *Elektron. Obrab. Mater.* **1982**(1), 87.
66. A. L. Sergeev, *Elektron. Obrab. Mater.* **1983**(6), 50.
67. O. P. Smirnov and E. M. Balandin, *Elektron. Obrab. Mater.* **1979**(1), 85.
68. O. P. Smirnov, E. M. Balandin, and V. S. Kikhno, *Elektron. Obrab. Mater.* **1982**(1), 46.
69. E. M. Balandin, V. S. Kikhno, and N. V. Vlasyuk, *Khim. Tekhnol. Vody* **4**(2), 129 (1982).
70. F. L. Cross and J. L. Tessitore, *Munic. Ind. Waste Annu. Madison Waste Conf.* 7th, University of Wisconsin Ext., Madison, Wisconsin, 1984, pp. 100–107.
71. A. D. Venderevskii, V. N. Syrovatko, and T. P. Pershina, *Bum. Prom-st.* **1976**(8), 25.
72. V. G. Selivanov, V. P. Svitel'skii, V. G. Ryumin, and N. V. Samborskii, *Bum. Prom-st.* **1976**(8), 24.
73. R. C. Clayton and J. G. Noble, *Effluent Water Treat. J.* **14** (Refocus, August), 21 (1974).
74. V. V. Otletov and Yu. A. Kovalenko, Soviet Patent 710,986 (1980); *Byull. Izobret.* **1980**(3), 94 (*Chem. Abstr.* **93**, 31352).
75. M. A. Kraizman, I. Z. Eifer, and I. G. Shimko, *Khim. Volokna* **15**(4), 6 (1973).
76. M. A. Kraizman, G. N. Filippov, I. Z. Eifer, I. G. Shimko, E. A. Shrader, A. G. Gol'man, and R. M. Zekel', *Khim. Volokna* **17**(5), 37 (1975).
77. S. M. Shifrin, I. G. Krasnoborod'ko, E. S. Svetashova, L. N. Gubanov, and R. S. Safin, *Tekst. Prom-st.* (*Moscow*) **1976**(8), 76.
78. K. C. Käuffer, *Melliand Textilber.* **65**, 855 (1984).
79. V. B. Chebanov, A. A. Mamakov, and L. B. Fainshtain, *Elektron. Obrab. Mater.* **1972**(5), 8.
80. D. P. Avetisyan, A. S. Tarkhanyan, and L. N. Safaryan, *Khim. Tekhnol. Vody* **6**, 345 (1984).
81. J. R. Backhurst and K. A. Matis, *J. Chem. Technol. Biotechnol.* **31**, 431 (1981).
82. F. Barrett, *Water Pollut. Control* **74**(1), 59 (1975).
83. Q. H. McKenna, H. H. Helber, L. M. Carrell, and R. F. Tobias, Electrochemical Flotation Concept for Removing Oil from Water, AD 760 056, NTIS, Springfield, Virginia (1973), 120 pp.
84. V. N. Anapol'skii, V. M. Rogov, and Yu. V. Davidyuk, *Khim. Tekhnol. Vody* **5**, 528 (1984).
85. Mitsugawa Kogyo K. K., *Jpn. Patent* 60 110,391 (1985).
86. P. Costaz, J. Miquel, and R. Reinbold, *Water Res.* **17**, 255 (1983).
87. P. Musquere and I. Richy, *Water Supply* **1**(2/3), SS8-1 (1983).
88. O. M. Frank, O. M. Murashov, and I. P. Mochalov, *Elektron. Obrab. Mater.* **1982**(1), 51.
89. G. S. Kucherenko and E. A. Golovash, *Elektron. Obrab. Mater.* **1979**(6), 49.
90. V. I. Mikhailov, V. S. Dyubchenko, and A. V. Teterin, *Gidroliz. Lesokhim. Prom-st.* **1982**(6), 10.
91. P. Hogan, A. T. Kuhn, and B. A. Wills, *Camborne Sch. Mines J.* **76**, 48 (1976).
92. P. Hogan, A. T. Kuhn, and J. F. Turner, *Trans. Inst. Min. Metall., Sect. C* **88**(June), C83 (1979).
93. A. M. Romanov, A. A. Mamakov, and L. P. Nasen'ka, *Elektron. Obrab. Mater.* **1974**(3), 50.
94. D. Wong, F. H. Cocks, and J. Giner, Development of Electrochemical Methods for the Enhancement of Flotation Extraction with Special Reference to Chromium Ores, PB 297 052, NTIS, Springfield, Virginia (1978), 31 pp.
95. Yu. A. Syasin, L. M. Tyrina, T. A. Kaidalova, and A. F. Morozov, *Elektron. Obrab. Mater.* **1982**(4), 44; **1982**(5), 42.
96. V. A. Glembotskii, A. A. Mamakov, and V. N. Sorokina, *Elektron. Obrab. Mater.* **1973**(6), 46.
97. A. M. Romanov, V. I. Zelentsov, V. N. Sorokina, and L. P. Nasen'ka, *Elektron. Obrab. Mater.* **1974**(6), 62.

98. R. Sh. Shafeev, V. A. Chanturiya, M. A. Sal'nikov, R. I. Sturua, and V. P. Bezrodnykh, *Elektron. Obrab. Mater.* **1971**(1), 45.

99. V. A. Bocharov and V. P. Sapozhnikov, *Tsvetn. Met.* (*Moscow*) **1977**(7), 80.

100. B. R. Lazarenko, V. A. Glembotskii, A. A. Mamakov, and M. I. Avvakumov, *Elektron. Obrab. Mater.* **1969**(6), 50.

101. A. M. Romanov, E. S. Nenno, V. E. Nenno, and A. A. Mamakov, *Elektron. Obrab. Mater.* **1973**(1), 54.

102. V. A. Glembotskii, I. I. Grazhdantsev, A. A. Mamakov, V. E. Nenno, E. S. Nenno, N. D. Postoenko, and A. M. Romanov, *Elektron. Obrab. Mater.* **1976**(3), 49.

103. G. V. Rao, F. U. Schneider, and H. Hoberg, *Erzmetall.* **40**, 183 (1987).

104. R. D. Gott and J. M. Lafferty, *Ind. Water Eng.* **15**(2), 6 (1978).

105. R. K. Alekseeva and A. P. Seligerskaya, *Tsvetn. Met.* (*Moscow*) **1977**(5), 53.

106. K. Zmudzinski and Z. Konaszynska, *Przem. Chem.* **62**, 524 (1983).

107. V. I. Golovanov, N. Ya. Kovarskii, and I. S. Pryazhevskaya, *Zh. Prikl. Khim.* **49**, 788 (1976).

108. K. A. Matis, *Water Pollut. Control* **79**(1), 136 (1980).

109. V. P. Shvedov and M. F. Yakushev, *Radiokhimiya* **12**, 871 (1970).

110. V. P. Shvedov and M. F. Yakushev, *Radiokhimiya* **12**, 876 (1970).

111. V. P. Shvedov and M. F. Yakushev, *Radiokhimiya* **15**, 428 (1973).

112. C. Manohar, V. K. Kelkar, and J. V. Yakhmi, *J. Colloid Interface Sci.* **89**, 54 (1982).

113. J. Gnieser, *Wasser Luft Betr.* **21**, 343 (1977).

114. Z. L. Faingol'd, V. F. Karpukhin, E. V. Zav'yalova, V. B. Nikolaev, A. R. Yakubova, and A. S. Memorskaya, *Antibiot. Med. Biotekhnol.* **32**, 346 (1987).

115. Z. L. Faingol'd, E. V. Zav'yalova, V. F. Karpukhin, and T. A. Ivankova, *Antibiot. Med. Biotekhnol.* **32**, 120 (1987).

116. A. M. Romanov, *Elektron. Obrab. Mater.* **1985**(6), 43.

117. A. M. Gol'man and A. A. Lavrinenko, *Elektron. Obrab. Mater.* **1987**(1), 74.

118. J. O'M. Bockris, in: *Comprehensive Treatise on Electrochemistry*, Vol. 3 (J. O'M. Bockris, B. E. Conway, E. Yeager, and R. E. White, eds.), pp. 1–38, Plenum Press, New York (1981).

119. E. P. Okun and B. M. Matov, *Dev. Food Sci.* **9** (Food Ind. Environ.), 69 (1984).

120. L. M. Safley, R. L. Vetter, and L. D. Smith, *Poult. Sci.* **66**, 941 (1987).

3

Importance of Electrical Double Layer in Technological Applications

Zlata Kovac and Carlos J. Sambucetti

1. INTRODUCTION

Electrochemical technology is applied to a large number of products in the modern electronics industry.[1-3] Various manufacturing processes leading to important commercial implementations in the computer industry are based on the study, understanding, and direct utilization of double-layer phenomena. The objective of this chapter is to describe in some detail how double-layer theory is being utilized for the development of two rather different products.

An electrical double layer is formed when two phases come into contact and at least one of them contains mobile charges. It is associated with a distribution of charges in the interfacial region, which differs from that in the bulk of either phase. The metal/electrolyte solution interface is probably the most important electrochemical system. The reactions taking place at this interface such as electron, proton, and metal-ion transfer contribute to important branches of electrochemistry such as electrodeposition, electrochemical etching and machining, and corrosion and passivation of metals and their alloys. The nonmetal/solution interface is important in stabilization of colloidal systems, electrophoretic processes, catalysis for activation and metallization of nonconductive substrates, and many other processes.[4]

Figure 1 shows the well-known model of the electrical double layer at the metal/aqueous electrolyte interface according to Bockris, Devanathan, and Müller.[5] As can be seen from Fig. 1,[5,6] specifically adsorbed anions and water molecules are in direct contact with a metal surface. They form the so-called inner compact or Helmholtz layer.[7] Beside anions, large cations, such as Cs^+[8] and quaternary ammonium compounds,[9] which are not hydrated, are specifically adsorbed and are part of this layer. Hydrated cations are separated from the metal by water molecules. They form the outer compact layer. The potential drop across this layer is linear. A model for the electrical double layer is that of a parallel-plate capacitor. Assuming straight lines, the metal surface forms one plane of a capacitor, an imaginary line going through the centers of specifically adsorbed anions defines the second plane, that is, the inner Helmholtz plane, and the third plane goes through the centers of hydrated cations, that is, the outer Helmholtz or Gouy plane. Beyond the compact layer is a diffuse layer where the distribution of the ions is affected by their electrostatic forces as well as their thermal motion. Potential decay is exponential across this layer, and its thickness is given by the

Zlata Kovac • IBM Research, Almaden Research Center, San Jose, California 95120-6099. *Carlos J. Sambucetti* • IBM Research, T. J. Watson Research Center, Yorktown Heights, New York 10598.

Electrochemistry in Transition, edited by Oliver J. Murphy *et al.* Plenum Press, New York, 1992.

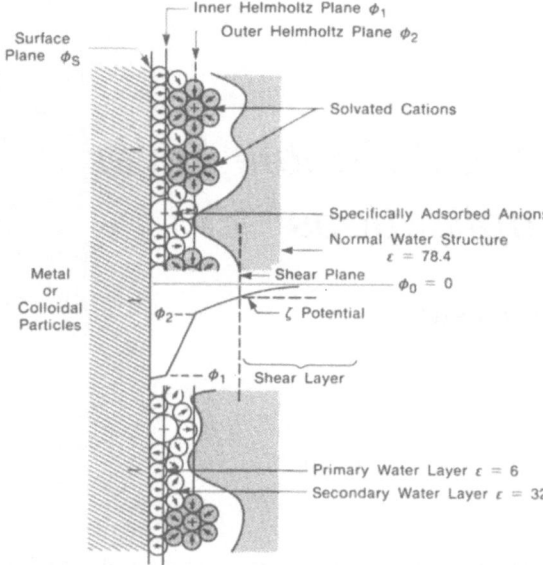

FIGURE 1. A model of the double layer at metal or colloidal particle/solution interfaces from Ref. 5. Schematic representation of potential versus distance change superimposed in the center from Ref. 6.

Debye length of the solution. Our knowledge of double-layer structure is mainly based on experimental data for easily accessible Hg/solution interface (electrocapillary and capacity) measurements. Capacity measurements can also be carried out on the solid metals. They reveal that the Helmholtz capacity strongly depends on both the nature of the metal and the solvent.[10,11]

New *in situ* spectroscopic techniques now available, such as surface extended X-ray adsorption fine structure (SEXAFS) and grazing incident X-ray scattering (GIXS), provide a more detailed and unambiguous picture of double-layer structure on the solid metal/electrolyte interface under potential control. These measurements, for example, showed the geometric arrangement of atoms during underpotential deposition and the orientation of polyatomic anions on a metal, for example, SO_4^{2-} or CH_3COO^-, which are adsorbed through oxygen atoms,[12] (Fig. 2).

Scanning tunneling microscopy (STM) in solution[13] is being used to study the metal side of the double layer on an atomic scale under potential control. STM will certainly contribute to our understanding of the metal/solution interface under charge transfer conditions, making possible the visualization of electrodeposition and corrosion processes.

New theories and models such as "jellium/hard-sphere electrolyte" are emerging.[14,15] They consider the contribution of the metal electrons, which penetrate slightly and "spill over" into the solution side of the double layer. The combination of experimental data obtained using the new techniques with more refined theories will give further insight into double-layer structure and mechanism.

Colloidal particles in solution have an ionic charge which is due to dissociation of surface groups and/or preferential adsorption, and a double layer, as shown in Fig. 1, is formed around each particle. When the particles move through a solution, they carry their double layer with them, which extends up to the so-called plane of shear. The potential drop through the diffuse layer up to the hydrodynamic plane of shear is called the electrokinetic or zeta potential, ζ, which is crucial for the stability of colloids. According to the theories of Derjaguin and Landau, and Verwey and Overbeek (now frequently termed the DLVO theory), colloidal stability is achieved when electrostatic repulsion, determined by the value of the zeta potential,

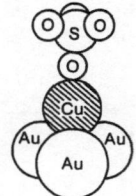

FIGURE 2. Surface structure in plating solution from Ref. 12, showing geometrical arrangement and orientation of cations and polyatomic anions during underpotential deposition.

exceeds van der Waals attraction. If the colloidal particles possess a magnetic moment, then an additional term due to the magnetic attraction between two spheres must be added to the DLVO equation. For a given particle size, it was possible to determine[16] the values of ζ-potential which gave stable magnetic inks.

The present chapter reviews the utilization of double-layer phenomena in two particular cases:

(1) Catalytic activation/metallization of nonconductive substrates for the fabrication of printed circuit boards in the computer industry.
(2) Formation of stable magnetic colloidal fluid for displays and ink jet printing.

2. CATALYTIC ACTIVATION/METALLIZATION OF PRINTED CIRCUIT BOARDS

The printed circuit boards used in computers are highly complex systems which form a communication network for the transmission of electrical signals between the outside world and silicon memory and logic chips attached to the ceramic chip carriers on the board.[2,3] The manufacture of these multilayer panels involves various processes from many disciplines of material sciences: electrochemistry, surface science, laser chemistry and physics, photochemistry, polymer chemistry and physics, and others. The main processing steps are (i) high-pressure and high-temperature lamination of glass fiber sheets with epoxy resin to form the mechanical structure of a board; (ii) photolithography to define the circuit pattern; (iii) catalytic activation and electroless Cu plating of the electrical circuit; and (iv) laser and mechanical drilling to fabricate vias and plated-through holes (PTH) which vertically connect different layers of signal and power planes. In such a way, large boards are formed containing kilometers of electrolessly plated Cu wires with thousands of vias and PTH.

Figure 3 shows an isometric view of the cross section of a multilayer printed circuit panel. Details of the inner wired planes and the plated-through holes connecting the various panel levels can be seen. In these double-sided and multilayer structures, all fine lines and through-hole metallization is accomplished by catalysis and electroless plating of the dielectric laminate, usually epoxy–glass composite. Electroless plating is a heterogeneous reaction where two soluble reactants interact on the surface, at the surface–solution interface, and plating occurs selectively only at the surface sites "seeded" with a catalyst, usually Pd metal, although other metals such as Pt, Au, and Cu (autocatalysis) can also be used[17]:

$$Cu^{2+} + 2H_2CO + 4OH^- \xrightarrow{Pd} Cu^0 + H_2 + 2HCOO^- + 2H_2O \tag{1}$$

In these solutions, the copper ions are complexed with an organic ligand in order to increase their stability in the highly alkaline solution. The reducing agent is usually formaledehyde, although others such as hypophosphite and dimethylaminoborane (DMAB) have also been used. Therefore, seeding of the dielectric laminate is essential to provide catalytic sites on

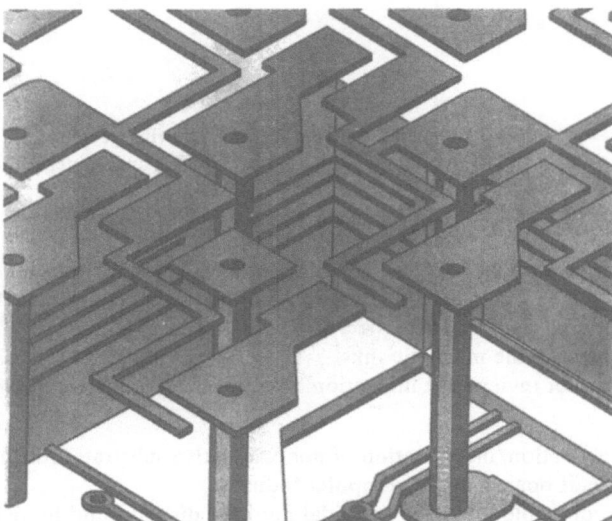

FIGURE 3. Isometric view of a cross section of a multilayer printed circuit board, showing interconnections between layers by plated-through holes (PTH).

the substrate for the autodeposition and selective line circuitization. Thus, the process of full electroless plating or electrolytic plating is as follows:

$$\text{Seeding} + \text{Electroless plating} \rightarrow \text{Conductive substrate}$$

To initiate plating, the catalyst has to be sufficiently active to bring about the anodic oxidation of the reducing agent, for example, HCHO in the case of Cu plating, or NaH_2PO_2 or DMAB in the case of other metals. Palladium-based catalysts are the catalysts of choice although other noble metals, for example, Pt, are potentially suitable. The technology of Pd-based catalytic seeding has undergone a considerable evolution, starting from the two-step stannous chloride and palladium chloride sequential immersion with a rinsing step in between[18-22] and progressing to a single-step Pd/Sn colloidal system.[23-29] This latter system is widely used today in multilayer circuit board manufacturing, usually accompanied by a solution "acceleration" process[23,24] to remove excess tin.

To obtain high quality and reliability in the performance of multilayer boards, it is essential that the copper electroless layer deposited on the glass–epoxy laminate surface has excellent adhesion to the substrate and forms a continuous, impervious, ductile metallurgical layer. We developed an improved process by (a) the selection and electrokinetic characterization of the catalyst system for seeding and (b) the understanding of its double-layer interactions with the substrate.

2.1. The Colloid Catalyst: Double-Layer Phenomena

For our studies, we selected the well-known Pd–Sn colloid. The preparation of this colloid is described in the literature.[23-28] In general, it involves mixing $PdCl_2$ with a large excess of $SnCl_2$ in concentrated HCl solution (see Fig. 5). Considerable controversy existed initially as to whether the Pd–Sn catalysts are true solutions or colloids. However, the colloidal nature was established by Mössbauer spectroscopy,[30-32] by ultracentrifugation studies of

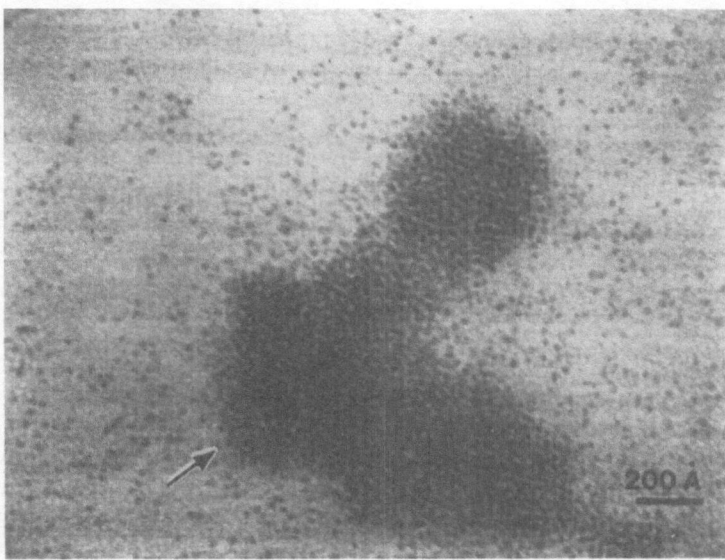

FIGURE 4. STEM high-magnification image of a Pd–Sn colloid dispersion, showing mainly monodisperse, 20–40 Å particles.

the catalyst solutions,[33] and by Rutherford ion backscattering.[34] We confirmed and characterized the colloid structure of this system in cooperation with the Central Research Colloidal Laboratory at Du Pont.[35] Samples were characterized by transmission electron microscopy, using vacuum generators HB5 STEM.

The colloidal characterization procedure was as follows. A drop of colloidal suspension was placed on a microscope grid, and most of the liquid was drawn away by air drying. Several areas on each specimen grid were photographed at high magnification (0.5×10^6). The elemental compositions of selected particles in each area were determined by X-ray emission microanalysis. With this combination of imaging and analysis, Pd–Sn particles were distinguished from Sn particles, and the predominant structures of the Pd–Sn particles were identified.[35]

Figure 4 is an image of a sample showing the predominant microstructure: uniform, monodisperse 20–40 Å-diameter Pd–Sn particles. The core consists of Pd particles about 1–2-μm diameters with a small amount of Sn.

Electrokinetic measurements on suspensions of this colloidal system indicated a value of the zeta potential of -170 ± 20 mV. This is also consistent with the fact that the colloid is stabilized by adsorption of Sn^{2+} ions and a surrounding layer, consisting predominantly of Cl^- ions (Cohen model).[21,22] Therefore, the key of the stability of this colloid is the surface-coulombic repulsion between interacting double layers.

2.2. Electrokinetic Interaction between Seed Colloid and Substrates

2.2.1. Experimentation and Testing

Systematic experiments were carried out to determine the electrochemical affinity between the negatively charged Pd–Sn colloid and a number of specially prepared substrates: glass, glass–epoxy composites, and Mylar. A dilute solution of a standard negative silica colloid

(0.01% Ludox Du Pont silica colloid) was used in the comparative tests. Normally unreactive and lyophobic glass or glass–epoxy surfaces were immersed in the processing solution of the Pd–Sn colloid. The resulting Pd film coverage on glass surfaces was practically zero. On epoxy, the amount of Pd found was about 0.5 $\mu g/cm^2$. This surface concentration of the catalyst is ineffective in achieving good plating initiation in a formaldehyde electroless bath.

If the glass–epoxy substrates are first immersed in the negative 0.01% Ludox solution and rinsed, the surface becomes very lyophilic and wettable due to the adsorption of silanol groups:

$$\text{Glass} + \text{Silica colloid} \rightarrow \ \ \diagdown_{\diagup}\!\!Si\!\diagup^{\displaystyle O^-}_{\diagdown O^-} \qquad \text{(silanol adsorbed)} \qquad (2)$$

All these surface samples gave zero Pd coverage upon immersion in the Pd–Sn seed colloid. However, if the initially strongly negative charged surface, covered by silanol groups, is first immersed in a positively charged, cationic surfactant, which is specifically adsorbed, and is then treated with the Pd–Sn colloid, an increased Pd coverage results. If the cationic surfactant is multifunctional in nature (see below), the amount of seed coverage is of the order of 5–8 μg Pd/cm^2. Evidently, the fundamental phenomenon here is the formation of a bilayer on the substrate surface. Initially, an adsorbed monolayer forms due to electrostatic attraction between the negative silanol groups (see Eq. 2) and the positive cationic surfactant groups, $(NR_4)^+$. Next, these groups attract the negative Pd–Sn colloidal particles, significantly increasing the seed coverage of the glass–epoxy surface:

$$\left(=Si\diagup^{\displaystyle O^-}_{\diagdown O^-}\right)[n(NR_4)]^{n+}(Pd\text{–}Sn)^- \text{ (adsorbed)} \qquad (3)$$

These findings are very significant because glass–epoxy printed circuit boards exhibit many areas, especially on the surface of drilled-through holes, where glass and other negative type surfaces must be treated with the Pd–Sn colloid before electroless plating. The seed coverage in these untreated areas is normally inefficient due to the electrokinetic repulsion between the negative charges on the surface and those of the seed colloid.

2.2.2. Seeding by Electrostatic Principles

The metallization of nonconductors by catalysis and electroless plating depends on the energy levels. For example, to plate on the surface of a drilled hole on epoxy, a certain energy level is required. This can be considered to be given by the sum of the catalytic energy stored at the wall surface (Pd catalyst surface coverage) plus the activity or driving energy of the electroless bath. If the catalyst concentration at the PTH surface is low (smooth glass–epoxy substrate with poor seed coverage), then the plating bath has to operate at high activity, and, even so, it is difficult to avoid voids and PTH unplated areas. Furthermore, a high-energy electroless bath will produce abnormal growth on the adjacent Cu lines, where no catalyst is required. Thus, open areas on plated-through holes, foldover areas, and striation are often the result of poor seeding at glass–epoxy exposed areas.

The findings discussed in the previous section allowed us to develop an "electrostatic seed process" based on electrokinetic properties of the interacting, modified surfaces and the seed colloid. It consisted in pretreatment of the surface with an aqueous solution of a multifunctional cationic copolymer.[4] The positive-charge functionality was provided by many tetraalkylammonium groups attached to an acrylic polymer chain. This created multiple

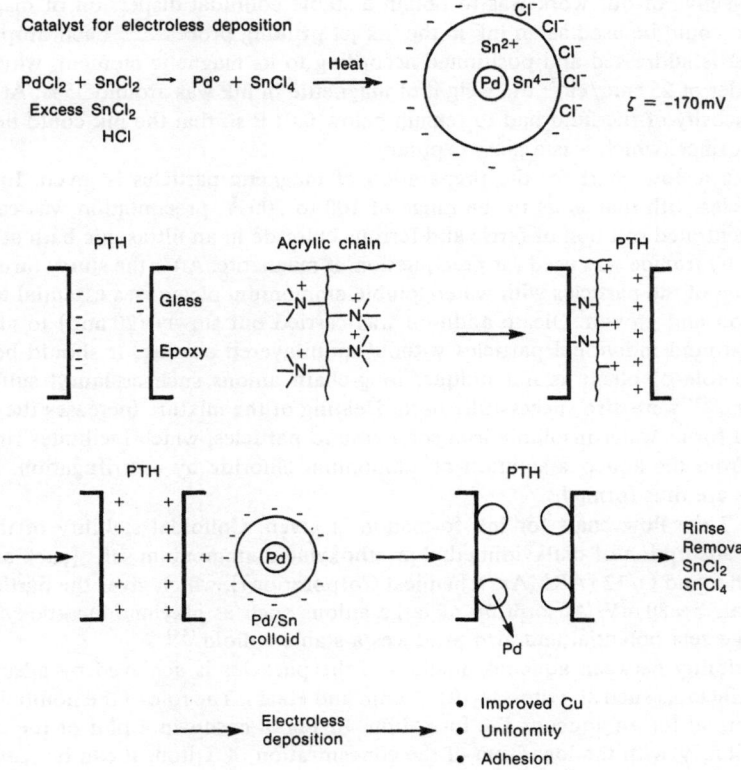

FIGURE 5. Flow chart for the process of metallization of glass–epoxy PTH surfaces.

anchoring sites, yielding excellent seed uniformity and a very active catalytic surface for the electroless plating solution. The sequence process steps is illustrated in Fig. 5.

3. FORMATION OF STABLE MAGNETIC COLLOIDAL DISPERSIONS

Ferromagnetic particles of colloidal dimensions are required for formation of magnetic fluids. They are dispersed in a carrier liquid to form stable colloidal dispersions. Such a dispersion can be made to move by a magnetic field gradient; this property is utilized in producing displays,[36] ink jet printers,[37,38] or rotary seals.[39] Single-domain magnetic particles can be made by grinding or by precipitation. Papell[40] and Kaiser and Miskolczy[41] prepared magnetic fluids by ball mill grinding of magnetite in kerosene or water, using surfactants as the dispersing agents. Since only a small amount of the energy in grinding is used to break up aggregates into smaller particles, the rest going into heat transfer, it is a slow and inefficient process, which required 1000 hours. Precipitation is fast and suitable for large-batch processing. Khalafalla and Reimers[42] precipitated magnetite from ferrous and ferric chloride with ammonium hydroxide; the magnetite thus formed was coated with ammonium oleate. Addition of kerosene to this mixture with heating and stirring resulted in phase separation. Fe_3O_4 coated with hydrophobic oleate went into the kerosene phase, and water-soluble NH_4Cl remained in the water phase.

The objective of our work was to obtain a stable colloidal dispersion of magnetite in water, which could be used as an ink in the ink jet printing process.[43] Each drop of liquid in the ink jet is addressed and positioned according to its magnetic moment, which had to be of the order of 25 emu/cm^3; the weight of magnetite in ink was around 35%. At the same time, the viscosity of the fluid had to remain below 0.01 P so that the ink could be injected through an orifice, which was a glass capillary.

In Fig. 6 a flow chart for the preparation of magnetic particles is given. In order to obtain particles with diameters in the range of 100 to 200 Å, precipitation was carried out from a concentrated solution of ferric and ferrous chloride in an ultrasonic bath at 10–25°C. Ammonium hydroxide was used for precipitation of magnetite. After the slurry turned black, *in situ* coating of the particles with water-soluble ammonium oleate was essential to prevent agglomeration and growth. Oleate addition was carried out slowly (20 min) to allow slow adsorption around individual particles without multilayered coating. It should be pointed out that the role of oleate is not unique; long-chain anions such as lauryl sulfate[44] or dodecylamine[45] were also successfully used. Heating of the mixture increases the magnetic moment and forms water-insoluble iron soap around particles, which facilitates rinsing and separation from the aqueous solution of ammonium chloride by centrifugation. Magnetic ink particles are thus formed.

In Fig. 7 the flow chart for ink formation is given. Colloidal stability of this ink is obtained by adsorption of dialkyldimethyl or ethoxylated ammonium salts [such as Arquad 2H-75 or Ethoquad O/12 (ARMAK Chemical Corporation)], which gives the particles large zeta potential, $>+80$ mV. Adsorption of large anions such as alkylaminocarboxylate gives particles large zeta potential and also produces a stable colloid.[44]

Compatibility between aqueous media and the particles is achieved by adsorption of nonionic surfactants such as Triton N-101 (Rohm and Haas). The role of the nonionic wetting agent is essential for an aqueous ink formation. In Fig. 8 is shown a plot of the change in surface tension, γ, with the logarithm of the concentration of Triton. It can be seen that the critical micelle concentration, cmc, is the same for water and Triton only or with addition

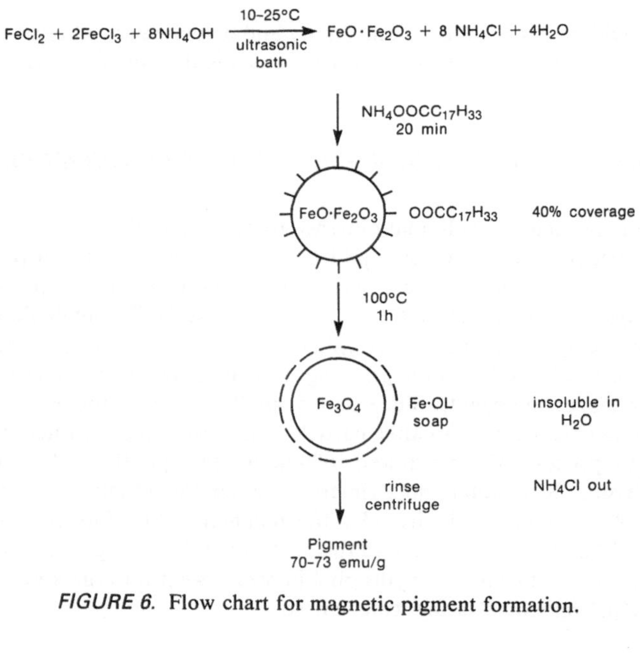

FIGURE 6. Flow chart for magnetic pigment formation.

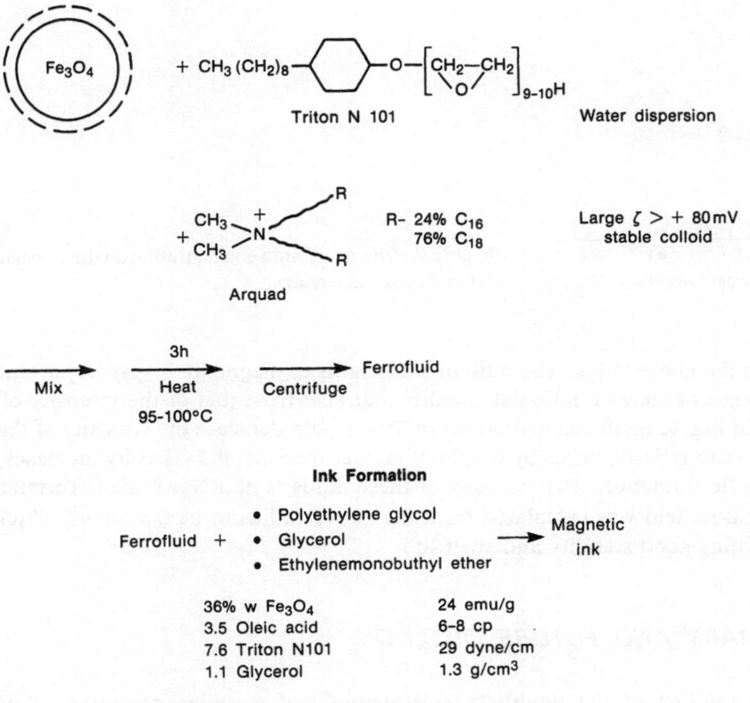

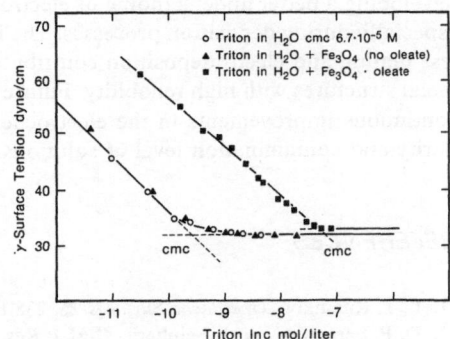

Fe_3O_4 + $CH_3(CH_2)_8$—⬡—O—$\left[CH_2-CH_2 \atop O\right]_{9-10}H$

Triton N 101   Water dispersion

CH_3—$\overset{+}{N}$—R   R– 24% C_{16} Large $\zeta > +80\,mV$
CH_3 R   76% C_{18} stable colloid

Arquad

Mix → Heat $\overset{3h}{95-100°C}$ → Centrifuge → Ferrofluid

Ink Formation

Ferrofluid +
- Polyethylene glycol
- Glycerol
- Ethylenemonobuthyl ether
→ Magnetic ink

36% w Fe_3O_4	24 emu/g
3.5 Oleic acid	6–8 cp
7.6 Triton N101	29 dyne/cm
1.1 Glycerol	1.3 g/cm³

FIGURE 7. Flow chart for ferrofluid formation.

of uncoated freshly precipitated Fe_3O_4, indicating no adsorption. When oleate-coated Fe_3O_4 was added, the cmc shifted by a few orders of magnitude due to adsorption. On the basis of the principle of like to like in colloidal systems, Triton is probably absorbed with the alkyl end of the molecule oriented toward the oleate-coated particles, and the (CH_2-CH_2-O) groups oriented toward the solution, forming a hydration shell around the particles. Sedimentation velocity in the centrifuge (16,000 rpm) was found to be equal to $2 \times 10^{-11}\,s^{-1}$, from which the effective radius of the sedimenting particles was found to be 135 Å. Transmission electron micrographs gave the average radius of Fe_3O_4 as 55 ± 18 Å, which leaves some 60–80 Å for the length of the surfactants and the radiation of the hydration shell. The fully stretched length of Triton N-101 is 27.8 Å, and that of Arquad 2H-75 is 28.6 Å, which

FIGURE 8. Change in surface tension with natural logarithm of Triton concentration without and with magnetite particles present.

o Triton in H_2O cmc $6.7 \cdot 10^{-5}$ M
▲ Triton in H_2O + Fe_3O_4 (no oleate)
■ Triton in H_2O + Fe_3O_4 · oleate

γ–Surface Tension dyne/cm

Triton lnc mol/liter

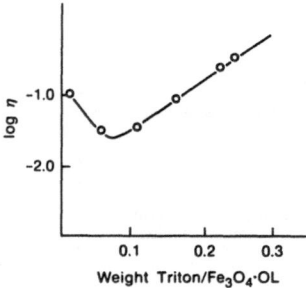

FIGURE 9. Effect of Triton concentration on the viscosity of oleate-coated Fe_3O_4 dispersions.

agrees with the above value. The ratio of surfactants to magnetite is very important, not only from the point of view of colloidal stability, but also from that of the viscosity of the fluid. As shown in Fig. 9, small concentrations of Triton help decrease the viscosity of the colloidal system. At Triton/Fe_3O_4 ratios by weight of greater than 0.1, the viscosity increases, probably due to micelle formation. The viscosity of these fluids is non-Newtonian. Terminal velocity in a gravitation field was calculated from the above radius to be 0.5 cm/yr, which gave the magnetic fluids good stability and shelf life.

4. SUMMARY AND FUTURE OUTLOOK

Understanding of the double-layer structure and specific adsorption of anions and quaternary ammonium cations has been successfully applied in the development of new products for the electronics industry. Novel techniques are now being rapidly developed bringing new experimental data and theories. *In situ* structural determination of the metal/solution interface under potential control is now possible. Scanning tunneling and atomic force microscopy are revealing the structure of metals and insulators in solution, air, and ultrahigh vacuum on the atomic scale. The configuration of adsorbed molecules on surfaces is visible now.

Electroless plating and electrodeposition of metals and alloys with low capital equipment cost, large batch size, and tailored properties of the deposited films will remain dominant in electronic industrial applications. High-rate deposition and dissolution processes will need further development. More work is required on the modeling and the theory of current distribution during electroplating, electropolishing, and machining. Better process controls and techniques to characterize finished products are needed.

Frumkin and Bockris stressed the usefulness of high-purity electrochemical systems in developing a better understanding of electrode/solution interfaces and the role of impurities, especially during deposition processes. The industrial application of this approach to electroless plating and electrodeposition contributed to the ability to manufacture smaller, thinner metal structures with high reliability. Future miniaturization of electronic devices will require continuous improvements in the electrochemical processes as well as in the control of the purity and contamination level of solutions.

REFERENCES

1. L. T. Rmankiw, *Oberfläche-Surfaces* **25**, 238 (1984).
2. D. P. Seraphim and I. Feinberg, *IBM J. Res. Dev.* **25**, 617 (1981).

3. D. P. Seraphim, R. Lasky, and C. Yu-Li, *Principles of Electronic Packaging*, McGraw-Hill, New York, (1989).
4. C. J. Sambucetti, J. Varsic, and W. Laboy, in: *Proceedings of the Symposium on Electrochemical Technology in Electronics*, Honolulu, Hawaii, Proceedings Vol. 88-23, pp. 59-62, The Electrochemical Society, Pennington, New Jersey (1987).
5. J. O.'M Bockris, M. A. V. Devanathan, and K. Müller, *Proc. Roy. Soc. (London)*, Ser. A **274**, 55 (1963).
6. W. N. Hansen, *J. Electroanal. Chem.* **150**, 133 (1983).
7. D. C. Grahame, *Chemistry* **41**, 441 (1947).
8. H. Wroblowa, Z. Kovac, and J. O.'M. Bockris, *Trans. Faraday Soc.* **61**, 1523 (1965).
9. M. A. V. Devanathan and J. M. Fernando, *Trans. Faraday Soc.* **58**, 368 (1962).
10. D. C. Grahame, *J. Am. Chem. Soc.* **76**, 4819 (1954).
11. G. Valette, *J. Electroanal. Chem.* **122**, 285 (1981).
12. J. G. Gordon II, O. R. Melroy, and L. Blum, in: *Diffusion at Interfaces: Microscopic Concepts* (M. Grunze, H. J. Kreuzer, and J. J. Weimer, eds.), *Springer Ser. Surf. Sci.* **12**, 172 (1988).
13. R. G. Sonnenfeld, J. Schneir, and P. K. Hansma, in: *Modern Aspects of Electrochemistry*, No. 21 (R. E. White, J. O.'M. Bockris, and B. E. Conway, eds.), pp. 1-28, Plenum Press, New York (1990).
14. J. P. Badiali, M. L. Rosenberg, F. Vericat, and L. Blum, *J. Electroanal. Chem.* **158**, 253 (1983).
15. W. Schmickler and D. Henderson, *J. Chem. Phys.* **80**, 3381 (1984); **85**, 1650 (1986).
16. Z. Kovac and C. J. Sambucetti, in: *Colloids and Surfaces in Reprographic Technology* (M. Hair and M. D. Croucher, eds.), ACS Symposium Series 200, p. 543, American Chemical Society, Washington, D.C. (1982).
17. M. Paunovic and I. Ohno (eds.), *Proceedings of the Symposium on Electroless Deposition of Metals*, Proceedings Vol. 88-12, The Electrochemical Society, Pennington, New Jersey (1988).
18. W. Goldie, *Metallic Coating of Plastic*, Vol. 1, Electrochemical Publications Ltd., Middlesex, England (1968).
19. R. Sarc, *J. Electrochem. Soc.* **117**, 864 (1970).
20. M. Schlesinger and J. Kisel, in: *Proceedings of the Symposium on Electroless Deposition of Metals* (M. Paunovic and I. Ohno, eds.), p. 100, Proceedings Vol. 88-12, The Electrochemical Society, Pennington, New Jersey (1988).
21. R. L. Cohen, J. F. D'Amico, and K. W. West, *J. Electrochem. Soc.* **118**, 2042 (1971).
22. R. L. Cohen and K. W. West, *J. Electrochem. Soc.* **119**, 433 (1972).
23. C. R. Shipley, Jr., U.S. Patent 3,011,920 (1961).
24. J. Horkans, K. Jim, C. McGrath, and L. T. Romankiw, *J. Electrochem. Soc.* **134**, 300 (1987).
25. R. J. Zeblisky, U.S. Patent 3,682,671 (1972).
26. J. Horkans and C. J. Sambucetti, *IBM J. Res. Dev.* **28**, 690 (1984).
27. E. J. M. O'Sullivan, J. Horkans, J. R. White, and J. M. Roldan, *IBM J. Res. Dev.* **32**, 591 (1988).
28. R. I. Jensen, J. P. Cummings, and H. Vora, *Proceedings of 34th Electronic Components Conference*, p. 73 Institute of Electrical and Electronics Engineers, New York (1984).
29. F. A. Lowenheim (ed.), *Modern Electroplating*, Chapter 19, McGraw-Hill, New York, (1978).
30. R. L. Cohen and K. W. West, *J. Electrochem. Soc.* **120**, 502 (1973).
31. R. L. Cohen and R. L. Meeks, *J. Colloid Interface Sci.* **55**, 156 (1976).
32. R. L. Cohenand K. W. West, *Chem. Phys. Lett.* **16**, 128 (1972).
33. E. Matijevic, A. M. Poskanzer, and P. Zuman, *Plating* **62**, 958 (1975).
34. R. L. Meek, *J. Electrochem. Soc.* **122**, 1177 (1975).
35. U. Chaudhry, Du Pont Central Research Colloidal Laboratory, private communication (1985).
36. L. T. Romankiw, M. Slusarczuk, and D. Thompson, U.S. Patent 3,972,595 (1976).
37. G. J. Fan and R. A. Toupin, U.S. Patent 3,805,272 (1974); 3,916,419 (1975).
38. G. J. Fan, W. Hurley, D. C. W. Lo, and J. W. Mitchell, U.S. Patent 4,068,240 (1978).
39. R. E. Rosensweig, U.S. Patent 3,612,584 (1971); 3,620,630 (1971); 3,734,578 (1973).
40. S. S. Pappel, U.S. Patent 3,215,572 (1965).
41. R. Kaiser and G. Miskolczy, *J. Appl. Phys.* **41**, 1064 (1970).
42. S. E. Khalafalla and G. W. Reimers, U.S. Patent 3,843,540 (1974).
43. Z. Kovac and B. A. Gardineer, U.S. Patent 3,990,981 (1976).
44. Z. Kovac and C. J. Sambucetti, U.S. Patent 4,107,063 (1978).
45. S. E. Khalafalla and G. W. Reimers, *IEEE Trans. Magn.* **MAG-16**, 178 (1980).

4

Electrochromism

M. Ahsan Habib

1. INTRODUCTION

Electrochromism is a phenomenon of reversible change in optical properties produced electrochemically. The color change of the electrochromic materials can be best demonstrated by making thin films of these materials on transparent conductive surfaces. An electrochromic electrode would be capable of variable light transmission by electrochemical oxidation or reduction. When the polarity of the applied potential is reversed, the direction of the optical density change is altered. Thus, an electrochromic panel may function as a shutter with variable aperture that can be electrochemically operated. Recent advances in optical switching technology have stimulated research and development of electrochromic materials for "smart windows," the name usually used for multilayer window glazings that act as variable-absorbance or variable-reflectance light apertures. These materials have potential for architectural, automotive, and many other applications. They provide control of radiant solar energy transmission, resulting in the reduction of energy consumption.

Most research on electrochromic (EC) materials has focused on display applications, with high switching speed (>1 kHz) being the major objective.[1] Liquid crystal (LC) films, which rely on light scattering to control transmission electrically, usually have very fast switching speed.[2] In an LC film, the effective refractive index of the LC droplet generally differs from that of the encapsulating medium; therefore, the film scatters light and appears cloudy. When voltage is applied across the film, the LC molecules within the droplet tend to align along the applied field. If the LC and the encapsulating medium are properly chosen, the refractive indices of droplets and encapsulant in this aligned state nearly coincide, light scattering is significantly reduced, and the film becomes transparent.[2] This orientation change with the application of the electric field being very fast, the switching speed of the LC device is usually >1 kHz and, hence, is most suitable for display applications. On the other hand, electrochromism is not a physical effect produced by the applied electric field. The depth of coloration and the resulting transmittance change depend on the quantity and rate of ionic charge injection into (or depletion from) the electrochromic layer. Specifically, the change in optical density is directly proportional to this charge until the film is completely oxidized or reduced or some side reaction occurs. The response is of the order of hertz instead of kilohertz. This slow response time is the principal reason why electrochromic materials are not widely used for display applications. However, for the control of solar energy transmission

M. Ahsan Habib • Department of Physical Chemistry, General Motors Research Laboratories, Warren, Michigan 48090.

Electrochemistry in Transition, edited by Oliver J. Murphy *et al.* Plenum Press, New York, 1992.

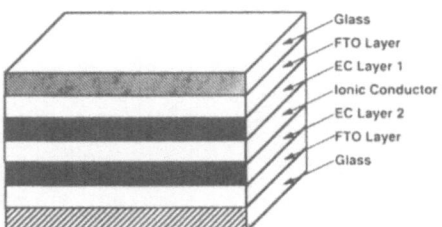

Glass
FTO Layer
EC Layer 1
Ionic Conductor
EC Layer 2
FTO Layer
Glass

FIGURE 1. Schematic diagram of an electrochromic cell with electrochromic films on both the electrodes. The electrochromic (EC) layers 1 and 2 color in a complementary way; that is, one is colored when it is oxidized, and the other when reduced. Thus, both the electrodes color at the same time and become transparent at the same time. FTO layer refers to the transparent fluorine-doped tin oxide conducting layer.

through "smart windows," a response time of the order of hertz appears acceptable. Thus, recently a frenzy of research activity has taken place in the field of electrochromics.[3] A typical configuration of an electrochromic system is shown in Fig. 1.

Although research on electrochromic materials is not new, the basic understanding of various aspects of electrochromism is still incomplete. In this chapter, we will discuss various types of electrochromic materials and mechanisms of electrochromic phenomena in some of the well-known systems.

2. TYPES OF ELECTROCHROMIC MATERIALS

A great variety of physical and chemical phenomena are involved in bringing about a variable radiative change in certain solids or liquids, which may be organic or inorganic materials. Classification of these materials into different categories according to the physical phenomena involved is an important step toward the understanding of these materials.

In general, electrochromic materials are classified into three categories: (a) transition-metal oxides, (b) organic polymers, and (c) inorganic complexes. In each class, some of the materials develop absorption properties upon reduction and hence are known as cathodically coloring materials. Some materials develop absorption bands upon oxidation and are known as anodically coloring materials.

3. TRANSITION-METAL OXIDES

3.1. Cathodically Coloring Oxides

Well-known cathodically coloring oxides include WO_3,[3-5] MoO_3,[6] V_2O_5,[7] Nb_2O_5, and TiO_2.[8] Some heteropoly compounds of W and Mo such as phosphotungstic acid[9] and silicomolybdate[10] which contain the units of oxides of these metals also show electrochromic behavior. Amorphous V_2O_5 changes from yellow to blue upon reduction, which results in the insertion of Li^+ ions[7] as follows:

$$V_2O_5 + xLi^+ + xe^- \rightleftharpoons Li_xV_2O_5 \tag{1}$$

For Nb_2O_5, which exhibits electrochromism similar to that of V_2O_5 and WO_3, the suggested mechanism is[8]

$$Nb_2O_5 + 2H^+ + 2e^- \rightleftharpoons Nb_2O_3(OH)_2 \text{ or } 2Nb_2O_2 \cdot H_2O \tag{2}$$

The electrochromic reaction of amorphous TiO_2 is (8)

$$TiO_2 + H^+ + e^- \rightleftharpoons TiOOH \tag{3}$$

MoO_3 follows the same mechanism as that discussed below for the electrochromic phenomena of WO_3 (Eq. 4). Among the transition-metal oxides, WO_3 is the most widely studied material, 'and we summarize below the electrochemical properties of this material.

WO_3 Electrochromics

Tungsten trioxide is essentially colorless, but when hydrogen or alkali-metal atoms are inserted into the interstitial sites of this oxide, it absorbs light and has a blue color. In the late 1960s and early 1970s, it became increasingly clear that the reduction of WO_3 involves the formation of a tungsten bronze following the insertion of ions[11]:

$$WO_3 + xM^+ + xe^- \rightleftharpoons M_xWO_3 \qquad (4)$$

where $M^+ = H^+$, Li^+, K^+, or Na^+. The mechanism given by Eq. (4) is the well-accepted-mechanism for WO_3 electrochromism today. Cyclic voltammograms of a WO_3 film in a solution of propylene carbonate containing Li^+ ions are shown in Fig. 2. The main features of these voltammograms are the large anodic peaks at potentials cathodic to 0.0 V versus SCE and the cathodic currents which continue to increase up to -1 V versus SCE. On a Pt surface, however, distinct cathodic peaks are observed in aqueous acidic solution.[11] Both amorphous and polycrystalline WO_3 films undergo electrochromic phenomena, with perform-ance improving with a decrease in crystal size.[12] The value of x in M_xWO_3 usually varies between 0 and 0.5 and depends on the density and crystallinity of the films, which varies for different methods of preparation.[13] Although the phenomenon of bronze formation was known for many years, Deb and Choparian[6,14] were the first to apply the concept for the construction of an optically switchable layered device in the late 1960s. Optical measurements yield further information about the electronic energy levels as a function of the x value. A typical example of optical density variation upon electrochemical reduction of WO_3 is shown in Fig. 3.[5] The optical absorbance of the film in the entire visible region of the spectra (400–850 nm) increases with increased amount of injected cathodic charge. The coloration process starts at the WO_3/electrolyte interface, since the dynamics of coloration are limited by the diffusion of protons, and advances toward the substrate from the electrolyte side (Fig. 4).[15] From an *in situ* measurement, Ohtsuka *et al.* [16] have reported that during the coloration process, H_2O molecules also are incorporated into the H_xWO_3 structure.

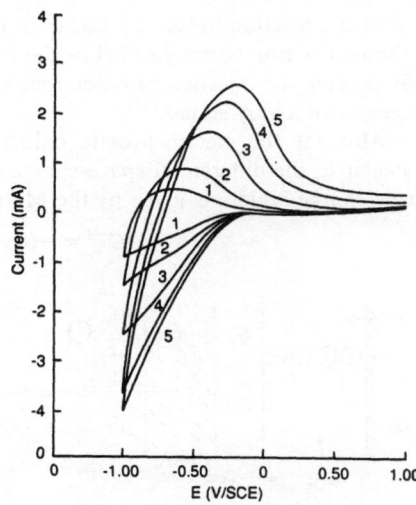

FIGURE 2. Cyclic voltammograms of a tungsten trioxide film with various scan rates: (1) 50 mV/s, (2) 100 mV/s, (3) 250 mV/s, (4) 500 mV/s, (5) 750 mV/s. Solution was propylene carbonate with 1 mol/dm³ $LiClO_4$.[5]

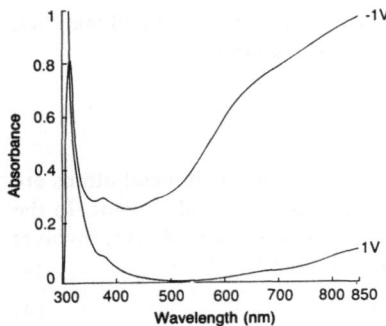

FIGURE 3. Absorbance spectra of a tungsten trioxide film in the colored (-1 V versus SCE) and in the bleached state (1 V versus SCE). The solution was propylene carbonate with 1 mol/dm^3 LiClO$_4$.[5]

The properties of both the sodium and lithium tungsten bronzes are very similar to those of H$_x$WO$_3$.[3] Na$_x$WO$_3$ shows a change from light green to deep blue.[17] The electrochromic response appears to be an order of magnitude slower than that of H$_x$WO$_3$.[5] McHardy and Bockris[11] investigated the electrochromic effect of Na$_x$WO$_3$ bronze by performing reflectance spectroscopic measurements during potential sweeps. Absorbance by the bronze electrode was shown to increase with the increase of the electrode potential in the cathodic direction. They[11] also showed that as soon as a crystal of sodium tungsten bronze is immersed in oxygen- (or air-) saturated acid solution, Na$^+$ ions are lost from the surface by a corrosion mechanism:

$$O_2 + 4H^+ + 4e^- \rightarrow 2H_2O \tag{5}$$

$$Na_xWO_3 \rightarrow WO_3 + xNa^+ + xe^- \tag{6}$$

The anodic dissolution of sodium proceeds at a rate which diminishes over a period of days, roughly in proportion to the logarithm of the time. The aging process is accelerated during cycling of the electrode between anodic and cathodic potentials. The decay in anodic partial current is manifested during a steady-state test both as an increasing net current for oxygen reduction at cathodic overpotentials and as an extension of the oxygen evolution Tafel line to lower current densities at anodic overpotentials. Both the dissolution reaction and the anodic evolution of oxygen can be stimulated by light, which creates electron–hole pairs in the electrode surface; the mechanism of stimulation entails hole capture,

$$2H_2O + 4h^+ \rightarrow 4H^+ + O_2 \tag{7}$$

Unlike the reaction in Eq. (6), however, this process can be reversed, and the corresponding cathodic reaction occurs just below the theoretical oxygen electrode potential. The oxidation may involve the injection of oxide ions into the bronze surface, but the evidence for such a process is not very sound.[18]

Most of the electrochromic oxides possess semiconductor properties. The relation between C, the differential space-charge capacitance, and ΔE, the potential drop across the space-charge region, is given by the Mott–Schottky equation[11]:

$$C^{-2} = -(8\pi/en_d\varepsilon)[(\Delta E - kT)/e] \tag{8}$$

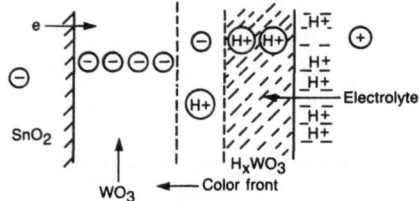

FIGURE 4. Schematic representation of the interfacial region during coloration of a WO$_3$ film.[15]

where e is the electronic charge, n_d is the donor concentration, and ε is the dielectric constant. The Mott–Schottky plot for Na_xWO_3 reported by Bockris et al.[11] is shown in Fig. 5. The donor concentration, n_d, was of the order of 10^{21} cm^{-3}, which corresponds to an x value of <0.1. This value is lower than that observed for H_xWO_3[12] and probably reflects incomplete ionization of the Na^+ donor ions.[19]

For Li_xWO_3, it has been argued that the presence of Li^+ in place of H^+ extends its lifetime. H_xWO_3 in acid electrolyte tends to discolor with time.[20] The importance of the granular structure of WO_3 is vital to the electrochromic response of the films. The angular grain boundaries of polycrystalline WO_3 thin films permit rapid diffusion of discharged lithium ions throughout the grain boundary manifold, followed by slower diffusion into the grains. The diffusion of Li^+ ions into the grains is a highly nonlinear process, being field assisted by the electric field associated with the electron space charge in each grain. From potentiodynamic measurements, Habib and Glueck[5] have reported the diffusion coefficient of Li^+ ions in WO_3 films in contact with propylene carbonate solution containing $LiClO_4$ to be 3×10^{-11} cm^2/s. This value may be compared with 2.8×10^{-11} cm^2/s, found by Ho et al.[21] by AC impedance techniques, and 2×10^{-11} and 5×10^{-9} cm^2/s, reported by Green[22] and Mahapatra,[23] respectively, and based on potentiostatic measurements.

3.2. Anodically Coloring Oxides

The anodically coloring oxides include IrO_2,[24] Rh_2O_3,[25] and NiO_x or $Ni(OH)_2$.[26] Hydrated iridium oxide is transparent in its reduced form and turns blue-black upon oxidation when hydroxide ions are inserted into the film[24]:

$$Ir(OH)_3 + OH^- + h^+ \rightleftharpoons IrO_2 \cdot H_2O \tag{9}$$

Hydrous Rh_2O_3 can be transformed from its initial reduced yellow state to a dark green or dark brown oxidized state according to the following reaction[25]:

$$\tfrac{1}{2}Rh_2O_3 + OH^- + h^+ \rightleftharpoons RhO_2 + \tfrac{1}{2}H_2O \tag{10}$$

Nickel Hydroxides

The electrochromic properties of nickel hydroxides and related nickel oxide materials have been studied quite extensively in recent years.[26] Nickel hydroxide films acquire an intense dark-brownish color upon oxidation,[26]

$$Ni(OH)_2 + OH^- \rightleftharpoons NiOOH + H_2O + e^- \tag{11}$$

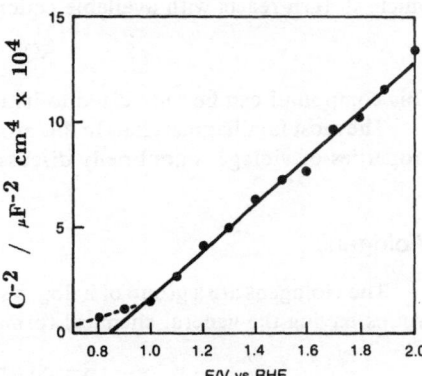

FIGURE 5. Mott–Schottky plot of a sodium tungsten bronze (Na_xWO_3) electrode in an acidic solution (0.1 mol/dm^3 $HClO_4$).[11]

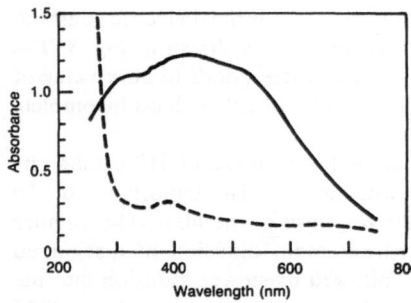

FIGURE 6. Absorbance spectra of reduced and oxidized nickel hydroxide films. Dashed curve indicates spectrum of a reduced film with a thickness of about 1 μm. Solid curve indicates spectrum of an oxidized film with a thickness of about 150 nm.[27]

The UV-visible spectra in the colored and bleached states, as reported by Carpenter and Corrigan,[27] are shown in Fig. 6. The reduced film is essentially transparent throughout the visible region. The transmittance of the oxidized (colored) form ranged between 30% and 50% in the visible region due to a strong and broad absorption band centered around 500 nm and is associated with photoelectrochemical activity[27] and changes in the electronic conductivity[28] accompanying the redox reaction. Thus, the coloration is likely due to a band-to-band transition in the oxidized material.[28] The durability of the films was found to be improved when Y^{3+}, Ce^{3+}, and La^{3+} ions were incorporated in the films.[29]

4. ORGANIC ELECTROCHROMICS

4.1. Cathodically Coloring Organics

A large number of organic compounds develop absorption bands in the visible region of the spectrum upon reduction. Most of the compounds in this class exhibit electrochromic phenomena by reversible electrodeposition. This group of materials includes viologens,[30] pyridines,[31] ortho-tolidines,[32] and anthraquinones.[33] Several derivatives of pyridine, e.g., Co-pyridinoporphyrazine[31] films, become violet upon electrochemical reduction. ortho-Tolidine ($C_{14}H_{16}N_2$; 4,4'-diamine-3,3'-dimethylbiheptyl) forms a colored precipitate with anions when in the monovalent cationic state.[32] Upon oxidation, it turns whitish or reddish in color. Anthraquinones (AQ)[33] form colored salts with cations when in the monovalent anionic state. The AQ molecule forms an anion radical upon reduction,

$$AQ + e^- \rightleftharpoons AQ^- \tag{12}$$

which, in turn reacts with available cations, M^+, giving the colored compound, MAQ,

$$AQ^- + M^+ \rightleftharpoons MAQ \tag{13}$$

This compound can be reoxidized to its neutral, colorless form, AQ.

The most familiar materials in this group are the viologen compounds. The electrochromic properties of viologens are briefly discussed below.

Viologens

The viologens are a group of halogen salts of organic compounds containing dipyridinium groups having the general chemical formula

$$[R-NC_5H_4-C_5H_4N-R]X_2$$

where R may be an alkyl, cycloalkyl, or other organic substitute, and X is a halogen. The name "viologen" was given by Michaelis and Hill[34] to the above 4,4'-dipyridinium compound because of the deep-blue–purple color upon reduction. These compounds are almost colorless in the oxidized state. Since the redox potentials of viologens are independent of pH,[35] these compounds are routinely used as oxidation–reduction indicators. The simplest form of viologen, methyl viologen (MV), clearly exhibits a two-step reduction at −0.68 and −1.07 V versus SCE, respectively.[35] The first reduction step at −0.68 V is quite reversible while that at −1.07 V is irreversible. The reduction wave at the first potential, E_1, may be attributed to the reversible one-electron reduction and that at E_2 is due to one more electron reduction reaction[36]:

$$R-N^+C_5H_4-C_5H_4N^+-R + e^- \rightleftharpoons R-N^+C_5H_4-C_5H_4N-R + e^-$$

Colorless   Colored

$$\rightleftharpoons R-NC_5H_4=C_5H_4N-R$$

Colored

The absorption spectrum of MV^+ has two main absorption peaks at 390 and 605 nm. Strojek et $al.$[37] reported the change of absorbance at the two absorption peaks as a function of time during electrochemical reduction. If the redox reactions are diffusion-controlled, then they can be diagnosed by the variation of current, i, with time, t, as expressed by the following equation[38]:

$$i = (nFAD_0^{1/2}C_0)/(\pi^{1/2}t^{1/2}) \tag{14}$$

where F, A, D_0, and C_0 are the Faraday constant, the area of the electrode, the diffusion coefficient of the ions, and the initial concentration of the reactants, respectively. The integral form of Eq. (14) from $t = 0$ gives the cumulative charge passed:

$$q = \int i\, dt = (2nFAD_0^{1/2}C_0t^{1/2})/(\pi^{1/2}) \tag{15}$$

In terms of the amount of charge, q, required to bring about the change in absorbance,

$$\Delta Absorbance = \alpha q/nFA \tag{16}$$

Substituting q from Eq. (15) in Eq. (16), the change in absorbance with time may be obtained from

$$\Delta Absorbance = (2\alpha C_0 D_0^{1/2}t^{1/2})/(\pi^{1/2}) \tag{17}$$

Strojek et $al.$[37] found that the absorbance is proportional to $t^{1/2}$, indicating that the electroredox reaction of the MV^+ film is diffusion-controlled (Eq. 17). The diffusion coefficient of H^+ ions was found to be 9.5×10^{-6} cm^2/s.

Both the absorption band at 390 nm and that of 605 nm of the reduced MV appear during both the one-electron and the two-electron reduction process. Therefore, these absorption bands are characteristic of the absorption of MV^+ free radicals. Moreover, the slopes of the absorbance versus $t^{1/2}$ plots for both absorption bands at the second electron reduction potential are double those at the first reduction potential, indicating the following disproportionate type of reaction[1]:

$$MV^{2+} + e^- \rightarrow MV^+ \qquad \text{at } E_1 \tag{18}$$

$$MV^+ + e^- \rightarrow MV \qquad \text{at } E_2 \tag{19}$$

$$MV + MV^{2+} \rightarrow 2MV^+ \tag{20}$$

A similar phenomenon has also been observed with diheptylviologen dibromide in aqueous solution.[39] For benzyl viologen halides mixed with polymerized viologen dibromide,[40] colors range from a whitish blue to reddish purple. Various other mixed viologens such as tetramethylene[bis-4(ethylpyridine-4'-yl)pyridinium] perchlorate[41] have also been found to possess good electrochromic properties. Bookbinder and Wrighton[30] made a derivative of a viologen dibromide salt with a silane compound, resulting in a good surface adhesive viologen film. This redox system was found to be a durable surface electrochromic material that exhibited fast response and coloration per unit of charge expected for a viologen-based chromophore in a variety of aqueous electrolyte systems.

4.2. Anodically Coloring Organics

The anodically coloring organics are mostly polymers which undergo changes in their optical properties upon oxidation. Thin films of these materials are usually prepared by electropolymerization on a conducting surface. Some of the materials in this group are tetrathiofulvalines,[42] pyrazolines,[43] polythiophenes,[44] phthalocyanines,[45] and polyaniline.[46, 47] Tetrathiofulvalines and pyrazoline polymer-coated electrodes can be oxidized from light yellow to green to purple.[42,43] Lutetium diphthalocyanine, $HLuPc_2$, is green in its neutral form and changes to orange, red, or brown upon oxidation and to blue upon reduction.[45] A brief discussion of the electrochromic properties of polyaniline, which have attracted significant attention in recent years, is given below.

Polyaniline

Polyaniline (PANI) in its uncontrolled and uncharacterized form has been the subject of several studies for a long time. Recently, interest in polyaniline has been revived in connection with its application as a conductive polymeric electrode in rechargeable batteries,[48] diodes or transistors,[49] and electrochromic devices.[47] Upon oxidation, a PANI film exhibits a green to blue color, and upon reduction it becomes nearly transparent with a slight yellow-green tint (Fig. 7). Chiang and MacDiarmid[47] studied the electrochemical oxidation and reduction of polyaniline in aqueous electrolytes of varying pH values and inferred that two classes of redox processes occur at different potentials which differ from each other by the extent to which the processes are accompanied by deprotonation (during oxidation) and protonation (during reduction). The conductivity of the emaraldine salt of polyaniline lies in the metallic regime, and Chiang and MacDiarmid[47] proposed a symmetric conjugated structure having extensive charge delocalization. This structure resulted from the doping formation of an organic polymer salt rather than oxidation, which occurs in the p-doping of other conducting polymer systems. Konig and Schultze[50] studied the detailed kinetics of polyaniline formation and redox processes. The polymerization of aniline proceeds in accordance with the following reaction:

$$2nC_6H_5-NH_2 \rightarrow (-C_6H_4-NH-C_6H_4-NH-)_n + 2nH^+ + 2ne^- \qquad (21)$$

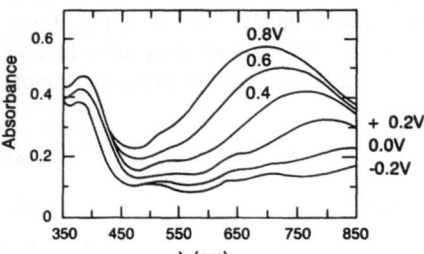

FIGURE 7. In situ absorbance spectra of a polyaniline film on a fluorine-doped SnO_2-coated glass substrate at various potentials.[46]

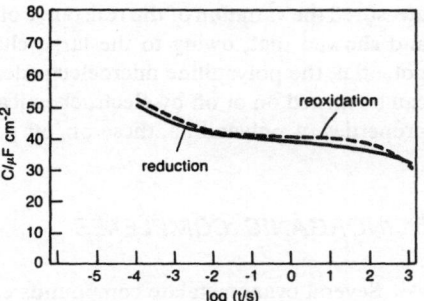

FIGURE 8. Variation of the polyaniline electrode capacity, C, during reduction from 1 to 0 V versus RHE (——) and reoxidation from 0 to 1 V (- - -) in 0.5 mol/dm³ H_2SO_4 (aqueous).[50]

which consumes about 1.2 electrons per molecule since, in addition to the deprotonation, anions are inserted into the polymer film. During reduction of the polymerized film, protons are inserted and anions are removed while in the reoxidation process the reverse processes occur. As shown in Fig. 8, the electrode capacities during reduction and reoxidation are almost independent of time and are quite high (40 μF cm² at 1 ms < t < 100 s). Thus, the reduction and reoxidation follow the same reaction pattern and are explained by an almost constant motion of the reaction front through the polymer film.[50]

Habib and Maheswari[46] investigated the structural changes of the PANI films by an *in situ* Fourier transform infrared (FTIR) spectroelectrochemical technique. Coloring (oxidation) and bleaching of the film were shown to be associated with the deprotonation and protonation of the film, respectively. The N—H stretching bond around 2970 cm⁻¹ decreased or increased as the film was gradually oxidized or reduced, respectively (Fig. 9). These results signify the breaking and formation of the N—H bond during the oxidation and reduction processes as indicated by the following mechanism:

$$(-C_6H_4-NH-C_6H_4-NH-)_n \rightleftharpoons (-C_6H_4-N=C_6H_4=N-)_n + 2nH^+ + 2ne^- \quad (22)$$

From an electrogravimetric measurement, Habib[51] has also shown that the oxidized polyaniline surface is more hydrophilic than the reduced surface. Wrighton and co-workers[49]

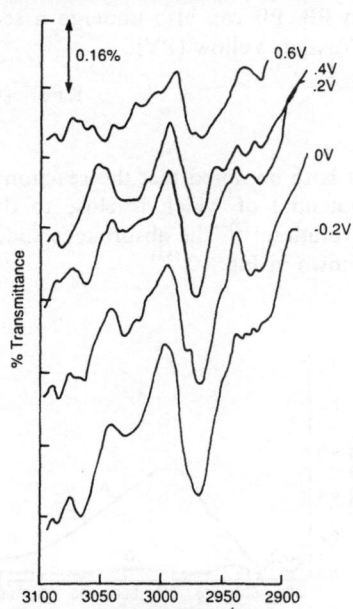

FIGURE 9. *In situ* FTIR spectra of a polyaniline film in the N—H stretching region as a function of applied potential (versus SCE).[46]

measured the variation of the resistance of polyaniline films as a function of electrode potential and showed that, owing to the large change in resistance upon change in electrochemical potential, the polyaniline microelectrodes can function as diodes and transistors in that they can be turned on or off by electrochemical oxidation or reduction. Due to the electrochromic properties of polyaniline, these on/off positions can be optically visualized.

5. INORGANIC COMPLEXES

Several cyanometalate compounds exhibit electrochromic properties. These compounds have the general formula $(M_1^X)_k[M_2^Y(CN)_m]_n \cdot H_2O$, where M_1 and M_2 are metal atoms, X and Y are valencies, and k, m, and n denote stoichiometry. One of the most durable electrochromic materials of this group is Prussian Blue (PB), nominally $KFe[Fe(CN)_6]$.

Prussian Blue (PB)

The electrochromic transition of PB, from blue in the oxidized state to transparent in the reduced state, occurs reversibly and with good efficiency.[52] Two types of PB have been identified: "soluble" PB, $KFeFe(CN)_6$, and "insoluble" PB, $Fe_4[Fe(CN)_6]_3$, both of which generally contain an indeterminate amount of water in the interstices of their nominally cubic lattice structures. (The terms "soluble" and "insoluble" reflect the relative ease of peptization rather than the actual solubility.) Both forms of PB can be reversibly reduced by an injection of electrons accompanied by the intercalation of alkali-metal ions, M^+, to yield the transparent Everitt's salt (ES). This electrochromic reaction for the soluble form of PB is given by

$$MFeFe(CN)_6 + M^+ + e^- \rightleftharpoons M_2FeFe(CN)_6$$

$$\text{Prussian Blue} \qquad\qquad \text{Everitt's salt} \qquad\qquad\qquad (23)$$

When M^+ is a potassium ion, PB can be repeatedly reduced to Everitt's salt and reoxidized to PB. PB can also undergo a second electrochromic reaction in which it is oxidized to Prussian Yellow (PY):

$$KFeFe(CN)_6 \rightleftharpoons FeFe(CN)_6 + K^+ + e^- \qquad\qquad (24)$$

$$\text{Prussian Yellow}$$

It is to be noted that the reaction in Eq. (23) is more reversible than that in Eq. (24), the potential of which is close to the decomposition potential of PY and that of oxygen evolution.[52] The absorbance spectra of PB in the bleached and in the colored states are shown in Fig. 10.[53]

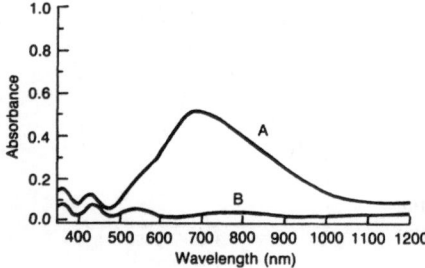

FIGURE 10. *In situ* UV-vis-NIR spectra of a Prussian Blue film in an aqueous solution containing 1 mol/dm³ KCl (pH 4). (A) Colored state at 0.6 V versus SCE; (B) clear state at −0.2 V versus SCE.[53]

Rajan and Neff[54] studied the kinetics of PB electrochromic reactions. From a potentiodynamic study, they found that peak currents and peak separations increase linearly with scan rate at low scan rates. At intermediate scan rates there is a gradual change to square root dependence of both peak current and peak separation. At very high scan rates, or under pulse conditions, one must take account of chemical potential gradients within the film.[54] Lundgren and Murray[55] found no elecctrolyte anion dependence on the redox potential. Neff and co-workers[54,56] formulated the transport behavior in terms of the Nernst–Planck equation. Itaya et al.[52] have studied the short-time behavior of PB films and have found large migration effects for the reduction of PB but not for the reoxidation of ES. This may lead to charge imbalance during long-term cycling of the film. However, Prussian Blue is one of the very few electrochromic materials which have been known to undergo up to 10^6 cycles.[57]

6. SUMMARY AND CONCLUSIONS

The state of the art in electrochromic materials has been briefly reviewed. These materials have potential for architectural and automotive applications as they provide control of radiant solar energy transmission, resulting in the reduction of energy consumption. The main features of the electrochromic characteristics of the most well-known groups of materials have been described. The properties of some well-known materials such as WO_3, $Ni(OH)_2$, viologens, polyaniline, and Prussian Blue have been discussed with particular reference to the mechanism of electrochromic reactions. For WO_3, the long-term reversibility of the change in its optical density associated with a reversible structural change has been established. The cycle life of nickel hydroxide electrochromics has been shown to improve when certain ions such as La^{3+}, Ce^{3+}, and Y^{3+} are incorporated in the films. The mechanism of polyaniline electrochromism is reasonably well understood. Long-term cycle life and UV stability are still unclear. For Prussian Blue, although the cycle life is quite high, further research is needed for a basic understanding in terms of the charge imbalance during coloration and bleaching.

ACKNOWLEDGMENTS

Thanks are due to Drs. Don MacArthur and Mike Carpenter for several stimulating discussions.

REFERENCES

1. I. F. Chang, *Non-Emissive Electro-optic Displays*, p. 155, Plenum Press, New York (1976).
2. K. Nassau, *The Physics and Chemistry of Color*, p. 320, John Wiley & Sons, New York (1983).
3. C. Lampert, *Sol. Energy Mater.* **11**, 1 (1984).
4. S. K. Deb, *Phil. Mag.* **27**, 801 (1973).
5. M. A. Habib and D. Glueck, *Sol. Energy Mater.* **18**, 127 (1989).
6. S. K. Deb and J. A. Choporian, *J. Appl. Phys.* **37**, 4818 (1966).
7. S. F. Cogan, N. M. Nguyen, S. J. Perrotti, and D. Rauh, *Proc. SPIE* **1016**, 57 (1988).
8. C. K. Dyer and J. S. Lech, *J. Electrochem. Soc.* **135**, 23 (1978).
9. S. P. Maheswari and M. A. Habib, *Sol. Energy Mater.* **18**, 75 (1988).
10. S. K. Mahapatra, G. D. Boyd, F. G. Storz, and S. Wagner, *J. Electrochem. Soc.* **126**, 805 (1979).
11. J. McHardy and J. O'M. Bockris, *J. Electrochem. Soc.* **120**, 53 (1973).
12. R. S. Crandall, P. J. Wojtowicz, and B. W. Faughnan, *Solid State Commun.* **18**, 1409 (1976).

13. M. Green and K. Kang, *Displays* **9**, 166 (1988).
14. S. K. Deb, *Appl. Opt.* **192** (*Suppl.* 3), 1969 (1969); *Proc. Roy. Soc.* (*London*), *Ser. A* **306**, 211 (1968).
15. B. W. Faughnan, R. S. Crandall, and P. M. Heyman, *RCA Rev.* **36**, 177 (1975).
16. T. Ohtsuka, N. Goto, N. Sato, and K. Kunimatsu, *Ber. Bunsenges. Phys. Chem.* **91**, 313 (1987).
17. J. P. Randin, A. K. Vijh, and A. B. Chughtai, *J. Electrochem. Soc.* **120**, 1174 (1973).
18. J. O'M. Bockris and J. McHardy, *J. Electrochem. Soc.* **120**, 61 (1973).
19. J. F. Dewald, *J. Phys. Chem. Solids* **14**, 155 (1960).
20. M. Campagna, G. K. Wertheim, H. R. Shanks, F. Zumsteg, and E. Banks, *Phys. Rev. Lett.* **34**, 738 (1975).
21. C. Ho, I. D. Raistrick, and R. A. Huggins, *J. Electrochem. Soc.* **127**, 343 (1980).
22. M. Green, *Thin Solid Films* **50**, 145 (1978).
23. S. K. Mahapatra, *J. Electrochem. Soc.* **125**, 284 (1978).
24. J. D. E. McIntyre, S. Basu, W. F. Peck, Jr., W. L. Brown, and W. M. Augustyniak, *Phys. Rev.* B **25**, 7242 (1982).
25. S. Gottesfeld, *J. Electrochem. Soc.* **227**, 272 (1980).
26. M. K. Carpenter, R. S. Conell, and D. A. Corrigan, *Sol. Energy Mater.* **16**, 333 (1987).
27. M. K. Carpenter and D. A. Corrigan, *J. Electrochem. Soc.* **136**, 1022 (1989).
28. M. J. Natan, D. Belanger, M. K. Carpenter, and M. S. Wrighton, *J. Phys. Chem.* **91**, 1834 (1987).
29. D. A. Corrigan and M. K. Carpenter, in: *Large Area Chromogenics: Materials and Devices for Transmittance Control* (C. G. Graqvist and C. M. Lampert, eds.), SPIE, Bellingham, Washington (1990).
30. D. C. Bookbinder and M. S. Wrighton, *J. Electrochem. Soc.* **130**, 1080 (1983).
31. M. Yamana, N. Nashiwazaki, M. Yamamoto, and T. Nakano, *Displays* **9**, 190 (1988).
32. I. F. Chang, B. L. Gilbert, and T. I. Sun, *J. Electrochem. Soc.* **122**, 955 (1975).
33. L. G. Van Vitert, G. J. Zydzik, S. Singh, and I. Camlidel, *Appl. Phys. Lett.* **36**, 109 (1980).
34. L. Michaelis and E. S. Hill, *J. Am. Chem. Soc.* **55**, 1481 (1983).
35. R. M. Elofson and R. L. Edsberg, *Can. J. Chem.* **35**, 646 (1957).
36. M. Ito and T. Kuwana, *J. Electroanal. Chem.* **32**, 415 (1971).
37. J. W. Strojek, G. A. Gruver, and T. Kuwana, *Anal. Chem.* **41**, 481 (1969).
38. J. O'M. Bockris and A. K. N. Reddy, *Modern Electrochemistry*, Vol. 1, Plenum Press, New York (1970).
39. T. Kawata, M. Yamamota, M. Yamana, M. M. Jajima, and N. Nakano, *Jpn. J. Appl. Phys.* **14**, 725 (1975).
40. G. G. Barna and J. G. Fish, *J. Electrochem. Soc.* **128**, 1290 (1981).
41. J. Bruinink, C. G. A. Kregting, and J. J. Panjee, *J. Electrochem. Soc.* **124**, 1854 (1977).
42. K. F. Kaufman, A. H. Schoeder, E. M. Engler, S. R. Kramer, and J. Q. Chambers, *J. Am. Chem. Soc.* **102**, 483 (1980).
43. F. B. Kaufman, A. H. Schroeder, E. M. Engler, and V. V. Patel, *Appl. Phys. Lett.* **36**, 422 (1980).
44. A. O. Patil, A. J. Heeger, and F. Wudl, *J. Am. Chem. Soc.* **88**, 183 (1988).
45. N. M. Nicholson and F. A. Pizzarello, *J. Electrochem. Soc.* **126**, 1490 (1979).
46. M. A. Habib and S. P. Maheswari, *J. Electrochem. Soc.* **136**, 1050 (1989).
47. J.-C. Ching and A. G. MacDiarmid, *Synth. Met.* **13**, 193 (1986).
48. A. Kitani, J. Yano, and K. Sasaki, *J. Electroanal. Chem.* **209**, 227 (1986).
49. E. W. Paul, A. J. Ricco, and M. S. Wrighton, *J. Phys. Chem.* **89**, 1441 (1985).
50. U. Konig and J. W. Schultze, *J. Electroanal. Chem.* **242**, 243 (1988).
51. M. A. Habib, *Langmuir* **4**, 1302 (1988).
52. K. Itaya, I. Uchida, and V. D. Neff, *Acc. Chem. Res.* **19**, 162 (1986).
53. M. A. Habib, S. P. Maheswari, and M. K. Carpenter, *J. Appl. Electrochem.* **21**, 203 (1991).
54. K. P. Rajan and V. D. Neff, *J. Phys. Chem.* **86**, 4361 (1982).
55. C. A. Lundgren and R. W. Murray, *Inorg. Chem.* **27**, 933 (1988).
56. D. Ellis, M. Eckhoff, and V. D. Neff, *J. Phys. Chem.* **85**, 1225 (1981).
57. K. Itaya, I. Uchida, and S. Toshima, *New Mater. New Processes* **2**, 508 (1983).

Spectroscopic Studies

5

Ellipsometry at Metal–Electrolyte Interfaces

W. Visscher

1. INTRODUCTION

Ellipsometry was first introduced by P. Drude in 1890. This technique is based on the changes in the polarization state of light upon reflection at a substrate.

The method is in particular suitable for the measurement of thin films which are present at a substrate. The great advantages of ellipsometry are (i) it provides *in situ* detection, (ii) it is a nondestructive method, (iii) its sensitivity makes it applicable to the measurement of submonolayer films, and (iv) it can be employed to monitor film growth with time, which can yield information about the initial stages of the formation, the nature of the layer, and eventual changes in properties during the growth process.

In this chapter some examples will be given of the measurement of thin films by ellipsometry in the investigation of the electrodeposition of metallic films, viz., deposition of copper on platinum and of nickel on carbon, and the growth of an anodic oxide layer on a nickel–cobalt alloy.

2. COPPER DEPOSITION ON PLATINUM

The phenomenon of underpotential—as well as bulk—deposition was investigated for copper on platinum. All optical experiments were carried out in a cylindrical Teflon cell with quartz windows arranged for an angle of incidence of 70° at the substrate. The ellipsometer was the automatic Rudolph RR2200, equipped with a tungsten iodine light source and a monochromatic filter for 546.1 nm. The electrochemical equipment consisted of a Wenking potentiostat POS 73, controlled by a PAR model 175 Universal Programmer. The electrolyte was $1M$ $H_2SO_4 + xM$ $CuSO_4$, prepared from AnalaR chemicals and deaerated with N_2; the reference electrode was mercury-mercurous sulfate (MSE; +0.65 V versus RHE), and the counter electrode a Pt foil. The substrate was polycrystalline Pt, area 0.5 cm^2, polished with alumina (0.05 μm).

W. Visscher • Laboratory for Inorganic Chemistry and Catalysis, Faculty of Chemical Technology, Eindhoven University of Technology, 5600 MB Eindhoven, The Netherlands.

Electrochemistry in Transition, edited by Oliver J. Murphy *et al.* Plenum Press, New York, 1992.

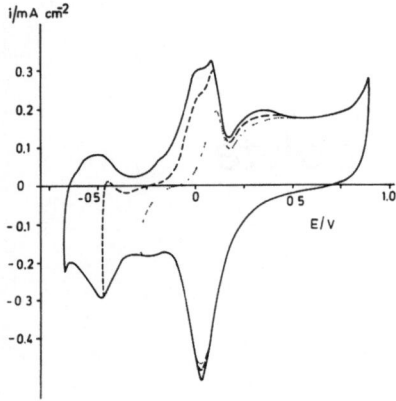

FIGURE 1. Voltammogram of Pt in $0.5M$ $H_2SO_4 + 6 \times 10^{-5}M$ CuSO$_4$; scan rate: 200 mV/s.

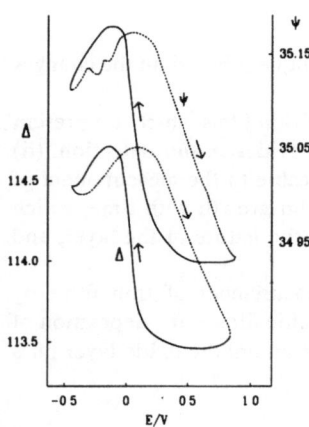

FIGURE 2. Change of Δ and ψ in potential range -0.5 to $+0.9$ V versus MSE; scan rate: 20 mV/s; electrolyte: $0.5M$ $H_2SO_4 + 6 \times 10^{-5}M$ CuSO$_4$.

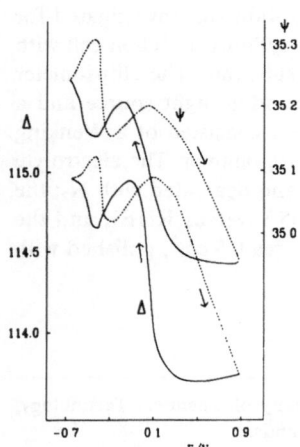

FIGURE 3. Change of Δ and ψ in potential range -0.7 to $+0.9$ V versus MSE; other conditions as in Fig. 2.

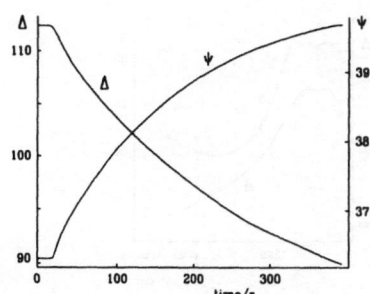

FIGURE 4. Ellipsometric time curves for Cu deposition at $E = -0.6$ V versus MSE; electrolyte: $0.5M$ $H_2SO_4 + 6 \times 10^{-3} M$ $CuSO_4$.

The voltammogram of the Cu/Pt system is shown in Fig. 1 for three different potential ranges. The underpotential deposition (upd) is clearly manifested by the anodic peaks at 0.0 and +0.8 V; the bulk deposition begins at -0.45 V. The optical response is given in Figs. 2 and 3. In the upd range a distinct change of the optical parameters Δ and ψ is observed (Fig. 2). For the upd layer a coulometric charge corresponding to $\theta = 0.6$ was obtained. With $r(Cu) = 0.128$ nm, this represents a thickness of 0.15 nm, and the refractive index was found to be $N = 2.5 - 3.3i$.

With increasing negative potential, Δ and ψ pass through a minimum at -0.45 V (Fig. 3), and around $E = -0.65$ to -0.70 V, Δ and ψ are about equal to the values for the bare substrate (at $E = 0.1$ V). Coulometrically, a charge equivalent to $\theta = 1$ is reached at about -0.7 V, so it must be concluded that a rearrangement of the surface atoms takes place before the actual bulk deposition can start. It has been discussed[1-4] whether or not a complete monolayer is formed before deposition of the bulk metal begins. The ellipsometric results indicate that the deposition of the bulk metal is preceded by a reorientation at the surface which takes place when the coverage reaches about a monolayer. It is therefore not surprising that the refractive index of the upd layer differs from the bulk value of copper.[5,6] A structural change in the adsorbate layer at $\theta = 0.6$ was also concluded by Kolb and Kotz[1] on the basis of reflectance spectroscopy.

The growth of the multilayer deposit was further studied at constant potential or current. Figure 4 gives the Δ versus time and ψ versus time curves during deposition at $E = -0.6$ V versus MSE. These data are replotted in Fig. 5 as a ψ versus Δ graph, together with calculated ψ versus Δ curves for films with refractive index $n = 0.65$, and $k = 2.70$ and 2.80. The close agreement between the experimental points and the calculated curves implies that copper is deposited as a homogeneous layer with bulk properties.

This result is typical for deposition at $E = -0.6$ V versus MSE. At $E = -0.50$ V and $E = -0.45$ V or at low current, the ellipsometric data are different: a typical result is depicted in Fig. 6a for $i = 0.05$ mA/cm². It was noted that the potential did not drop below -0.5 V. The ψ versus Δ plot (Fig. 6b) indicates that during deposition under these conditions the optical properties of the film change continuously. Only after a charge of about 60 mC/cm²

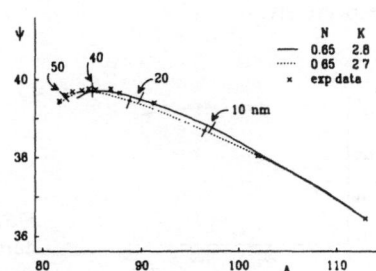

FIGURE 5. ψ-Δ plot of data of Fig. 4. Solid lines are calculated curves for refractive index $n = 0.65$; $k = 2.7$ and 2.8.

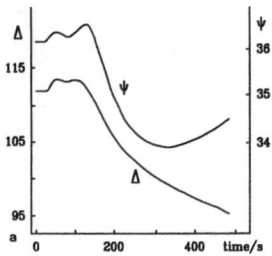

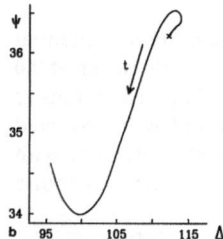

FIGURE 6. (a) Ellipsometric change with time for Cu deposition at $i = 0.05$ mA/cm^2; electrolyte as in Fig. 4. (b) Corresponding ψ–Δ plot.

had been passed through were the bulk values of copper reached. The reason for this difference in behavior can be attributed to a different nucleation and growth process. At high overpotential, instantaneously a large number of nuclei are deposited, which then grow three-dimensionally, so that a homogeneous deposit is obtained. At lower overpotentials, a smaller number of nuclei are formed, with the result that the film is gradually filled with metallic deposit; therefore, the composition of the layer changes with time.

A detailed analysis of the nucleation and growth process is provided by the current–time response during a potential step.[7,8] This method provides the ability to distinguish between progressive and instantaneous nucleation with two- and three-dimensional growth.

This technique was combined with ellipsometry for the study of the deposition of nickel on vitreous carbon.[9]

3. NICKEL DEPOSITION ON CARBON

The vitreous carbon, area 0.2 cm^2, was mechanically polished with alumina down to 0.05 μm and then cleaned ultrasonically in double-distilled water. The deposition was carried out in 0.1M NiSO$_4$ + 0.6M NaCl + 0.58M H$_3$BO$_3$, prepared from AnalaR chemicals. After the carbon electrode had been mounted, the potential was set at -0.12 V versus SCE, and then a potential step to -0.90 V was applied.

The current and optical responses to this potential step are given in Figs. 7 and 8, respectively.

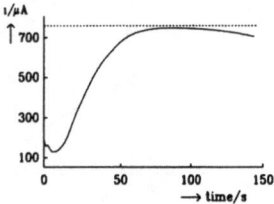

FIGURE 7. Nickel deposition on vitreous carbon. Current transient during potential step -0.12 to -0.90 V versus SCE is shown; electrolyte: 0.1M NiSO$_4$ + 0.6M NaCl + 0.58M H$_3$BO$_3$.

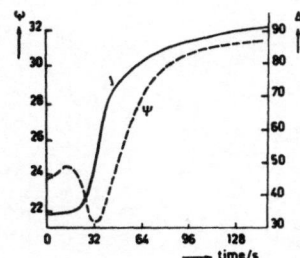

FIGURE 8. Δ and ψ transients during nickel deposition.

The electrocrystallization process involves nucleation and three-dimensional growth of the centers. The initial deposit is therefore not smooth, and its topography changes with time. For the optical evaluation, it will be supposed that this system can be described by a substrate/film/electrolyte medium model, where the film is represented by a layer of thickness, d, with an effective refractive index (N_{eff}) which is determined by the deposited metal and the electrolyte:

$$N_{eff} = wN_{dep} + (1 - w)N_{med}$$

where w, the volume fraction of the deposit, is time-dependent and determined by the growth model.

If it is assumed that right circular cones are formed, then their growth with time can be presented schematically as pictured in Fig. 9. The characteristic parameters are the outward growth rate (k') and a parameter related to the parallel growth rate (k) and the nucleation rate. These can be evaluated from the current–time plot.

N_{eff} can thus be calculated as a function of time, and from this the Δ and ψ transients are obtained.[9] This procedure was performed for circular cones as well as for hemispherical centers. Figure 10 gives the calculated curve for a hemispherical model, which was found to give the best agreement with the experimental results of Fig. 8. The corresponding variation of the refractive index with time is given in Fig. 11.

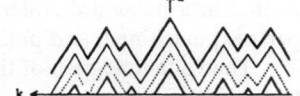

FIGURE 9. Schematic model of the growth of right circular cones at four different times.

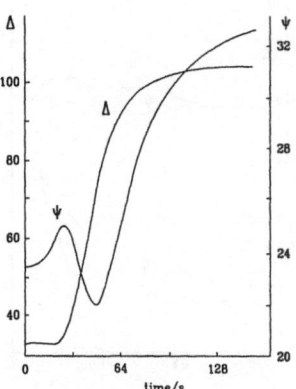

FIGURE 10. Calculated Δ and ψ transients for hemispherical model, with a growth rate of 0.3 nm/s and $N(\text{Ni}) = 2.0 - 4.0i$.

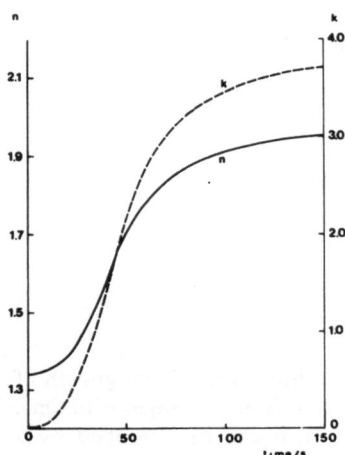

FIGURE 11. Change of the variation of n and k with time for data of Fig. 10.

The shape of the Δ and ψ transients depends on the optical density of the film. It was found that the ψ transient is particularly influenced by the growth model. Thus, ellipsometry can provide additional information about the topography of the deposit.

4. ANODIC OXIDATION OF NICKEL–COBALT ALLOY

As an illustration of the formation of an oxide layer, the growth of an anodic film on a nickel–cobalt alloy will be discussed. The oxide film of Ni–Co $(1:2)$ alloy is of interest in view of the excellent catalytic properties of the spinel oxide $NiCo_2O_4$ for electrochemical water oxidation.

The oxidation was investigated in $5M$ KOH electrolyte, prepared from analytical grade chemicals, and the reference electrode was an Hg, HgO electrode ($+0.926$ V versus RHE); the potentials are quoted versus RHE. The nickel–cobalt alloy was prepared from a $1:2$ mixture of nickel (99.8%) and cobalt (99.99%) powder. This was pressed into a pellet and melted in a flame arc under an argon atmosphere. The resulting button was embedded in perspex, machined, and polished to a flat surface ($0.4\,cm^2$).

The voltammogram of the anodic behavior during a positive potential scan from -0.075 to $+1.425$ V is given in Fig. 12, together with the simultaneously recorded ellipsometric curves.

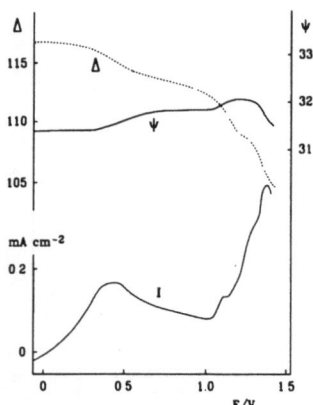

FIGURE 12. Current and ellipsometric behavior of Ni-Co $(1:2)$ alloy during potential scan at $20\,mV/s$ in $5M$ KOH.

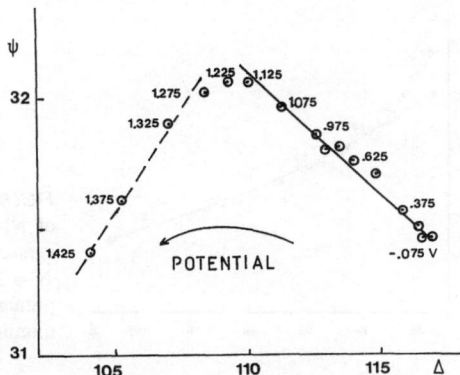

FIGURE 13. ψ–Δ plot of data of Fig. 12. The points refer to potential values. The solid line is the curve for a film with $N = 2.3 - 0.1i$; the dashed line is that for $N = 2.9 - 2.1i$.

Replotting the optical data in a ψ versus Δ graph (Fig. 13) shows that the oxidation can be considered to proceed in two stages: first, an increase of ψ, and thereafter a decrease, while Δ continuously decreases. The changeover occurs around $E = 1.25$ V versus RHE, that is, where the voltammogram indicates the onset of another oxidation step. The drawn line is the best fit to the results during the first oxidation step; this implies that the first film grows at constant refractive index, $N = 2.3 - 0.1i$. Also, the data for the second oxidation appear to lie along a straight line. Several models were tried to fit the results, such as conversion of film 1 to another film or growth of a second film under film 1 or on top of film 1, and the best agreement was obtained with a model in which the second film is formed on top of the first film.[10] The dashed line gives the best fit with $N = 2.9 - 2.1i$. In this experiment the second film reaches a thickness of 2.4 nm, and the first film 3.4 nm.

A remarkable feature of this alloy is its change in electrocatalytic properties for the oxygen evolution with different pretreatments: At the freshly polished alloy a Tafel slope of 40 mV is obtained; when the electrode is subjected to multiple cycling between −0.075 and +1.425 V, the overpotential decreases (Fig. 14), whereas after preanodization at 1.8 V a

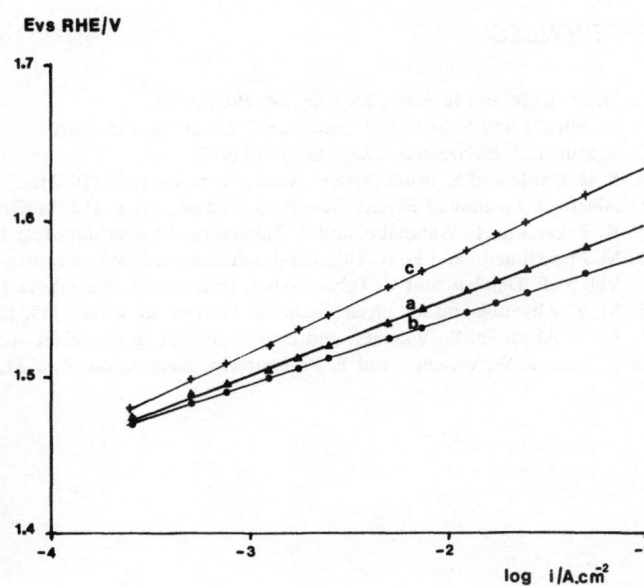

FIGURE 14. Effect of pretreatment on the Tafel plot for oxygen evolution at Ni–Co (1:2) alloy in $5M$ KOH (IR corrected). (a) Freshly polished alloy; (b) after prolonged cycling (22 h); (c) after prolonged oxidation (10 h) at 1.8 V.

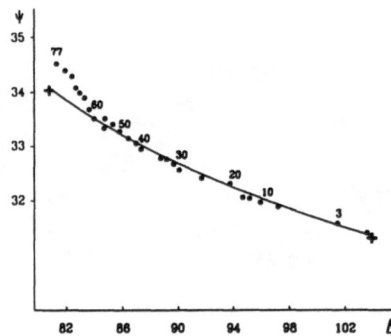

FIGURE 15. Change of Δ and ψ during repeated cycling of Ni–Co (1:2) alloy; the data were measured at 1.425 V versus RHE. The solid line is the calculated curve for $N = 2.3 - 0.1i$ with thickness increasing up to 17.5 nm. The points refer to data measured at 1.425 V after the indicated number of cycles.

higher overpotential and a slope of 60 mV are found. This difference in behavior must be caused by a change in the oxide layer, and therefore ellipsometric measurements were carried out during prolonged cycling. It was found that with cycling Δ and ψ keep changing (Fig. 15). This points to a continuous growth of an oxide layer. The best agreement with the experimental results is obtained when it is assumed that this cycling treatment promotes the growth of film 1 while film 2 does not increase. The solid line in Fig. 15 is a calculated curve for the homogeneous growth of film 1 ($N = 2.3 - 0.1i$).

If, however, the anodic layer is formed by constant-potential oxidation at 1.8 V, its catalytic properties are less (cf. Fig. 14, curve c), and from the optical results it follows that now the growth of the first film is limited. Hence, during the oxidation of Ni–Co (1:2) alloy a two-layer oxide film is formed; the first film appears to be poorly conducting (low k), whereas the second film, on top, has a higher conductivity. This difference in conductivity implies that if a spinel-like oxide layer is formed, it must be the upper film. It is, however, not possible to deduce the composition of the oxide layer from these results. It can also be argued that Ni and Co oxides are formed from which Co may dissolve depending on the oxidation mode.[10]

REFERENCES

1. D. M. Kolb and R. Kotz, *Surf. Sci.* **64**, 698 (1977).
2. N. Furuya and S. Motoo, *J. Electroanal. Chem.* **72**, 165 (1976).
3. S. Stucki, *J. Electroanal. Chem.* **80**, 375 (1977).
4. S. H. Cadle and S. Bruckenstein, *Anal. Chem.* **43**, 1858 (1971).
5. *American Institute of Physics Handbook*, 2nd ed., pp. 6–114, McGraw-Hill, New York (1963).
6. K. Takamura, F. Watanabe, and T. Takamura, *Electrochim. Acta* **26**, 979 (1981).
7. M. Fleischmann and H. R. Thirsk, in: *Advances in Electrochemistry and Electrochemical Engineering*, Vol. 3 (P. Delahay and C. Tobias, eds.), Interscience, New York (1963).
8. M. Y. Abyaneh and M. Fleischmann, *J. Electroanal. Chem.* **119**, 187 (1981).
9. M. Y. Abyaneh, W. Visscher, and E. Barendrecht, *Electrochim. Acta* **28**, 285 (1983).
10. J. Haenen, W. Visscher, and E. Barendrecht, *Electrochim. Acta* **31**, 1541 (1986).

6

Combined Ellipsometric and Reflectance Measurements for Characterization of Films Formed on Electrodes

Woon-kie Paik and Duckhwan Lee

1. INTRODUCTION

The conventional form of ellipsometry has difficulty in characterizing a light-absorbing film, because there are only two equations for three unknown properties of the film at a fixed angle of incidence and at fixed wavelength. The three unknown properties are the real and imaginary parts of the dielectric constant and the thickness of the film.[1] In order to overcome this difficulty, supplementary measurements are necessary. However, since these supplementary measurements are often nonoptical or nonsimultaneous with the optical measurements, they are not suitable for *in situ* monitoring of the formation of thin films on electrode surfaces. The *reflectance–ellipsometry* technique, or *three-parameter ellipsometry* as it is often called, in which the ellipsometric parameters and reflectance are simultaneously measured, has proved to be applicable without such difficulties for thin films formed on electrode surfaces.[2-5] With this technique, *transient reflectance–ellipsometry* measurements become possible and can be used to characterize the films as functions of time during film formation and during the early stage of growth and change of the films.

In this chapter, we present some results of transient reflectance–ellipsometric measurements on anodic films. We also discuss the precautionary considerations concerning the precision and error with regard to the experimental situations for the individual systems studied by the method.

2. MEASUREMENTS

The principle of the reflectance–ellipsometry technique is described in the literature.[2-6] Unlike in the classical form of ellipsometry, in which only two optical parameters, Δ and Ψ, are measured, three optical parameters, usually reflectance R, Δ, and Ψ, are measured in the

Woon-kie Paik and Duckhwan Lee ● Department of Chemistry, Sogang University, Seoul 121-742, Korea.

Electrochemistry in Transition, edited by Oliver J. Murphy *et al.* Plenum Press, New York, 1992.

reflectance–ellipsometry technique. Alternatively, two components of R, R_p and R_s, can be measured in place of R and Ψ. In the studies reported here, the set of R, Δ, and Ψ is measured for films formed on electrode metal surfaces. Actually, the Δ and Ψ values for the film-free reference surface, Δ_0 and Ψ_0, are measured first, from which the optical constants of the bulk metal are calculated and hence R_0 is obtained. Then the relative change in R is followed by recording the intensity of the reflected light relative to the incident light. The changes in Δ and Ψ are simultaneously recorded. The conditions for formation and growth of the films were controlled electrochemically.

An automatic ellipsometer built in this laboratory[7,8] to suit the requirements of the transient reflectance–ellipsometry technique was used for most of the results presented here except those that were made before 1982. The instrument is a self-nulling ellipsometer using two faraday modulator-rotators with the extra capability of recording the ratio of incident and reflected light.

3. FILM PROPERTIES AS FUNCTIONS OF TIME

By means of the reflectance–ellipsometry technique, anodic films on several metal electrodes have been characterized. Table 1 is a partial summary of the results obtained by one of the present authors and co-workers from steady-state measurements since the method was conceived in the early 1970s. It can be seen from the table that, for these extremely thin films, with thicknesses of the order of nanometers, the optical constants (n and k) and the thickness of the films could be determined with reasonable accuracy without auxiliary measurements or assumptions other than the use of the model of a single uniform film. The three simultaneous equations for Δ, Ψ and R, expressed in terms of three unknown film properties (n, k, τ), can be solved numerically, for example, by the Newton–Raphson method. All films have small, but nonzero, values of the imaginary part of the optical constant, $\tilde{n} = n - ik$. For this reason, conventional ellipsometry dealing with two parameters only could not have been applicable for determination of the three unknowns in a single measurement.

The reflectance–ellipsometry technique is suitable in particular for transient measurements on films in the formation or initial growth stages because all three optical signals can be recorded simultaneously with an automatic ellipsometer equipped with an intensity recording device.

In Table 2 are summarized the results of such measurements, from which the changing values of the thickness and the optical constants could be determined as functions of time. Time resolution in the subsecond range has been possible, but a much shorter response time should be possible with improvements in the instrumentation. From the table it can be seen

TABLE 1

Thickness (τ) and Optical Constants (n, k) of Surface Films on Electrodes Determined by Reflectance–Ellipsometry Technique

Metal/solution	λ (nm)	n	k	τ (nm)	Conditions	Reference
Co/borate, pH 8.3	545.6	2.6	0.4	1.3	−0.08 V vs. SCE	2
Pt/0.05M H$_2$SO$_4$	546.1	2.8	1.7	0.2–0.5	Anodic, steady	6
Ni/sulfate, pH 11.0	632.8	2.61	0.46	1.55	1 min, 0.23 V vs. SCE	7
Ni/sulfate, pH 3.21	632.8	1.94	0.35	1.23	1 min, 0.5 V vs. SCE	8
Fe/borate, pH 8.4	632.8	2.59	0.42	2.25	1 min, 0.0 V vs. SCE	9

TABLE 2

Transient Values of the Optical Constants (n, k) and Thickness (τ) of Surface Films Formed on Electrodes Determined by the Reflectance–Ellipsometry Technique

Metal/solution	Potential	λ (nm)	Time	n	k	τ (Å)
Ni/pH 3.21[a]	500 mV vs. SCE	632.8	0.2 s	2.1	0.4	5.5
			1 s	1.80	0.28	11.3
			1 min	1.94	0.35	12.3
Ni/pH 1.81[a]	520 mV vs. SCE	632.8	0.2 s	1.9	0.4	7.5
			1 s	1.70	0.23	13.7
			1 min	1.85	0.34	14.0
Ni/0.1M NaOH[b]	558 mV vs. SHE	546.1	10 s	2.36	0.55	15.0
			10 min	2.31	0.40	17.1
Ni/pH 13[c]	128 mV vs. SHE	632.8	1 s	2.41	0.7	9.2
			10 s	2.56	0.70	10.1
			1 min	3.01	0.54	9.6
Ni/pH 11[c]	228 mV vs. SHE	632.8	1 s	2.1	0.2	7.0
			10 s	2.44	0.43	10.7
			1 min	2.61	0.46	15.5
Fe/pH 8.4[d]	0 mV vs. SCE	632.8	1 s	2.4	0.7	
			10 s	2.56	0.62	20.5
			1 min	2.59	0.42	22.5
Fe/pH 13[d]	−200 mV vs. SCE	632.8	1 s	2.8	0.3	
			10 s	2.83	0.37	15.2
			1 min	2.85	0.28	17.2

[a] Ref. 8.
[b] Ref. 10.
[c] Ref. 7.
[d] Ref. 9.

that, although n and k change to an extent which is beyond the error range of the determinations, there is no abrupt change in the optical constants which might suggest a phase transformation in the films during the passive film growth on ion or nickel.

4. ERROR AND THE ANGLE OF INCIDENCE

For a successful application of the reflectance–ellipsometry technique, an evaluation of the error of measurement and the precision of determination has to be made. This is particularly important for extremely thin films forming on electrodes such as the anodic films on noble metals and the passive films forming on ferrous metals. Although the three equations involved are independent from each other, their numerical near-dependency can be serious. This problem is compounded by the fact that the ellipsometry signals, particularly the changes in Ψ and R, can be very small. Simulation calculations have been made for the optical signal changes for typical electrode films using the known thickness and optical constants and for hypothetical variations of the optical constants and thickness. Figures 1–3 show such calculated changes in the optical signals as functions of the angle of incidence of the light beam in the ambient medium. The results shown in the figures are for the passive film formed on

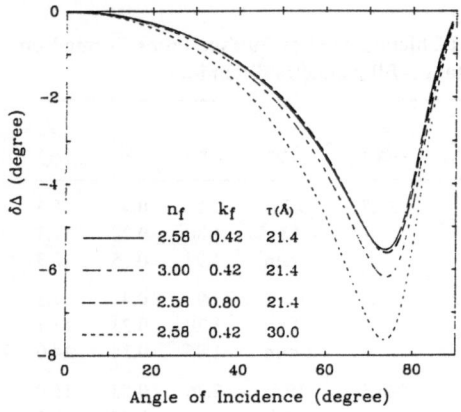

FIGURE 1. Calculated changes in Δ due to the passive film formed on an iron electrode in borate buffer solution of pH 8.4 at the electrode potential of 0 V versus SCE as a function of angle of incidence. The thickness and the optical constants used for the calculation of the solid line are obtained from the actual reflectance–ellipsometry measurements at one angle of incidence. Effects of hypothetical changes in the film properties and thickness indicated on the figure are shown by the broken lines. The wavelength used is λ = 632.8 nm.

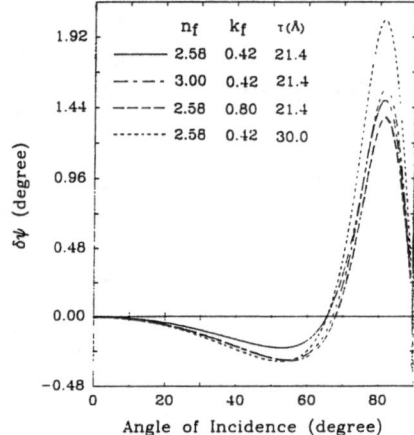

FIGURE 2. Same as Fig. 1 for changes in Ψ.

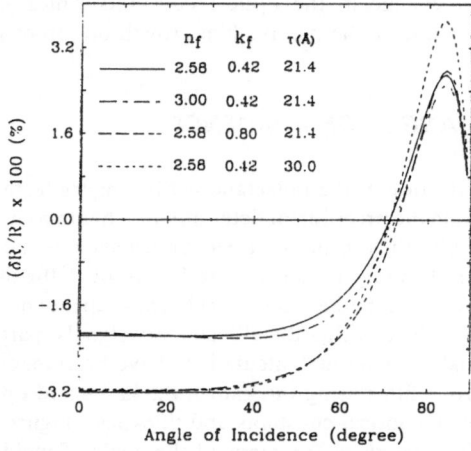

FIGURE 3. Same as Fig. 1 for changes in relative reflectance R.

iron, but the general features are typical of many thin films on metal electrodes. The change
in Δ is the largest (Fig. 1) at the angle of incidence near 73°, which is called the *principal
angle of incidence*. Conventional ellipsometry in which only two parameters are measured is
usually performed at angles near this principal angle of incidence. This angle is, however, a
very poor choice for the reflectance–ellipsometry technique since the reflectance hardly
changes with film formation at this angle (Fig. 2). A visual inspection of the figures reveals
that at angles between 50° and 60°, not only do the reflectance and Ψ change significantly,
but also the changes in Δ are of good magnitude. Another factor to be taken into consideration
is whether each of the film properties affects the optical signals in a distinctive way. This is
related to the precision of determination using the three simultaneous equations. Although
the three equations for Δ, Ψ, and R are independent, determination of the film properties
by numerical solution of the three simultaneous equations may not be precise. Numerical
solution employs the first-order approximation relating the variations in the film properties
(n, k, τ) to the changes in the optical signals (Δ, Ψ, R),

$$\delta\Delta = \left(\frac{\partial\Delta}{\partial n}\right)_{k,\tau} \delta n + \left(\frac{\partial\Delta}{\partial k}\right)_{n,\tau} \delta k + \left(\frac{\partial\Delta}{\partial \tau}\right)_{k,n} \delta\tau$$

$$\delta\Psi = \left(\frac{\partial\Psi}{\partial n}\right)_{k,\tau} \delta n + \left(\frac{\partial\Psi}{\partial k}\right)_{n,\tau} \delta k + \left(\frac{\partial\Psi}{\partial \tau}\right)_{k,n} \delta\tau$$

$$\delta R = \left(\frac{\partial R}{\partial n}\right)_{k,\tau} \partial n + \left(\frac{\partial R}{\partial k}\right)_{n,\tau} \delta k + \left(\frac{\partial R}{\partial \tau}\right)_{k,n} \delta\tau$$

According to Cramer's rule, the simultaneous equations have a solution if the determinant
of the coefficients

$$\begin{vmatrix} \left(\dfrac{\partial\Delta}{\partial n}\right)_{k,\tau} & \left(\dfrac{\partial\Delta}{\partial k}\right)_{n,\tau} & \left(\dfrac{\partial\Delta}{\partial \tau}\right)_{k,n} \\[2em] \left(\dfrac{\partial\Psi}{\partial n}\right)_{k,\tau} & \left(\dfrac{\partial\Psi}{\partial k}\right)_{n,\tau} & \left(\dfrac{\partial\Psi}{\partial \tau}\right)_{k,n} \\[2em] \left(\dfrac{\partial R}{\partial n}\right)_{k,\tau} & \left(\dfrac{\partial R}{\partial k}\right)_{n,\tau} & \left(\dfrac{\partial R}{\partial \tau}\right)_{k,n} \end{vmatrix}$$

does not vanish. If the determinant is very small, the errors propagate through the calculation
to result in large uncertainties in the final results. This is the situation when the optical signal
changes with each of n, k, and τ are not distinctive, because the determinant becomes very
small when any two columns or any two rows of the determinant are proportional to each
other, for example,

$$\left(\frac{\partial\Delta}{\partial \tau}\right) \bigg/ \left(\frac{\partial\Delta}{\partial n}\right) \approx \left(\frac{\partial\Psi}{\partial \tau}\right) \bigg/ \left(\frac{\partial\Psi}{\partial n}\right) \approx \left(\frac{\partial R}{\partial \tau}\right) \bigg/ \left(\frac{\partial R}{\partial n}\right)$$

The calculated magnitudes of the determinant of the coefficients are listed in Tables 3–5 for
films on nickel and platinum, respectively, as functions of the angle of incidence along with
the magnitudes of the optical signals. The calculated magnitudes of the determinants for
anodic films on iron, nickel, and platinum are shown in Fig. 4 as functions of the angle of
incidence. A strong dependence of the determinant on the angle is noticed, especially for
angles greater than the principal angle of incidence. With the platinum surface film, for which
the optical signals are small due to the thinness of the film, the problem becomes more serious
because of the extremely small value of the determinant over the whole range of angles of
incidence. This mandates careful preliminary considerations for the experiment with platinum.

TABLE 3

Dependence of Magnitudes of the Optical Signals and of the Determinant of the Coefficients on the Angle of Incidence for Passive Films on Nickel[a]

Angle (deg)	$-\delta\Delta$ (deg)	$-\delta\Psi$ (deg)	$-\delta R/R$ (%)	Determinant
40	0.54	0.112	0.911	0.00405
50	0.89	0.151	0.889	0.00867
55	1.11	0.256	0.827	0.0115
60	1.36	0.140	0.710	0.0142
70	1.79	0.012	0.236	0.0158
80	1.36	−0.247	−0.442	0.00631

[a] $n = 1.733$, $k = 0.274$, $\tau = 1.55$ nm, $\lambda = 632.8$ nm, film formed in solution of pH 1.81.

TABLE 4

Dependence of Magnitudes of the Optical Signals and of the Determinant of the Coefficients on the Angle of Incidence for Anodic Film on Platinum[a]

Angle (deg)	$-\delta\Delta$ (deg)	$-\delta\Psi$ (deg)	$-\delta R/R$ (%)	Determinant
40	0.266	0.078	0.983	0.000347
50	0.443	0.118	0.961	0.000758
60	0.695	0.150	0.877	0.00130
70	1.008	0.127	0.634	0.00161
80	1.001	−0.018	0.125	0.000813

[a] $n = 3.02$, $k = 1.45$, $\tau = 0.376$ nm, $\lambda = 546.1$ nm; film formed in $0.05M$ H_2SO_4.

The maxima in the determinant value also appear near the principal angle of incidence. Therefore, in the case of platinum an angle of incidence near the principal angle seems to be a better choice at the expense of sensitivity in R. Angles immediately beyond the principal angle also give optical signals comparable to or better than those obtained at angles 10° to 20° below the principal angle. However, at these angles the curves in the figure are very steep, and hence only a small error in the incidence angle may result in a large error in the final results.

TABLE 5

Magnitudes of the Optical Signal Changes and of the Determinant of the Coefficients in the Newton–Raphson Solutions of the Reflectance-Ellipsometry Equations for the Surface Films

Metal/solution	Potential (mV vs. SCE)	λ (nm)	Angle (deg)	$-\delta\Delta$ (deg)	$-\delta\Psi$ (deg)	$-\delta R/R$ (%)	Determinant
Co/pH 8.3	−80	546	58.10	2.23	0.17	1.46	0.0200
Pt/0.05M H_2SO_4		546	69.00	0.98	0.135	0.67	0.00161
Ni/pH 1.81	520	632.8	55.00	1.17	0.152	0.90	0.0116
Ni/pH 3.21	500	632.8	55.00	1.11	0.125	0.85	0.0100
Fe/pH 8.4	0.0	632.8	60.00	3.41	0.168	1.45	0.0260
Fe/pH 13	−200	632.8	60.00	2.98	0.133	1.39	0.0165

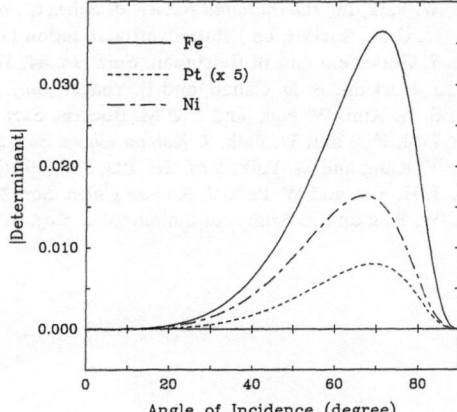

FIGURE 4. Calculated magnitudes of the determinant of coefficients of the three reflectance-ellipsometry equations as functions of angle of incidence. Calculations are made for anodic films on iron, nickel, and platinum.

The changes in Ψ are steep with changes in the angle of incidence near the principal angle of incidence (Fig. 2). Therefore, any small error in the incidence angle due to a misalignment of the instrument can result in large errors in the calculated final result. This is one more reason to avoid incidence angles near the principal angle. The choices of angle of incidence made for the systems studied are listed in Table 5. Also shown in the table are the magnitudes of the optical signals and of the determinant of the coefficients, all of which are factors to be taken into consideration for the proper choice of angle of incidence. As a rule, angles between 10° and 15° below the principal angle of incidence seem to be a good choice in the reflectance–ellipsometry technique.

5. SUMMARY

1. The reflectance–ellipsometry technique (three-parameter ellipsometry) is suitable for studying thin films formed on electrodes.
2. The reflectance–ellipsometry technique with an automatic instrument is suitable for transient measurements of film formation and in early stages of film growth.
3. Considerations about the problems of error lead to recognition of the importance of proper choice of the angle of incidence in the reflectance–ellipsometry technique.

ACKNOWLEDGMENT

The financial support from the Korean Science and Engineering Foundation given to one of the authors (W.P.) is gratefully acknowledged.

REFERENCES

1. F. L. McCrackin and J. P. Colson, in: *Ellipsometry in the Measurement of Surfaces and Thin Films* (E. Passaglia, R. R. Stromberg, and J. Kruger, eds.) pp. 61–80, National Bureau of Standards Miscellaneous Publication 256, Washington, D.C. (1964).
2. W. Paik and J. O'M. Bockris, *Surf. Sci.* **28**, 61 (1971).

3. W. Paik, in: *International Review of Science, Physical Chemistry Series One*, Vol. 6, *Electrochemistry* (J. O'M. Bockris, ed.) Butterworths, London (1973).
4. S. Gottesfeld and B. Reichman, *Surf. Sci.* **44**, 377 (1974).
5. J. Horkans, B. D. Cahan, and E. Yeager, *Surf. Sci.* **46**, 1 (1974).
6. S. H. Kim, W. Paik, and J. O'M. Bockris, *Surf. Sci.* **33**, 617 (1972).
7. D. J. Kim and W. Paik, *J. Korean Chem. Soc.* **26**, 369 (1982).
8. Y. Kang and W. Paik, *Surf. Sci.* **182**, 257 (1987).
9. I. H. Yeo and W. Paik, *J. Korean Chem. Soc.* **28**, 271 (1984).
10. W. Paik and Z. Szklarska-Smialowska, *Surf. Sci.* **96**, 401 (1980).

In Situ *SNIFTIRS and Radiotracer Study of Adsorption on Platinum:* CF_3SO_3H *and* H_3PO_4

Piotr Zelenay

1. INTRODUCTION

Methods for the study of adsorption extend at present well beyond the "classical" electrochemical techniques used for many years in measurements of both molecular and ionic adsorption on a liquid mercury electrode. The latter techniques, which basically rely upon inherently indirect measurements of one electrical variable (such as current, charge, or capacitance) as a function of another (e.g., electrode potential), prove insufficient when nonideally polarizable solid electrodes are used for adsorption studies. Quite frequently, many complex electrochemical reactions accompany adsorption on the catalytic surface of a solid, thereby making it impossible to perform any accurate adsorption measurements by purely electrochemical means. Carefully selected adsorbate-sensitive *in situ* techniques must thus be employed to measure the adsorption process among a variety of other electrode reactions which usually occur on the surface of a given electrode. The vibrational spectroscopic probes, that is, the infrared and Raman spectroscopies, available now to electrochemical adsorption studies (for a review, see Bewick and Pons[11]), can provide deep molecular-level insight into layers formed on solid electrodes and greatly facilitate investigation of the adsorption mechanism and formulation of a coherent adsorption model.

Although the vibrational spectroscopies have undoubtedly been of great service to adsorption studies, they have continuously suffered from inability to provide unambiguous information on the amount of species adsorbed on the surface. When used in combination with some reliable quantitative methods, the infrared and Raman spectroscopies become powerful tools for adsorption studies on solid electrodes. Recent reports[2-5] have shown that merging the vibrational spectroscopies and various radiotracer techniques, probably the best methods of quantitative determination of adsorption on solids,[6] is of particular advantage to surface electrochemical studies.

Piotr Zelenay ● Department of Chemistry, Warsaw University, 02-089 Warsaw, Poland.

Electrochemistry in Transition, edited by Oliver J. Murphy *et al.* Plenum Press, New York, 1992.

2. INFRARED AND RADIOTRACER STUDY OF CF₃SO₃H AND H₃PO₄ ADSORPTION ON PLATINUM

2.1. Methods

The IR spectra of adsorbed trifluoromethanesulfonic acid (CF_3SO_3H) and phosphoric acid (H_3PO_4) were recorded by using surface normalized interfacial Fourier transform infrared spectroscopy (SNIFTIRS) according to the procedure described elsewhere.[1,2,7] In this method, the reflectivity change of a carefully polished platinum electrode, defined as ΔR, was measured as a function of adsorption potential. It was normalized with respect to the infrared reflectance spectrum taken at the potential where adsorption was judged to be negligible (R_0). The normalized reflectivity change ($\Delta R/R_0$), equal to the IR absorbance of the adsorbed monolayer, was then plotted as a function of the electrode potential and compared with radiotracer data.

The radiotracer experiment was carried out[4] on the electrolytically deposited platinum electrodes of slightly roughened surface (roughness factor of about 60). The surface concentration of adsorbed CF_3SO_3H (^{14}C labeled) and H_3PO_4 (^{32}P labeled) was calculated from the formula derived by Wroblowa and Green[8]:

$$\Gamma = [(N_{\text{total}} - N_{\text{backg}})/N_{\text{backg}}]c/(\mu \cdot r) \tag{1}$$

Here, Γ is the surface concentration (mol/cm²), N_{total} and N_{backg} are the counts measured in the adsorption and no-adsorption situation, respectively, c is the solution concentration (mol/cm³), μ is the linear coefficient of absorption of β^- radiation in the solution phase (cm⁻¹), and r is the roughness factor measured as a ratio of the real to the geometric surface area. The values of μ for β^- radiation emitted by ^{14}C and ^{32}P were calculated from the semiempirical equations given by Libby[9] and were 314 and 8.3 cm⁻¹, respectively.

2.2. Potential Dependence of Adsorption

In Fig. 1, differential ($\Delta R/R_0$) spectra of a platinum surface in a $2 \times 10^{-2}M$ aqueous solution of CF_3SO_3H are shown at 0.00, 0.40, and 1.00 V with respect to the normal hydrogen electrode (NHE). The wide absorption band in the region from 1400 to 950 cm⁻¹ consists of

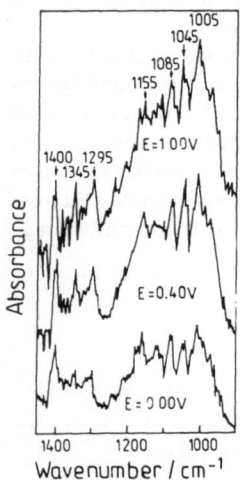

FIGURE 1. SNIFTIRS spectra of adsorbed $CF_3SO_3^-$ on platinum at various potentials[4]; $c = 2 \times 10^{-2}M$, $T = 298$ K. Reprinted by permission of the publisher, The Electrochemical Society, Inc.

several rather poorly developed peaks. Their wave numbers can be compared with the IR frequency data published by Balicheva et al.[10] for CF_3SO_3H dissolved in water. The peaks observed at 1045 and 1400 cm^{-1} are attributed to the symmetric and asymmetric stretching vibration of the S—O bonds, respectively, whereas the peaks at 1155 and 1295 cm^{-1} are assigned to the C—F bond stretch.[10]

In order to minimize the possible effect of an impurity upon a given peak, the entire band area (1400-950 cm^{-1}) was taken as a measure of $CF_3SO_3^-$ adsorption. The maximum band area at 0.80 V was calibrated with respect to the maximum surface concentration of adsorbed trifluoromethanesulfonic ion, as determined by the radiotracer method. The surface concentration data were then converted into surface coverage values (θ) using the calculated value of the area occupied by one ion (17.6 Å^2). As shown in Fig. 2, the calculated surface coverage increases with potential up to 0.80 V, then remains constant up to ca. 1.00 V, dropping slightly when the potential becomes more anodic. The latter observation is in disagreement with the radiotrace result, which shows a sharp increase of Γ at highly positive potentials.

A possible explanation of the discrepancy between SNIFTIRS and radiotracer data is in terms of a surface chemical reaction leading to the formation of fluorosubstituted organic sulfate:

$$2CF_3SO_3^- + 2PtO \rightarrow (CF_3)_2SO_4 + SO_4^{2-} + 2Pt \qquad (2)$$

or perfluorinated ethane (Kolbe-type reaction):

$$2CF_3SO_3^- \rightleftharpoons 2CF_3SO_3 + 2e^- \rightarrow C_2F_6 + 2SO_3 \qquad (3)$$

If the products of reactions (2) and (3) remain on the electrode surface, an increase in the radiotracer signal is expected due to presence of two carbon atoms in each molecule. At the same time, the SNIFTIRS signal should drop as a result of gradual removal of $CF_3SO_3^-$ from the adsorbed layer.

The IR spectra of H_3PO_4 (Fig. 3), recorded in $6.2 \times 10^{-2}M$ aqueous solution, show an absorbance peak at 1074 cm^{-1}. It is merely visible at 0.20 V but increases with potential and reaches its maximum intensity at about 0.80 V. The peaks in close proximity to that at 1074 cm^{-1} behave in the same way and hence are of the same origin as the main one. It can easily be ascertained from the infrared data[11] that the peak at 1074 cm^{-1} is due to the P—O bond stretching vibration.

The potential dependence of H_3PO_4 adsorption, obtained in a manner similar to that used in the study of $CF_3SO_3^-$ adsorption, is given in Fig. 4. The θ values for phosphoric acid adsorption were calculated using 19.4 Å^2 as the cross-sectional area of the adsorbed H_3PO_4 molecule. The very high maximum surface coverage of 0.9 corresponds to the surface concentration of 8×10^{-10} mol/cm^2 which was measured radiometrically at 0.80 V.

It should be noted that almost perfect agreement between SNIFTIRS and radiotracer adsorption data is found in the case of phosphoric acid adsorption.

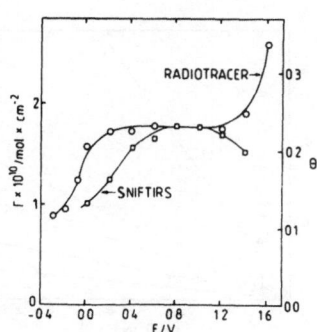

FIGURE 2. Potential dependence of $CF_3SO_3^-$ adsorption on platinum[4]; $c = 2 \times 10^{-2}M$, $T = 298$ K. Reprinted by permission of the publisher, The Electrochemical Society, Inc.

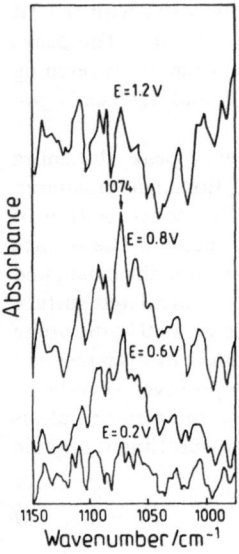

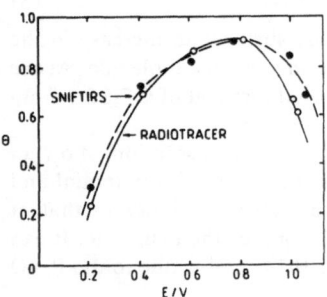

FIGURE 3. SNIFTIRS spectra of adsorbed H_3PO_4 on platinum at various potentials[2]; $c = 6.2 \times 10^{-2} M$, $T = 298$ K. Reprinted by permission of the publisher, The Electrochemical Society, Inc.

FIGURE 4. Potential dependence of H_3PO_4 adsorption on platinum as measured by SNIFTIRS[2] and the ratiotracer method[4]; $c = 6.2 \times 10^{-2} M$, $T = 298$ K. Selected data reprinted by permission of the publishers, The Electrochemical Society, Inc., and the American Chemical Society.

2.3. Effect of Solution Concentration and Temperature

The amount of adsorbed $CF_3SO_3^-$ sharply decreases with an increase in temperature (Fig. 5). The concentration dependence of θ is consistent with the Bockris–Swinkels solvent displacement model,[12] namely:

$$[\theta/(1-\theta)^n]\{[\theta + n(1-\theta)]^{n-1}/n^n\} = K(c/c_w) \tag{4}$$

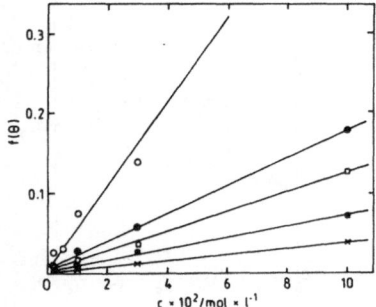

FIGURE 5. Adsorption of $CF_3SO_3^-$ at 0.80 V as a function of concentration and temperature[4]: ○, 298 K; ●, 323 K; □, 333 K; ■, 343 K; ×, 353 K. The ordinate $f(\theta)$ is a configurational term in the Bockris–Swinkels isotherm (see text). Reprinted by permission of the American Chemical Society.

Here, c and c_w are the concentrations of solute and solvent, respectively, n is the number of solvent (water) molecules displaced from the surface by one adsorbing solute, and K is the equilibrium constant for adsorption.

The best fit of the model to the experimental results was obtained for $n = 3$. Somewhat unexpectedly, no repulsive interaction is detected within the adsorbed layer; this may be caused by the relatively low electrode coverage with trifluoromethanesulfonic ions (ca. 0.35 at maximum; Fig. 2).

Phosphoric acid adsorption also decreases with an increase in temperature and with a decrease in solution concentration (Fig. 6). As in the case of $CF_3SO_3^-$ adsorption, the qualitative behavior of the relation between surface coverage and solution concentration is independent of temperature. Several adsorption isotherms were tested in order to find the best consistency with the experimental results, including those of Langmuir, Frumkin, Temkin, and Flory–Huggins. The best fit was found, however, for the Bockris–Swinkels isotherm ($n = 3$, the same value as for $CF_3SO_3^-$) after dispersive interaction within the layer of adsorbed H_3PO_4 had been accounted for.[4] A modified version of the Bockris–Swinkels isotherm, thus obtained, is as follows:

$$f(\theta) \exp(-A \cdot \theta^3/RT) = K \cdot (c/c_w) \tag{5}$$

Here, $f(\theta)$ is the configurational term on the left-hand side of Eq. (4), and A is a term based upon lateral dispersive interaction between molecules of the adsorbate.[13] The value of A was found to be 3.9 kJ/mol.

It should be stressed that the number of water molecules displaced upon both $CF_3SO_3^-$ and H_3PO_4 adsorption on platinum corresponds very well to the difference in size between each adsorbate and water.

2.4. Evaluation of Thermodynamic Data

In Fig. 7, the $\Delta G^0/T$ values, calculated from the slopes of the straight lines in Figs. 5 and 6, are given for adsorption of the trifluoromethanesulfonic ion and the phosphoric acid molecule, respectively. For both adsorption systems investigated in this work, the standard Gibbs energy of adsorption is negative at all temperatures (Table 1). The standard enthalpy of adsorption, ΔH^0, and the standard entropy of adsorption, ΔS^0, for $CF_3SO_3^-$ and H_3PO_4, calculated from the slope and from the intercept of the plots in Fig. 7, are also given in Table 1. For the sake of comparison, all thermodynamic data presented in this table for $E = 0.80$ V were recalculated in such a way[4,14,15] that they refer to common standard-state conditions of $\theta^0 = 0.5$ and $c^0 = 1$ mol/liter. It is worth noting that both the enthalpy and entropy of H_ePO_4 adsorption on platinum are more than two times higher (more negative) than those found for $CF_3SO_3^-$.

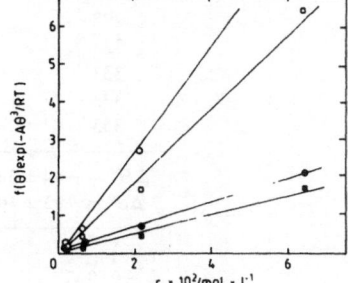

FIGURE 6. Adsorption of H_3PO_4 at 0.80 V as a function of concentration and temperature[4]: ○, 298 K; □, 308 K; ●, 318 K; ■, 328 K. Reprinted by permission of the American Chemical Society.

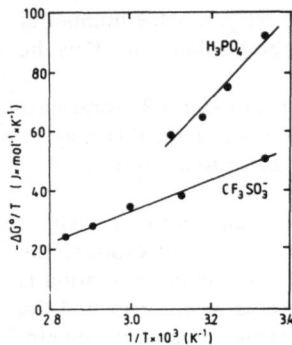

FIGURE 7. Temperature dependence of the standard Gibbs energy of adsorption for $CF_3SO_3^-$ and H_3PO_4 ($E = 0.80$ V).

2.5. State of the Adsorbed Species

The surprising lack of measurable repulsive interaction in the adsorbed $CF_3SO_3^-$ layer may only be explained if the lateral forces between adsorbed ions are diminished as a result of "screening" of their negative charges. This may happen when the charged group of the ion, that is, the SO_3^- group, is located away from the electrode surface. In such a case, the charged part of the $CF_3SO_3^-$ ion would be placed in an environment of high dielectric constant, ε, comparable to that for bulk water ($\varepsilon \approx 80$), rather than in the molecular layer adjacent to the surface, where the dielectric constant is estimated[16] to be around 6.

Strong support of the latter hypothesis is provided[4] when the applicability of the Bockris–Devanathan–Müller (BDM) theory[17] to $CF_3SO_3^-$ adsorption on platinum is tested. The results of such a test not only confirm the applicability of the BDM model in this case, but also indicate that the dielectric constant for lateral interaction in the adsorbed $CF_2SO_3^-$ layer is equal to ca. 50.[4] The latter value is consistent with a vertical orientation of trifluoromethanesulfonic ion with the SO_3^- group located ca. 6 Å away from the surface and the CF_3 group in contact with platinum.

TABLE 1
Comparison between Thermodynamic Data for $CF_3SO_3^-$
and H_3PO_4 Adsorption on Platinum at the Standard State
of $\theta^\circ = 0.5$ and $c^\circ = 1$ mol/liter[a, b].

CF$_3$SO$_3^-$		H$_3$PO$_4$	
T (K)	ΔG° (KJ/mol)	T (K)	ΔG° (KJ/mol)
298	−15.6	298	−28.0
323	−13.6	308	−24.1
333	−13.0	318	−21.6
343	−11.7	328	−21.7
353	−10.2	—	—

$\Delta H^\circ = -44 \pm 3$ kJ/mol $\Delta H^\circ = -92 \pm 20$ kJ/mol
$\Delta S^\circ = -93 \pm 10$ J/mol · K $\Delta S^\circ = -219 \pm 63$ J/mol · K

[a] Ref. 14.
[b] $E = 0.80$ V.

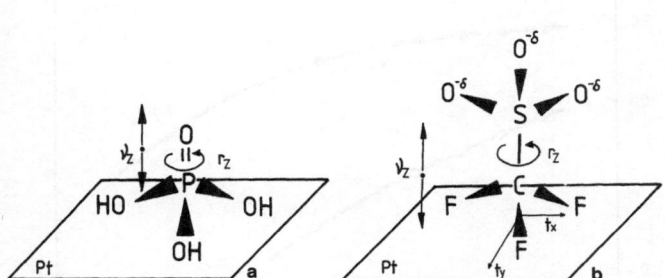

FIGURE 8. Orientation at the surface and motions executed by adsorbed H_3PO_4 (a) and $CF_3SO_3^-$ (b); ν_z denotes vibration along the z-axis (which is chosen to be perpendicular to the surface), r_z refers to restricted rotation about the z-axis, and t_x, t_y denote translation along the x- and the y-axis, respectively.

Further, the strongly hydrophobic nature of the CF_3 group, on the one hand, and the hydrophilic behavior of the SO_3^- group (formation of hydrogen bonds with water), on the other, should stabilize this orientation of adsorbed $CF_3SO_3^-$.

Since adsorption of both $CF_3SO_3^-$ and H_3PO_4 on the Pt electrode is accompanied by the displacement of water molecules, the total entropy of adsorption, $\Delta S°$, is the sum of the change in the entropy associated with adsorption of either $CF_3SO_3^-$ or H_3PO_4 and that associated with the desorption of water from the surface:

$$\Delta S° = (S°_{ads,i} - S°_{sol,i}) + n(S°_{sol,w} - S°_{ads,w}) \tag{6}$$

where n is the number of water molecules displaced, the subscripts ads and sol denote the adsorbed and the dissolved state, respectively, and the subscripts i and w refer to adsorbing species and water, respectively. Each of the entropy terms on the right-hand side of Eq. (6) consists of several partial entropies associated with various motions of respective molecules in either the adsorbed or the dissolved state. Of these partial entropies, only entropies of translation, restricted rotation (including libration), and vibration considerably contribute to the total entropy value. Treating the solution as a quasicrystalline solid and using relatively simple partition functions for energies of the aforementioned motions, the total entropies of adsorption can be calculated[4] for different *assumed* states of adsorbed $CF_3SO_3^-$ and H_3PO_4. The calculated total entropy values which give best agreement with experment (for experimental entropy values, see Table 1) are -90 J/(mol · K) and -232 J/(mol · K), for $CF_3SO_3^-$ and H_3PO_4, respectively. In the case of the trifluoromethanesulfonic ion, the best correlation between calculation and experiment is reached when the adsorbed ion retains two degrees of translational freedom, rotates about the axis perpendicular to the surface, and vibrates along the same axis. In the most likely model for phosphoric acid adsorption, the molecule has no translational degree of freedom but is allowed to vibrate and rotate about the axis perpendicular to the platinum surface.

The most probable orientations of $CF_3SO_3^-$ and H_3PO_4 on the platinum electrode surface, together with the motions which the adsorbed entities are allowed to execute, are shown in Fig. 8.

3. RELEVANCE OF ADSORPTION BEHAVIOR TO FUEL CELL PEFORMANCE

The performance of a fuel cell is strongly affected by the electrolyte used. Certain undesirable characteristics of orthophosphoric acid[18,19] have stimulated growing interest in alternative acid fuel cell electrolytes. Of several electrolytes investigated so far, mainly perfluorinated organic acids, only trifluoromethanesulfonic acid appears stable enough to be a likely substitute for H_3PO_4 in acid fuel cells.

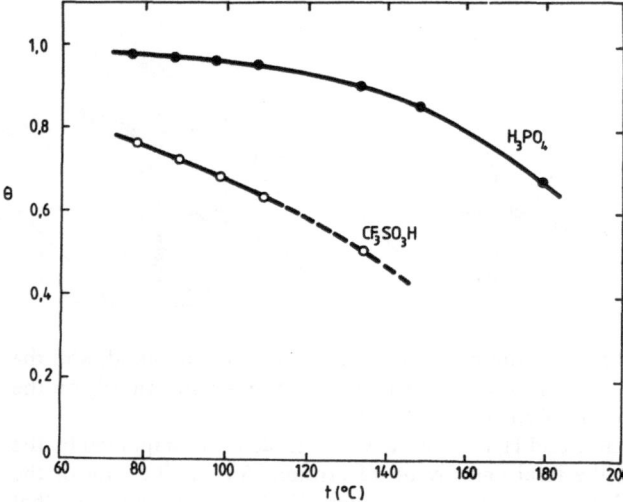

FIGURE 9. Extrapolation of adsorption data to high temperatures and concentrations[14]: O, $5M$ CF_3SO_3H; ●, 98% H_3PO_4. Reprinted by permission of the publisher, The Electrochemical Society, Inc.

One important reason for differences in fuel cell performance can be adsorption of the electrolyte species on a platinum cathode, which should result in lowering of the rate of oxygen reduction (the reaction which usually determines the overall fuel cell performance). As originally suggested by Appleby and Baker,[20] particularly promising rates of oxygen reduction in trifluoromethanesulfonic acid, as compared with phosphoric acid, could be related to low specific adsorption of $CF_3SO_3^-$.

Measurements of adsorption in the region of acid fuel cell practice encounter serious difficulties. The operating conditions for the phosphoric acid fuel cell are 180°C and an acid concentration of 98%, whereas those expected for trifluoromethanesulfonic acid are ca. 100°C and $5M$, respectively. Thus, a meaningful comparison of the two acids can be made after extrapolation of the present results to these temperatures and concentrations. The results of such an extrapolation, which can be made by using the thermodynamic data from Table 1, are shown in Fig. 9.[14] The estimated coverages of $CF_3SO_3^-$ and H_3PO_4 at 0.80 V, under conditions resembling those for fuel cell operation, are much closer to each other than at low temperatures and concentrations. The expected coverage by $CF_3SO_3^-$ in $5M$ aqueous solution and at 100°C would be about 0.7, the same as the coverage estimated for H_3PO_4 in 98% solution and at 180°C.

Since the areas available for oxygen reduction in both media are similar, adsorption of the electrolyte species should have little effect on the performance of the respective fuel cells. It has already been shown experimentally[21] that both oxygen solubility and diffusivity, along with the basic rate constants for oxygen reduction, play more important roles in oxygen reduction in CF_3SO_3H and H_3PO_4 than adsorption of the corresponding electrolyte species.

REFERENCES

1. A. Bewick and S. Pons, in: *Advances in Infrared and Raman Spectroscopy*, Vol. 12 (R. J. H. Clark and R. E. Hester, eds.), pp. 1-62, Wiley Heyden, New York 1(985).
2. M. A. Habib and J. O'M. Bockris, *J. Electrochem. Soc.* **130**, 2510 (1983).
3. M. A. Habib and J. O'M. Bockris, *J. Electrochem. Soc.* **132**, 108 (1985).
4. P. Zelenay, M. A. Habib, and J. O'M. Bockris, *Langmuir* **2**, 393 (1986).

5. D. S. Corrigan, E. K. Krauskopf, L. M. Rice, A. Wieckowski, and M. J. Weaver, *J. Phys. Chem.* **92,** 1596 (1988).
6. P. Zelenay and A. Wieckowski, in: *Electrochemical Interfaces: Modern Techniques for In-Situ Interface Characterization* (H. D. Abruña, ed.), Chapter 9, VCH Publishers, New York (1991).
7. M. A. Habib and J. O'M. Bockris, *J. Electroanal. Chem.* **180,** 287 (1984).
8. H. Wroblowa and M. Green, *Electrochim. Acta* **8,** 679 (1963).
9. W. F. Libby, *Phys. Rev.* **103,** 1900 (1956).
10. T. G. Balicheva, V. I. Ligus, and Yu. Ya. Fialkov, *Russ. J. Inorg. Chem.* (*Engl. Transl.*) **18,** 1701 (1973).
11. K. Nakamoto, *Infrared and Raman Spectra of Inorganic and Coordination Compounds,* 3rd ed. John Wiley & Sons, New York (1978).
12. J. O'M. Bockris and D. A. J. Swinkels, *J. Electrochem. Soc.* **111,** 736 (1964).
13. E. Blomgren and J. O'M. Bockris, *J. Phys. Chem.* **63,** 1475 (1959).
14. P. Zelenay, B. R. Scharifker, J. O'M. Bockris, and D. Gervasio, *J. Electrochem. Soc.* **133,** 2262 (1986).
15. B. R. Scharifker and P. Zelenay, *Acta Cient. Venez.* **39,** 315 (1988).
16. W. R. Smythe, *Static and Dynamic Electricity,* 3rd ed., McGraw-Hill, New York (1968).
17. J. O'M. Bockris, M. A. V. Devanathan, and K. Müller, *Proc. Roy. Soc.* (*London*) *Ser.* A **274,** 55 (1963).
18. A. J. Appleby, *J. Electrochem. Soc.* **117,** 328 (1970).
19. H. R. Kunz and G. A. Graver, *J. Electrochem. Soc.* **122,** 1279 (1975).
20. A. J. Appleby and B. S. Baker, *J. Electrochem. Soc.* **125,** 404 (1978).
21. B. R. Scharifker, P. Zelenay, and K. O'M. Bockris, *J. Electrochem. Soc.* **134,** 2714 (1987).

Electrode Kinetics and Electrocatalysis

8

Transition-Metal Oxide Electrocatalysts for O_2 Electrodes: The Pyrochlores

Jai Prakash, Donald Tryk, Wesley Aldred, and Ernest Yeager

1. INTRODUCTION

There has been wide interest in the search for bifunctional oxygen electrocatalysts which can reversibly or nearly reversibly catalyze both the reduction and the generation of O_2. There are two possible approaches: (1) use of a single bifunctional electrocatalyst which promotes both reactions; and (2) use of separate electrocatalysts for the two reactions within one electrode. Both approaches have been tried by various groups. The use of separate catalysts for these two functions provides a much wider range of materials for consideration. In this chapter, however, the first approach will be emphasized, with the focus on the transition-metal pyrochlore oxides, particularly the lead ruthenate pyrochlore and related materials. These pyrochlores have metallic conductivity, can be prepared in very high area forms, and also show high electrocatalytic activity for both O_2 reduction and generation.[1-4] The nature of the electrocatalysis is also very much dependent on the surface electronic properties of the catalyst, which in turn are dependent to some extent on the bulk properties.

This chapter will review some of the results obtained by various research groups, including the authors', for the lead, bismuth, yttrium, and rare-earth ruthenates concerning their electronic, electrochemical, and electrocatalytic properties, especially for O_2 reduction and generation. Areas to be described include the use of ionomer membranes to stabilize pyrochlore-based gas-fed electrodes, use of high-area carbon as a support material in gas-fed electrodes, and the optimization of the structure of self-supporting gas-fed electrodes.

Much of the effort in the authors' laboratory has been focused on the stoichiometric lead ruthenate $Pb_2Ru_2O_{7-y}$ as a model compound because of its relative simplicity. Additional advantages are that its electronic conductivity is higher than that of the lead-rich compounds,[5] and its stability in concentrated alkaline solution is also higher.[2,3] Horowitz et al.[2,3] have concluded that the catalytic properties for the lead ruthenates are not strongly dependent upon composition but simply depend upon the surface area. They have favored the lead-rich compounds because they can be prepared more readily in high-area form,[2,3,6-8] although they have also been able to prepare the stoichiometric compound in high area form (150 m^2/g;

Jai Prakash • Chemical Technology, Argonne National Laboratory, Argonne, Illinois 60439 and Case Center for Electrochemical Sciences and the Chemistry Department, Case Western Reserve University, Cleveland, Ohio 44106. *Donald Tryk, Wesley Aldred, and Ernest Yeager* • Case Center for Electrochemical Sciences and the Chemistry Department, Case Western Reserve University, Cleveland, Ohio 44106.

Electrochemistry in Transition, edited by Oliver J. Murphy *et al.* Plenum Press, New York, 1992.

Ref. 8). The authors' research group has also been able to prepare the stoichiometric compound in relatively high-area forms.

2. STRUCTURAL, PHYSICAL, AND CHEMICAL PROPERTIES

2.1. General

The pyrochlores can be represented by the formula $A_2B_2O_6O'$, where A is typically a rare earth or an element such as Tl, Pb, or Bi, B is typically a transition or post-transition metal, and O and O' are crystallographically distinct types of oxygen. Vacancies can occur at the O' sites, and other anions such as fluoride can also be substituted at these sites. Thus, there are three different types of sites, A, B, and O', in which substitutions can be made as long as the ionic radii are appropriate and charge neutrality is maintained. Consequently, a large number of different pyrochlores exist with a wide range of physical properties.[9] The $A_2B_2O_7$ type of pyrochlore has been referred to as the stoichiometric or normal type as opposed to the defect pyrochlores, for which vacancies can exist, both in A and O'. With A cations such as Tl^+, Pb^{2+}, and Bi^{3+}, O' vacancies can occur, resulting in the formula $A_2B_2O_{7-x}$. The O' vacancies are thought to be central to the electronic and electrocatalytic properties of these pyrochlores (see, e.g., Refs. 10 and 11), and will be discussed in Section 2.3.

Another variant of the pyrochlore structure is one in which part of the B sites are substituted with A-type cations. This occurs, for example, with lead- and bismuth-based pyrochlores. These Pb-rich and Bi-rich compounds, particularly ruthenates and iridates, have been investigated by Horowitz and co-workers for their electrocatalytic properties[1-3] as well as their precisely controllable resistivities.[5]

2.2. Preparation

Randall and Ward[12] were the first to report the synthesis of Pb–Ru and Pb–Ir oxides with the pyrochlore structure. These compounds actually had Pb/Ru and Pb/Ir molar ratios of 2:1 and were thus lead-rich. Subsequently, the stoichiometric lead ruthenate $Pb_2Ru_2O_{7-y}$ and lead iridate $Pb_2Ir_2O_{7-y}$ were prepared by Sleight[13] and Horowitz et al.[10] These syntheses were traditional solid-state-type ceramic preparations. In order to promote the diffusion of ions during the synthesis and thereby approach a homogeneous composition, the materials were subjected to prolonged high-temperature heat treatments. In this process, however, the surface areas were considerably decreased. Chemically bound water would also tend to be driven off. Materials prepared in this way tend to exhibit relatively low catalytic activities. The full catalytic activity of the pyrochlores was not realized until Horowitz et al.[6,7] reported the synthesis of lead and bismuth ruthenates with quite high surface areas, ranging from 50 to 200 m^2/g. They used an alkaline aqueous solution both as a means of reacting the appropriate metal cations by precipitation and subsequently as a medium for the crystallization of the precipitate. The presence of oxygen in the solution was used to control the oxidation state of the cations. The important factors controlling the alkaline solution synthesis, particularly the composition (i.e., A cation substitution at B cation sites) and surface area, are (1) the solubilities of the A and B cations, which are often in complex anionic forms; (2) pH; (3) temperature; and (4) effective redox potential of the solution environment. The high solubility of lead species relative to ruthenium species, especially with increasing pH and temperature, must be taken into account.[3,6,7] Thus, an excess of lead salt must be used in many cases. To favor the formation of the stoichiometric lead ruthenate $Pb_2Ru_2O_{7-y}$, the concentration of lead in solution should be kept low by replacing the alkaline solution medium periodically with fresh solution.[8] If, however, there is insufficient lead to form the stoichiometric pyrochlore, an impurity phase of RuO_2 can form. The formation of RuO_2 was

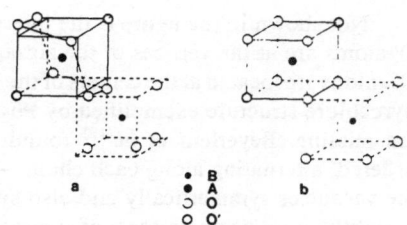

FIGURE 1. Changes in the shapes of the coordination polyhedra of the A cations (e.g., Pb^{2+}) and B cations (e.g., $Ru^{4.5+}$) with the oxygen positional parameter x_O (48f oxygens in the space group $Fd3m$) in the $A_2B_2O_6O'$ structure: (a) $x_O = 0.3125$ (origin at the B cation); (b) $x_O = 0.375$. (Taken from Subramanian et al.[9])

also found to be favored when the pH, temperature, and O_2 partial pressure were all lower than optimum for the pyrochlore.[14]

The relatively low synthesis temperature (70–90°C) used in the alkaline solution technique yields pyrochlores with quite high surface areas.[6–8] Horowitz et al. have been able to prepare crystalline $Pb_2Ru_2O_{7-y}$ with a surface area of 35 m^2/g even without further heat treatment. More recent preparative efforts using the same techniques in this laboratory have achieved crystalline material with a surface area of 105 m^2/g.[15]

2.3. Crystal Structure

There are several different ways of describing the pyrochlore structure. As pointed out by several authors,[9,16,17] this is partly due to the fact that the coordination polyhedra of O^{2-} anions surrounding the A and B cations change shape as a function of the variable positional parameter x_O for the oxygens in the BO_6 polyhedra (Fig. 1). At one extreme, x_O can in principle take the value 0.3125 (origin at the B cation) or $0.75 - 0.3125 = 0.4375$ (origin at the A cation), in which case the BO_6 polyhedra are regular octahedra. At the other extreme, x_O can take the value 0.375 (either A or B origin), in which case the BO_6 polyhedra are highly distorted octahedra, but the AO_8 polyhedra are regular cubes. In the latter case, the structure is similar to the body-centered cubic fluorite structure.[10] The practical range for x_O is actually 0.305 to 0.355.[18] The x_O parameter for $Pb_2Ru_2O_{6.5}$ has been established to be 0.3232 (B origin) or 0.4268 (A origin) by Beyerlein et al.[19] This value is 83% of the way toward the regular octahedron extreme. Thus, the description based on BO_6 octahedra is appropriate for this material.

The BO_6 octahedra are corner-shared and are in a diamondlike network.[16] Figure 2 shows two tetrahedral clusters of such a network. From this figure it can be seen that the B—O—B angle is not 180° as it would be in an ideal perovskite structure. Instead it is 135° (141° for regular octahedra), which affects the overlap of the B and O atomic orbitals that form the metal-like d band[10,11,20] and also affects the orientations of the d-like orbitals on the surface.

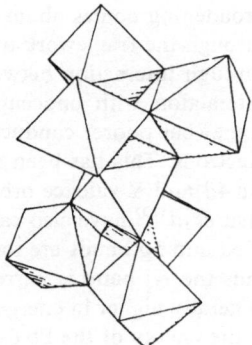

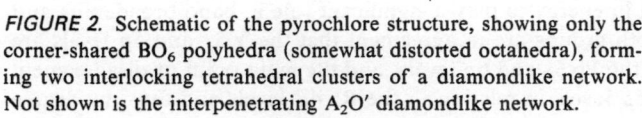

FIGURE 2. Schematic of the pyrochlore structure, showing only the corner-shared BO_6 polyhedra (somewhat distorted octahedra), forming two interlocking tetrahedral clusters of a diamondlike network. Not shown is the interpenetrating A_2O' diamondlike network.

Not shown in the figure is the interpenetrating diamondlike A_2O' network, in which the O' atoms are at the vertices of the zigzag $-O'-A-O'-A-$ chains[18] as in cuprite.[16] The O' anions are located at the centers of the large voids in the B_2O_6 network. In the anion-deficient pyrochlore structure exemplified by $Pb_2Ru_2O_{6.5}$ (or $Pb_2Ru_2O_6O'_{0.5}\square_{0.5}$), half of the O' atoms are missing. Beyerlein *et al.*[19] found using neutron diffraction that these vacancies are ordered, alternating along each chain: $-A-O'-A-\square-$. The Pb atoms are shifted toward the vacancies symmetrically and also symmetrically away from the O' anions. This leads to a slightly more complex form of symmetry, because in the ideal pyrochlore structure, which belongs to the $Fd3m$ space group, the A cation and O' anion positions are fixed, and the O' vacancies are assumed to be randomized. With the ordering of the vacancies and the associated movement of the A cations, the structure then belongs to the $F\bar{4}3m$ space group.[19]

Goodenough and co-workers[10,11] have proposed that the Pb $6s^2$ electron pairs are drawn into the O' vacancies so as to shield the repulsive interactions between the A cations. Such electrostatic shielding of the A^{2+} ions by negatively charged O' vacancies is needed to explain the stability of the defect pyrochlore structure relative to that of the perovskites $ARuO_3$, where A = Ca or Sr. Delocalization of the Pb lone-pair electrons into the O' vacancies, involving $6s$-$6p$ hybridization, was proposed,[10,11] leading to the concept of vacancy-mediated or "trap-mediated" Pb—Pb bonding. Alternatively, a shifting of charge toward the O' vacancies could occur while retaining the lone electron pair ($6s$-$6p$ hybrid) bound to the Pb^{2+} ions, without significant Pb—Pb bonding, as proposed by Beyerlein *et al.*[19]

Another explanation has been proposed for the stabilization of the anion-deficient pyrochlore structure by Sleight,[21] who stated that the pyrochlore structure allows the oxygen anions to have strong covalent bonds to four nearest neighbors in a nearly tetrahedral arrangement. This can only occur for A and B cations that are not strongly electropositive. This argument is able to explain the existence of the pyrochlore $Ag_2Sb_2O_6$, in which oxygen vacancy-mediated bonds could not occur due to the lack of lone-pair electrons.[22]

2.4. Physical Properties

As noted by several authors, the electronic and magnetic properties of the pyrochlores vary over a wide range (see Ref. 9). Even among the ruthenate pyrochlores, Goodenough and co-workers[11] pointed out that the electronic conductivities range from semiconducting to metallically conducting and the magnetic behavior from moderately temperature-dependent paramagnetism, due to localized electrons, to very low-level, essentially temperature-independent (Pauli-type) paramagnetism due to delocalized conduction electrons in metallic conductors (see Table 1). For example, lead ruthenate has very low resistivity together with a positive thermal coefficient of resistance (TCR) and Pauli-type paramagnetism. These metal-like properties have been rationalized by both Cox *et al.*[11] and Hsu *et al.*[23] as being due to broadening of the partially filled Ru $4d\,t_{2g}$ band. Cox *et al.*[11] proposed that the band broadening comes about due to interaction of the Ru $4d$ orbitals with the Pb $6s$ orbitals through the framework oxygens, while Hsu *et al.*[23] considered that the band is broadened through interaction between Ru $4d$ and O $2p$ orbitals, with further broadening due to an interaction with unoccupied Pb $6p$ orbitals. Yttrium ruthenate, on the other hand, is a somewhat poorer conductor and exhibits paramagnetic behavior due to localized unpaired electrons. This has been rationalized by Cox *et al.*[11] as being due to competition between Ru $4d$ and Y valence orbitals for oxygen p electrons, thus narrowing the conduction band. Hsu *et al.*[23] have also carried out band calculations for $Y_2Ru_2O_7$, which indicated that the Y $5s$ and $5p$ bands are too high in energy to play a significant role in band broadening and thus the t_{2g} band is narrow. Both groups are in agreement that the Y $5s$ and $5p$ levels are generally higher in energy than the Pb $6s$ and $6p$ levels, and the main point of disagreement is the energy of the Pb $6s$ and $6p$ levels in relation to the Ru $4d$ levels and thus the degree of mixing.

TABLE 1
Electrical and Magnetic Properties of Selected Ruthenate Pyrochlores

Compound	Sample type	$\rho_{298}{}^a$ (Ω cm)	TCR^b (ppm)	Magnetic behavior	References
$Pb_2Ru_2O_{6.5}$	sinter	2.7×10^{-4}	—	Pauli paramagnetic	10
	sinter	5×10^{-4}	—	—	13
	sinter	2.0×10^{-3}	—	—	24
	sinter	3.0×10^{-4}	—	—	25
	powder	1.2×10^{-3}	+1700	—	5
	sinter	4.7×10^{-4}	+2770	—	26
$Y_2Ru_2O_7$	sinter	2.7×10^{-3}	—	—	25
	powder	—	—	paramagnetic	27, 28
$Nd_2Ru_2O_7$	single crystal	2.0	negative	—	20
	sinter	4.2×10^{-3}	—	—	25
	powder	—	—	paramagnetic	27
$Dy_2Ru_2O_7$	sinter	4.4×10^{-3}	—	—	25
	powder	—	—	paramagnetic	27

^{a} Resistivity at 298 K.
^{b} Thermal coefficient of resistivity, in 10^{-6} K K^{-1}.

The rare-earth ruthenates, for example, $Nd_2Ru_2O_7$ and $Dy_2Ru_2O_7$, have electrical and magnetic behavior similar to that of $Y_2Ru_2O_7$ (Table 1). There is a somewhat puzzling discrepancy, however, between the resistivities reported for $Nd_2Ru_2O_7$ single-crystal[20] and presumably polycrystalline samples.[27] The difference is more than the expected scatter seen, for example, for the lead ruthenate resistivities (Table 1) and is much greater than the difference expected on the basis of single-crystal versus polycrystalline character. For example, such a difference is less than an order of magnitude for $Bi_2Ru_2O_{7-y}$.[20] In any case, on the basis of their magnetic behavior, the yttrium and rare-earth pyrochlores are not expected to be metallic conductors.

Both Cox et al.[11] and Hsu et al.[23] showed that there is a relatively high DOS at E_F for $Pb_2Ru_2O_{6.5}$ as indicated by UPS. Cox et al. showed that $Y_2Ru_2O_7$ has an insignificant DOS at E_F. In their band calculations, Hsu et al. also found that $Y_2Ru_2O_7$ has a very small Y DOS at E_F.

As will be discussed later, $Y_2Ru_2O_7$ as well as $Nd_2Ru_2O_7$ and $Dy_2Ru_2O_7$ have very low catalytic activities for O_2 reduction and generation, while $Pb_2Ru_2O_{7-y}$ is very active for both reactions. With the data available at present, it is tempting to speculate that the differences in catalytic activity are related to the differences in DOS at E_F.

3. ELECTROCHEMICAL AND ELECTROCATALYTIC PROPERTIES

3.1. Electrochemical Properties

The intrinsic electrochemical behavior of the stoichiometric lead ruthenate pyrochlore $Pb_2Ru_2O_{7-y}$ in alkaline solution is somewhat complicated and is still not completely understood. Results from cyclic voltammetry (CV) have been presented by several groups,[29-31] including the authors',[4] with varying interpretations. An additional complicating factor is that the shapes of the CV curves are probably somewhat dependent upon the details of the oxide preparation and electrode fabrication as well as other factors such as potential cycling history.

The electrochemical behavior has some similarities with that of thermally prepared RuO_2, in that the charge under the voltammetric curve is probably due, at least in part, to a series of redox processes involving multiple electron transfers to the Ru cations.[32,33] The lead in the pyrochlore may also undergo redox processes. The CV charge for the pyrochlore is only slightly dependent upon the potnetial sweep rate, at least for low-area samples. This is different from the behavior of the RuO_2 films, which can have substantial sweep rate dependence.[33,34]

The anodic sweep in the CV for a thin PTFE-bonded layer of $Pb_2Ru_2O_{7-y}$ exhibits a series of small peaks apparently superimposed on a gradually increasing background (Fig. 3). The total anodic charge over the potential range -0.68 to $+0.52$ V versus Hg/HgO was approximately 2.1×10^4 C/mol of Ru (0.22 faradays per mole of Ru). The cathodic charge was very similar, 2.2×10^4 C/mol of Ru (0.24 faradays per mole of Ru), indicating negligible amounts of excess anodic charge, which can arise due to O_2 generation and electrodissolution. An average surface coverage of both Ru and Pb cations on the oxide particles might be considered to be $\sim 7 \times 10^{-10}$ mol/cm² [7.3×10^{-10} mol/cm², averaged for (111) and (222) surfaces, and 6.3×10^{-10} for the (100) surface] or 6.8×10^{-5} C/cm² for each Ru or Pb cation, assuming one electron transfer, while the experimental charge was 3.5×10^{-3} C/cm², a factor of ~ 50 higher. The possible redox processes in this potential range include (see Refs. 35–39 concerning Ru and Refs. 40–42 concerning Pb):

$$Pb^{2+} \xrightarrow{-2e} Pb^{4+}$$

$$Ru^{3+} \xrightarrow{-4e} Ru^{7+}$$

for a total of up to six electrons. Thus, the charge developed might involve more than nine layers of cations, which are spaced ~ 2.56 Å apart in the [100] direction.

The small superimposed peaks are thought to involve discrete redox processes occurring at the surface of the oxide particles. For example, the charge under the unresolved peaks c and d totaled $\sim 1.4 \times 10^{-4}$ C/cm² (true area), which corresponds to approximately two electrons or approximately one electron for each peak. The areas under the other peaks cannot be measured precisely but are also in the one-electron range. These peaks unfortunately cannot be unequivocally assigned to particular redox processes due to the uncertainty in the hydration of the surface. The voltammetry in acid solution and at intermediate pH may aid

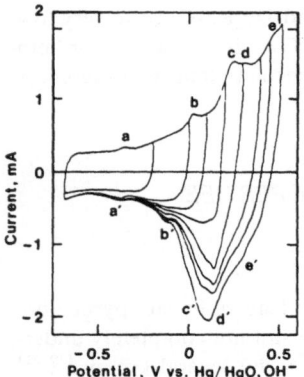

FIGURE 3. Cyclic voltammetry for $Pb_2Ru_2O_{7-y}$ pyrochlore in the form of a thin PTFE-bonded porous coating on the basal plane of a pyrolytic graphite disk in N_2-saturated $1.0M$ KOH at 22°C. The pyrochlore content of the coating was 4.0 mg/cm², and the Teflon T30B content was 0.2 mg/cm²; electrode area was 0.196 cm^{-2}; sweep rate was 20 mV/s^1.

in the interpretation of these peaks. Preliminary work in the authors' laboratory has shown that there are peak potential shifts with pH over the range ~12 to 14, but these shifts may partially be complicated by local pH shifts within the porous layer due to lack of buffer capacity.

Goodenough et al.[31] have recently presented the voltammetry of $Pb_2Ru_2O_{7-y}$ in acid solution, which shows the presence of three redox couples. These were assigned as follows: $E_0 \simeq +0.33$ V (SHE) to Ru^{2+}/Ru^{3+}; $E_0 \simeq +1.14$ V (SHE) to Ru^{3+}/Ru^{4+}; and $E_0 \approx +1.38$ V (SHE) to Ru^{4+}/Ru^{5+}. At pH 14 (assuming 0.06 V per pH unit), the +1.14 V peak would be expected to shift to ~+0.31 V and the +1.38 V peak to ~+0.55 V. These potentials are in reasonably good agreement with E_0 values obtained in our voltammetry of +0.32 V (SHE) and +0.48 V (SHE) (Fig. 3). Different samples of $Pb_2Ru_2O_{7-y}$ have also shown deviations from these values, so that the ranges are +0.28–+0.32 V and +0.46–+0.48 V.

The +0.04 V (SHE) peak (Fig. 3) occurs in the same region as the onset of O₂ reduction, but this peak does not occur in all $Pb_2Ru_2O_{7-y}$ samples. Thus, the possible involvement of discrete redox couples in the O₂ reduction is still in question.

In the future the use of in situ spectroscopic techniques may shed further light on the assignment of specific voltammetric peaks as well as the overall process, which may involve some of the bulk structure of the oxide. Mössbauer spectroscopy could be a particularly powerful technique in identifying the ruthenium valency and spin states. Another technique that could be quite powerful in this regard is X-ray absorption near-edge spectroscopy (XANES).

3.2. Electrocatalysis—General Aspects

Although the principal emphasis in this chapter is on the O₂ reduction and generation reactions, it is worth noting that the lead and bismuth ruthenate pyrochlores have electrocatalytic activity for other reactions. The first documented use of pyrochlores of this type as electrocatalysts was by Welch in 1974[43] for chlorine generation. He also recognized their possible application for a wide range of other electrolytic processes in which dimensionally stable anodes are needed. Most of the other examples of electrocatalysis involve organic oxidation reactions, including oxidation of alkenes,[3,7,30] monosaccharides and lignins,[44] alcohols,[3,7,45] and ketones.[7] The remainder of this chapter will be devoted to the oxygen reduction and generation reactions. These are of special interest due to their involvement in fuel cells, energy storage systems, and industrial electrolysis.

3.3. O₂ Reduction

The kinetic analysis for O₂ reduction was carried out using a graphite disk–gold ring rotating electrode with a thin Teflon (PTFE)-bonded coating of $Pb_2Ru_2O_{7-y}$ on the disk.[4] O₂ transport with such electrodes involves transport through the bulk solution to the porous electrode and transport within the electrode in gas-filled pores. The effectiveness of the O₂ transport within the electrode is very great. The active catalyst layer is only of the order of a few microns in thickness. The standard treatments of the rotating disk and ring-disk electrode techniques must be used with caution, however.

The disk and ring currents at several rotation rates are shown in Fig. 4. The plots of (current)$^{-1}$ versus (rotation rate)$^{-1/2}$ based on the data in Fig. 4 were linear (Fig. 5). The slopes (inversely related to the Levich B coefficient) can be used to determine the overall number of electrons transferred, n. The apparent n value, ~3.8, was relatively independent of rotation rate and potential over the potential range −0.04 to −0.1 V versus Hg/HgO, OH⁻. The ring currents for peroxide oxidation were essentially zero for the first ~100 mV after the

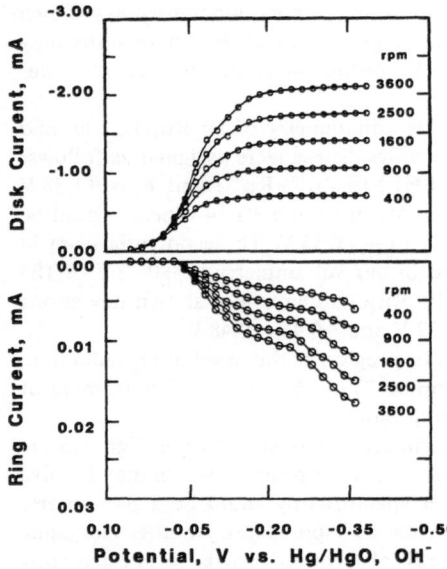

FIGURE 4. Rotating ring-disk polarization curves (steady state) for O_2 reduction on a thin PTFE-bonded porous coating of $Pb_2Ru_2O_{7-y}$ in O_2-saturated $1.0M$ KOH at 22°C. The composition of the coating was as shown in Fig. 3. Electrode area was 0.45 cm²; collection efficiency was 0.177; rotation rates were as shown.

onset of O_2 reduction. Thereafter, they rose to a maximum of ~5% of the disk current (after correction for the collection efficiency). Unfortunately, however, the porous character of the coating precludes a confident quantitiative analysis of the ring data. This is because peroxide can be produced in the catalyst pores and then be decomposed at the pore walls before having an opportunity to diffuse into the bulk solution. The transport of peroxide out of the porous electrode is likely to be very restricted because of slow diffusion within the active catalyst layer for a solution-phase component. Thus, it remains open to question whether the pyrochlore is catalyzing the direct four-electron reduction of O_2 or is just a very efficient peroxide decomposer. Experiments carried out in the authors' laboratory indicate that at least the latter is true. Further research must be done with nonporous forms of the catalyst in order to settle this question.

A mass-transport-correlated Tafel plot (not shown) based on the data of Fig. 4 exhibited a linear portion over the potential range +0.04 to −0.10 V versus Hg/HgO with a slope of −0.063 V/decade. This slope could indicate (a) a rate-determining first electron transfer step involving an adsorbed product such as OH^- with Temkin adsorption behavior or (b) a

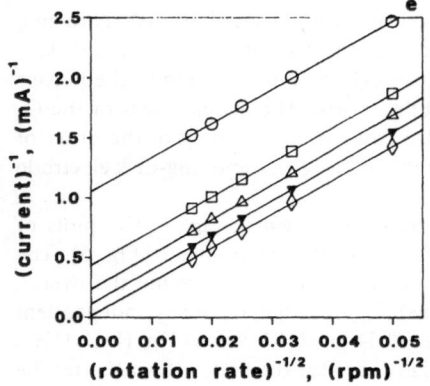

FIGURE 5. (current)$^{-1}$ versus (rotation rate)$^{-1/2}$ for O_2 reduction on a rotating disk electrode with a thin Teflon-bonded coating of $Pb_2Ru_2O_{7-y}$, based on the data of Fig. 4: (a) −0.4 V; (b) −0.1 V; (c) −0.08 V; (d) −0.06 V; (3) −0.04 V versus Hg/HgO, OH^-.

rate-determining chemical step following a fast first electron transfer step. A redox mediator mechanism would be an example of the latter.

The effect of pH variation (11.9 to 13.9) on the O_2 reduction was examined, and the reaction order in OH^- concentration was found to be -0.50 ± 0.02. The slopes of the mass-transport-corrected Tafel plots on which this reaction order was based were nearly identical, providing evidence that the reaction mechanism remained the same over this pH range.

The reaction order of -0.5 together with the -0.06 V/decade Tafel slope is consistent with the data of Egdell et al.,[29] which was explained as follows. First, there is proposed to be a fast outer-sphere reduction of O_2 to O_2^- followed by protonation to yield physisorbed HO_2. This then displaces a surface OH^- in the rate-determining step. An electrostatic argument was invoked to explain the -0.5 reaction order for OH^-.

In more recent work of Goodenough et al. at pHs extending from acid to base,[31] the mechanism for O_2 reduction on $Pb_2Ru_2O_{7-y}$ at pH > 2 was proposed to involve the following:

$$O_2 + e^- \rightarrow (O_2^-)_{ads} \tag{1}$$

$$Pb^{2+}OH^- + (O_2^-)_{ads} \rightarrow Pb^{2+}O_2^- + OH^- \qquad (rds) \tag{2}$$

$$Pb^{2+}O_2^- + 2H_2O + 3e^- \rightarrow Pb^{2+}OH^- + 3OH^- \tag{3}$$

Surface charge density versus pH measurements indicated that the oxygens in the Ru_2O_6 framework are not protonated at higher pH and thus are not labile enough to undergo exchange with O_2^-.[31] At pH values below 2, the Ru-bound oxygens become protonated, and the Ru^{3+} cations can then take part in a step analogous to the reaction in Eq. (2).[31] Steps such as this (e.g., Eq. (2)), not involving charge transfer, are consistent with the observed Tafel slope (~ -0.06 V/decade). The reaction in Eq. (3) is consistent with a lack of peroxide generation. The -0.5 reaction order for OH^- remains to be explained, however.

In order to fully elucidate the mechanisms for both O_2 reduction and generation on $Pb_2Ru_2O_{7-y}$, other factors may have to be taken into account, such as the specific steric-electronic environment on the surface of the oxide. In this regard, it should be helpful to examine the structure of the various surfaces both theoretically and experimentally in future work.

As a first step in this direction, one could look at the surface just as a simple termination of the bulk structure, realizing that this may be a gross oversimplification. A characteristic feature of both the (111) and (100) faces of $Pb_2Ru_2O_{7-y}$ is the zigzag chains of RuO_6 polyhedra, with the Ru—O—Ru angle being 135° (Fig. 6). The octahedrally bound oxygens which point up and out of the surface are either inclined toward or away from each other. In the former case, these oxygens would ideally be about 2.8 Å from each other, assuming no relaxation of the structure at the surface. This can also be seen in more detail in Ref. 46. The zigzag nature of the ideal pyrochlore surface is unique and distinguishes it from both the perovskite (e.g., $SrRuO_3$) and the rutile (e.g., RuO_2) structures. The proximity of these oxygens allows one to envision mechanisms in which breakage of the O—O bond in O_2 could be assisted by either these closely situated oxygens or by the Ru cations themselves. The details of such interactions remain to be worked out but would have to take into account the lack of protonation of the surface at high pH, as pointed out by Goodenough et al.[31]

3.4. O_2 Generation

The mechanism of the O_2 generation reaction on $Pb_2Ru_2O_{7-y}$ in alkaline solution also remains to be worked out in detail in the light of the surface structures. The mechanism that has been proposed by our research group[4] involves adjacent Me—OH sites (probably

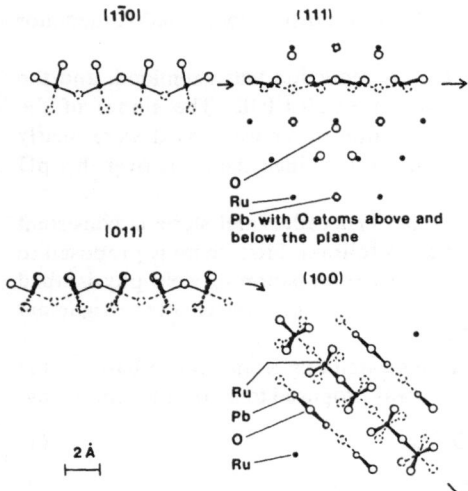

FIGURE 6. Projections of the (111) and (100) surfaces of the $Pb_2Ru_2O_{7-y}$ pyrochlore (right-hand side) and the zigzag chains of corner-shared RuO_6 polyhedra in the indicated directions across these surfaces (left-hand side). Dashed lines show atoms and bonds lying below the surface.

Ru—OH), and, with the surface structure in mind, it is possible to imagine how this might work on the (111) surface (Fig. 7). The fact that the Ru cations are probably already six-coordinate, as shown in **I**, means that the first step involves the formation of a seven-coordinate Ru species; such species are known to exist as intermediates in some ligand substitution reactions.[47] Goodenough et al.,[31] on the other hand, have proposed a six-coordinate $Me^{4+}OO^{2-}$ intermediate (Me = Ir or Ru), which could be written alternatively as $Me^{4+}(O—O)^{2-}$.

In order to gain further understanding of the surface structure and how O_2, H_2O, and OH^- interact with it, it will be necessary to use some of the surface analytical techniques such as LEED, XPS, AES, SIMS, and STM. The preparation of single-crystal surfaces of pyrochlore oxides is not expected to be straightforward but should be well within the realm of possibility. One example of the expected problems is the fact that the (111) faces have

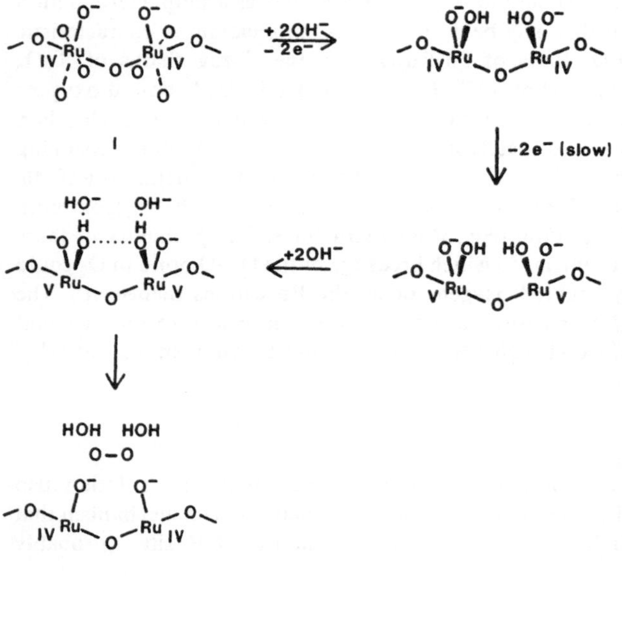

FIGURE 7. Proposed mechanism for O_2 generation on the (111) face of $Pb_2Ru_2O_{7-y}$ in alkaline solution, showing a side-on view of two corner-shared RuO_6 polyhedra in the zigzag chain shown in Fig. 6. In **I**, the three oxygens lying behind, below, and in front of a Ru(IV) cation are shown but are omitted thereafter for clarity.

75% of the surface cations being Ru and 25% Pb, while the situation is reversed on the (222) faces. With the use of special etching techniques, it may be possible to handle such problems.

3.5. Gas-Fed Electrode Measurements

The performance of porous gas-fed O_2 electrodes based on Pb–Ru pyrochlores has been shown to be quite good in alkaline solution by several groups.[2-4,31] The O_2 generation performance has also been shown to be good.[2-4] One drawback to the use of these pyrochlores as electrocatalysts is their non-negligible equilibrium solubility in concentrated alkaline solution, which increases with more positive potentials.[2,3]

Both the electrocatalytic activity and the stability can be modified through the partial or complete substitution of the Pb and/or Ru cations. For example, as noted earlier, Pb can be replaced by Y or rare-earth elements such as Nd or Dy. The resulting materials are paramagnetic semiconductors. The performance in both O_2 reduction and generation was found to be quite poor, which can be related either to their poor conductivity or to a low DOS at E_F. To some extent, the poor performance was also a result of their low surface areas.

Part of the Ru can be substituted with Pb to form the "lead-rich" pyrochlores discussed earlier. The performance of gas-fed electrodes based on these compounds has been found to be superior to that of the stoichiometric pyrochlore $Pb_2Ru_2O_{7-y}$ but is related to the fact that the lead-rich compounds can be prepared in higher area form.[2,3] In our experience, however, the O_2 reduction performance of gas-fed electrodes made from the lead-rich compounds degrades much faster than that of the electrodes made from the stoichiometric compound, especially when anodic polarization measurements are also made.

Part of the Ru can be substituted with Ir, and this is expected to modify the long-term stability. The performance on both O_2 reduction and generation of gas-fed electrodes based on such materials showed small improvements,[4] but the long-term stability has not been tested. There is some indication that partial substitution of the Ru with Sb can markedly improve the stability of Pb–Ru pyrochlores, especially in the more positive potential range.[48]

Other factors are also operative in determining stability. Such factors include the crystallinity and the presence of impurity phases such as RuO_2 and PbO. One can determine the importance of the crystallinity as long as one realizes that the crystallinity of a given sample cannot be improved without decreasing the BET surface area.

An alternative approach to stabilizing the pyrochlore involves the use of anion-exchange polymers, both as a replacement for the liquid electrolyte phase within the porous electrode and as a discrete overlayer on the electrolyte side of the electrode.[49] Very encouraging results have been obtained using both techniques.[4] The diffusion of the ruthenate ion RuO_4^{2-} out of the porous electrode has been monitored spectrophotometrically during anodic polarization measurements, and a partially fluorinated anion-exchange membrane (RAI 4035, RAI Corp., Hauppage, New York) was found to retard this process significantly.[50] The ion diffuses through the membrane relatively slowly, probably due to size and electrostatic effects, although this needs further investigation.

A high-area carbon support matrix has been used for some of the gas-fed electrodes fabricated in our laboratory. This has usually been an acetylene black (Shawinigan Black, Chevron Chemical Co., Olefins and Derivatives Division, Houston, Texas; BET surface area, ~80 m^2/g). This has been used mainly because it facilitates the fabrication of the Teflon-bonded porous electrodes. Although this carbon is comparatively oxidation-resistant in the O_2 generation mode,[51] other carbons have been developed that are even more corrosion-resistant,[52] such as that developed by Ross and Sattler.[53] This carbon can also be used in gas-fed electrodes in conjunction with pyrochlores, with little change in performance compared to that obtained with Shawinigan black.

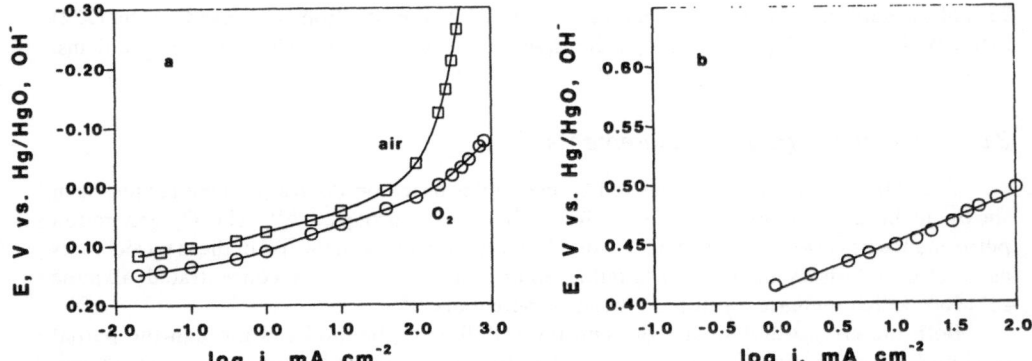

FIGURE 8. Polarization curves for O_2 reduction (a) and O_2 generation (b) with gas-fed (1 atm) electrodes based on $Pb_2Ru_2O_{7-y}$ pyrochlore in 5.5M KOH at 25°C. The electrode in (a) contained 83.3 mg/cm^2 pyrochlore and 27.8 mg/cm^2 Teflon T30B and was heat-treated at 330°C for 2 h in flowing helium. Ammonium bicarbonate (18.8 mg/cm^2) was added as a pore former before the heat treatment. The electrode in (b) was made with a similar composition, except that 41.7 mg/cm^2 Teflon and 16.7 mg/cm^2 NH_4HCO_3 were used.

Under conditions of high temperature and O_2 pressure, such as would be used in high-performance alkaline fuel cells, it may be necessary to completely avoid the use of carbon matrices, however. Techniques have been developed in our laboratory to fabricate "self-supported" gas-fed electrodes based on $Pb_2Ru_2O_{7-y}$ and Teflon T30B.[15] The short-term performance for O_2 reduction and generation of such electrodes was very good (Fig. 8). Within the statistical accuracy of the measurement, the O_2 reduction current was first order with respect to the O_2 partial pressure over the range 1 to 10 mA/cm^2. Deviations from linearity of the polarization curve in the low-current-density range are probably associated with (1) surface redox couples on the pyrochlore and/or (2) rate-determining peroxide decomposition within the porous electrode. Deviations in the high-current-density range are most likely due to ohmic and/or mass-transport limitations. Further improvements in electrode structures and fabrication techniques are needed.

The polarization curve for O_2 generation (Fig. 8b) was linear, with a Tafel slope very close to +0.04 V/decade. This is consistent with a rate-determining second electron transfer step, as shown in the proposed mechanism in Fig. 7.

4. CONCLUSIONS AND RECOMMENDATIONS

The lead–ruthenium pyrochlores are promising electrocatalysts for O_2 reduction and generation in alkaline media. Their high catalytic activity may be associated with the relatively high density of electronic states at the Fermi level. Mechanisms by which these materials catalyze the O_2 reduction and generation reactions have been proposed but are far from certain. Specific ways in which the surface structures may be involved have also been proposed. Further fundamental research is needed to relate the electrocatalytic activity of the pyrochlores to their surface electronic and structural properties.

The lead–ruthenium pyrochlores have significant solubility in alkaline solution in the O_2 generation potential range, but the use of anion-conducting polymers, both as membrane overlayers and as an integral component of the gas-diffusion electrode, promises to alleviate this problem. Further improvements are also needed in gas-diffusion electrode fabrication techniques in order to achieve higher performance and stability.

ACKNOWLEDGMENTS

This research was supported by grants from the U.S. Department of Energy through a subcontract with Lawrence Berkeley National Laboratory, from the NASA Lewis Research Center, from the U.S. Air Force, Wright-Patterson AFB, through a subcontract with NASA-Lewis Research Center, and from the Office of Naval Research. Helpful comments from Professor J. B. Goodenough are gratefully acknowledged.

Note added in proof. Professor J. B. Goodenough has kindly provided the authors with a preprint of an article which discusses the stabilization of Pb-Ru pyrochlores against dissolution in an acid environment using Dow perfluorosulfonic acid polymer (J.-M. Zen, R. Manoharan and J. B. Goodenough, *J. Appl. Electrochem.*, in press).

REFERENCES

1. H. S. Horowitz, J. M. Longo, and J. I. Haberman, U.S. Patent 4,124,539 (1978).
2. H. S. Horowitz, J. M. Longo, and H. H. Horowitz, *J. Electrochem. Soc.* **130**, 1851 (1983).
3. H. S. Horowitz, J. M. Longo, H. H. Horowitz, and J. T. Lewandowski, in: *Solid State Chemistry in Catalysis* (R. K. Grasselli and J. F. Brazdil, eds.), ACS Symposium Series 279, pp. 143–163, American Chemical Society, Washington, D.C. (1985).
4. J. Prakash, D. Tryk, and E. Yeager, *J. Power Sources* **29**, 413 (1990).
5. R. A. Beyerlein, H. S. Horowitz, and J. M. Longo, *J. Solid State Chem.* **72**, 2 (1988).
6. H. S. Horowitz, J. M. Longo, and J. T. Lewandowski, U.S. Patent 4,129,525 (1978).
7. H. H. Horowitz, H. S. Horowitz, and J. M. Longo, in: *Electrocatalysis* (W. E. O'Grady, P. N. Ross, and F. G. Will, eds.), pp. 285–290, The Electrochemical Society, Pennington, New Jersey (1982).
8. H. S. Horowitz, J. M. Longo, and J. T. Lewandowski, U.S. Patent 4,176,094 (1979).
9. M. A. Subramanian, G. Aravamudan, and G. V. Subba Rao, *Prog. Solid State Chem.* **15**, 55 (1983).
10. J. M. Longo, P. M. Raccah, and J. B. Goodenough, *Mater. Res. Bull.* **4**, 191 (1969).
11. P. A. Cox, R. G. Egdell, J. B. Goodenough, A. Hamnett, and C. C. Naish, *J. Phys. C: Solid State Phys.* **16**, 6221 (1983).
12. J. J. Randall and R. Ward, *J. Am. Chem. Soc.* **81**, 2629 (1959).
13. A. W. Sleight, *Mater. Res. Bull.* **6**, 775 (1971).
14. H. S. Horowitz, J. M. Longo, and J. T. Lewandowski, U.S. Patent 4,203,871 (1980).
15. M. Shingler, W. Aldred, D. Tryk, and E. Yeager, Final Report: Catalysts for Ultrahigh Current Density Oxygen Cathodes for Space Fuel Cell Applications, Contract No. NAG3-694 with NASA—Lewis Research Center, prepared by Case Center for Electrochemical Sciences, Case Western Reserve University, May, 1990.
16. A. F. Wells, *Structural Inorganic Chemistry*, 5th ed., pp. 129, 258, Clarendon Press, Oxford (1984).
17. R. A. McCauley, *J. Appl. Phys.* **51**, 290 (1980).
18. A. W. Sleight, *Inorg. Chem.* **7**, 1704 (1968).
19. R. A. Beyerlein, H. S. Horowitz, J. M. Longo, M. E. Leonowicz, J. D. Jorgensen, and F. J. Rotella, *J. Solid State Chem.* **51**, 253 (1984).
20. A. W. Sleight and R. J. Bouchard, in: *Solid State Chemistry*, NBS Special Publication 364 (R. S. Roth and S. J. Schneider, eds.), pp. 227–232, National Bureau of Standards, U.S. Dept. of Commerce, Washington, D.C. (1972).
21. A. W. Sleight, *Mater. Res. Bull.* **4**, 377 (1969).
22. R. J. Bouchard and J. L. Gillson, *Mater. Res. Bull.* **6**, 669 (1971).
23. W. Y. Hsu, R. V. Kasowski, T. Miller, and T. C. Chiang, *Appl. Phys. Lett.* **52**, 792 (1988).
24. P. R. van Loan, *Ceram. Bull.* **51**, 231 (1972).
25. V. B. Lazarev and I. S. Shlaplygin, *Russ. J. Inorg. Chem.*, **23**, 163 (1978).
26. G. Mayer-von Kurthy, W. Wischert, R. Kiemel, S. Kemmler-Sack, R. Gross, and R. P. Huebener, *J. Solid State Chem.* **79**, 34 (1989).

27. R. Aleonard, E. F. Bertaut, M. C. Montmory, and R. Pauthenet, *J. Appl. Phys.* **33**(Suppl.), 1205 (1962).
28. J. Rosset and D. K. Ray, *J. Chem. Phys.* **37**, 1017 (1962).
29. R. G. Egdell, J. B. Goodenough, A. Hamnett, and C. C. Naish, *J. Chem. Soc., Faraday Trans. 1*, **79**, 893 (1983).
30. J. A. R. van Veen, J. M. van der Eijk, R. de Ruiter, and S. Huizinga, *Electrochim. Acta* **33**, 51 (1988).
31. J. B. Goodenough, R. Manoharan, and M. Paranthaman, *J. Am. Chem. Soc.* **112**, 2076 (1990).
32. L. D. Burke and O. J. Murphy, *J. Electroanal. Chem.* **109**, 199 (1980).
33. A. Ardizzone, G. Fregonara, and S. Trassatti, *Electrochim. Acta* **35**, 263 (1990).
34. L. D. Burke and O. J. Murphy, *J. Electroanal. Chem.* **96**, 19 (1979).
35. M. Pourbaix, *Atlas of Electrochemical Equilibria in Aqueous Solutions*, pp. 343–349, Pergamon, Oxford (1966).
36. J. F. Llopis and I. M. Tordesillas, in: *Encyclopedia of Electrochemistry of the Elements*, Vol. VI (A. J. Bard, ed.) pp. 277–298, Marcel Dekker, New York (1976).
37. K. W. Lam, K. E. Johnson, and D. G. Lee, *J. Electrochem. Soc.* **125**, 1069 (1978).
38. L. D. Burke and D. P. Whelan, *J. Electroanal. Chem.* **103**, 179 (1987).
39. F. Colom, in: *Standard Potentials in Aqueous Solution* (A. J. Bard, R. Parsons, and J. Jordan, eds.), pp. 413–427, Marcel Dekker, New York (1985).
40. M. Pourbaix, *Atlas of Electrochemical Equilibria in Aqueous Solutions*, pp. 485–492, Pergamon, Oxford, (1966).
41. T. F. Sharpe, in: *Encyclopedia of Electrochemistry of the Elements*, Vol. I (A. J. Bard, ed.), pp. 235–347, Marcel Dekker, New York (1973).
42. Z. Galus, in: *Standard Potentials in Aqueous Solution* (A. J. Bard, R. Parsons, and J. Jordan, eds.), pp. 220–235, Marcel Dekker, New York (1985).
43. C. N. Welch, U.S. Patent 3,801,490 (1974).
44. M. R. St. John, U.S. Patent 4,395,316 (1983).
45. T. R. Felthouse, *J. Am. Chem. Soc.* **109**, 7566 (1987).
46. O. Knop, F. Brisse, and L. Castelliz, *Can. J. Chem.* **43**, 2812 (1965).
47. P. C. Ford, J. R. Kuempel, and H. Taube, *Inorg. Chem.* **7**, 1976 (1968).
48. C. Cha, Wuhan University, Wuhan, People's Republic of China, personal communication.
49. M. S. Hossain, D. Tryk, and A. Gordon, Extended Abstracts, 171st Meeting of the Electrochemical Society, Philadelphia, May 1987, pp. 466–467.
50. J. Prakash, D. Tryk, W. Aldred, and E. Yeager, unpublished results.
51. L. B. Berk and D. Zuckerbrod, in: *The Electrochemistry of Carbon* (S. Sarangapani, J. R. Akridge, and B. Schumm, eds.), pp. 238–250, The Electrochemical Society, Pennington, New Jersey (1984).
52. D. Tryk, W. Aldred, and E. Yeager, in: *The Electrochemistry of Carbon* (S. Sarangapani, J. R. Akridge, and B. Schumm, eds.), pp. 192–220, The Electrochemical Society, Pennington, New Jersey (1984).
53. P. N. Ross and M. Sattler, *J. Electrochem. Soc.* **135**, 1464 (1988).

9

Progress in the Studies of Oxygen Reduction during the Last Thirty Years

A. Damjanovic

1. INTRODUCTION

The oxygen reduction reaction, along with the hydrogen reaction, stands among the most important reactions in electrochemistry both from the practical/economical and theoretical point of view. Whereas meaningful kinetic and mechanistic studies of the hydrogen reaction can be traced back to the 1930s,[1] not much work had been done on the oxygen reduction reaction prior to 1960. At that time, kinetic–mechanistic studies in general, as opposed to those based on the equilibrium state and thermodynamics, were still, in a sense, in the embryonic stage. The first systematic analysis of the kinetics of the oxygen reduction appears to have been initiated around 1960 in Bockris's laboratory at the University of Pennsylvania with the aim of determining the mechanisms of the reduction and to gain an insight into the catalysis and factors affecting the catalysis at different metals.[2-4] A program was set up at that time to study this reaction on noble metal electrodes and their alloys, specifically on Pt, Pd, and Rh, as well as on their alloys with Au.[4] In this early time, the accent was placed on Pt, and three major observations made regarding the kinetics of the reduction and the nature of the electrode surfaces proved to be keystones and provided the basis for much of the later and more detailed studies of this reaction, not only on Pt, but on other electrodes as well.

First, there is a sharp difference in the rates of the reduction at a given electrode potential depending on the Pt electrode pretreatment prior to measurements. This is illustrated in Fig. 1 with the early E–log i data for an electrode that was anodically pretreated and an electrode which was cathodically pretreated prior to the measurements.[3, 5] Clearly, below about 1.0 V versus RHE the rates at the same potential at the prereduced electrode are significantly higher compared to those at the anodically pretreated electrode.

The second, and the most significant, observation was that the kinetics at the prereduced electrodes follows a rate law completely different from that at the anodically pretreated electrodes. The Tafel slope at the prereduced electrodes is close to $-2.3 \times 2RT/F$, whereas at the anodically pretreated electrodes it is $-2.3 \times 2RT/F$.

The third observation in this early period was that between 0.7 and 1.0 V versus RHE, the Pt electrode is covered by some kind of *adsorbed* oxygen species, other than water.[2]

A. Damjanovic • Allied-Signal Incorporated, Corporate Technology, Morristown, New Jersey 07962.

Electrochemistry in Transition, edited by Oliver J. Murphy *et al.* Plenum Press, New York, 1992.

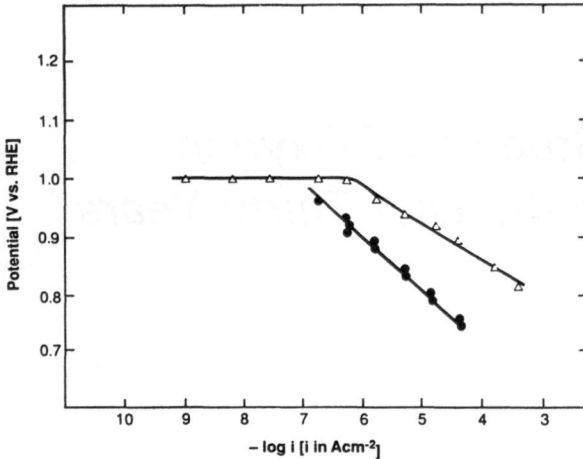

FIGURE 1. The early data showing E–log i relations for oxygen reduction at prereduced ($\triangle$) and preanodized ($\bullet$) Pt electrodes. (After Ref. 5.)

Above 1.0 V, an oxide phase forms in the sense that Pt atoms have left their regular position in the metal lattice to form more intimate multidirectional bonds with oxygen species, presumed at that time to be discharged OH.[6] Oxygen species in the adsorbed stage behave reversibly, and their coverage changed linearly with electrode potential. The maximum coverage reached at 1.0 V is below that corresponding to a full monolayer. It corresponds to a fractional coverage of only about 0.35 calculated on the basis of OH as the adsorbed species.[2] Coverage above about 1.0 V is not a function of electrode potential. In general, at any potential, or applied anodic current, it increases with time and can reach coverages equivalent to a few monolayers of an oxide film, that is, film thicknesses of about 6–15 Å.[6] A strong hysteresis is observed—both coulometrically and ellipsometrically[7,8]—in the coverage or film thickness. Due to this hysteresis, oxides that are formed at higher potentials reduce at potentials much lower than the potentials at which they are thermodynamically stable. Of course, the reduction is a time-related process, and the oxide film will gradually reduce at any potential below, say, 0.95 V versus RHE.

Oxygen reduction at the anodically pretreated Pt electrodes occurs, therefore, on an oxide-covered electrode, and, because of the hysteresis, the reduction of oxygen could be extended even below 1.0 V versus RHE. However, below 1.0 V only quasi-steady-state measurements are possible, and, as the oxide films are gradually reduced, the rates of the oxygen reduction at a given potential increase with time in an irregular fashion. Because of this, a variety of E–log i data were reported in the early literature. Consequently, a study of the mechanisms of the reduction was virtually impossible. Only with fairly fast measurements was it possible to establish on the preanodized, oxide-covered electrodes fairly linear Tafel relations with the slope of −120 mV (but see below).

In contrast to the case with oxide-film-covered electrodes, the kinetics at the prereduced electrodes proved to be reproducible and to represent a "true" steady state. Tafel lines can be retraced both in the process of increasing and decreasing the applied current. The kinetics is characterized by the Tafel slope of −60 mV and this showed that the mechanism at the prereduced, oxide-free electrodes is quite different from that at the anodically pretreated, oxide-film-covered electrodes. As the current density at the oxide-free electrodes is decreased below about 3×10^{-7} A/cm^2, corresponding to the electrode potential close to 0.98 V versus RHE, the potential does not increase further (see Fig. 1). It assumes the value of the rest potential, E_r, otherwise observed at the prereduced electrode in oxygen-saturated solutions when no current is applied to the electrode.[2,3,9] If an electrode is brought to a potential above 1.0 V, for example, by applying a constant potential or in a potential sweep, an anodic current will appear and an oxide film will form even in oxygen-saturated solutions. However, formation of an oxide film at or close to E_r is not a primary reason why in the galvanostatic measurements the potential at the prereduced electrodes cannot be extended beyond E_r. The nature of the processes controlling the rest potential, E_r, as well as the current density down to which the Tafel relationships extend, will be discussed below.

Because Pt catalysts are oxide-free in fuel cell practice, the major effort in the early studies of the kinetics and mechanism of oxygen reduction was placed at the prereduced, oxide-free electrodes. Only much later, the study of oxygen reduction was extended to the oxide-film-covered electrodes in order to obtain knowledge regarding the mechanisms of charge transfer, both in the anodic and cathodic direction, across thin, electronically non-conductive surface-oxide films.

The work initiated in the Bockris laboratory in Philadelphia was subsequently transplanted and expanded by his students and co-workers in other laboratories. Selected work from the laboratory in Philadelphia, and from other laboratories, is included in this short overview of the oxygen reduction at various electrodes.

2. EARLY MECHANISTIC STUDY OF OXYGEN REDUCTION AT Pt ELECTRODES

Brusic was the first to examine in detail the kinetics of O_2 reduction at oxide-free Pt electrodes.[3] In Fig. 2, E–log i relationships are shown for the reduction in oxygen-saturated $HClO_4$ solutions of two pHs. In Fig. 3, data are shown for the reduction in $0.1 N$ $HClO_4$ at different partial pressures of oxygen. Based on these data, the following rate equation for the reduction was given:

$$i = kp_{O_2}a_{H^+}^{3/2} \exp\left(\frac{-FE}{RT}\right) \tag{1}$$

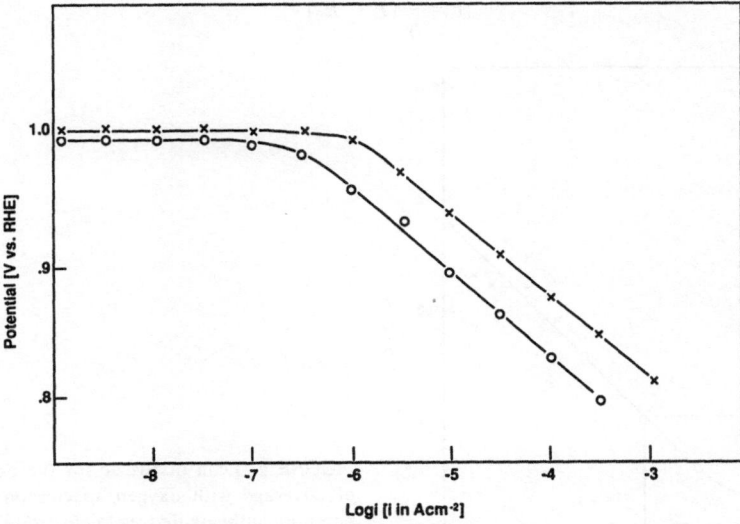

FIGURE 2. The early E–log i relations obtained by Brusic for oxygen reduction at oxide-free Pt electrodes in $HClO_4$ solutions. ×, pH 1; ○, pH 2.2.[3]

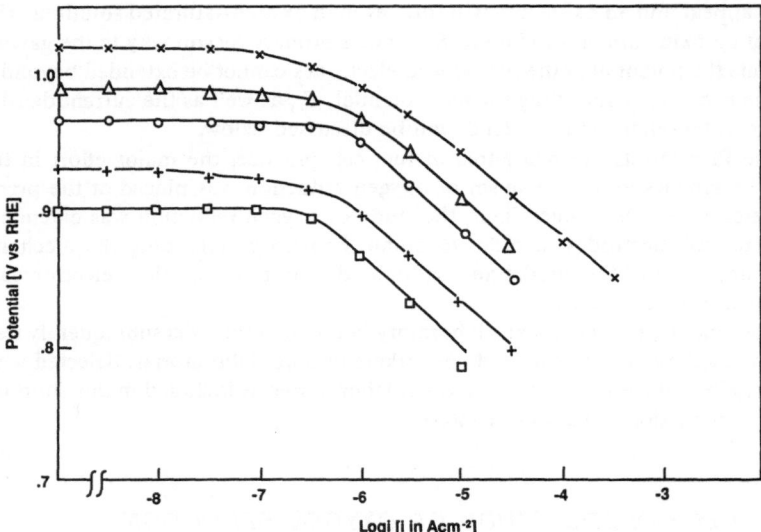

FIGURE 3. The early Tafel relations obtained by Brusic for oxygen reduction at Pt electrodes in $0.1\,N$ HClO$_4$ solution at different partial pressures of oxygen: ×, 1; △, 0.3; ○, 0.1; +, 0.03; □, 0.01 atm.[3]

where E is the electrode potential versus a pH-independent reference electrode, and other symbols have their usual significance.

The most characteristic feature of the observed kinetics is the fractional reaction order with respect to H⁺. To explain this reaction order, Brusic invoked the observation that in the potential region with the Tafel slope of $-60\,$mV the coverage, θ, with adsorbed oxygen species changes linearly with E and that the θ–E relationship depends on pH according to (see Fig. 4):

$$\theta = \frac{F}{\gamma}(E - E_0) \tag{2}$$

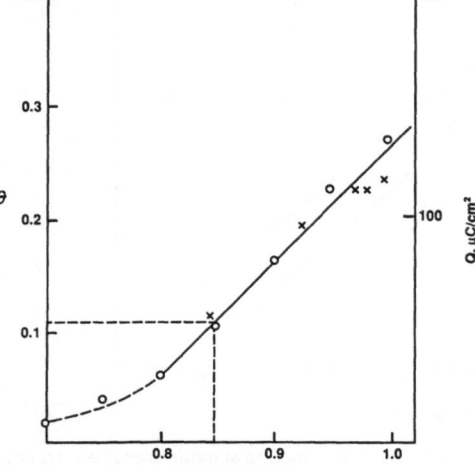

FIGURE 4. Data of Brusic for the dependence of coverage with oxygen species on electrode potential at the oxide-free Pt electrode. Coverage does not depend on whether the solution ($0.1\,N$ HClO$_4$) is saturated with N$_2$ (○) or O$_2$ (×).[3]

Here, $\gamma(\approx 23 \text{ kcal/equiv})$ is a constant that is experimentally available, and E_0 is the potential at which the θ–E dependence in a solution of a given pH extrapolates to $\theta = 0$. A detailed analysis of the θ–E dependences over the entire pH range was subsequently carried out by Sepa et al.[10] Their data are shown in Fig. 5. It was found that the E_0 versus a pH-independent reference electrode decreases by 60 mV as the pH increases by one unit,[3,10] i.e.,

$$E_0 = E_{0,\text{pH}=0} - \frac{2.3RT}{F}\text{pH} \tag{3}$$

where $E_{0,\text{pH}=0}$ is the extrapolated potential in a solution of pH 0.

Because fractional coverages with oxygen species, which are assumed to be intermediates in the oxygen reduction process,[3] have values between 0 and 0.35, Brusic suggested that the kinetics is controlled by the Temkin rather than Langmuirian adsorption conditions. In this case, the enthalpy of activation increases with coverage, and hence electrode potential, according to:

$$\Delta H_\theta^\ddagger = \Delta H_{E=E_0}^\ddagger + \beta\gamma\theta \tag{4}$$

$$= \Delta H_{E=E_0}^\ddagger + \beta F E + 2.3\beta RT\text{pH} \tag{5}$$

Here, β is the symmetry factor, and $\Delta H_{E=E_0}^\ddagger$ is the enthalpy of activation before the onset of the Tenkin condition, say, at $\theta \approx 0.05$.

According to the original analysis of Brusic, the first electron transfer is the rate-determining step (rds); e.g.,

$$S\cdots O_2 + H^+ + e^- \rightarrow SO_2H \tag{6}$$

Here, S stands for a surface site. For this step, the rate equation under the Temkin condition is given by

$$i \approx k'nF[H^+]p_{O_2}\exp\left(\frac{-\Delta H_\theta^\ddagger}{RT}\right)\exp\left(\frac{-\beta FE}{RT}\right) \tag{7}$$

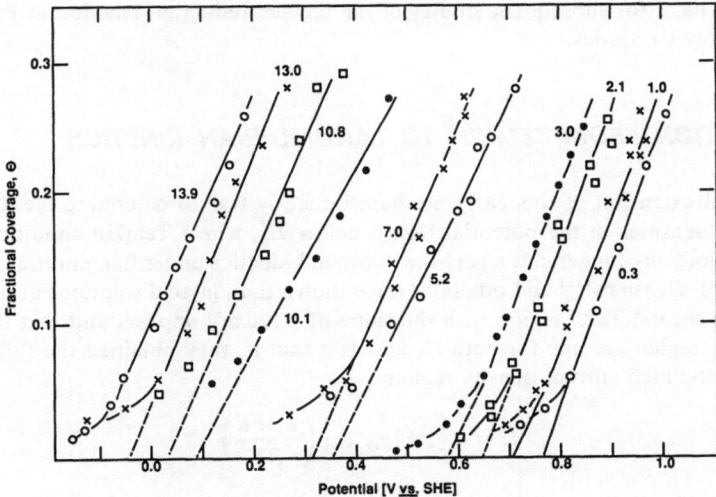

FIGURE 5. Dependence on the potential of the coverage with oxygen species at Pt in solutions of the pH values indicated.[10]

In this equation, the entropy of activation is included into the preexponential constant. Inserting Eq. (5) into this equation and rearranging, the following rate equation in obtained:

$$i \approx knFp_{O_2}[H^+]^{3/2} \exp\left(\frac{-FE}{RT}\right) \tag{8}$$

which is in accordance with the observed rate equation, including the pH and O_2 pressure dependences.

Thus, although the first electron transfer step is the rds, the Tafel slope is -60 mV, not -120 mV, as usually expected when the first charge transfer step is the rds. The most significant result of this analysis, however, is that under Temkin adsorption conditions the fractional reaction order is obtained. It arises because of the dependence on pH of θ at any potential versus a pH-independent reference electrode.

In alkaline solutions, Wong[11] found that the Tafel slopes are again close to -60 mV, but that the pH dependence, $(dE/d\text{pH})i$, is only -26 mV. This pH dependence corresponds to a fractional reaction order of $\frac{1}{2}$ with respect to H^+. With respect to oxygen, the rates are first order. Brusic suggested that in alkaline solutions too, the first electron transfer step was the rds under Temkin adsorption conditions. Then, for the step

$$O_2 + e^- \rightarrow O_2^- \tag{9}$$

and using the same relation for θ and $\Delta H^{\ddagger}$ as in acid solutions, she obtained the following rate equation:

$$i = kFp_{O_2}[H^+]^{1/2} \exp\left(\frac{-FE}{RT}\right) \tag{10}$$

which is in accordance with the observed kinetics in alkaline solutions.

Thus, although H^+ does not participate as a reactant in the rds, the fractional reaction order with respect to H^+ appears in the rate equation because of the dependence of θ on pH. As in acid solutions, the Tafel slope of -120 mV for the first charge transfer step is modified to -60 mV because of the dependence of θ on E.

These early studies of the kinetics and mechanisms of O_2 reduction at Pt electrodes provided the basis for subsequent studies of the oxygen reduction reaction at Pt and other metal and alloy electrodes.

3. TRANSITION FROM TEMKIN TO LANGMUIRIAN KINETICS

A logical extension of this early mechanistic study was to determine the $E-i$ data at high current densities in the potential region below E_0, where Temkin conditions are not expected to hold any longer and where the "normal" kinetics under Langmuirian conditions should prevail. Genshaw[12] and others[13] have shown that, in acid solutions at high current densities, the second Tafel region with the slope of -120 mV appears and that the reaction orders in this region are one for both O_2 and H^+; that is, they obtained the following rate equation for the high-current-density region:

$$i_h \approx k_h[H^+]p_{O_2} \exp\left(\frac{-\beta FE}{RT}\right) \tag{11}$$

This rate equation corresponds to the same reaction step as proposed for the low-current-density region in acid solutions. Thus, the first electron transfer step (Eq. 6) is the rds in the high-current-density region also, but now under Langmuirian adsorption conditions.

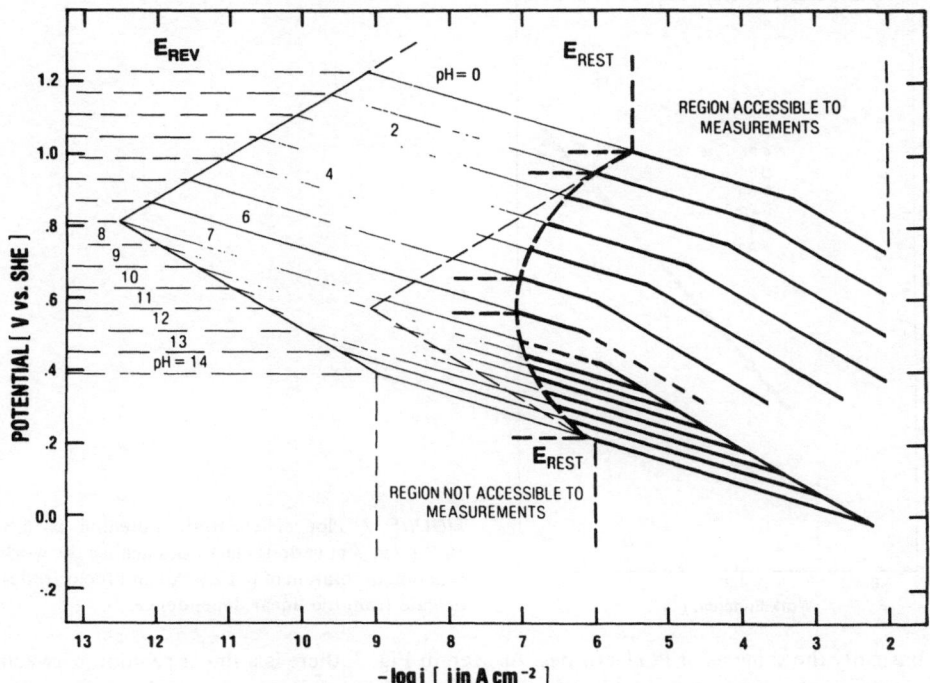

FIGURE 6. Composite idealized E–log i diagram for the oxygen reduction at Pt electrodes over the entire pH range. E_{rev} is the reversible potential, and E_{rest} is the rest potential in O_2-saturated solutions.[10]

Extensive studies of the reduction at Pt over the entire pH range and at various temperatures were carried out by Sepa and co-workers[10,14] at Belgrade University. Based on their studies, a composite E–log i diagram for the entire pH range is constructed. It is shown in Fig. 6. As seen in this idealized diagram, in alkaline solutions also, the second Tafel region with the slope -120 mV appears at high current densities. However, the pH dependence there is zero, that is, $(dE/d\text{pH})_i = 0$, and the rds is the same as in the low-current-density region with the Tafel slope of -60 mV. As expected, the family of the Tafel lines from the low-pH end meets the family of lines from the high-pH end close to pH 7. Because of this, the overall pH dependence from the acid to the alkaline end appears to be -60 mV. This diagram illustrates the danger of drawing conclusions regarding the pH dependence from a single measurement in acid and alkaline solution.

At low current densities E–log i lines do not extend all the way to the reversible potential. As shown in another study,[10] the reason for this is not the formation of a surface oxide film, which would then retard the reduction, but rather it is the anodic dissolution of Pt. The observed rest potentials, E_r, are in fact the corrosion potentials with the cathodic process being O_2 reduction and the anodic process Pt dissolution.

4. OXYGEN REDUCTION AT ELECTRODES OTHER THAN Pt

Similar kinetic studies were carried out with other noble metal and alloy electrodes.[4,10,15-19] The most catalytically active electrode is Pt followed by Pd, Rh, Au, and Ir. The kinetics and mechanisms at Pd and Rh, as well as at Pt- and Pd-rich alloys with Au,

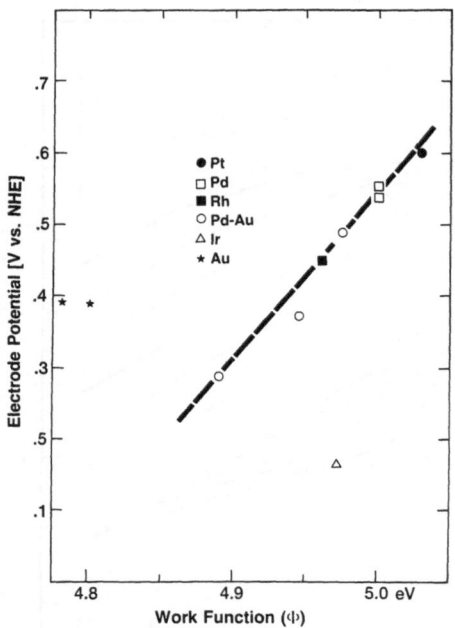

FIGURE 7. Plot of electrode potential at 5×10^{-4} A cm^{-2} at various electrodes against the work function in solution of pH 2.5. Au and Ir electrodes deviate from the linear dependence.

are basically the same as at Pt electrodes. As seen in Fig. 7, there is a linear relation between the potentials at a constant current density in the high-current-density region at these electrodes and the "preferred" work functions, Φ, that are suggested by Trasatti.[20] This is, however, only an indirect effect of Φ on the kinetics, perhaps due to an increase of the energy of adsorption of the products in the rds with increasing Φ. Ir deviates from the linear relation because the mechanism of the reduction at the oxide-covered Ir electrodes is different from that at the oxide-film-free Pt, Pd, and Rh electrodes. Au, although free from surface oxide films, deviates from the linear relation also for the same reason. It may be noted that similar linear dependences have been observed at these metal electrodes for other reactions and were related to bond strengths of reactants with the electrode surface.

5. DEPENDENCE OF THE TAFEL SLOPES ON TEMPERATURE

In Fig. 8, Tafel slopes, b, for the oxygen reduction at Pt in the high-current-density region are plotted against temperature, T, for a number of experiments.[21] The line in the

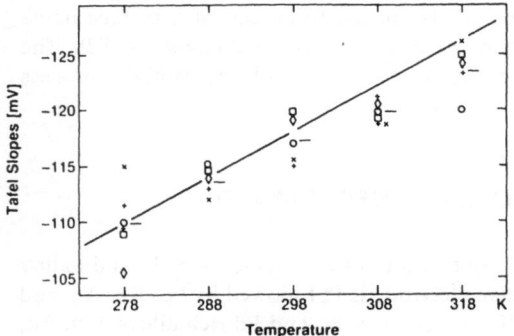

FIGURE 8. Tafel slopes for the oxygen reduction at Pt electrodes in the high-current-density region versus temperature: ×, pH 1.2; ○, pH 1.9; +, pH 2.7; (˙), pH 3.9; and ( ), pH 5.0. The average slopes from all solutions are indicated by —. These slopes deviate from the "ideal" dependence of $-2.3RT/\beta F$ with $\beta = \frac{1}{2}$, which is shown by the straight line.[21]

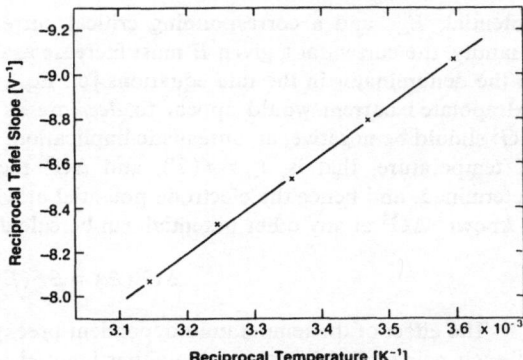

FIGURE 9. Plot of the reciprocal average Tafel slopes (the Lefat slopes) from Fig. 8 versus reciprocal temperature. From this plot, $\beta_H = 0.44$ and $\beta_S = 2.3 \times 10^{-4}$ K^{-1}.[21,27]

figure indicates the ideal dependence of b on T with the symmetry factor, β, being independent of temperature. It is seen that the average values of b do not follow this ideal dependence. This shows that a part of β varies with T; that is,

$$\beta = \beta_H + \beta_S T \tag{12}$$

Here, β_H is the enthalpic and β_S the entropic component of β.[22,23] Both β_H and β_S are evaluated from the Conway plot[24] of the reciprocal slope (the Lefat slope[25]) against $1/T$. In this plot (Fig. 9), the average Lefat slopes decrease linearly with $1/T$. Using the relation

$$\frac{1}{b} = -\frac{\beta_H F}{2.3RT} - \frac{\beta_S F}{2.3R} \tag{13}$$

β_H is found to be 0.44, and β_S 2.3×10^{-4} K^{-1}. Similar values are found for Pd and Rh electrodes. However, for Ir[26] and Au[19] electrodes, the Tafel slopes are independent of T; that is, for these electrodes $\beta_H \approx 0$ and $\beta = \beta_S T$ (see below). Thus, at Pt, Pd, and Rh, β is controlled mainly by the enthalpic component of the activation energy. At Au and Ir, however, it is controlled by the entropic component of the activation energy.

Since β_H and β_S are, respectively, given by[22,24]

$$\beta_H = \frac{d\Delta H^{\ddagger}(E)}{F\,dE} \tag{14}$$

and

$$\beta_S = -\frac{d\Delta S^{\ddagger}(E)}{F\,dE} \tag{15}$$

it follows that at Pt, Pd, and Rh electrodes the entropy of activation for the oxygen reduction changes little with electrode potential, the major contribution to the change of the free energy of activation, $\Delta G^{\ddagger}$, at any potential arising from the enthalpy of activation, which strongly depends on electrode potential. At Au and Ir electrodes, the enthalpy of activation changes little with the electrode potential, while the entropy of activation depends strongly on electrode potential. The significance of the temperature dependence of the Tafel slopes is discussed below.

6. ACTIVATION ENERGIES OBTAINED FROM THE INTERSECTION OF TAFEL LINES

When β is closely constant, that is, $\beta \approx \beta_H$ as in the case of O_2 reduction at Pt discussed above, Tafel lines at different temperatures intersect, upon extrapolation, at some critical potential, E_{cr}, and a corresponding critical current density, i_{cr}. Since $\Delta G^{\ddagger}$ is a positive quantity, the current at a given E must increase as temperature increases due to the T factor in the denominator in the rate equations [cf. Eq. (11)]. At a potential below E_{cr}, however, extrapolated current would appear to *decrease* as temperature increases. This means that $\Delta G^{\ddagger}$ should be negative, an unrealistic implication. At E_{cr}, the current becomes independent of temperature, that is, $i_{cr} \neq f(T)$, and $\Delta G^{\ddagger}$ therefore must be zero.[23,27] Once E_{cr} is determined, and hence the electrode potential at which the activation energy would be zero is known, $\Delta G^{\ddagger}$ at any other potential can be calculated from the relation

$$\Delta G^{\ddagger}(E) = \beta F(E - E_{cr}) \tag{16}$$

The effect of the temperature-dependent preexponential factor on the intersection point of each pair of Tafel lines can be either ignored in the narrow temperature range of most electrochemical experiments, or, if desired, a correction for it can be made.

Equation (16) can be written in the form

$$\Delta G^{\ddagger}(E) \approx (\beta_H + \beta_S T) F(E - E_{cr}) \tag{17}$$

For small values of β_S, the variation of E_{cr} with temperature is small and can be ignored. In this case, as an approximation, one can write[23,27]

$$\Delta H^{\ddagger}(E) \approx \beta_H (E - E_{cr}) \tag{18}$$

and

$$\Delta S^{\ddagger}(E) \approx \beta_S (E - E_{cr}) \tag{19}$$

Thus, when $\beta_S T \ll \beta_H$, $T\Delta S^{\ddagger}(E) \ll \Delta H^{\ddagger}(E)$. In the limit when $\beta_S = 0$, $\Delta S^{\ddagger}(E)$ is zero, that is, for $\beta = \text{const.} \neq f(T)$, only the enthalpy of activation, $\Delta H^{\ddagger}(E)$, controls the rate of an electrochemical reaction. In this case, the intersection point of the Tafel lines signifies the potential at which $\Delta H^{\ddagger}$, rather than $\Delta G^{\ddagger}$, is zero.

In Fig. 10, E–log i data are shown for the oxygen reduction at Pt in pH 1.9 and 3.8 solutions at two temperatures.[23, 27] Upon extrapolation, the Tafel lines obtained at different temperatures in the high-current-density region intersect at $E_{cr} \approx 0.33$ V versus RHE in pH 1.9 solution, and at $E_{cr} \approx 0.41$ V versus RHE in pH 3.8 solution. In Fig. 11, E_{cr} is plotted against

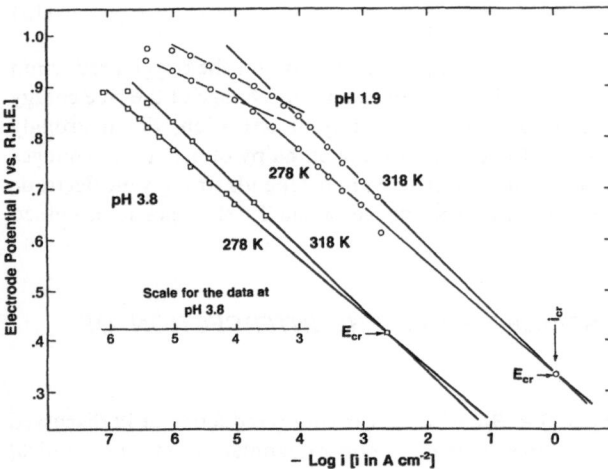

FIGURE 10. E–log i relations for the oxygen reduction at 278 and 318 K in pH 1.9 and pH 3.8 solutions. Upon extrapolation, the Tafel lines in the high-current-density region intersect at the critical potentials, E_{cr}, 0.33 and 0.41 V versus RHE for the pH 1.9 and the pH 3.8 solution, respectively. The intersection points indicate the potentials at which $\Delta H^{\ddagger}$ is zero.[27]

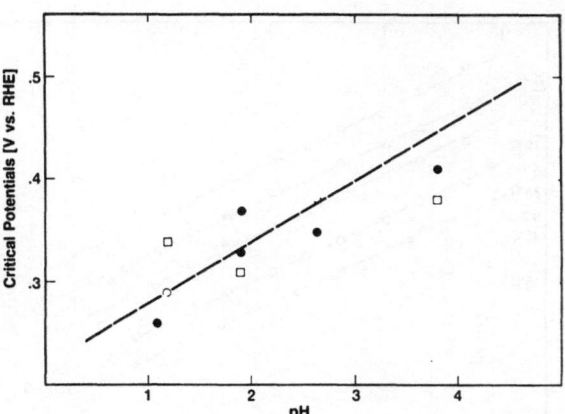

FIGURE 11. Plot of E_{cr} versus pH. ●, E_{cr} obtained from the intersection points, such as those in Fig. 10; ○, E_{cr} obtained by algebraic solution for the intersection points using $(E, \log i)$ data in the Tafel region with the -120 mV slope; □, E_{cr} evaluated from the Tafel region with the -60 mV slope. The dashed line represents the ideal dependence of $2.3RT/F$.[27]

pH for all experimental data available.[27] Full symbols represent direct readings of the intersection points. Open symbols are obtained by algebraic solution for the intersection points using $(\log i, E)$ data in the Tafel regions. Irrespective of the substantial scatter of the intersection points, it is seen that E_{cr} versus RHE increases as pH increases. The dashed line in the figure does not represent the best fit through the experimental points, but rather it indicates the ideal pH dependence of $2.3RT/F$, which will be discussed below. Note that with respect to SHE, E_{cr} is independent of pH.

With $E_{cr} = 0.33$ V versus RHE, in solutions of pH 1.9, $\Delta H^{\ddagger}(0.8)$ is calculated with $\beta = \frac{1}{2}$ to be ~ 0.23 eV (22.7 kJ/mol). This value compares satisfactorily with the values of 21 and 24 kJ/mol for $\Delta H^{\ddagger}(0.8)$ in the same solutions obtained from classical Arrhenius plots. With $\beta_H \approx 0.45$, as found experimentally, $\Delta H^{\ddagger}(E)$ is 25 kJ/mol.

The enthalpies of activation at 0.8 V versus RHE that are calculated from E_{cr} are given in Fig. 12. Although the scattering of experimental points is substantial, it is seen that the enthalpy decreases as pH increases. Note that the enthalpies of activation at a constant potential versus SHE are independent of pH. This is expected for the observed kinetics, which shows that, at a constant E versus SHE, $\Delta H^{\ddagger}(E) \neq f(\text{pH})$. Because of the relation in Eq. (18), E_{cr} versus SHE is then expected to be independent of pH.

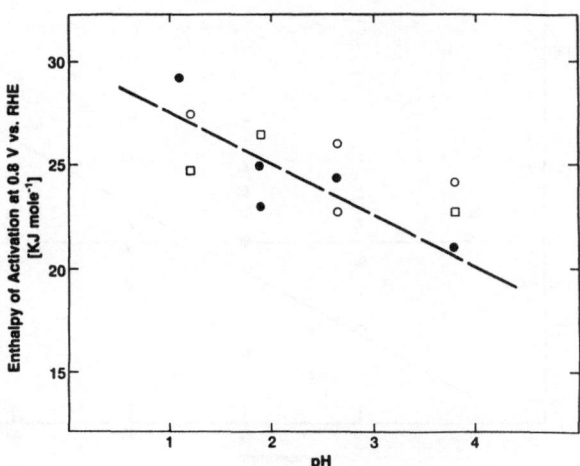

FIGURE 12. Plot of enthalpy of activation at 0.8 V versus RHE, $\Delta H^{\ddagger}(0.8)$, versus pH. These enthalpies are calculated from E_{cr} in Fig. 11. Symbols are the same as in Fig. 11.

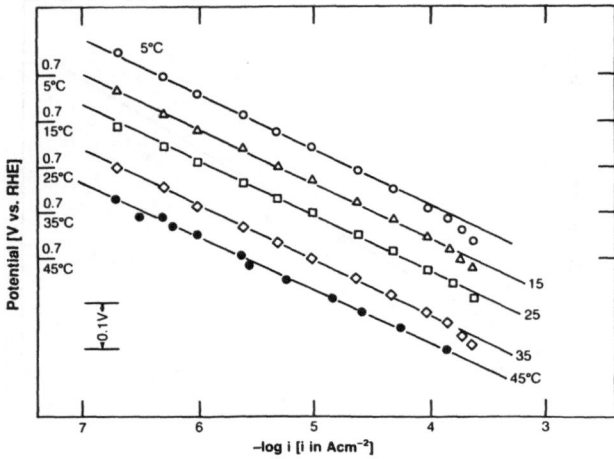

FIGURE 13. Steady state E-log i relations at Au electrodes at different temperatures in pH 3 (HClO$_4$) solution. Potential scales are given versus RHE at room temperature.[19]

7. OXYGEN REDUCTION AT Au ELECTRODES

In Fig. 13, steady-state E-log i data are shown for Au electrodes in pH 2.2 solutions at different temperatures.[19] Potentials refer to the reversible hydrogen electrode kept at room temperature in all experiments. It is clear that at all temperatures Tafel slopes are close to −120 mV. As seen in Fig. 14, Tafel slopes do not fit the conventional dependence of $2.3RT/\beta F$ with β as a constant. This conventional dependence is indicated in the figure by the straight line passing through $T = 0$ K. Instead, the slopes appear to be independent of temperature, and therefore $\beta = \beta_S T$. This behavior can be contrasted with that for the reduction at Pt, Pd, and Rh electrodes (see Fig. 8). The reduction at the gold electrode is pH independent. Thus, the mechanism of the reaction is also different from that at Pt electrodes. Evidently, the rds

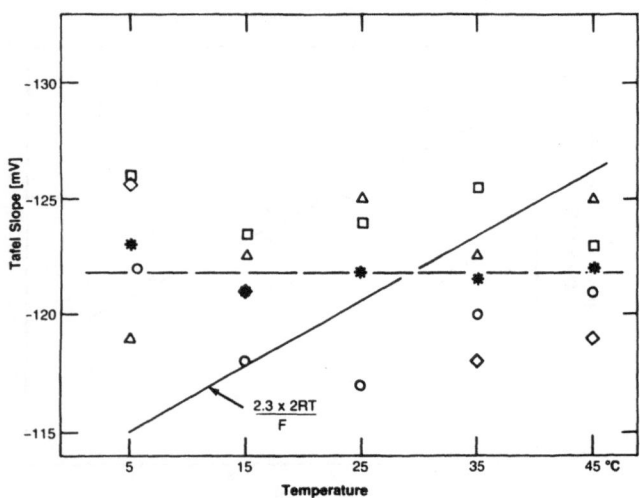

FIGURE 14. Plot of the Tafel slopes for the oxygen reduction at Au electrodes versus temperature. Full line indicates the ideal $2.3RT/\beta F$ dependence when $\beta = \frac{1}{2}$. Clearly, Tafel slopes are independent of T, i.e., $\beta = \beta_S T$.[19]

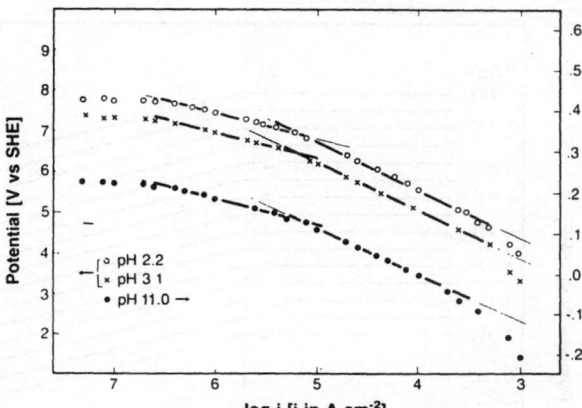

FIGURE 15. E–log i dependence for the oxygen reduction at Ir electrodes in pH 2.2 (○) and pH 3.1 solutions (×) (left-hand axis) and in pH 11.0 (●) solutions (right-hand axis). Oxygen under standard pressure.[18]

is the first discharge step without H^+ as the reactant in the step, i.e., $O_2 + e^- \to O_2^-$, even though in acid solutions.

8. OXYGEN REDUCTION AT Ir ELECTRODES

Steady-state E–log i relations at an Ir electrode in aqueous acid and alkaline solutions show two linear Tafel regions.[18,26] At low current densities, Tafel slopes are close to -60 mV, and at high current densities they are close to -120 mV (see Fig. 15). These behaviors are superficially similar to those at Pt, Pd, and Rh electrodes. The similarity, however, ceases when the pH dependence and temperature effect on the reduction at Ir electrodes is compared to that at Pt electrodes. Thus, the pH dependence in both current density regions, expressed as $(dE/d\text{pH})_i$, is -60 mV. Based on these data, and the observed dependence of the rates on oxygen partial pressure, the following rate equations are obtained for the low and the high current-density region, respectively[18]:

$$i_l = k_l[H_3O^+]p_{O_2}\exp\left(\frac{-FE}{RT}\right) \tag{20}$$

and

$$i_h = k_h[H_3O^+]^{1/2}p_{O_2}\exp\left(\frac{-FE}{2RT}\right) \tag{21}$$

In Fig. 16, the idealized composite diagram is sketched for the reduction at Ir electrodes over the entire pH range.[18] This diagram can be contrasted with the corresponding diagram for Pt electrodes (see Fig. 6). In the figure, the observed positions of the rest potentials, E_r, are shown. E_r evidently decreases by ca. $2.3RT/F$ as the pH increases by one unit. Again, this potential seems to be the corrosion potential.

Comparative analysis of the data at Ir electrodes with those at Pt and Au electrodes clearly shows that the kinetic behavior and the mechanism of the reduction at the Ir electrodes are quite different from those at the other electrodes discussed above.

9. MECHANISM OF THE REDUCTION AT Ir ELECTRODES

The observed kinetics in the low-current-density region in acid solutions could readily be comprehended in terms of a rate-determining chemical step that follows a first electron

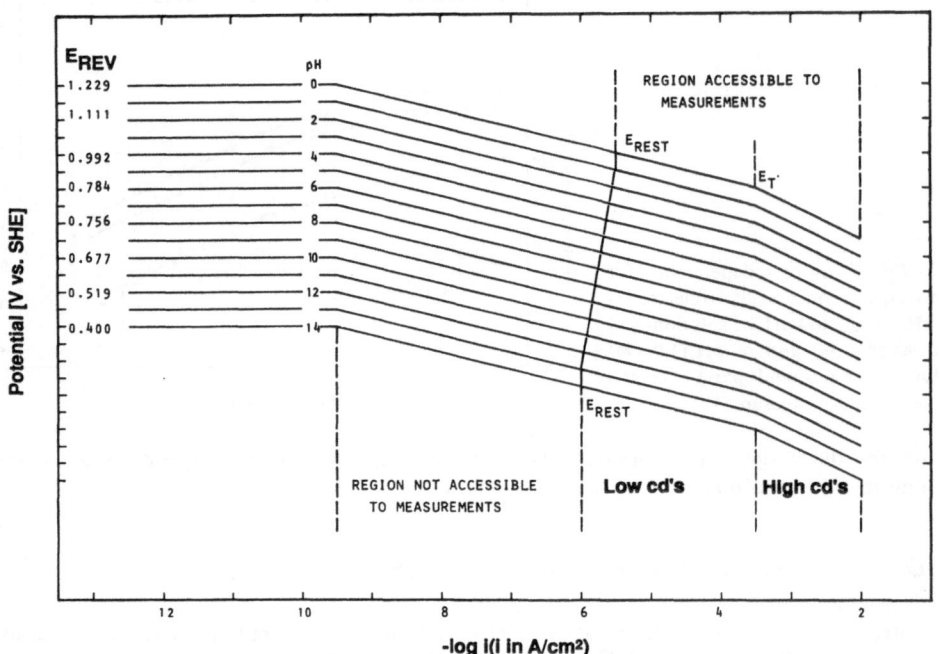

FIGURE 16. Ideal composite E-log i diagram for the oxygen reduction at Ir electrodes over the entire pH range. $p_{O_2} = 1$ atm. E_{rev} is the reversible potential, and E_{rest} is the rest potential above which no measurements of E-log i can be extended; E_T is the potential of the transition from one Tafel region to another; cd is current density.[18]

transfer step, for example, that in Eq. (6), under simple Langmuirian conditions. However, because of the diffusional limitation of H^+ it is difficult to see how this step can be the rds in alkaline solutions either under the Langmuirian or under the Temkin adsorption conditions. Further, the fractional reaction order of $\frac{1}{2}$ with respect to H^+ for the kinetics in the high-current-density region cannot be explained in terms of any classical kinetics analysis under either adsorption condition.

The fact that the E-log i region with the -60 mV slope at low current densities is followed, without any discontinuity, by the region with the -120-mV slope at high current densities is interpreted to indicate that the same reaction path is operative in both current density regions[28] and that either the rds in that path changes with electrode potential or the adsorption conditions in the same rds are altered at the point of the transition from one Tafel region to another, causing the change in the Tafel slope. The latter conditions are illustrated above with the change from Temkin to Langmuirian adsorption in the reduction at Pt electrodes. The kinetic data at Ir electrodes, however, cannot be accounted for by the change of the adsorption conditions.

In contrast to the Au and Pt electrodes, Ir electrodes in the potential region of O_2 reduction are covered by an oxide phase film.[18,26] Consequently, a specific effect of the oxide film on the kinetic behavior regarding O_2 reduction is to be expected.

It is suggested here that the fractional reaction order with respect to H^+, which is observed in the high-current-density region, can be explained by invoking the concept of heterogeneous acid–base equilibrium at the oxide surface/solution interface.[29] The same concept should then be compatible with the explanations of the kinetics in the low-current-density region.

The acid–base equilibrium at the oxide-film-covered electrodes in acid solutions can be represented by the equations[29]

$$>S-OH \leftrightarrow >S^+ + (OH^-)^* \tag{22}$$

and/or

$$>S-OH + H_2O \leftrightarrow >S-OH_2^+ + (OH^-)^* \tag{23}$$

In basic solutions, it can be represented by

$$=S-OH \leftrightarrow =S-O^- + (H^+)^* \tag{24}$$

Here, S stands for a surface site. As a consequence of these hydrations and dissociations, a potential difference, $\Delta\phi^*$, is established between the plane of adsorption of protons, or hydroxyl ions, and the bulk of the solution. In a simple way, this difference is controlled by the equilibrium that exists between the H^+, or OH^-, ions in the plane of adsorption and the bulk of the solution. It is given by[29]

$$\Delta\phi^* = \text{const.} - \frac{2.3RT}{F} pH \tag{25}$$

The asterisk here and in the above equilibria signifies position in the adsorption plane.

If it is taken that the O_2 molecules that are reacting in the first electron transfer step are located in the same adsorption planes as H^+ and/or OH^-, then only a part of the total electrode potential difference across the electrode/solution interface acts in the electron transfer process. For this process, that is, for

$$O_2 + e^- \rightarrow O_2^- \tag{26}$$

the rate equation can be written in the form

$$i_h = k'_h p_{O_2} \exp\left[\frac{-\beta F(E - \Delta\phi^*)}{RT}\right] \tag{27}$$

Here, E is the electrode potential versus a pH-independent reference electrode. Replacing $\Delta\phi^*$ with Eq. (25), the rate equation assumes the form

$$i_h = k_h p_{O_2}[H_S^+]^{1/2} \exp\left(\frac{-\beta FE}{RT}\right) \tag{28}$$

Here, $[H_S^+]$ represents the activity of H^+ in the bulk of the solution. It can be shown that this rate equation for the current in the high-current-density region holds over the entire pH range.

The kinetics in the low-current-density region is explained by a chemical rate-determining step which follows the first electron transfer step, for instance, the step

$$O_2^- + H_2O \rightarrow HO_2^- + OH \tag{29}$$

The rate of this step is given in the usual way by

$$i_l = k'_l[O_2^-] \tag{30}$$

The concentration of O_2^- is readily obtained from the quasi-equilibrium along the preceding first electron transfer step, which yields

$$i_l = k_l p_{O_2}[H_S^+] \exp\left(\frac{-FE}{RT}\right) \tag{31}$$

This rate equation, too, like that in Eq. (28), is valid over the entire pH range.

In these derivations it is assumed that $(H^+)^*$ and $(OH^-)^*$ behave as a part of the oxide phase and that their activity is constant, independent of electrode potential and pH. However, the activity of the product in the first step, O_2^-, which also is assumed to be located in the ϕ^* plane, is taken to vary with the electrode potential; that is, O_2^- is not considered to be a part of the solid state. Only with the assumption that $(H^+)^*$ and/or $(OH)^*$ are independent of pH do the acid–base equilibria lead to a $\Delta\phi^*$ which is pH dependent.

It may be noted that this potential difference between the plane of adsorption and the bulk of the solution is, at least formally, equivalent to the potential difference across the outer Helmholtz layer (OHL), which also was found, in oxygen evolution at Pt electrodes, to be independent of the change in the electrode potential but dependent on pH.[30] The adsorption plane of the acid–base equilibria then corresponds to the inner Helmholtz plane (IHP), in which the reactants are located.

Thus, with the concept of acid–base equilibria it is possible to explain the observed kinetics for O_2 reduction at Ir electrodes in both current density regions over the entire pH range. In particular, it is possible to account for the fractional reaction order with respect to H^+ observed in the high-current-density region.

10. OXYGEN REDUCTION AT OXIDE-FILM-COVERED Pt ELECTRODES

Recently, O_2 reduction at oxide-film-covered Pt electrodes has been examined under controlled thickness of the surface oxide films for each measured E–$\log i$ point.[31] In contrast to the previous data, two linear E–$\log i$ regions are observed. At low current densities, the Tafel slope is -60 mV, and at high current densities it is -120 mV. The pH dependence in both current density regions is given by $(dE/d\mathrm{pH})_i \approx -60$ mV. This dependence leads to the same reaction step as in the case of the Ir electrodes and to the fractional reaction order with respect to H^+ for the rates in the high-current-density region. Following a previous analysis of oxygen evolution at Pt,[30] it was suggested that the reactants are located in the IHP and that there is a potential difference across the OHL which is independent of electrode potential but varies with pH. It was found[30] that

$$^{IHP}\Delta^{OHP}\phi = 0.96 - \frac{2.3RT}{F}\mathrm{pH} \tag{32}$$

This potential difference is related to the onset of oxide film formation. It introduces the fractional reaction order into the rate equation at high current densities. Equally well, the observed reaction order at Pt electrodes can be related to the acid–base equilibrium in much the same way as was done above in the analysis of the kinetics at Ir electrodes. The only important difference between the oxide-film-covered Pt and Ir electrodes is that at Pt the rates depend exponentially on the oxide film thickness, whereas this does not seem to be the case with the rates at Ir electrodes. The reason for this is that the oxide film at Pt is an electronic insulator, and electrons tunnel through the film, resulting in the exponential diminution of the rates at a given potential with the film thickness.[30] It seems useful to examine O_2 reduction at other oxide-film-covered electrodes (e.g., Rh) and compare the catalytic activities at such electrodes, as was done for the O_2 reduction at the oxide-film-free electrodes (see Fig. 7) and for O_2 evolution at a number of electrodes.[32]

11. COMPARISON OF THE TAFEL SLOPES AT DIFFERENT ELECTRODES

In contrast to the Tafel slopes at oxide-free Pt, Pd, and Rh electrodes, those at Ir electrodes are unaffected by temperature (Fig. 17). Here too, as at Au electrodes, $\beta = \beta_S T$.

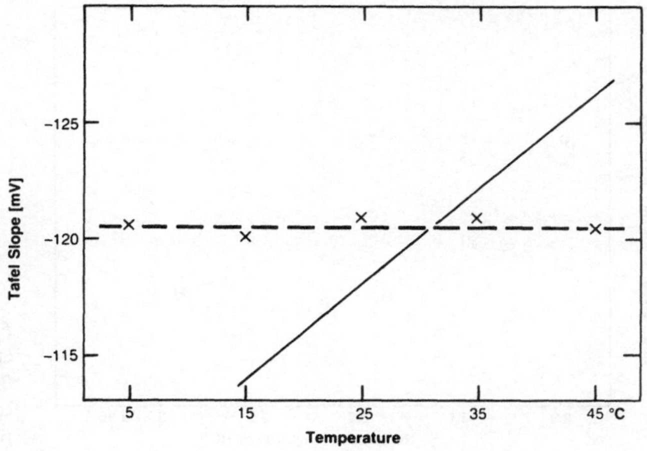

FIGURE 17. Average Tafel slopes from six measurements for the oxygen reduction at Ir electrodes in the high-current-density region in solutions of different pHs. Full line shows the ideal $2.3RT/\beta F$ dependence with $\beta = \frac{1}{2} = $ const. Tafel slopes are independent of T.[26]

At the present time, the reason for β being constant at some electrodes and varying strongly with temperature at other electrodes is not clear. It appears that, at least for oxygen reactions at oxide-covered electrodes both in the evolution[33] and reduction,[18,26] $\beta = \beta_S T$. At oxide-free electrodes, with the exception of Au, β is approximately constant. It may be noted that at Au electrodes, and perhaps at oxide-covered electrodes, there is only a weak interaction of reactants and/or reaction intermediates with the constituents of electrode surfaces.

12. THE SIGNIFICANCE OF THE ARRHENIUS PLOTS WHEN $\beta = \beta_S T$

Writing the rate equation for the first discharge step in the cathodic direction in the form

$$\ln i = k' - \frac{\Delta G_0^{\ddagger}}{2.3R} - \frac{\beta_S T F \Delta \phi}{RT}$$

$$= K - \frac{\Delta H_0^{\ddagger}}{RT} + \frac{\Delta S_0^{\ddagger}}{R} - \frac{\beta_S F E}{R} \tag{33}$$

it follows that the log i–$1/T$ plots at a constant E should be linear and, for different E, also parallel. Slopes of these plots,

$$\frac{d \log i}{d(1/T)} = -\frac{\Delta H_0^{\ddagger}}{2.3R} \tag{34}$$

should then yield the enthalpy of activation at the zero Galvani potential difference, $\Delta \phi = 0$.†
Nearly linear plots have been reported for the oxygen reduction at Au electrodes in a narrow temperature range. Recently, it was shown that for the oxygen evolution at Pt[33] over a wide temperature range (273–348 K) and for hydrogen evolution[34] in "frozen" systems (~50–300 K) such plots are not linear, although β for these reactions in $\beta_S T$. This is illustrated in Fig. 18 for the oxygen evolution reaction. Surprisingly, however, the plots of log i versus T are *linear* (see Fig. 19). This is not expected because differentiation of Eq. (33) yields

$$\left[\frac{d \log i}{dT} \right]_E = \frac{\Delta H_0^{\ddagger}}{RT^2} \tag{35}$$

---

† In view of the foregoing discussion of the reaction rates in terms of acid–base equilibrium, $\Delta \phi = 0$ refers to the zero electric potential difference between the bulk of the solution and the adsorption plane.

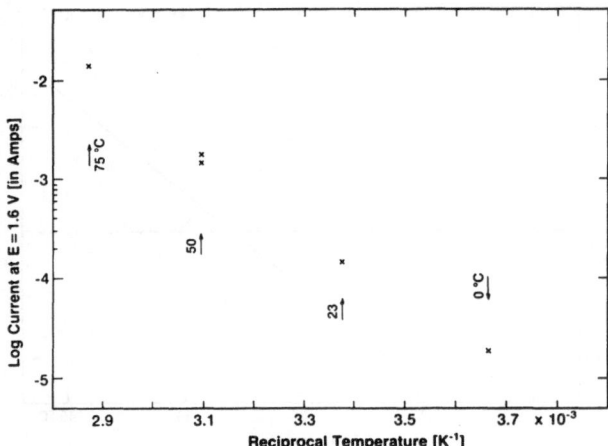

FIGURE 18. Arrhenius plot of log i at $E = 1.6$ V versus SCE against T^{-1} for O_2 evolution at Pt in H_2SO_4 solution, pH 1.5, for which $\beta = \beta_S T$.

Thus, not a linear but a very strong nonlinear dependence on T is expected. To account for the observed linearity, it is suggested that $\Delta H_0^{\ddagger}$ is small, if not zero, compared to some other factor or factors in the expression for the current density [see Eq. (33)], which then must increase *linearly* with temperature. It is further suggested that ΔS_0 changes linearly with temperature because the entropy in the ground state of the reacting species varies linearly with temperature.[35] Moreover, when the entropy term in the Gibbs activation energy predominates, and the entropy changes with temperature, $\beta_H = 0$ and $\beta = \beta_S T$. When the enthalpy term predominates, and it is independent of temperature, $\beta_H = $ const. and $\beta_S = 0$. This seems to be in accord with the view that $\beta_S T$ is associated with the change of entropy of activation with electrode potential [see Eq. (15)], whereas β_H is associated with the change of enthalpy of activation with electrode potential [see Eq. (14)]. A mixed dependence, $\beta = \beta_H + \beta_S T$, would then signify a combined effect of both entropy and enthalpy of activation on the reaction rate.

13. CONCLUSION

Further studies with accurate experimental data for the activation energies for O_2 reduction at oxide-free as well as at oxide-covered electrodes over a wide temperature range, combined with parallel studies of oxygen evolution, seem to be very desirable here. These

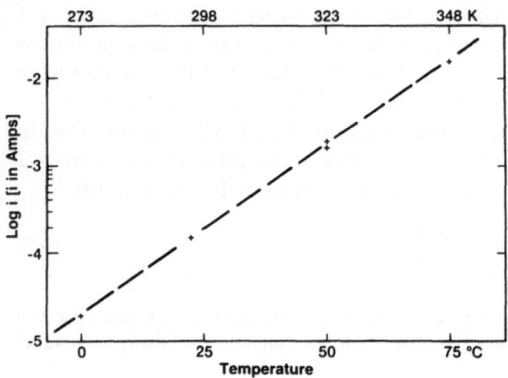

FIGURE 19. The same data as in Fig. 18 plotted against T. A linear relation over the entire temperature range is evident.

studies are expected to provide a clue as to the origin of the observed differences in the temperature behavior of symmetry factors at various electrodes. Such studies, in turn, will help to elucidate the relations between the β factors and activation energies.

It is seen from this short review that substantial progress has been made since the time of the initial studies in the laboratory of Professor Bockris at the University of Pennsylvania. This progress has been triggered by the sound foundation that was laid down by Professor Bockris and his colleagues at that early time. Subsequently, the work initiated in the laboratory of Professor Bockris was carried on in other laboratories, resulting in a substantial understanding of the mechanisms of oxygen reactions and paving the way for future research and studies.

ACKNOWLEDGMENT

Thanks are due to Drs. D. Narasimhan and R. Morris for their help and support with the preparation of the manuscript.

REFERENCES

1. A. N. Frumkin, in: *Advances in Electrochemistry and Electrochemical Engineering*, Vol. 1 (P. Delahay, ed.), p. 65, Interscience Publishers, New York (1961).
2. M. L. B. Rao, A. Damjanovic, and J. O'M. Bockris, *J. Phys. Chem.* **67**, 2508 (1968).
3. A. Damjanovic and V. Brusic, *Electrochim. Acta* **12**, 615 (1967).
4. A. Damjanovic and V. Brusic, *Electrochim. Acta* **12**, 1171 (1967).
5. A. Damjanovic and J. O'M. Bockris, *Electrochim. Acta* **11**, 376 (1966).
6. A. Damjanovic, A. T. Ward, B. Ulrick, and M. O'Jea, *J. Electrochem. Soc.* **122**, 471 (1975).
7. A. Damjanovic, A. T. Ward, and M. O'Jea, *J. Electrochem. Soc.* **121**, 1186 (1974).
8. M. Genshaw, A. K. N. Reddy, and J. O'M. Bockris, *J. Electroanal. Chem.* **8**, 406 (1964).
9. H. Wroblowa, M. L. B. Rao, A. Damjanovic, and J. O'M. Bockris, *J. Electroanal. Chem.* **15**, 139 (1967).
10. D. B. Sepa, M. V. Vojnovic, and A. Damjanovic, *Electrochim. Acta* **26**, 781 (1981).
11. M. Wong, in: Reversible Oxygen Electrodes, 5th Report ECOM U.S. Army Electronics Command, Fort Monmouth, New Jersey (1966) p. 1.
12. A. Damjanovic and M. A. Genshaw, *Electrochim. Acta* **15**, 1281 (1970).
13. A. Damjanovic, D. B. Sepa, and M. V. Vojnovic, *Electrochim. Acta* **24**, 887 (1979).
14. D. B. Sepa, M. V. Vojnovic, and A. Damjanovic, *Electrochim. Acta* **25**, 1491 (1980).
15. A. Damjanovic, V. Brusic, and J. O'M. Bockris, *J. Phys. Chem.* **71**, 2471 (1967).
16. L. M. Vracar, D. B. Sepa, and A. Damjanovic, *J. Electrochem. Soc.* **133**, 1835 (1986).
17. J. M. Martinovic, D. B. Sepa, M. V. Vojnovic, and A. Damjanovic, *Electrochim. Acta* **33**, 1267 (1988).
18. D. B. Sepa, M. V. Vojnovic, M. Stojanovic, and A. Damjanovic, *J. Electroanal. Chem.* **218**, 265 (1987).
19. D. B. Sepa, M. V. Vojnovic, L. M. Vracar, and A. Damjanovic, *Electrochim. Acta* **31**, 1105 (1986).
20. S. Trasatti, in: *Advances in Electrochemistry and Electrochemical Engineering*, Vol. 10 (H. Gerischer and C. W. Tobias, eds.), p. 213, John Wiley & Sons, New York (1977).
21. A. Damjanovic, D. B. Sepa, L. M. Vracar, and M. V. Vojnovic, *Ber. Bunsenges. Phys. Chem.* **90**, 1231 (1986).
22. B. E. Conway, D. F. Tessier, and D. P. Wilkinson, *J. Electroanal. Chem.* **199**, 249 (1986).
23. A. Damjanovic, B. E. Conway, and D. B. Sepa, *Ber. Bunsenges. Phys. Chem.* **93**, 510 (1989).
24. B. E. Conway, D. P. Wilkinson, and D. F. Tessier, *Ber. Bunsenges. Phys. Chem.* **91**, 484 (1987).
25. B. E. Conway, in: *Modern Aspects of Electrochemistry*, No. 16 (J. O'M. Bockris and B. E. Conway, eds.), p. 103, Plenum Press, New York (1985).
26. D. B. Sepa, M. V. Vojnovic, M. Stojanovic, and A. Damjanovic, *J. Electrochem. Soc.* **134**, 845 (1987).
27. A. Damjanovic and D. B. Sepa, *Electrochim. Acta*, in press.
28. A. Damjanovic, A. Dey, and J. O'M. Bockris, *J. Electrochem. Soc.* **113**, 739 (1966).
29. A. Daghetti, G. Lodi, and S. Trasatti, *Mater. Chem. Phys.* **8**, 1 (1983).

30. V. I. Birss and A. Damjanovic, *J. Electrochem. Soc.* **130**, 1694 (1983).
31. A. Damjanovic and P. G. Hudson, *J. Electrochem. Soc.* **135**, 2269 (1988).
32. S. Trasatti, *Electrochim. Acta* **29**, 1503 (1984).
33. A. Damjanovic, A. T. Walsh, and D. B. Sepa, *J. Phys. Chem.* **94**, 1967 (1990).
34. A. Damjanovic, J. O'M. Bockris, and J. C. Wass, private communication.
35. A. Damjanovic, V. I. Birss, and D. B. Sepa, in preparation.

10

Mechanism of Oxygen Reduction on Iron in Neutral Aqueous Solutions: Oxygen Chemisorption Model

Vladimir Jovancicevic

1. INTRODUCTION

Though the anodic reactions of metal and alloy dissolution[1-5] have been extensively studied in solutions of various pHs, the corresponding cathodic processes have received little attention in the neutral pH range.[6-8] Generally, under practical conditions, the cathodic reaction of hydrogen evolution in acid solutions and the oxygen reduction reaction in neutral and alkali solutions are the main cathodic depolarization reactions.

The principal difficulty considered to confront experimental work in neutral solutions is related to the local changes in pH in the vicinity of the double layer in the course of the electrode reactions. Thus, while the anodic reaction results in a decrease in pH, the cathodic reaction gives rise to an increase in pH relative to the bulk solution pH. The surface pH change, in turn, can influence the oxygen reduction reaction either directly through the reaction mechanism involving OH^- or indirectly via redox acid–base properties of the surface oxide layer.

In studying the kinetics and mechanisms of the oxygen reduction reaction, the reduction characteristics and the structure and composition of the surface-active sites of the oxide film on iron are of primary interest. Despite the large body of experimental work and theoretical rationalization, there are still considerable uncertainties about the detailed structure of the surface-active sites of iron oxides.

A few papers have been devoted to the study of O_2 reduction on Fe in neutral or alkaline solutions. Delahay[6] studied the reduction of O_2 at an iron electrode in near-neutral solutions. For potentials negative to -0.6 V versus NHE, the reaction was a four-electron process with little H_2O_2 detected. Haruyama et al.[7] proposed an electrochemical O_2 reduction mechanism at pH 8.4 in which the chemical step was the catalytic decomposition of adsorbed peroxide. Foroulis[8] examined the effect of rotating disk velocity and the dissolved oxygen concentration on the initial corrosion behavior of a rotating iron electrode in "pure" water. Fabjan et al.[9] studied the reduction of O_2 on iron in alkaline solution and found reaction orders with respect to O_2 and OH^- of $\frac{1}{2}$ and -1, respectively. A similar result was obtained by Calvo and

Vladimir Jovancicevic ● W. R. Grace & Co., Washington Research Center, Columbia, Maryland 21044.

Electrochemistry in Transition, edited by Oliver J. Murphy *et al.* Plenum Press, New York, 1992.

Schiffrin[10] in NaOH solutions, suggesting the Fe^{2+}-mediated formation of adsorbed peroxide species under Temkin conditions.

The present study was undertaken to gain information about the basic mechanism of O_2 reduction, including the pathway and rate-determining steps, on bare iron and passive iron in neutral borate buffer solution. The redox property of the passive film on iron and its role in the O_2 reduction reaction were investigated with respect to the Fe^{2+}/Fe^{3+} ratio and bound water.

2. EXPERIMENTAL TECHNIQUES

2.1. Rotating Disk-Ring Electrode

The details of the three-compartment cell with a rotating disk electrode assembly have been described elsewhere.[11] An iron–platinum rotating disk-ring electrode (RDRE) with collection efficiency of 0.37 was used for the detection of hydrogen peroxide. RDRE experiments were carried out by using a bipotentiostat (Pine Instrument) and a rotator (Pine Instrument). The reference electrode was a saturated calomel electrode (SCE); all potentials are referred to the normal hydrogen electrode (NHE).

The polarization V–log i curves for O_2 reduction on the iron disk electrode were obtained by means of the potential-step method following the reduction of the air-formed oxide layer at -0.74 V versus NHE. After introducing O_2 at a given partial pressure, the steady-state cathodic current was measured at various potentials with 50 mV interval. The corresponding anodic currents for H_2O_2 oxidation on platinum ring were measured at 1.0 V versus NHE under diffusion control.

Borate buffer solutions were prepared from triply distilled water and analytical-grade sodium tetraborate ($0.0375M$) and boric acid ($0.15M$), pH 8.4. The solution pH was varied by adjusting the pH with boric acid (pH 7.0) and sodium hydroxide (pH 9.8). Sodium sulfate as supporting electrolyte was used in some experiments.

2.2. Surface Spectroscopy

Slow, single-sweep voltammetry was used for the formation and reduction of the passive oxide layer as described previously.[12] In some experiments, the cathodic sweep was preceded by an anodic step at various potentials in the passive range. The thickness and the optical properties of the film were studied as a function of potential in the presence and absence of O_2 by using Fourier transform ellipsometry (FTE) (Rudolph Research 2000 FT).

The surface analysis of the passive layers formed at different anodic potentials was carried out with a Kratos X SAM-8000 X-ray photoelectron spectrometer (XPS) using a monochromatic Mg Kα X-ray source. Direct transfer of the electrode under N_2 atmosphere from the electrochemical cell to the XPS sample chamber was performed in order to reduce possible changes of the surface of the passive film.

3. PROPERTIES OF OXIDE FILM ON IRON

A first cyclic voltammogram of the formation of the passive layer in borate buffer solution is shown in Fig. 1. On the anodic sweep, the iron dissolution peak (I) at -0.4 V versus NHE, corresponding to the formation of a monolayer of iron oxide film, is followed by a broad anodic peak (II) at 0.1 V versus NHE. On the reverse sweep, a cathodic peak (II') at -0.35 V

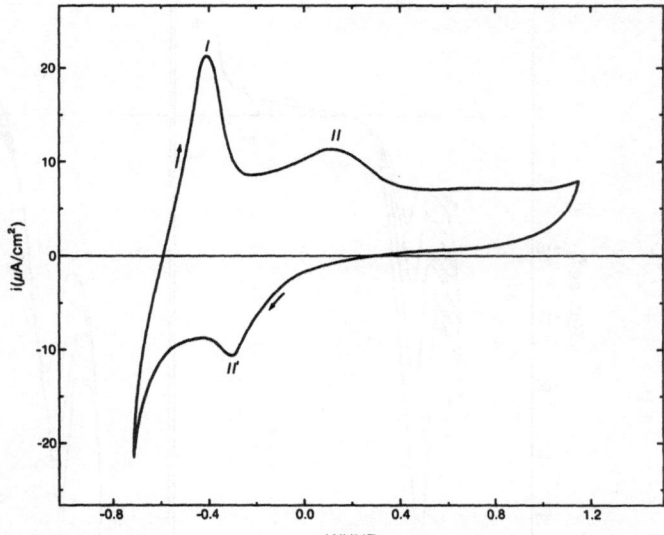

FIGURE 1. First cyclic voltammogram of the formation of passive film on iron in pH 8.4 borate buffer solution. $v = 5$ mV/s.

versus NHE can be observed. The increase of the current at more cathodic potentials is due to the simultaneous occurrence of the reduction of the passive film and the hydrogen evolution reaction. After complete reduction of the film at potentials less than or equal to -0.74 V versus NHE, the same reproducible voltammogram was obtained.

The effect of the sweep rate on the reduction of the passive layer formed at 1.1 V versus NHE is given in Fig. 2a. In the potential range between 1.1 and 0.1 V versus NHE, no reduction of the passive layer occurs. At more cathodic potentials, a relatively sharp peak is superimposed on a large background current. The higher the sweep rate, the smaller is the charge associated with the background current. The charge corresponding to the cathodic peak current (II') is independent of the sweep rate.

The effect of the anodic potential limit on the cathodic peak II', at a sweep rate of 5 mV/s, is shown in Fig. 2b. As the thickness increases with increasing potential, the reduction peak (II') current increases but the corresponding charge remains constant. The increase of the peak current is associated with higher background current. The thickness of the passive layer in the potential region from -0.5 to 0.25 V versus NHE, as obtained from ellipsometric data in the absence and in the presence of oxygen, is shown in Fig. 3. In the absence of O_2, an important feature of the curve is the quasiconstant thickness of the passive layer between -0.3 and -0.2 V versus NHE. The thicknesses obtained at the more positive potentials are 5–10 Å higher than those previously reported.[13,14]

A surprising difference in the thickness–potential curve relationship is observed in the presence of oxygen. Thus, the plateau in the thickness–potential curve is no longer noticeable, and the thickness increases linearly with potential, but at a lower rate than in the absence of oxygen.

The optical constants, n and k, of the passive film obtained by ellipsometry in the wavelength range from 4000 to 7000 Å are given in Fig. 4, for various potentials. Two distinct peaks in the extinction coefficient, k, are present at 5250 and 6000 Å. The peak intensity at 6000 Å decreases with potential until it becomes negligible at 0.2 V versus NHE, and that at 5250 Å also decreases with potential approaching its limiting value of 0.09 at the same potential. The background value of k for wavelengths longer than 5000 Å decreases correspondingly from 0.4 to 0 at 0.2 V versus NHE. The related spectra of refractive index, n, show a Lorentzian type of behavior with minima at the characteristic wavelengths.

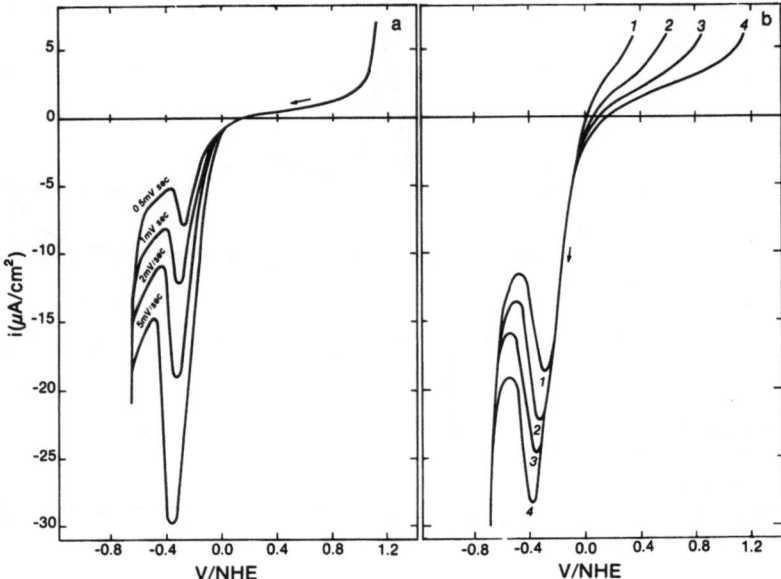

FIGURE 2. (a) Cathodic potentiodynamic curves of the passive layer formed in borate buffer solution (pH = 8.4) at 1.1 V versus NHE for the different sweep rates indicated. (b) Cathodic potentiodynamic curves of the passive layer formed in borate buffer solution (pH = 8.4) at different anodic potentials; $v = 5$ mV/s.

The X-ray photoelectron spectra of Fe $2p_{3/2}$ and O $1s$ for different potentials are shown in Fig. 5. The first large peak of Fe could be deconvoluted into two peaks at 708 and 710.5 V. The peak with a binding energy of 710.5 eV can be assigned to Fe^{3+}, while the peak at 708 eV is attributed to Fe^{2+}.[15,16] The peak of metallic iron appears at 706 eV. Two peaks for O $1s$ at 530.2 and 531.9 eV are attributed to O^{2-} and $OH^-(H_2O)$, respectively.

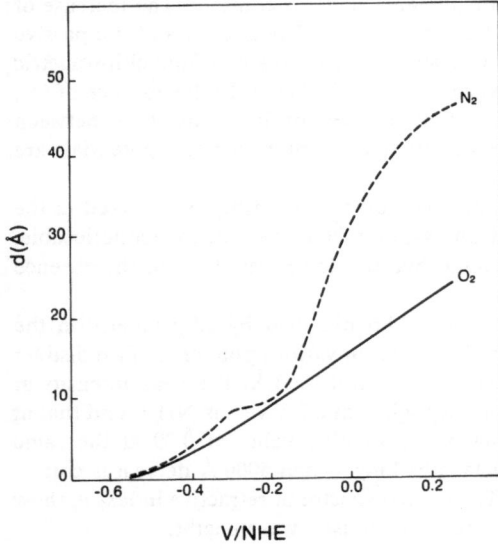

FIGURE 3. Thickness versus potential plots of the passive layer formed in borate buffer solution (pH = 8.4) in the presence (——) and absence (- - -) of O_2. $v = 5$ mV/s.

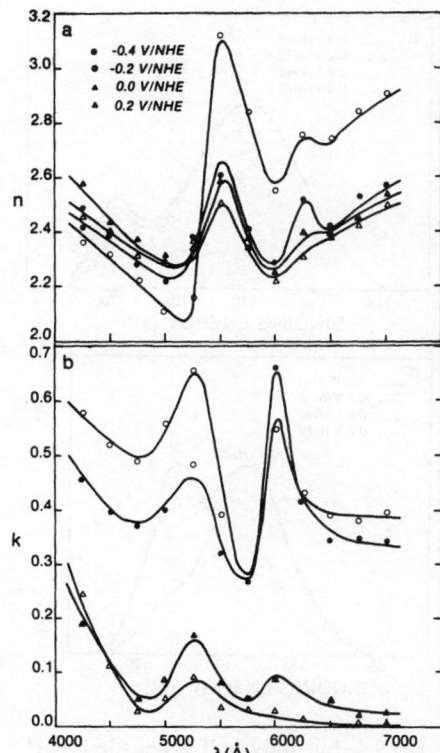

FIGURE 4. Potential dependence of refractive index (a) and extinction coefficient (b) of the passive layer for various wavelengths.

On deconvoluting the spectra, the ratios Fe^{2+}/Fe^{3+} and $OH^-(H_2O)/O^{2-}$ are obtained, and their variations with potential are given in Fig. 6. The ratios obtained for very thin films were more difficult to analyze due to continuing oxidation of the surface exposed to the solution after the loss of potential control. At more anodic potentials, the Fe^{2+}/Fe^{3+} ratio decreases substantially, approaching 0.2 at 0.2 V versus NHE. The $OH^-(H_2O)/O^{2-}$ ratio exhibits a minimum in the potential range corresponding to constant film thickness and maximum Fe^{2+}/Fe^{3+} ratio.

4. KINETICS OF OXYGEN REDUCTION

4.1. Tafel Plots and Reaction Orders

Potentiostatic Tafel plots for O_2 reduction at different pHs are shown in the potential range from 0.05 V (passive layer) to -0.7 V versus NHE (hydrogen evolution) in Fig. 7. The anodic current corresponding to the dissolution of iron (-0.4 V versus NHE) is not observed even at pH 7.0, though some dissolved iron could be detected using the rotating disk-ring electrode. The cathodic current in the potential range from -0.25 to -0.40 V versus NHE was not corrected for the anodic current since the latter was negligible compared to the former. Two Tafel slopes are obtained, one of -110 mV/decade at low current densities (passive layer) and the second of -120 mV/decade at high current densities (bare iron). The Tafel slope corresponding to the passive region extends over two decades of current density

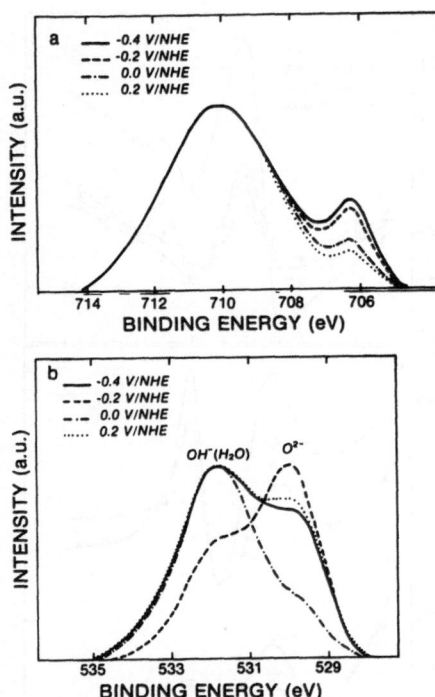

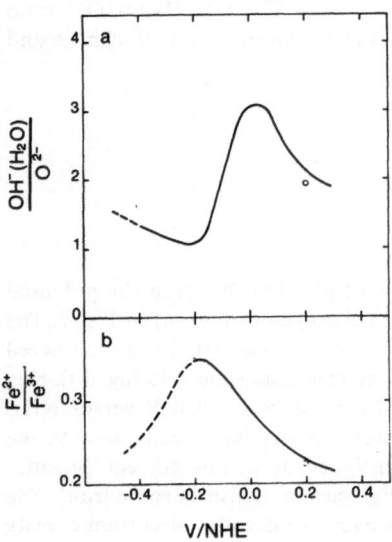

FIGURE 5. XPS data for Fe $2p_{3/2}$ (a) and O $1s$ (b) of the passive layer for different potentials.

at pH 9.4, but only half a decade at pH 7.0. The Tafel slope at high current densities was obtained using a rotating disk in order to correct for the diffusion component of the current.

For the first-order reaction with respect to O_2, the disk current density, i, is related to the rotation rate, ω, by the expression[17]

$$\frac{1}{i} = \frac{1}{i_k} + \frac{1}{B\omega^{1/2}} \tag{1}$$

FIGURE 6. $OH^-(H_2O)/O^{2-}$ (a) and Fe^{2+}/Fe^{3+} (b) ratios of the passive layer obtained from XPS data as a function of potential.

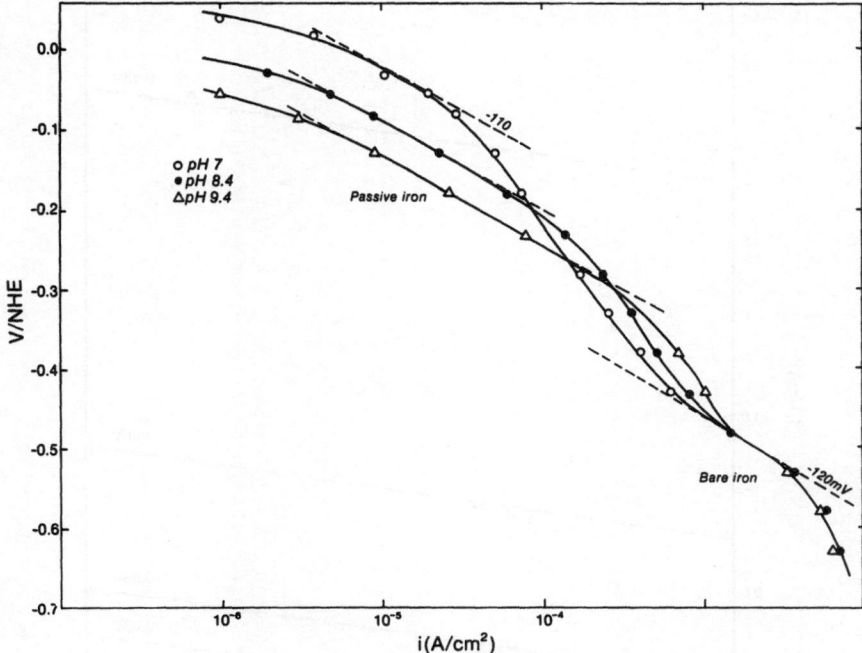

FIGURE 7. Tafel plots for O_2 reduction on iron in borate buffer solution at various pHs ($0.5M$ H_3BO_3, $0.0375M$ $Na_2B_4O_7 \cdot 10H_2O$). $P_{O_2} = 1$ atm.

where i_k is the kinetic current density for O_2 reduction, and B is related to the diffusion-limiting current (i_l) by the expression $i_l = B\omega^{1/2}$. B is given by

$$B = 0.2(D_{O_2})^{2/3} \nu^{-1/6} nFC_{O_2} \qquad (2)$$

where ν is called the kinematic viscosity.

Figure 8 shows a plot of $1/i$ versus $\omega^{-1/2}$ obtained from the polarization data for different potentials from -0.43 to -0.68 V versus NHE. The experimental value of B obtained from the slope of these plots is 0.14 mA/cm^2 rpm$^{1/2}$. This is in good agreement with the theoretical value of 0.13 mA/cm^2 rpm$^{1/2}$ calculated from Eq. (2) for $n = 4$, using data for the solubility and diffusion coefficient of O_2 in borate buffer solution.[18]

The intercepts of the parallel straight lines for various potentials with the $1/i$ axis yield the i_k values. These values are used to plot the V–$\log i$ curve at high current densities (Fig. 7). The slope of the Tafel plot of -120 mV/decade in this region extends to 10^{-2} A/cm^2 in the presence of $0.5M$ Na_2SO_4 as supporting electrolyte. Though the increase in the ionic strength of the solution has negligible effect on the kinetic reduction current, it significantly reduces the diffusion limiting current. This is due to the decrease in the solubility and diffusivity of O_2 by 50% and 25%, respectively, with the increase in electrolyte concentration.[18]

A series of Tafel plots were made from measurements in borate buffer solution at pHs ranging from 7.0 to 9.4 (Fig. 1). In the potential range -0.5 to -0.3 V versus NHE, the state of the iron electrode surface changes from a passive oxide layer at more positive potentials to bare iron at negative potentials. The polarization data for O_2 reduction in the Tafel regions indicate an apparent reaction order with respect to pH of -0.4 for passive iron and 0 for bare iron.

Figure 9 shows the O_2 reduction current as a function of oxygen partial pressure obtained on passive iron (-0.08 and -0.25 V versus NHE) and bare iron (-0.48 V versus NHE)

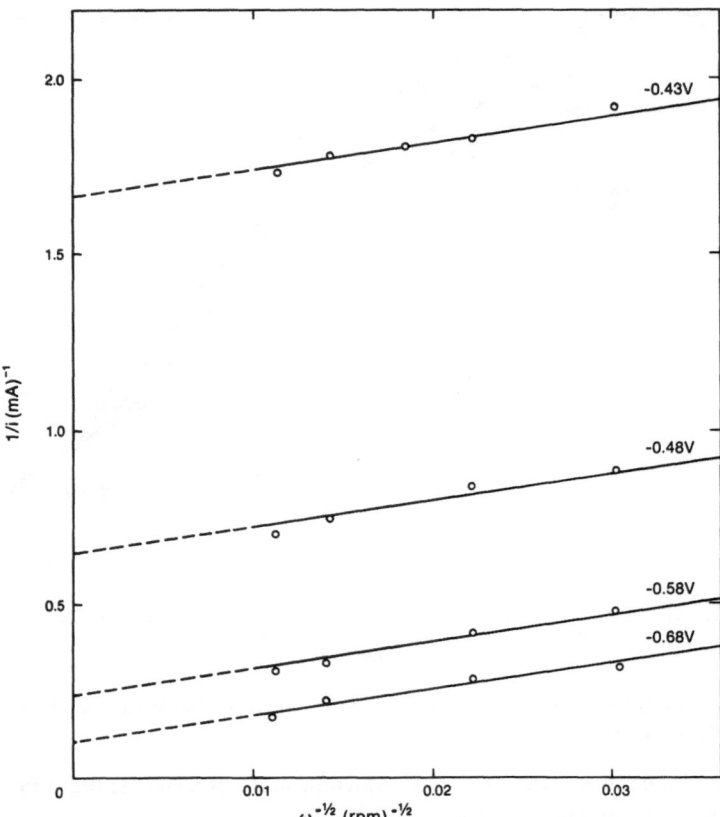

FIGURE 8. Plots of $1/i$ versus $\omega^{-1/2}$ for O_2 reduction on bare iron in borate buffer solution ($0.15M$ H_3BO_3, $0.0375M$ $Na_2B_4O_7 \cdot 10H_2O$) as a function of potential. $B = 0.14\,mA/cm^2\,rpm^{1/2}$.

surfaces. The reaction order for oxygen on both surfaces as 1. The gradual change of the gradient, $\delta \log i / \delta \log P_{O_2}$, at low O_2 partial pressure ($<0.2\,atm$) on the passive film is due to the contribution of the anodic current of iron dissolution to the total current. On iron, the $\log i$–$\log P_{O_2}$ plot yields two regions, both having a reaction order of unity.

4.2. Hydrogen Peroxide Formation

A rotating disk-ring electrode was used to determine whether hydrogen peroxide is produced as a product in the consecutive reaction pathway or as an intermediate in the parallel reaction pathway of the oxygen reduction. Figure 10 depicts the disk and ring currents as a function of the disk potential at 5000 rpm. The potential of the ring was held at 1.0 V versus NHE, where H_2O_2 oxidation is diffusion-controlled. Two anodic current peaks for H_2O_2 oxidation on the ring corresponding to different disk potential regions are associated with two different surface states of the electrode. The change from passive iron to bare iron is accompanied by a decrease in the ring current. The H_2 evolution reaction at more negative potentials has a similar effect on the ring current. The influence of the solution pH on the formation of H_2O_2 is shown in Fig. 11. As pH decreases, the ring current corresponding to the passive iron disk current increases, while that related to bare iron remains unchanged. This pH dependence of the ring current reflects the change of the disk current with pH.

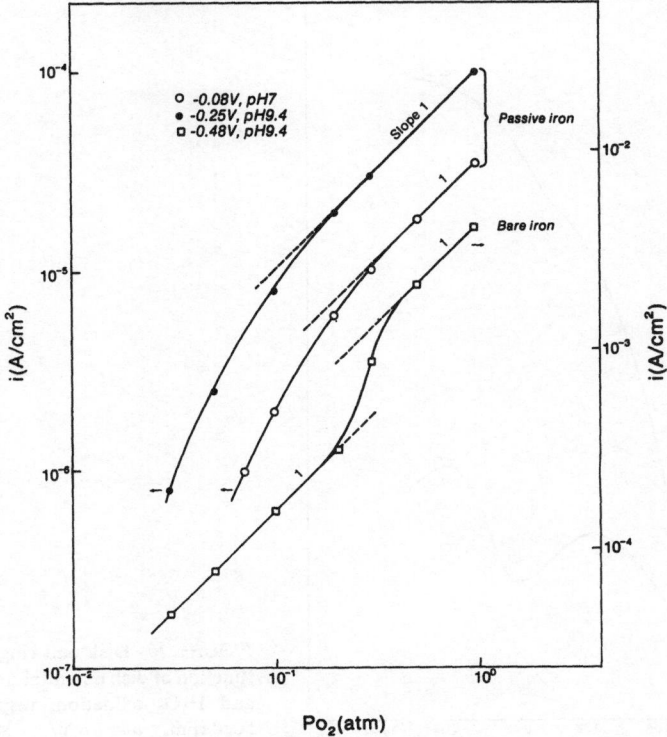

FIGURE 9. O_2 reduction as a function of oxygen partial pressure on passive iron (-0.08 and -0.25 V versus NHE) and bare iron (-0.48 V versus NHE).

5. DISK-RING ANALYSIS

Rotating disk-ring electrode data for O_2 reactions were analyzed considering a general serial–parallel reaction pathway.[19,20] In neutral solution, the reaction schemes can be represented as

$$O_{2(s)} \xrightarrow{\text{Diffusion}} O_2^* \underset{k_{-5}}{\overset{k_5}{\rightleftarrows}} O_{2(a)} \underset{k_{-2}(-2e^-)}{\overset{k_2(2e^-)}{\rightleftarrows}} H_2O_{2(a)} \xrightarrow{k_3(2e^-)} OH^-$$

with $k_1(4e^-)$ over the top from $O_{2(a)}$ to OH^-, k_4 and k_6, k_{-6} pathways to $H_2O_2^*$.

$$H_2O_2^*$$

$$\underline{\text{Disk}}$$
$$\text{Ring} \qquad H_2O_{2(s)} \xrightarrow{(-2e^-)} O_2$$

In this reaction scheme, the symbols (s), *, and (a) refer to the species in the bulk solution, in the solution adjacent to the electrode surface, and in the adsorbed state, respectively. The O_2 can be reduced to OH^- either directly without the intermediate formation of $H_2O_{2(a)}$ (k_1) or indirectly through the formation of $H_2O_{2(a)}$ (k_2, k_3) or to $H_2O_2^*$ (k_2, k_6). The

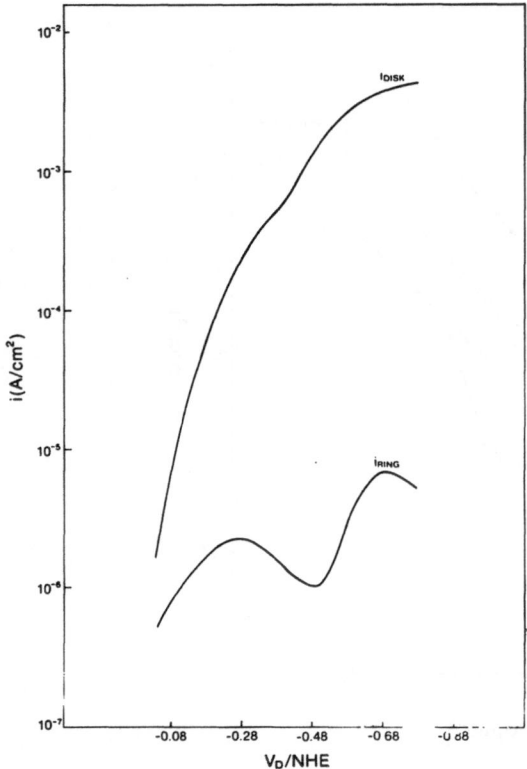

FIGURE 10. Disk and ring currents as a function of disk potential for O_2 reduction and H_2O_2 oxidation, respectively; $\omega = 5000$ rpm; $v = 1$ mV/s; $V_R = 1.0$ V vs. NHE; pH = 8.4.

adsorbed $H_2O_{2(a)}$ can also be oxidized back to O_2 (k_{-2}), catalytically decomposed (k_4), or desorbed (k_6) and diffused into the solution. The rate constants k_5, k_{-5} and k_6, k_{-6} are for adsorption and desorption of O_2 and H_2O_2, respectively.

The above reaction scheme predicts the relationship between the ratio of currents at the disk (I_D) and the ring (I_R) and the electrode rotation rate, ω, as

$$\frac{I_D}{I_R} = \frac{1}{N}\left(1 + \frac{2k_1}{k_2} + A\frac{1}{k_{-6}} + A\frac{k_6}{k_{-6}Z\omega^{1/2}}\right) \qquad (3)$$

where $A = (1 + 2k_1/k_2)(k_{-2} + k_3 + k_4) + (k_3 - k_{-2})$, $Z = 0.62D^{2/3}v^{-1/6}$, and N is the collection efficiency. Equation (3) provides essentially the same diagnostic criterion for the O_2 reduction pathway as a criterion first developed by Damjanovic et al.[21] and later presented in a more complete form by Wroblawa et al.[22]

The experimental disk-ring data for oxygen reduction on iron in borate buffer solution are represented by plotting I_D/I_R versus $\omega^{-1/2}$ in Figs. 12 and 13. Figure 12 shows the I_D/I_R plots obtained at less negative potentials for oxygen reduction on passive iron at pH 8.4. While the linear plots are independent of the rotation speed ω, the intercepts of the straight lines depend on the potential of the electrode. At about -0.08 V versus NHE, where the passive film is formed, the intercept becomes $1/N$; that is, there is no longer any contribution from the direct four-electrode reduction pathway. Figure 13 shows I_D/I_R plots for the bare iron electrode at various potentials. The slopes of the linear plots decrease with increasing negative potential, while the intercept remains almost constant at ~ 300.

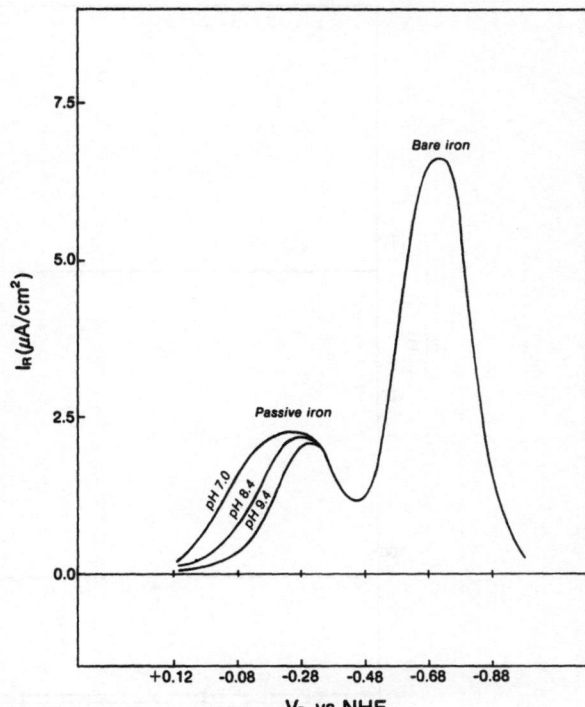

FIGURE 11. pH dependence of the
formation of H_2O_2 on iron.

6. MECHANISM OF OXYGEN REDUCTION

The kinetic data for O_2 reduction on passive iron and bare iron surfaces in borate buffer solution are summarized in Table 1. The criteria shown are used for the discussion of the mechanisms of oxygen reduction.

6.1. Passive Iron

6.1.1. Oxygen Reduction Pathway

It has been shown that the reaction path for O_2 reduction on passive iron leads to a two-electron transfer reaction, that is, the formation of H_2O_2. The exchange current density for H_2O_2 of about $1.2 \times 10^{-7} A/cm^2$ was obtained by extrapolating the Tafel plot to reversible standard potential (0.11 V versus NHE[23]). The depolarization reaction of the formation of H_2O_2 at the corrosion potential (0.05 V versus NHE) yields a corrosion current of $1.8 \times 10^{-6} A/cm^2$ (Table 1).

In Table 2, the most probable pathways and rate-determining steps for the formation of H_2O_2 are summarized. Three criteria for O_2 reduction on passive iron, namely, the Tafel slope of $2RT/F$, the reaction order with respect to oxygen of 1, and the reaction order with respect to $[OH^-]$ of -0.4, were compared with those obtained under Langmuir and Temkin conditions for different reaction pathways. It can be seen from the comparison that the only pathways which fit these criteria are pathways A and C (corrected for potential difference in the double layer) under Temkin and Langmuir conditions, respectively.

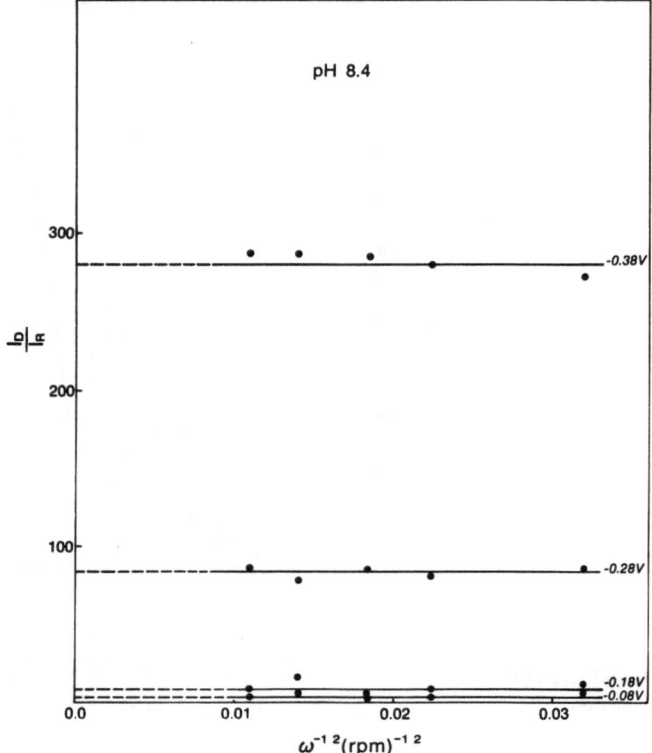

FIGURE 12. I_D/I_R versus $\omega^{-1/2}$ for O_2 reduction on passive layer at various potentials; pH = 8.4.

Pathway C, involving the formation of O_2H^+ on a passive layer, seems improbable due to the presence of predominantly Fe^{2+} active sites at cathodic potentials, which favors Fe^2-O_2 adduct formation instead.[24]

Mechanism A gives $2RT/F$ for the Tafel slope, chemisorption of O_2 as the rate-determining step, and a reaction order with respect to $[OH^-]$ of -0.5 for the Temkin adsorption isotherm. The reaction order with respect to O_2 is 1.

The observed Tafel slope of 110 mV/decade indicates that the potential drop across the passive layer is negligible in comparison with that in the Helmholtz double layer. The small increase in the thickness of the film with potential (Fig. 3) does not affect the Tafel slope.

In respect to the model where the adsorption of intermediate in the step after the rate-determining step follows a Temkin isotherm,[25] one can write

$$r\theta/RT = \log[OH^-] + FV/RT + \text{const.} \qquad (4)$$

where r and θ are the interaction parameter and surface coverage of the adsorbate, respectively.

The rate of the oxygen reduction is of the form

$$i = kP_{O_2} \exp(-\alpha FV/RT) \exp(-\beta \Delta G_{ads}/RT) \qquad (5)$$

where k is a rate constant, ΔG_{ads} is the standard free energy of adsorption of the O_{2ads} intermediate following the rate-determining step, and α and β are the transfer coefficient

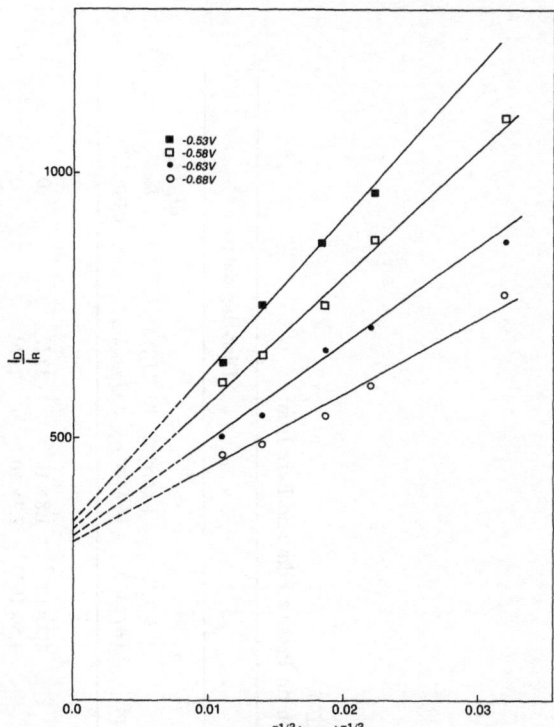

FIGURE 13. I_D/I_R versus $\omega^{-1/2}$ for O_2 reduction on bare iron at various potentials.

and symmetry factor, respectively. The free energy of adsorption is $r\theta$ dependent; that is,

$$\Delta G_{ads} = \Delta G_{ads}^0 + r\theta \tag{6}$$

where ΔG_{ads}^0 is the zero-coverage value. Then, Eq. (5) can be written as

$$i = k \cdot P_{O_2} \exp(-\alpha FV/RT) \exp(-\beta \Delta G_{ads}^\circ/RT) \exp(\log[OH^-]^{-\beta}\beta FV/RT) \tag{7}$$

Rearranging Eq. (7) gives

$$i = k' P_{O_2}[OH]^{-1/2} \exp[-(\alpha + \tfrac{1}{2})VF/RT] \tag{8}$$

where k' is another constant, and $\beta = \tfrac{1}{2}$. This gives a reaction order with respect to $[OH^-]$ and O_2 of -0.5 (compare to the experimental values of -0.4) and 1, respectively. The Tafel slope under Temkin conditions for $\alpha = 0$ (chemical rate-determining step preceding an equilibrium electron transfer step) is $2RT/F$.

The chemisorption of oxygen as the rate-controlling step, that is,

$$O_2 \rightarrow O_{2ads} \quad \text{(rds)} \tag{9}$$

has been considered in the oxygen reduction on catalytic surfaces.[26] Although the steps after the first rate-determining step cannot be ascertained, two possible sequences involving O_{2ads}^- and O_2H_{ads} as the adsorbed intermediate may be considered.

TABLE 1

Kinetic Parameters for O_2 Reduction on Passive Film and Bare Iron

Electrode surface	Disk data					Disk-ring data	
	$\delta V/\delta \log i$	$\delta \log i/\delta \log P_{O_2}$	$\delta \log i/\delta \text{pH}$	i_{0,O_2} (A/cm^2)	$i(\text{corr})$ (A/cm^2)	$\dfrac{(I_D)}{(I_R)}_{\omega\to\infty}$	$\dfrac{\delta I_D/I_R}{\delta\omega}$ (rpm)$^{1/2}$
Passive iron							
pH 7	−110	1	−0.4	1.2×10^{-7}	1.8×10^{-6}	$1/N$	0
pH 9	−110	1	−0.4	1.5×10^{-7}	2.2×10^{-6}	$1/N$	0
Bare iron							
pH 7	−120	1[a]	0	8×10^{-15}	1.2×10^{-3}	300	$1.5\times10^{4}\text{-}3.0\times10^{4}$
pH 9	−120	1[a]	0	1.5×10^{-13}	1.3×10^{-3}	300	$1.5\times10^{4}\text{-}3.0\times10^{4}$

[a] For $P_{O_2} < 0.2$ and $P_{O_2} > 0.3$.

TABLE 2
Proposed Mechanisms for Oxygen Reduction on Passive Iron

Mechanism[a]	Reaction pathway	Langmuir conditions						Temkin conditions		
		$\delta E/\delta \ln i$		$\delta \log i/\delta \log P_{O_2}$		$\delta \log i/\delta \mathrm{pH}$		$\delta E/\delta \ln i$	$\delta \log i/\delta \log P_{O_2}$	$\delta \log i/\delta \mathrm{pH}$
		$\theta \to 0$	$\theta \to 1$	$\theta \to 0$	$\theta \to 1$	$\theta \to 0$	$\theta \to 1$			
A	$O_2 \to O_{2\,ads}$	∞	—	1	—	0	—	$2RT/F$	1	-0.5
	$O_{2\,ads} + H_2O + e^- \to O_2H_{ads} + OH^-$	$2RT/F$	$2RT/F$	1	0	0	0	RT/F	1	-0.5
	$O_2H_{ads} + e^- \to O_2H^-$	$2RT/3F$	$2RT/3F$	1	0	-1	0	$2RT/F$	0	0
	$O_2H^- + H_2O \to H_2O_2 + OH^-$	$RT/2F$	—	1	0	-1	0	∞	-0.25	0
B	$O_2 + e^- \to P^-_{2\,ads}$	$2RT/F$	—	1	0	0	0	∞	1	0
	$O^-_{2\,ads} + H_2O \to O_2H_{ads} + OH^-$	RT/F	—	1	0	0	0	$2RT/F$	0	0
	$O_2H_{ads} + e^- \to O_2H^-_{ads}$	$2RT/3F$	—	1	0	-1	0	$2RT/F$	0	0
	$O_2H^-_{ads} + H_2O \to O_2H_2 + OH^-$	RT/F	—	1	0	-1	0	∞	-0.25	0
C	$O_2 + H_2O \to O_2H^+ + OH^-$	$2RT/F$	—	1	—	0	—	∞	1	0
	$O_2H^+ + e^- \to O_2H_{ads}$	$2RT/F$	$2RT/F$	1	0	-0.5	0	$2RT/F$	0	0
	$O_2H_{ads} + e^- \to O_2H^-_{ads}$	$2RT/3F$	$2RT/3F$	1	0	-1	0	$2RT/F$	0	0
	$O_2H^-_{ads} + H_2O \to H_2O_2 + OH^-$	RT/F	—	1	0	-1	0	∞	-0.25	0
D	$O_2 + H_2O + e^- \to O_2H_{ads} + OH^-$	$2RT/F$	0	1	0	0	0	$2RT/F$	1	0
	$O_2H_{ads} + e^- \to O_{ads} + OH^-$	$2RT/3F$	$2RT/F$	1	0	-1	0	$2RT/F$	0	0
	$O_{ads} + H_2O \to H_2O_2$	$RT/2F$	—	1	0	-1	0	∞	-0.25	0
E	$O_2 + e^- \to O^-_{2\,ads}$	$2RT/F$	—	1	0	0	0	$2RT/F$	1	0
	$O^-_{2\,ads} + H_2O + e^- \to O_2H_{ads} + OH^-$	$2RT/3F$	—	1	-1	-1	0	$2RT/3F$	1	-1
	$O_2H_{ads} + H_2O \to H_2O_2 + OH^-$	$RT/2F$	—	1	0	-1	0	∞	-0.25	0

[a] Pathway A, oxygen chemisorption path; B, Bogotskii's path; C, O_2H^+ adsorption path; D, oxygen-water discharge path; E, superoxide adsorption path.

The possibility of chemibonding of O_2^- and O_2H may be essentially the same from the molecular orbital theory standpoint. However, in aqueous solutions O_2^- has an added hydration energy of about -100 kcal/mol, which would result in a less adsorbed intermediate, and thus is less likely to be present in the pathway.

A likely mechanism for oxygen reduction on the passive layer following the rate-determining step could be described as

$$O_{2ads} + H_2O + e^- \rightarrow O_2H_{ads} + OH^- \tag{10}$$

$$O_2H_{ads} + e^- \rightarrow O_2H^- \tag{11}$$

$$O_2H^- + H_2O \rightarrow H_2O_2 + OH^- \tag{12}$$

where O_2H_{ads} follows a Temkin isotherm.

6.1.2. Oxygen Chemisorption

The bonding of two oxygen atoms in the O_2 molecule, according to valence bond theory, involves the resonating configurations

where the lobes and the circles represent oxygen $2p$ orbitals in the plane and out of the plane, respectively, and the dots indicate the number of electrons.[27] This results in a particularly large resonance effect in the ground state of O_2, with each of the two configurations shown above having two singly occupied orbitals, yielding both a spin singlet and a spin triplet, the ground state being the triplet.[28] In the triplet state, out of the total 119 kcal involved in the $O=O$ bond, about 47 kcal is associated with the σ bond, while 72 kcal involves π electrons. It is the process of decreasing the O_2 π resonance that represents the critical step in activating the O_2 bond.

The bonding of a radical, such as H, causes most of the O_2 π resonance to be lost, resulting in a weakened $H-O_2$,[27,28]

$$D(H-O_2) \approx 50 \text{ kcal}$$

The activation of O_2 by H yields, in terms of the reduced $O-O$ bond energy,

$$D(HO-O) \approx 66 \text{ kcal}$$

$$D(HO-OH) \approx 51 \text{ kcal}$$

Another way of reducing $O-O$ bond energy is through electron transfer, for example,

$$D(HO-O^-) \approx 63 \text{ kcal}$$

$$D(O-O^-) \approx 95 \text{ kcal}$$

Thus, atom transfer is more effective than electron transfer in weakening the $O-O$ bond.

The bonding of O_2 to iron oxide surfaces can be presented in terms of dynamic structural models of the surface that closely specify its electronic structure. Despite the considerable uncertainties and complexities of the detailed structure of the iron oxide surface, most current models lead to the idea that each surface Fe atom can be in a tetrahedral or an octahedral configuration, depending on its oxidation state, with one electron in a dangling bond orbital (Fe^{2+}),

The first step of bonding of O_2 leads to a peroxy radical (rate-determining step).

The calculation indicates that in the case of a silicon surface there would be a barrier to closing of the surface peroxy radical species to form a bridged peroxide.[29] The reduced surface species (superoxide) on iron oxide, however, could lead to bridged peroxide species, providing another surface-adjacent bonding site.

6.2. Bare Iron

6.2.1. Oxygen Reduction Pathway

The pathway of O_2 reduction on a bare iron surface has been elucidated from the rotating disk-ring data analysis (Table 1). The ratio of the rate of O_2 reduction without formation of H_2O_2 to the rate of O_2 reduction involving H_2O_2 as an intermediate, given by the intercept I_D/I_R of ~300 (Fig. 13), indicates that about 99% of the reduction current goes directly to form OH^-.

The four-electron pathway may proceed via formation of adsorbed peroxide or superoxide where these adsorbates do not lead to a solution-phase species. If the peroxide state is involved, the parallel reaction pathway with rate constant k would be different from the series pathway with rate constants k_2 and k_3 in terms of the activity of the adsorption sites. Thus, the adsorbed peroxide will be further reduced before any appreciable desorption takes place.

The pathway involving H_2O_2 as an intermediate leading to the solution-phase species occurs with a rate constant k_2 about 100 times lower on the less active sites. Further reduction of H_2O_2 appears to be little dependent on the potential (Fig. 13), with the rate constant k_3 calculated from the slope of the I_D/I_R versus $\omega^{-1/2}$ plots (0.5 cm/s) varying by less than 2 for a 0.15-V change in potential. This indicates either a process controlled by the diffusion of H_2O_2 to the sites where it can be reduced or a slow chemical step preceding the first electrochemical reduction process.

In order to determine the mechanism and rate-determining step for the four-electron reduction of O_2 on a bare iron surface, the kinetic criteria, Tafel slope, and reaction order with respect to O_2 obtained experimentally (Table 1) are compared with those derived from the most probable mechanistic schemes (Table 3). Such a comparison will be made in the next section, following a brief discussion of superoxide formation.

TABLE 3

Proposed Mechanisms for Oxygen Reduction on Bare Iron

Mechanism	Reaction pathway	$\delta E / \delta \ln i$	$\delta \log i / \delta \log P_{O_2}$	$\delta \log i / pH$
A	$O_{2ads} + e^- \rightarrow O_{ads} + O_{ads}^-$	$2RT/F$	1	0
	$O_{ads}^- + H_2O \rightarrow OH_{ads} + OH^-$	RT/F	1	0
	$O_{ads} + H_2O + e^- \rightarrow OH_{ads} + OH^-$	$2RT/3F$	1	0
	$OH_{ads} + e^- \rightarrow OH^-$	$2RT/3F$	1	−1
B	$O_2 + H_2O + e^- \rightarrow O_{ads} + OH_{ads} + OH^-$	$2RT/F$	1	0
	$O_{ads} + H_2O + e^- \rightarrow OH_{ads} + OH^-$	$2RT/3F$	1	−1
	$OH_{ads} + e^- \rightarrow OH^-$	$2RT/3F$	1	−2
C	$O_2 \rightarrow O_{ads} + O_{ads}$	—	1	0
	$O_{ads} + H_2O + e^- \rightarrow OH_{ads} + OH^-$	$2RT/F$	0.5	0
	$OH_{ads} + e^- \rightarrow OH^-$	$2RT/3F$	0.5	−1
D	$O_2 + e^- \rightarrow O_2^-$	$2RT/F$	1	0
	$O_2^- + H_2O \rightarrow OH_{ads} + O_{ads} + OH^-$	RT/F	1	−1
	$O_{ads} + H_2O + e^- \rightarrow OH_{ads} + OH^-$	$2RT/3F$	1	−1
	$OH_{ads} + e^- \rightarrow OH^-$	$2RT/3F$	1	−2
E	$O_2 + e^- \rightarrow O_2^-$	$2RT/F$	1	0
	$O_2^- + H_2O \rightarrow HO_{2ads} + OH^-$	RT/F	1	0
	$HO_{2ads} + H_2O + e^- \rightarrow 2OH_{ads} + OH^-$	RT/F	1	−1
	$OH_{ads} + e^- \rightarrow OH^-$	RT/F	0.5	−2
F	$O_2 + e^- \rightarrow O_2^-$	$2RT/F$	1	0
	$O_2^- + H_2O \rightarrow HO_{2ads} + OH^-$	RT/F	1	0
	$HO_{2ads} + H_2O + e^- \rightarrow H_2O_{2ads} + OH^-$	$2RT/3F$	1	−1
	$H_2O_{2ads} \rightarrow 2OH_{ads}$	$RT/2F$	1	−2
	$OH_{ads} + e^- \rightarrow OH^-$	$2RT/5F$	0.5	−2
G	$O_2 + H_2O + e^- \rightarrow O_2H_{ads} + OH^-$	$2RT/F$	1	0
	$O_2H_{ads} + H_2O + e^- \rightarrow 2OH_{ads} + OH^-$	$2RT/3F$	1	−1
	$OH_{ads} + e^- \rightarrow OH^-$	$RT/2F$	0.5	−2

6.2.2. Superoxide Formation

At cathodic potentials, the active iron surface site for O_2 adsorption can be visualized as

$$(13)$$

leading to the formation of bridged reduced superoxide species

$$(14)$$

with the cleavage of the O—O bond.

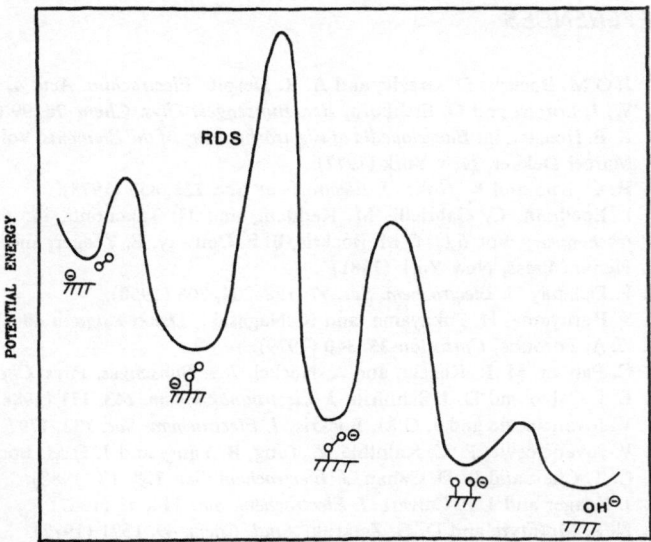

FIGURE 14. Potential energy profile for O_2 reduction on bare iron, involving absorption of reactant and product in the rate-determining step (rds).

The reaction sequence can proceed either through the electrolytic pathway involving bond breaking in the rate-determining step (mechanisms A and B in Table 3) or through the formation of superoxide or O_2H_{ads} in the rate-determining step (mechanisms D–G).

It can be seen by comparison of Tables 1 and 3 that the six pathways with the first step as the rate-determining step satisfy the criteria in Table 1. The potential energy profile for rate-determining electron transfer to O_2 is presented schematically in Fig. 14. Since the electrocatalytic pathway (mechanisms A and B) involves a high energy of activation for simultaneous charge transfer and bond breaking, the most probable mechanism would be

$$O_{2ads} + e^- \rightarrow O_{2ads}^- \tag{15}$$

followed by chemical (mechanisms D and F) or electrochemical (mechanism E) O—O bond cleavage.

7. CONCLUSION

The kinetics and mechanism of O_2 reduction on passive iron and bare iron in neutral aqueous solutions have been established. The major findings are as follows:

- The rates of O_2 reduction, i_{0,O_2}, on passive iron and bare iron are 1.2×10^{-7} and 8×10^{-15} A/cm^2, respectively, at pH 7.0.
- The reaction order with respect to pH is −0.4 on the passive layer and zero on bare iron. With respect to O_2, the reaction order on both surfaces is 1.
- The O_2 reduction proceeds through a four-electron pathway with little H_2O_2 as an intermediate on the bare iron and a two-electron pathway with the formation of H_2O_2 as the product on the passive layer.
- The rate-determining step on bare iron involves the formation of O_2^- followed by chemical or electrochemical O—O bond cleavage, while on the passive layer it is O_2 chemisorption under Temkin conditions followed by H atom and electron transfer successively.

REFERENCES

1. J. O'M. Bockris, D. Dražić, and A. R. Despić, *Electrochim. Acta* **4**, 325 (1961).
2. W. J. Lorenz and G. Eichkorn, *Ber. Bunsenges. Phys. Chem.* **70**, 99 (1966).
3. K. E. Heusler, in: *Encyclopedia of Electrochemistry of the Elements*, Vol. 9 (A. J. Bard, ed.), pp. 299–353, Marcel Dekker, New York (1977).
4. H. C. Kuo and K. Nobe, *J. Electrochem. Soc.* **125**, 853 (1978).
5. I. Epelboin, C. Gabrielli, M. Keddam, and H. Takenouti, in: *Comprehensive Treatise of Electrochemistry*, Vol. 5 (J. O'M. Bockris, B. E. Conway, E. Yeager, and R. E. White, eds.), pp. 151–192, Plenum Press, New York (1981).
6. P. Delahay, *J. Electrochem. Soc.* **97**, 198–204, 205 (1950).
7. S. Haruyama, H. Fukayama, and K. Nagasaki, *Denki Kagaku* **40**(9), 637 (1972).
8. Z. A. Foroulis, *Corrosion* **35**, 340 (1979).
9. C. Fabjan, M. R. Kazehi, and A. Neckel, *Ber. Bunsenges. Phys. Chem.* **84**, 1026 (1980).
10. E. J. Calvo and D. J. Schiffrin, *J. Electroanal. Chem.* **243**, 171 (1988).
11. V. Jovancicevic and J. O'M. Bockris, *J. Electrochem. Soc.* **133**, 1797 (1986).
12. V. Jovancicevic, R. C. Kainthla, Z. Tang, B. Yang, and J. O'M. Bockris, *Langmuir* **3**, 388 (1987).
13. C. T. Chen and B. D. Cahan, *J. Electrochem. Soc.* **129**, 17 (1982).
14. J. Kruger and J. P. Calvert, *J. Electrochem. Soc.* **114**, 43 (1967).
15. N. S. McIntyre and D. G. Zetaruk, *Anal. Chem.* **49**, 1521 (1977).
16. A. J. McEvoy and W. Gissler, *Thin Solid Films* **83**, L165 (1981).
17. V. G. Levich, *Physicochemical Hydrodynamics*, Prentice-Hall, Englewood Cliffs, New Jersey (1962).
18. V. Jovancicevic, P. Zelenay, and B. R. Scharifker, *Electrochim. Acta* **32**, 1553 (1987).
19. V. S. Bagotskii, M. R. Tarasevich, and V. Yu Filinovskii, *Electrokhimiya* **5**, 1218 (1969); **8**, 84–87 (1972).
20. J. Appleby and M. Savy, *J. Electroanal. Chem.* **92**, 15 (1978).
21. A. Damjanovic, M. A. Genshaw, and J. O'M. Bockris, *J. Chem. Phys.* **45**, 4057 (1966).
22. H. S. Wroblowa, Y. C. Fan, and G. Razumney, *J. Electroanal. Chem.* **69**, 195 (1976).
23. J. P. Hoare, in: *Encyclopedia of Electrochemistry of the Elements*, Vol. 2 (A. J. Bard, ed.), pp. 192–351, Marcel Dekker, New York (1974).
24. J. E. Newton and M. B. Hall, *Inorg. Chem.* **23**, 4627 (1984).
25. A. J. Appleby, in: *Comprehensive Treatise of Electrochemistry*, Vol. 7 (J. O'M. Bockris, B. E. Conway, E. Yeager, S. U. M. Khan, and R. E. White, eds.), pp. 173–239, Plenum Press, New York (1983).
26. M. R. Tarasevich, A. Sadkowski, and E. Yeager, in: *Comprehensive Treatise of Electrochemistry*, Vol. 7 (J. O'M. Bockris, E. Yeager, and R. E. White, eds.), pp. 301–398, Plenum Press, New York (1983).
27. B. J. Moss and W. A. Goddard III, *J. Chem. Phys.* **63**, 3523 (1975).
28. W. A. Goddard III, J. J. Low, B. Olafson, A. Redondo, Y. Zeiri, M. L. Steigerwald, E. A. Carter, J. N. Allison, and R. Chang, *Proceedings of the Symposium on The Chemistry and Physics of Electrocatalysis* (J. D. E. McIntyre, M. J. Weaver, and E. B. Yeager, eds.), Proceedings Vol. 84-12, pp. 63–95, The Electrochemical Society, Pennington, New Jersey (1984).
29. A. Redondo, W. A. Goddard III, C. A. Swarts, and T. C. McGill, *J. Vac. Sci. Technol.* **19**(3), 498 (1981).

11

Rechargeable Manganese Oxide Electrodes and Cells

Halina S. Wroblowa

1. INTRODUCTION

Rechargeable batteries are indispensable as subsidiary power plants or as auxiliaries in a large segment of industrial products. The necessity of their further development increases with the rapid expansion of industries using throwaway primary cells and with the concern about energetic and environmental problems. Primary batteries contribute to a considerable energy waste (the energy expended in manufacturing may exceed ten times the energy supplied by the battery during its lifetime), valuable material waste, and waste disposal problems. Secondary batteries presently in use might have to be replaced if the use of toxic lead and cadmium were to be prohibited.

The development of rechargeable manganese oxide/zinc batteries might considerably ameliorate the problems mentioned above. The system is presently the most widely used one among primaries. The Leclanché batteries, still predominant in the Third World countries, have been largely displaced in the West by alkaline cells. The higher price of the latter is compensated by a much better performance due to a unique set of properties which include, in addition to the low cost and abundance of reactants, a capability of high discharge rates for prolonged times, long shelf life, and stable performance over a wide temperature range. Their drawback is a poor rechargeability. Worldwide effort to make the system rechargeable has continued since the end of the 19th century.

2. HISTORICAL RETROSPECT

Manganese dioxide was known already in ancient Egypt and Greece, where it was used to modify the color of glass. Pliny described a natural MnO_2 ore some two thousand years ago. It was first used in the positive electrode of the Volta pile in 1799 by J. W. Ritter.[1] About 60 years later, Leclanché patented a wet primary MnO_2/Zn cell with ammonium chloride electrolyte and founded a battery factory in Brussels.[2] In a short time, his original cell design was turned inside out by replacing the negative zinc rod with a zinc can which

Halina S. Wroblowa • Ford Motor Company, Research Staff, Dearborn, Michigan 48121. *Present address*: 5924 Dunmore Drive, West Bloomfield, Michigan 48322.

Electrochemistry in Transition, edited by Oliver J. Murphy *et al.* Plenum Press, New York, 1992.

served both as an anode and the cell container. In the 1880s, the French company Leclanché began to manufacture dry cells, whose invention is attributed to Gassner,[3] although the apothecary C. H. Wolf in Blankenese described a rechargeable dry cell in 1884[2]—three years prior to the issue of Gassner's patent. The first wet alkaline manganese dioxide/tin cell "regenerable by an electric current" was patented in 1882 in Germany by Leuchs.[4] A U.S. patent for a wet manganese dioxide/zinc alkaline cell was granted in 1903 to Yai.[5] A dry alkaline cell was first reported in the period 1912–1914 by Aschenbach in a series of German and U.S. patents.[6]

In spite of this relatively early entrance of alkaline cells into the literature, their commercialization had to wait some 50 years until Herbert patented[7] the "crown" cell in the early 1950s. His work paved the way to the first alkaline batteries marketed by P. R. Mallory for use in transistor radios.

Since then, the introduction of large-area zinc electrodes and voluminous work leading to the development of positive electrodes with high reactivity, coupled with growing demands of the electronics industries, led to the emergence of a several billion dollars MnO_2/Zn cell market. The unique performance characteristics make the alkaline MnO_2 battery almost universally applicable, and the incentive of converting it into a rechargeable system makes it extremely appealing. The incentives are connected not only with energetics, material savings, and environmental reasons, but also with the possibility of entering the vast (presently over $6 billion/year) market of secondary batteries, dominated now by more expensive and more toxic lead acid (automotive SLI, specialized traction, emergency power, utilities, telephone, and remote and portable equipment) and nickel/cadmium (aircraft, communication, photographic, electronic, and other portable equipment) systems. Successful development of rechargeable MnO_2-based batteries is of interest for almost all of the present applications of both primary and secondary batteries. In this respect, alkaline MnO_2/Zn cells have the greatest promise. Although partially discharged dry Leclanché cells can be recharged, the success of the process depends on a strictly defined set of conditions[2] whose control is virtually impossible in common practice. The rechargeability of the presently existing alkaline cells is also limited to conditions of shallow discharge; however, its cycling regime is much less demanding.

The first, relatively inexpensive, rechargeable MnO_2/Zn battery, designed for portable TV use, appeared for a short time on the market in the 1970s. The main reason for its lack of commercial success was a very short cycle life (<30 cycles) coupled with inconvenience to the user: the depth of discharge had to be very closely controlled to prevent the total loss of rechargeable properties. The continued effort did not bring much success, the primary reason being the inherent irreversible behavior of manganese dioxide electrodes. Until the early 1980s the lifetime of the alkaline MnO_2 electrode was limited to some hundred recharge cycles (Fig. 1) if the depth of discharge never exceeded ~25% of the theoretical one-electron capacity.[8] A single discharge to $MnO_{1.5}$ made the electrode totally inactive. A vast effort to

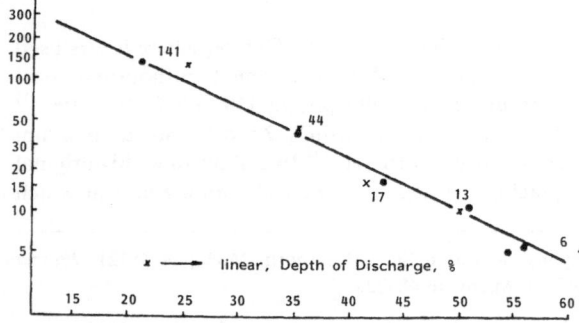

FIGURE 1. The number of discharge–charge cycles as a function of the depth of discharge (% of the theoretical one-electron capacity) for electrolytic γ-MnO_2.[8]

overcome these limitations was directed at understanding and correlating the irreversible behavior with electrochemical processes in the cell.

3. MECHANISTIC ASPECTS

Fundamental studies of the structural, physicochemical, and redox properties of MnO_2, which have continued since the 1960s, have provided a large amount of information concerning alkaline MnO_2 electrodes; however, even now, neither the reaction mechanism nor the factors affecting the rechargeable behavior are clearly understood. The uncertainties and contradictions that exist in the literature are largely due to the great variety and complexity of crystallographic forms (over 30 varieties of MnO_2 have been described) and poor, often overlapping X-ray diffraction (XRD) patterns (the given species may need to be present in 50–100% concentration to be detected[9]). The electrochemical behavior of manganese dioxide depends on its crystal structure and structural disorder, stoichiometry, presence of impurities, electrical conductivity, and physical properties.[8,10-12] Standardization of MnO_2 in the form of International Common Samples (ICS)[13] introduced a reference for the study, evaluation, and comparison of various results, thus diminishing the confusion arising from the use of different materials by different authors.

The 1966 English edition of the Pourbaix Atlas[14] still contained a reversible Mn^{4+}/Mn^{2+} electrode reaction. However, it was already realized at that time that the electrode could be recharged only if the depth of discharge was limited to a fraction of the one-electron capacity. Some researchers blamed the loss of rechargeability on the diffusion of Zn ions to the positive electrode and the formation there of nonreactive hydrohetaerolite[15]—a mixed compound containing trivalent manganese oxide. Although in the presence of zincate ions $Mn_2O_3 \cdot ZnO$ is indeed formed at a sufficient depth of discharge, MnO_2 electrodes were shown in the 1960s to be inherently nonrechargeable when reduced (in absence of Zn or Zn ions) below the $MnO_{1.7}$ composition. Discharge experiments coupled with X-ray diffraction analysis led Bell and Huber[2] to suggest that at this composition, which they identified with MnO(OH), a homogeneous reduction is replaced by a heterogeneous reaction leading to $MnO_{1.47}$ (identified as γ-Mn_2O_3). Boden et al.,[16] who carried out similar experiments in the absence of zinc, reported the initial reduction to consist in assimilation of protons into the γ-MnO_2 lattice, which collapses at the $MnO_{1.7}$ composition into an amorphous phase.

Earlier work[17-21] led Kozawa and co-workers[22,23] to the formulation of the mechanism for the first reduction step of γ-MnO_2 in alkaline electrolytes. According to this generally accepted hypothesis, the first step consists in the insertion of an electron and a proton into the MnO_2 lattice (Fig. 2). Both manganese and oxygen ions are thought to remain in their sites, while the reduction, which progresses by movement of protons and electrons, yields a trivalent Mn species initially isostructural with MnO_2. It has been repeatedly pointed out by Kozawa[24] that at least the initial stage of reduction of β- and γ-MnO_2 proceeds in a single

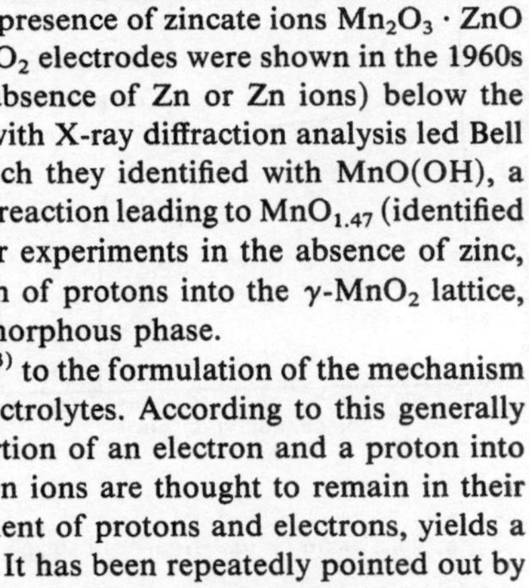

FIGURE 2. Schematic representation of the initial discharge mechanism. Movement of electrons (-----→) and protons (——→) is indicated.[24]

solid phase. The confirmation of this can be found, for instance in the open-circuit potential, which exhibits single-phase Nernstian behavior (Fig. 3). In principle, this reaction is reversible. However, the replacement of tetravalent manganese and oxygen ions by ions with larger radii—trivalent manganese and hydroxyl ion, respectively—results in a lattice dilation which progresses with increasing depth of discharge.[16,25,26] The dilation was macroscopically demonstrated[8] by the expansion and contraction of the MnO$_2$ electrode during shallow discharge and recharge half-cycles, respectively (Fig. 4). The progressive loss of capacity with cycling was interpreted in terms of a mechanical disintegration of the electrode followed by its increased resistance. The same authors[8] have also shown that mechanically restrained MnO$_2$ electrodes indeed performed better than their free unrestrained counterparts (cf. Fig. 4). Attempts to limit practical rechargeable cells to the required shallow depth of discharge included voltage-controlled discharge cutoff, use of low electrolyte concentration, which forces zinc into early passivation, or limitation of the cell capacity by the zinc anode. The first solution is inconvenient for the user, and the latter limits cell life on the negative side as well: the alkaline zinc electrode can survive prolonged cycling only in the presence of a two- to fourfold excess of zinc (this delays the shape change and the loss of the active Zn surface).

The progressive dilation of the lattice during reduction leads to a highly strained structure which collapses at the MnO$_{1.7}$ composition into more stable and less reactive compounds.[16,25,26] This collapse may contribute to the limitation of the depth of discharge required for the rechargeable behavior.

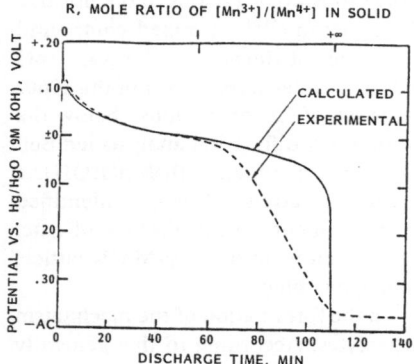

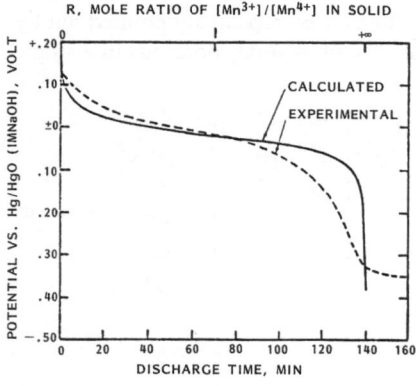

FIGURE 3. Open-circuit potential versus log [Mn^{3+}]/[Mn^{4+}] relation in 9M KOH (*top*) and 1M NaOH (*bottom*) at 23°C.[22]

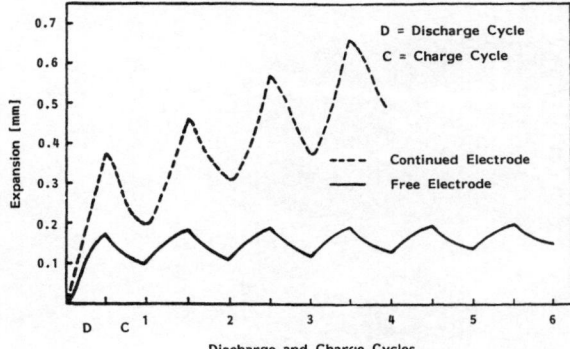

FIGURE 4. The expansion and contraction of MnO_2–graphite disks during cycling.[8]

Subsequent reduction steps, which lead eventually to divalent manganese species, are even less understood, complex heterogeneous processes involving, apart from solid phases, dissolved manganese ions of lower valence.[27] According to Boden *et al.*[16] the amorphous phase formed below the $MnO_{1.7}$ composition continues to assimilate protons until saturation, whereupon it recrystallizes to nonreducible Mn_3O_4. At low potentials the amorphous phase was reported to crystallize directly to $Mn(OH)_2$.

The possibility of reoxidation of the products of deep discharge is limited to a few cycles, with rapidly diminishing electrode capacity. The observed irreversibility has been variously ascribed to the loss of surface conductivity[28,29]; increased internal resistance of reaction intermediates and/or products[25-27]; and formation of Mn_2O_3[30] or hausmannite, Mn_3O_4.[31] According to McBreen,[31] the reversible first electron reduction leads to a collapse of the original lattice to amorphous MnOOH, which then is reduced to $Mn(OH)_2$ with accompanying cementation. Partial reoxidation of the divalent hydroxide leads to the progressive accumulation of nonrechargeable hausmannite (Fig. 5).

4. THE RECHARGEABLE MANGANESE OXIDE ELECTRODE

A breakthrough in the field occurred in the early 1980s when the first truly rechargeable "chemically modified" MnO_2 materials which had no constraints connected with the depth of discharge were described.[32] Their development was based on consideration of the redox mechanisms proposed in the literature. These implied that the rechargeability might be attained if the reactant had an open structure which could be maintained throughout the

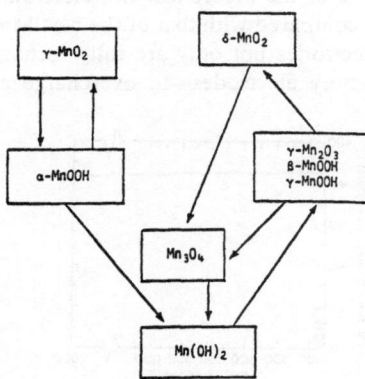

FIGURE 5. Schematic representation of the reaction paths for the reduction and oxidation of manganese oxides in $7M$ KOH.[9]

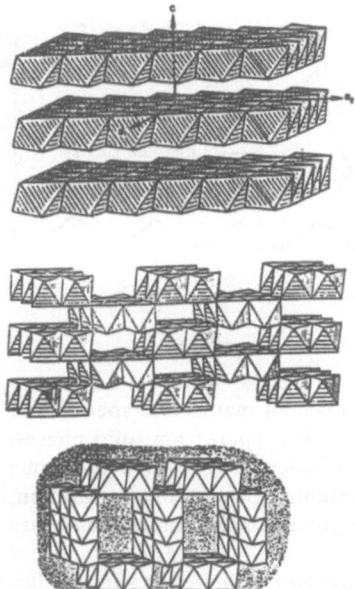

FIGURE 6. Examples of MnO_2 structures: *top*, layers of edge-sharing octahedra of a phyllomanganate; *middle*, γ-MnO_2, *bottom*, romanechite.[34]

redox cycle, thus improving the mobility of protons in the lattice, inhibiting pronounced volumetric changes, and preventing the collapse of the original structure to more stable, denser, and less electroactive and/or less conductive compounds. This premise led in turn to the synthesis of birnessites[33]—complexes of tetravalent manganese, which belong to the group of phyllomanganates. Their structure has been described as infinite two-dimensional sheets of edge-shared $[Mn^{IV}O_6]$ octahedra[34,35] separated by ~7 or 10 Å (Fig. 6). This structure may contain layers of water and hydroxyl ions bonded to the octahedral layers by foreign metal ions.[34,35] The openness of the lattice and the possibility of stabilizing it by insertion of foreign cations seemed suitable to fulfill the demands of the working hypothesis.

A half-cell study of electrodes containing phyllomanganates of various types incorporating a number of foreign cations inserted during synthesis[33,36] showed that Ag, Al, Ca, Ce, Cu, K, La, Na, Sn, Y, and Zn ions had no effect on rechargeability. Barium and antimony ions resulted in some improvement, incomparable, however, with the effects of the introduction into the birnessite lattice of trivalent Bi or divalent Pb. Bismuth and lead birnessites can be cycled electrochemically for thousands of cycles[37,38] at reactant utilization reaching over 90% of the theoretical two-electron capacity. In Fig. 7 the behavior of modified electrodes is compared with that of the best MnO_2 electrodes produced heretofore. The modified MnO_2 electrodes not only are fully rechargeable but also are insensitive—unlike the majority of battery electrodes—to overcharge and overdischarge, to changes of temperature, and to

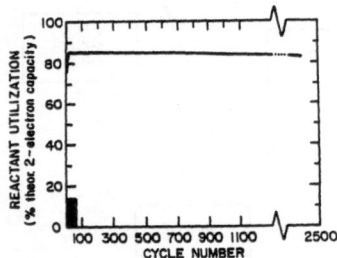

FIGURE 7. Fraction of the theoretical two-electron capacity of Bi-birnessite electrode as a function of the cycle number. Black rectangle indicates performance of best electrolytic γ-MnO_2 electrodes.

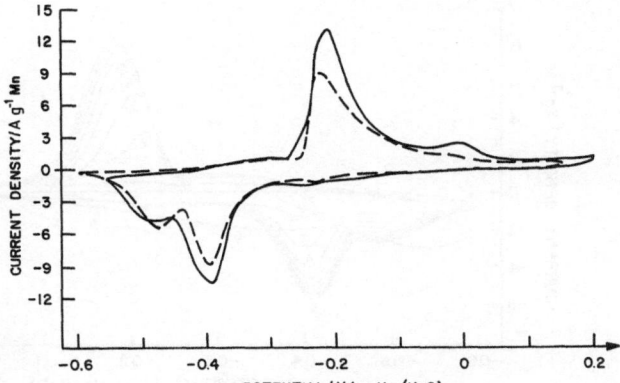

FIGURE 8. Cyclic voltammograms of chemically modified materials: ——, Bi-birnessite; – – –, Pb-birnessite.[36]

dehydration, are capable of extremely high current drains, and have a very long shelf life. Cyclic voltammograms of chemically modified MnO_2 electrodes, shown in Fig. 8, resemble quantitatively those obtainable for a few cycles with γ- and β-MnO_2.[36,39]

Similar results were obtained by "physically modifying" various manganese oxides by admixing to them Bi_2O_3 or PbO.[37,38] Somewhat less improved rechargeability was obtained using bismuth and lead sulfides. Admixture of oxides of other metals was ineffective, as in the case of synthetic birnessites containing metal ions other than Bi^{3+} and Pb^{2+}. Doping with titanium ions was reported[40] to improve rechargeability of MnO_2, albeit within the regime of shallow depths of discharge.

Physically modified materials differ from synthetic ones in their behavior during the initial cycles. As shown in Figs. 9 and 10, the electrode capacity develops with cycling to achieve within the first 10–30 cycles capacities comparable to those of chemically modified electrodes. Thereafter, their cycling behavior is similar to that of the chemically modified electrodes. The reactant utilization depends on the manganese oxide material used in the mixture (Fig. 11).† Best results are obtained with Bi_2O_3-doped γ- and δ-MnO_2 with mixed chain–tunnel and layered structures, respectively.

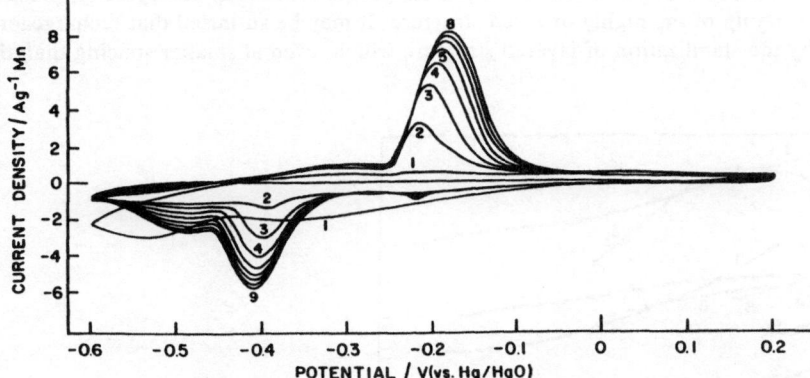

FIGURE 9. Initial cyclic voltammograms of a γ-MnO_2 + Bi_2O_3 electrode in 9M KOH. Cycle numbers are indicated on curves.[39]

† The relatively short cycle life shown in Fig. 11 results from the design of the unoptimized laboratory cell (used for comparative purposes only), from which soluble tri- and divalent manganese ions were permitted to escape.

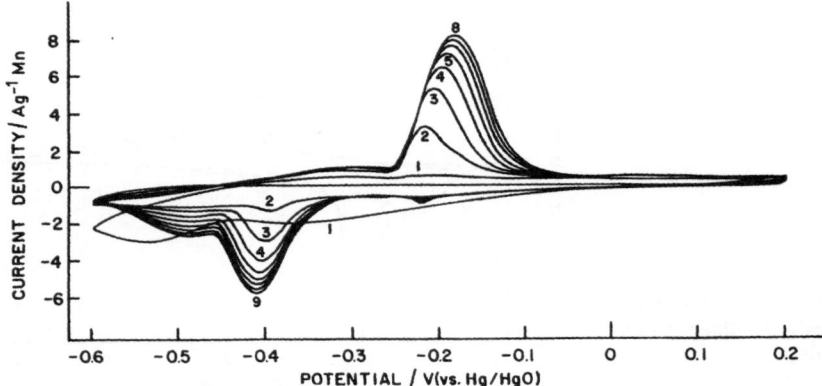

FIGURE 10. Initial cyclic voltammograms of a γ-MnO_2 + PbO electrode in $9M$ KOH. Cycle numbers are indicated on curves.[39]

Factors affecting the rechargeability of manganese oxides remain unclear. In spite of the success in attaining rechargeability by following the original premise linking rechargeable behavior with the presence of a highly open lattice, a host of evidence was collected by Ford Research workers[36-38] which lends itself to alternative interpretations of the effects of foreign ions. As discussed below, the original premise of facilitating movements of protons in the open layered lattice can be as well argued for as the effects of Bi and Pb on conductive, fractal, or "catalytic" properties, the latter connected with the possible steering of the reaction path away from producing irreversible intermediates.

The lattice-stabilizing effect of bismuth and lead cations lacks a direct confirmation owing to the low crystallinity of materials present at intermediate depths of discharge and charge. The XRD spectra show that the Bi- or Pb-modified lattice oscillates between that of birnessite ($d = 7$ Å) in the fully oxidized state and that of pyrochroite (a layered manganous hydroxide with $d = 4.7$ Å) in the fully reduced state.[36] Even these structures can be identified only for the first few tens of cycles owing to the progressive X-ray amorphicity, presumably due to tearing up of the highly oriented structure. It may be surmised that rechargeability is ensured by the stabilization of layered structure which, even at smaller spacing than that of

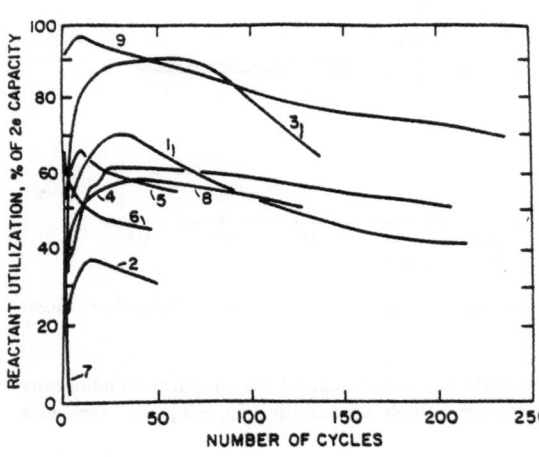

FIGURE 11. Reactant utilization as a function of the cycle number: (1) and (3) ICS #1 + Bi_2O_3; (2) ICS #1 + PbO; (4) ICS #6 + Bi_2O_3; (5) ICS #7 + Bi_2O_3; (6) Ghana ore + Bi_2O_3; (7) ICS #1; (8) ICS #1 + Bi_2O_3 + PbO; (9) Australian vernadite ore + Bi_2O_3.[39]

·birnessite, is sufficient to prevent structural changes hindering movement of protons and leading to the appearance of irreversible intermediates. The layered structure is maintained throughout the redox process by Bi or Pb ions held tenaciously in their original positions while Mn ions move between the birnessite and pyrochroite lattice. The immobility of bismuth ions seems to be confirmed by cyclic voltammetry. The overdischarge of the modified electrodes below -0.56 V (versus Hg/HgO reference), at which the reduction to divalent Mn^{2+} is completed, results in the peaks of quantitative Bi^{3+}/Bi reduction, without any change in the surface content of Bi. Subsequent recharge produces corresponding peaks of Bi oxidation preceding the anodic voltammogram of progressive manganese hydroxide oxidation. Neither this half-cycle nor ensuing cyclic voltammograms differ from those corresponding to nonoverdischarged electrodes.[37-39]

However, the structural interpretation of the reversible behavior may be contested in view of the rechargeability attained by physical admixing of bismuth and lead oxides or sulfides to a variety of manganese oxides (cf. Fig. 11), regardless of their chain, tunnel, or layer structure (cf. Fig. 6). This conclusion seems to be strengthened further by the fact that the rechargeable behavior and reactant utilization are hardly affected by the Bi:Mn ratio in the mixture (Fig. 12), at least within the ~1 to 26 at. % range. It is easier to envisage dopant levels of the order of 1 at. % affecting conductive properties and/or reaction pathways than having lattice stabilizing effects. The latter mechanisms seem to be indicated by the fact that the synthesis of birnessite using the ion-exchange method[36] (in which the first step carried out in the absence of Bi or Pb ions leads to Na-birnessite) must be carried out at low temperature, since at ambient temperature hausmannite is the major product. Products obtained using the coprecipitation method,[36] which requires the presence of Bi or Pb from the beginning of synthesis, contain—even if produced at room temperature—no hausmannite (the compound often cited in the literature[9,16] as the irreversible species, whose accumulation makes manganese dioxide electrodes nonrechargeable).

On the other hand, physical mixtures require several conditioning cycles (cf. Figs. 9 and 10) to attain capacity comparable to that of synthetic bismuth or lead birnessites. Thus, it is conceivable that electrochemical cycling in the presence of Bi or Pb ions results in the formation of a compound with an open structure resembling that of birnessites. The amorphicity of electrochemically cycled species prevents a direct confirmation or refutation of this hypothesis. Indirect evidence is found in the XRD data for cycled mixture electrodes. Bismuth trioxide can be detected in discharged samples if the electrodes are prepared by admixing to manganese oxide an excessive amount of bismuth trioxide (excessive with respect to the stoichiometric Bi:Mn ratio in synthetic birnessites). However, no Bi_2O_3 (or PbO) lines are detected after the first conditioning cycles if the Bi:Mn ratio in physical mixtures is kept close to the stoichiometry of the respective birnessite complex (7-10% at. % Bi:Mn).

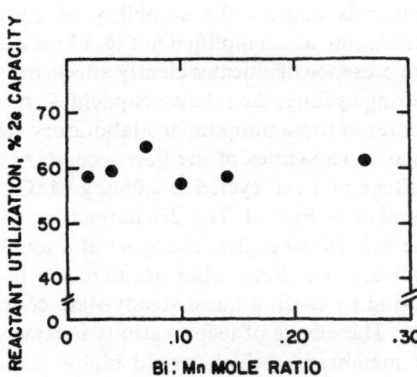

FIGURE 12. Reactant utilization in the 60th cycle as a function of Bi:Mn mole ratio.[39]

The specificity of the effect limited to Bi and Pb cations can also be ambiguously interpreted: the structure-stabilizing effects might be expected also for other cations with relatively large ionic radius and high valence. The only property uniquely shared by Bi^{3+} and Pb^{2+} cations is the same ionic radius, 1.2 Å. Unless it is incidental, a demand for a specific ionic radius value may enable incorporation of the ion into a different vacancy position from those of other ions. This in turn may be connected with structure-stabilizing factors.

5. CELLS CONTAINING RECHARGEABLE MANGANESE OXIDE ELECTRODES

In alkaline aqueous electroytes the highest cell voltage can be achieved by coupling manganese dioxide electrodes with zinc. Unfortunately, zinc is highly soluble in alkali, in which the concentration of zincate ions can reach over $2M$ (in $9M$ KOH) under electro-chemical cycling conditions. The presence of zincate ions in the electrolyte shortens the cell life, leading to the well-known phenomena of shape change, densification, and/or dendritic growth. The failure to produce a viable Ni/Zn electric vehicle battery was due to these phenomena, which limit, at present, the cycle life of conventional Ni/Zn cells to 200–300 cycles. Other applications are less demanding. Thus, for example, commercial secondary dry cells would not require a cycle life of more than 200 cycles. Lead/acid systems, used primarily as automotive SLI batteries, undergo, as a rule, shallow discharges only. Under these conditions, the zinc electrode does not deteriorate in the fashion described above and may last—if not abused (i.e. undergoing frequent deep discharge)— for the life of the automobile.

However, apart from these inherent problems of the alkaline zinc electrode, the latter is known to affect the performance of the conventional manganese dioxide materials. Migration and diffusion of zincate ions to the positive electrode leads to the formation of electrochemically inactive hetaerolite. The same was shown to occur in modified Mn-oxide/Zn cells.[41-43] However, unlike cells with the conventional MnO_2 materials, the cells with modified MnO_2 electrodes can be cycled for some two hundred cycles before the capacity would drop to 15–20% of the theoretical two-electron capacity. Apparently, the rate of formation of hetaerolite is inhibited to some extent by the structure of modified manganese oxides and/or by the presence of bismuth cations. The pejorative effects decrease with lowered zincate ion concentration and almost completely disappear if the latter drops to $\sim 0.1 M$.[39] The better performance of modified MnO_2 electrodes at lower zincate ion concentration is ascribed to the lower rate of the transport of zincate ions to the positive electrode and, correspondingly, to the reduced rate of hetaerolite formation.

Attempts to suppress the zincate ion concentration by addition of calcium oxide were unsuccessful,[44] since the zincate levels were lowered only to $0.5M$. A number of organic materials depress the solubility of zincate ions to lower levels thus increasing reactant utilization, as exemplified in Fig. 13 for $Zn(OH)_4^{2-}$ concentration lowered to $<0.2M$. Although the presence of zincates clearly affects the reactant utilization of the modified MnO_2 electrodes during cycling, the relative capacities (ratio of the actual discharge capacity to the theoretical value) of these unoptimized laboratory cells were better, even after 200 cycles, than the initial relative capacities of the best secondary MnO_2/Zn cells described in the literature. The cell voltage of a cell cycled at 0.05A/g MnO_2 is shown as a function of discharge time and cycle number in Fig. 14. The discharge time to 0.8 V cutoff voltage decreases most rapidly within the first 30–50 cycles, changing at a much slower rate for the next 200 cycles. This seems to indicate that hetaerolite accumulates in the electrode primarily within the initial cycling period to attain a quasi-steady-state concentration in the electrode.

The effects of using various separators are also shown in Fig. 13. It is clear that the use of membranes which would highly impede the transport of zincate ions to the modified

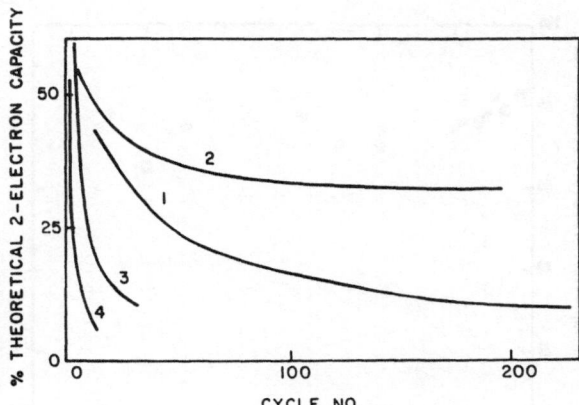

FIGURE 13. Reactant utilization in modified MnO_2/Zn cell as a function of cycle number: (1) $9M$ KOH, Daramic separator; (2) 50% MeOH in $9M$ KOH, Daramic separator; (3) $9M$ KOH, Nafion separator; (4) $9M$ KOH, Celgard separator.[43]

manganese oxide electrodes can further improve the capacity of MnO_2/Zn cells cycled for prolonged times. Data described above were obtained under extremely demanding conditions of continuous, fast charges and discharges. A cycling regime which allows much better results to be obtained has already been devised.[44]

The advantages of alkaline Fe/modified MnO_2 cells lie in their cycling longevity and inexpensive, abundant reactants. (The sintered iron electrodes have been reported to be capable of some 3000 cycles in alkaline electrolytes.) The main disadvantage is the relatively low cell voltage (0.6–0.8 V) under operating conditions (comparable to those of fuel cells). A battery requires a larger number of cells connected in series; therefore, the system would be best suited for stationary applications where a large footprint and maintenance requirements can be compensated by the low cost and long cycle life of the storage battery. The low overvoltage of hydrogen evolution on iron causes gassing and requires periodic water replenishment. This disadvantage, common to all secondary iron batteries (e.g., nickel/iron), has been somewhat alleviated by the development of systems for single-point watering in multicell batteries and for flame retardation[45] in case of malfunction and hydrogen accumulation.

The performance of a cell containing modified MnO_2 and iron electrodes is shown in Fig. 15.[43] As expected, the presence of the iron electrode (which, in spite of its dissolution/precipitation redox mechanism, can be considered, similarly to nickel oxide or graphite,

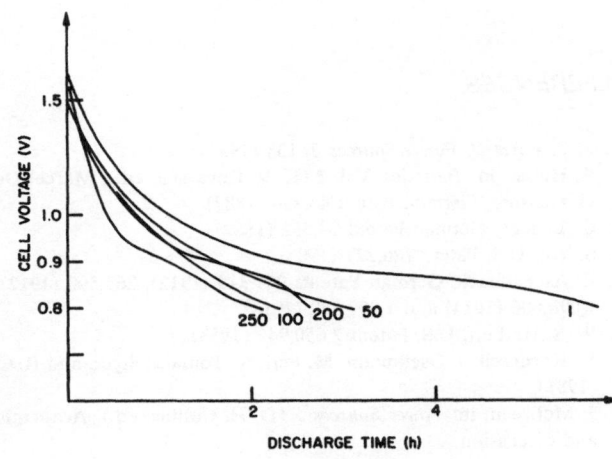

FIGURE 14. Cell voltage of a modified MnO_2/Zn cell as a function of discharge time and cycle number, indicated on the curves. Electrolyte: 50% MeOH in $9M$ KOH.[43]

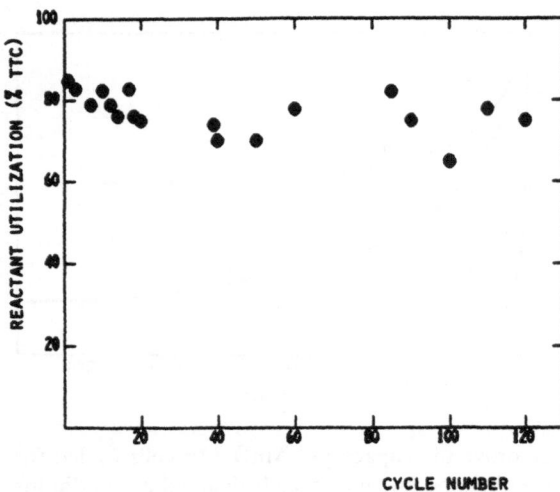

FIGURE 15. Reactant utilization versus cycle number for a modified $MnO_2/KOH/$sintered Fe cell.[43]

as "insoluble"), has no pejorative effects on the performance, rechargeability, and capacity of the modified manganese dioxide electrodes.

Modified MnO_2 materials can be coupled in aqueous alkaline cells with cadmium.[46] The operating voltage is not much better than in the case of iron; however, gassing is prevented, and voltaic and coulombic energy efficiencies are higher. The disadvantage of cadmium is its toxicity.

In nonaqueous electrolytes, modified MnO_2 materials can be coupled with lithium, and probably sodium, electrodes to form rechargeable systems.[46] The advantage of these cells lies in their high operating voltages, unattainable in aqueous systems, and potentially high energy and power densities. Their development may lead to batteries suitable for both conventional and special-purpose applications.

The applied aspects of the recent breakthrough in the area of rechargeable manganese oxide electrode and cells are self-evident. The scientific aspects need to be resolved in that the nature of factors impeding and imparting rechargeable behavior remains an open question since the results of diagnostic experiments lend themselves to ambiguous interpretation. Apart from being of purely academic interest, the resolution of this problem should help further developments of rechargeable batteries, so much needed in the future.

REFERENCES

1. K. J. Euler, *J. Power Sources* **8**, 133 (1982).
2. R. Huber, in: *Batteries*, Vol. 1 (K. V. Kordesch, ed.), Marcel Dekker, New York (1974).
3. H. Gassner, German Patent 45,250 (1887).
4. G. Leuchs, German Patent 24,552 (1882).
5. S. Yai, U.S. Patent 746,227 (1903).
6. E. Aschenbach, German Patents 261,319 (1912), 265,590 (1912), and 279,911 (1913); U.S. Patents, 1,098,606 (1914) and 1,090,372 (1914).
7. W. S. Herbert, U.S. Patent 2,650,945 (1953).
8. K. Kordesch, J. Gsellmann, M. Peri, K. Tomantschger, and R. Chemelli, *Electrochim. Acta* **26**, 1495 (1981).
9. J. McBreen, in: *Power Sources 5* (D. H. Collins, ed.), Academic Press, London (1975), paper #31 and discussion.

10. W. C. Vosburgh, *J. Electrochem. Soc.* **106**, 839 (1959),
11. R. Giovanoli, Proceedings of the MnO_2 Symposium, Tokyo, 1980, Vol. 2, paper #7.
12. J. P. Brenet, J. P. Chevillot, and J. Brenet, *Schweitzer Archiven* **1**, 10 (1960).
13. A. Kozawa and R. A. Powers, Proceedings of the MnO_2 Symposium, Cleveland, Ohio, 1975, Vol. 1, p. 4.
14. M. Pourbaix, *Atlas d'Equilibres Electrochimiques*, Gauthier-Villars (1963); English ed., Pergamon Press, New York (1966).
15. K. Miyazaki, *J. Electroanal. Chem.* **21**, 414 (1969); *J. Electrochem. Soc.* **116**, 1469 (1969); *J. Electrochem. Soc.* **117**, 821 (1970).
16. D. Boden, C. J. Venuto, D. Wisler, and R. B. Wylie, *J. Electrochem. Soc.* **114**, 415 (1967).
17. J. P. Brenet, in: *Proceedings of the 8th CITCE, Madrid*, 1956, p. 394, Butterworths, London (1958).
18. J. Coleman, *Trans. Electrochem. Soc.* **90**, 545 (1946).
19. D. T. Ferrel and W. C. Vosburgh, *J. Electrochem. Soc.* **98**, 334 (1951).
20. R. S. Johnson and W. C. Vosburgh, *J. Electrochem. Soc.* **100**, 471 (1953).
21. A. B. Scott, *J. Electrochem. Soc.* **107**, 941 (1960).
22. A. Kozawa and J. F. Yaeger, *J. Electrochem. Soc.* **112**, 959 (1965).
23. A. Kozawa and R. A. Powers, *J. Electrochem. Soc.* **113**, 870 (1966).
24. A. Kozawa, Proceedings of the MnO_2 Symposium, Tokyo, 1980, Vol. 2, paper #22.
25. J. P. Brenet and S. Ghosh, *Electrochim. Acta* **7**, 449 (1962).
26. G. S. Bell and R. Huber, *J. Electrochem. Soc.* **111**, 1 (1964).
27. A. Kozawa, T. Kalnoki-Kis, and J. F. Yaeger, *J. Electrochem. Soc.* **113**, 405 (1966).
28. J. P. Brenet, in: *Batteries*, Vol. II, (D. H. Collins, ed.), p. 245 (1964).
29. D. Boden, C. J. Venuto, D. Wisler, and R. B. Wylie, *J. Electrochem. Soc.* **115**, 333 (1968).
30. H. Y. Kang and C. C. Liang, *J. Electrochem. Soc.* **115**, 6 (1968).
31. J. McBreen, *Electrochim. Acta* **28**, 221 (1975).
32. H. S. Wroblowa, N. Gupta and Y. F. Yao, Proceedings of the 3rd MnO_2 Symposium, Graz, 1985, Vol. II, p. 57.
33. Y. F. Yao, U.S. Patent 4,520,005 (1985).
34. R. G. Burns and V. M. Burns, Proceedings of the MnO_2 Symposium, Tokyo, 1980, Vol. II, paper #6.
35. R. Giovanoli, Proceedings of the MnO_2 Symposium, Tokyo, 1980, Vol. II, paper #7.
36. Y. F. Yao, N. Gupta, and H. S. Wroblowa, *Extended Abstracts*, Vol. 85-2, The Electrochemical Society, Pennington, New Jersey (1985); *J. Electroanal. Chem.* **223**, 107 (1987).
37. H. S. Wroblowa and N. K. Gupta, unpublished results.
38. F. Chouaib, O. Cauquil, and M. Lamache, *Electrochim. Acta* **26**, 325 (1981).
39. H. S. Wroblowa and N. Gupta, *J. Electroanal. Chem.* **238**, 93 (1987).
40. J. Gsellmann, W. Harer, K. Holzleitner, and K. Kordesch, Extended Abstracts, Vol. 84-2, p. 116, The Electrochemical Society, Pennington, New Jersey (1984).
41. H. S. Wroblowa, J. T. Kummer, M. Dzieciuch, and N. Gupta, U.S. Patent 4,451,543 (1984).
42. M. Dzieciuch and H. S. Wroblowa, Extended Abstracts, Vol. 86-2, The Electrochemical Society, Pennington, New Jersey (1986).
43. M. A. Dzieciuch, N. Gupta, and H. S. Wroblowa, *J. Electrochem. Soc.* **135**, 2415 (1988).
44. H. Wroblowa and N. K. Gupta, patent pending.
45. B. D. Edwards and M. F. Mangan, Proceedings Electric and Hybrid Vehicles, Systems Assessment Seminar, Gainesville, Florida, paper #313, Dec. 13-16, 1983.
46. H. S. Wroblowa and M. Dzieciuch, unpublished results.

12

A Direction of Study of Electrocatalysis in Anodic O₂ Evolution through Characterization of Chemisorption Behavior of Intermediates

B. E. Conway

1. INTRODUCTION

1.1. Nature of Electrocatalysis and Role of Adsorption of Intermediates

Electrocatalysis is manifested when it is found that the electrochemical rate constant for an electrode process, standardized with respect to some reference potential (often the thermodynamic reversible potential for the same process), depends on the chemical nature of the electrode metal, the physical state of the electrode surface, the crystal orientation of single-crystal surfaces, or on, for example, alloying effects. One of the first definitions and first instances of recognition of the phenomenon of electrocatalysis, along these lines, is to be found in the important, but little quoted, paper of Busing and Kauzmann.[1] Electrocatalysis is thus the phenomenon by which the rate or rate constant of the *rate-determining step* is enhanced at the catalytic surface in relation to the kinetics at a less catalytic one; of course, the rate-determining step in a given multistep process may also be changed from one catalyst to another. It is perhaps necessary to emphasize that, in electrode kinetics, there can be no *non*catalyzed reference processes since all electrode reactions must proceed heterogeneously at some kind of surface acting as an electron source or electron sink. Also, the reaction mechanism may be found to be dependent on the above factors; in special cases, for a given reactant, even the reaction pathway, for example, in electrochemical reduction of ketones or alkyl halides or electrochemical oxidation of aliphatic acids (the Kolbe and Hofer–Moest reactions), may also depend on those factors.

The wide variation of rate constants, for a given reaction, with the electrode material is a commonly observed effect in electrocatalysis, for example, in the electrochemical H_2(HER)[2,3] and O_2 (OER) evolution reactions,[3,4] and is attributed to (a) variation of extents of coverage, θ, of the electrode surface by chemisorbed reaction intermediates, determined by their standard Gibbs energies of adsorption,[5] and (b) a dependence of the

---

B. E. Conway • Chemistry Department, University of Ottawa, Ottawa, Ontario, Canada K1N 6N5.

Electrochemistry in Transition, edited by Oliver J. Murphy *et al.* Plenum Press, New York, 1992.

Gibbs energy of activation upon electrode material, also related to the energy of chemisorption of intermediates.[6,7] These relations to chemisorption energies can often be represented by so-called "volcano" plots of the kinetic parameter (standard rate constant or exchange current density, i_0) versus the standard Gibbs energy of adsorption of the kinetically involved intermediate.

The chemisorption factor in electrocatalysis is also the leading basis for *selectivity*[8] in electrochemical reactions involving adsorption, as it is in regular heterogeneous catalysis. Selectivity in electrode processes is, of course, most clearly manifested in reactions involving small molecules, for example, their electrochemical oxidation or reduction including hydrogenation and dimerization. In reactions involving production of H_2 or O_2, no chemical selectivity can arise, but the i_0 values depend widely on electrode material.[2-5] In O_2 reduction, however, selectivity in relation to electrocatalyst material is observed, leading to distinction between two-electron and four-electron reduction of O_2 to peroxide or water, respectively, a topic that has received much attention in the literature, especially in the work of Yeager and co-workers.[9]

Electrocatalytic reactions are of two principal types: (a) those which proceed by electron transfer to or from a molecule or ion, producing a chemisorbed species on the electrode surface, which then, with further steps, forms a stable molecule (e.g. H_2, O_2, or Cl_2) through a heterogeneous electrochemical or chemical recombination step; and (b) reactions that involve an initial dissociative, or associative, chemisorption step, as with H_2, CH_3OH, and C_2H_4 oxidation or O_2 reduction, followed by electrochemical charge-transfer steps involving the initially formed chemisorbed intermediates.

In the relatively simple reactions of H_2 and O_2 evolution at electrodes, it is surprising that, until recently (e.g., cf. Refs. 10–13), the important matter of relating the kinetic and mechanistic behavior of the processes to extents of chemisorption of the respective chemisorbed intermediates has been little studied, although this matter has received considerable attention in the literature on theoretical aspects of kinetics and mechanisms of electrode reactions, especially in the works of Bockris,[2,14] Butler,[6] and Parsons[5] and in some papers of the present author.[12,13,15]

The particular aspects of chemisorption behavior of intermediates that are of interest are (a) the dependence of coverage, θ, on potential, V; (b) the corresponding adsorption pseudocapacitance, C_ϕ; (c) the dependence of these quantities for a given reaction on electrode material; and (d) the dependence of C_ϕ and θ on V in relation to the Tafel slope for the process, $b = dV/d \log i$.

It is important to make a distinction here between chemisorbed species that are formed in underpotential deposition processes (Upd species), and thus do not take part in an overall continuous faradaic reaction at appreciable current densities, and those species that are chemisorbed as the kinetically significant intermediates† in a faradaic reaction taking place at appreciable continuous current density and corresponding overpotential [referred to as overpotential-deposited (opd) species]. In some cases, the upd and opd species can be chemically identical, as with H at noble metals such as Pt, Pd, and Rh, but the state of chemisorption must be different: for example, at Pt, the surface is already fully covered by chemisorbed upd H at the H_2 reversible potential, so the further chemisorption of opd H at finite cathodic currents and overpotentials must take place on, or within, this completed upd film of adsorbed H. Related questions arise in the OER; anodic O_2 evolution and deposition of the OH and O intermediates[14] in this process take place on a chemisorbed oxygen film (a surface oxide) at the noble metals and on thicker oxide films at baser metals.[13]

---

† It is interesting to calculate the "turnover" rates at each catalytic site, as often evaluated for heterogeneously catalyzed processes; they are 6 for 1 mA/cm² and 6000 for 1 A/cm² at Pt, and thus of course can be controlled by the current density.

These "pre-reaction" states of electrode surfaces for the HER and especially the OER, through oxide film formation, are obviously of major significance for the faradaic electrocatalytic processes that go on, on these surfaces, at finite net current densities at significant or appreciable overvoltages.

One of the first attempts to characterize the adsorption behavior of the H intermediate in the HER at significant cathodic current densities (i) was that by Bockris et al.,[16] using a double-pulse galvanostatic procedure. On base metals, difficulties arise with any faradaic stripping procedure due, for example, to reoxidation of evolved H_2 (in the HER case) and surface oxidation of the metal depending on the excursion of potential allowed or adventitiously arising in the transient. This method was applied to determination of H adsorption at Ni by Devanathan and Selvaratnam,[17] but extremely large apparent coverages ($\theta_H \gg 1$) were found at high overpotentials. A method involving fewer problems of interpretation was described by Gerischer and Mehl[18] for investigation of chemisorbed H at H_2 cathodes and was based on ac impedance measurements and has been extensively developed since the time of that publication. More recently, a method based on interpretation of open-circuit potential relaxation transients recorded over a logarithmically wide time range (5–7 decades), following interruption of polarization currents at various controlled potentials, has been successfully applied[11-13] and is free of some of the difficulties of current-pulse methods owing to the fact that on open circuit, for all (over)potentials above the reversible potential, no back reaction of the product can take place and electrode surface oxidation is avoided except under conditions where open-circuit corrosion could take place in the case of very base metals in water.

In this chapter, we describe experiments on the OER at Pt electrodes in acid and alkaline solutions, with particular reference to the evaluation of the behavior of surface intermediate states in the reaction. Such states, as we discuss later, are to be identified with chemisorbed OH and O species, together with high oxidation states of metal ions of the electrode material in the surface of the oxide film on which the OER takes place. Such oxidation states can act as mediators in discharge steps of the OER and must also be counted as intermediates in the overall reaction schemes for anodic O_2 evolution.

1.2. Some Previous Work on the OER at Pt

A general scheme of formal reaction pathways in the OER was given by Bockris,[14] with diagnostic criteria for distinction of mechanisms, especially the Tafel slopes, reaction orders, and related pH effects on the kinetics. This paper was also of significance on account of the use, for the first time in electrode kinetics, of the Christiansen method for dealing with kinetics of multistep reaction pathways.

As with the HER at Pt, there is considerable interest in the OER from the points of view of the role of surface intermediate states in the reaction mechanism[13,15,19-21] and also the influence of oxide-film thickness.[22] For satisfactory experimental studies, the latter factor must be known or otherwise controlled, as has been recognized in various recent works on anodic O_2 and Cl_2 evolution kinetics.[22-24] It is well known that, at Pt, the oxide film itself is relatively stable down to much less positive potentials than those required for O_2 evolution and, in fact, is not reduced until potentials near, or in, the upd H potential range for Pt are approached, as is seen from typical cyclic voltammograms illustrated in Fig. 1. There is hence a necessity to distinguish between the oxide film at Pt anodes and the OER intermediate states at its surface. This is also indicated by the recent results of Willsau et al.,[25] who found by means of O-labeling experiments that gaseous O_2, electrolytically generated at Pt from water, does not contain significant quantities of O atoms from the oxide film itself. However, this conclusion is at variance with that of Ref. 26.

The importance of the oxide film at Pt anodes on which the OER proceeds was recognized by Schultze and co-workers[22,23] and treated in terms of electron tunneling,[22,27] in particular

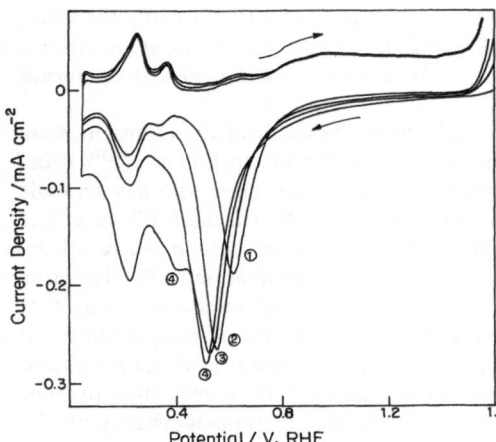

FIGURE 1. Cyclic-voltammetry plots for formation and reduction of surface oxide at Pt for various conditions of pre-formation of the oxide film in 1 mol/dm^3 aqueous NaOH at 298 K: (1) anodic limit, $V_a = 1.60$ V, holding time, $t_h = 60$ min; (2) $V_a = 1.90$ V, $t_h = 100$ min; (3) $V_a = 2.10$ V, $t_h = 100$ min; (4) $V_a = 2.34$ V, $t_h = 100$ min. Sweep rate: 25 mV/s.

by Birss and Damjanovic[28] in relation to the origin of the "$\frac{3}{2}$" reaction order in [OH$^-$] in acid solution associated with Tafel slope values of 60 and 120 mV per decade of current density. A "dual-barrier" model was considered,[28] which provided an explanation of this behavior. However, the essential information for examination of electrocatalysis in the OER in terms of the adsorption behavior of the intermediate states in the OER reaction pathway is still lacking and will be provided by the results of the work described in this chapter for the case of Pt in alkaline solution.

For the OER, we proposed in an earlier paper[19] that the intermediate states are not only discharged OH or O[17] but also higher oxidation states of the metal ions in the surface of the oxide film, for examples at Ni.O.OH, for which it was shown[19] how such states can act as mediators in the OER; this mechanism was later applied to the Cl$_2$ evolution reaction[29] and in other work on the OER.[21,28,29] In fact, changes of Tafel slope for the OER are associated with the potentials for change of oxidation state of oxide films.[21]

1.3. Reaction Pathways in the OER

Detailed possibilities for the reaction pathways in the OER were examined by Bockris,[14] and criteria for their distinction and for identification of rate-controlling steps within them were given. Here we illustrate the most likely pathways that may be involved in the OER at oxidized Pt electrodes, as treated in the present work. The discharge steps are written with OH$^-$ as the oxygen source for alkaline pHs, but obviously equivalent steps can be written when H$_2$O is the source. However, the surface state of oxidized Pt at given O$_2$ overpotentials can be different in acid and alkaline solutions depending on the "acidity" of the surface, especially in higher oxidation states, and activation energies of the respective steps can differ between acid and alkaline conditions.

In the process of O$_2$ evolution, the pathways are more complex than for the HER but the following, among others, have been proposed:

$$OH^- + M.ox. \rightarrow M.ox.OH + e^- \tag{1}$$

$$OH^- + M.ox.OH \rightarrow M.ox.O + e^- + H_2O \tag{2}$$

$$2M.ox.O \rightarrow 2M.ox. + O_2 \tag{3}$$

Other pathways are possible, such as:

$$2M.ox.OH \rightarrow M.ox.O + H_2O + M.ox. \tag{4}$$

or

$$2M.ox.OH \rightarrow \begin{matrix} M.ox. \\ M.ox. \end{matrix} O + H_2O \tag{5}$$

and a pathway involving an adsorbed peroxo species has been suggested.[14] In the above equations, M.ox. represents the surface of an oxide or oxide film on which the reaction steps proceed.

2. EXPERIMENTAL

2.1. General Procedure

The general experimental procedure, involving digital recording of potential-relaxation transients over a time range of ca. six decades, from microseconds to seconds and digital plotting of Tafel relations using a computer-controlled potentiostat, was as described previously.[30,31]

2.2. Preformation of Pt Surface Oxide Film

As in other studies of anodic processes at Pt electrodes,[22–24,28] it is important that the experiments on the OER and potential relaxation be conducted on oxidized Pt surfaces at well-defined, controlled potentials and/or where a definite thickness or state of the oxide film has been previously established by polarization for a controlled period of time to potentials above the highest to which the electrode will be taken in the kinetic measurements so that further changes do not take place during the kinetic experiment itself. Such a procedure was adopted in the present work, as in other work on the OER at Pt[22,28] and in our work on the Cl_2 evolution reaction at Pt.[24] The stability of such "preoxidized" surfaces is demonstrated by well-known cyclic voltammetry behavior which shows that oxide films on Pt formed at high potentials (up to 1.8–2.2 V E_{H_2} in the present work) do not become reduced until potentials near those for the H upd region are attained (cf. Fig. 1). Therefore, the preformed oxide film itself, on which the OER is caused to take place, does not suffer reduction during the potential-relaxation transients, but there is an adjustment of surface density of intermediates on the film during the open-circuit potential-relaxation transient as the potential declines spontaneously toward the O_2 reversible potential.

2.3. Solutions

Aqueous solutions of NaOH (1 mol/dm³), made up from NaOH recrystallized at low temperature from twice-distilled water, were used as the electrolyte at 298 K. Pyrodistilled water, which is required for upd and cathodic HER studies, is found to be unnecessary for anodic OER experiments. Solutions in the working electrode compartment were saturated with O_2 bubbled through the solution while purified N_2 was bubbled in the counter electrode compartment.

Aqueous sulfuric acid solutions (0.5 mol/dm³) were prepared from BDH Aristar grade H_2SO_4 in water doubly distilled from alkaline permanganate (pyrodistilled water is not required for anodic polarization studies at Pt). Experiments were conducted at 298 K.

2.4. Reference Electrodes

A Hg/HgO reference electrode was used in the same experimental NaOH solution but maintained in a compartment separated from the Pt working electrode by a closed, wetted stopcock. For the H_2SO_4 solution, a Hg/Hg_2SO_4 electrode was used. An H_2Pt electrode was not used in order to avoid diffusion of traces of H_2 to the Pt anode, which can give rise to depolarization effects in sensitive potential-relaxation experiments. Potentials recorded in this paper are converted to the RHE scale (denoted by E_{H_2}) for the same $1\ mol/dm^3$ NaOH solution, as also checked in separate experiments.

2.5. Pt Working Electrode

High-purity grade Pt (99.9%) wires from Johnson Matthey Company were used as the working electrodes. After preliminary cleaning and sealing into a soft-glass tube, the electrode was subjected to ca. 10 anodic/cathodic potential-sweep cycles at $50\ mV/s^1$ between 0.05 and $1.2\ V\ E_{H_2}$ until the cyclic voltammogram corresponded to that for a very clean Pt surface. The charge for upd H accommodation provided the real area of the electrode, in the usual way.

2.6. Cyclic Voltammetry

Cyclic or linear-sweep voltammetry was also performed in the ususal way to determine the state of, and charge for, oxide film formation (see Section 2.2) at various potentials and for controlled times of anodic polarization (see Fig. 1).

3. RESULTS AND DISCUSSION

3.1. Basis of the Potential-Relaxation Method

It is first necessary to outline the basis of the potential-relaxation procedure used in this work to derive information on the surface intermediate states in the OER at Pt.

In recent papers,[32,33] we have proposed approaches based on potential-relaxation analysis for providing information on the kinetically involved adsorbed intermediates in electrocatalytic reactions. Details have been given previously,[30,32] and a full theoretical discussion of the significance of potential-relaxation measurements was presented recently in the paper of Harrington and Conway.[33] In the potential-relaxation method, it is the open-circuit potential-relaxation transients themselves, recorded following interruption of polarizing current, that provide the necessary time-dependent information (as also may be obtained from ac impedance[18]) complementary to that from the steady-state polarization characteristic; these two kinds of information together enable the adsorption behavior of the surface intermediate states that are kinetically involved in the principal reaction pathway to be quantitatively determined. The information on the kinetically involved intermediate states in the reaction is derived in the form of the potential-dependent adsorption pseudocapacitance[23,24] of the overpotential-deposited (opd) species in the steady states prior to interruptions of polarizing currents. We emphasize that the nature and states of such opd species will generally be different from those of upd species generated at Pt at lower potentials, <1.23 V versus RHE.

The basic equation representing the potential decay rate, dV/dt, is

$$-C(dV/dt) = i_0 \exp(\alpha VF/RT) \equiv i_{ss}(V) \qquad (6)$$

where V is the measured electrode potential referred to a reference electrode, t is the time, i_0 is the exchange current density of the process, α is its transfer coefficient, and i_{ss} is the initial steady-state current density. C is the net interfacial capacitance, which is characterized mainly by the adsorption pseudocapacitance, C_ϕ,[34,35] for the process when appreciable potential-dependent surface coverages, θ, of the intermediate states of the reaction arise over the potential range of interest in the kinetic study of the reaction. Formally, $C = C_\phi + C_{dl}$, where C_{dl} is the interfacial double-layer capacitance.

As was discussed previously,[33] three types of adsorption pseudocapacitance of the opd species are to be distinguished: (i) a steady-state (ss) pseudocapacitance defined, as in the work of Gileadi and Conway,[34] by

$$C_{\phi,ss} = q(d\theta_{ss}/dV_{ss}) \tag{7}$$

(ii) a transient pseudocapacitance[33]

$$C_{\phi,t} = q(d\theta/dt)/(dV/dt) \tag{8}$$

and (iii) an operational pseudocapacitance,[33] $C_{\phi,0}$, as measured here through use of Eq. (6) with an experimentally recorded Tafel polarization characteristic $i_{ss}(V)$ [in Eq. (9)] and the experimental potential-decay transient, $V(t)$; that is [cf. Eq. (6)],

$$C_{\phi,0} = -\frac{i_{ss}(V)}{dV(t)/dt} \tag{9}$$

In Eqs. (6), (7), and (8), q signifies the formal charge required to complete a monolayer of the kinetically significant intermediate species involved in the reaction pathway. (Note, this charge q is not necessarily identical with the charges for monolayer surface oxidation of Pt by OH or O since here the species of kinetic significance are new states on or at the already oxidized Pt surface that are participating in the anodic formation of O₂.) Here we apply the potential-relaxation method to the study of the surface intermediate states in the OER from alkaline and acid solutions at Pt and compare the results.

3.2. Experimental Information Obtained

The following experimental information was obtained and processed as described in Section 3.1 above.

(i) Potentiostatically determined Tafel relations, as in Fig. 2, were digitally recorded at a Pt electrode bearing a "preformed" oxide film grown at a potential of 2.3 V E_{H_2} for 600 s, that is, at a potential higher than the highest anodic potential at which polarization measurements were made in electrode-kinetic OER runs or from which potential-relaxation transients were recorded. The points shown in Fig. 2 are for the descending direction of potential change below 2.3 V E_{H_2} and represent the reproducible curves for the OER at the "preoxidized"

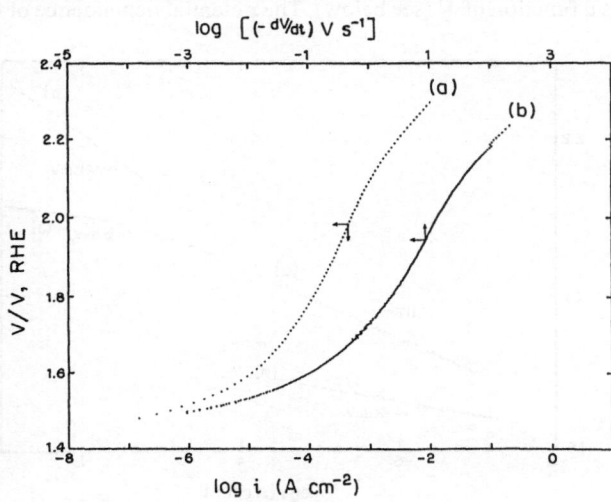

FIGURE 2. (a) Potentiostatically determined Tafel polarization line for the O₂ evolution reaction at Pt in 1 mol/dm³ aqueous NaOH at 298 K. Data for descending and ascending changes of anodic potential, from 2.25 V to 1.47 V versus RHE. (b) Plot of $\log(-dV/dt)$ from 2.00 V, versus RHE, determined from the potential relaxation transient taken at the same electrode.

Pt anode surface. The technique used here was similar to that described previously for the Cl₂ evolution reaction.[24]

In alkaline solution, the Tafel relations (curve b, Fig. 3) obtained consist of three regions: at low overpotentials, a low-slope region having $b \simeq 46$ mV; at high potentials, above 2.0 V, a high-slope region with $b \simeq 148$ mV, and a transition region in between, covering a span of ca. 0.35 V. This behavior notably contrasts with that observed in acid (0.5 mol/dm³ H₂SO₄) (curve a, Fig. 3) under otherwise similar conditions of oxide film formation, where the Tafel relation at potentials less than 1.85 V versus RHE has a slope of ca. 120 mV while above 1.9 V versus RHE the slope is ca. 57 mV. This change, corresponding formally to kinetics determined by two alternative parallel reactions, has been interpreted as due to onset of "resonance tunneling"[22,27] or mediation of the OER by higher oxidation states[19,39,40] of Pt in the oxide film surface,[36,37] as was first proposed by Conway and Bourgault[19] for the OER at oxidized Ni. While the forms of the two Tafel relations in Fig. 3 are qualitatively different, it may be significant that the transition in slope at higher overpotentials occurs at a potential, V_{tr}, which is almost the same (relative to the H₂ electrode in each solution) in acid and alkaline solution, viz., ca. 1.85 V versus RHE.

(ii) Potential-relaxation transients, as plotted out from the digital data acquisition system over the time range of microseconds to seconds, are shown in log time in Fig. 4. The same data subjected to a computer-programed differentiation procedure to obtain $\log(-dV/dt)$ as a function of voltage are plotted in Fig. 5. Another $\log(-dV/dt)$ versus V plot, from a transient taken from 2.23 V (near the upper limit of the Tafel line a in Fig. 2), is shown as curve b in Fig. 2 to provide comparison with the Tafel relation (cf. Fig. 2, curve a) over a comparable potential range.

The relation between the shapes or the slopes of the V versus $\log(-dV/dt)$ profiles and those of the Tafel relations (curves a and b of Fig. 2) can be understood through inspection of Eq. (6), which indicates that

$$\ln(-dV/dt) = \ln(i_0/C) + \alpha VF/RT \tag{10}$$

that is, $\ln(-dV/dt)$ as a function of V has the same slope, $\alpha F/RT$, as the Tafel relation only if C is independent of potential. Generally, when $C_{\phi,0}$ is appreciable and thence[34,35] potential-dependent, $\ln i_0/C$ in Eq. (10) must also be potential-dependent so that the plots of $\ln(-dV/dt)$ versus V will generally not be identical in form to the Tafel relation $\ln i$ versus V, as in fact is the case for the results in Fig. 2. However, generally from Eq. (6) it is seen that $\ln i(V) - \ln(-dV/dt) = \ln C(V)$, and it is this relation that enables C to be evaluated as a function of V (see below). The potential dependence of $\ln i_0/C$ in Eq. (10) arises because

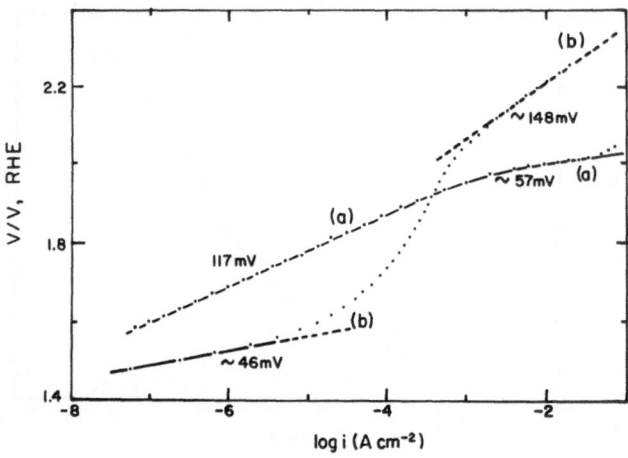

FIGURE 3. Comparison of Tafel relations for anodic O₂ evolution on preoxidized Pt electrodes in mol/dm³ aqueous NaOH (b) and 0.5 mol/dm³ aqueous H₂SO₄ (a) at 298 K (the latter from Ref. 26).

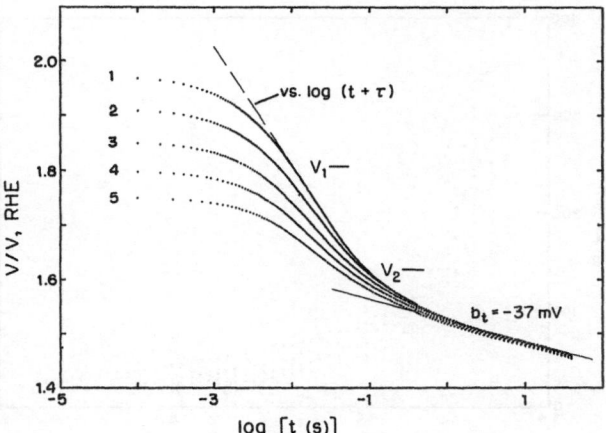

FIGURE 4. Potential-relaxation transients in log time, t, for the OER polarization at Pt in 1 mol/dm³ aqueous NaOH from five initial polarization potentials: (1) 1.98 V; (2) 1.91 V; (3) 1.85 V; (4) 1.80 V; (5) 1.75 V. Dashed line shows the corresponding plot for curve 1 versus $\log(t + \tau)$.

the coverage factors θ or $1 - \theta$ for the electroactive intermediate(s) are expected to be potential-dependent, limitingly exponential in $\pm VF/RT$[34,35]; tests of such behavior are shown later.

It is found that the logarithmic potential decay plots of Fig. 3 determined from different anodic potentials do not become superimposable when properly converted to plots in $\log(t + \tau)$ (cf. Refs. 30, 32, and 38), where τ is the integration constant[38] for Eq. (6), determined appropriately for each initial current density. In the case of the HER at Ni,[30] they do, as is expected. The difference here must be due to different states of the oxide surface involving the adsorbed intermediates that are generated at the Pt oxide film at different potentials.

3.3. Capacitance Behavior and Adsorption of Intermediates

We have referred, in Eqs. (7), (8), and (9), to the three distinguishable types of pseudocapacitance associated with faradaically generated adsorbed intermediates. The operational pseudocapacitance, $C_{\phi,0}$ (Eq. 9), which is experimentally accessible, is calculated from the experimental data points for $\log i(V)$ and dV/dt at various values of V. It must be mentioned that derivation of the $C_{\phi,0}$ behavior relies in part, and necessarily, on the $i(V)$ characteristic itself [right-hand side of Eq. (6)]; that is, evaluation of $C_{\phi,0}$ (V) from $V(t)$ transients requires the complementary and independently determined experimental data

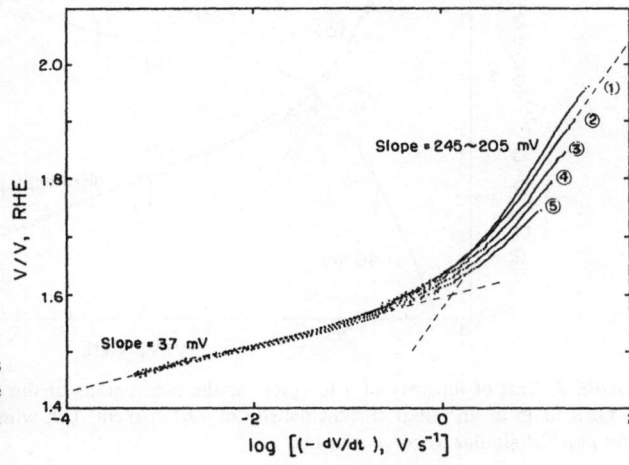

FIGURE 5. Log $(-dV/dt)$ plots versus electrode potential from the respective potential-relaxation transients of Fig. 4.

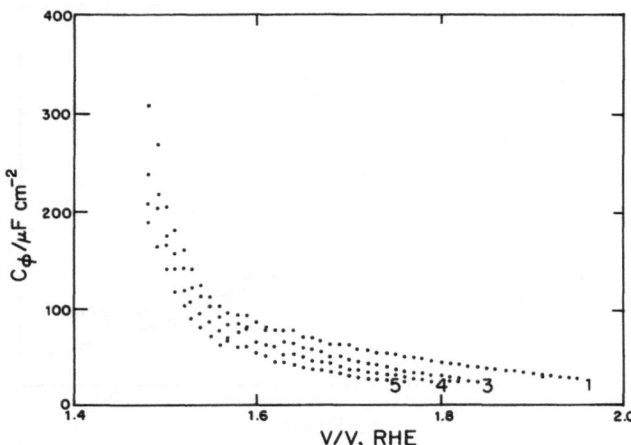

FIGURE 6. Plots of the operational pseudocapacitance, $C_{\phi,O}$ for the OER at Pt as a function of electrode potential for the more anodic range of potentials, derived from the potential-relaxation transients of Fig. 4, curves 1, 3, 4, and 5.

provided by the potential-relaxation transients, which enable dV/dt to be accurately derived as a function of V from the digitally recorded $V(t)$ transients, ca. 1200 points per volt, using a differentiating subroutine previously checked for accuracy on an analytic algebraic function.

From the V versus $\log i$ and the V versus $\log t$ plots in Figs. 3 and 4, we thus obtain the $C_{\phi,0}$ versus V profiles of Fig. 6, corresponding to curves 1, 3, 4, and 5 of Fig. 4. These curves, especially 1 and 3, taken from higher anodic potentials, are evidently composed of two distinguishable regions with an inflection around 1.85 V E_{H_2}, corresponding to the upper inflection region in the Tafel line (Fig. 3, curve a). This is seen more clearly when $\log C_{\phi,0}$ is plotted versus V and compared with $\log i$ versus V, drawn on the same graph (Fig. 7); the $\log C_{\phi,0}$ versus V plots are best represented by two, or possibly three, linear regions having slopes dependent on the potential from which the potential relaxation was initiated, as listed in Table 1. These plots give, for the first time on an experimental basis, a clear idea of how changes of Tafel slope, b, for an electrode process are intimately connected with the changing conditions of adsorption of the intermediate(s) and the related potential dependence of

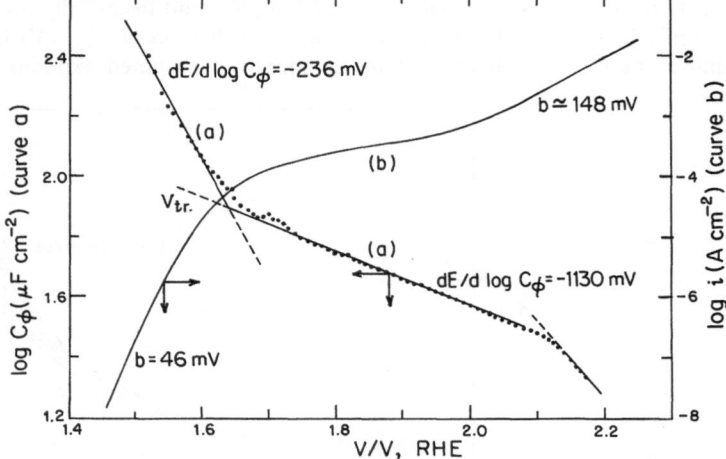

FIGURE 7. Test of linearity of a log plot for the two regions of the $C_{\phi,O}$ versus potential relation for the OER at Pt at an initial current density of 1.61 mA/cm^2 (a), with the relation superimposed on a Tafel plot (b) similar to curve a in Fig. 2.

TABLE 1
Values of the Derivative $-dV/d \log C_{\phi,0}$

Curve no. in Fig. 8	Initial potential (V)	$-dV/d \log C_{\phi,0}(V)$	
		Low V	High V
1	1.97	0.19	0.74
3	1.85	0.19	0.51
4	1.80	0.15	0.45
5	1.74	0.14	0.42

coverage by the kinetically significant intermediate states (to be discussed further below) in the OER at a Pt surface, measured here in terms of the operational pseudocapacitance, $C_{\phi,0}$.

In the case of electrochemical desorption steps, the relation between b and the potential dependence of coverage by adsorbed intermediates is clear and is given by

$$b^{-1} = d(\ln \theta)/dV + \beta F/RT \tag{11}$$

where the first term on the right-hand side is a derivative of the electrochemical adsorption isotherm, and the second term arises from the potential dependence of the electron transfer rate. In a limiting case, for low θ, Eq. (11) gives rise to the well-known result[2] $b^{-1} = (1 + \beta)F/RT$, or $2.3b = 42$ mV at $T = 298$ K.

In terms of the pseudocapacitance $C_\phi = q_1 \, d\theta/dV$, Eq. (11) becomes

$$b^{-1} = \frac{1}{\theta} \cdot \frac{C}{q} + \beta F/RT \tag{12}$$

and θ is related to $\int C(V) \, dV$ over the relevant potential range.

For limiting coverage conditions, it is expected[34,35] that $C_{\phi,0}$ will be an exponential function of V of positive or negative argument depending on whether $\theta \to 0$ or $\theta \to 1$; that is, $\log C_{\phi,0}$ can be linear in V. It is evident that although the relations between V and $\log C_{\phi,0}$ can be represented quite well by straight lines (Fig. 8), the slopes are very different from the limiting value of a. -59 mV expected in terms of a simple pseudocapacitance relation for a one-electron electroactive species approaching full coverage (cf. Refs. 34 and 35).

The fact that the slopes $dV/d \log C_{\phi,0}$ depend on the initial polarization current density or potential (Fig. 8) suggests that some characteristic state of the Pt oxide surface is developed

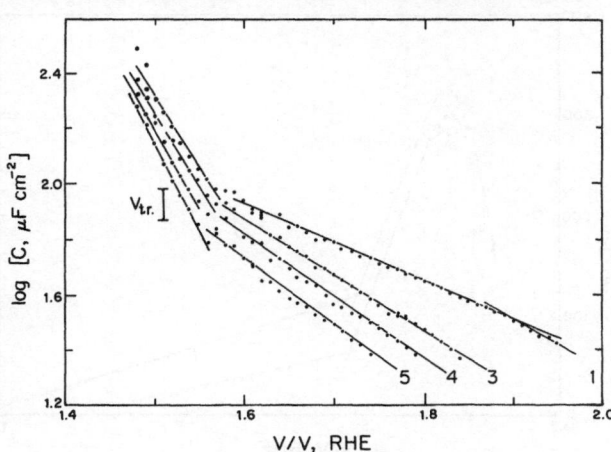

FIGURE 8. Logarithmic plots of the operational capacitance $C_{\phi,0}$ for the OER at Pt in 1 mol/dm³ aqueous NaOH derived from four of the potential-relaxation transients of Fig. 4, curves 1, 3, 4, and 5.

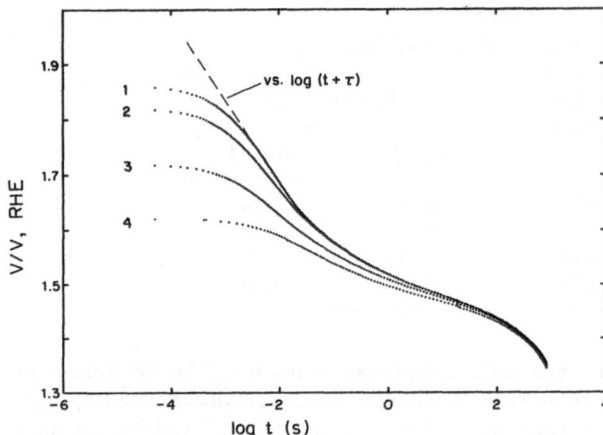

FIGURE 9. Potential-relaxation transients, as in Fig. 4, but for lower ranges of potential: from (1) 1.86 V; (2) 1.82 V; (3) 1.72 V, and (4) 1.62 V, down to 1.34 V E_{H_2}.

during polarization at each oxygen evolution current density. For example, this may be due to the degree to which higher oxidation states of $Pt^{(39,40)}$ are developed in the oxide film's surface during O_2 evolution, even though a definite overall extent of film growth was established in each series of experiments by "preformation" of the oxide film for controlled time and potential before the polarization and current-interruption experiments were performed.

Qualitatively, the only other factor likely to lead to a changing electrocatalytic behavior of the surface oxide, as potential relaxation transients are taken from successively lower potentials, is the continuing reconstruction of the quasi-three-dimensional surface oxide that can take place in time, without any further growth (i.e., without further increase in charge for reduction). This could cause a time-dependent change of the actual physical and/or chemical constitution of the catalytically effective surface in contact with the electrolyte.

It was of interest to establish if $C_{\phi,0}$ exhibited any maximum at low overpotentials as was found for H in the H_2 evolution reaction at Ni and Ni/Mo composite electrodes.[16] Potential-relaxation transients, as in Fig. 9, were accordingly recorded down to lower potentials and gave the $C_{\phi,0}$ versus V profiles of Fig. 10 for two initial current densities, each exhibiting the expected maximum (cf. Refs. 33 and 34), which arises in this case at around 1.43 ± 0.05 V E_{H_2}.

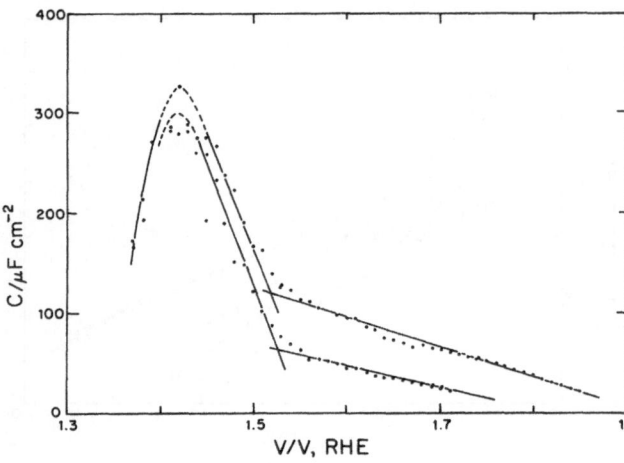

FIGURE 10. Plots of the operational pseudocapacitance $C_{\phi,O}$ for the OER at Pt as in Fig. 8 but for the lower ranges of potential covered in the potential-relaxation transients of Fig. 9. Note the development of the maxima in $C_{\phi,O}$ at ca. 1.42 V E_{H_2}. Data for two polarization current densities of 0.63 and 1.52 mA/cm² in the steeply rising part of curve a in Fig. 2.

The total charges under the $C_{\phi,0}(V)$ versus V profiles over the potential range 1.46 to 1.8 through to 1.99 V (Fig. 6) lie between 48 and 56 $\mu C/cm^2$,† corresponding formally to about 25% site occupancy relative to the original Pt metal surface, based on the H accommodation and one electron per site (see below). This corresponds, of course, formally to every other lattice position on the oxide film in each direction on, say, a two-dimensional (100) surface being occupied by an intermediate in the OER at the highest potential. These figures seem reasonable bearing in mind that undischarged OH^- ions are also resident in the double layer adjacent to the surface of the Pt oxide film.

One of the interesting aspects of the results presented here (Fig. 7 and Table 1) is that the slopes of the two regions of the V versus $\log C_{\phi,0}$ plots have remarkable values: at lower potentials in curve a of Fig. 7 $dV/d \log C_{\phi,0} \cong -236$ mV, while the longer region of curve a has a slope of ca. -1130 mV. Data for other polarization conditions were given in Table 1. The limiting slopes of curves of $\log C_{\phi,0}$ versus V, for low and high coverage, in the absence of lateral interaction effects are expected[34,35] to be $+59$ and -59 mV, respectively, very different from the observed values in Fig. 7 or Table 1. Introduction of a large lateral repulsive interaction parameter, $g = 12RT$, in the adsorption isotherm (cf. Refs. 34 and 35) is required to account for such a high slope as -1130 mV, and, even then, the calculated relation for $\log C_{\phi,0}$ versus V is not as linear as that actually observed in Fig. 7 or 8. Use of an isotherm of the form[41]

$$\frac{\theta}{1-\theta} = Kc \exp(-\tfrac{3}{2}g\theta^{1/2}) \exp VF/RT \tag{13}$$

corresponding to lateral repulsion of surface or image dipoles experiencing a pairwise interaction energy $U(\theta) = g\theta^{3/2}$ gives rise to an adsorption pseudocapacitance for an intermediate produced in a quasiequilibrium step, given by

$$C(\theta) = \frac{qF}{RT} \cdot \frac{\theta(1-\theta)}{1 + \tfrac{3}{2}g\theta^{1/2}(1-\theta)} \tag{14}$$

This does not improve the situation, as the direction of asymmetry of the corresponding curve of $C(V)$ versus V about its maximum is opposite to that observed in Fig. 7 or 10, and a lengthy linear logarithmic region of the kind found in Fig. 7 cannot be derived.

It does not seem that the unusual results for the potential dependence of $\log C_{\phi,0}$ are likely to be the result of a flaw in the employment of the potential-relaxation method or analysis of the results therefrom since (a) previously we have found good agreement between results for the capacitance behavior of intermediates in the OER at Co oxide electrodes[15] independently determined by means of potential-relaxation and ac impedance measurements; (b) over the potential range for appreciable currents for H_2 evolution at Au, for which H chemisorption is very weak, the potential-relaxation method gives only double-layer capacitance values (26–38 $\mu F/cm^2$) (corresponding, as expected for Au, to no significant adsorption pseudocapacitance for H); and (c) the experimental behavior found previously for the HER at Ni and Pt[12,30] by means of the potential-relaxation method can be well simulated in terms of kinetic equations[33] without reference to any arbitrary equivalent circuits for the interfacial processes.

An alternative origin for large values of the slope $dV/d \log C_{\phi,0}$ would be if the electroactive intermediates were deposited or generated with transfer of only a small partial

---

† We have considered whether this linear region in Fig. 6 could be due to double-layer capacitance, but the charges involved over the whole potential range referred to here appear to be too large for this to be the case; integrally, they would amount to ca. 150 $\mu C/V$.

charge, γe,† so that the limiting slopes would be 2.3 $RT/\gamma F$. However, the values of γ required to account for the observed slopes of ca. -236 and -1130 mV would, on such a basis, have to be ca. 0.25 and 0.05 (2), respectively, and other values would arise corresponding respectively to the $dV/d \log C_{\phi,0}$ data of Table 1. For the case $\gamma = 0.05$, the charge passed over the linear log region of slope -1130 mV would be ca. 9.5 $\mu C/cm^2$, which would correspond to a particle number density equivalent in charge to 9.5/0.05, that is, 190 $\mu C/cm^2$ if a full $1e$ charge were passed per particle. Such a figure is interestingly close to a change of surface density of the intermediate involved equivalent to monolayer coverage. For the region of higher slope at lower potentials, the charge passed would be ca. 20 $\mu C/cm^2$, and with $\gamma = 0.25$, as above, this would correspond to a change in coverage by the species involved in that process equivalent to ca. 40% of a monolayer.

3.4. Comparison with Behavior in Acid Solution

As was shown in Fig. 3, the forms of Tafel relations for the OER at Pt in alkaline and acid solutions are quite disparate, although an inflection occurs in both cases at a potential V_{tr} around 1.85 V, but the Tafel slopes above V_{tr} are quite different, as they also are in the lower potential range, 46 and 117 mV (see Fig. 3). In the acid case, the change of slope was attributed by Schultze and Haga[22] to onset of "resonance tunneling."

A significant difference in the reaction mechanism of the OER between acid and alkaline solutions, involving a difference in the activation energy, arises first from the difference in the source of OH or O species: H_2O at acid pHs and OH^-_{aq} in alkali. The overall energy difference for these processes corresponds to the energy of ionization of water, viz., 73 kJ/mol in ΔG^0; that is, discharge of OH from H_2O is expected to be kinetically more difficult than from OH^-_{aq}. (Relative to the H_2 electrode, the overall energy for discharge of O_2 is, of course, the same, independent of pH.) This seems to be consistent with the difference in the Tafel slopes at low potentials in alkaline and acid solutions, which have values of 46 and 117 mV, respectively; the first of these values could correspond (cf. Refs. 14, 28, and 42) to a rate-determining step such as

$$M_{ox}.OH + OH^- \rightarrow M_{ox}.O + H_2O + e^- \tag{15}$$

for alkaline solution having a slope of ca. 46 mV [$2.3RT/(1 + \beta)F$ with $\beta \cong 0.5$; cf. curve b of Fig. 3], with the prior discharge step, $M_{ox} + OH^- \rightarrow M_{ox}.OH + e^-$, having a greater rate constant, in the usual way, while, in acid, the first step, $M_{ox} + H_2O \rightarrow M_{ox}.OH + H^+ + e^-$, could account for the 117-mV slope of curve a in Fig. 3. This low-slope region does not, however, pass continuously into a region of higher slope, $2.3RT/\beta F$, expected in the simple analysis of a step such as that in Eq. (15), but passes through an extended inflection (Fig. 3) over which the V versus $\log C_{\phi,0}$ relation (Figs. 7 and 8) has the unusually high (negative) slope referred to earlier, before the upper linear Tafel region of high slope, 148 mV, is reached (curve a, Fig. 3).

We have emphasized earlier in this chapter, and elsewhere,[13,24] that electrocatalysis in anodic reactions taking place from aqueous solutions must depend on the state, including the oxidation state, of the oxide film [M_{ox} in eq. (15)] that is generated at the metal electrode rather than on the properties of the substrate metal itself. In the case of the OER proceeding at oxide films on transition-metal anodes, the intermediate states on and in the oxide film surface are to be identified not only with adsorbed, discharged OH· and O· species (cf. Ref. 14) but also with high oxidation states of the metal ions of the oxide film (here probably

---

† Or course, if formal charges of less than one electron are passed per intermediate state generated in the reaction pathway, correspondingly greater charge must be passed in the subsequent steps, with $4F/mol$ required for overall molecular O_2 formation.

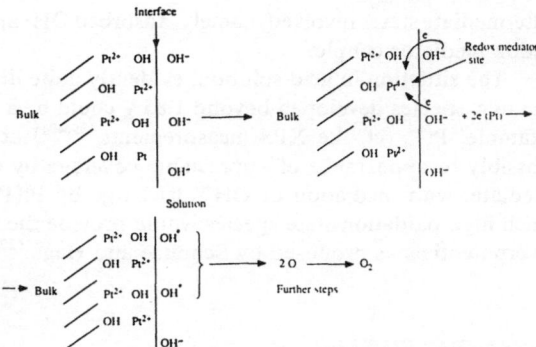

FIGURE 11. Involvement of higher oxidation states of platinum in the oxide film as mediators in the oxygen evolution mechanism.

Pt^{2+} and Pt^{4+}; see Fig. 11) acting as a mediator couple[19-21] in the discharge of OH^-, or of $OH\cdot$ from H_2O, depending on pH. The third stage of the process in Fig. 11 regenerates the initial Pt^{2+} sites on the oxide, thus providing a continuous mediation mechanism in charge transfer. A related mediator scheme involving O^{2-} species in the oxide film's bulk and surface structure, substituting some of the OH^-, together with the Pt^{4+} and chemisorbed $O\cdot$ as the kinetically involved intermediate states, could obviously be written.

An indication of the substantial difference in the surface-chemical behavior of the intermediate surface states that are involved in the OER on oxide films at Pt in alkaline and acid solutions is given by the curves of log $C_{\phi,0}$ versus V in Fig. 12. In alkaline solution, log $C_{\phi,0}$ versus V plot show the higher and the lower negative dV/d log $C_{\phi,0}$ values indicated on Fig. 12, whereas, for acid solution, the negative slope changes to a positive one beyond about 1.85 V, at which potential the Tafel slope change arises.

This implies, for the acid-solution case, the appearance of a new electroactive species beyond 1.85 V, increasing in surface concentration with potential from some initially low value (corresponding to the slope 57 mV in Fig. 2, curve a). We have seen from the present results that this behavior of the OER is not exhibited in alkaline solution, and the log $C_{\phi,0}$ versus V relation maintains a high negative dV/d log $C_{\phi,0}$ value until the further linear Tafel region (slope = 148 mV, curve b, Fig. 3) appears at high potentials. The negative slopes of the V versus log $C_{\phi,0}$ relations (except for potentials below that for the maximum in Fig. 10) imply increasing coverage toward a limitingly high (saturation) value of the coverage by the

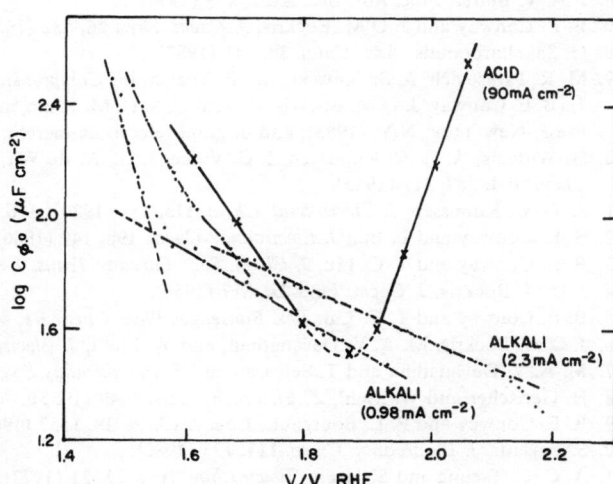

FIGURE 12. Comparison of plots of log $C_{\phi,0}$ versus V for Pt in acid (Ref. 36) and alkaline (present work) solutions at 298 K.

intermediate states involved, namely, adsorbed OH· and O· species and probably the Pt^{2+}/Pt^{4+} redox mediator couple.

The situation in acid solution, evidently quite different from that in alkali, suggests that the new species developed beyond 1.85 V could be a higher state of oxidation of Pt ions, for example, Pt^{4+} (cf. the XPS measurements,[39,40]) at the oxide-film surface, corresponding possibly to appearance of appreciable coverage by O rather than OH species as the intermediate, with mediation of OH^- discharge by Pt(IV) cations, for example, as in Fig. 11. Such high-oxidation-state species would provide the "resonance tunneling" pathway at high overpotentials as proposed by Schultze and Haga.[22]

ACKNOWLEDGMENTS

Grateful acknowledgment is made to the Natural Sciences and Engineering Research Council of Canada for support of this work on a Strategic Grant. We also thank Dr. L. Bai of this laboratory for helpful discussions. The results described in this paper were largely obtained by Mr. T. C. Liu in our laboratory and are specially acknowledged. They are now published in more detail in regular journals (Refs. 36 and 37).

The author also acknowledges this pleasant opportunity of presenting this paper in honor of Professor J. O'M. Bockris, with whom he worked as one of the first research students in the first electrochemistry group of Professor Bockris at Imperial College, London, in the period 1946-1949. He remembers those years with affection, and for the intellectual and scientific stimulation John Bockris provided to all of his co-workers.

REFERENCES

1. W. R. Busing and W. Kauzmann, *J. Chem. Phys.* **20**, 1129 (1952).
2. J. O'M. Bockris, in: *Modern Aspects of Electrochemistry*, No. 1 (J. O'M. Bockris, ed.), Chapter 4, Butterworths, London (1954).
3. P. Ruetschi and P. Delahay, *J. Chem. Phys.* **23**, 195 (1955).
4. P. Ruetschi and P. Delahay, *J. Chem. Phys.* **24**, 556 (1956).
5. R. Parsons, *Trans. Faraday Soc.* **54**, 1053 (1958).
6. J. A. V. Butler, *Proc. Roy. Soc.* **A157**, 423 (1936).
7. B. E. Conway and J. O'M. Bockris, *J. Chem. Phys.* **26**, 532 (1957).
8. G. Sakellaropoulis, *Adv. Catal.* **30**, 341 (1987).
9. M. R. Tarasevich, A. Sadkowski, and E. Yeager, in: *Comprehensive Treatise of Electrochemistry*, Vol. 7, (B. E. Conway, J. O'M. Bockris, E. Yeager, S. U. M. Khan, and R. E. White, eds.), p. 301, Plenum Press, New York, N.Y. (1983), and original references therein.
10. H. Willems, A. G. C. Kobussen, I. C. Vinke, J. H. W. de Wit, and G. H. J. Broers, *J. Electroanal. Chem.* **194**, 287, 317 (1985).
11. A. G. C. Kobussen, *J. Electroanal. Chem.* **115**, 131 (1980); **126**, 199 (1981).
12. B. E. Conway and L. Bai, *J. Electroanal. Chem.* **198**, 149 (1986).
13. B. E. Conway and T. C. Liu, *J. Chem. Soc., Faraday Trans. 1* **83**, 1063 (1987).
14. J. O'M. Bockris, *J. Chem. Phys.* **24**, 817 (1956).
15. B. E. Conway and T. C. Liu, *Ber. Bunsenges Phys. Chem.* **91**, 461 (1987).
16. J. O'M. Bockris, M. A. V. Devanathan, and W. Mehl, *J. Electroanal. Chem.* **1**, 143 (1959).
17. M. A. V. Devanathan and T. Selvaratnam, *Trans. Faraday Soc.* **56**, 1820 (1960).
18. H. Gerischer and W. Mehl, *Z. Elektrochem.* **59**, 1049 (1955).
19. B. E. Conway and P. L. Bourgault, *Can. J. Chem.* **38**, 1557 (1960); **40**, 1690 (1962).
20. S. Trasatti, *J. Electroanal. Chem.* **111**, 125 (1980).
21. A. C. C. Tseung and S. Jasem, *Electrochim. Acta* **22**, 31 (1977); *J. Electrochem. Soc.* **22**, 31 (1977).

22. J. W. Schultze and M. Haga, *Z. Phys. Chem.*, *N.F.* **104**, 73 (1977).
23. K. J. Vetter and J. W. Schultze, *J. Electroanal. Chem.* **34**, 131; 141 (1973).
24. S. Roscoe and B. E. Conway, *J. Electroanal. Chem.* **224**, 163 (1987).
25. J. Willsau, O. Wolter, and J. Heitbaum, *J. Electroanal. Chem.* **195**, 299 (1985).
26. K. I. Rozenthal and V. I. Veselovskii, *Dolk. Akad. Nauk U.S.S.R.* **111**, 647 (1956).
27. W. Schmickler and J. W. Schultze, *Z. Phys. Chem.*, *N.F.* **110**, 277 (1978).
28. V. Birss and A. Damjanovic, *J. Electrochem. Soc.* **130**, 1688 (1983); **134**, 113 (1987); **133**, 1621 (1986).
29. M. Tamura, *Electrochim. Acta* **23**, 9 (1978); **24**, 993 (1979); see also *Denki Kagaku* **48**, 173 (1980).
30. B. E. Conway and L. Bai, *J. Chem. Soc., Faraday Trans. 1* **81**, 1841 (1985).
31. B. E. Conway and T. C. Liu, *Ber. Bunsenges, Phys. Chem.* **91**, 461 (1987).
32. B. E. Conway, L. Bai, and D. F. Tessier, *J. Electroanal. Chem.* **161**, 39 (1984).
33. D. A. Harrington and B. E. Conway, *J. Electroanal. Chem.* **21**, 1 (1987).
34. E. Gileadi and B. E. Conway, *J. Chem. Phys.* **39**, 3420 (1963).
35. B. E. Conway and E. Gileadi, *Trans. Faraday Soc.* **58**, 2493 (1962).
36. B. E. Conway and T. C. Liu, *Proc. Roy. Soc.* (*London*) **A429**, 375 (1990).
37. B. E. Conway and T. C. Liu, *Langmuir* **6**, 268 (1990).
38. H. B. Morley and F. E. W. Wetmore, *Can. J. Chem.* **34**, 359 (1956).
39. G. C. Allen, P. M. Tucker, A. Capon, and R. Parsons, *J. Electroanal. Chem.* **50**, 335 (1974).
40. T. Dickinson, A. Povey, and P. M. A. Sherwood, *J. Chem. Soc., Faraday Trans. 1* **71**, 298 (1975).
41. B. E. Conway, E. Gileadi, and M. Dzieciuch, *Electrochim. Acta* **8**, 143 (1963).
42. J. O'M. Bockris and A. K. M. S. Huq, *Proc. Roy. Soc.* (*London*) *Ser.* A **237**, 277 (1956).

13

Quantum-Mechanical Formalisms of Electron Transfer Reactions at Electrode–Electrolyte Interfaces

Shahed U. M. Khan

1. INTRODUCTION

A quantum-mechanical theory of electron transfer processes at a metal electrode/electrolyte interface was originated by Gurney[1] in 1931. According to this theory, the electrochemical electron transfer reaction at an electrode metal involves the tunneling of electrons across the interfacial barrier to the activated ions in solution. This approach of Gurney was later named *molecular theory* and was further developed by Butler,[2] Gerischer,[3, 4] Christov,[5-7] Bockris and co-workers,[8-12] Schmickler,[13, 14] Khan and co-workers,[15-19] and Ovchinnikov and Benderskii.[20]

Another approach to the electron transfer reaction is connected with the work of Weiss[21] and Libby,[22] who suggested that a considerable part of the Franck–Condon barrier, that is, the activation barrier in solution, is due to continuum solvent polarization. This approach, which came to be named *continuum theory*, was further developed by Platzmann and Frank,[23] Kubo and Toyozawa,[24] Marcus,[25, 26] Levich and Dogonadze,[27, 28] and Dogonadze *et al.*[29, 30] Platzmann and Frank[23] introduced the concept of radiationless transition of electrons, and later it was developed by Lax[31] and Parker[32] for the process of charge transfer in liquids. Kubo and Toyozawa introduced the concept of polarons in the theory of radiationless transition of electrons. Detailed reviews on continuum models of electron transfer processes have been presented by Dogonadze[33] and also by Ulstrup.[34]

This chapter will focus mainly on recent developments of quantum-mechanical formalisms for molecular models of electron transfer reactions at electrode/electrolyte interfaces.

2. MOLECULAR MODELS

Gurney[1] put forward a theory of the radiationless electron transfer process from a metal electrode to ions (e.g., protons) in solution. According to this theory, an activated

Shahed U. M. Khan • Department of Chemistry, Duquesne University, Pittsburgh, Pennsylvania 15282.

Electrochemistry in Transition, edited by Oliver J. Murphy *et al.* Plenum Press, New York, 1992.

proton in the solution (originating due to molecular motion of the ion-solvent bond) is considered to be on the saddle point of the reaction hypersurface. When an electronic energy of the activated protons equals the energy of an occupied metal electronic level, radiationless electron transfer to an ion (e.g., a proton) in the interface may result via electron tunneling through the nonfluctuating interfacial barrier. The Born-Oppenheimer (Franck-Condon) principle is assumed to hold for the nuclei; that is, their movement is neglected during the short period of the electron transfer process. The theory then envisages a first-order reaction rate for electron or proton transfer reactions.

2.1. An Expression of Current Density

On the basis of Gurney's model,[1] an expression for the cathodic current density at a metal electrode/solution interface in terms of various probabilities can be obtained[35]:

$$i_c = e_0 \delta C_i Q^{-1} \int n(E)D(E)\nu_a(E)T(E)\, dE \tag{1}$$

where e_0 is the electronic charge, δ is the distance between the metal surface and the reaction plane at the interface, c_i is the concentration of reactant ions in solution, $n(E)$ is the occupation number of electrons in the metal at any energy state E, $D(E)$ is the density of electronic states in the metal electrode, $T(E)$ is the transition probability of electrons across the interfacial barrier, $\nu_a(E)$ is the frequency of activation of reacting ions in solution, which is coupled with the frequency of electron transition from the metal to the activated species in solution, and Q is the normalization function for the total number of electrons in the various energy states in the metal and can be expressed as[35]

$$Q = n(E)D(E)\, dE$$

$$= A \int_{E_F}^{\infty} E^{1/2}\{\exp[(E - E_F)/kT] + 1\}^{-1}\, dE \tag{2}$$

where A is a constant in the density of states function, $D(E)$. The electron occupation number, $n(E)$, in Eq. (1) is expressed as

$$n(E) = \{\exp[(E - E_F)/kT] - 1\}^{-1} \tag{3}$$

where E_F is the Fermi energy in the metal electrode. The density of states of electrons in the metal at any energy state can be expressed as

$$D(E) = AE^{1/2} = [4\pi(2m/h^2)^{3/2}]E^{1/2} \tag{4}$$

where m is the mass of the electron, and h is Planck's constant. The integral in Eq. (2) has been solved using the occupation number, that is, the Fermi function, as an exponential function for $E > kT$ and obtained as

$$Q = (2/h^3)(2\pi mkT)^{3/2} \tag{5}$$

2.2. The Frequency of Activation, $\nu_a(E)$

2.2.1. Theory

Recently, a model of frequency of activation of ions in solution has been given[35, 38] in terms of the phonon-vibron coupling concept using time-dependent perturbation theory.[38] According to the phonon-vibron coupling (PVC) model, an ion in solution can receive energy in a condensed medium due to transfer of energy from the phonons produced by the surrounding solvent oscillators to the ion-solvent bond vibrations (vibrons) in the inner sphere (see Fig. 1).

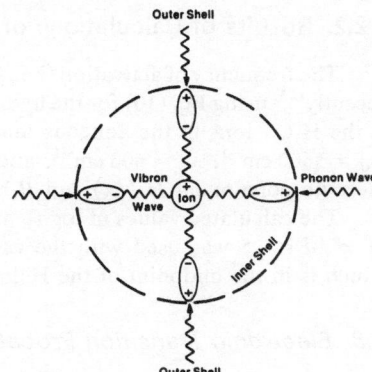

FIGURE 1. Schematic representation of phonon–vibron coupling (PVC) model of activation.

The frequency at which the ion–solvent bond or ion–ligand bond oscillator in an initial state, ψ_i, interacts with phonons and becomes excited to an excited (e.g., activated state) vibrational state, ψ_f, can be expressed in terms of Fermi's golden rule of time-dependent perturbation theory as[35]

$$\nu_a(E) = (2\pi/\hbar)|\langle\psi_f|H_{\text{int}}|\psi_i\rangle|^2\rho(E_f) \tag{6}$$

where $\hbar = h/2\pi$, and $\rho(E_f)$ is the density of the final excited state having energy $E = E_f$, the final state energy, and can be expressed as $\rho(E_f) = 1/h\omega_{if}$, where ω_{if} is the frequency of the ion–solvent bond. In Eq. (6) the interaction Hamiltonian, H_{int}, involves the interaction potential V_1 and the phonon occupation number n_q, which can be expressed as[35]

$$n_q = [\exp(h\omega_q/kT) - 1]^{-1} \tag{7}$$

where ω_q is the phonon frequency.

Taking ψ_i and ψ_f in Eq. (6) as harmonic oscillator wave functions, one obtains[35]

$$\nu_a = (\pi\omega_q/\hbar m v_q^2\omega_{if})|\langle\psi_f|V_1|\psi_i\rangle|^2[\exp(\hbar\omega_q/kT) - 1]^{-1} \tag{8}$$

The energy must be conserved[36] such that one can write

$$\hbar\omega_q = \Delta E_{if} \tag{9}$$

Considering the ion–solvent bond as a one-dimensional harmonic oscillator, and then using the normalized harmonic oscillator wave function for $n = 0$ as the initial state wave function ψ_i and that for $n = 1$ as the final state wave function ψ_f, one can express Eq. (8) as

$$\nu_a = C\left|\int_0^\infty H_n(\xi x)\exp(-\xi^2 x^2)k_v x^2\,dx\right|^2 \{\exp[-\Delta G^\ddagger - (x^* e_0\eta_c/d)]/kT - 1\}^{-1} \tag{10}$$

where $H_n(\xi x)$ is the Hermite polynomial of order n, $\Delta E_{if} = (\Delta G^\ddagger - x^* e_0\eta_c/d)$, where $\Delta G^\ddagger$ is the free energy of activation, η_c is the cathodic overpotential, and k_v is the force constant of the ion–dipole bond vibrator.

In Eq. (10) x^* is the position of the activated complex on the reaction hypersurface from the electrode surface in the electrical double layer of width d, and

$$C = \pi B^2\omega_q/2m v_q^2\hbar\omega_{if} \tag{11}$$

with

$$B = \xi/(2^n n!\pi^{1/2})^{1/4} \tag{12}$$

$$\xi^2 = m\omega_{if}/\hbar \tag{13}$$

and v_q is the velocity of phonon waves.

2.2.2. Results of Calculations of Frequency of Activation

The frequency of activation (i.e., the probability of activation per unit time) was calculated recently[35] using Eq. (10) for the hydrogen evolution reaction, for which the activating species is the H_3O^+ ions in the aqueous medium. The frequency of the H^+—H_2O bond vibration, $\omega_{if} = 3600$ cm^{-1}, $\omega_q = 600$ cm^{-1}, and $\Delta G^\ddagger = 0.9$ eV[37] at an Hg electrode were used. The Hermite polynomial $H_n(\xi x)$ and B have been evaluated for $n = 1$.

The calculated values of log ν_a at different overpotentials are plotted in Fig. 2. The value of $x^*/d = 0.5$ was used with the assumption that the activated complex is situated at x^* which is in the midpoint of the Helmholtz double layer of width d.

2.3. Electronic Transition Probability, T(E)

In terms of time-dependent perturbation theory, the electronic transition probability, $T(E)$, can be expressed as[38,39]

$$T(E) = (2\pi\tau/\hbar)|\langle \psi_f|V(r)|\psi_i\rangle|^2\rho(E_f)$$
$$= (2\pi\tau/\hbar)|M|^2\rho(E_f) \tag{14}$$

where ψ_i and ψ_f are, respectively, the initial and the final state wave function of the electron, $\rho(E_f)$ is the density of electronic states in the final state in the acceptor metal ion for the cathodic process or in the metal for the anodic process, and τ is the transition time of the electron across the interface.

For a model calculation of the transition probability, $T(E)$, it was considered that an electron transfers from an available Fe^{2+} ion in solution to a metal electrode having equal energy. The following matrix element was derived for the above process using the plane wave (with an effective mass approximation) in the metal as[15,38]

$$M = (4\pi/L_0^{3/2})\left[\sum_{m=-2}^{2} Y_{2m}(k)(-1)\int_0^\infty r^2\,dr\,j_2(kr)V(r)\psi_d(r)\right.$$
$$+ A\sum_{m=-3}^{3} Y_{3m}(k)(-i)\int_0^\infty r^2\,dr\,j_3(kr)V(r)\psi_f(r)$$
$$\left.+ B'\sum_{m=-1}^{1} Y_{1m}(k)(i)\int_0^\infty r^2\,dr\,j_1(kr)V(r)\psi_p(r)\right] \tag{15}$$

Here, $Y_{lm}(k)$ represents the spherical harmonics; the $j_l(kr)$ are the Bessel functions; ψ_p, ψ_d, and ψ_f are, respectively, the p, d, and f orbitals in the Fe^{2+} ions in solution; $L_0 = 2\pi$; and the terms A and B' are expressed as

$$A = 3\xi'|M_1|/\sqrt{35}(E_d^0 - E_f^0) \tag{16}$$
$$B' = 2\xi'|M_2|/\sqrt{15}(E_d^0 - E_p^0) \tag{17}$$

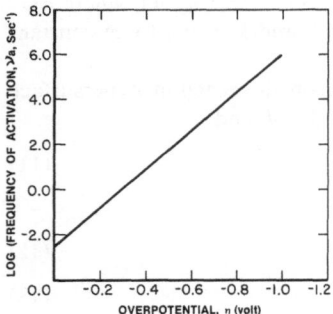

FIGURE 2. Plot of the logarithm of the frequency of activation as a function of overpotential, η_c.

where

$$V(r) = 6e_0\mu\cos\theta/|R - r|^2 - \xi'/|R - r|^2 \tag{18}$$

where e_0 is the electronic charge, μ is the dipole moment of the ligand or solvent molecule, $|R - r|$ is the radial vector from the dipole to the $3d$ electron, θ is the angle between $|R - r|$ and the dipole moment vector, μ, and $\xi' = 6e_0\mu\cos\theta$. In Eq. (16),

$$M_1 = [(2l + 1)/2R]\sum_l (2l + 1)\begin{pmatrix} 3 & l & 3 \\ 0 & 0 & 0 \end{pmatrix}^2 P_l(\cos\Omega)r^2\,dr\,\psi_f(r)Q_l[(R^2 + r^2/2Rr)] \tag{19}$$

In Eq. (17), M_2 is similar to M_1 with the $3j$ symbol of $\begin{pmatrix} 1 & l & 1 \\ 0 & 0 & 0 \end{pmatrix}$ and in the integral ψ_f is replaced by ψ_p. P_l and Q_l are the associated Legendre and Legendre functions, respectively. E_p^0, E_d^0 and E_f^0 are the ground state energies of p, d, and f state wave functions, respectively, in the Fe^{2+} ion.

In Eq. (14) the density of states of the electron is taken as that in the metal electrode level at the reversible potential (where electron transfer rates are equal for both directions) and is expressed as[38]

$$\rho(E_f) = mk_f L_0^3/2\pi^2 h^2 \tag{20}$$

where k_f is the wave number of electrons corresponding to the Fermi level energy at the reversible potential, $E_{f,\text{rev}}$.

For the typical value of $\tau = 1.66 \times 10^{-16}$ s, the transition probability at the reversible potential of the Fe^{3+}/Fe^{2+} system becomes[15,38]

$$T(E) = 7.2 \times 10^{-3} \tag{21}$$

One can also determine $T(E)$ at other potentials.

2.4. Free Energy of Activation

One needs the value of the free energy of activation, $\Delta G^{\ddagger}$, in Eq. (10), and this can be expressed as

$$\Delta G^{\ddagger} = \Delta G^{\ddagger}_{\text{inner}} + \Delta G^{\ddagger}_{\text{outer}} = (R_{\text{inner}} + R_{\text{outer}})/4 \tag{22}$$

where $\Delta G^{\ddagger}_{\text{inner}}$ and $\Delta G^{\ddagger}_{\text{outer}}$ are equal to the inner-sphere reorganization energy, $R_{\text{inner}}/4$, and the outer-sphere reorganization energy, $R_{\text{outer}}/4$, respectively.

2.4.1. Inner-Sphere Reorganization Energy

2.4.1.1. George–Griffith Model. According to the general George–Griffith model, the inner-sphere reorganization energy for one type of ions in solution can be expressed as[40]

$$R_i = n[f_z f_{z+1}/(f_z + f_{z+1})]\Delta q^2 \tag{23}$$

where n is the number of ligands attached to the central ion, and Δq is the equilibrium bond length difference between the oxidized and the reduced ion. It should be noted that the values of the force constants f_z and f_{z+1} are obtained from vibrational spectroscopic data using the following equation:

$$f_i = 4\pi^2 c^2 \omega_i^2 (m_L/N) \tag{24}$$

where ω_i is the observed symmetric stretching frequency of the ith ion–ligand bond (i represents either the z or the $z + 1$ state), c is the velocity of light, m_L is the molecular weight of the ligand, and N is Avogadro's number.

However, the George–Griffith model requires a knowledge of experimental stretching frequency (ω_i) data, the paucity of which severely limits its frequent utility. To get around this difficulty, one can replace f_i values by the corresponding theoretical force constant of the reorganized state, $f(z, r_{z+1})$, which is obtained by using ion–dipole orbiting models[41,42] and also using the INDO/2 molecular orbital method.[19]

2.4.1.2. Ion–Dipole Orbiting Models. The ion–dipole orbiting potential of the reorganized state, $V(z, r_{z+1})$, in terms of its nuclear charge z, and nuclear configuration, r_{z+1} (i.e., ion–dipole separation), can be expressed as[16-18]

$$-V(z, r_{z+1}) = (ze^2\alpha/2r_{z+1}^4) + (ze\mu/r_{z+1}^2)\langle\cos\theta\rangle \tag{25}$$

where e is the electronic charge, α is the molecular polarizability of the ligand, θ is the angle between the dipole direction and r_{z+1}, and $\langle\cos\theta\rangle$ is the average value of $\cos\theta$ for a given r_{z+1}. The force constant of the reorganized state, $f(z, r_{z+1})$, is then obtained from Eq. (25) using

$$f(z, r_{z+1}) = -\partial^2[V(z, r_{z+1})]/\partial r_{z+1}^2 \tag{26a}$$

Note that the force constant for the individual ion type having electronic charge z and nuclear configuration r_z can be expressed as

$$f(z, r_z) = -\partial^2[V(z, r_z)]/\partial r_z^2 \tag{26b}$$

where $V(z, r_z)$ can be computed from Eq. (25) by replacing r_{z+1} by r_z. Similarly, the force constant $f(z + 1, r_{z+1})$ can be obtained.

The final expressions of $f(z, r_{z+1})$ and $f(z, r_z)$ depend on the dipole orientation model used for obtaining $\langle\cos\theta\rangle$. Finally, the expression for R_i is obtained from

$$R_i = (n/2)f(z, r_{z+1})\Delta q^2 \tag{27}$$

Various ion–dipole orientation models are available, and two methods of obtaining $\langle\cos\theta\rangle$ which generate the best results are given below.

(a) Locked Dipole Orientation (LDO) Model. Earlier attempts to calculate the average value of $\cos\theta$, $\langle\cos\theta\rangle$, were made by Moran and Hamill,[41] who found that the dipoles approximately lock onto the ion, and consequently $\langle\cos\theta\rangle = 1$ is obtained. This model is applicable mainly for dipoles or ligands of high dipole moments.

(b) Improved Average Dipole Orientation (IADO) Model. In the improved average dipole orientation (IADO) model,[41,42] the average value of $\cos\theta$ is given by

$$\langle\cos\theta\rangle = (\tfrac{1}{3})[3 - (2^{5/2}r_{z+1})(E(\infty)/ze\mu)^{1/2}] \tag{28}$$

$E(\infty)$ is the value of $E(r_{z+1}, \theta)$ at infinite separation such that it becomes equal to kT.

Hence, using this value of $\langle\cos\theta\rangle$ from the IADO model in Eq. (25), one obtains the force constant using Eq. (26a) as

$$f(z, r_{z+1}) = 10ze^2\alpha/r_{z+1}^6 + 6ze\mu/r_{z+1}^6 - 3.772(ze\mu kT)^{1/2}/r_{z+1}^3 \tag{29}$$

The calculation of the inner-sphere reorganization energy of various ions was made by using Eqs. (27) and (29). The calculated values of the force constant and the reorganization energy for hexaaquo redox complexes are given in Tables 1 and 2, respectively. Importantly, this model provides a simple method of computing R_{inner}, and it does not require the rarely available vibrational spectroscopic data.

2.4.1.3. Intermediate Neglect of Differential Overlap (INDO/2) Molecular Orbital (MO) Method. The INDO/2 molecular orbital method has been applied recently to compute the inner-sphere reorganization energy of a few transition-metal ions in solution.[19] The force constants and inner-sphere reorganization energies calculated using the quantum-chemical INDO/2 MO method and the classical LDO and IADO methods and the corresponding experimental values are summarized in Tables 1 and 2, respectively.

TABLE 1

Force Constants of Metal-Ion–Ligand Bonds of Hexaaquo Complexes

Complex	Force constant (mdyn/Å)			
	$f_{i,w}$ (expt) [a]	$f_{i,MO}$ [b]	$f_{i,LDO}$ [c]	$f_{i,IADO}$ [d]
$[V(H_2O)_6]^{2+}$	1.60	1.16	0.92	0.85
$[Cr(H_2O)_6]^{2+}$	1.60	1.54	1.01	0.93
$[Mn(H_2O)_6]^{2+}$	1.65	1.56	1.10	1.02
$[Fe(H_2O)_6]^{2+}$	1.60	1.72	1.22	1.12
$[Co(H_2O)_6]^{2+}$	1.60	2.03	1.28	1.15
$[V(H_2O)_6]^{3+}$	2.44	1.67	2.17	2.02
$[Cr(H_2O)_6]^{3+}$	2.54	2.40	2.34	2.18
$[Mn(H_2O)_6]^{3+}$	2.54	2.15	2.74	2.55
$[Fe(H_2O)_6]^{3+}$	2.54	2.16	2.74	2.55
$[Co(H_2O)_6]^{3+}$	2.54	2.43	2.97	2.76

[a] From Eq. (24) using experimental spectroscopic frequency (ω) data.
[b] Results obtained using quantum-chemical INDO/2 molecular orbital method.
[c] Results from classical locked dipole orientation (LDO) model using Eq. (26b).
[d] Results from classical improved average dipole orientation (IADO) model using Eq. (26b).

2.4.2. Outer-Sphere Reorganization Energy

An expression for the outer-sphere reorganization energy was given by Marcus[25, 26] using nonequilibrium continuum solvent polarization. Assuming the central ion to be a spherical conductor, the expression of the reorganization energy was obtained as[26]

$$R_{outer} = [(ne)^2/2](1/a - 1/r)(1/D_{op} - 1/D_s) \qquad (30)$$

where ne is the number of electrons transferred, a is the radius of the ion including the first solvation shell, r is the distance from the center of the ion to the center of its image in the electrode, and D_{op} and D_s are, respectively, the optical and the static dielectric constant of the solvent medium.

TABLE 2

Inner-Sphere Reorganization Energy, R_i, for Electron Transfer Reaction of Hexaaquo Complexes

Complex	r_i (Å) [a]		$R_i^{(calc)}$ (eV)			$R_i^{(expt)}$ (eV)
	r_z	r_{z+1}	$R_{i,MO}$ [b]	$R_{i,LDO}$ [c]	$R_{i,IADO}$ [d]	$R_{i,\omega}$ [e]
$[V(H_2O)_6]^{2+}$	2.26	2.07	1.23	0.87	0.80	1.30
$[Cr(H_2O)_6]^{2+}$	2.22	2.04	1.14	0.79	0.85	1.19
$[Mn(H_2O)_6]^{2+}$	2.18	1.97	1.35	1.17	1.09	1.50
$[Fe(H_2O)_6]^{2+}$	2.14	1.98	0.92	0.81	0.75	0.94
$[Co(H_2O)_6]^{2+}$	2.12	1.95	1.20	0.97	0.88	1.06

[a] The equilibrium metal-ion–ligand bond lengths obtained using experimental crystallographic radii.
[b] Calculated from Eq. (23) using force constant of INDO/2 MO method, $f_{i,MO}$, and experimental values of $\Delta q(= r_z - r_{z+1}$, where r_z and r_{z+1} are given in columns 2 and 3, respectively).
[c] Calculated from Eq. (23) using force constant of locked dipole orientation (LDO) model and experimental values of Δq.
[d] Calculated from Eq. (23) using force constant of improved average dipole orientation (IADO) model and experimental values of Δq.
[e] Calculated from Eq. (23) using experimental values of force constant $f_{i,\omega}$ and experimental values of Δq.

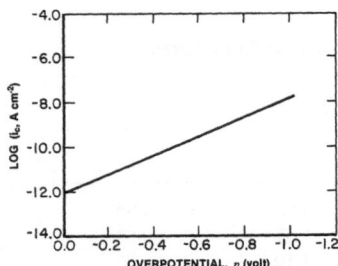

FIGURE 3. Plot of the logarithm of the current density as a function of overpotential, η_c.

2.5. Current Densities from Molecular Models

The current densities can be computed from Eq. (1) with the use of Eqs. (3)-(5) and (10)-(29). Calculated values of the current densities obtained by considering $T(E) \simeq 1$ and $\Delta G^{\ddagger} = \Delta G^{\ddagger}_{\text{expt}}$ are given in Fig. 3 for the hydrogen evolution reaction at a metal electrode.

3. ELECTRON TRANSFER REACTIONS AT A BIOLOGICAL-MEMBRANE-COVERED ELECTRODE

Recently, theoretical expressions of cathodic current density have been put forward[43] for the electron transfer process at a biological-membrane-covered electrode.[44,45] For this, two possible mechanisms of electron transfer processes through a protein-containing biological membrane to redox species in solution were considered. In one, an electron transport due to both diffusion and drift across the conduction band of the protein (embedded in the membrane) to the membrane/solution interface (where an electron transfer to redox ion in solution occurs) was taken into account. In the second model, both direct and resonance tunneling of electrons across the biomembrane layer were considered.

3.1. Transport Model

In the transport model an expression for the cathodic current density, i_c, was given in terms of the electronic charge, e_0, the charge transfer rate constant, k_{ct}, and the density of electrons in the semiconducting protein, N_D, as[43]

$$i_c(\text{transport}) = e_0[k_{ct}/(k_{ct} + YD_e)]D_e N_D Y \exp(-e_0|V_s|/kT) \qquad (31)$$

The charge transfer rate constant, k_{ct}, in Eq. (31) was expressed as[43]

$$k_{ct} = S_{th}(N_a/N_t)P_T(E_c) \exp[-(E_a - e_0\beta|\eta_c| - E_c)/kT] \qquad (32)$$

where S_{th} is the thermal velocity of the electron in the semiconducting protein; $P_T(E_c)$ is the electron tunneling probability across the interfacial barrier at the membrane/solution interface, which we assume, for simplicity, to be approximately unity; E_a is the activation energy of species in solution, β is the electrochemical symmetry factor,[38] and η_c is the electrochemical overpotential[38]; N_a is the number of electron acceptor species per unit area of the reaction plane [e.g., the outer Helmholtz plane (OHP)]; N_t is the total number of sites per unit area of the reaction plane; and E_c is the energy of an electron at the conduction band edge at the membrane/solution interface. In Eq. (31), D_e is the diffusion coefficient of the electron in the membrane, and

$$Y = e_0|V_s|/dkT \qquad (33)$$

where d is the thickness of the membrane, and $|V_s|$ is the absolute value of the surface potential at the protein/solution interface. Also in Eq. (31),

$$N_D = N_m \exp[-(E_C^0 - E_F)/kT] \tag{34}$$

where N_m is the number density of conduction electrons in the metal, E_C^0 is the bottom of the conduction band edge at the metal/membrane interface, and E_F is the Fermi level energy of the metal (see Fig. 4).

3.2. Direct Tunneling Model

For the direct tunneling model, the electron is considered to tunnel from the metal surface through the membrane to acceptor species at the membrane/solution interface (Fig. 4). The protein-containing biological membrane acts as a barrier. The current density for the direct tunneling of electrons through such a barrier can be expressed as[43]

$$i_c(\text{direct}) = e_0(2m^*/\pi^2 h^3)(N_a/N_i) \int E\{1 + \exp[(E - E_F)/kT]\}^{-1} P_T(E) D(E, e_0 \eta_c) \, dE \tag{35}$$

where e_0 is the electronic charge, m^* is the effective mass of the electron, E is the energy of the outcoming electron, and E_F is the Fermi energy of the electron in the metal.

In terms of the Wentzel, Kramer, and Brillouin (WKB) approximation, the direct tunneling probability, $P_T(E)$, across a membrane of thickness d can be expressed as[43]

$$P_T(E) = \exp\{-\tfrac{4}{3}(2m^*/\hbar^2)^{1/2}[d/(U_R - U_L)][(U_R - E)^{3/2} - (U_L - E)^{3/2}]\} \tag{36}$$

where U_R is the barrier height in the right hand side and U_L is the barrier height in the left hand side (see Figs. 4 and 5). The distribution function of electronic states in the species in solution can be expressed as[43]

$$D(E, e_0 \eta_c) = \exp[-(E_a - e_0 \beta |\eta_c| - E)/kT] \tag{37}$$

where E_a is the activation energy for the redox species in solution, β is the electrochemical symmetry factor,[38] having values $0 < \beta < 1$, and $|\eta_c|$ is the absolute value of the cathodic overpotential.[38]

3.3. Resonance Tunneling Model

In the resonance tunneling model, the semiconducting protein in the biological membrane is assumed to be uniformly doped with impurities (e.g., proton or organic species in aqueous solution). The energy levels of these impurity states are assumed to lie at a fixed distance

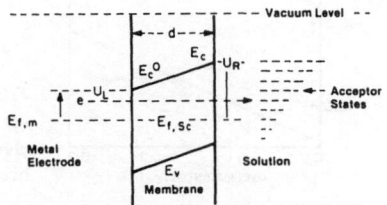

FIGURE 4. Schematic diagram of the energetics of electrons at metal electrode/semiconducting-protein-containing bio-membrane/electrolyte interface to represent the direct tunneling of electrons.

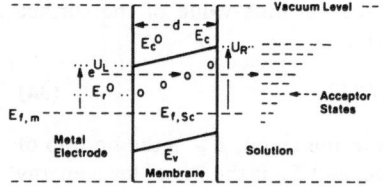

FIGURE 5. Schematic diagram of the energetics of electrons at a metal electrode/semiconducting-protein-containing biomembrane/electrolyte interface to represent the resonance tunneling of electrons.

below the bottom of the conduction band (Fig. 5). The resonance current density can be expressed as[43,46-48]

$$i_c(\text{resonance}) = e_0(2m^*/\pi^2h^3)(N/N_t)\int_0^\infty E_r\{1$$
$$+ \exp[(E_r - E_F)/kT]\}^{-1}R(E_r)D(E_r, e_0\eta_c)\, dE_r \qquad (38)$$

where $R(E_r)$ is the resonance tunneling factor via a resonance energy level, E_r, and can be expressed as[44,46-48]

$$R(E_r) = R_1(E_r)R_2(E_r)/[R_1(E_r) + R_2(E_r)] \qquad (39)$$

where

$$R_1(E_r) = \exp\{-(4/3a)(2m^*/\hbar^2)^{1/2}[(U_L - E_r + al)^{3/2} - (U_L - E_r)^{3/2}]\} \qquad (40)$$

$$R_2(E_r) = \exp\{-(4/3a)(2m^*/\hbar^2)^{1/2}[(U_R - E_r)^{3/2} - (U_L + al - E_r)^{3/2}]\} \qquad (41)$$

Note that the distance of the impurity state, l, from the metal surface is dependent on the resonance energy state, E_r, by the relation

$$l = (E_r - E_r^0)/a \qquad (42)$$

where E_r^0 is the position of the resonance level at the surface of the metal, and the slope of the linear barrier is

$$a = (U_R - U_L)/d \qquad (43)$$

where U_R is the barrier height in the membrane/solution interface, and U_L is that in the metal/membrane interface (Fig. 5).

The results of numerical calculations of the current densities/overpotential dependences are presented in Fig. 6.

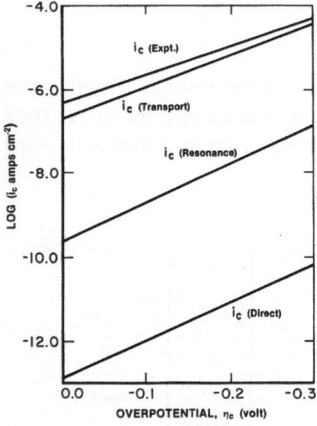

FIGURE 6. Current density–overpotential dependences obtained for transport, direct, and resonance tunneling processes at a metal electrode covered with a 100-Å-thick protein-containing biomembrane and their comparison with experimental results.[44, 45]

The above results reveal that (1) electrons can in fact transfer through the protein-containing biological membrane and react with the redox species at the membrane/solution interface, and (2) two pathways for electron transfer across the biomembrane are possible, that is, either the transport or the resonance tunneling mechanism.

The transport mechanism will be the most viable path if the conduction band edge of the protein at the metal/membrane interface is very close to the Fermi level of the metal.

4. CONCLUDING REMARKS

From the time of Gurney (1931) to date, the quantum-mechanical theories of electron transfer processes in solutions and at electrodes in terms of molecular models have progressed slowly to a reasonable extent. In particular, attempts have been made to compute the frequency of activation of ions in a condensed medium in terms of the phonon–vibron coupling (PVC) model using time-dependent perturbation theory. Alternative methods of computation of the inner-sphere reorganization energy of ions in solution using the classical improved dipole orientation model (IADO) and quantum-chemical INDO/2 MO methods have been developed. Progress has been made on the development of quantum-mechanical theory of direct and resonance tunneling of electrons across an oxide-covered and biomembrane-covered electrode.

However, though considerable attention has been given to the nonmolecular continuum theory of outer-sphere electron transfer reactions, few attempts have been made to develop the quantum-mechanical theory of bond-breaking electron transfer reactions at electrodes that involve adsorption and thereby electrocatalysis.

Though some attempts have been made to do quantum-mechanical computation of transmission coefficients for outer-sphere electron transfer reactions at electrodes[15] and in solution,[49,50] it is essential to carry out similar computations for the bond-breaking reactions that constitute the major electrochemical reactions.

REFERENCES

1. R. W. Gurney, *Proc. Roy. Soc. (London)*, *Ser. A* **134**, 137 (1931).
2. J. A. V. Butler, *Proc. Roy. Soc. (London)*, *Ser. A* **157**, 423 (1936).
3. H. Gerischer, *Z. Phys. Chem.* **26**, 223 (1960); **29**, 325 (1960).
4. H. Gerischer, *Z. Phys. Chem.* **27**, 48 (1961).
5. S. G. Christov, *Electrochim. Acta* **4**, 306 (1961).
6. S. G. Christov, *Electrochim. Acta* **9**, 575 (1964).
7. S. G. Christov, *J. Res. Inst. Catal. Hokkaido Univ.* **16**, 169 (1968).
8. J. O'M. Bockris and D. B. Matthews, *Proc. Roy. Soc. (London)*, *Ser. A* **292**, 479 (1966).
9. J. O'M. Bockris and D. B. Matthews, *J. Phys. Chem.* **44**, 298 (1966).
10. J. O'M. Bockris and S. Srinivasan, *J. Electrochem. Soc.* **111**, 853 (1964).
11. J. O'M. Bockris, S. Srinivasan, and D. B. Matthews, *Discuss. Faraday Soc.* **39**, 239 (1965).
12. J. O'M. Bockris and D. B. Matthews, *Electrochim. Acta* **11**, 143 (1966).
13. W. Schmickler, *J. Electroanal. Chem.* **82**, 65 (1977).
14. W. Schmickler, *J. Electroanal. Chem.* **84**, 203 (1977).
15. S. U. M. Khan, P. Wright, and J. O'M. Bockris, *Electrokhimiya* **13**, 914 (1977).
16. M. S. Tunuli and S. U. M. Khan, *J. Phys. Chem.* **89**, 4667 (1985).
17. M. S. Tunuli and S. U. M. Khan, *Trans. Faraday Soc.* **82**, 2911 (1986).
18. M. S. Tunuli and S. U. M. Khan, *J. Phys. Chem.* **91**, 3474 (1987).
19. Z. Zhou and S. U. M. Khan, *J. Phys. Chem.* **93**, 5292 (1989).
20. A. Ovchinnikov and V. A. Benderskii, *J. Electroanal. Chem.* **100**, 563 (1979).

21. J. Weiss, *J. Phys. Chem.* **19,** 1066 (1951).
22. W. Libby, *J. Phys. Chem.* **56,** 863 (1952).
23. R. Platzmann and J. Frank, *Z. Phys.* **138,** 411 (1958).
24. R. Kubo and Y. Toyozawa, *Prog. Theor. Phys.* **138,** 411 (1955).
25. R. A. Marcus, *J. Chem. Phys.* **24,** 966 (1956); **26,** 867 (1957).
26. R. A. Marcus, *J. Chem. Phys.* **38,** 1353, 1858 (1963); **43,** 679 (1965).
27. V. G. Levich and R. R. Dogonadze, *Dokl. Akad. Nauk SSSR* **124,** 123 (1959).
28. V. G. Levich and R. R. Dogonadze, *Dokl. Akad. Nauk SSSR* **133,** 158 (1960); *Collect. Czech. Chem. Commun.* **26,** 193 (1961).
29. R. R. Dogonadze, A. M. Kuznetsov, and V. G. Levich, *Electrochim. Acta* **13,** 1025 (1968).
30. R. R. Dogonadze, in: *Reactions of Molecules at Electrodes* (N. S. Hush, ed.), Chapter 3, Wiley Interscience, New York (1971).
31. M. Lax, *J. Chem. Phys.* **20,** 1752 (1952).
32. S. I. Parker, *Investigations of Electronic Theories of Crystals* (Russian ed.), Fizmatgiz, Moscow (1951).
33. R. R. Dogonadze, in: *Comprehensive Treatise of Electrochemistry, Vol. 7* (B. E. Conway, J. O'M. Bockris, E. Yeager, S. U. M. Khan, and R. E. White, eds.), Chapter 1, Plenum Press, New York (1983).
34. J. Ulstrup, *Charge Transfer Processes in Condensed Media*, Springer, Berlin (1979).
35. S. U. M. Khan, *Appl. Phys. Commun.* **4,** 149 (1984).
36. B. Goodman, A. W. Lawson, and L. I. Schiff, *Phys. Rev.* **71,** 191 (1946).
37. J. O'M. Bockris and D. B. Matthews, *Electrochim. Acta* **11,** 143 (1966).
38. J. O'M. Bockris and S. U. M. Khan, *Quantum Electrochemistry*, Chapter 4, Plenum Press, New York (1979).
39. E. Merzbacher, *Quantum Mechanics*, John Wiley & Sons, New York (1970).
40. P. George and J. S. Griffith, in: *The Enzymes*, Vol. 1, (P. D. Boyer, H. Lardy, and K. Myrback, eds.), p. 347, Academic Press, New York (1956).
41. T. F. Moran and W. H. Hamill, *J. Chem. Phys.* **39,** 1413 (1963).
42. D. R. Bates, *Chem. Phys. Lett.* **82,** 396 (1981).
43. S. U. M. Khan, *J. Phys. Chem.* **92,** 2541 (1988).
44. M. A. Habib and J. O'M. Bockris, *J. Bioelectricity* **3,** 247 (1984).
45. M. A. Habib and J. O'M. Bockris, *J. Biophys.* **14,** 31 (1986).
46. W. Schmickler, *J. Electroanal. Chem.* **83,** 387 (1977).
47. W. Schmickler and J. Ulstrup, *J. Chem. Phys.* **19,** 217 (1977).
48. S. U. M. Khan and W. Schmickler, *J. Electroanal. Chem.* **134,** 167 (1982).
49. M. D. Newton, *Int. J. Quantum. Chem. Quantum Chem. Symp.* **14,** 363 (1980); *Chem. Rev.* **91,** 767 (1991).
50. S. U. M. Khan and Z. Y. Zhou, *J. Chem. Phys.*, **93,** 8808 (1990).

IV

Photoelectrochemistry

14

Novel Approaches for the Study of Surface Structure and Reactivity of Semiconductor Electrodes

Kohei Uosaki, Per Carlsson, Hideaki Kita, and Bertil Holmström

1. INTRODUCTION

Photoelectrochemical reactions at semiconductor electrodes have been intensively studied during the last 15 years in relation to solar energy conversion.[1] The photoexcited minority carrier plays a major role in photoelectrochemical reactions, and the lifetime of the minority carrier is controlled by the potential and charge distribution within the semiconductor and the surface recombination velocity.[2] Reactivities of solids for heterogeneous reactions, including electrochemical reactions, are strongly affected by the surface structures of the solids. Moreover, the reactivity of a solid for a particular reaction often changes while the reaction proceeds, and it is generally believed that this change is related to changes in the surface structure of the solid during the reaction. Thus, to understand the detailed mechanism of a photoelectrochemical reaction at a semiconductor electrode, it is essential to know the surface structure of the electrode, the potential and charge distribution at the semiconductor/electrolyte interface, and the potential dependence of the surface recombination velocity.

In this study, we applied photoluminescence (PL) measurements[3-6] to probe the potential and charge distribution at the semiconductor/electrolyte interface and to determine the surface recombination velocity and the scanning laser spot (SLS) technique[7] and scanning tunneling microscopy (STM)[8] to obtain information on the surface structure with micron (μm) and atomic (nm) resolution, respectively.

2. EXPERIMENTAL

2.1. Materials

Zn-doped p-GaAs single crystals (4.66×10^{18} cm^{-3}, Morgan Semiconductor), Si-doped n-GaAs single crystals (4×10^{18}–8×10^{18} cm^{-3} or 0.084×10^{18} cm^{-3}, Laser Diode), and p-InSe

Kohei Uosaki and Hideaki Kita ● Department of Chemistry, Faculty of Science, Hokkaido University, Sapporo 060, Japan.   *Per Carlsson and Bertil Holmström* ● Department of Physical Chemistry, Chalmers University of Technology/Göteborg University, Göteborg, S-41296, Sweden.

Electrochemistry in Transition, edited by Oliver J. Murphy *et al.* Plenum Press, New York, 1992.

(donated by Dr. A. Chevy) were used in this study. An ohmic contact was obtained using an In–Zn alloy for p-GaAs and p-InSe and using In for n-GaAs. The GaAs surface was etched in HNO_3–HCl (1:1) before each experiment. The electrolyte solution was prepared using reagent-grade chemicals and water purified by a Milli-Q water purification system. Pt treatment of InSe and GaAs was carried out by dipping the electrode into 20mM H_2PtCl_6 for 5 min and 15 s, respectively, followed by rinsing with water.

2.2. Electrochemical Measurements

An ordinary two-compartment, three-electrode cell with a Pyrex window was used for the measurements. The counter and reference electrodes were a Pt foil and an Ag/AgCl electrode, respectively. The electrode potential was controlled by using a potentiostat (Nikko Keisoku, NPGS-301S). An external potential was provided either by a function generator (Nikko Keisoku, NFG-3) or by a programmable function generator (Hokuto Denko, HB-501) which was controlled by a personal computer (NEC, PC-8801 mkII) via a GP-IB interface. Values of current and potential were collected by the personal computer via a 12-bit A/D converter and stored on a floppy disk.

2.3. PL Measurements

A block diagram for a typical luminescence measurement is shown in Fig. 1. A 500-W Xe lamp was used as an excitation light source. Monochromatic light was obtained by passing the light through an appropriate interference filter. The light intensity was kept constant by adjusting the lamp current to avoid the possible effect of excitation light intensity on the PL behavior. The luminescence intensity was monitored by using a GaAs-based photomultiplier tube (PMT; Hamamatsu Photonics, R636) which has a flat response between 300 and 800 nm. To improve the S/N ratio, the excitation light was chopped by using a light chopper (NF Electronics, CH-353), and the PMT signal intensity was determined by using a photon counter (NF Electronics, PC-545A). The personal computer was used to control the photon counter and for the data acquisition.

2.4. SLS System

The experimental configuration for the SLS system[7] is shown in Fig. 2. The system consists of a single-mode optical fiber which is tuned to 632 nm and has a core diameter of 4 μm and a Selfoc micro lens (Nippon Sheet Glass). The micro lens is cylindrical with a

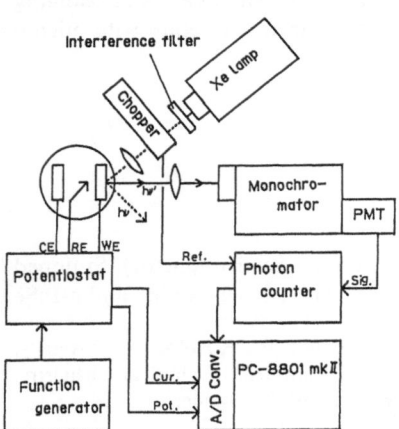

FIGURE 1. Block diagram for electrochemical and photo-luminescence measurements.

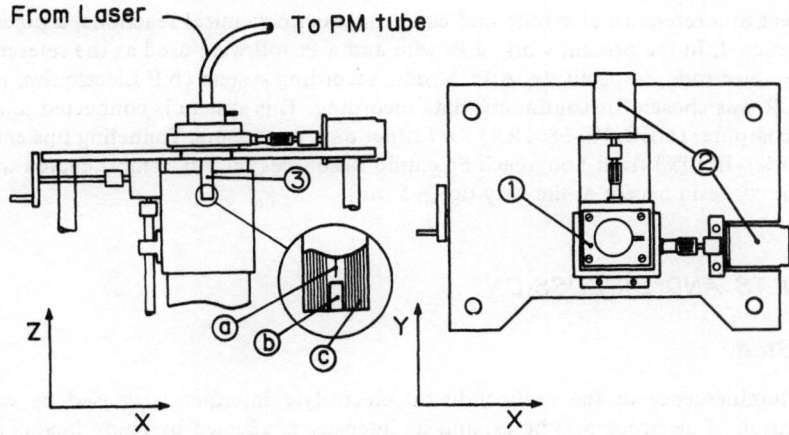

From Laser — **To PM tube**

FIGURE 2. Block diagram of SLS system. (1) X-Y coordinate table. (2) Step motors. (3) Laser spot probe with single-mode optical fiber (a), Selfoc micro lens (b), and assembly of optical fibers (c).

graded index of refraction which decreases with the square of the radial distance from the optical axis. The laser spot on the electrode is approximately a 1:1 image of the fiber core, and the smallest spot diameter obtained was 4.15 μm. A bundle ($d = 8.5$ mm) of fibers is arranged around the micro lens in order to collect the reflected light. The X-Y scan was accomplished using a coordinate table equipped with step motors, and the minimum step width was 0.5 μm. The data acquisition and control was handled by a personal computer (NEC, PC-8801 MR). The light from a He–Ne laser (NEC, 5 mW) was chopped, and the photocurrent and PMT output of the reflected light were amplified with a lock-in amplifier (NF Electronics, LI-574).

2.5. STM System

A block diagram of the STM system[8] is shown in Fig. 3. The STM used in this study was a NanoScope I from Digital Instruments. To control the potential of the tip and a sample

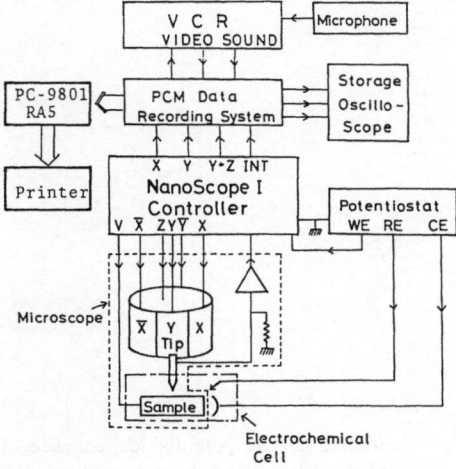

FIGURE 3. Block diagram of STM system.

with respect to a reference electrode and carry out electrochemical reactions, a potentiostat was constructed. In the present work, a Pt wire and a Pt foil were used as the reference and the counter electrode, respectively. A PCM data recording system (NF Electronics, RP-880) with a VCR was chosen for continuous data recording. This system is connected to a 32-bit personal computer (NEC, PC-9801RA) for further data processing. Tunneling tips employed were Pt(90%)–Ir(10%) from Longreach Scientific Resources ($\phi = 0.25$ mm) coated with soft glass or epoxy resin except at the very tip ($\sim 5 \ \mu$m).

3. RESULTS AND DISCUSSION

3.1. PL Study

The luminescence at the semiconductor/electrolyte interface is caused by radiative recombination of electrons and holes, and its intensity is affected by many factors such as the concentration of electrons in the conduction band and that of holes in the valence band, the electric field within semiconductor, the surface recombination velocity, and the minority carrier lifetime. Thus, the study of the luminescent properties provides much useful information on the electrochemical characteristics of semiconductor/electrolyte interfaces.[9-13]

Figure 4 shows the potential dependence of dark current, photocurrent, and PL intensity at etched and Pt-treated p-GaAs electrodes in $1 M$ NaOH. While dark current was increased significantly, photocurrent was decreased by the Pt treatment. Similar results were reported for Ru-treated GaAs.[14] The potential dependence of the PL intensity of the Pt-treated p-GaAs electrode was smaller than that of the etched electrode. At the etched p-GaAs electrode, large hysteresis was observed in the PL–potential relation if the anodic limit was too positive, but hysteresis was very small at Pt-treated p-GaAs even when the anodic limit was relatively positive. Figure 5 shows the potential dependence of the PL intensity of the etched and Pt-treated p-GaAs electrodes in $1 M$ NaOH measured with several excitation wavelengths. At both electrodes, the potential dependence of PL intensities increased as the excitation wavelengths became shorter. Hysteresis was observed in the PL intensity–potential relations

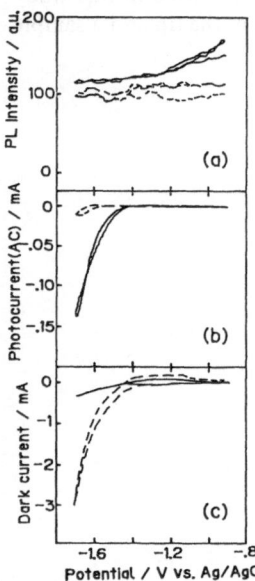

FIGURE 4. The potential dependence of PL intensity (a), photocurrent (b), and dark current (c) at etched (——) and Pt-treated (– – –) p-GaAs electrodes in $1 M$ NaOH.[6]

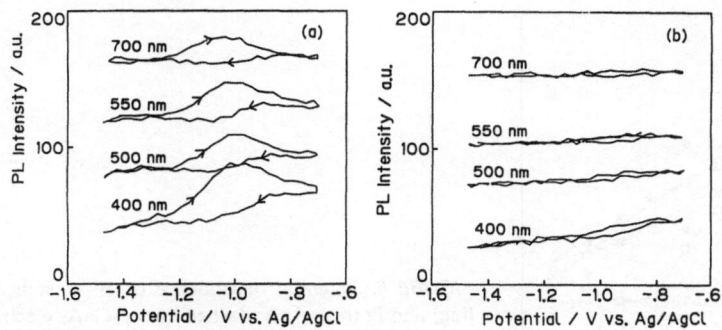

FIGURE 5. Potential dependence of PL intensity of etched (a) and Pt-treated (b) p-GaAs electrodes in $1M$ NaOH measured with several excitation wavelengths monitored at 870 nm.[6]

at etched p-GaAs, but almost reversible behavior was obtained at the Pt-treated electrode. These results suggest that while the surface nature of the etched p-GaAs changed during the potential sweep, that of the Pt-treated electrode was relatively constant.

The space-charge layer widths (W) and surface recombination velocities (S_{rv}) can be determined by analyzing the results obtained above as follows. The potential dependence of PL intensity under reverse bias can be generally explained in terms of a "dead layer" model.[15,16] The model assumes that electron–hole pairs formed within the region where the electric field exists in a semiconductor are swept apart so rapidly that they do not recombine radiatively. This nonemissive region, which is almost equivalent to the space-charge layer, is called the "dead layer" in this model. According to this model, the potential dependence of PL intensity is related to the potential dependence of the thickness of the dead layer. This model is valid only when the surface recombination velocity does not depend on the electrode potential. A more advanced model, proposed by Mettler,[17] takes into account not only the thickness of the space-charge layer but also the surface recombination velocity. According to Mettler's model, the PL intensity, I_{PL}, is given by[17]

$$I_{PL} = K\{\alpha_e L/[(\alpha_e L)^2 - 1]\}\{[(S_{rv} + \alpha_e L)/(S_{rv} + 1)(\alpha_p L + 1)] - [1/(\alpha_e + \alpha_p)L]\}$$
$$\times \exp[-(\alpha_e + \alpha_p)W] \tag{1}$$

where K is a constant containing the integral quantum efficiency and geometrical factor, L is the diffusion length of the minority carrier, α_e and α_p are the absorption coefficients at the wavelength of the excitation light and the emitted light, respectively, W is the space-charge layer width, and S_{rv} is the reduced surface recombination velocity, defined as

$$S_{rv} = S_v(\tau/L) \tag{2}$$

where S_v is a virtual surface recombination velocity, and τ is the lifetime of the minority carrier. W and S_{rv} can be determined by fitting the PL intensity obtained experimentally at various excitation wavelengths, that is, at various penetration depths, α_e^{-1}, to Eq. (1). Literature values are used for the absorption coefficients at given wavelengths.[18] L was assumed as 4 μm, and the justification of this value was provided by Johnson et al.[19] The fitting was carried out using SAS (Statistical Analysis System) at Hokkaido University Computing Center. The potential dependence of W and S_{rv} at etched and Pt-treated p-GaAs thus obtained is shown in Figs. 6 and 7, respectively. While large hysteresis was observed in the potential dependence of W and S_{rv} at etched p-GaAs, almost no hysteresis was observed at Pt-treated p-GaAs.

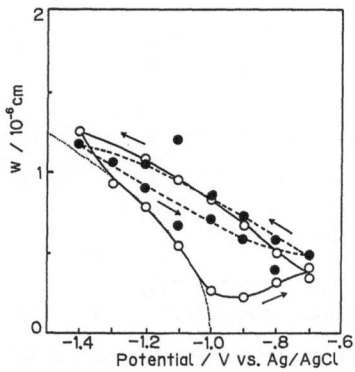

FIGURE 6. Potential dependence of W at etched (O, solid line) and Pt treated (●, dashed line) p-GaAs electrodes in $1M$ NaOH. Dotted line shows calculated values based on Eq. (3), assuming $V_{fb} = -1.0$ V.[6]

The position of the band edge is one of the most important properties in controlling potential distribution and kinetics at semiconductor electrodes, and the possibility of a shift in the flat-band potential (FBP), that is, a band-edge shift, under illumination has been suggested.[20] The results presented in Fig. 7 support the FBP shift. When the electrode potential was swept from −0.7 V to more negative potentials, W gradually increased. On the return (anodic) sweep, W decreased as the potential became positive, showed a minimum at −1.0 V, and then gradually increased. At a given potential, the PL intensity on the cathodic sweep is much larger than that on the anodic sweep. Thus, there exists a large hysteresis. The FBP is determined as ca. −1.0 V in $1M$ NaOH based on the potential dependence of W on the anodic sweep. The potential dependence of the space-charge layer width calculated using Eq. (3), which gives the value of W based on a simple Schottky junction model with the assumption that $V_{fb} = -1.0$ V, is shown as a dotted line in Fig. 6.

$$W = (2\varepsilon\varepsilon_0 \Delta\phi_{sc}/eN)^{1/2} \tag{3}$$

Here ε is the dielectric constant of the semiconductor, ε_0 is the permittivity, N is the carrier concentration, and $\Delta\phi_{sc}$ is the barrier height. The calculated values agree well with experimental results obtained during the anodic sweep, supporting that the FBP under illumination is −1.0 V. On the other hand, the FBP determined by impedance measurements in the dark is about −0.35 V. The difference between the value in the dark and that under illumination means that the FBP shifts in the negative direction by 0.6 V under illumination.

Thus, the potential dependence of the PL intensity can be explained by using the band diagram shown in Fig. 8. When the potential is swept positively, the band bending and W in p-GaAs decrease and become almost zero around V_{fb}^*, which is the FBP under illumination and is −1.0 V in $1M$ NaOH. The FBP shifts gradually toward the value in the dark, V_{fb}, which is −0.35 V in $1M$ NaOH, even under illumination if the potential is relatively positive. The band bending, W, and, therefore, the PL intensity increase with time although the potential is swept to more positive values. The present results also confirm the results of Johnson et al.[19] as S_{rv} also changes around −1.0 V, suggesting a change in the nature of the

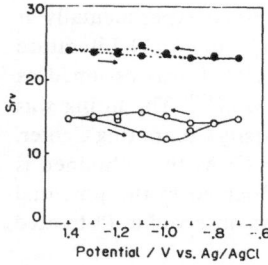

FIGURE 7. Potential dependence of S_{rv} at etched (O) and Pt-treated (●) p-GaAs electrodes in $1M$ NaOH.[6].

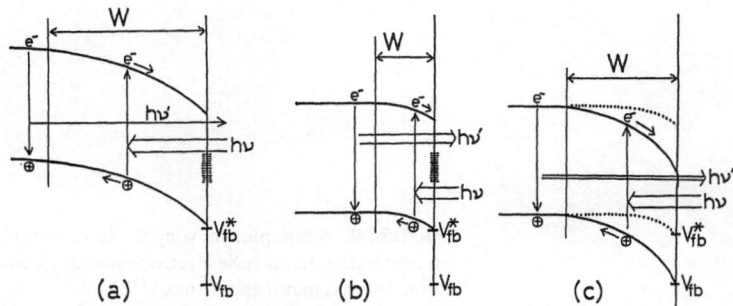

FIGURE 8. Band diagrams of p-GaAs/$1M$ NaOH solution interface under illumination: (a) at a relatively negative potential; (b) at a relatively positive potential which is close to V_{fb}; (c) at the same potential as in (b) but after a certain time. V_{fb}^* is the flat-band potential under illumination and is ca. -1.0 V. V_{fb} in the dark is -0.35 V.[6]

surface in this potential region. Thus, this surface change causes both the FBP shift and the S_{rv} change. If the potential sweep is reversed after the band-edge shift, the PL intensity at a given potential should be smaller than that on the positive sweep, since the space-charge layer is thicker on the negative sweep than on the positive sweep. This leads to the hysteresis observed in the PL–potential relations. The band edge should shift again in the negative potential direction while the photoelectrochemical hydrogen evolution reaction (HER) takes place. Although further study is needed to understand this surface change in more detail, the most probable reaction involved is formation and reduction of the oxide layer on GaAs in this potential region.[21] Thus, the steeper increase in the PL intensity with increase of positive potential at the etched electrode is considered to be due to the change of the surface nature of GaAs, possibly electrochemical reduction of the GaAs surface itself under illumination, leading to the band-edge shift and the change in the surface recombination velocity.

At the Pt-treated p-GaAs electrode, however, almost no hysteresis was observed in the potential dependence of W and S_{rv} as shown in Figs. 6 and 7, respectively. The Pt treatment seems to prevent the surface from changing its nature, suggesting that the HER takes place at Pt sites introduced by the Pt treatment. The Pt treatment increased the HER rate in the dark quite significantly but decreased the photocurrent. These results can be explained either by the introduction of a surface state which acts as a surface recombination center, as we proposed before, or by the formation of a leaky barrier, that is, band-edge unpinning. The results presented in Fig. 7 clearly demonstrate that the S_{rv} value at the Pt-treated surface is larger than that at the etched surface, confirming the introduction of recombination centers by Pt treatment. We have also demonstrated before that electroluminescence caused by electron injection from adsorbed hydrogen at p-GaAs is quenched by Ru treatment and suggested that the noble metal treatment introduces recombination centers.[22] The potential dependence of W at Pt-treated p-GaAs shows that the FBP of this electrode in $1M$ NaOH is -0.35 V both in the dark and under illumination; this is the same as the value for the etched electrode in the dark. Actually, the potential dependence of the PL intensity at Pt-treated p-GaAs can be explained very well by the simple dead layer model with $V_{fb} = -0.35$ V. This means that the band edge of the Pt-treated electrode is fixed at the position of etched p-GaAs in the dark. Thus, the increase of dark current and the decrease of photocurrent by Pt treatment are due to the introduction of a surface state which acts as a surface recombination center. Impedance analysis showed that the density of surface states introduced by Pt treatment is as high as 2.45×10^{13} cm^{-2}. It must be noted here that this value is almost equivalent to the surface concentration of Pt determined by X-ray photoelectron spectroscopy (XPS).

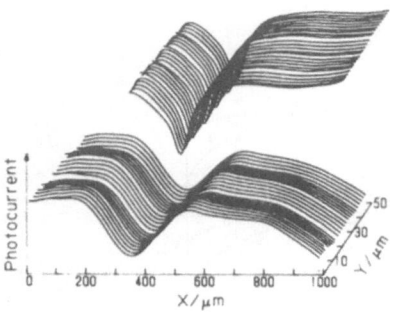

FIGURE 9. A line plot showing the laser spot photocurrent across a step on an InSe electrode before (lower lines) and after Pt treatment (upper lines).[7]

3.2. SLS Study

The presence of scratches, grain boundaries, and, in layered compounds, steps is known to reduce the photocurrent as these defects induce recombination. The scanning laser spot (SLS) technique permits laterally resolved studies of the photoelectrochemical properties of semiconductors, and it is possible to visualize the effect of these defects on the photocurrent.[23-30] Since InSe is a layered compound, the characteristic surface morphology is steps between adjacent layers. This type of compound is especially well suited for SLS studies as ordered surface structures are very easy to prepare by lifting off layers with sticking tape. The laser spot (LS) photocurrent which was observed when the spot was swept over a step is shown in Fig. 9. The upper line scan shows the photocurrent after Pt dip treatment, and the lower that before Pt treatment. The step influences the photocurrent over a distance that is much larger than the dimension of the step and the light spot. After the Pt treatment, the drop in the photocurrent becomes sharper and more localized at the step. These results can be explained by considering the formation of a charged semisphere in the semiconductor upon illumination as shown in Fig. 10. Some of the photoexcited minority carriers, electrons in this case, diffuse away from the excited spot before being transferred to the electrolyte solution. If a recombination center, such as a step, exists within the semisphere, the number of electrons that recombine within the semiconductor increases, and, therefore, the number of electrons which are transferred to the electrolyte solution, that is, the photocurrent, decreases. Thus, the resolution of this measurement is determined neither by the spot size nor by the dimension of the steps but by the size of the semisphere. The size of the semisphere is controlled by the diffusion coefficient and the lifetime of the photogenerated minority carrier. Therefore, if the rate of electron transfer to the electrolyte solution increases, the size of the semisphere decreases and resolution is improved. It is a well-known fact that Pt

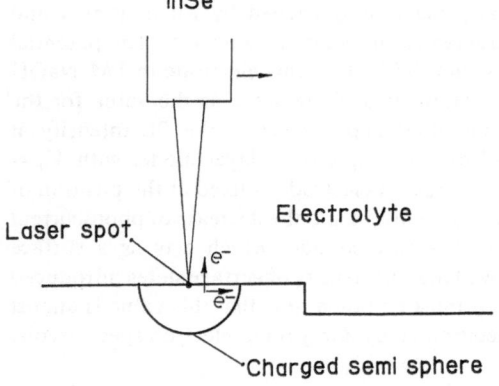

FIGURE 10. A model showing the generation of a charged semisphere within InSe upon illumination of laser spot.

treatment improves the photocathodic current on p-type semiconductors by its catalytic effect on the hydrogen evolution reaction.[31-33] As Pt treatment increases the photocathodic charge transfer reaction rate, the charged sphere diameter decreases and the resolution increases. More quantitative treatment is given elsewhere.[34]

The SLS probe with an array of optical fibers around the central emitting fiber makes it possible to record both the reflected light intensity and the photocurrent at the same time. With this optical arrangement, it is possible to observe light which has been reflected by irregularities, diffracted, or emitted as photoluminescence. Figure 11 shows an example of the SLS image of the reflected light intensity and laser spot photocurrent at an InSe electrode which was dip treated in H_2PtCl_6 solution. In this case, it was confirmed that the light detected was not photoluminescence because no light was detected if a long-pass filter ($\lambda > 690$ nm) was placed in front of the PMT. The electrode surface was first observed by an optical microscope, and the increased reflected light image in Fig. 11 is associated with a folded surface layer which rises like a hill over the surrounding surface. The two main peaks in the reflected light image are due to reflection at the slopes of the hill, and the minimum between them is the reflectivity at the top of the hill. In the left part of the reflected light image in Fig. 11, there is a sharp hump in the intensity, which is related to a step. The increase in the photocurrent might be related to the fact that the light spot covers a larger area at the slopes of the hill. Another possible explanation is uneven distribution of Pt. We have studied this effect by putting a drop, approximately 1 mm in diameter, of the Pt solution for 5 min on the electrode surface. The electrode was rinsed with water and analyzed with SLS. We observed much higher photocurrent at the area which was covered with the Pt drop compared to the surface which was untreated. Further experiments are required to explain the reason for the increased photocurrent observed in Fig. 11.

3.3. STM Study

As mentioned above, the resolution obtained in SLS measurements is on the order of microns, at best, and information on the surface structure with much higher resolution is required in many cases. The recently developed scanning tunneling microscope (STM) seems to be most suitable for this purpose. The real space image of conductive surfaces with atomic scale resolution can be obtained by STM with relative ease, and, more importantly, measurements by STM are possible in various media including electrolyte solutions.[35] *In situ* observation of an electrode surface by STM was first reported by Sonnenfeld and co-workers

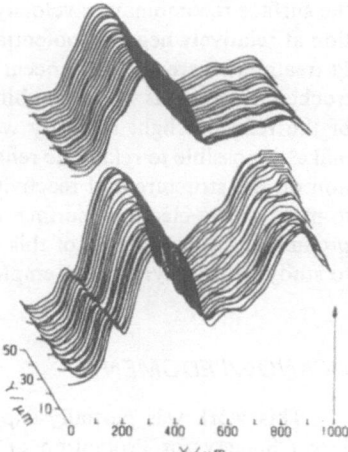

FIGURE 11. The reflected light (lower lines) and laser spot photocurrent (upper lines) from the same area.[7]

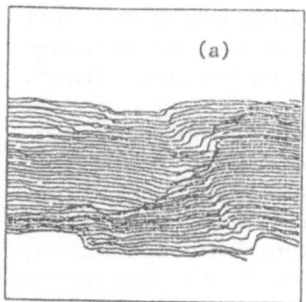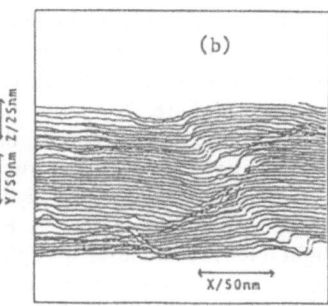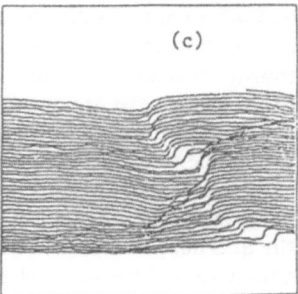

FIGURE 12. STM images of n-GaAs in 2 mM HClO$_4$ under illumination: (a) just after reaction started; (b) after 20 s; (c) after 50 s.

for Ag and Au deposition on highly oriented pyrolytic graphite (HOPG).[36,37] The first STM investigation of a semiconductor electrode surface under potentiostatic conditions was carried out by Itaya and Tomita.[38] They used TiO$_2$ as an electrode and found that the STM image could not be obtained when the electrode was under reverse-bias condition. We experienced a similar situation at the n-GaAs surface but were able to obtain STM images if reverse bias was relatively large. This may be due to the smaller energy gap of GaAs and/or the existence of midgap surface states acting as a mediator for electron tunneling.[39] We were also able to obtain STM images of the irradiated surface of n-GaAs under anodic bias continuously in 2mM HClO$_4$ while photoanodic current (10–20 μA/0.07 cm^2) due to dissolution was flowing. As shown in Fig. 12, flat portions and steps can be seen.[40] Although a large structural change was not observed because of the relatively short measuring time, it is clear that the hilly area observed at the beginning (Fig. 12a) became smaller after 20 s (Fig. 12b) and disappeared after 50 s (Fig. 12c). More detailed study is under way.

4. CONCLUSION

In this study, three nonelectrochemical methods were employed to investigate semiconductor/electrolyte interfaces. PL measurements were carried out to probe the potential and charge distribution at semiconductor/electrolyte interfaces and the potential dependence of the surface recombination velocity. At a p-GaAs electrode, a band-edge shift under illumination at relatively negative potential and an increase in the surface recombination velocity by Pt treatment were shown to occur. A new fiber-optical concept for SLS studies of photoelectrochemical systems which combines laser spot photocurrent measurement with measurement of the reflected light intensity was presented. It was demonstrated that this combination makes it possible to relate the reflectivity to photocurrent and provides much useful information on the structure and reactivity of semiconductor electrodes. Finally, STM was applied to monitor the electrode surface *in situ* at much higher resolution. Although the results are preliminary, the potential of this technique is clearly demonstrated. We are now attempting to study various systems by employing these techniques simultaneously.

ACKNOWLEDGMENTS

This work was partially supported by The Mazda Foundation's Research Grant, the Japan Society for Promotion of Science, the Swedish Board for Technical Development

(STU), and the International Scientific Research Program (Joint Research 01044005) of the Ministry of Education, Science, and Culture, Japan.

REFERENCES

1. A. Hamnett, in: *Comprehensive Chemical Kinetics*, Vol. 27 (R. G. Compton, ed.), Chapter 2, Elsevier, Amsterdam (1987).
2. K. Uosaki and H. Kita, in: *Modern Aspects of Electrochemistry*, No. 18 (R. E. White, J. O'M. Bockris, and B. E. Conway, eds.), Chapter 1, Plenum Press, New York (1986).
3. S. Kaneko, K. Uosaki, and H. Kita, *J. Phys. Chem.* **90,** 6654 (1986).
4. S. Kaneko, K. Uosaki, and H. Kita, *Chem. Lett.* **1986,** 1951.
5. K. Uosaki, Y. Shigematsu, and H. Kita, *Chem. Lett.* **1988,** 1815.
6. K. Uosaki, Y. Shigematsu, S. Kaneko, and H. Kita, *J. Phys. Chem.*, **93,** 6521 (1989).
7. P. Carlsson, B. Holmstrom, K. Uosaki, and H. Kita, *Appl. Phys. Lett.* **53,** 965 (1988).
8. K. Uosaki and H. Kita, *J. Electroanal. Chem.* **259,** 301 (1989).
9. K. H. Beckmann and R. Memming, *J. Electrochem. Soc.* **116,** 368 (1969).
10. H. H. Streckert, J. Tong, and A. B. Ellis, *J. Am. Chem. Soc.* **104,** 581 (1982).
11. Y. Nakato, A. Tsumura, and H. Tsubomura, *Chem. Phys. Lett.* **85,** 387 (1984).
12. W. S. Hobson and A. B. Ellis, *J. Appl. Phys.* **54,** 5956 (1984).
13. A. A. Burk, P. B. Johnson, W. S. Hobson, and A. B. Ellis, *J. Appl. Phys.* **59,** 1621 (1986).
14. K. Uosaki and H. Kita, *Chem. Lett.* **1984,** 953.
15. D. B. Wittry and D. F. Kyser, *J. Appl. Phys.* **38,** 375 (1967).
16. R. E. Hollingsworth and J. R. Sites, *J. Appl. Phys.* **57,** 5357 (1982).
17. K. Mettler, *Appl. Phys.* **12,** 75 (1977).
18. E. D. Palik, in: *Handbook of Optical Constants of Solids* (E. D. Palik, ed.), p. 429, Academic Press, Orlando (1985).
19. P. B. Johnson, C. S. McMillan, A. B. Ellis, and W. S. Hobson, *J. Appl. Phys.* **62,** 4903 (1987).
20. J. J. Kelly and R. Memming, *J. Electrochem. Soc.* **129,** 730 (1982).
21. H. Gerischer and I. Mattes, *Z. Phys. Chem.*, *N.F.* **49,** 112 (1966).
22. K. Uosaki and H. Kita, *J. Am. Chem. Soc.* **108,** 4294 (1986).
23. T. E. Furtak, D. C. Canfield, and B. A. Parkinson, *J. Appl. Phys.* **51,** 6018 (1980).
24. M. A. Butler, *J. Electrochem. Soc.* **130,** 2358 (1983).
25. M. A. Butler, *J. Electrochem. Soc.* **131,** 2158 (1984).
26. C. A. Kavassalis, D. H. Longendorfer, R. A. Le Lievre, and R. D. Rauh, *Mater. Res. Soc. Symp. Proc.* **29,** 151 (1984).
27. G. Vercruysse, W. Rigole, and W. P. Gomes, *Sol. Energy Mater.* **12,** 157 (1985).
28. P. S. Tyler, M. R. Kozlowski, W. H. Smyrl, and R. T. Atanasoki, *J. Electroanal. Chem.* **237,** 295 (1987).
29. D. Shukla, T. Wines, and U. Stimming, *J. Electrochem. Soc.* **134,** 2086 (1987).
30. P. Carlsson and B. Halmstrom, VIth International Conference on Photochemical Conversion and Storage of Solar Energy, Paris, 1987, D-7.
31. A. Heller, E. Aharon-Shalom, W. A. Bonner, and B. Miller, *J. Am. Chem. Soc.* **104,** 6942 (1982).
32. R. N. Dominey, N. S. Lewis, J. A. Bruce, D. C. Bookbinder, and M. S. Wrighton, *J. Am. Chem. Soc.* **104,** 467 (1982).
33. K. Uosaki, S. Kaneko, H. Kita, and A. Chevy, *Bull. Chem. Soc. Jpn.* **59,** 599 (1986).
34. S. Eriksson, P. Carlsson, B. Holmström, and K. Uosaki, *J. Appl. Phys.* **69,** 2324 (1991).
35. P. D. Hansma and J. Tersoff, *J. Appl. Phys.* **61,** R1 (1987).
36. R. Sonnenfeld and B. C. Schardt, *Appl. Phys. Lett.* **49,** 1172 (1986).
37. B. Drake, R. Sonnenfeld, J. Schneir, and P. K. Hansma, *Surf. Sci.* **181,** 92 (1987).
38. K. Itaya and E. Tomita, *Chem. Lett.* **1989,** 285.
39. P. Carlsson, B. Holmström, H. Kita, and K. Uosaki, *Surf. Sci.* **237,** 280 (1990).
40. K. Uosaki andd H. Kita, *Denki Kagaku*, **57,** 1213 (1989).

15

Photoelectrocatalysis

Marek Szklarczyk

1. INTRODUCTION

Photoelectrochemistry is one of the oldest branches of electrochemistry. The development of this field can be divided into three periods. The first period started with Edmond Becquerel's observation in 1859 of the flow of current at any illuminated silver electrode immersed in chloride solution.[1] The next period began in 1955, with the work of Brattain and Garrett on electrochemical photopotential.[2] In that period several experimental and theoretical papers were published. Dewald was the first to study the electrochemical behavior of wide-band-gap semiconductors,[3] and Green formulated the first $i = i(E)$ dependency for photoelectrodes.[4] Gerischer published several papers on the thermodynamics, chemical stability, and behavior of different redox systems at the semiconductor/solution interface.[5,6] Myamlin and Pleskov gave the first systematic description of the electrochemistry of semiconducting materials.[7] Memming published work on the capacitance of semiconductor electrodes.[8] In 1960 Williams pointed out that semiconductor electrodes can be used in practical devices.[9] This period in the history of photoelectrochemistry ended at the turning point in the sixties and seventies, when Fujishima and Honda demonstrated the possibility of self-driven water splitting in photoelectrochemical cells.[10] The possibility of cheap production of hydrogen fuel greatly increased interest in the field of photoelectrochemistry. In the last decade further new applications of photoelectrochemistry have been developed, for example, reduction of CO_2,[11] oxidation of H_2S,[12] and redox reaction of $HCOOH$.[13-17] In the middle of the eighties, a theoretical interpretation of the observation of quantum yields exceeding unity[18-21] was proposed.[21,22] In the next section of this chapter, results from the investigation of the dependence of the hydrogen evolution reaction (HER) on the state of the semiconductor surface will be discussed. The model explaining the dependence of the rate of the HER on surface properties, that is, a model of photoelectrocatalysis, will be given. In the last part of this chapter, results on photoelectroreduction of formic acid will be presented.

2. PHOTOCATALYSIS OF HYDROGEN EVOLUTION REACTION

The generation of hydrogen by photoirradiation of the semiconductor/solution interface has been the subject of many papers.[20,21,23-31] The aim of all these studies was to increase

Marek Szklarczyk • Department of Chemistry, Warsaw University, 02–089 Warsaw, Poland.

Electrochemistry in Transition, edited by Oliver J. Murphy *et al.* Plenum Press, New York, 1992.

TABLE 1
Etching Procedures

i-E curve[a]	Etching baths	n[b]	$\log i_0$ (A/cm^2)
A	HF, 15 min; water rinse, 1 min	1.46	−7.4
B	HNO$_3$/HF (7:2), 1 min; water rinse, 1 min	1.45	−5.9
C	HNO$_3$/HF (7:2), 1 min; HF, 1 min; water rinse, 1 min	1.45	−6.8
D	HNO$_3$/HCl/HF (1:3:2), 15 min; water rinse, 1 min	1.81	−5.4
E	HNO$_3$/HCl/HF (1:3:2), 15 min; HF, 15 min; water rinse, 1 min	2.49	−4.2

[a] See Fig. 1.
[b] n, Refractive index detected ellipsometrically[26] for particular surface after carrying out etching procedure.

the rate of the HER, which was achieved by changing the surface properties of the semiconductor electrode (e.g., by etching)[31] or by deposition of a catalyst (e.g., metal) on the semiconductor electrode.[20,21,25–30]

The models proposed to explain the effect of change in semiconductor electrode properties on its photoactivity can be divided into two groups. The first group of models relate changes in surface properties to changes of properties of the semiconductor material (e.g., band bending, surface states).[26–28,32] In these models the rate-determining step (rds) of the HER is considered to occur within the semiconductor material. In the second group of models, the influence of changes in surface properties on the electrochemical double layer, that is, the electrode/solution interface, is taken into account, and the rds of the HER is considered to be the electrochemical reaction.[29–31]

2.1. Influence of Surface Etching on the Photoactivity of p-Si Photocathode

The pronounced influence of the various modes of surface preparation of p-Si (Table 1) on the HER is seen in Fig. 1.

The chemical structure of the electrode surface after etching was determined by ellipsometry, X-ray photoelectron spectroscopy (XPS), secondary-ion mass spectrometry, and ion-scattering spectroscopy (ISS).[31–33] It was proposed that the structure of the surface changes from $(SiO_2)_n$ for samples A–C (Table 1), through $(SiO)_n$ for samples D and E, to $(SiOH)_n$ for the sample labeled "after hydrogenation" (Fig. 2).

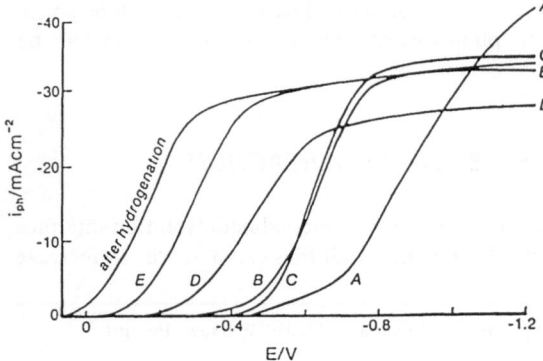

FIGURE 1. Photocurrent–potential relations for p-Si electrode following the etching treatments listed in Table 1. The run described as "after hydrogenation" represents the effect of continuing potential sweeps at 200 mV/s in the −0.04–0.7 V range (50 mW/cm^2 Xe light, 0.5 M H$_2$SO$_4$.)[31]

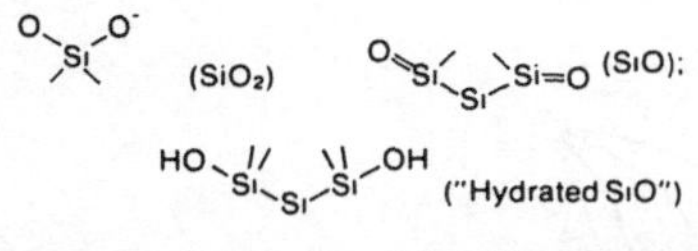

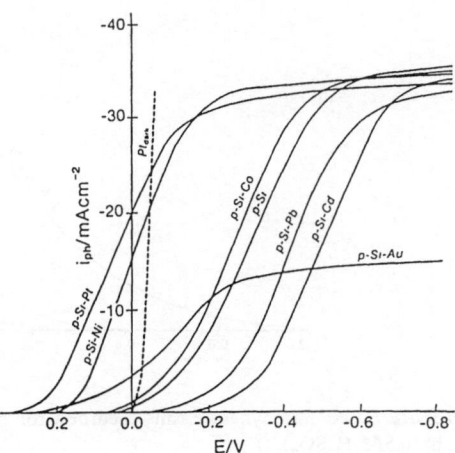

FIGURE 2. Surface compounds produced by different etching procedures.[31]

Thus, the major changes which are seen in Fig. 1 would depend upon the valency of surface atoms available for bonding the discharged proton. The surface reduction process leads to the most photoelectrochemically active electrode (cf. Fig. 1, dependence "after hydrogenation," Fig. 2 "SiOH" structure).

In Table 1 are listed the refractive indices and log i_0 values for the electrodes studied. It indicates that the electrochemical parameter log i_0 of p-Si varies with the surface structure of the p-Si electrode. Furthermore, the calculated Tafel slopes for the i_{ph}-E plots shown in Fig. 1 varies from 112 mV to 170 mV.[31]

These slopes for the photoelectrochemical current are close to those obtained for metals (112 mV) evolving H_2 in the dark with discharge of H_3O^+ as the rate-determining step.[34] This is consistent with the suggestion made above. Hence, the dependence shown in Fig. 1 supports the view that an electrochemical surface step is rate-determining in photoelectrochemical hydrogen evolution on these electrodes.

2.2. Influence of Metal Deposits on the Rate of Hydrogen Photoevolution

The possibility of electrocatalysis in the photoelectrochemical evolution of hydrogen at the photocathode surface was studied by several authors.[26-32] It was observed that deposition of some metals in an islet form caused an increase in the rate of evolution while others decreased the rate of the HER compared to that on naked semiconductor electrodes (Figs. 3-5).

According to Dominey et al.,[28] effects of metal additions to semiconductor surfaces on the photocurrent are connected with the properties of the semiconductor/solution interface. The charged Pt aggregates are not regarded as a factor in determining the efficiency of H_2 photoevolution. Alternatively, it has been proposed that changes in the Si–SiO$_x$ interface are significant, and no changes in the rate of hydrogen photoevolution should be observed from one catalyst to another.[28] However, this latter prediction has not been confirmed. In fact, deposition of different metals displaces the photocurrent–potential relation by up to 0.8 V for metals ranging from Cd to Pt (Figs. 3-5).

FIGURE 3. Potentiodynamic runs recorded for p-Si and p-Si-Me electrodes (50 mW/cm^2 Xe light, 0.5M H$_2$SO$_4$).[20,30]

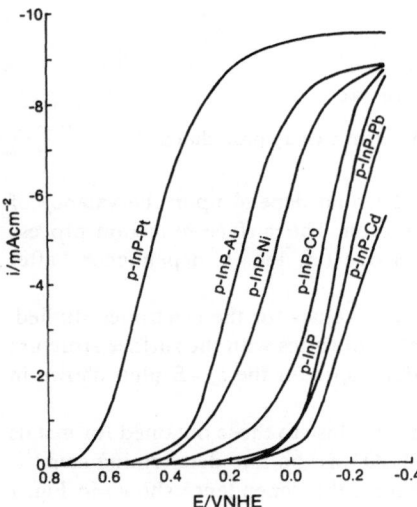

FIGURE 4. Photocurrent-potential relations for p-InP and p-InP-Me electrodes (50 mW/cm^2 Xe light, $0.5M$ H$_2$SO$_4$).[21]

Heller *et al.*[27] observed impressive effects, particularly with noble metals on InP.[26,27] These workers developed a model based upon the concept that the rate of the photoelectrochemical reaction is mainly determined by the field inside the semiconductor (Schottky barrier model).

Thus, if the deposited metal has a work function lower than that of the semiconductor, the Fermi level of the latter will be lowered, and so the potential gradient within the solid will increase, leading to an (on a Schottky barrier model) increase in the current density. (Here, the charge transfer reaction at the metal/solution interface is taken as "fast," that is, not rate-controlling.) It follows that, taking the metal–solution work functions of the metals used in the present experiments, the lowest photocurrent would be obtained with Pt (which should show the smallest shift, that is, the smallest ΔE) and the biggest ΔE would be obtained with Cd. The expected shifts are opposite to the results experimentally observed (Figs. 3–5 and Refs. 27 and 29–32). To be able to sustain this theory, there would have to be changes

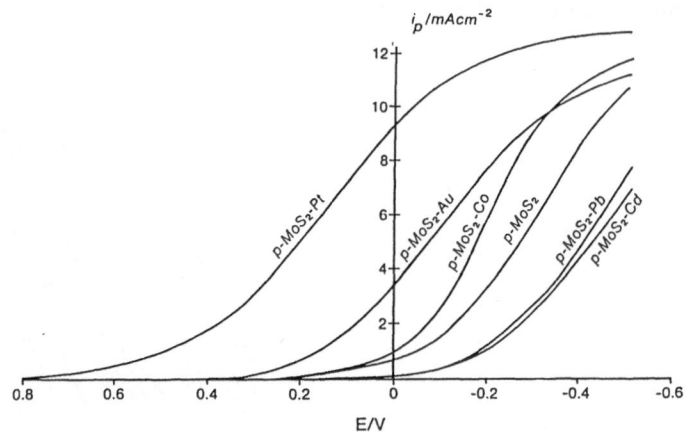

FIGURE 5. Potentiodynamic runs recorded for p-MoS$_2$ and p-MoS$_2$-Me electrodes (50 mW/cm^2 Xe light, $0.5M$ H$_2$SO$_4$).[30]

in the work function caused by H which would need to be large and negative (e.g., 1 eV) for Pt and highly positive for Cd. The effect of H on the work function of metals is given in Table 2. However, according to the requirements of this theory, the changes should be several times larger than those shown in Table 2.

A different interpretation was made by Butler and Ginley,[32] who observed an acceleration in hydrogen evolution on p-GaP in the presence of Ru, Pt, Rh, Bi, and Pb ions, which were adsorbed on the semiconductor surface. The results of such experiments are similar to those reported here (Figs. 3–5). It was observed that when the adsorbed metal ion increased the rate of photoevolution of hydrogen, it introduced surface states in the band gap of the semiconductor. Lead did not introduce surface states, but its effect on hydrogen evolution was not given.[32] Such an explanation of the effect of adsorbed metal ions is tenable but lacks, for example, a demonstrated correlation to the band edge and the potential at which the ions begin to adsorb significantly.[32]

Another explanation can be given in terms of the hole–electron recombination velocity. Thus, it has been shown by Wilson[40] that the photocurrent density can be written in the form

$$i_{ph} = \frac{k_{ct}}{k_{ct} + k_{sr}} \text{[function of transport]} \tag{1}$$

where k_{ct} is the velocity constant for charge transfer at the semiconductor/solution interface, and k_{sr} is a similar quantity for surface recombination. In the exponential region of the i_{ph}–E relation, it may be that k_{ct} can be much smaller than k_{sr}. Then, i_{ph} at a semiconductor with metal islets may be modified by changes in k_{sr}. Wilson presented relations obtained by changing k_{st} (Fig. 2 in Ref. 40) which resemble those shown in Figs. 3–5. However, if this model is applicable to the results reported here, the variation of k_{sr} would have to be closely related to the electrochemical properties of metals, for example, the exchange current density for hydrogen evolution on massive metals. There seems little to connect these two phenomena because of the observation of the dependence of the shift in the i_{ph}–E relation, ΔE, caused by metal deposition on $\log i_0$ of the metals (Fig. 6).

The results shown in Fig. 6 appear to be more consistent with the interpretation given by Szklarczyk and Bockris.[20,21,29,30] According to this model, namely, the photoelectrocatalytic model, charge transfer across the solution/semiconductor interface may be assumed to be the rate-determining step, and the properties of the deposited catalysts then determine the reaction rate, in agreement with the view proposed earlier.[25]

TABLE 2
Value of the Work Function for Metals in Vacuum
(ϕ_v) and Solution (ϕ_s) and Changes Caused by
Hydrogen Adsorption ($\Delta\phi_H$)[a]

Metal	ϕ_v (eV)	ϕ_s (eV)	$\Delta\phi_H$ (eV)
Cd	3.0	4.1	
Pb	4.0	4.2	
Co	4.9	4.7	+0.33
Au	5.2	4.8	+0.2
Ni	5.1	4.7	+0.4
Pt	5.5	5.0	+0.1
Ru	4.7	4.8	+0.4
Rh	4.9	5.0	+0.3

[a] Refs. 29 and 37–41.

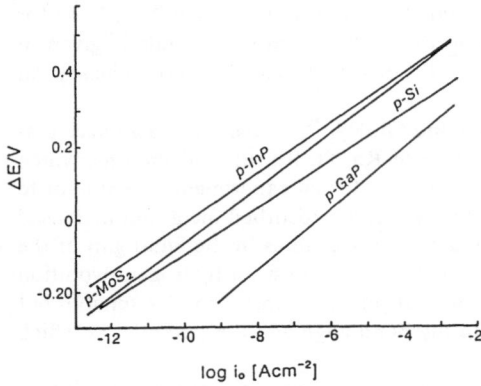

FIGURE 6. The potential shift, ΔE, due to the presence of metals on semiconductor electrodes as a function of log i_0, the exchange current density for hydrogen evolution on massive metals. Lines for p-InP, p-MoS$_2$, p-Si, and p-GaP obtained from data in Refs. 20, 21, 30, and 25, respectively.

Thus, let it be assumed that, under conditions of constant illumination, the variation of the photocurrent density with the constitution of the metal islets depends principally on the dark kinetics of hydrogen evolution on the islets. In this case the photoelectrode potential would have two contributions, one from the photopotential and the other from the dark potential, at the current density used.

Let E_{ph} depend only on the semiconductor substrate, Sc, and be constant for all conditions of metal islet concentration, thus depending only on the conditions of illumination, which are constant.

The dark potential, E_D, which is being defined as independent of the conditions of illumination for a given i_{ph}, is given by

$$E_D = E_{Rev,H_2} + \eta_{H_2} \tag{2}$$

where η_{H_2} is the overpotential for hydrogen evolution at i_{ph}. Thus,

$$E_D = E_{Rev,H_2} + \frac{RT}{\alpha F} \ln \frac{(i_0)_{M_i}}{i_{ph}} \tag{3}$$

where $RT/\alpha F$ has its usual meaning in electrode kinetics, and i_0 is the exchange current density for hydrogen evolution on a specific metal, M_i, in the dark. Then,

$$E_{M_i} = E_{ph} + E_{Rev,H_2} + \frac{RT}{\alpha_{M_i} F} \ln \frac{(i_0)_{M_i}}{i_{ph}} \tag{4}$$

Supposing that

$$\alpha_{M_1} = \alpha_{M_2} = \alpha_{M_i} = \alpha_{Sc} = \alpha \tag{5}$$

then

$$E_{M_i} - E_{Sc} = \frac{RT}{\alpha F} \ln \frac{(i_0)_{M_i}}{(i_0)_{Sc}} = \Delta E_{ph} \tag{6}$$

$$\frac{\Delta E_{ph}}{d[\Delta \log (i_0)_{M_i}]} = 2.303 \frac{RT}{\alpha F} = \frac{0.058}{\alpha} \tag{7}$$

Equation (7) then relates the quality observed under illumination, ΔE_{ph}, with the dark property of metals, log i_0.

Correspondingly, it is possible to calculate on the basis of the photoelectrocatalytic model the shifts in ΔE among pairs of p-Sc–M couples.[30] Pairs of such correlations are calculated in Table 3 and compared with experimental observations. There is a reasonable degree of correlation between the observed and predicted results.

TABLE 3
Correlation between Experimental and Calculated Shifts of i_p-E Dependence

System 1	System 2	Calculated value (mV)	Experimentally observed value (mV)
p-InP-Pt	p-InP-Ni	220	300
p-MoS$_2$-Pt	p-MoS$_2$-Au	320	270
p-InP-Pt	p-InP-Co	410	440
p-Si-Pt	p-Si-Co	410	280
p-MoS$_2$-Pt	p-MoS$_2$-Co	410	360
p-MoS$_2$-Pt	p-MoS$_2$-Cd	900	650
p-Si-Au	p-Si-Co	90	90
p-MoS$_2$-Au	p-MoS$_2$-Co	90	80
p-InP-Au	p-InP-Cd	590	430
p-Si-Au	p-Si-Cd	590	300
p-MoS$_2$-Au	p-MoS$_2$-Cd	580	390

Correlations have been made between the work function of the metal and effect of the metal on photoelectrochemical kinetics (Fig. 7).

The observed linear dependence of ΔE on $\phi_{\mathrm{Me_{sol}}}$ (Fig. 7) requires interpretation in terms of the electrocatalytic model. The shift of the i_p-E curve should be a function of the hydrogen overpotential at a constant current density, and this is proportional to log i_0. However, the values of log i_0 increase with decreasing M—H bond strength for hydrogen evolution in the dark on most metals in acid solution. Hence, if M—H decreases with increasing work function, there is a qualitative rationalization of the observations of Fig. 6. The requisite dependence is shown in Fig. 8.

A corresponding relation of log i_0 to ϕ (log $i_0 = A\phi + B$) has been given by Trasatti.[41]

The photoelectrocatalytic model suggests that electrons are emitted from surfaces consisting only of the islet area, for otherwise the dependence on the dark characteristics of the islets would not be dominant. This trend continues, surprisingly, for islets which have i_0 lower than that of the semiconductor, where one would have expected the hydrogen evolution to have occurred via the Sc. However, this expectation neglects local effects of the metal islets on the field gradient in the semiconductor: there is likely to be a large increase of dV/dx near the metal. Thus, near a metal islet, within the semiconductor the local photoelectron flow is greater than that in other areas, although the interfacial reaction velocities are

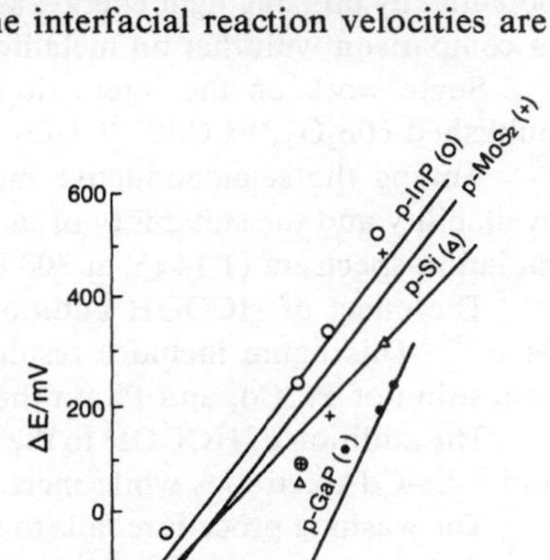

FIGURE 7. The dependence of the anodic shift, ΔE, on the electron work function of the metal in solution for various p-type semiconductor metal systems.

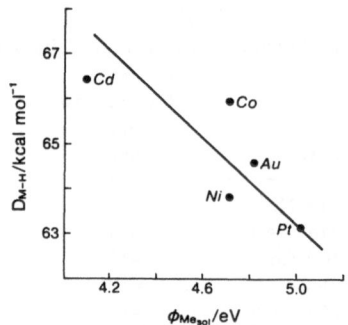

FIGURE 8. The dependence of the heat of hydrogen adsorption, D_{M-H}, on the work function in solution.

greater in the Sc areas than in those for which $(i_0)_M > (i_0)_{Sc}$. The semiconductor itself will be relatively denuded of electrons because of the potential gradient near the metal.

One way of distinguishing which process determines the hydrogen photoevolution rate is to study the same process in dark, majority charge carriers. Thus, in both cases—photoevolution and dark evolution—the surface is the same but the origin of the electrons differs: in one case the orgin is hole–electron pair formation by light excitation; in the other, it is thermal activation from donor levels. The results of such experiments are in excellent agreement for the two situations.[42]

These results[42] seem to be crucial in determining the rate-determining step in the photoevolution of hydrogen on p-type semiconductors with metal additions. Were this to be electron–hole pair formation, then (as it cannot be this in the n-type dark reaction) one would expect a different trend in the effect of metal catalysts on the observed currents in photoevolution as compared to the dark-evolution case.

3. INTERACTION OF HCOOH WITH SILICON ELECTRODE

The reduction of formic acid appears to be the rate-determining step in the electrochemical reduction of carbon dioxide to methanol ($CO_2 + 2H_2O \rightarrow CH_3OH + \frac{3}{2}O_2$, $E^0 = -1.198$ V) on metallic electrodes.[43,44] The use of semiconducting electrodes can decrease the reduction potential by utilizing light energy. Moreover, a different product distribution can be expected in comparison with that on metallic electrodes.

Some work on the interaction of HCOOH with semiconductor electrodes has been published (Fe_2O_3,[13] CdS[14] TiO_2[15]).

Among the semiconductive materials, silicon appears to be promising because of its availability and the suitability of its band gap, which matches reasonably well with the solar radiation spectrum (1.14 eV at 300 K).

The effect of HCOOH addition on i_{ph}–E curves for p-Si electrodes is shown in Fig. 9a–d.[16] This figure includes results obtained after a washing procedure.[16] The effect of deposition of Pt, Cd, and Pb on the silicon surface is also shown.

The addition of HCOOH to the solution causes a decrease in the current for p-Si, p-Si–Pt, and p-Si–Cd electrodes while increasing the current for the p-Si–Pb electrode.

The washing procedure fails to restore the initial i–E curve characteristic of semiconductors that were never in contact with HCOOH (Fig. 9a–d, run "0"), which indicates that irreversible chemisorption of either HCOOH or some species derived from it occurs at the electrodes.[16]

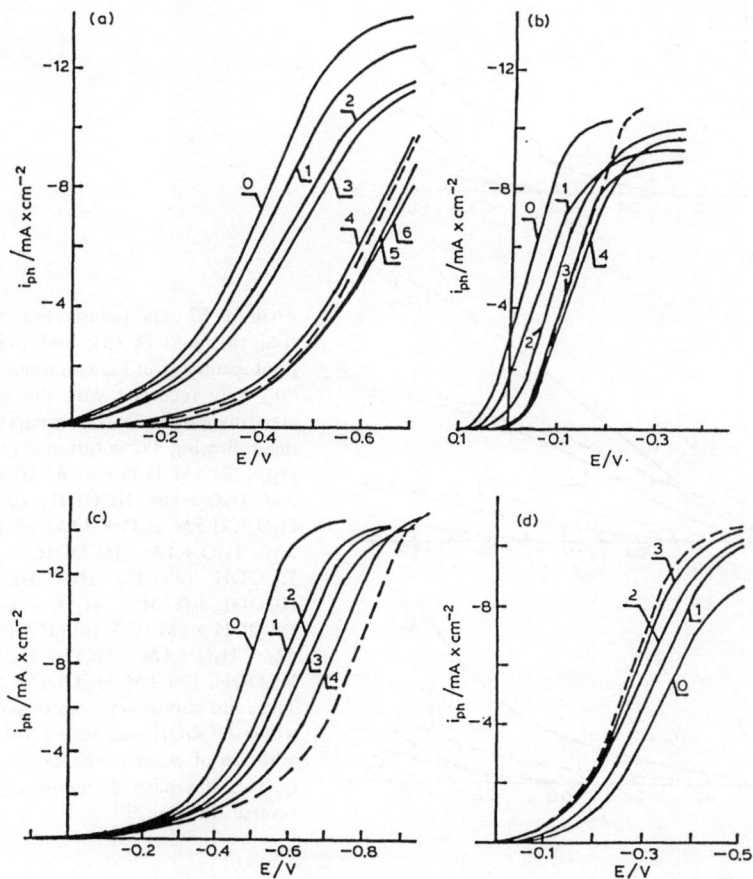

FIGURE 9. Influence of the HCOOH concentration on the i_{ph}-E curves.[16] Curves "0" denote the polarization curves for the supporting electrolyte only. Other curves were recorded in the presence of HCOOH. The dashed curves represent the polarization curves after the washing procedure (see text). (a) p-Si: (1) 0.005M, (2) 0.01M, (3) 0.03M, (4) 0.05M, (5) 0.10M, (6) 0.15M; (b) p-Si-Pt: (1) 0.01M, (2) 0.05M, (3) 0.10M, (4) 0.30M; (c) p-Si-Cd: (1) 0.05M, (2) 0.10M, (3) 0.30M, (4) 0.50M; (d) p-Si-Pb: (1) 0.05M, (2) 0.10M, (3) 0.30M.[17]

In order to determine the role of water molecules in the HCOOH reduction process, Szklarczyk *et al.* studied the process in an organic solvent, propylene carbonate (PC).[16] The i_{ph}-E dependencies for different combinations of HCOOH, H_2O, and PC are shown in Fig. 10a–c. Similar effects of H_2O or HCOOH addition on the polarization curves for different p-Si-Me electrodes was observed without illumination.[16] It should be pointed out that although the effects are similar in both experiments, that is, in the absence and in the presence of light, the origin of charge carriers has to be different.

The effect of HCOOH can be characterized by the shift of the polarization curve, that is, its ΔE value. If $\Delta E > 0$, as in the case of p-Si-Pb, the rate of hydrogen photoevolution increases and/or HCOOH reduction occurs. If $\Delta E < 0$, as in the case of p-Si-Pt and p-Si-Cd, the photoevolution of hydrogen is inhibited, probably due to poisoning of the electrode surface by the chemisorbed products of HCOOH. In this case the decrease in the photocurrent should be proportional to the surface coverage by the adsorbed species. However, it is possible

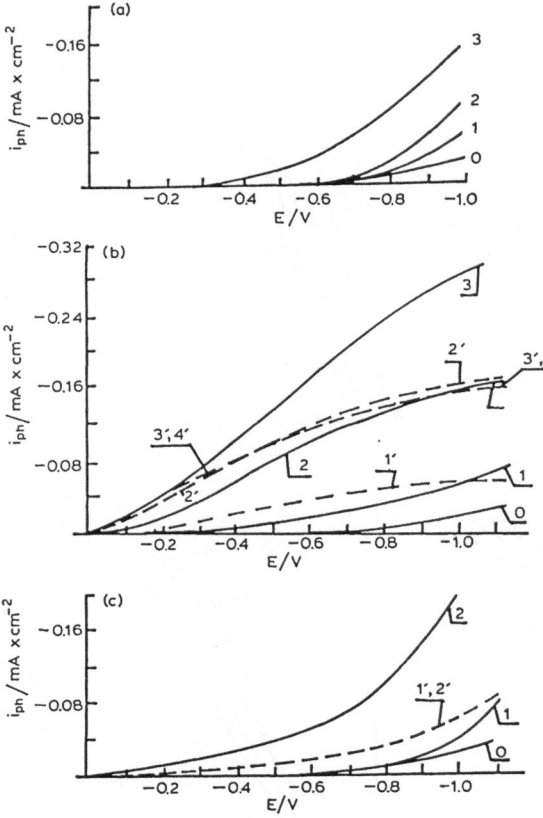

FIGURE 10. The polarization curves for
p-Si (a), p-Si–Pt (b), and p-Si–Cd (c)
photocathodes in PC solution.[16] Curves
"0" were recorded with the supporting
electrolyte only. The other curves refer to
the following PC solutions: (a) (1) 5M
H_2O, (2) 5M H_2O + 0.1M HCOOH, (3)
5M H_2O + 1M HCOOH; (b) (1) 5M
H_2O, (2) 5M H_2O + 0.1M HCOOH, (3)
5M H_2O + 1M HCOOH, (1′) 0.1M
HCOOH, (2′) 1M HCOOH, (3′) 1M
HCOOH + 0.1M H_2O, (4′) 1M
HCOOH + 1M H_2O; (c) (1) 5M H_2O, (2)
5M H_2O + 1M HCOOH, (1′) 1M
HCOOH, (2′) 1M HCOOH + 5M H_2O.
The solid curves are from experiments in
which HCOOH was added following the
addition of water to the supporting electrolyte; the dashed curves are for the
reverse situation.[17]

that suppression of the hydrogen evolution may be balanced partially by the photoreduction
of HCOOH.

The results of the experiments presented in Fig. 9a–d show that the species derived from
HCOOH chemisorbed on p-Si and p-Si–Pt electrodes are similar. It can be assumed that
there is no direct reduction of HCOOH from the bulk of solution and that the decrease in
the photocurrent is due to an increasing poisoning of the electrode surface with increase in
the bulk concentration of HCOOH.

The polarization curve for p-Si–Cd (Fig. 9c) after the washing procedure differs markedly
from the curves for p-Si and p-Si–Pt (Fig. 9a and 9b). In the former case, the current is
observed to decrease further on removal of HCOOH from the bulk of solution (notice the
difference between the runs indicated by line 4 and the dashed line in Fig. 9c). The only
explanation is that the net current is the sum of the hydrogen evolution photocurrent and
the current resulting from the direct reduction of HCOOH from the bulk of the solution on
the surface partially covered with chemisorbed products derived from HCOOH.

For the p-Si–Pb electrode (Fig. 9d), a shift of the polarization curve in the positive range
of potentials is observed in the presence of HCOOH. This means that the current increases
with increasing HCOOH concentration owing to the reduction of HCOOH. However, after
the washing procedure, no decrease in the current is observed. No satisfactory explanation
could be offered, except for the suggestion that the presence of adsorbed organic species on
the electrode surface accelerates the hydrogen evolution.

The influence of deposited metals on the kinetics of processes on semiconductor electrodes can be considered in terms of the change in energy levels of the semiconductor or of

the change in the double layer's electrochemical properties. In the first case the variations in the Fermi level energy as well as band bending of the semiconductor by the deposited metals should be taken into account. In particular, the work function of the deposited metals affects both values. The work functions of the metals studied follow the order $\phi_{Pt} > E_{F,Si} > \phi_{Pb} > \phi_{Cd}$ ($E_{F,Si}$ is the Fermi level of the Si electrode). Hence, the band bending should follow the reverse order. As the rate of the electrode processes is proportional to the band bending,[18] the shift of the i_{ph}–E curves, ΔE, should be positive for p-Si–Cd and p-Si–Pb and negative for p-Si–Pt. One can see (Fig. 9) that this is not the case.

It was proposed[16] that metal deposited on a semiconductor introduces local energy levels allowing electrons to flow from the semiconductor to the metal island prior to being transferred to species in the solution.[20] In this case electrochemical double-layer effects (e.g., exchange current density for the reaction considered) at the metal/solution interface seem to determine the mechanism and kinetics of the processes investigated.

The interaction of HCOOH and H_2O molecules with the p-Si electrode differs markedly (Fig. 10). It is evident that H_2O molecules replace PC molecules adsorbed on the electrode surface. The subsequent addition of HCOOH (H_2O/HCOOH system) causes a further increase in the photocurrent. It is likely that the adsorbed hydrogen, originating from the water discharge, reacts with the formic acid molecule on the electrode surface.[23] When HCOOH solution is introduced into the supporting electrolyte before water is added (HCOOH/H_2O system), no increase in the current due to water reduction is observed. This means that the adsorbed molecules of formic acid and/or the products of chemisorption are bound more strongly to the surface than the water molecules and prevent the discharge of H_2O on the electrode. Therefore, the surface concentration of hydrogen is lower and the reduction of HCOOH is slower than in the H_2O/HCOOH system.

Surprisingly, similar effects were observed for nonilluminated p-type silicon electrodes.[16] During illumination the photoexcited electrons of the conduction band (e_{ph}^-) are the charge carriers, which is not the case in the "dark" experiments. Taking into account the band position of p-Si[45] and the high negative potential of the electrode, the formation of an inversion layer is possible. In the inversion state, a direct flow of electrons from the valence band (e_{VB}^-) to the interface is possible, which can explain the behavior of p-Si–Pt electrode as a "dark" cathode.

The overall reactions of HCOOH reduction on the electrodes studied can be assumed to be as follows:

$$HCOOH + 2H_{ads} \rightarrow HCOH + H_2O \qquad (0.96 \text{ kJ/mol}, 0.0 \text{ V}) \qquad (8)$$

$$HCOOH + 4H_{ads} \rightarrow CH_3OH + H_2O \qquad (57.37 \text{ kJ/mol}, 0.15 \text{ V}) \qquad (9)$$

$$HCOOH + 6H_{ads} \rightarrow CH_4 + 2H_2O \qquad (179.07 \text{ kJ/mol}, 0.31 \text{ V}) \qquad (10)$$

The following sequences are proposed for the reaction of formic acid on the p-Si–Me electrodes:

$$(11)$$

$$(12)$$

$$HCOH + H_2O; \quad CH_3OH + H_2O; \quad CH_4 + H_2O$$

Reaction (11) should be neglected for the $HCOOH/H_2O$ system in PC solution. The electrons of the valence band are involved only in the PC solution.

4. CONCLUSIONS

1. When metals are deposited in islet form on a semiconductor electrode so that they cover a fraction of the semiconductor surface, the photocurrent-potential relation is shifted by an amount ΔE that is characteristic of the nature of the metal. This ΔE can be correlated to the exchange current density for hydrogen evolution in the dark on the bulk form of the metals concerned.
2. The linear dependence of ΔE on i_0 indicates that, in the presence of metal islets, the rate of the overall process of photoelectrochemical evolution of H_2 is determined by a charge transfer step at the metal/solution interface.
3. Deposition of metals on a semiconductor electrode strongly influences interaction of organic molecules with the electrode surface. The nature of the change shows the dependence of processes on the properties of the electrochemical double layer.
4. It is proposed that in photoelectrochemical processes involving an adsorption step, the rate of reaction depends on the process taking place in the electrochemical double layer and not within the bulk of the semiconducting electrode.

ACKNOWLEDGMENT

    The author wishes to thank Mrs. Babli Kapur for her help in the editing of this text and the Welch Foundation for support of this work carried out in the Department of Chemistry, Texas A&M University.

REFERENCES

1. E. Becquerel, *C. R. Acad. Sci.* **9**, 561 (1839).
2. W. H. Brattain and C. G. B. Garrett, *Bell Syst. Tech. J.* **34**, 129 (1955).
3. J. F. Dewald, in: *Semiconductors* (N. B. Hannay, ed.), Reinhold, New York (1957).
4. M. Green, in: *Modern Aspects of Electrochemistry*, No. 2 (J. O'M. Bockris, ed.), p. 343, Butterworths, London (1959).

5. H. Gerischer, in: *Advances in Electrochemistry and Electrochemical Engineering* Vol. 1 (P. Delahay, ed.), p. 139, Interscience, New York (1961).
6. H. Gerischer, in: *Physical Chemistry, An Advanced Treatise*, Vol. 9 (H. Eyring, ed.), p. 463, Academic Press, New York (1970).
7. W. A. Myamlin and J. V. Pleskov, *Electrochemistry of Semiconductors*, Nauka, Moscow (1965).
8. R. Memming, *J. Electrochem. Soc.* **116**, 785 (1969).
9. R. Williams, *J. Chem. Phys.* **32**, 1505 (1960).
10. A. Fujishim and K. Honda, *Nature* **238**, 37 (1972).
11. M. Halmann, *Nature* **275**, 115 (1978).
12. M. Gratzel (ed.), *Energy Resources through Photochemistry and Catalysis*, p. 25, Academic Press, New York (1983).
13. J. H. Kennedy and D. Dünwald, *J. Electrochem. Soc.* **130**, 2013 (1983).
14. M. Matsamura, M. Hiramoto, T. Iehara, and H. Tsubomura, *J. Phys. Chem.* **88**, 248 (1989).
15. M. H. Miles, A. N. Fletcher, G. E. Manis, and L. O. Spreer, *J. Electroanal. Chem.* **190**, 157 (1985).
16. M. Szklarczyk, J. Sobkowski, and J. Pacocha, *J. Electroanal. Chem.* **215**, 307 (1986).
17. M. Szklarczyk, *Electrochim. Acta* **32**, 1257 (1987).
18. F. Beck and H. Gerischer, *Z. Elektrochem.*, **63**, 500 (1969).
19. H. Tamura, H. Yoneyama, C. Iwakura, H. Sakamoto, and S. Murakami, *J. Electroanal. Chem.* **80**, 357 (1977).
20. M. Szklarczyk and J. O'M. Bockris, *J. Phys. Chem.* **88**, 1808 (1984).
21. M. Szklarczyk and J. O'M. Bockris, *J. Phys. Chem.* **88**, 5241 (1984).
22. M. Szklarczyk and R. E. Allen, *Appl. Phys. Lett.* **49**, 1028 (1986).
23. Y. Nakato, S. Tonomura, and H. Tsubomura, *Ber. Bunsenges, Phys. Chem.* **80**, 1289 (1976).
24. K. Ohashi, J. McCann, and J. O'M. Bockris, *Energy Res. Abstr.* **1**, 259 (1977).
25. W. Kautek, J. Gobrecht, and H. Gerischer, *Ber. Bunsenges, Phys. Chem.* **84**, 1034 (1980).
26. A. Heller and R. G. Vadimsky, *Phys. Rev. Lett.* **46**, 1153 (1981).
27. A. Heller, E. Aharon-Shalom, W. A. Bonner, and B. Miller, *J. Am. Chem. Soc.* **104**, 6942 (1982).
28. R. N. Dominey, N. S. Lewis, J. A. Bruce, D. C. Bookbinder, and M. S. Wrighton, *J. Am. Chem. Soc.* **104**, 467 (1982).
29. M. Szklarczyk and J. O'M. Bockris *Appl. Phys. Lett.* **2**, 1035 (1983).
30. M. Szklarczyk and J. O'M. Bockris, *Int. J. Hydrogen Energy* **9**, 831 (1984).
31. M. Szklarczyk, J. O'M. Bockris, V. Brusic, and G. Sparrow, *Int. J. Hydrogen Energy* **9**, 707 (1984).
32. M.A. Butler and D. S. Ginley, Extended Abstracts, 163rd Meeting of the Electrochemical Society, San Francisco, May 8–13, 1983, Abstract 723.
33. M. Szklarczyk, A. Q. Contractor, J. O'M. Bockris, V. Y. Young, L. A. Bernard and G. Sparrow, *Sol. Energy Mater.* **11**, 1051 (1984).
34. H. Kita *J. Electrochem. Soc.* **113**, 1095 (1966).
35. V. V. Gorodetskii and B. E. Nieuwenhuys, *Surf. Sci.* **108**, 225 (1981).
36. G. F. Voronina, L. A. Larin, and T. V. Kalish, *Elektrokhimiya* **14**, 297 (1978).
37. F. C. Tompkins, in: *The Solid–Gas Interface*, Vol. 2 (E. A. Flood, ed.), p. 765, Marcel Dekker, New York (1967).
38. R. V. Culver and F. C. Tompkins in: *Advances in Catalysis and Related Subjects*, Vol. XI (D. D. Eley, P. W. Selwood, and P. B. Weisz, eds.), p. 67, Academic Press, London (1959).
39. L. Whalley, B. J. Davis, and L. Moss, *Trans. Faraday Soc.* **66**, 3143 (1970).
40. R. H. Wilson, *J. Appl. Phys.* **48**, 4292 (1977).
41. S. Trasatti, *J. Electroanal. Chem.* **39**, 163 (1972).
42. A. Q. Contractor, M. Szklarczyk, and J. O'M. Bockris, *J. Electroanal. Chem.* **157**, 175 (1983).
43. P. G. Russel, N. Kovac, S. Srinivasan, and M. Steinberg, *J. Electrochem. Soc.* **124**, 1329 (1977).
44. A. Czerwinski and J. Sobkowski, *J. Electroanal. Chem.* **59**, 41 (1975).
45. P. A. Kohl, S. N. Frank, and A. J. Bard, *J. Electrochem. Soc.* **124**, 225 (1977).

16

An Application of Optical Waveguides to Electrochemical and Photoelectrochemical Processes

Kiminori Itoh and Akira Fujishima

1. INTRODUCTION

An optical waveguide (OWG) consists of a substrate with a transparent thin layer (a few microns in thickness) on the top, through which layer a light wave can propagate. Since the light is confined to a small surface region of the OWG, the electric field associated with the light wave is very large at the surface of the OWG. Consequently, light absorption occurs significantly when suitable species, for example, dyes, are adsorbed onto the OWG surface. Thus, optical absorption by these species will be greatly enhanced by using OWGs.[1,2]

We have applied the OWG technique to electrochemical and photoelectrochemical reactions for the first time.[3] The sensitivity of the optical detection with the OWGs appeared to be more than 100 times higher than that with conventional optical methods. Moreover, since the OWG method is suitable for fast transient measurements, we can apply, for instance, flash photolysis to semiconductor systems by using suitable OWGs. Thus, we can avoid several difficulties encountered with semiconductor systems, namely, semiconductor electrodes give only small optical changes associated with surface reactions, and semiconductor particles largely scatter light. We present here some results on electrochemical reduction of an organic dye, Methylene Blue, and photoelectrochemical deposition of silver from aqueous solution. Furthermore, numerical analyses of multilayered OWGs were carried out to clarify basic characteristics of these OWG systems.

2. EXPERIMENTAL

2.1. Glass OWGs, OWGs Coated with TiO$_2$ Powder, and OWG Electrodes

Figure 1 depicts the structure of the OWGs used in this study. Slab-type glass optical waveguides (Fig. 1a) were prepared by immersing a glass slide into molten KNO$_3$ for 0.5–4 h.[4]

Kiminori Itoh • Institute of Environmental Science and Technology, Yokohama National University, Yokohama 240, Japan.   *Akira Fujishima* • Department of Synthetic Chemistry, Faculty of Engineering, The University of Tokyo, Tokyo 113, Japan.

Electrochemistry in Transition, edited by Oliver J. Murphy *et al.* Plenum Press, New York, 1992.

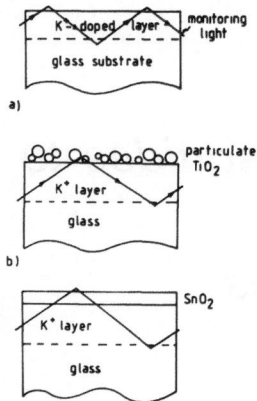

FIGURE 1. Schematic diagrams of the structure of OWGs used in the present study: (a) glass OWG with K⁺-doped layer; (b) glass OWG covered with particulate TiO_2; (c) OWG electrode—glass OWG coated with conductive SnO_2.

For photochemical measurements, one side of the glass OWG thus prepared was brought into contact with an aqueous suspension of TiO_2 fine powder (Japan Aerosil P-25). The suspension was removed after several minutes, and only TiO_2 adsorbed onto the OWG surface remained (Fig. 1b). In order to use the OWGs in electrochemical measurements, the surface of the OWGs should be electrically conductive; this type of OWG can be called an OWG electrode (Fig. 1c). A thin film of SnO_2 doped with Sb was coated onto the glass OWG samples by using a spray pyrolysis technique.

2.2. Characterization of the OWGs

The OWGs were mounted on a turntable in order to make a laser beam (He–Ne laser, 633 nm) incouple and outcouple with the waveguide layer by rotating the table. For this purpose, two prisms or two holographic gratings are attached to the surface of the OWG. The sensitivity of the OWGs in absorption measurements was estimated using an adsorbed dye, Methylene Blue (MB). Absorption spectra were taken for MB adsorbed from an aqueous solution onto glass surfaces. The optical density (OD) at 633 nm was, for example 0.003 for a $2 \times 10^{-5} M$ solution. The OD values thus obtained were used as standard values in a rough estimation of the OWG sensitivity. The peak intensity of the guided light was measured with and without MB adsorbed, and OD changes due to MB adsorption were estimated.

2.3. Photochemical and Electrochemical Measurements with OWGs

Figure 2 shows a schematic diagram of the experimental apparatus for photochemical and electrochemical measurements with OWGs. The intensity of the guided light was

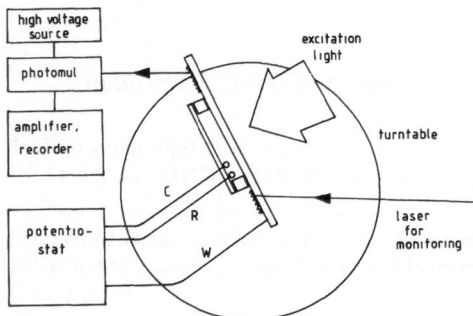

FIGURE 2. Schematic diagram of experimental setup for electrochemical and photochemical measurements with an OWG.

monitored with a photomultiplier and was recorded with an oscilloscope or a transient memory. The front side of the OWG was in contact with a solution in a reservoir, and the OWG was irradiated from the back side if necessary. The light sources used were an N_2 laser (pulse width = 5 ns, wavelength = 337 nm) and a high-pressure Hg lamp. For the electrochemical measurements, a Pt counter electrode and an Ag wire were used.

2.4. Photocatalytic Reduction of Silver Ions at TiO₂ Particles

It is known that photoexcited TiO_2 powder reduces metal ions in solution, causing deposition of metal clusters or islands onto semiconductor surfaces (see, for instance, Ref. 5). These deposition processes are typical photocatalytic reactions and are useful for recovering noble metals in solution[5] or for preparing highly dispersed catalysts.[6] Of these reactions, we employed here silver deposition because it has been studied most widely as an efficient reaction for electron trapping.

2.5. Electrochemical Reduction of MB at the OWG Electrodes

The intensity of the guided light (I_{OWG}) is greatly decreased as a result of adsorption of MB molecules onto the OWG surface (cf. Fig. 3). I_{OWG} decreased in a similar manner at the OWG electrodes (glass OWG/SnO₂ system). Therefore, I_{OWG} should increase when the MB molecules adsorbed at the OWG surface are reduced to give the colorless leuco form.[7] Thus, we employed the electrochemical MB reduction as a model reaction to demonstrate the function of the OWG electrodes.

3. RESULTS AND DISCUSSION

3.1. Characteristics of the OWG

Figure 3 shows how the monitoring light is introduced into the OWG layer by rotating the turntable. At the peaks shown in the figure, the intensity of the guided light is so large that we could see traces of the guided light on the OWG surfaces. Glass OWG A gave four peaks within one third of a degree, and glass OWG B gave one peak. The thickness of the OWG layer in these samples was estimated to be ca. 8 μm for OWG A and ca. 2 μm for OWG B from the relation between the thickness and the number of modes or by using the inverse WKB method.[8]

The OWG sensitivity for the single-mode OWG (OWG B) was ca. 150/(3.3-cm optical path) and was larger than that for the multimode OWG (OWG A). This is consistent with

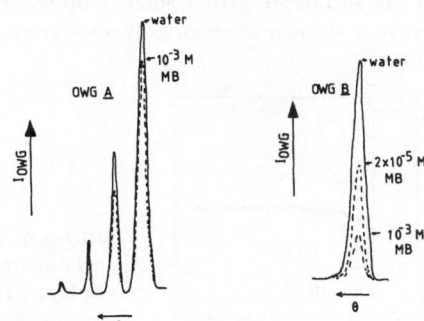

FIGURE 3. OWG characteristics: intensity of guided light as a function of the angle of the turntable. OWG A: Glass OWG doped with K^+ by immersion into molten KNO₃ for 4 h; OWG B: immersion time = 0.5 h.

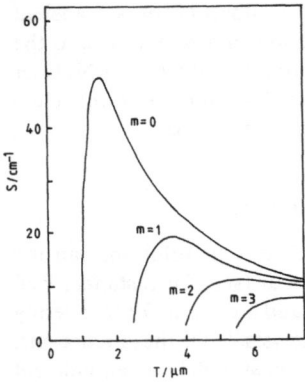

FIGURE 4. Theoretically calculated dependence of the OWG sensitivity on the OWG thickness.

numerical analyses based on a Gaussian distribution of the refractive index $n(x)$ in the OWG layer; that is,

$$n(x) = n(\infty) + [n(0) - n(\infty)] \exp[-(x/T)^2] \tag{1}$$

where $n(\infty)$ is refractive index of the substrate, and $n(0)$ is that of the OWG surface. We can change the value of T, the thickness of the OWG layer, experimentally by changing the duration of immersion of the glass substrate into the molten salt. An example of such a calculation is shown in Fig. 4, where $n(\infty) = 1.51$, $n(0) = 1.518$, and $n = 1.00$ for $x < 0$ (air). For $T < 2.5 \ \mu m$, the number of modes is one, and the OWG sensitivity has the highest values. The details have been decribed elsewhere.[9]

3.2. Photocatalytic Reduction of Silver Ions at TiO₂ Particles

An OWG covered with TiO_2 powder was mounted on the cell shown in Fig. 2, and the solution reservoir was filled with an aqueous solution of $AgNO_3$ (1mM). Figure 5 shows the change in I_{OWG} due to light irradiation (high-pressure Hg lamp, duration = 1 s). I_{OWG} immediately decreased upon irradiation, slowly increased after the irradiation was stopped, and reached a steady-state value. Irradiation for 10 s made I_{OWG} almost zero, and the surface of the OWG became brown. The magnitude of the change in I_{OWG} due to the irradiation was enhanced by adding EtOH (30 vol %) to the solution.

The above results clearly show that the glass OWG/particulate TiO_2 system is very effective in monitoring the photoinduced reaction taking place at semiconductor surfaces. The increase in I_{OWG} after the light irradiation was stopped shows either that the metallic silver, once deposited onto TiO_2, was oxidized, or that the deposited metallic silver rearranged at the surface to give a smaller optical density. The former explanation is more plausible because oxygen is reportedly produced through a reaction between water and remaining holes.[5]

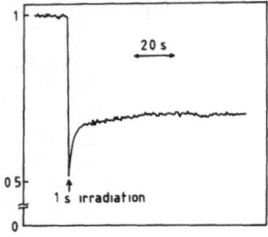

FIGURE 5. Change in the intensity of the guided light (I_{OWG}) associated with irradiation. A glass OWG covered with particulate TiO_2 was employed, and the solution contained a 1mM $AgNO_3$ aqueous solution. The light source was a high-pressure Hg lamp (duration 1 s).

We next applied laser flash photolysis to this system, that is, silver deposition at glass OWG/particulate TiO_2, in order to study the characteristics of the OWG method in fast transient measurements. Figure 6 shows typical transient behavior of I_{OWG} in this system. A fast decrease in I_{OWG} with an onset time of ca. 50 μs was observed when the OWG system was irradiated with light pulses from an N_2 laser. This onset time was reduced to ca. 10 μs when the concentration of $AgNO_3$ was increased to 10mM. These results show that the transient behavior indeed represents the process of silver deposition and is not due to artifacts arising from heat generation at the OWG surface or mismatching of the guided light. In fact, the surface of the OWG gradually changed to a brownish color during continual pulsed irradiation (5 Hz), and I_{OWG} was decreased by ca. 50% after 32 light pulses.

3.3. Electrochemical Reduction of MB on OWG Electrodes

OWG electrodes should have sufficiently low resistivity for electrochemical measurements. The resistivity of SnO_2 films deposited onto the glass OWG decreases with an increase in concentration of the dopant (Sb, in this study). However, we should take into account that free electrons and color centers generated in the films in the doping process absorb the guided light. Thus, we made OWG electrodes with the optimum concentration of Sb (2.5 wt. %). The sheet resistance of these OWGs was ca. 500 Ω, and attenuation of the guided light was sufficiently small.

The OWG electrode was brought into contact with an aqueous solution of MB (1mM), and I_{OWG} was monitored as a function of the electrode potential. A cyclic voltammogram was recorded at the same time. As shown in Fig. 7, I_{OWG} increased when the electrode was cathodically polarized to reduce MB and decreased upon anodic polarization, which gave rise to reoxidation of MB.

Figure 8 shows results obtained for a wider range of electrode potentials and a smaller MB concentration. One can see that MB seemingly gave colorless substances at highly anodic potentials as a result of extensive oxidation and that the I_{OWG} versus potential curve is tilted. This kind of tilt can also be observed in electroreflectance measurements and has been attributed to dependence of the refractive index at the electrode surface on the electrode potential.[10] We can see also a curious hysteresis in the I_{OWG} versus potential curve. We cannot explain this hysteresis in a conventional manner, and it probably comes from an iR drop due to the relatively large resistance of the SnO_2 film.

3.4. Analyses of Multilayered OWGs

We have demonstrated in this chapter that the OWG method is useful in electrochemical and photoelectrochemical measurements by using simple OWG systems. We are now constructing various types of OWGs to overcome the problems that arise in carrying out more sophisticated measurements. For instance, it would be of interest to deposit photoactive

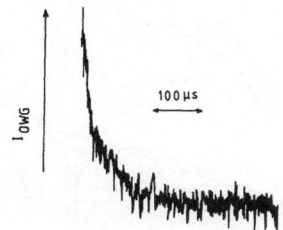

FIGURE 6. Change in the intensity of guided light due to pulsed irradiation. The conditions are the same as those in Fig. 5 except the light source was an N_2 pulsed laser.

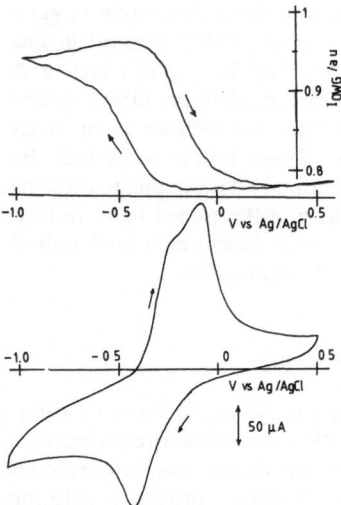

FIGURE 7. Change in the intensity of the guided light as a function of electrode potential for an OWG electrode (a glass OWG coated with conductive SnO_2) in contact with a $1mM$ Methylene Blue aqueous solution. A cyclic voltammogram is also shown.

semiconductor films on OWG electrodes in order to observe photoelectrode reactions on the OWGs. Although we can make OWGs very conveniently with a K^+-doped layer, it was rather difficult to construct, for example, glass $OWG/SnO_2/TiO_2$ OWG systems. This is presumably because the K^+ ions diffused during the deposition process, which requires a temperature of ca. 400°C. However, this difficulty is not a fundamental one. We can employ OWGs that are much more thermally stable,[4] and the semiconductor films can be deposited by other methods.[11]

Besides the practical problems involved in making such multilayered OWG's, we should consider that the OWG characteristics of the multilayered systems change very sensitively with changes in OWG parameters. We have performed model calculations on SnO_2/K^+-doped glass OWG systems. Figure 9 shows the dependence of the OWG sensitivity on the thickness (T_1) of the top SnO_2 ($n = 2.0$) layer. Parameters for the glass OWG layer (Gaussian type) used are $n(\infty) = 1.51$, $n(0) = 1.518$, and $T = 4.0$ μm [cf. Eq. (1)]. The numbers indicated on the figure are mode numbers, which are equal to the numbers of nodes associated with the light wave propagating in the OWG layer. One can see that the OWG sensitivity changes very much and oscillates from 10 to 1000 as the thickness of the SnO_2 layer changes. Thus, we should optimize the design of the OWG systems by using suitable values of the OWG parameters. For instance, the thickness of the top layer should be controlled precisely to avoid too high or too low sensitivity.

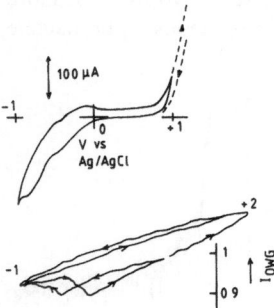

FIGURE 8. Change in the intensity of the guided light as a function of electrode potential. Conditions are the same as those in Fig. 7 except for the potential range.

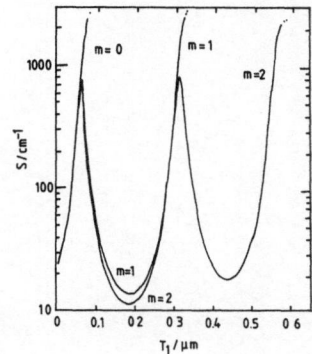

FIGURE 9. Dependence of the OWG sensitivity on the top-layer thickness in SnO_2/K^+-doped glass systems.

REFERENCES

1. J. D. Swalen, M. Tacke, R. Santo, K. E. Rieckhoff, and J. Fischer, *Helv. Chim. Acta* **61**, 960 (1978).
2. Z. Z. Ho, W. M. K. P. Wijekoon, E. W. Koenig, and W. M. Hetherington III, *J. Phys. Chem.* **91**, 757 (1987).
3. K. Itoh and A. Fujishima, *J. Am. Chem. Soc.* **110**, 6267 (1988); *J. Phys. Chem.* **92**, 7043 (1988).
4. V. Neuman, O. Parriaux, and L. M. Walpita, *Electron. Lett.* **15**, 704 (1979).
5. R. Baba, R. Konda, A. Fujishima, and K. Honda, *Chem. Lett.* **1986**, 1307, and references therein.
6. S. Sato, private communication.
7. H. Yoneyama, Y. Toyoguchi, and H. Tamura, *J. Phys. Chem.* **76**, 3460 (1972).
8. J. M. White and P. F. Heidrich, *Appl. Opt.* **15**, 151 (1976).
9. K. Itoh and M. Murabayashi, in: *Trends in Physical Chemistry*, *1* (Council of Scientific Research Integration, eds.), p. 179, Research Trends, Indiana (1991).
10. J. D. E. McIntyre and D. M. Kolb, *Symp. Faraday Soc.* **4**, 99 (1970).
11. K. Heuberger and W. Lukosz, *Appl. Opt.* **25**, 1499 (1986).

17

Electrons, Interfaces, and Societies in the 21st Century

J. O'M. Bockris

1. INTRODUCTION

Electrochemistry is still, in a sense, a little known subject. When people talk about the fuel cells which give the energy aboard space vehicles, they talk about "an electronic device." One of the difficulties electrochemistry has is, where to fit in. On the fundamental side it has contributions from so many areas. There comes the quantum mechanics of interfacial electron transfer, the surface chemistry, the consequences of the presence of very strong interfacial fields, the structure of the surface of metals, and the solid-state chemistry of semiconductors. Then, there is the huge area of biology subject to interpretation in terms of electrochemical mechanisms.

On the applied side, one may sometimes be tempted to see too much as part of electrochemistry. This is because the *surface* of most objects is the locus of many happenings at them, certainly in respect to their longevity, but also in respect to many of their functions. In practice, surfaces are covered with moisture and constitute an electrochemical interface. Many things depend on this, the most obvious being the energy conversion devices, those involved in electrical energy storage, the production of many metals, biosyntheses, and much of analytical chemistry.

It is interesting to try to get some kind of an idea of what fraction of chemistry is really electrochemistry, and one measure of this is the relative size of the meetings of the American Chemical Society, 5000 to 10,000, and those of the Electrochemical Society, 1000 to 2000 people. This sounds about right. The dollar value of the products of the electrochemical industry is around one-quarter that of those of the chemical industry. Something of this conclusion was reached in a 1966 article by Wenglowski in *Modern Aspects of Electrochemistry.*[1]

On the other hand, the number of persons who are trained in electrochemistry is discrepant with this. The number of Ph.D.s produced in chemistry in the United States per year is in the region of 1000, and that in electrochemistry is nearer 10 than 100.

There are some deeper aspects of electrochemistry which should be brought out before we decide where to put it. It is not a part of chemistry exactly, but "another kind of chemistry."

J. O'M. Bockris • Department of Chemistry, Texas A&M University, College Station, Texas 77843. This chapter is an updated text of the lecture given in Toronto in June 1988.

Electrochemistry in Transition, edited by Oliver J. Murphy *et al.* Plenum Press, New York, 1992.

Toronto, June, 1988

John O'Mara Bockris, 1990

Imperial College of Science and Technology, University of London, South Kensington, London, 1947
Front row: A. Azzam; R. Parsons; JOMB; J. F. Herringshaw; B. E. Conway; H. Egan. *Back row*: J.
Bowler-Reed; J. Tomlinson; H. Rosenberg; E. Potter; J. F. Wetterholm; M. Fleischmann.

There are remarkable, and little realized, basic differences between electrochemical reactions
and chemical reactions. For one thing, at a fundamental level, the electrochemical ones do
not produce heat. The ΔH is zero, and what heat they produce is due to polarization effects
or I^2R heating. Combustion carried out electrochemically is "cold," as Justi and Winsel said
many years ago.[2]

There is another fact about electrochemical reactions which makes them differ fundamentally from chemical reactions—namely, they do not take place by means of collisions
between the reactants. Thus, "the mechanism of chemical reactions" is generally presented
in terms of collisions between highly energized molecules of the reactants. On the other hand,
in electrochemical reactions there are collisions between each reactant and a catalyst–electrode
interface, but the reactants (e.g., C_2H_4 and O_2) may be separated by centimeters.

Thus, electrochemistry is *another* chemistry. It is the other side of the penny to thermal
chemistry, and it is perhaps best called electric chemistry. In this view, chemistry is divided
into two parts: the thermal part and the electrical part.

In a big sense, the changes in society reflect the changes implicitly suggested here in
how to look at electrochemistry. Since the beginning of the Industrial Age—let us associate
that with Newcommen's steam engine, in 1712—there has been stress upon heat and fire (i.e.,
explosions) as the basis of technology, and it is obvious that we are now (because of the
difficulties of atmospheric pollution) seeing the *end of the age of fire* and approaching an
age in which we shall live on light as the basic fuel, collected from the sun (eventually in
space) and distributed in the forms of electricity and/or hydrogen. When it is desired, heat
will be produced from electricity.

In the sketch that I am going to give here, I shall, of course, stress my own work because
this chapter is drawn from an awards lecture. However, I think that the examples I have
chosen illustrate the theme by which I have titled the chapter.

2. THE EARLY DAYS

This account is about electrons, interfaces, and society and not primarily about my life,
but, nevertheless, it is probably a good beginning to note that I entered Imperial College as

a Research Student in 1943 to work with H. J. T. Ellingham, who is known for the free-energy-temperature plots called Ellingham diagrams and for his book, coauthored with A. T. Allmand, *The Principles of Applied Electrochemistry*.[3]

The Imperial College of Science and Technology, in South Kensington, London, was a grand place to work in the sense that it contained, in 1943, along with that at Cambridge University, the leading chemical school in England, housed in splendidly modern-looking buildings (amazingly, built in 1851). They were remarkable to view both on a good summer's day and when lit up at night (of course, after the end of World War II).

I spent two years of World War II at Imperial College, and remained another eight years. I was on "The Roof Party"†—the team which was always assembled in a maintenance room on the first floor of the physics and chemistry buildings, ready to man the roof at the sign of an air raid, warn of fire bombs (and put them out when they struck us) or bombers appearing to be near enough for emergency action (dropping down under the desk, etc.). When I was supposed to be "on" and had to be present in the department all night, it seemed pointless to sit with the others in the "maintenance room" and so I worked down in my laboratory and, when the air raid sirens did not go off (in 1944, that was seldom), slept on a mattress on the floor and continued carrying on my measurements, which at this time involved a reading every 30 minutes (alarm clock).

Around 1947

Left to right: Stan Ignatowitz; Brian Conway; Roger Parsons; Hanna Rosenberg; Ed Potter; Ahmed Azzam; Harold Egan, John Tomlinson.

---

† One had to be *asked* to join the Roof Party—associated with the membership of which was the privilege of living in the hostel on Prince Consort Road, where life was remarkably free from any suggestion of interference with one's own personal predilections. Now, there was an initiation test. Mine was conceived by a research student in inorganic chemistry called Con. "So you'll be on, then," he said breezily. "Well, before we can accept you finally, we'll just have a trip around the roof. Alright?" It was raining that fine London drizzle and I was wearing leather shoes. But I was 20 and there was a war on. The first part was alright—and I only slipped twice and didn't really fall over. Then, he came to the end of the building. There was, of course, another building, only about five or six feet maybe, but several hundred feet above the street. He jumped. I jumped. "You'll do," he said.

My first research student, Stanley Ignatowitz, was brought into my room somewhat hesitantly by Harry J. Emeléus, the great inorganic chemist (later Professor of Chemistry at Cambridge). Ignatowitz had just been released from being a prisoner of war in Russia, where he had been put to work in a salt mine and used in place of a donkey, to turn a rotor working a grinder. He was probably 40 years of age at the time. I put him to work upon determining what in those days was called "the deposition potentials" of the tetraalkylammonium salts. The resulting paper has, perhaps fortunately, been mislaid and has been dropped out of my publication list.

Harold Eagan was number two. He was a man whom I had known at Acton Technical College, where I taught third- and fourth-year physical chemistry when I was a graduate student. Harold was a man of enormous dynamism, optimistic and smiling, and carried out this work on salting out with great panache. He later attained the exalted position of "The Government Chemist," the official who supervised chemical test procedures for the British government.

Roger Parsons was number three. In those early days he proved to be an extremely effective experimentalist. It was remarkable to see him arrive in the morning, turn to the coat hook, and deposit a coat thereon with one hand while turning on the flow of hydrogen gas with the other. Parsons was thermodynamically oriented compared with Conway, who was definitely a mechanist from the beginning, and he was always very *confident* of his intellectual positions—even though these changed with fair frequency. Roger Parsons' subject was hydrogen evolution kinetics, and he specialized in the kinetics in nonaqueous solutions.

Brian Conway came along fourthly—around the same time as Parsons—and he worked on several things, but in particular the effect of trace poisons on electrode kinetics. Conway's work contributed greatly to our ability to make reproducible measurements on solid electrodes. I found Brian, even in those days, the soul of sincerity and had, perhaps, the best discussions with him.

Ahmed Azzam, from Cairo, "turned up" one day—there having been no previous correspondence—and said that he had decided to work with me. After I had persuaded him

On a Picnic, 1947

Left to right: Ed Potter (hydrogen kinetics theory); Roger Parsons (hydrogen in metals and $d\alpha/dT$); Hanna Rosenberg (protons in nonaqueous solutions); Ahmed Azzam (hydrogen evolution at 100 amps/cm^2); G. Wetterholm (hydrogen electrodes in the nA/cm^2 region).

(under protest) to take an informal examination, he settled down to work on hydrogen evolution at very high current densities.

Of course, anyone reading this will understand that in the late 1940s electrochemical technique was a trifle less complex than it can be now, particularly on the electronic side. Cathode-ray oscilloscopes were a rarity, and having one was regarded as a great privilege. The simple potentiometer was the essential instrument, and constant-current measurements were the norm.

Nevertheless, in spite of these simple techniques, these were heady days in the chemistry department at Imperial College, and the bottom corridor, where, by 1948, ten people were working in earnest on electrochemical research, and another one, Martin Fleischmann, was a corridor away, working with J. F. Herringshaw, and frequently within our own group, both for scientific discussions and social activities.†

I think the best part of those days, as I recall them, was the discussions, many of which we used to have walking two-by-two up and down the lengthy corridors at Imperial College. Probably, if one could resuscitate them now, the fare would seem rather bland, but, of course, in those days it was brilliantly exciting, and as the participants involved Parsons, Conway, Fleischmann, and myself, as well as Tomlinson, Azzam, MacKenzie, Potter, and Watson, it has sometimes been called by others "the golden age" (for it was near the beginning of the postwar era in which electrode kinetics grew so rapidly).

I think I should mention here the influence on us of the Russian group in Moscow under Frumkin. Before I got my Ph.D., I had understood that *Acta Physicochimica* of the U.S.S.R. was the place to find descriptions of high-standard fundamental electrochemical measurements. Here it was that I first saw the long trains used to purify gases, the preelectrolysis, the cells gas tight against the atmosphere, the constant stress upon purification. Nor was it easy to find the Russian journals, many of which did not reach us until after the end of World War II (though I think there was a certain leakage via Portugal). Saturday morning was my time for going down to the Chemical Society eagerly to read the latest that I could get. We certainly owed a good deal to Frumkin although the "relations" were so cold that when Parsons and I first tried to get a visa to make a visit to Moscow (about 1948) we were told, "No, there is no hotel accommodation in Moscow."

I have mentioned the phrase "the golden age," referring to the bright intellectual atmosphere which existed in the electrochemistry group at Imperial College. However, I might say that the phrase could be interpreted by those who remember the days to have a double meaning because we were all rather young, you see—and mostly unmarried. The average age during the nine years of my tenure of the lectureship at Imperial College was 26. I was, on the whole, some two to three years older than my students. Not only group members came on the picnics—there would have been too many males. Those others (particularly) from the library may have been attracted by the golden hue. Picnics in the country, visits to the Isle of Wight, and other such—purely social—activities in the group led to much human interaction, which, in retrospect, had, on the whole, happy consequences.

But, of course, that is all part of a longer story.

Lastly, perhaps I should say that in those days I was regarded as a bit of a slave driver. What of it? Another way of looking at being a slave driver if you have a bunch of young students who are two to three years younger than you is that you create a lot of verve in the group. We got on fast at Imperial College around 1950 (see Table 1).

---

† Martin's penchant for the rapid production of a string of differential equations to cover almost any topic was already apparent well before he got his Ph.D. His somewhat turgid style and refusal to talk down to simpler minds was often a cause for visitors who came to my room with stupid questions to recall that there was a somewhat earlier train from nearby Victoria Station than they had firstly considered. It saved my time.

Picnic at Freshwater Boating, 1949
Left to right: JOMB; Josephine Mertens (library); John Tomlinson (liquid silicates); Hanna Rosenberg; M. G. Fouad (electropolishing); Boatman.

3. GENERAL CONTRIBUTIONS

There are 60 items in my publication list here, and compression is therefore difficult, particularly considering the diversity of the material. I have decided to mention only four items.

The first is an article "Electrochemistry: The Underdeveloped Science."[4] This article was written in the heyday of the 1960s when support for electrochemistry from NASA (under

Isle of Wight Bicycles, 1949
Left to right: Josephine Mertens; John Tomlinson; Ahmed Azzam; JOMB; Hanna Rosenberg.

Picnic at Freshwater 1949
Left to right: Hanna Rosenberg; B. E. Conway; JOMB; Josephine Mertens.

Off for the day! Around 1953
Front row: I. A. Ammar (with racquet) (kinetics of hydrogen evolution on WOMO); John Parry-Jones (friction and potential); Mrs. Dorothy Lowe; Mrs. Linda Bockris; Josephine Mertens. *Back row*: A. K. M. S. Huq; Joan Digby; Dennis Lowe; JOMB.

TABLE 1
People Obtaining Their Ph.D. Degrees in the Chemistry Department at
Imperial College after Studying with J. O'M. Bockris (1945–1957)

Year	Name	Thesis title
1948	Roger Parsons	Studies in the Kinetics of Electrode Processes
1949	Brian E. Conway	Studies in Hydrogen Overpotential
1949	Harold Eagan	Salting Out
1951	A. E. Davies	Studies in the Electrochemistry of Silicate Melts: Transport
1950	Alec Libermann[a]	Physical Chemistry of Iron and Steel
1950	J. W. Tomlinson	The Electrochemistry of Molten Silicates
1951	A. M. Azzam	Kinetics of Hydrogen Evolution at High Current Densities
1951	J. W. Bowler-Reed	On Salting Out and Salting In
1951	Lucy Oldfield	Redox Reactions with Benzidines
1951	E. C. Potter	The Hydrogen Evolution Reaction: On Nickel and Alkaline Solution: The Stoichiometric Number
1951	Hanna Rosenberg	Proton Conductivity in Non-Aqueous Solution
1951	S. A. Spratt[b]	The Physical Chemistry of Iron and Steel
1952	R. G. H. Watson	Hydrogen Evolution in Alkaline Solutions: Mechanism Studies
1952	J. W. Evans[b]	The Physical Chemistry of Iron and Steel
1952	Molly Gleiser[b]	The Dissolution of Oxygen in Very Pure Iron
1952	I. A. Ammar	Studies in Electrolytic and Dielectric Polarization
1953	D. C. Lowe	The Measurement of Viscosity at High Temperatures—Molten Silicates
1953	J. D. MacKenzie	Viscous Flow in Liquid Silicates
1953	D. F. Parsons	A Magneto Chemical Study of Free Radicals
1954	G. W. Mellors	Electric Transport in Liquid Silicates and Borates
1955	Eric Sheldon[c]	Investigations on the Mechanism of Hydrogen Evolution at Various Metal Cathodes of Intermediate Overpotential
1955	D. L. Hill[a]	Studies in Titanium Chemistry
1955	Alina Borucka[b]	Diffusion Studies in Molten Salts
1955	M. R. S. Heynes	The Partial Molar Volume of Silica and Liquid Silicates
1955	A. K. M. S. Huq	Oxygen Evolution Kinetics with Special Reference to the Reversible Oxygen Potential
1957	Douglas Inman[a]	Studies on the Electrochemistry of Fused Salt Systems
1957	I. A. Menzies[a]	Studies in Titanium Chemistry

[a] Supervision shared with G. J. Hills.
[b] Supervision shared with J. A. Kitchener.
[c] External student, Acton Technical College.

the direction of the irrepressible Ernst Cohn) was great,† in connection with the development of electrocatalysts for fuel cells in space vehicles. The future of electrochemical technology looked particularly bright in those days—just as it seems so again today because of the pressures of air pollution.

The second is a general article called "Overpotential."[5] This was an article which I wrote in one day in the Dorchester Hotel in London when confined there is a snowstorm, around Christmas of 1970. It is an article which has been useful as an amusing introduction to a concept which not all mature physical chemists of that time had bothered to understand.

Then, of course, there is Bockris and Reddy, *Modern Electrochemistry*, published by Plenum in 1970,[6] with corrected editions through the 1970s. This is, without doubt, a very well known work.

Just as Bockris and Reddy is well known, my next example is very little known. It is called "The Great Academician" and is an account of my personal impressions of Alexander N. Frumkin.[7] The article has recently been republished by the Russians in a book on the life of the remarkable Frumkin, who, in my view, was (with his approximately 400 collaborators) the greatest contributor to the development of fundamental electrochemistry in the first half of the 20th century.

Lastly, I wish to mention the *Comprehensive Treatise of Electrochemistry*,[8] which I initiated and for which I was fortunately able to attract the coeditorship of Conway and Yeager and the assistant editorship of S. Srinivasan, Ralph White, Y. Chizmadzhev, and P. Sarangapani. The treatise covers all electrochemistry, and I suppose that we should try for a new edition around about 2000, to be published about 2005. One hopes that someone will take a new look at everything every 15 to 20 years.

4. NEW METHODS OF INVESTIGATION (1947–1990, 40 PUBLICATIONS)

I have published many papers in this category (about one a year) so that the papers mentioned have had to be picked rather arbitrarily: I decided to mention five papers from work of particular difficulty and interest.

The first is the viscometer that Dennis Lowe[9] built in 1952 and which was used to measure viscosities (largely for liquid silicates) up to 2000°C (Fig. 1). Viscometers for measurements at these sorts of temperatures are very interesting. One has to guard against, for example, the growth of electronic conductivity in alumina tubing. Silica evaporates and dissociates to silicon monoxide—this kind of thing. The diagram shows the complex apparatus. Viscometers for the 2000°C range are not greatly changed in 1990.

Nanosecond time measurements would be the second choice because they were made in my laboratory for the first time in 1959,[10] due largely to the electronic skills of Eric Blomgren. Such very-high-frequency measurements involve inductances (for the inductive impedance is ωL). These contributions would have nothing to do with the electrode reaction under examination, but would be entirely artifacts of the circuit. They can be balanced by placing a counter current wire near that causing the false inductive impedance. It was possible by the utilization of the nanosecond time domains to measure particularly high exchange current densities in molten salts.

---

† However, the article is also seen by some as having added to the overpotential in the electrochemical cell. In praising the work of some, I left out the work of others—the latter did not agree with my choice. They appear to have been reacting ever since.

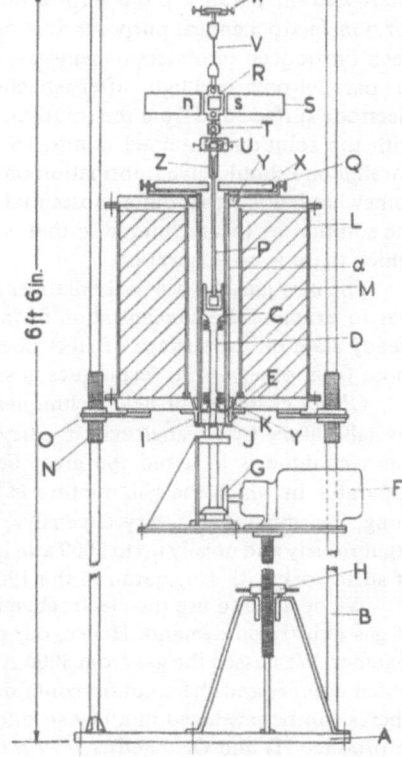

FIGURE 1. General view of viscometer. A, Base plate; B, main supports; C, 2-in. internal diameter recrystallized alumina tube; D, crucible support; E, self-centering chuck; F, $\frac{1}{4}$ horsepower motor; G, bevel gears; H, capstan jack; I, chuck housing; J, chick backing plate; K, grease nipple; L, brass furnace casing; M, lamp black furnace packing; N, threaded sleeves; O, furnace support rings; P, graphite center spindle; Q, D-plates housing; R, Mg–Al coil former; S, Alcomax II magnet; T, galvanometer wire; U, total reflecting prism; V, Be–Cu suspension wire; W, inner cylinder centering prosm; X, water-cooled D-plates; Y, furnace element contact cones; Z, Tufnol collar; α, molybdenum furnace element.

My third choice would be the radioactive technique published by Blomgren and myself in 1960.[11] Although there had been earlier work on radio trace measurements at electrodes, they had been concerned with the deposition of polonium on the electrode itself and were not measurements of adsorption. Blomgren and I devised a Geiger counter, one end of which was covered with a nylon membrane on which very thin layers of platinum were deposited. The surface of the Geiger counter-electrode-membrane device was lowered toward the solution containing the radioative material (thiourea). As the electrode approached the surface of the solution, through the air, the frequency of registration of clicks on the Geiger counter monitor increased slowly, but when the electrode contacted the solution there was a very large increase, which corresponded to the adsorbed material on the surface. The Geiger counter registration was then compared with a similar one where a known amount of radioactive material was introduced in drops onto the surface of the electrode and allowed to evaporate.

It was possible to follow the adsorption of thiourea as a function of potential, and this was the first potential-controlled radioactive adsorption measurement, the model for much subsequent work done both in my own laboratory but particularly in that of Kazarinov† in the Frumkin Institute and by Horanyi in Budapest.

Two other techniques are of much more recent vintage.[12] One is a polarization modulation modification of the Fourier transform infrared (FTIR) techniques. This, in its SNIFTIRS form, was taught to Ahsan Habib by Stan Pons, then at the University of Edmonton in Canada. The modification which Habib made involved the introduction of a polarizer which

† Dr. Kazarinov worked in my laboratory at the University of Pennsylvania for about a year around 1963.

alternated the parallel to the vertical phases of the polarization of light. This had been done for nonelectrochemical purposes at Proctor and Gamble, but it was the first time that it had been introduced into electrochemistry. If one accepts Greenler's theorem at its face value, the parallel-polarized light, after reflection from the electrode, contains information from the electrode surface and from the solution, but the vertically polarized component interacts only with the solution-contained bonds, so subtraction of the vertical signal strength from the parallel one should give information only from the surface. K. Chandrasekaran then showed somewhat later[13] that quite substantial amounts of information *were* being picked up† from the solution by this technique so that we began to subtract a quantity taken at a potential at which there is no adsorption.

The new parallel–vertical polarizer technique enabled us to do many things. One of them was to examine the concentration of intermediates for methanol oxidation reactions in the steady state on the surface of electrodes. A comparison of results from this technique with those from other *in situ* techniques is shown in Fig. 2.

Of my examples of new techniques here, I would like to take the recent work done in my laboratory by Jovancicevic in carrying on pulse-pressure measurements.[14] The idea of this technique is to avoid the great awkwardness of trying to work at 2000 atm with an apparatus in which the gas mixture is mechanically *pumped* up to the pressure.‡ For one thing, the apparatus is very expensive, takes up most of the laboratory, and beats its way lugubriously and noisily up to 2000 atm in, for example, eight hours. I had plenty of experience of such work with Nagajaran in the 1960s.

We decided to use the electrochemist's ability to cause the production of large amounts of gas quickly on demand. Hence, our pressure vessel (Fig. 3) has two long electrodes close together. We passed the gas from 3000 A into a 0.5-cm³ space. The gas pressed on a membrane which compressed the solution containing the cell and its electrodes. One thousand atmospheres can be produced in a few seconds for 1% of the cost. The only thing is, it is not wise to produce H_2 and O_2 together.§[14] An O_2 or H_2 producing system is possible.

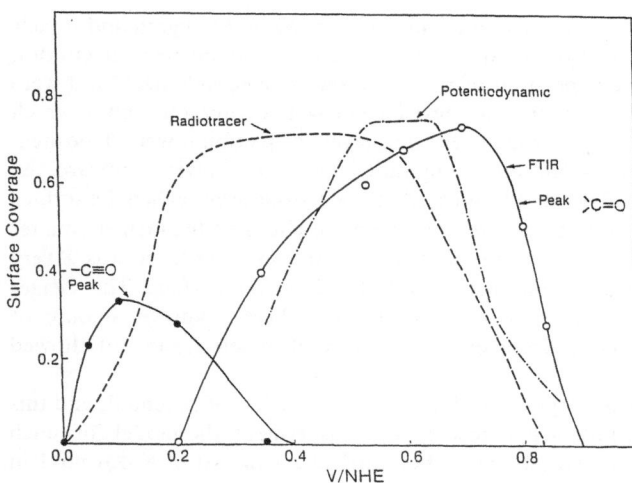

FIGURE 2. Comparison of various *in situ* techniques for monitoring the chemisorption of methanol on platinum.

† Chandrasekaran and I did not challenge Greenler's work. He applied his concepts to the gas phase, and there the concentration of bands (compared with those on the surface) is negligible. This is not so in solution.

‡ One must also fear the dissolved gas in the system from the approach.

§ Although Jovancicevic proceeded to do this with impunity for many months, Enayetullah, his successor, was less fortunate. He refused to enter the laboratory again.

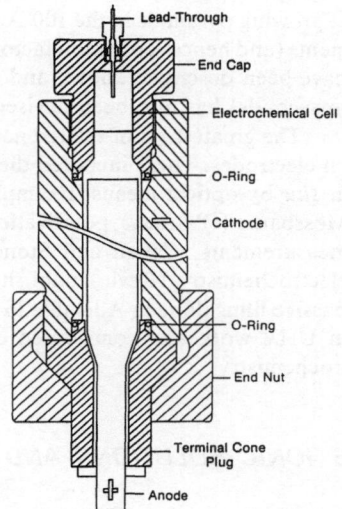

FIGURE 3. A cross section through the high-pressure apparatus.

This technique should be the father of a new technique for doing high-pressure physical chemistry, pulsed and steady state.

These brief sketches do not get to the heart of the attitude toward experimentation in my laboratory, which has always been to think it out and devise a new approach made in the workshop rather than to buy expensive equipment.

Thus, several new techniques for 2000°C work were initiated in the 1950s with Kitchener and Tomlinson.[15] Magnetochemical advances were made with D. F. Parsons.[16] Several different new methods for measuring diffusion coefficients were devised including some for making measurements at high T and high D together[17]; measurements of fugacity and partial molar volume of H in metals were carried out with McBreen.[18] *In situ* observations

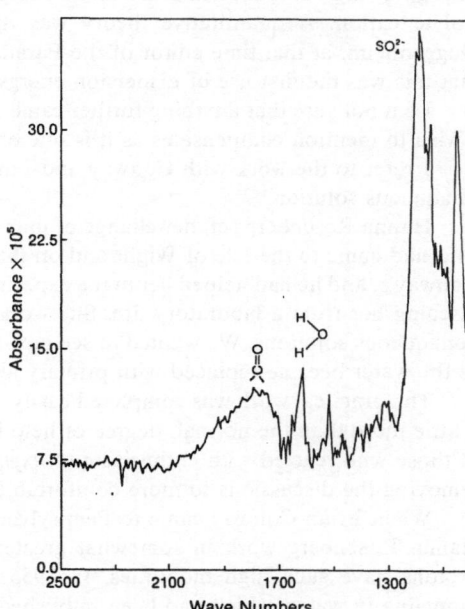

FIGURE 4. IR spectrum of $H_2C = O$, H_2O, and SO_4^{2-} on Pt electrode.

of growing pyramids in the 100-Å range were achieved with Damjanovic.[19] Tritium measurements (and hence separation factors) were made in 1963.[20] The new pressure pulse techniques have been described above, and a temperature pulse technique (heats η, activity measured in seconds) has also been devised with Velev.

The greatest *set* of related new techniques were spectroscopic and optical measurements on electrodes. These included the initiation of the modern trend to study electrode surfaces *in situ* by optical means. Examples include radiotracer (Blomgren), ellipsometry (Reddy), Mössbauer (O'Grady), polarization modulation IR (Habib) (Fig. 4), and surface conductance measurements (Cahan and Stoner). The first ultra-high-vacuum (UHV) measurements in electrochemistry (Revie) published in a refereed paper were the Auger measurements of passive films done in Adelaide in collaboration with Bruce Baker, who had much experience in UHV work and contributed enormously to its little recognized introduction into electrochemistry.

5. IONIC SOLUTIONS AND LIQUIDS

5.1. Ionic Solutions (1958–1984, 32 Publications)

When I became a research supervisor at the tender age of 22, it seemed clear that I ought to be doing some research on ionic solutions as well as on electrode processing, and I started out with Harold Eagan, who wanted to continue the study of the effect of ions on the solubility of organics, whiche had been doing earlier at Acton Technical College.

Of the two papers I want to mention in this section, the first contains the theory of salting-in.[21] John Bowler-Reed had made a series of studies on the tetraalkylammonium salts and their effect on the solubility of benzoic acid. We found that, surprisingly, these large cations caused an increase and not a decrease of solubility, as had been expected.[22] Dispersion energy equations were quite familiar to me in 1950, and it was immediately clear that the large radius of the alkylammonium cations meant a large polarizability, which in turn gave large attractive intermolecular forces between the organic molecule and the electrolyte cation. A quantitative theory was the consequence, and the very critical E. A. Guggenheim, at that time editor of the Faraday Society, actually wrote a note pointing out that this was the first use of dispersion energy equations applied to electrolyte theory.

I am not sure that anything further came of this work, but, if not, then the second paper I wish to mention compensates as it is one of the most influential papers I have published.

I refer to the work with Conway and Linton on the mechanism of proton conductivity in aqueous solution.[23]

Hanna Rosenberg (cf. her change of name to Linton, described below) was one of those who had come to the Isle of Wight and on the picnics. She had been a close friend of Brian Conway's, and he had helped her in the experimental part of her Ph.D. thesis, on one occasion rescuing her from a laboratory fire. She worked with me largely on proton conductivity in nonaqueous solutions. We wanted to see how far the excess conductivity of the proton lasted as the water became replaced with primary alcohols of increasingly large size.

The practical work was completed easily, but Ms. Rosenberg had to have a fair, perhaps a little more than the normal, degree of help in writing her thesis. I found that she was one of those who reacted with enthusiasm to *explicit encouragement*, which we accomplished by removing the discussions to more comfortable locales than that of the chairs at my desk.

When Brian Conway came to Pennsylvania, we got down to tackling the theory of the Hanna Rosenberg work in somewhat greater detail. We reexamined the concept of why protons have such high mobilities. In 1955 the theory of Bernal and Fowler on proton tunneling in water, which had been published in the first edition of the *Journal of Chemical*

Physics in 1933, was still fresh in mind. Protons tunneled from H_3O^+ to the next water molecule: this was the reason for the high mobility.

Conway soon worked out that while this view was stimulating, it was also naive: the predicted velocity was 100 times higher than that actually observed, and there must be another rate-controlling step than the tunneling.

Many dozens of diagrams later (and not a few contortionate bodily immitations of various ensembles of waters in changing juxtapositions), we came to the conclusion that the step holding up passage of the too easily tunneling proton was the rotation of the water molecules under the influence of the field of an approaching proton. After a proton had passed from one molecule to another, the molecule had to "turn back," offering an orbital to another proton. This turning back took some time, and calculations showed that, for normal concentrations of protons, there would have to be a significant wait before the water turned back to offer a vacant orbital (Fig. 5).

We then went into what for the time was *Journal of Chemical Physics* standard mathematics and did, indeed, derive some quite nice equations for the effect of the local field of the approaching proton on the water molecule, showing how the water molecule rotated in time.

The paper was published in the names of Conway, Bockris, and Linton,[23] not Rosenberg, because by this time Ms. Rosenberg (whom many considered to be very beautiful) had decided to become a successful film actress and had changed her name to Hedda Linton.†

Much of the Discussion part of the proton transport paper was devised in the warm Philadelphia evenings of 1955, for we were at that time without the blessing of air conditioning. Sometimes we would take to the streets, get our 2:00 A.M. coffees from the local stand, and come back to continue the discussions.

Some months after we had finished the proton transport paper, and while one morning at my house in Philadelphia, Conway and I realized that our mechanism threw light on the perplexing problem of why it is that the mobility of protons is so much greater in ice than it is in water. Intuitively, of course, one would think that the OH_2 in ice would rotate more slowly (because more bound) than in liquid water. However, in fact, as had been determined much earlier by Bradley, the mobility of protons in ice is ~100 times greater than that in water.[24]

It occurred to Conway and me that the point lay in the *concentration* of protons. Protons are far less concentrated in pure ice than in pure water so that the average time between the arrival of one proton and the next in a given water molecule was much greater in ice than in an aqueous solution of any reasonable concentration. Now, the water molecules could indeed get back into the appropriate position (to give a hand to the oncoming proton) themselves in time without having to wait for the protons to come and turn them back again. So the original conditions of the Bernal–Fowler theory ("pure tunneling control") would be much nearer to being fulfilled. We rushed off to the university to write the note and send it to *Journal of Chemical Physics*.[25]

All this seemed rather a halcyon tale, but there lay ahead a summer storm. We were attacked by Eigen on the basis that some of the integrals in our paper had been evaluated classically and not quantally.[26] It was easy to show that it made no difference.[27] However, we were disturbed by the publication of a very long paper by Eigen and DeMayer[28] *qualitatively* proposing a theory startling similar to that which we had published, claiming, of course, that our own theory was wrong because of the classical calculation of some integrals.

The story was saddened by the fact that Eigen had visited us during the writing of the 1956 *Journal of Chemical Physics* paper. Since that time, the story of protons in solution has often been discussed and revised. A good review of this has been given by Erdey-Gruz in

---

† The romantic Miss Rosenberg considered her personal fate similar to a mixture of those of Hedda Gabler (of Ibsen's play) and Cathy Linton (of Brontë's *Wuthering Heights*). I concur with her evaluation.

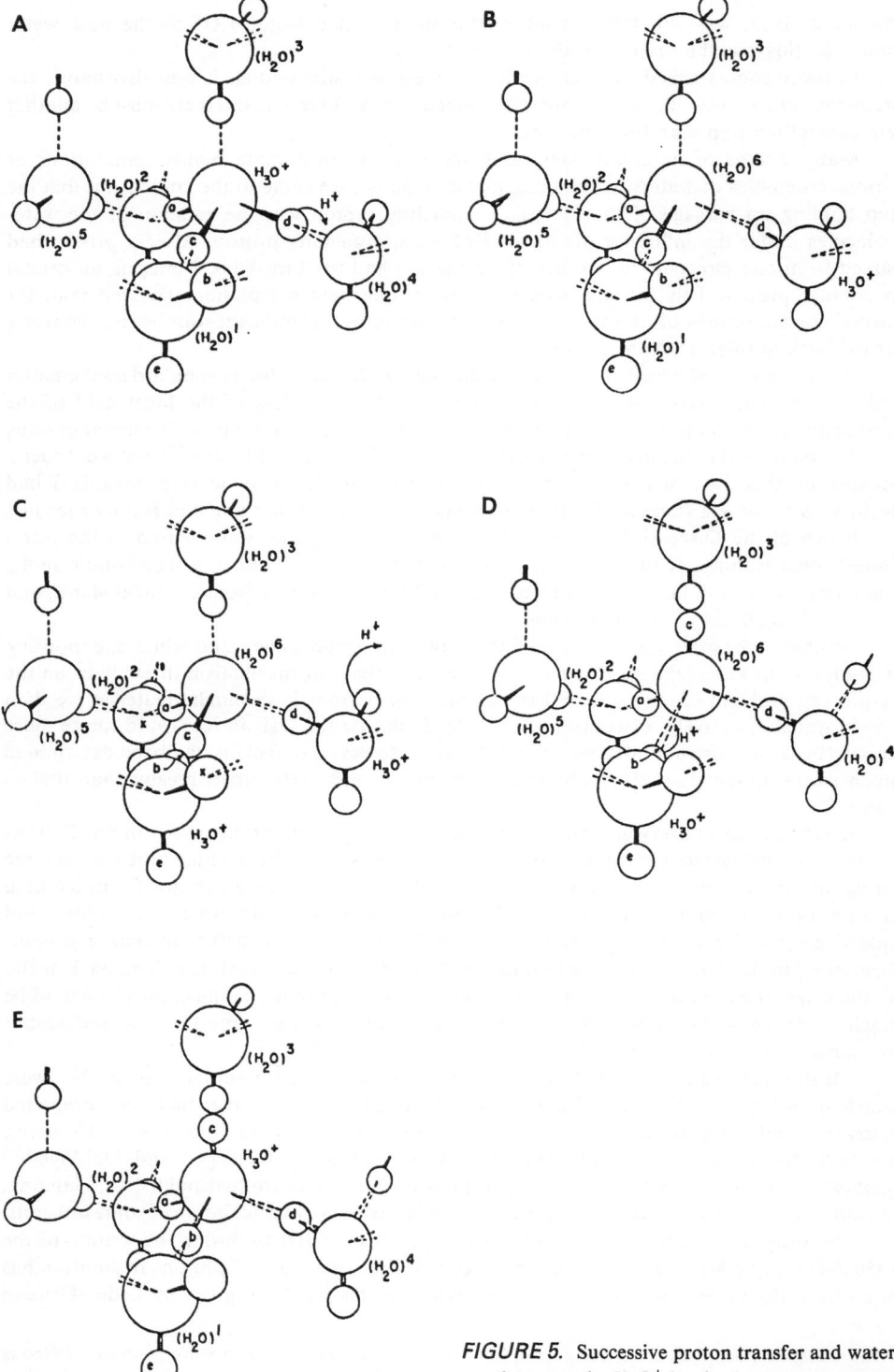

FIGURE 5. Successive proton transfer and water rotation near the H_3O^+ ion in the water structure.

Number 12 of *Modern Aspects of Electrochemistry*. Although some authors have tried to see things a little differently, the basic picture has remained based on the Bernal–Fowler tunneling, with rate determination spelled out in terms of the change of configuration of the water molecule, as effected by the approaching proton.

Our 1956 model also gives the mobility ratio, u_{H^+}/u_{D^+}, correctly and—not to be forgotten—allowed us both to see the long-term significance of Ms. Rosenberg's curves.

5.2. Ionic Liquids (1948–1984), (34 publications)

The unusual term which is used for study under the above heading is molten salts. I do not use this term because some of the systems referred to are not salts.

These researcher were carried out after I had about eight years' experience of physical chemistry at high temperatures (around 2000°C). From the point of view of technique, it seemed a comedown to work at 1000°C but was much easier, to say the least.

My selection here must include the study made by beautiful Alina Borucka, who worked with Joe Kitchener and me in the later London days and remained behind with Kitchener when I went to the United States in 1953.†

University of Pennsylvania Around 1957

Front row: Cathy Jesch (electrocapillary); Erik Blomgren (radiotracer method); Branko Lovsrececk (rectification by ionic junctions); Jacqueline Diaz (germanium dendrites); Michio Enyo ($Cu^{2+} + e^- \to Cu^+$); Evan Crooks (conductance of molten salts). *Back row*: Asa Despić ($Ag \to Ag^+ + e^-$ and theory of β); Jim Barton (dendrite growth); Mr. Stephens (manager); Austen Angel (diffusion in molten salt mixtures); Franscisco Colum (H dissolution from Fe); George Myers (18-years administrative assistant); Mr. Ritchie (general assistance).

† Poor Alina! Her beauty was counteracted by her sharp tongue and proud Polish intellectual ways. This mitigated against her ability to attract a husband, which irked her considerably, but worse was to come, for her overcritical ways and overuse of words such as "fool" and "idiot" in respect to her collaborators made it difficult for her to hold a job in the United States. After having been appointed group leader at Allied Chemicals, and duly fired for being imperious, she fell rapidly. In 1983, she was found at Kennedy Airport, a bag lady, with no money, who would like to get back home. (After Poland and prisoner-of-war camp in Russia, her home had been London.) She was taken to the hospital by the police and died from cirrhosis of the liver, her body being taken care of by the Polish Community.

The first account here must concern the Nernst–Einstein equation, that product of the greatest electrochemist and the greatest physicist of the 20th century, respectively, a phenomenological relation that connects equivalent conductivity to diffusion coefficient. For all intents and purposes, it *should* be true. Alina Borucka and I were curious, however, whether the relationship would stand up to a test made by radiotracer determination of D^+ and D^-, which we were able to make using ^{24}Na and ^{36}Cl. The sum of the diffusion coefficients multiplied by appropriate dimension-correcting factors was some 10 to 20% less than the equivalent conductance. This was an interesting challenge: a phenomenologically deduced relation "cannot be wrong."

At this time (around 1956) I knew Carl Wagner at MIT. Alina was writing her thesis in London and I was corresponding with her from Philadelphia. During a conversation with Wagner, he sketched out on an envelope how it might be possible for there to be a mechanism for diffusion which would not be relevant for conductance: ion pairs could hop together, and then they would contribute to the diffusion but not to the conductance.†[29] Hence, the sum of the diffusion coefficients could be greater than the conductance. The idea is shown in Fig. 6, and the paper was published in the prestigious *Proceedings of the Royal Society.*‡

Nolan Richards came to me from Harry Bloom in New Zealand. (Bloom and I had been contemporaneous graduate students at Imperial College.) What Richards did was to pair up with Ed Brauer, an excellent electronics engineer of the time, and invent a method involving Lissajous figures for presenting the velocity of sound in molten salts. From there, a well-known formula gives compressibility. From this one can obtain numerous things, including the free volume of the salt.[30]

A magnificent set of experimental data was delivered by Nolan Richards. Bloom, who happened to be present at the time in my laboratory, then suggested to us that we try applying a little-known theory by Furth.

Furth had published a version of the concept of holes in molten salts, a model popularized by Henry Eyring. The point about Furth's version, however, was that the holes were distributed in size. The theory was more sophisticated than what I had seen published from Eyring. However, the paper was little known because Furth published it in the *Proceedings* of the obscure but prestigious Cambridge Philosophical Society.[31] We were able to calculate from the theory the compressibility and expansivity of molten salts. The agreement was good, as shown in Table 2.

The degree of agreement is the more impressive in view of the fact that there are no adjustable constants or scaling factors in the theory concerned.

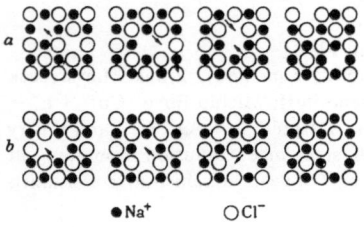

● Na⁺   ○ Cl⁻

FIGURE 6. (a) Formation, rotation, and dissociation of a coupled vacancy. (b) Diffusion of a coupled vacancy.

---

† The combative Alina later contested the fact that I had suggested the idea as a result of my conversation with Wagner and claimed that it was entirely her own concept.

‡ The Alina Borucka radiotracer work was an interesting example of how one can obtain money for one thing but happen to do another with agreement of the sponsors. Thus, Alina Borucka had started out with *money obtained by her* because she wanted to work on carbonate fuel cells, a love for most of her lovelorn life. It was desirable to find out the transport number of the carbonate ion. She actually came to work with me to do this, but I pointed out to her that carbonates were difficult to do at first and suggested that we try something simpler.

TABLE 2
Comparison of Calculated Isothermal Compressibilities and Expansivities with the Experimental Values

Salt	Temperature (°C)	$10^{12}\beta$ (calc.) (cm²/dyn¹)	$10^{12}\beta$ (obs.) (cm²/dyn¹)	$10^4\alpha$ (calc.) (°C⁻¹)	$10^4\alpha$ (obs.) (°C⁻¹)
LiCl	614	20.8	19.4	—	—
	800	30.5	24.7	4.0	3.0
	900	39.1	28.6	—	—
	1000	47.9	33.0	4.0	3.3
NaCl	800	27.8	28.7	3.3	3.6
	900	36.9	22.8	—	—
	1000	49.6	40.0	3.6	3.9
KCl	772	27.6	36.2	—	—
	800	30.2	38.4	2.8	3.9
	900	42.3	45.7	—	—
	1000	56.7	54.7	3.2	4.2
CsCl	642	20.2	38.0	—	—
	800	39.0	51.8	4.2	4.1
	900	57.0	62.6	—	—
	1000	72.5	76.3	4.4	4.5
CdCl₂	600	31.0	29.8	2.7	2.4
	700	37.3	33.1	2.7	2.4
	800	46.9	36.9	2.9	2.5
NaBr	747	32.8	31.6	—	—
	800	40.1	33.6	3.2	3.1
	900	47.4	38.6	—	—
	1000	59.5	44.9	4.0	3.4
KBr	735	28.0	39.8	2.5	3.7
	800	34.5	43.8	2.6	3.8
	900	51.3	52.1	—	—
	1000	69.9	62.1	3.0	4.1
CsBr	636	56.3	49.1	4.0	4.4
	800	76.0	67.1	4.2	4.6
	900	97.5	82.7	—	—
	1000	130.5	103.1	4.4	5.1
NaI	651	29.1	37.8	5.3	3.6
	800	43.7	47.3	2.5	3.8
	900	74.1	55.6	—	—
	1000	106.1	65.6	3.9	4.2
KI	683	40.3	46.7	3.4	3.8
	800	55.4	59.9	3.3	4.1
	900	81.9	72.0	—	—
	1000	124.1	87.3	3.5	4.5
LiNO₃	300	38	19.6	—	—
	400	56	23.4	11.0	3.2
	500	66	28.9	9.3	3.3
NaNO₃	400	26	21.6	8.5	3.8
	500	47	26.8	7.2	3.9
KNO₃	400	45	23.4	9.0	4.0
	500	57	29.4	7.5	4.1

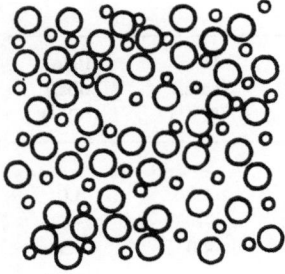

FIGURE 7. The hole model with randomly located and variable-sized holes in the liquid.

This work should be given more attention than it has gotten. People have gone away from making *models*, and it will be seen by comparing diagrams of the structure obtained by Monte Carlo calculations (the best known is by Woodcock and Singer) that it compares well with the model derived in 1957! (see Fig. 7).

Next came a piece of work which still worries me: the discovery with Leonard Nannis of a formula which relates the heats of activation for transport processes in liquids to their melting points.[32] The formula runs:

$$\Delta E^x_{\text{Transport}} = 3.7RT_{\text{mp}}$$

where T_{mp} is the melting point, and $\Delta E^x_{\text{Transport}}$ is the heat of activation for diffusion, viscous flow, and conductance (Fig. 8).

Why? Here difficulties arise because the diagram shows that the equation applies to liquid rare gases, organic liquids which are not associated, molten salts, and molten metals. It applies over a range of a few degrees above absolute zero to 2000 K! Thus, a similar mode of transport in all these very discrepant types of systems is implied.

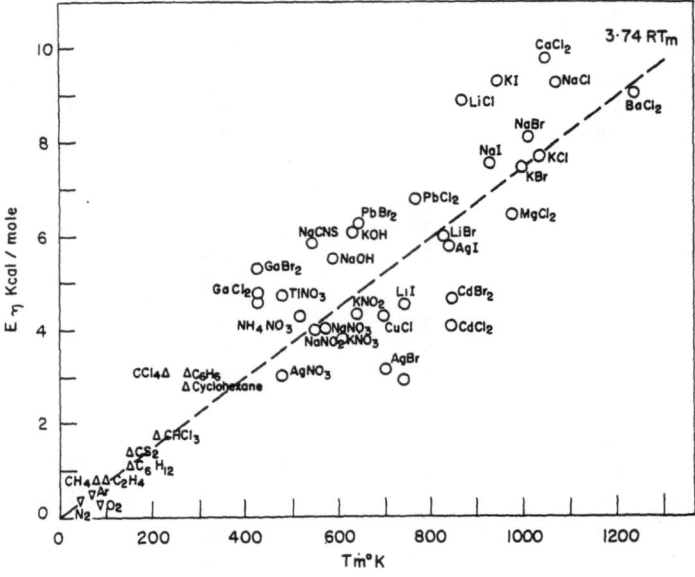

FIGURE 8. The dependence of the experimental energy of activation for viscous flow on the melting point.

Toshiko Emi came to me from the Kawasaki Steel Company near Tokyo, and it was he who eventually gave a very approximate and phenomenological theory which nevertheless, perhaps, contains the essence of the situation.[33] The modeling Emi gave was one of holes in liquids, which gave a fair numerical account of the heat of activation for transport at constant pressure in simple liquids, based on the somewhat tacit assumption that the functional dependence of σ (surface tension) on r (radius of hole) does not affect the integration; that is, σ does not depend on r.

Until the mid 1960s I had concentrated upon ionic liquids which are simple (a relief to the complexity of liquid silicates, borates, and phosphates). My colleagues tried to shift me out of this attitude and into molten salts which contain complex ions. However, a complex ion in a molten salt might be a difficult thing to define. I therefore set about to prove the existence of clearly bonded aggregates in molten salts. I did this with two approaches. The first was with Tanaka.[34] This work produced the first Raman measurements made in molten salts and marked the first time we used a laser (1963). We were able to make measurements of the polarization states of the ions as well as the frequency and to detect thereby $CdCl_3^-$ as well as other complexes.

The study I liked most, however, in respect to complexes in molten salts was that carried out by Douglas Inman.[35] He not only detected and identified the complexes in molten salts by quarter time potential measurements but also made measurements of their lifetimes, finding in the same cadmium chloride system a lifetime in the region of 10^{-2} s by utilizing galvanostatic measurements, where the delay times were interpreted in terms of the lifetime of the complex as yielding ions to take part in the electrode process.[36]

As time went on into the 1960s, discussions in this field because more connected with the fundamental movements of ions. Two views existed. One, corresponding to the whole view, saw ions moving into vacancies when they turned up in the flickering pattern of holes adventitiously occurring in the liquid. The other view saw a kind of "struggling through the crowd" mode of movement, in which each movement was far less than that of a radius.

I tried to solve this with Nagajaran[37] by setting up a piece of equipment (slow and cumbersome, but effective) in which we measured the diffusion coefficient of Na^+ in sodium nitrate at different pressures up to 2000 atm at 1000°C. If one knows the variation of the diffusion coefficient with temperature at various pressures, it is easy to obtain the variation with temperature at constant volume as no new holes can form with increasing temperature at constant volume. Under these conditions, $[\partial \ln D/\partial(1/T)]_v$ gives you the heat of activation for jumping into the holes (E_{jump}) whereas the normal heat of activation obtained at constant pressure gives you the sum of $E_{jump} + E_{hole}$ formed.

The results here were rather interesting: it turned out that the normally observed constant-pressure heat was $E_{hole formed}$ (i.e., the Furth model was correct), and the E_{jump} was small.

Of course, this model does not apply to all liquids, but if the mode of transport varies with the liquid (excluding, of course, the associated liquids)† how is it that the Nannis formula $\Delta D^x = 3.7 RT_{mp}$ applies to them all?

On the electrode process side, I have contributed little in the molten state. However, there was an interesting study with Douglas Inman and Grahame Hills on the deposition of titanium, uranium, and thorium, an early exhibition of the contribution of electrochemistry to the separation of components from nuclear waste. This marked the beginning of my interest in dendrites. We found that there were three conditions under which these materials deposited.

---

† There the work carried out by Dennis Lowe[38] and Doug MacKenzie[39] for viscosity in silicates shows that here (for the concentrations of less than 10% of metal oxide) one is breaking bonds and that is the rate-determining step. I suspect this is the case for water, too, just as we found it to be so for molten silica, which bears resemblance to water in its structure.

At low current densities, they deposited easily and there were no dendrites formed, the deposit being entirely the transition metals. But then came an intermediate current density, and when one examined the electrode data, one could see bits of dendritic crystals sticking out from the electrode. Finally, at higher current densities the transition-metal crystal formed passed to what is called in industry "fines," that is, a powdery type of deposit, but at the same time, one could see that it contained alkali metal. This could be removed by immersing the cathode with its product, taken out from the molten salt, into water. Here, then, the transition-metal cation exceeded its limiting current, alkali metal began to deposit, and the fines were a secondary product of chemical reactions between the titanium, uranium, and thorium ions and sodium diffusing out into the diffusion layer, and reacting homogeneously.

6. THE STRUCTURE OF MOLTEN SILICATES AND GLASSES (1948–1958, 14 PUBLICATIONS)

Although this work was done in the 1950s, there have been few repeated studies, partly due to the inhibitions which some physical chemists feel about work at 2000°C. The work also illustrates my penchant for getting associated with a variety of new things and taking an independent look at them. Thus, when Sir Charles Goodeve and F. D. Richardson of the British Iron and Steel Research Association called upon me one day at Imperial College in 1947 (I was 24 years old) to ask me if I would lead some work upon the nature of slags, I readily agreed (the financial terms were good) but directly they had gone went quickly to the library to find out what slags are.

In 1947, when this work began, concepts on the structure of slags were as described in great detail in a massive book on steel making by Eitel.[40] The corresponding structure with glasses was that which had been originated by Zachariasen.[41]

Slags (which are complex mixtures of several metal oxides dissolved in SiO_2 and made molten at temperatures in the region of 1650°C) were described by Eitel as consisting of complex *molecular compounds*, in which no ions were present but metal cations such as calcium were bound covalently to lengthy silicon-containing molecules. Stress was placed on precise (but imaginary) molecular entities in the slag in terms of "activity calculations," which were prevalent in the thermodynamic attitudes of the 1930s and 1940s.

I was joined in the work supported by the British Iron and Steel Research Association by Joe Kitchener, a worthy physical chemist in the chemistry department at Imperial College who later transferred to work in the metallurgy department.

I at once began to think that the compounds suggested in the German literature were unlikely, and the obvious way to test for the presence of free ions was to measure conductivity. Conductance measurements at 1650°C and above (we eventually went to 2400°C) are not easy to make. Kitchener and I took a trip around the British steel industry and learned a lot of things about the technique of those days in making measurements at 1600°C and above.

We were soon able to knock out the old ideas and show that slags were, indeed, as one might expect, ionic liquids. The electric conductance of silicate melts was first shown to be very high by my first graduate student, Stan Ignatowicz, and the inimitable John Tomlinson, who later became vice-chancellor at Wellington University in New Zealand.[42–44] We managed to devise ceramic cells whereby the transport number of the various entities in the slag could be measured and found that the cation transport number was always one.

The most remarkable measurements in this series were carried out by Dennis Lowe, who contrived the rotating disk viscometer working up to 2000°C.[45] (See Fig. 1). The structure of the polyanions shown below, which proved the basis of a new detailed structure for slags (cf. the glasses discussed below), was based primarily upon the relationship which Lowe was able to establish for the heat of activation of the flow of silicates as a function of the silicon/oxygen ratio. We interpreted the rapid fall of E_η with increasing M_xO_y, followed by a flat portion, in which E_η^x depends little on composition, as providing evidence for a small flow unit. It seemed inconsistent with ideas suggested for glasses by Zachariasen, for the latter involved three-dimensional structures of glasses (hence slags) at low Si/O ratios.

The climax of the experimental work was the measurement of the viscosity of SiO_2 itself (Fig. 9).[46]

However, the major success here was the devising of a series of anions to correspond to the structures. Some of these are illustrated in Fig. 10.[46,47]

The ideas that polyanions of silicate and accompanying metal cations were the whole basis of the structure of slags was not supported by work later carried out by Jack L. White, (who had come from California to London to work with me). He measured the density of the silicate melts as a function of composition and calculated the partial molar volume.[48,49] The trouble was that, as shown in Fig. 11, the partial molar volume of the silicate melts was little dependent on composition. It was suggested, therefore, that there existed in the structure a number of *molecules* of silica themselves, which would then come together and form small groups which we called "icebergs" (cf. the presence of individual H_2O molecules at higher temperatures in water). Thus, the *predominant* anion structure was that of the SiO_4O_{12} anion, but to make up the appropriate stoichiometry there would be some pure silica in the form of the so-called "iceberg" groups of SiO_2, "frozen silica."

All this was done without any concept that it had application to glasses, and I was surprised therefore, traveling one day in the bullet train in Japan, about 1965, when I was located by an enthusiastic Japanese scientist who had boarded the train at an intermediate stop to find me. He told me that the work that we had led on the structure of molten slags in the 1950s had had a significant effect on the structure of glasses and jumped up from his seat to bow to me!

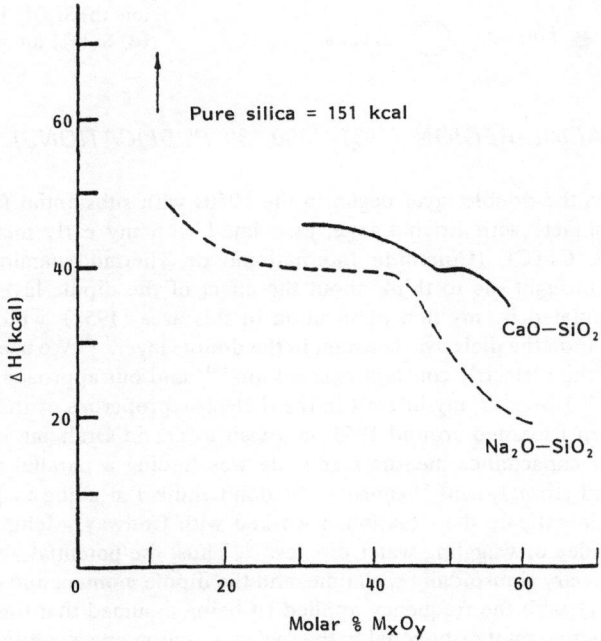

FIGURE 9. The heat of activation for viscous flow of silica.

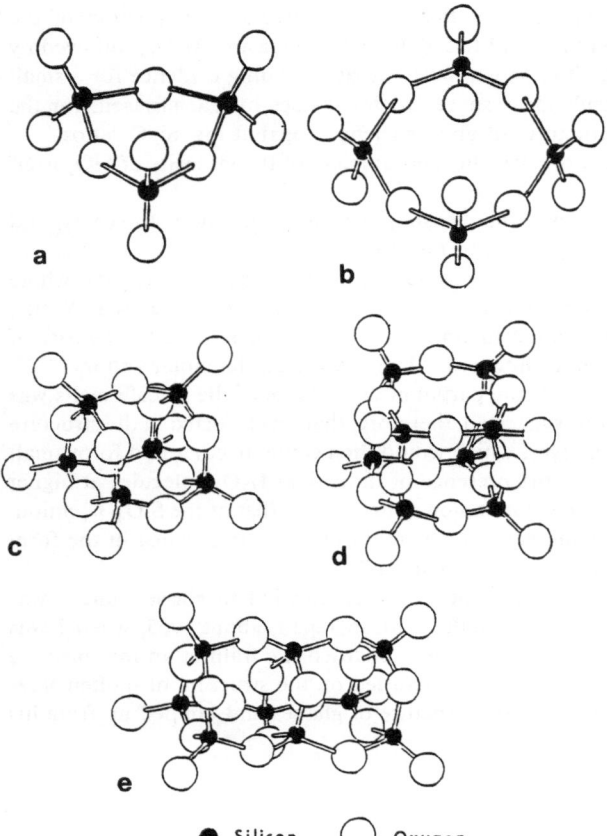

FIGURE 10. Suggested discrete ions in liquid silicates: (a) $Si_3O_9^{6-}$ ion, (b) $Si_4O_{12}^{8-}$ ion, (c) $Si_6O_{15}^{6-}$ ion, (d) $Si_8O_{20}^{8-}$ ion, (e) $Si_9O_{21}^{6-}$ ion.

7. THE INTERFACIAL REGION (1951–1990, 38 PUBLICATIONS)

My forays into the double layer began in the 1950s with substantial faith in the Stern theory, but my contacts with Erich Lange, associated with my early membership of the forerunner of ISE, CITCE (Committé International de Thermodynamique et Cinétique Electrochimique), brought me to think about the effect of the dipole layer at the surface. This was also stimulated by my first publication in this area (1951), which was that with Conway and Ammar on the dielectric constant in the double layer.[50] We used Webb's theory of the variation of the dielectric constant near an ion,[51] and our approach was superseded by that of Booth.[52] However, my interest in the dielectric properties of the inner layer had been aroused. When I learned around 1955 on a visit to David Grahame in Amherst, that, in his double-layer capacitance measurements, he was finding a parallel resistance in his bridge which varied strongly with *frequency* ("I don't think I'm going to publish this"), I was stimulated to investigate the situation. I worked with Conway, Mehl, and Young and we got to the first idea of waggling water dipoles.[53] Thus, the potential applied across the double layer would vary sinusoidally with time, and the dipole moment and hence the dipole potential would vary with the frequency applied (it being assumed that the dipole moment of the adsorbed water was not orthogonal to the surface), thus giving a conductive component in the double layer.

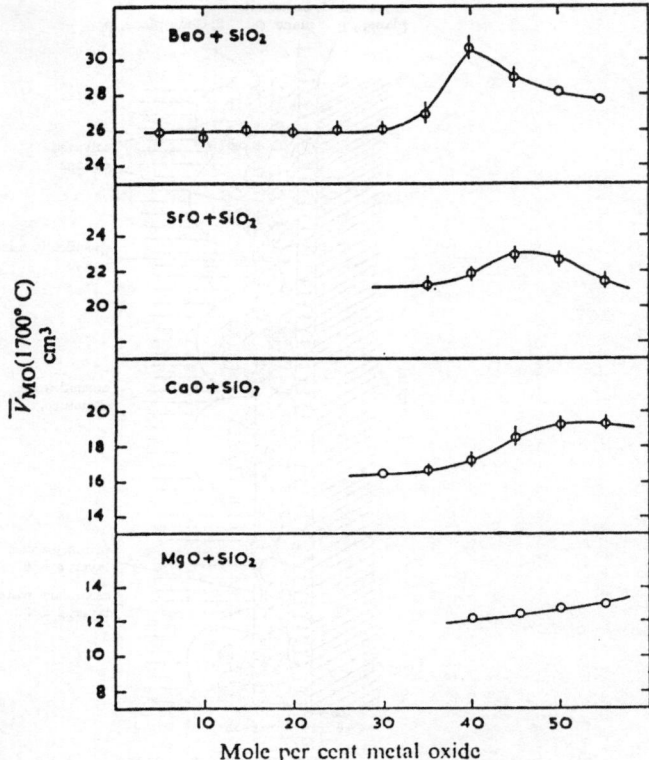

Partial molar volumes of metal oxides as a function of composition.

By 1959, we had gotten together with Blomgren[54] and proposed the first isotherm for adsorption at an electrode which took into account water displacement. Russian electrochemists are fond of quoting this as the Blomgren–Bockris isotherm, and, although it took into account dispersion forces between adsorbed molecules, it is too simple to have much application without modification. However, it was an interesting forerunner of the isotherms deduced later with Swinkels,[55] with Devanathan and Müller,[56] and with Zelenay and Habib.[57]

7.1. The Bockris, Devanathan, and Müller Paper

The Bockris, Devanathan, and Müller paper was a good example of a number of favorable circumstances converging to give a good result. We had the brilliant Müller, who was able to use electric calculating machines (much available), electronic computers (sparingly available), and his own extraordinary arithmetic talents to do the calculations. Devanathan had a long experience of double-layer work, based on his Ph.D. with Parsons at Imperial College, and I was keen on my water dipoles and their contributions to the double layer.† The resulting paper, "the BDM paper," was for a long time a most quoted paper in a section of the *Frequency of Quotation Abstracts*, and its picture of the double layer remains today the one most frequently drawn (see Fig. 12).

† I devised after discussions with Parsons and Grahame a version of the Stern theory but with χ_{Dipole}, but, unfortunately, this was the year in which Piontelli's printers went on strike and *Comptes Rendus de CITCE* of 1952 (in which would have been this modified Stern) was never published.

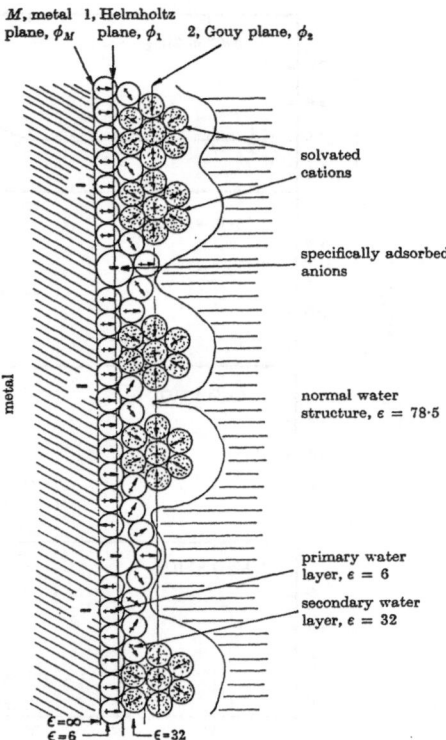

FIGURE 12. The BDM model.

The tan *h* relation between χ_{Dipole} and the electrode charge I derived from the analogy to ferromagnetic theory, where the direction of north–south and south–north dipoles aligns under the field (Bitter[58]).

Devanathan made the major contribution to the specific adsorption discussion, and one part of the BDM paper was essentially an update of work which Devanathan had published earlier.[59] However, the paper gave rise to a comprehensive isotherm based on single imaging and the first molecular theory of adsorption of molecules on the surface of electrodes.†

The BDM isotherm gave, for the first time, a rationalization of the relation of the heat of adsorption to coverage which was not that of Tempkin. It is difficult to derive Tempkin's

---

† However, we must be most careful to quote the work of colleagues who published at the same time, or very nearly. It was not until the first Australian meeting of 1963 that I met Ross Macdonald and learned that he and Barlow[60] had had a theory in which they had looked to water dipoles on the surface of electrodes. The theory which they put forward at this meeting was disturbingly similar to that of the water molecule model of Bockris, Devanathan, and Müller. Since this meeting I learned to respect and learn much from Ross Macdonald.

Mott and Watts-Tobin[61] published a paper which involved water dipoles before that of BDM (although Watts-Tobin was sent by Mott to study in my laboratory for three weeks during the writing of the paper). Those who have not compared the two papers have the opinion that BDM is a sophisticated second approximation to Watts-Tobin and Mott. However, the latter put their water dipoles in parallel with the ionic contributions to the potential difference, whereas in BDM they are in series. This, of course, means that as the water capacitance grows, it contributes *less* to the double layer; in Mott and Watts-Tobin, it contributes *more*. This radical difference has strong implications when one considers the origin of the capacitance hump in the C_{DL}-q_M relation.

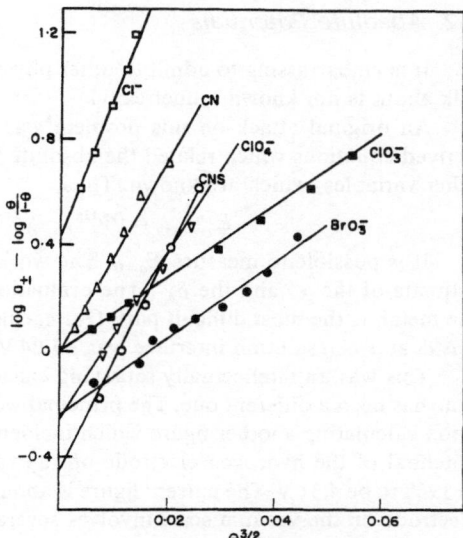

FIGURE 13. The plot of $\log a_{\pm} - \log[\theta/(1-\theta)]$ versus $\theta^{3/2}$.

linear dependence of the heat of adsorption on coverage, and the single imaging model gives $\Delta H_{\theta} = \Delta H_0 - A\theta^{3/2}$ (Fig. 13).

There were two follow-ups to BDM. The first was BGM,[62] by Bockris, Gileadi, and Müller. This is about the organic adsorption side of BDM, but it takes into account interaction between the organic molecules on the surface. Figure 14 shows the various directions with molecules on the surface as treated in BGM's paper.

Another follow-up was the Bockris–Swinkels isotherm.[55] This isotherm was developed following the Flory–Huggins principles of the entropy of mixing on the surface and has proved useful in treating the adsorption of phosphoric acid on platinum.[57]

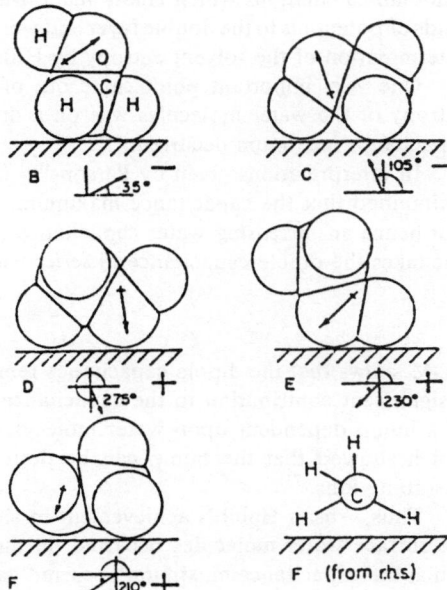

FIGURE 14. Orientations of adsorbed methanol molecules. The inclination of the dipole to the double layer is indicated.

7.2. Absolute Potentials

It is embarrassing to admit to other physical chemists that the potential electrochemists talk about is not known numerically.

An original attack on this problem was made by Argade and myself in 1968.[63] We derived equations which related the absolute Galvani potential difference at the interface to other variables, which are known. Thus,

$$E_{q=0,\mathrm{Hg}} = (\Phi^{\mathrm{Hg}}) - \chi^s + \delta\chi^{\mathrm{Hg}} - \Delta^s\phi_{\mathrm{ref}} + \mu_e/F$$

It is possible to measure $E_{q=0}$. The work function, Φ, is known, and one can make an estimate of the χ^s and the $\delta\chi^s$. The evaluation of μ_e^M, the Fermi energy of the electron in the metal, is the most difficult part. Our caculated value for the Galvani potential difference $^M\Delta^s\phi$ at a Na/solution interface was -1.24 V.

This was an intellectually satisfying calculation. The course of developments since that time has been a different one. The principal worker here has been Trasatti.[14] He has insisted upon calculating another figure which incidentally fell out of our calculations. I refer to the potential of the hydrogen electrode on the vacuum scale, which was calculated by Argade in 1968 to be 4.31 V. The current figure is about 4.5 V. However, the potential of the hydrogen electrode on the vacuum scale involves several other interfaces, and I prefer to think of the calculation of the metal/solution interface as the one which we primarily mean when we talk about an absolute potential (though, of course, it is true that, in practical cases, one has to look into the whole cell).

7.3. Remarkable Contributions by Ahsan Habib

Looking back on contributions to the structure of the interfacial region, I see M. A. V. Devanathan and Ahsan Habib as the two most outstanding scientists who collaborated with me in this field. In particular, Habib did so much in such a short time.

Firstly, he dealt with the question of the configurational entropy of the double layer.[65] There was some earlier work on this by Conway and Gordon,[66] but it was far less comprehensive than calculations which Habib made which were related to the total contribution of all kinds of potentials to the double layer and were quantitatively correlated with the experimental determination of the solvent entropy by Hills and Payne.[67]

One very important point came out of this. It was shown that the maximum of the entropy of the water molecules was on a different side of the pzc from that on which the capacitance maximum occurred.

In interpretations given by Parsons[68] (and by many other workers[69,70]), it is usually maintained that the capacitance maximum, or hump, was due to water dipole *orientation*, and hence an *increasing* water capacitance (for a parallel capacitance model). In BDM, if one takes the dipole capacitance *in series* with the ionic capacitance,

$$\frac{1}{C_{\mathrm{obs}}} = \frac{1}{C_{\mathrm{ion}}} + \frac{1}{C_{\mathrm{dipole}}}$$

BDM shows that the dipole capacitance remains still too big even at the pzc for it to make a significant contribution to the capacitance of the ionic double layer. Hence, the concept of a hump dependent upon water molecules is not likely. Later, Habib derived equations which showed that the hump can be derived as a function of image repulsion between adsorbing ions.[71]

Thus, Ahsan Habib's achievement in showing that the point of maximum disorder of the surface water molecules occurred on the side of the double layer opposite to that on which the capacitance maximum occurred gave substantial support to a model of the hump in terms of anion adsorption (Fig. 15).

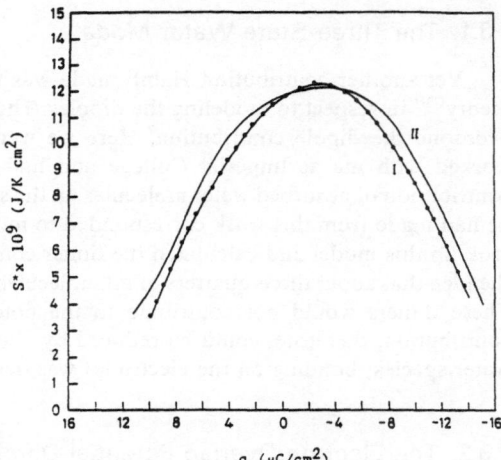

FIGURE 15. The solvent excess entropy as a function of electrode charge. (I) Experimental and (II) theoretical curves for 0.1M NaCl solution.

This work on entropy was only the beginning of Habib's work, and he went on to produce a detailed isotherm with two improvements over that of BDM. It treated the differences between single and multiple imaging. BDM had used a single-imaging situation. However, Levine et al.[72] had suggested that multiple imaging would be necessary and that the difference between the single- and multiple-imaging models would be large. Habib[73] showed, however, that this was only so because a sharp boundary had been assumed between the diffuse and the Helmholtz layer, and, directly one took into account the gradual nature of this boundary, the imaging fell off rather like a reduced reflection of light from a broken mirror.

Thus, the Habib isotherm[74] was a single-energy isotherm, but it also took into account a neglected contribution, the effect of dispersion forces between ions on the surface for higher charges (though cf. Blomgren and Bockris[54]). It is then possible quantitatively to interpret not only the hump on the capacitance–potential graph but also the fact that the capacitance then turns around again and heads upward when the charge is made positive to that of the hump (see Table 3).

TABLE 3

Calculated and Experimental Surface Coverage (θ) at Which Capacitance Hump and Capacitance Minimum Occur[a]

Ion	θ_{hump}		θ_{min}	
	Exptl.	Calc.	Exptl.	Calc.
Cl^-	0.07	0.070	0.100	0.310
Br^-	0.10	0.075	0.135	0.320
I^-	0.12	0.082	0.192	0.335
ClO_3^-	0.07	0.100	0.110	0.270
BrO_3^-	0.12	0.100	0.145	0.270
NO_3^-	0.10	0.100	0.256	0.305
ClO_4^-	0.11	0.095	0.240	0.275
SCN^-	0.07	0.065	0.198	0.285

[a] Data taken from ref. [7.25].

7.3.1. The Three-State Water Model

Yet another contribution Habib made was to work out an improvement on the BDM theory[80] in respect to modeling the dipoles. Thus, by the 1970s it was clear that BDM had overdone the dipole contribution. Here we went back to Law's thesis. J. T. Law[81] had worked with me at Imperial College and had carried out determinations of the dipole contribution of adsorbed water molecules on the surface of mercury. The entropy calculations he had made from this work corresponded to molecules as dimers on the surface. Habib[80] took up this model and calculated the dimer concentration on the surface, coming out with the idea that about three-quarters of all molecules on the surface would be in the dimer form. These dimers would not contribute to the potential difference at the interface, and the contribution, therefore, could be reduced by about four times. This kind of model (several water species; bonding on the electrode) was developed in the 1980s by Giudelli.

7.3.2. The Electron Overlap Potential Difference

The last contribution made by Ahsan Habib concerned the so-called electron overlap potential difference (pd)[75]. Thus, a simplification had earlier been made in talking about the surface potential. This had been put forward as depending on the dipole pd at the surface, following some remarks made originally by Lange and Miscenko.[76] Habib and I, however, were clearly aware of the fact—later exploited by Schmickler and Henderson[77,78]—that there would be an electron overlap potential difference at the surface. In fact, in Chapter 1 of Bockris and Reddy[79] there is an order-of-magnitude calculation of electron overspill, given as an example of the importance of quantum-mechanical considerations in the double layer.

Habib made a calculation of the electron overlap potential difference by calculating the Fermi energy of the solution and then utilizing the equations of Bockris and Argade[63] with a model value of χ_{dipole} with the experimental pzc values to obtain χ_e for the interface.

8. NUCLEAR ELECTROCHEMISTRY (1989–1990, 6 PUBLICATIONS)

It is now a part of scientific history that Fleischmann and Pons were the first to present evidence (March 23, 1989) which seemed to show that nuclear reactions could be produced at an electrochemical interface.[82]

I was particularly stimulated to do something about this because Fleischmann had at one time (around 1947) asked to do a Ph.D. with me while I was at Imperial College. I refused him because I was full up at the time, but he worked with J. F. Herringshaw, a fellow lecturer at Imperial College, and was often associated socially and scientifically with my own research group, that of Herringshaw having unit size. I had known him, therefore, for more than 40 years and had sporadic intellectual contacts with him all this time.

I talked to Fleischmann in Southampton a day or two after the announcement had been made, and he was good enough to give me some hints as to how to start the experimental work.

There has, of course, been controversy about the work, and very many people have come out with the statement that the alleged effects are, in fact, errors of measurement. Among the people who have worked with me at Texas A&M on this topic are Nigel Packham, Ramesh Kainthla, and Omo Velev, with the recent addition of Zoran Minevski.

We found tritium in copious amounts in our solutions not long after we began to make electrolyses of hydrogen and oxygen in lithium deuteroxide solutions using palladium cathodes and nickel anodes (Fig. 16).[83]

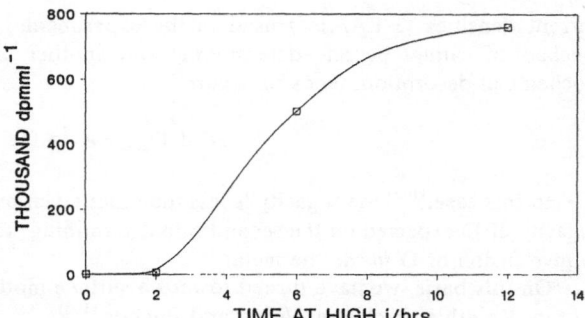

FIGURE 16. Production of tritium in cell A7 as a function of time (ordinate gives hours).

It seems that a key difference between the workers who say they found nothing[84-87] and the workers (now amounting to many[83,88-107]) who say they found something is the time which they wait for the electrolysis to occur. One has to electrolyze the solution for 1–10 weeks before the effects switch on. Some electrodes do not switch on in 10 weeks.

There seems to be a correlation between small size and effectiveness: as electrodes get smaller, they tend to evolve heat (on a per unit volume basis) better, and they tend to give a larger amount of tritium (though the two events are only weakly coupled, if at all).[89] If we restrict our observations to electrodes which are not larger than 1 mm in diameter, it has been possible to show that tritium (10^4 more than background, or greater) can be produced with a probability of 70%. [However, this must not be interpreted to indicate that one could go into a laboratory and get tritium *on a given day* with 70% of the electrodes. The 70% includes some electrodes which have been electrolyzed for 10 weeks (Fig. 17).[108]]

It is obvious that new hypotheses are necessary to explain these strange facts.

The original hypothesis of Fleischmann and Pons was that the key point relied upon the fugacity of the hydrogen inside the palladium, insofar as they could put up the basic formula:

$$f_{H_2} = p_{H_2} e^{-x\eta F/RT}$$

This assumes that the rate-determining step is $D_{ads} + D_{ads} \rightarrow D_2$.

Then, x turns out to be 2, and as η has numerical values of about -0.5 V, large internal fugacities are predicted. The Fleischmann–Pons concept is that as a result of such fugacities the deuterium atoms are pressed close together (fugacities of 10^{20} atm!).

However, the rate-determining step of atom recombination which gives rise to the *very* high fugacities only occurs when the coverage of the electrode surface is small ($\theta_H \ll 1$). Although θ_D has not yet been measured in the case of deuterium on palladium, at the high

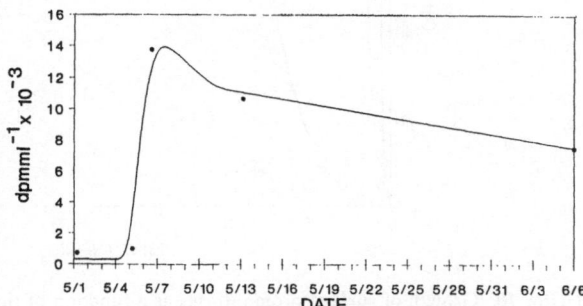

FIGURE 17. Production of tritium in cell A2 as a function of time (ordinate gives date).

current densities (~1 A/cm^2) used in the experiments, $\theta_D \rightarrow 1$, and this recombinational mechanism cannot be rate determining, and another one, that of rate-determining electrochemical desorption, tends to occur:

$$D^+ + D_{ads} + e^- \rightarrow D_2$$

In this case,[109] the fugacity is less than that given by the equation; indeed, the internal fugacity of D expected on the second rate-determining step (10^5–10^6 atm) would be unlikely to give fusion of D *inside* the metal.

On this basis, we have turned toward a surface model. It has been outlined in a paper by Lin, Kainthla, Packham, Velev, and Bockris.[108]

The model which we are putting forward depends upon the production of unusual promontories, which are supposed, in our view, to stick out from the surface. The long time which it takes for these dendrites to form is given by an expression derived by Popov and Maksimovic,[110] and it can be shown that several weeks' latency period is reasonable. After this time, the dendrites would come out from behind the diffusion layer on the surface, strong electric fields would form at their tips, and these fields would give rise to dielectric breakdown and gas layers around the dendritic tips (Fig. 18). Across these gas layers, deuterium ions would be accelerated in such a fashion that collision would occur with the surface of the palladium and give rise to the fusion reaction. (The D for the gas phase would collide with an adsorbed D, the order of magnitude of the energy of collision being about 1000 eV.)

This theory does explain the order of magnitude of the velocity at which tritium forms on the $D + D \rightarrow T + H$ reaction. However, there are other difficulties which beset the theory of electrochemically confined fusion at this time. For example, it does not predict the correct n/T ratio. In another view, the very high fugacity in voids is shown to bring about conditions whereby the particles D^+ in the plasma would lose their charges, and therefore be able to fuse.

Even if the theory that we put forward in the first year of work on this topic has some validity, there are many things which are not understood in respect to the nuclear reactions

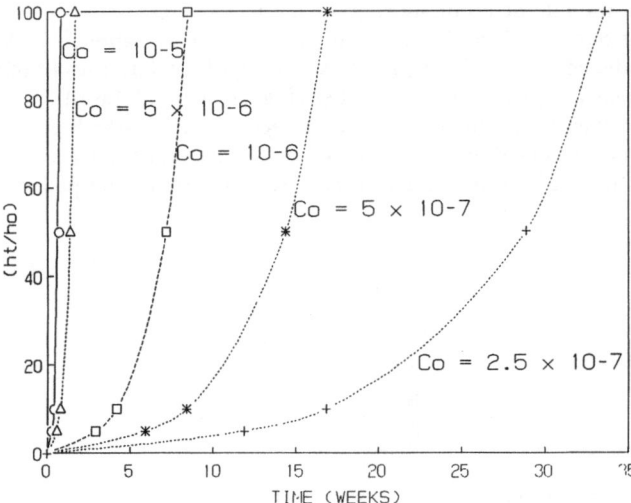

FIGURE 18. Growth of surface promontories as a function of time and concetration of bulk impurities.

which may be going on in these electrochemical cases. For example, we have little degree of correlation between the heat which should be produced from the tritium reaction and the heat which is seen (about 20% more than the classical heat put out by an electrochemical reaction).† Thus, at the present early time the suggestion is that there may be two processes, one giving rise to the tritium, and another, perhaps internal, which gives rise to the heat.

At the time of writing (1992), the atmosphere of doubt and indeed outright rejection of the phenomenon is somewhat relieved because observations of tritium formation at more than 10^3 times background have been verified by Wolf at Texas A&M, Scott at Oak Ridge, Storms at Los Alamos, Schoessow and Wethington at Gainesville, Ramirez in Mexico City, Adzoc and Yeager at Case Western, and (very extensively) Iyengar and a very large group working at the Bhabha Research Center in India and Menlove at Los Alamos. (However, note that in all cases the verifications made are irregular, irreproducible, and often take large numbers of runs and time in order to observe them.) Correspondingly, the excess heat has been reported by more than 50 groups.

Thus, an early retrospect here would say that there is no doubt that nuclear reactions at electrode surfaces have been discovered, and for this a great deal of credit should go to the original thinking of Fleischmann and Pons. Whether their discovery will lead in 20 years to grand technological advances, and even an energy source using fusion, is, of course, beyond our abilities to predict. However, it does appear that it is no longer reasonable for people to argue that no nuclear phenomena at electrodes exist.

University of Pennsylvania, 1965

Many now well-known electrochemists are in this group. *Front Row*: Miss Chiu (ellipsometry and adsorption); secretary; Asa Despić (BDD mechanism of dissolution); JOMB; Boris Cahan (porius electrode theory and many technical devices); Marvin Genshaw (oxygen reduction kinetics). Third row contains Bill O'Grady (Mössbauer spectra) and Vlasta Brusic (passive layers); fifth row contains George Myers; sixth row, P. K. Supramanyan (H in metals), Glenn Stoner (organic reaction mechanisms), Zoltan Nagy (electrocrystallization), and John McHardy (tungsten bronzes).

† Though a few workers have observed larger bursts, over 200% more than the classical heat.[82,90,93]

The above brief vignette concerns the electrochemical side. There are also the recent Japanese experiments in which glow discharge produces very large amounts of neutrons between electrodes of Pd.[105]

9. QUANTUM ELECTROCHEMISTRY (1948–1990, 48 PUBLICATIONS)

My work on quantum electrochemistry has often been connected with a basic disagreement which I have with the formulations made in this subject by R. A. Marcus, early on in my career.

Marcus is the most well known contributor to quantum electrochemistry, and his excellent reputation in this area was founded upon a paper published in 1956.[112]

Basically, the Marcus idea was to express the energy of ions undergoing electron transfer in terms of the optical-time-domain changes of solvation energy in an outer shell of solvent and assumed that the only influence on charge transfer was expressed by the Born equation.

At the time I first came into contact with this paper, I did not think it was to be taken in a serious vein, and hence did not respond to it. It seemed obvious that the energy of the solvation shell was not adequately represented by the Born equation,† and therefore any basic theory of electrode processes based on such calculations must have very much a club foot. (I calculated the error to be ~2 eV whereas the heats of activation measured were ~0.5 eV.)

Later on (c. 1985), I sharpened up my opposition to the use of Born's equation (duplicated by all who have followed Marcus).

In considering solvation, Born worked not in terms of the appropriate ion–water *interaction* energy but in terms of idealized charged metallic conductors. The energy *in vacuo* would be $q^2/2C$, where q is the charge on the sphere, and C is its capacitance. In water, a dielectric constant would reduce the energy from that in vacuum.

I concluded that it was fundamentally wrong to calculate a salt energy, for surely one must be concerned with the *energy of interaction* of the ion with water.

However, I spied a more fundamental inconsistency in Born's equation. Thus, in calculating the *experimental* value of solvation energies the assumption made is that the ions in the gas phase are an infinite distance apart and have zero potential energy.

In the Born equation, however, a different reference state (i.e., not zero in vacuum) is taken—the self-energy of the ion in the gas phase. Hence, the theoretical value from the Born calculation could not be directly compared with the experimental value and would significantly overrepresent the outer-sphere energy.

Nevertheless, there will be a component of energy in the outer sphere corresponding to electronic polarization of the water dielectric in the optical time domain (when the charge changes). I represent this by calculating the change of the polarization energy, $\frac{1}{2}\alpha_{H_2O}X_r^2$, for the energy of the water at a distance r from the iron. By allowing for the geometry of the spherical column and integrating to infinity, one may calculate the optical level contribution without using the Born equation.‡

---

† It is only fair to state that solvation entered my thesis discussion of 1945, and so I was familiar with the concepts modern at the time (Born's equation was a 1920 one!). I was also familiar with the earlier quantum-mechanical approach by Gurney and by Butler. Here, nearest-neighbor interaction had been discussed.

‡ The fact that I stress my disagreement with the basis of the Marcus approach must not be taken to imply a lack of recognition of the contributions which this scientist has made to physical chemistry in the United States. My knowledge of Rudy Marcus in the 1950s involved several friendly personal meetings with a good deal of verbal and intellectual combat.

Now, in furtherance of my interest in showing that the basic Marcus view was wrong, I published a work with Mathews and Khan in 1973,[113] in which we had gathered the rates of oxidation–reduction reactions for about 50 systems in solution and compared "experimental" values ΔF^{ox} with the theoretical values which arose from the basic Marcus solvent fluctuation view. This graph is shown in Fig. 19, where the discrepancy between the predictions of the original Marcus view and that of reality is brought out.

Having illustrated the fact that the original solvent fluctuation view gave results so far from experiment, the question was, of course, how we could calculate the correct values. In this respect, Paul Delahay[114] came to the rescue with a clever method for obtaining the total reorientation energy. It consisted essentially of measuring the photoeffects between an electrode kept just outside the solution and ions in the solution.

Delahay's conclusion on his first evaluation of his results[115] was that the outer-sphere contributions exceeded those from the inner shell. However, Khan and I[116] investigated Delahay's calculations and found that he had, unfortunately, not included the image energy of the ions in contact with the electrodes so that his calculations were incorrect. When we made them correctly, as accepted by *Chemical Physics Letters*, we found that, for the five systems in which we examined the division of the activation energy for redox electrode processes, it was something of the order of 50/50 between the solvent fluctuation and the inner-shell approach with some ions going more toward the vibrational side.

The work which Khan and I had been doing in this area since 1974 was now seen to be having an effect. Thus, the most prolific authors favoring the original view† of Marcus (solvent fluctuation only) had been the large Russian group in quantum studies in Moscow led by Levich with the able assistance of Dogonadze and Kuznetsov. Although they had been keen in following the Marcus view (solvent fluctuations only) up to the point at which Marcus added $-\lambda_i$, the energy of the inner shell, we begin to find that Russian visitors explained that this had all been due to Levich's evil influence (Levich having now emigrated to Israel and the United States), and they were now fully in agreement with the vibrational contribution being most important!

So much for an example of the effects of politics and peer pressure on electrochemistry.‡

In the 1980s we were much occupied with photoelectrochemistry, and this brought us to the work of Gerischer,[119] who had undoubtedly been the leading publisher in the associated area of semiconductors (although the first paper on semiconductor electrode kinetics was published by Mino Green in 1989). I had been puzzled over the years by the fact that

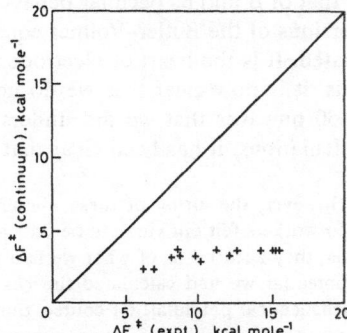

FIGURE 19. Plot of $\Delta F^{\ddagger}$ (continuum) against $\Delta F^{\ddagger}$ (experimental) for homogeneous electron transfer reactions in solutions involving reactants with water and ammonia molecules as ligands (correlation coefficient = 0.41).

† Marcus himself introduced an inner-shell term in 1965.[117]

‡ One member of the Russian group remains wedded to the pan-solvent concept, and this is Krishtalik,[118] who has published much on this view in his recent book.

particularly American (but also other) workers put a line on the solution side of their diagram showing that there was a "Fermi level" of the redox material in solution. This was supposed to be its redox potential on the vacuum scale. I remembered the original paper by Gerischer[120] in 1960 in which he had first introduced the idea of a Fermi level of electrons in solution. I had read a footnote in German in the paper which explained that the relationship $E_{Fermi} = -nFV_{redox}(vac)$ was only valid at zero charge and when there was no contribution from the surface potential.

It seemed to me that Gerischer had forgotten his own paper because he supported the identification of the redox potential with the Fermi energy in the solution, though, of course, it is never true that the surface potential is zero even at the potential of zero charge.

One day, when Khan and I were at Brookhaven on a visit, sitting in the lounge in the little house where we were lodged, we were playing with the equations of Gerischer's deduction and found a quite general argument showing that the redox potential on the absolute scale was equal to the *chemical* potential of the electron in the solution (not the electrochemical potential).

Unless one is familar with this field, one cannot feel the pulse of realization (the Japanese might call it a satori!) which went through us at this time. We had, of course, discovered what was wrong with equating the redox potential and the Fermi level energy. Thus, the Fermi level in an electrode is the *electrochemical* (not chemical) potential of electrons in the electrode in equilibrium with ions in the solution. The statements that were being made in the literature had left out the surface potential of the solution, and as this would not be negligible, having a value of perhaps several tenths of a volt, they could be significantly in error. We wrote up a note at once and published it in *Applied Physics Letters*.[121]

At the same time we wondered whether the concept of a Fermi energy in solution was a concept worth having. We came to the conclusion that it was not. I had made a study of the deduction of Fermi's law in writing *Modern Electrochemistry* with Reddy, and I knew that one of the postulates made in the deduction was that the electrons were to be fully mobile. It is obvious that electrons in redox ions are not mobile in the sense meant; that is, they cannot be communalized with the solution. For this reason, the idea of a *Fermi* energy in solution is not a sensible one, and it is better to deal with the ground state of the energy of the ions in solution, which can be calculated from the Born–Haber-type cycles as shown in earlier work.†[122]

Before I end this brief account of headlines in our quantum electrochemical work (which included, incidentally, the writing with Khan of a 500-page book, *Quantum Electrochemistry*,[124] published by Plenum in 1979), there is one more problem to discuss. It is that of β and its peculiar behavior. The β to which I refer is the symmetry factor in simple versions of the Butler–Volmer equation. The importance of this quantity cannot be overestimated. It is the heart of electrode kinetics, that is, the heart of electrochemistry. In spite of this, it is now clear that we do not understand it very well! We have thought from about 1950 onwards that we did understand β but since about 1985, on the basis of Conway's calculations, it has been clear that our understanding is very limited.

---

† However, the airing of these matters raised a hornet's nest in Germany where Gerischer and his co-workers felt our views to be critical of them.[123] They devised an excellent technique for defeating us; they accused us of what we had implicitly accused them. They said that in calculating the redox potential we had calculated the *chemical* potential in the solution, whereas it should be the electrochemical potential! Of course, this was just exactly our prior point but they turned it around in a way good enough to convince the unsuspecting reader.

However, we explained our position back in other notes, and, after a couple of exchanges in the literature, it cleared up and I do not think that anybody since that time has continued with the idea that the redox potential in the solution on a vacuum scale is equal to the Fermi level. As Gerischer had said in his original paper, it is only this in those very special circumstances that the surface potential is zero and the charge on the electrode has also been reduced to that for the pzc.

This is for two reasons.

Firstly, all the theories of electrode kinetics, beginning, for example, with that of Marcus, would show that β should vary with potential. On the other hand, the evidence to date shows that β remains constant[125] over considerable regions of potential difference. In the case of hydrogen and oxygen, these extend to 1 V, about the maximum potential variation of most electrode kinetic experiments.[126] Even with redox reactions, β is very constant over about 0.5 to 0.6 V, the maximum range examined.[127] This is quite inconsistent with the theory and demands some explanation.[128]

However, there is another aspect of the behavior of β which is anomalous, and that is the one to which Conway has drawn attention: β varies with temperature. It ought not to.[129]

Conway's explanation was to resuscitate a suggestion made by John Agar in 1947 at the Manchester meeting of the Faraday Society, concerning the work of Stout, who, with the oxidation of the azide, had found that β was dependent upon temperature.

Agar suggested that this meant that part of the potential dependence of the rate of the electrode reaction must imply that the entropy of activation, not only the potential energy, depended on potential.

Thus,

$$i = \frac{hKT}{h} e_A \, e^{-\Delta H^{0\ddagger}/RT} \, e^{\Delta S^{0\ddagger}/R} \, e^{-\beta\eta F/RT}$$

$$\frac{\partial \ln i}{\partial \eta} = \frac{1}{R} \frac{\partial \Delta S^{0\ddagger}}{\partial \eta} - \frac{\beta F}{RT} = \frac{-\beta_{ob} F}{RT}$$

$$\beta_{ob} = \beta - \frac{T}{F} \frac{\partial \Delta S}{\partial \eta}$$

Our own view on this was published with Gochev in 1986. Gochev[130] made a detailed examination of the ways in which β might vary with temperature, but when we added it all up, we could only come to about a tenth of the values which are experimentally obtained.

In more recent times, Conway and I have been speculating together about this and wonder whether the culprit is perhaps the variation of the transmission coefficient with potential. Thus,

$$i = \frac{\kappa kT}{h} e_0 \, e^{-\Delta H^{0\ddagger}/RT} \, e^{\Delta S^{0\ddagger}/R} \, e^{-\beta\eta F/RT}$$

$$\frac{\partial \ln i}{\partial \eta} = \frac{\partial \ln \kappa}{\partial \eta} + \frac{1}{R} \frac{\partial \Delta S^{0\ddagger}}{\partial \eta} - \frac{\beta F}{RT}$$

$$\beta_{ob} = \beta - \frac{T}{F} \frac{\partial \Delta S^{0\ddagger}}{\partial \eta} - \frac{RT}{F} \frac{\partial \ln \kappa}{\partial \eta}$$

$$K = \exp\left[-2 \int_a^0 \sqrt{(V - \eta - E)} \, dx \right]$$

$$\frac{\partial \ln K}{\partial \eta} = -\frac{2\partial}{\partial \eta} \left[\int_a^0 \sqrt{(V - \eta - E)} \, dx \right]$$

Some rough calculations show that, if the tunneling probability of protons is taken in a Gamow way, orders of magnitude turn out well. However, it is far too early to say that this is the explanation of the variation of β with temperature. [It would, however, explain one thing—the large $\partial\beta/\partial T$ seems to occur only for atom transfer reactions (where the Gamow term is far from 1), and not for electron transfer reactions.]

10. SPECTROSCOPIC STUDIES (1963–1990, 45 PUBLICATIONS)

Chemistry has become "spectroscopy" could be a description of a strong trend in physical chemistry since the early 1960s, and the studies which we have made in my own laboratory have been no exception. In fact, the studies we made on ellipsometry in 1962[131] were the earliest of an electrode/solution interface *in situ* under potential control. Here, one must bring out the strong role of A. K. Reddy. Nevertheless, it was not Reddy who began my interest in ellipsometry, which commenced in a visit to Trondheim, Norway, in 1952. There, I met Winterbottom, who, with Tronstad,† was the true originator of the application of ellipsometry to passivity, though in the dry. [Of course, ellipsometry itself goes back to Lord Rayleigh (1890).] A chance meeting with Lawrence Young at an Electrochemical Society meeting in the early 1960s gave me Young's agreement that ellipsometry might well be the best way to look at passive films in solution, and after that I asked Reddy, who had been trained at Imperial College in the chemical engineering department (and whose knowledge of electrochemistry at this time was not great), to concentrate upon developing ellipsometry for the electrochemist.

The amount of material in this section is large, and I can only give extremely brief descriptions of the highest peaks. One was the work with Reddy and Genshaw in first determining the extent of oxide film formation on platinum.[132] So much work has been done on this subject since then (particularly by B. E. Conway and Halina Kozlowska) that it does not always come out that by 1964 we had got evidence that a lattice film in $1 M$ H_2SO_4 was visible at about 0.8 on the normal hydrogen electrode (NHE) with adsorption of oxygen before this.[133]

The other great character who contributed to ellipsometry was Boris Cahan.‡ We utilized ellipsometry a great deal in looking at film growth, in both the initial stages and the later passivation stages on iron[134] (Fig. 20). Genshaw and Chiu[135,136] were the first to show that ellipsometry could give rise to some information about the adsorption of anions on surfaces.§

Then came a considerable contribution to ellipsometry. It was made by W. K. Paik from Korea. Paik was working on ellipsometry as a two-parameter method, giving the ellipsometry quantities Δ and Ψ.[137] When the film is absorptive, there are three unknowns, the thickness, the refractive index, and the absorption coefficient. Hitherto, the method had been to use an auxiliary method, for example, coulometry, to obtain one of these quantities, and then the other two could be obtained from ellipsometry. On the other hand, there were difficulties with this approach because one had to use a different film and a different setup for the other measurement, and, as with most surface things, reproducibility was difficult to obtain. What Paik did was to work out equations which showed that, with the same angle of incidence, the measurement of the reflectivity of the parallel component of the light from the surface

---

† Tronstad was one of Norway's ski troops in the war against Germany and was killed in 1940.

‡ The influence of Boris Cahan on my laboratory in the 1960s and early 1970s was all-pervasive. Boris was employed, after he got his Ph.D., to write a book on electronics in electrochemistry (this was never finished), but also the other half of his endeavors was to help the students design apparatus and equipment. In this endeavor his efforts were superb and fruitful, particularly in the progress made in transient ellipsometry. It is difficult to overestimate Boris Cahan's effect on my laboratory which, as far as the scientific part was concerned, was extremely good.

§ I am aware that there are some who think that it is not an effective method to employ. This is, at the moment, under reinvestigation in my laboratory, and our most recent thinking is that it is an acceptable tool.

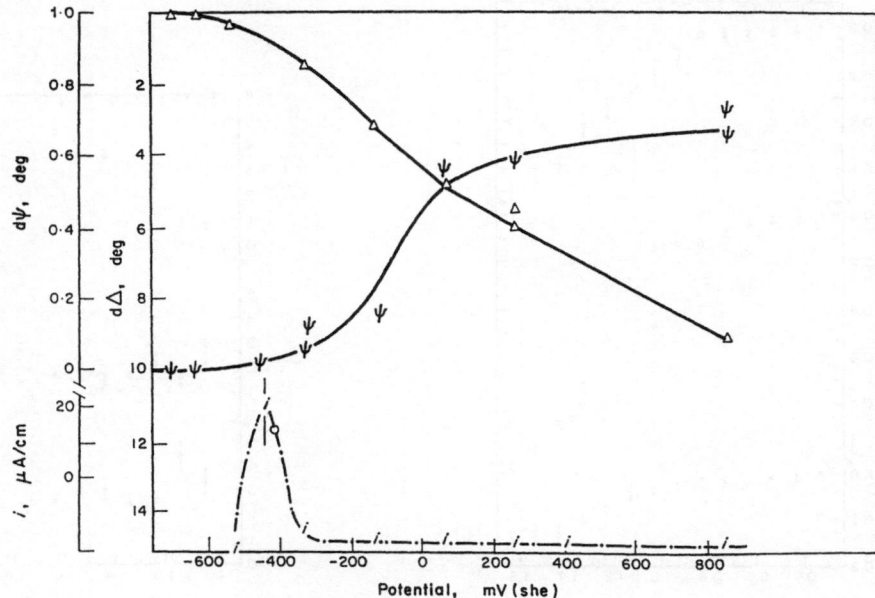

FIGURE 20. Δ, ψ, and i as functions of potential. For each observation, the electrode was oxidized in a single, fast potential sweep. Stirred solution.

gave information which could be added to the ellipsometric information Δ and Ψ and thus allow information on the thickness, refractive index, and absorption coefficient to be obtained (Fig. 21).[138]

Paik also did some pioneering work on plasmon effects. This was the first introduction of plasmon considerations into electrochemistry.† Paik determined ellipsometric parameters as a function of wavelength. If one was well away from the plasmon frequency, everything was alright, but each system had to be investigated for interference from this source.[139] Paik applied his methods and concepts to the study of platinum in an exemplary study.[140]

Down in Australia I was fortunate to become professor in a department in which surface chemistry was already the center of the research program. The leader in this program was Bruce Baker. It was he who helped us in the first Auger work in examining the passive film on iron, showing that the ratio of 0 to Fe was 2 and showing how this ratio decreased to 1.6 when chloride was added to the solution, work which was led by the inimitable Winston Revie[141,142] (Fig. 22). Then, back in the United States at Texas A&M and energized by Oliver Murphy's drive, we were able (with T. E. Pou, David Cocke, and Gene Sparrow) to get secondary-ion mass spectrometric (SIMS) evidence on passive layers. SIMS, of course, gives very direct evidence on what is present, and our main purpose was to find out whether our idea—that water was the main constituent which made iron passive—could be ascertained by SIMS (Fig. 23). The results were positive.[143]

Parallel studies on breakdown, too, were led by Pou and involved Vanessa Young and L. L. Tongsen, as well as Oliver Murphy,[144] but the decisive studies on the breakdown with chloride ions were the radiotracer studies in collaboration with Mizuno[145] (from Hokkaido, the home of Enyo, Kita, and Uosaki). Here we found that the chloride did indeed adsorb,

† In fact, rather dramatically so, because of the attack made on us by Wilf Hansen at the Electrochemical Society meeting at Atlantic City in 1969, where Hansen suggested that most of the ellipsometric effects in electrochemistry were due to plasmonic excitation.

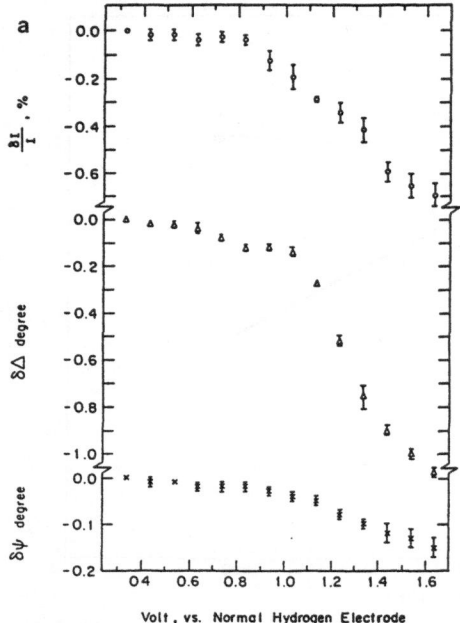
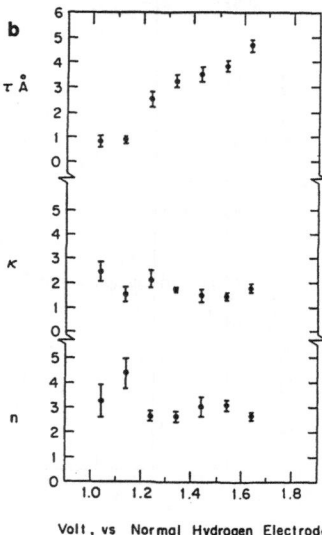

FIGURE 21. (A) The changes in reflectivity (○), Δ (△), and ψ (×) for a platinum electrode in 0.005M H_2SO_4. Wavelength, 5461 Å; angle of incidence, 69.00°. (B) The calculated values of the thickness (τ) and the real and imaginary parts of the refractive index of the film (n and k) at various potentials.

but, at a certain concentration and potential, it starting absorbing and diffusing into the passive layer, and, under these conditions, a rapid rise in current between the metal and the solution through the passive film occured, signifying breakdown of the passive layer, probably by dehydration (Fig. 24).

The studies of passive iron have continued, in particular with the masterful Jovancicevic.[146] A spectroscopic ellipsometric study was made on the passive film, identifying the regions of the various oxides and their contribution to the passive layer.

The application of FTIR spectroscopy in electrochemistry was certainly pioneered by others outside our own laboratory, first of all by workers in Austria[147] and then by Allen Bewick and Stanley Pons.[148] The latter brought this work to Edmonton in Canada, where he taught the technique in a workshop to Ahsan Habib. The redoubtable Habib, author of so many papers on the double layer, brought this work to Texas A&M and founded the technique of employing FTIR spectroscopy to study electrodes here around 1983. We made an important study of the orientation of water on platinum as a function of potential.[149]

Among papers which should be mentioned here are particularly the *in situ* spectroscopic investigations of adsorbed intermediate radicals, with the CO_2^{-}[150] (Fig. 25).

There is no doubt that FTIR spectroscopy has been something like a tour de force with us, and I quote the first publication in which hydrogen has been detected spectroscopically other than on platinum, showing the adsorption of hydrogen on iron and its dependence upon potential, etc., work carried out by Jose Carbajal as a graduate student under the supervision of B. R. Scharifker and with the help of K. Chandrasekaran[151] (Fig. 26).

Of all the *in situ* methods, FTIR spectroscopy is the best, and the only thing that I want is a much faster FTIR measurement time so that transient radical behavior can be seen.

In this series of works we are in the middle of the game, so to speak, and certainly other publications from our laboratory will be coming out of this area.

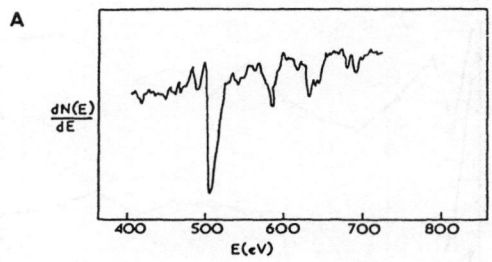

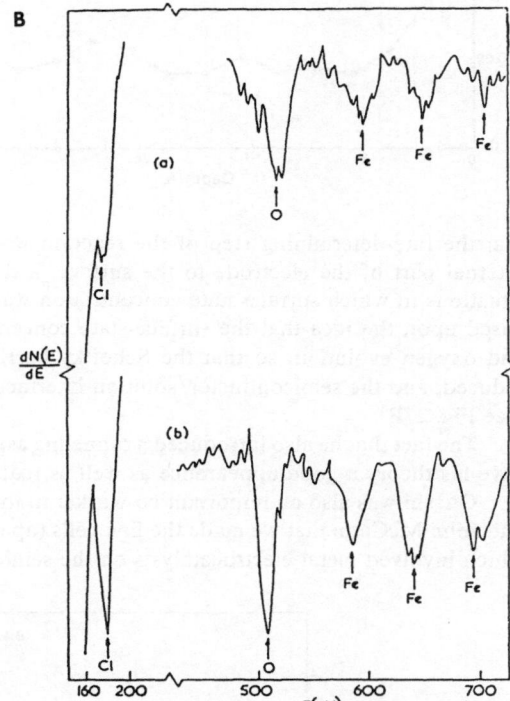

FIGURE 22. (A) Auger spectrum of evaporated film on iron after passivation in borate buffer solution, pH = 8.1, at 0.30 V versus SHE for 10 min. (B) Auger spectra of iron after potential control for 10 min in 1.0N KCl at 0.30 V (a) and 0.51 V (b) versus SHE.

11. PHOTOELECTROCHEMISTRY (1976–1990, 65 PUBLICATIONS)

Fujishima and Honda[152,153] made a remarkable contribution to electrochemistry when they showed that it was possible to decompose water by shining light upon titanium oxide, coupled to a platinum electrode.

Uosaki (whose nickname among his colleagues at the Flinders University was "the locomotive") slammed through a great deal of work on photoelectrochemistry, including the use of dual cells, both the cathode and the anode being photoelectrodes.[154–157] However, the most important part of his work was that which was published in 1978 in the *Journal of the Electrochemical Society* and which gave the first theory of photoelectrochemistry which was not based upon the idea of a Schottky barrier.[158]

Thus, until the work of Uosaki, photoelectrochemical theories had assumed that the situation in the semiconductor was dominated by a substantial Schottky barrier and, in fact,

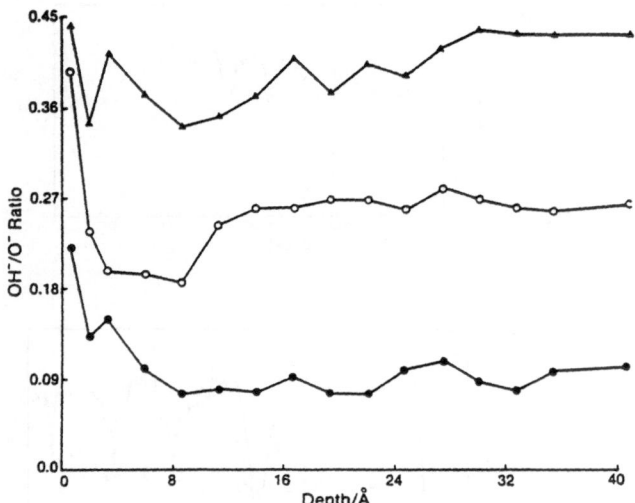

FIGURE 23. Variation of the OH$^-$/O$^-$ ratio, obtained from SIMS negative-ion spectra, with depth into the passive films: ○ and ●, pure iron electrodes passivated at 0.065 V and 0.65 V versus NHE in borate buffer, respectively; ▲, 80 Ni–20 Fe alloy (at.%) electrode passivated at 0.085 V versus NHE in borate buffer. The time for passivation was 40 min in each case.

that the rate-determining step of the reaction was the transport of the electrons from the internal part of the electrode to the surface, a theory which was reasonable in respect to situations in which surface-state concentration was low (see Fig. 27A). Uosaki's theory was based upon the idea that the surface-state concentration was high in the case of hydrogen and oxygen evolution, so that the Schottky barrier was effectively missing, or very much reduced, and the semiconductor/solution interface became more like that involving a metal (see Fig. 27B).

The fact that he also introduced a tunneling aspect and quantum-mechanical terminology gave his theory a good appearance as well as reality.

Ohashi was also an important co-worker in the Australian days because it was with him and John McCann that we made the first cells (apart from those earlier made by Tsubomura) which involved metal electrocatalysts on the semiconductor surfaces.[159]

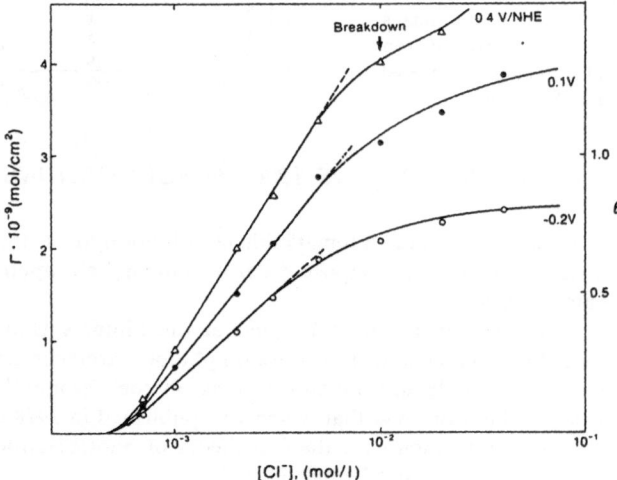

FIGURE 24. Concentration dependence of adsorption and absorption of chloride for different potentials (borate buffer, pH = 8.4).

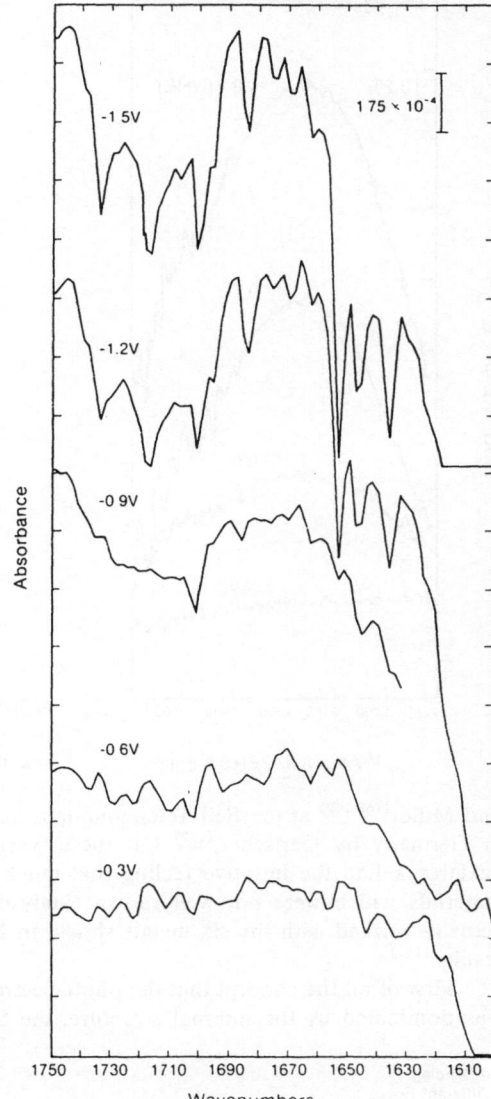

FIGURE 25. Absorption spectrum of CO_2^- radical absorbed on platinum in acetonitrile containing $0.4M$ $LiClO_4$.

When I came back to the United States in 1978, the only piece of apparatus which I brought with me was a light source. Vino Guruswamy, who had come from Sri Lanka to work with me in Australia, followed me back to the United States and was for some time the primary worker in photoelectrochemistry in my laboratory. She was able to demonstrate the photoelectrochemical production of chlorine from seawater.[160]

As this account is one in which space allows only the headlines, I will next jump to the work of Szklarczyk, which is directly associated with the concept of photoelectrocatalysis.[161-167] Szklarczyk, an extremely careful and meticulous experimental worker, had visited us from Poland on several occasions. On one of these occasions, he became the first to make a detailed study of the effect of traces of metals of quite different electrocatalytic properties on photoelectrodes. Earlier work had been carried out by Heller

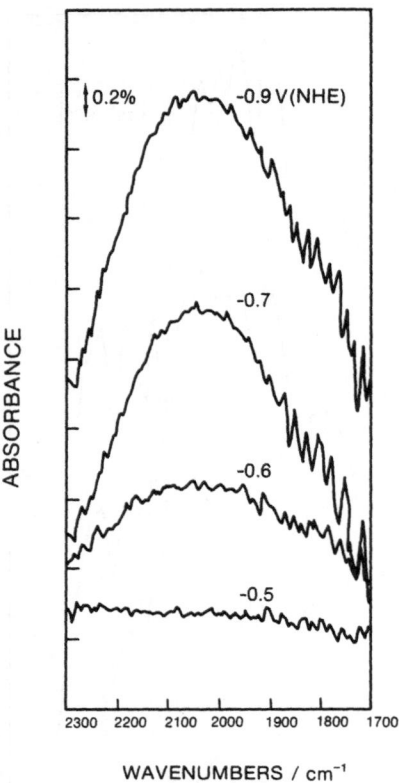

FIGURE 26. Differential IR spectra in absorbance of Fe—H vibration at different potentials in H_2O borate buffer solution.

and Miller[168,169] at the Bell Telephone labs as well as in Japan by Tsubomura[170,171] and in Germany by Gerischer,[172] but these workers had used only noble metal catalysts. Szklarczyk had the intuitive feeling that much could be gained by studying the effect of materials which were not well-known catalysts for the hydrogen evolution reaction. He actually worked with the six metals shown in Fig. 28 and came up with three remarkable results.[161]

First of all the concept that the photoelectrochemistry of hydrogen evolution on silicon was dominated by the internal structure, the Schottky barrier, was difficult to bring into

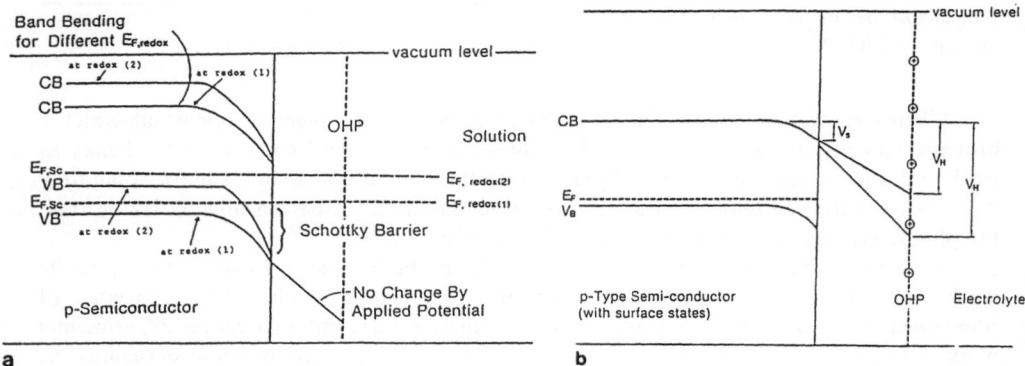

FIGURE 27. (A) Schottky barrier model at the interface. (B) Schematic diagram of p-type semi-conductor/solution interface in presence of high density of surface states.

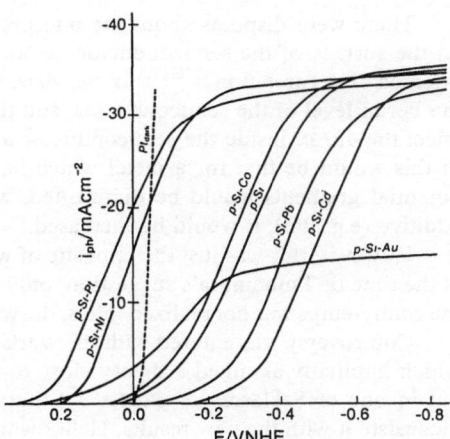

FIGURE 28. Potentiostatic runs on p-Si-M electrodes (50 mW/cm² Xe light, 0.5M H₂SO₄).

agreement with the result that etching changed the rate constant for the photoelectrochemical
evolution of hydrogen by many orders of magnitude. In fact, hydrogen evolution on a sample
of so-called silicon (in reality, of course, silicon covered with an oxide layer) was very slow,
but as the etching increased (ellipsometry showed that the oxide decreased), the film was
reduced and the velocity of hydrogen evolution greatly increased.[173]

Szklarczyk then showed the most remarkable result of all, namely, that the catalytic
effects of the various metals which he used could be related precisely to the exchange current
density for hydrogen evolution *in the dark* on the metal (Fig. 29). This result led to a clear
conclusion: the rate-determining step for the photoelectrochemical evolution was, then, the
thermal evolution of hydrogen on the metal spots on the surface of the electrode. The
semiconductor was acting as a light-absorbing and electron-producing box, but the rate-
determining step occurred on the surface.

Szklarczyk's graph of log i_0 against the shift of the i_p-potential curve was shown at the
Gordon Conference of 1983. This Gordon Conference was the West Coast one, in which
much of the clientele leans toward electro-analytical chemistry, and, perhaps for this reason,
the significance of the graph was not readily understood (electroanalytical chemists are less
familiar with the terminology of i_0 as the exchange current density). Szklarczyk's graph, so
full of implications, then projected on the screen was met with a blank silence.

A third result arose from Szklarczyk's work, and that is that when the metal on the
surface was lead or some other poor catalyst for hydrogen evolution, the photocurrent still
went via the metal and not via the surrounding silicon—there were far more silicon sites than
lead sites, of course—implying a pinch effect in which electron gradients inside the semiconductor were attracted toward the metal on the surface.

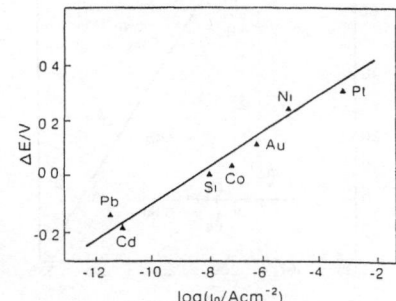

FIGURE 29. Dependence of the potential shift, ΔE, on the
log of exchange current densities, i_0, at various metal
electrodes.

There were disputes about the interpretation of the results and of the effect of metals on the surface of the semiconductor. At an earlier time, Tsubomura had given a "physics-oriented" interpretation.[170,171] In his view, the metal on the surface of the electrode affected the Fermi level of the semiconductor, and this, in turn, for a Schottky barrier model, would affect the $\partial v/\partial x$ inside the semiconductor and therefore the rate of the reaction. The result of this would be that for a metal which had a large work function, such as platinum, the potential gradients would be diminished, whereas for a small work function of the metal additive (e.g., Pb), it would be increased.

However, this was just the opposite of what had been observed. Because in the literature at the time of Tsubomura's suggestion, only work on platinum and noble metals was known, the contretemps was not realized. Thus, the wisdom of Szklarczyk's suspicions were confirmed.

Controversy was entered with the work of Heller and Miller, who had published papers which implicitly assumed a theory close to that of Tsubomura.[168,169] When it was pointed out in one of Szklarczyk's papers[167] that the concepts of the Heller–Miller papers were inconsistent with the new results, Heller wanted to "have these statements withdrawn." Two letters were exchanged in the *Journal of Physical Chemistry*,[174,175] and we had to compare Heller–Miller statements precisely with the facts.[174] Eventually, the editor of the journal asked us to communicate privately!

If there had been any further doubt about the mechanism of the effect of metals on photoelectrochemical currents, this was removed by work carried out by Contractor,[176] who measured the effect of metal particles present on the surface of TiO_2. Here, the electrochemical reaction was oxygen evolution, and a most informative result was obtained: in acid solutions (noble metal catalysts) there was no difference in the effect for any of the metals. In alkaline solution there were large differences, which turned out to show a linear dependence on the metal–oxygen bond strength (Fig. 30). The interpretation was clear. In the case of the acid solutions, the rate-determining step was the discharge of water. Titanium forms the strongest bond to water so the metal islets on part of the surfaces were unimportant to the velocity of the rate-determining step. In alkaline solutions, the rate-determining step would be the removal of OH from the surface of the electrode, and the added catalyst particles, on which M—OH bonding would be less strong than that with Ti, would be the desorption pathway. As desorption was the rate-determining step, the effect of the metal on oxygen evolution could be understood. The photoelectrocatalytic view was supported by these results. A new subfield, photoelectrocatalysis, had been begun.

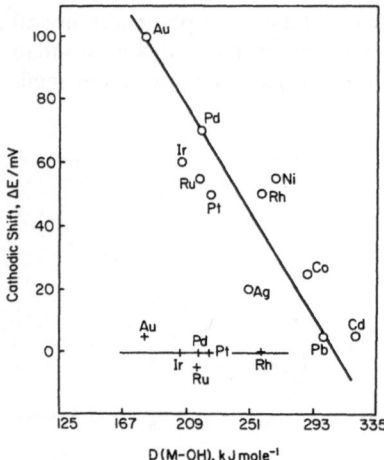

FIGURE 30. Dependence of the cathodic displacement, ΔE, as a function of bond energy, $D(M—OH)$.

While all this was going on, Shahed Khan was working on the theory of photoelec-trocatalytic reactions and produced a very detailed theory, published in the *Journal of Physical Chemistry.*[177] Basically, his model is that of Uosaki but calculated at a much deeper level. The Khan paper probably represents the ultimate in the theory of photoelectrochemistry for non-Schottky-barrier (i.e., high surface state) situations.

By 1984 it was felt that the situation with photoelectrochemistry and the decomposition of water was insufficiently well theorized in respect to conclusions as to what cells to build. Ramesh Kainthla settled down to apply the Khan theory and work out conditions which would give rise to optimal cell design.[178] He came to a nasty conclusion: In order to maximize the photoelectrochemical efficiency, one needed anode materials having flat-band potentials which in fact did not exist. In oxides one would have to use materials which are not oxides, and, of course, oxygen evolution would promptly form a blocking oxide layer.

The solution which Kainthla devised was based upon work which he had carried out earlier in India[179] at the Indian Institute of Technology and involved depositing various types of thin oxide films (i.e., transparent conducting oxides).

Thus, the cells which were made by Kainthla[180] were based on the optimization formulas which he had derived.[178] Thus, indium phosphide was used as a cathode, and gallium arsenide as an anode (Fig. 31). However, he decorated the indium phosphide with platinum, and the gallium arsenide was protected by manganese dioxide with a palladium catalyst on the outside of the film. Such cells gave rise to 8.2% efficiency at room temperatures for photoevolution of hydrogen and oxygen.†

Last to be mentioned in the summary of photoelectrochemical highlights is the Murphy cell.[181] This involves a different kind of approach to the decomposition of water, which is, in fact, photovoltaic electrolysis. One takes an $n-p$ junction consisting of, say, gallium arsenide and plates the n side of it with Pt, the p side being exposed to light. The other electrode is the same $n-p$ junction of gallium arsenide, but now the p side is plated with RuO_2, and the n side is exposed to light (Fig. 32).

These two $n-p$ junctions are then placed in a cell, and each electrode is exposed to light. The $n-p$ junction of the electrode coated with platinum then evolves hydrogen, and the $n-p$ junction of the electrode coated with ruthenium oxide evolves oxygen.

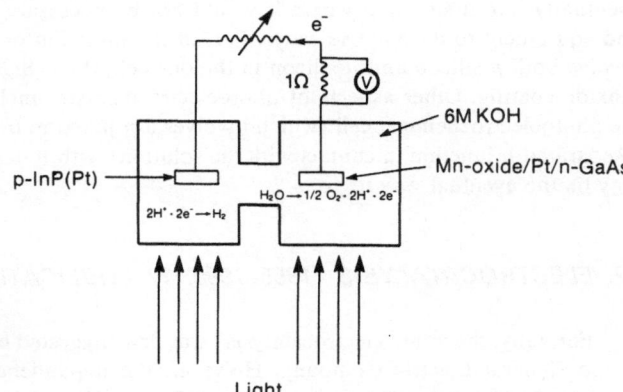

FIGURE 31. Schematic diagram of the self-driven photo-electrochemical cell for water electrolysis.

---

† At the SERI workshop on hydrogen production in Denver in December 1990, an attempt was made by Nozik *et al.* to discredit this result by saying that the high efficiency was obtained because the underlying gallium oxide was being dissolved. On the other hand, the cells can run for hundreds of hours, and when a calculation was made taking the rate of dissolution into account, it was shown that, were the current to be coming predominantly from dissolution, the electrode would be completely dissolved.

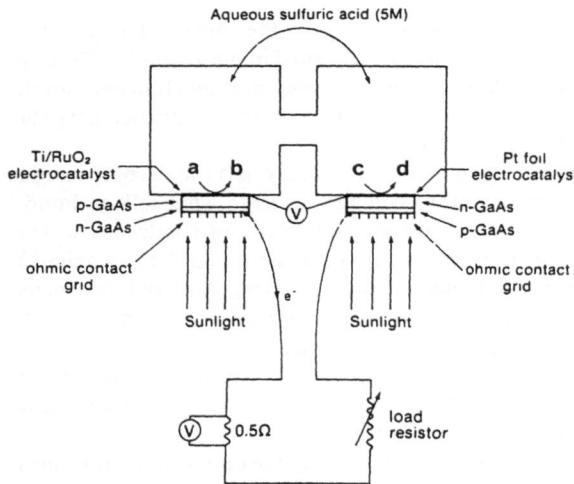

FIGURE 32. Diagrammatic sketch of electrochemical photovoltaic cell capable of splitting water into hydrogen and oxygen gases using only solar energy as input. (a) $H_2O \rightarrow$ (b) $\frac{1}{2}O_2 + 2H^+ + 2e$; $2H^+ + 2e \rightarrow$ (d) H_2.

The 9% efficiency obtained by using the Murphy cell indicates its practical usefulness. The concept of the publication was related to work which we are now carrying out, for one of the correlatives of this approach is to utilize silicon and silicon alloys to form a series of n–p junctions involving the pin concept. If the energy gaps are optimized by alloying, several of these cells together will give potentials sufficient to decompose water. Theoretically, the efficiency should be above 20%.

Before finishing this section, it is relevant to ask "Does work on photoelectrochemistry have a future for the decomposition of water?" Of course, the rival is dry photovoltaics, followed by water electrolyzers. There is much to be said for this second concept because there is so much work going on in photovoltaics. One might think that the tiny research support of photoelectrochemistry could not compete with the advances made in photovoltaics. However, there is a way to go in the photoelectrochemical situation which has the fundamental advantage of being able to evolve hydrogen and oxygen in the one cell, which could translate eventually into a situation where it would not be necessary to have two sets of apparatus and equipment to decompose water. Part of the direction of the work here is to attempt to involve both p-silicon and n-silicon in the one cell, the n-Si, of course, having a manganese dioxide coating. Other aspects of photoelectrochemistry include the concept of combining the photoelectrochemical cell (which involves a p junction in contract with the solution and a separated n junction in contact with the solution) with n–p cells backing it. Such dual cells may be the eventual way to go.

12. ELECTROCATALYSIS (1955–1990, 17 PUBLICATIONS)

Formally, the term "electrocatalysis" was first suggested by Grubb,[182] a fuel cell worker in the General Electric Company. However, the dependence of the reaction rate on the substrate had been studied by Bowden and Rideal[183] and by Baars[184] since 1928. The first review of "electrocatalysis" was published by Wroblowa and myself in 1964.[185]

The only papers I wish to mention are the ones concerning sodium tungsten bronzes. It was thought in 1967 that the bronzes themselves were catalysts for oxygen reduction which could be compared with Pt.[186] However, by 1972 it was realized that the result was a mirage.[187,188] Traces of platinum were found on the bronze surface! Although this discovery was certainly a big letdown, interesting results were gained because the activity of the traces of platinum on the tungsten bronzes was found to be about 100 times more per unit area than when the platinum was in bulk form.

This was the first of a series of works, some by Bagotzskii and co-workers,[189] which showed that trace elements on the surface of materials have an activity greater than their bulk values.

13. REACTIONS INVOLVING HYDROGEN AND OXYGEN (1946–1988, 64 PUBLICATIONS)

This area involves my Ph.D. thesis and is one in which I have kept up some work until the present time. It is possible to see now what has been of lasting importance.

The first paper I mention is that of Brian Conway from his Ph.D. thesis (1949): "The Effects of Catalytic Poisons at Platinum and Nickel."[190] We were the first to describe quantitatively what a marked effect very small concentrations of poisons can have on the rate of electrode processes. At that time we had difficulty in reproducing electrode kinetic results, particularly for solid electrodes. We thought that we had better try to find out whether the cause was impurities or irreproducible surfaces. We used preelectrolysis to get a steady value of the potential at constant current and then added tiny amounts of poison. We found that an effect began at 10^{-10} moles per litre of As_2O_3!

Another paper of lasting significance was that on the kinetics of hydrogen evolution in MeOH at low temperatures with Roger Parsons and Hanna Rosenberg.[191] published in 1951. This paper has particular relevance now because (as Brian Conway has pointed out so recently) the variation of α with temperature has become a hot topic in electrode kinetics. Figure 33 shows the peculiar nature of the effect.

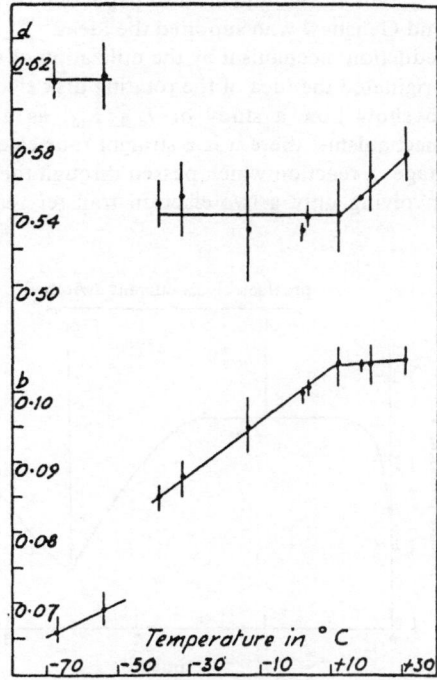

FIGURE 33. Variation of α and β with temperature.

Interestingly enough, the theory which Parsons suggested in 1951 is not so different from a more sophisticated version deduced by Gochev in 1986. The thickness of the double layer is supposed to vary with temperature. However, the correct numerical results can only be obtained if impurities are adsorbed and then this adsorption varies with temperature.

Then there were the two papers with Potter in 1952, but the one in the *Journal of the Electrochemical Society*[192] had material of more lasting significance because it gave the first formulation of several basic electrode kinetic sequences and derived the first equations for the decay of potential after switching off the current.

A. K. M. S. Huq, a Pakistani student who later became the Director of the Atomic Energy Commission in his country, worked on the mechanism of the evolution of oxygen on platinum at very low current densities at the time I was leaving London for Philadelphia. His paper[193] is an alpha one because it describes the experimental determination of the reversible potential for oxygen for the first time (Fig. 34). Many had tried to find the potential and failed because i was $\sim 10^{-10}$ A/cm^2 so that, in the region of the reversible potential, the currents were sensitive to tiny traces of competitors. We found by a detailed study of both anodic and cathodic preelectrolysis, and with enormous care, that we were able to work at 0.01 nA/cm^2 and reach 1.24 V at $i = 0$! (The condition was finally achieved in Philadelphia, in competition with that city's highly chlorinated water.)

After I had been at Pennsylvania for a year, Brian Conway joined me. We worked (*inter alia*) on the relation between the exchange current density and the heat of adsorption.[194] I had spotted two classes of metals by analysis of work function—overpotential data.[195] We were fascinated to find two groups of electrocatalysts, one in which the i_0 work function increased as the heat of adsorption increased, and the other (containing the transition metals) in which it decreased with increasing heat of adsorption (Fig. 35). Similar results were found with plots of log i_0 against ΔH_{H-M}. One rate-determining step involving adsorption, and one involving desorption! So simple sounding now, because the result, along with the volcano curve, has passed into the basic lore of electrode kinetics.

Not till 1967 did the next paper of lasting significance come in this area. It was Damjanovic and Genshaw who supplied the ideas.[196] The paper concerned determination of the oxygen reduction mechanism by the utilization of the rotating disk electrode with ring. Frumkin had originated the idea of the rotating disk electrode as a device to replace stirring. Our part was to show how a study of I_{disk}/I_{ring} as a function of $1/\omega^{1/2}$ could give information on mechanisms: there was a straight four-electron reduction without passing through the H_2O_2 stage; a reaction which passed through the H_2O_2 stage, but evolved O_2 finally; and reactions involving only a two-electron transfer reaction to H_2O_2 (Fig. 36). Halina Wroblowa and

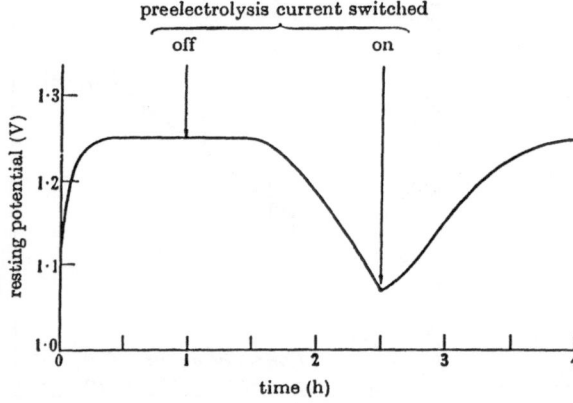

FIGURE 34. Effect of preelectrolysis on the resulting potential.

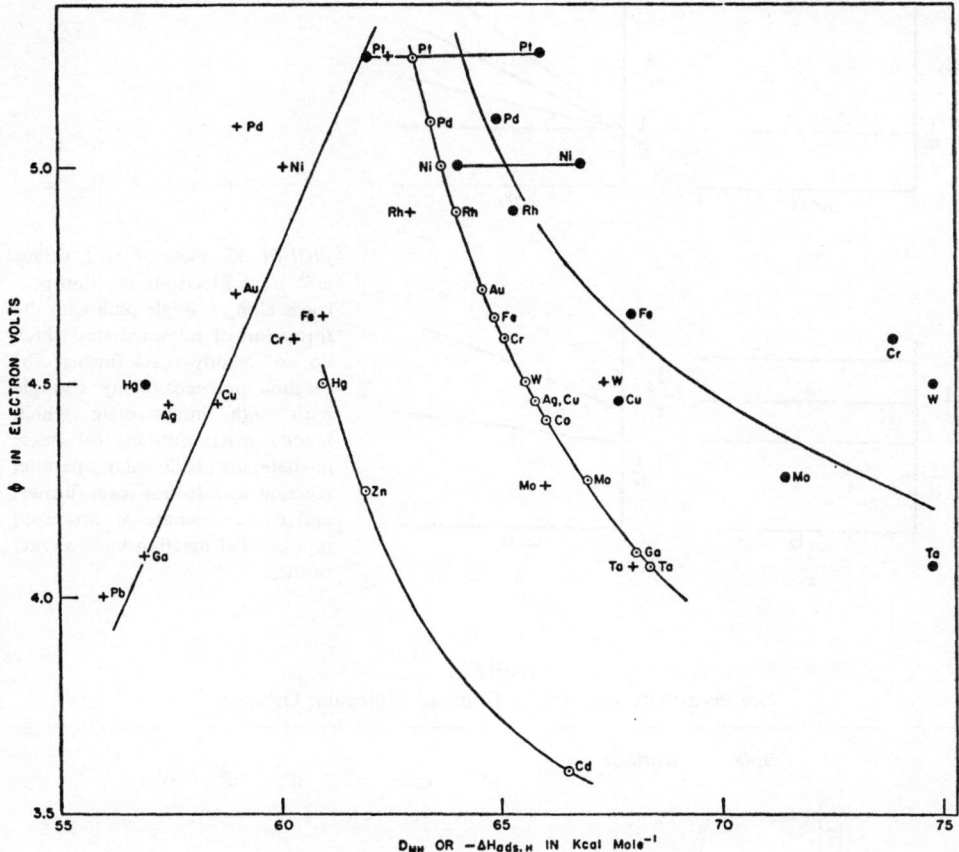

FIGURE 35. Calculated and experimental values of D_{MH} or $-\Delta H_{ads}$ as a function of electronic work function, ϕ.

others have improved our analysis,[197,198] but this kind of analysis has become a standard part of the armory of electrode kinetics.

We have made several studies on oxygen in the 1980s. Taka Ottagawa worked on the mechanism of oxygen evolution on perovskites.[199,200] This work concerned a series of mechanisms that are difficult to realize. At first sight, it is difficult to see how H_2O or OH discharge onto an oxide and form intermediates.

Luckily (and with assitance from Michael Hall of Texas A&M), Ottagawa was able to work out the likely OH–perovskite bonding in terms of molecular orbital theory (see Table 4), and he found that the mechanism of the reactions was as follows (Fig. 37):

Step a:

$$M^z + OH^- \rightleftarrows M^z{-}OH + e^-$$

Step b:

$$M^z{-}OH + OH^- \rightarrow M^z \cdots H_2O_2 + e^-$$

Step c:

$$(H_2O_2)_{phys} + OH^- \rightleftarrows (HO_2^-)_{phys} + H_2O$$

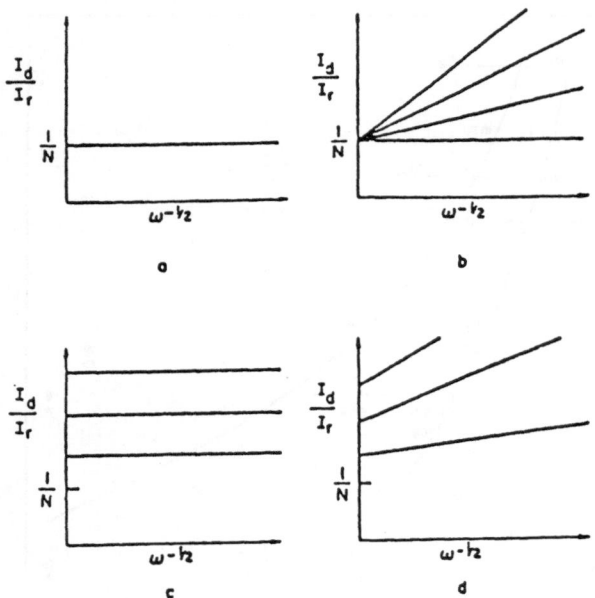

FIGURE 36. Plots of I_d/I_r versus $\omega^{-1/2}$. (a) Electrode reaction proceeds along a single path with the formation of intermediates which do not readily react further; (b) reaction proceeds along a single path with intermediate which readily reacts further; (c) intermediates are produced in a parallel reaction and do not react further; and (d) intermediates are produced in a parallel reaction but do react further.

TABLE 4
OH-Perovskite Bonding in Terms of Molecular Orbitals[a]

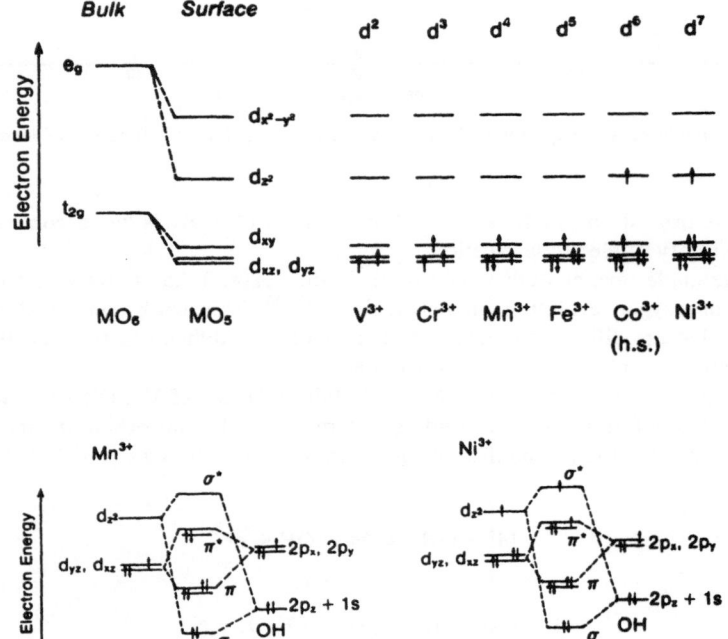

[a] This table depicts the d-electron configuration of transition-metal ions at the surface of perovskites (top); MO diagrams for the M^z—OH bonding at the surface of perovskites; manganites and nickelates (bottom).

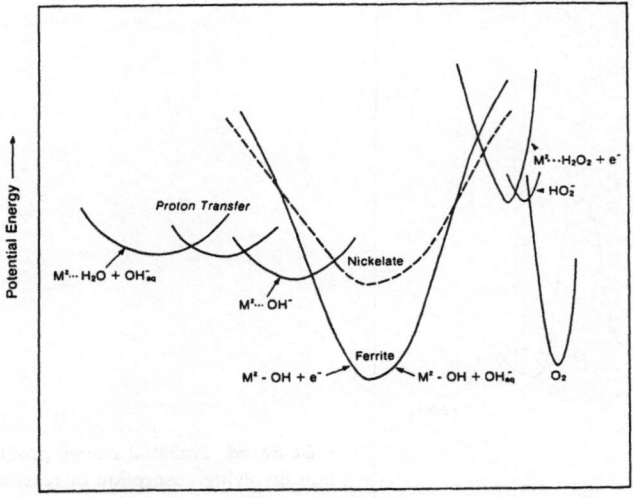

FIGURE 37. Potential energy diagrams for a rate-determining OH desorption mechanism.

Step d:

$$(H_2O_2)_{phys} + (HO_2^-)_{phys} \rightleftarrows H_2O + OH^- + O_2$$

where the subscript "phys" represents physical adsorption.

Our more recent paper in the hydrogen and oxygen group is that in 1986 with Jovancicevic on the mechanism of oxygen reduction on iron in *neutral* solutions.[201] This was ONR (Office of Naval Research)-supported work on O_2 reduction in *neutral* solutions, where few studies have been made but where most corrosion reactions take place. As many of these corrosion reactions are depolarized by oxygen reduction, it is important to find the mechanism of this well-known reaction under these conditions.

Thus, surprisingly the rate of oxygen reduction on passive iron is greater than that on bare iron. On the latter, oxygen reduction proceeds via a four-electron pathway with little H_2O_2 as intermediate, but on passive iron the reaction turns out to be a two-electron pathway with the final formation of H_2O_2. On bare iron the rate-determining step is the formation of O_2^- whereas, on the passive iron, oxygen chemisorption occurs under Tempkin conditions and is rate determining. (See Fig. 38.)

In our model, adsorption of intermediate occurs in the step after the rate-determining step and follows a Tempkin isotherm. One can write

$$r\theta/RT = \log[OH^-] + FV/RT + \text{const.}$$

where r is an interaction parameter for the adsorbate.

The rate of the oxygen reduction can be written in the form

$$i = kP_{O_2} \exp(-\alpha FV/RT) \exp(-\beta \Delta G_{ads}/RT)$$

where k is a rate constant, and ΔG_{ads} is the standard free energy of adsorption of intermediate following the rate-determining step. The free energy of adsorption is affected as a function of coverage, θ. Assuming that the $r\theta$ term affecting ΔG_{ads} is the same as that in the isotherm, that is,

$$\Delta G_{ads} = \Delta G_{ads}^0 + r\theta$$

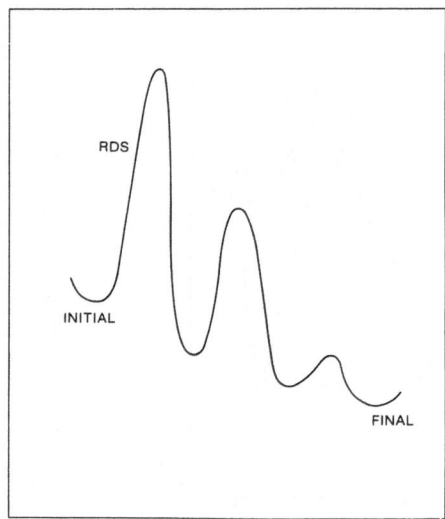

FIGURE 38. Potential energy profile for O_2 reduction involving adsorption of reactants and product in rate-determining step.

where ΔG^0_{ads} is the zero-coverage value, then

$$i = kP_{O_2} \exp(-\alpha FV/RT) \exp(-\beta \Delta G^0/RT) \exp(\log[OH^-]^{-\beta} - \beta FV/RT)$$

Rearranging gives

$$i = k'P_{O_2}[OH^-]^{-1/2} \exp[-(\alpha + \tfrac{1}{2})VF/RT]$$

where k' is another constant, and $\beta = \tfrac{1}{2}$. This explains an expected reaction order with respect to OH^- of -0.5. The label slope is correct for, under Tempkin conditions for this mechanism, $\alpha = 0$.

Flinders University, Adelaide, Australia, 1975

Front row: Sue (secretary); Shahed Khan (quantum mechanical); JOMB; Ashan Habib (double layer); a later rejected graduate student. *Back row*: Erik Stromberg (electrochemical mining); Koheii Uosaki (photoelectrochemical theory); John Pezy (technician); Harvey Flitt (H in metals); John Bourne (corrosion); John MacCann (photoelectrochemistry); T. Ohashi (water splitting).

The 60 or so papers I have published in this area have been a source of joy to me. Hydrogen and oxygen are involved in the most fundamental electrochemical reactions, and by studying them we can learn more about electrochemistry than by studying the simple one-step redox reactions which are so much preferred by many of our electroanalytically inclined colleagues.

14. ENERGY (1967–1990, 21 PUBLICATIONS)

The connection of the present group to fuel cells and energy conversion started in the 1950s with the connection to Tom Bacon, who had his famous laboratory in a disused airport in Cambridge. Bacon hired Rex Watson, who had worked with me in 1950–52 on the mechanism of hydrogen evolution from alkaline solutions. Watson was the electrochemist in the Bacon fuel cell work. Let us never, in any way, detract from Bacon's enormous persistence and the very large effect he has had upon the creation of the fuel cell. However, an ex-member of our group was in there. The famous 1955 fuel cell, the genesis of the fuel cells in NASA space vehicles, was run by Rex Watson in Bacon's laboratory.

My relationship with Tom Bacon has continued undiminished over 40 years, and, as I write these words, I am about to visit him in Little Shelford, near Cambridge, in his 85th year.

I shall pick out just four papers, those which seem right now to have the most chance of having a long-term influence.

The first of these would be the paper by Srinivasan and Hurwitz[202] in which the electrode kinetics of porous electrodes was formulated. This paper was the first to take account of activation control, ohmic influences, and, of course, diffusion all together. Then, the finite-contact-angle meniscus aspects from Cahan's thesis[203] were important in considerations which had been first made by Carl Wagner (working as a consultant for the United Aircraft Company in the early 1960s) as to the location of the electrochemical activity in a pore (Figs. 39 and 40). These papers led to attempts to locate the catalyst where the activity is and revealed the possibility of increasing the power per unit apparent area if the electrolyte was pumped through the electrode. The book *Fuel Cells: Their Electrochemistry* was published about this time with Srinivasan.[204]

Then two other papers of modern time which may have lasting value may be mentioned. One is a paper with Appleby in which we took up the concept of electrochemical overpotential

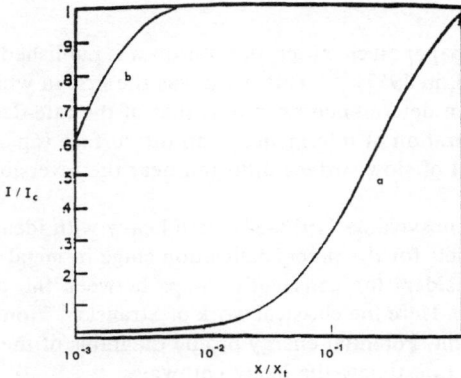

FIGURE 39. Current distribution along the length of the pores. $i_0 = 10^{-9}$ A/cm^2. (a) $\eta = 0.1$ V; (b) $\eta = 0.65$ V.

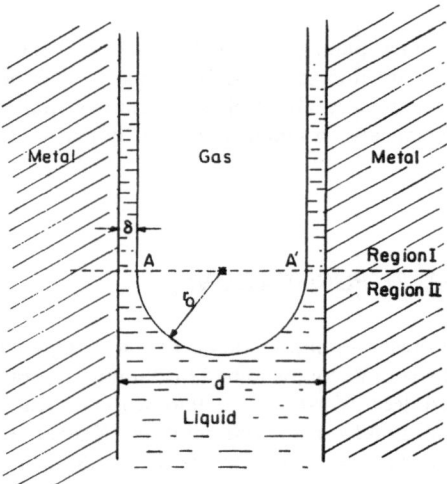

FIGURE 40. Cylindrical meniscus with a thin film.

and applied it to reactions in general.[205] Most people refer to the energy needed to carry out a reaction as the free energy for the concentration ratios concerned. However, there is also an "over free energy" which will increase with the rate of reaction and must be added to the thermodynamic free energy. This kind of consideration is little carried out in calculations of the economics of running chemical reactions.

Lastly, then for this collection, came the paper with Ghoroghchian on homopolar generators.[206] This began with an idea of Felix Gutmann, who pointed out that Faraday had suggested the use of homopolar generators and that little had been done about it after his time! One spins a disk in a high magnetic field, and the induced Emf gives rise to a potential difference across the disk. The spinning disk is immersed in the solution, and all then turns inside the magnetic field exerted between the north and south poles outside the spinning electrode. Hydrogen and oxygen are evolved, one from the center of the wheel and one from the external part.

The homopolar generator gives rise to high currents at low voltages—what electrolysis needs—but there are difficulties in practice unless one goes to the high magnetic fields available at liquid-hydrogen temperatures. Perhaps, high-temperature superconductors will make the ideas here more practical.

15. ELECTROCRYSTALLIZATION (1956–1973, 38 PUBLICATIONS)

My first noteworthy paper on electrocrystallization was published with one of Gerischer's students, Wolfgang Mehl, in 1957.[207] This work was the first in which the rate constant of the *fast* reaction had been determkned as well as that of the rate-determining step, and the first in which the concentration of intermediates on the surface (apart from H and O on Pt) was determined. A model of slow surface diffusion near the reversible potential was shown to fit the facts.

The early days at Pennsylvania (1953–58) were heavy with ideas. One line (with Brian Conway) concerned models for the precrystallization stage in metal deposition.[208]

The paper cited considers the consecutive steps between the charge transfer and the building into the kink site. Here the classical work of Stranski[209] on the point of repeatable growth was a starting point. Potential energy profile diagrams of the type first employed by Butler were used heavily to estimate the likely pathways.

The results from this paper have been the basis of several approaches to individual steps in the precrystallization phase. Adions, not adatoms, were at first formed on the surface. The deposition sites were planes, not kinks or edges. Surface diffusion occurred to edges and then to kinks and would tend to be rate determining near the reversible potential.

Afterwards, the sequence would repeat, and a spiral, the basic growth mechanism for a single crystal, would begin to wind up[210] (Fig. 41).

Another result which arose for the first time in these calculations was the idea that only *one* electron could participate in a given step. The result came out of the considerations of potential energy curves for the metal deposition, but on the visit of Nevil Mott to our lab in 1963, to discuss Watts-Tobin's work on the double layer, he fully supported the result on quantum-mechanical grounds.

Another paper,[211] which is the basis of a whole school of thought and much present activity in Europe, is that concerned with the electrodissolution of iron. This reaction has a special place in corrosion.

The Bockris, Despić, and Dražić (BDD) mechanism for the deposition and the dissolution of iron was worked out after surprising results were obtained: there is a pH dependency of the exchange current density. Thus, the mechanism was as follows:

$$Fe + H_2O \rightleftarrows FeOH_{ads} + H^+ + e^-$$

$$FeOH_{ads}\, rds \rightarrow FeOH^+ + e^-$$

$$FeOH^+ + H^+ \rightleftarrows Fe^{2+} + H_2O$$

However, a corresponding mechanism was suggested by Heussler. This mechanism (which was being competed against the BDD mechanism for almost 30 years) is as follows:

$$Fe + FeOH_{ads} \rightleftarrows [Fe(FeOH_{ads})]$$

$$[Fe(FeOH_{ads})] + OH^-\, rds \rightarrow FeOH^+ + FeOH_{ads} + 2\,e^-$$

$$FeOH^+ + H^+ \rightleftarrows Fe^{2+} + H_2O$$

FIGURE 41. Spiral growth in the electrocrystallization of Cu with pulsed current. [H. von Fischer, *Electrocrystallisation*, Springer-Verlag (1954).]

It is not appropriate here to go into the pros and cons of these two mechanisms. This has been done very thoroughly in a recent article in *Modern Aspects of Electrochemistry* by Dražić.[212] One unfortunate aspect of Heussler's mechanism is that it involves a two-electron transfer as a rate-determining step.

The papers mentioned so far concern the growth of submonolayers (in potential deposition). However, as time went on into the 1960s, considerations were given to more complex effects, and one interesting mechanism was worked out in detail. This 1961 by Barton and myself concerned the electrolytic growth of dendrites.[213]

Here the first work was done on microelectrodes, the practical application of which occurred later on a large scale. Dendrites would grow, preferentially to other mechanisms around them, if the tip of the growing pyramid, at the base of the dendrite, had a rate of curvature less than that of δ for the planar electrode base. The enhanced rate would be δ/r_{tip}, and, as r_{tip} was ~ 100 Å, the multiplier effect is $\sim 10^4$ in unstirred solutions.

This idea of Barton and myself has been taken up by others in discussions of dendritic growth. For example, the formation of snowflakes and the recrystallization of metals, such as iron, from the molten state are both dendritic. Barton and myself showed that the involvement of Kelvin's equation was essential. Thus, a drop of small radius, r, has an enhanced rate of evaporation according to:

$$\frac{p_r}{p^\infty} = e^{\gamma V/rkT}$$

This law was modified to be applied to electrochemical dissolution. As the radius of curvature grew smaller, into the 100-Å range, the dissolution velocity (which competed with the deposition current) grew larger than it would have been at a planar surface. Such a mechanism led to the stabilizing of growth at a maximum rate and optimal radius (Figs. 42a and 42b).

Another paper[214] concerned the basics of stress corrosion. A machine was used such that stress could be applied at a chosen rate, R. The strain was applied to a wire surrounded at intervals by cathodes, and it was possible to measure the rate of dissolution as a function of the rate of change of stress.

The results are shown in Fig. 43. At least for iron and molybdenum, there is a critical strain rate at which there is a sudden increase of the rate of dissolution by a factor of 10–100. Despić and Raicheff connected these happenings with the fact that, on these metals, slip planes are the high-index planes.[214] However (cf. Damjanovic), these have an i_0 value ~ 10 times greater than those of the low-index planes (see Fig. 44).

These more detailed studies of the dendritic growth and distribution of rates under changing stress as planes emerge from the metal interior brought us to the theory of morphology. Here, I was lucky once more in my co-worker because I had Zoltan Nagy (already trained as an electrochemist in Hungary), who combined with Dražić to make a mechanistic study of the morphology of zinc deposition from alkaline solution.[215] It was found possible to interpret in terms of a detailed model the whys and whats of the morphology.

At low overpotential (50 meV), epitaxial layer type growth was observed; the width of the macrosteps increased linearly with time. A theory explaining the linear step widening was proposed, which can also account for the observed effect of substrate orientation. At 100 mV, "boulders" were observed, not all of which were epitaxial with the substrate. They were suggested to originate from nucleation. The boulder density per square centimeter first sharply increased and then slowly decreased with time. A statistical calculation for this was given, based on a model in which large boulders consume the smaller ones. With further deposition, a small fraction (ca. 0.1%) of the boulders develop into dendrites, their number

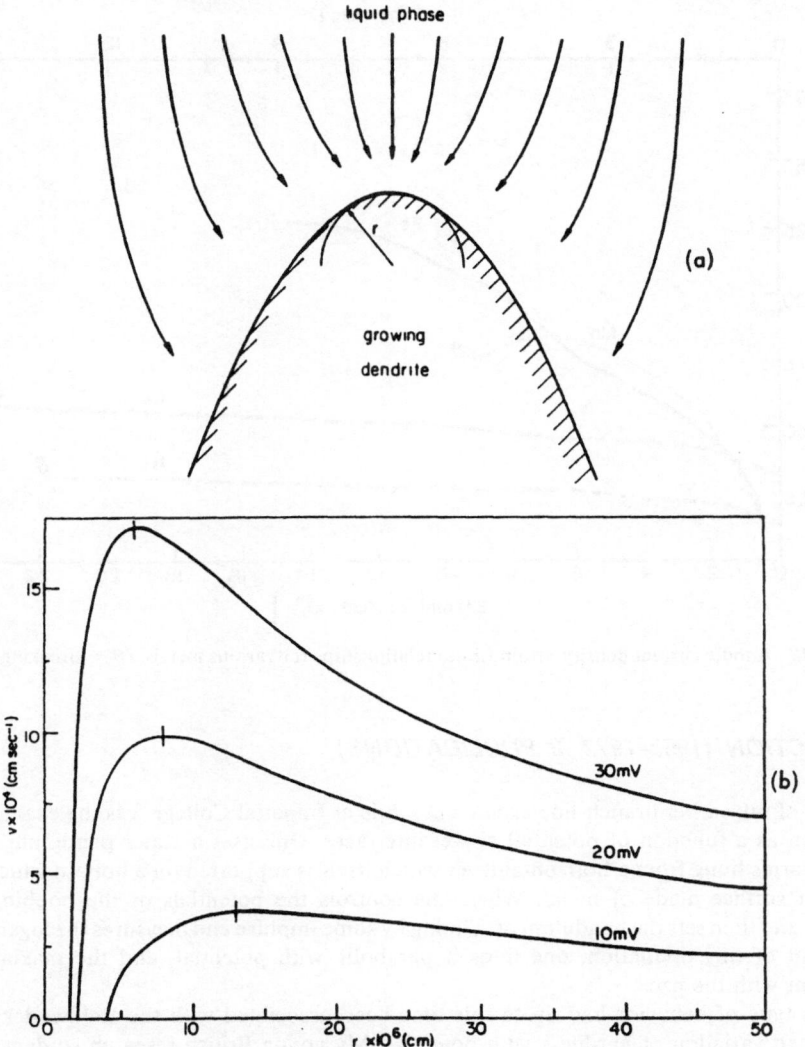

FIGURE 42. (a) Parabolic and spherical models of a dendrite tip. (b) Theoretical curves showing relation between tip radius and velocity for dendritic growth at various overpotentials ($T = 581$ K; $D_{Ag^+} = 1.4 \times 10^{-5}$ cm^2/s; $c_{Ag^+} = 5 \times 10^{-4}$ mol/cm^3; $i_0 = 50$ A/cm^2; $\gamma = 2 \times 10^5$ erg/cm^2). [J. Barton and J. O'M. Bockris, *Proc. Roy. Soc.* (*London*), *Ser. A* **268**, 485 (1962).]

being limited by the available total current. Experiments were carried out to differentiate between overpotential and current control of morphology, and overpotential was found to be the critical variable.

These studies (which included a number of Nomaski interferometric measurements of gliding plane and a good deal of work on bunching) were a substantial contribution to the understanding of the basis of the deposition and the retrocrystallization of metals and were largely led by Damjanovic. They have been followed up mainly in the Soviet Union, but little in the United States, where the publication of metal deposition research tends to be inhibited by considerations of commercial secrecy.

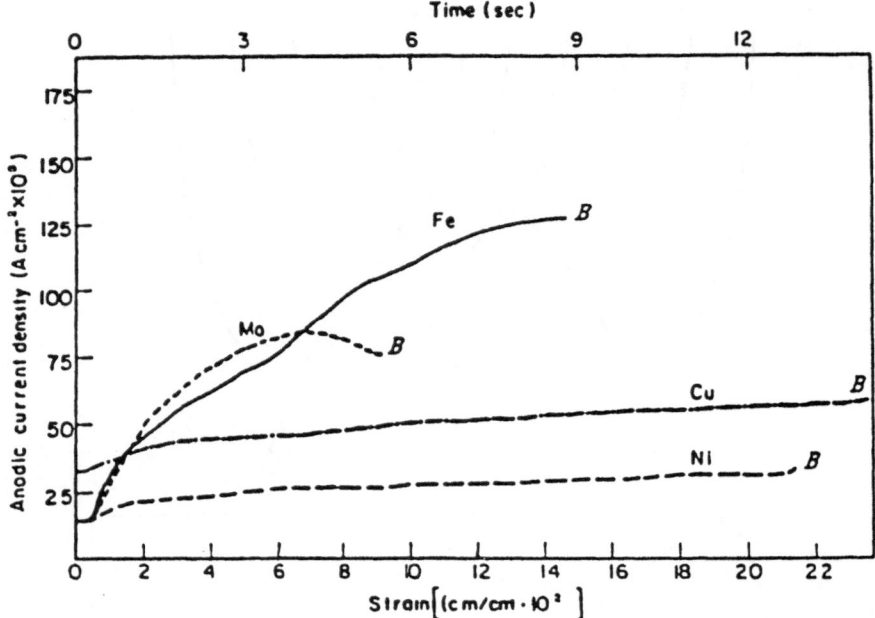

FIGURE 43. Anodic current density–strain (time) relationships for various metals. (β = Breaking point.)

16. FRICTION (1953–1972, 3 PUBLICATIONS)

One of my earlier branch-line excursions while at Imperial College was the examination of friction as a function of potential at wet interfaces. One uses a Kater pendulum. It has two side arms hung from a horizontal lever which itself is supported on a bobbin which rests on a wet surface made of metal. When one controls the potentials of the bobbin/metal interface and then sets the pendulum oscillating by some impulse and measures the logarithmic decrement of this oscillation, one finds it parabolic with potential, and the maximum is coincident with the pzc.

This type of behavior had up to this time been associated with the Rehbender effect, the claimed variation of hardness with potential. My young British research student of the

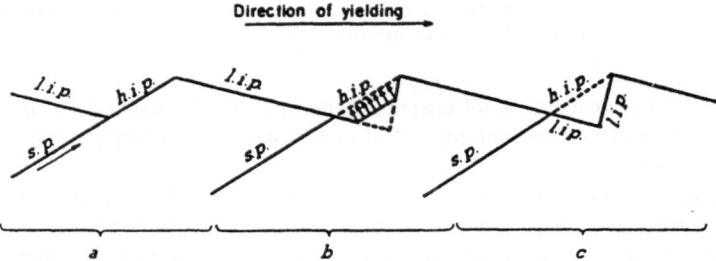

FIGURE 44. Schematic representation of development and preferential dissolution of the high-index planes. s.p., Slip planes; h.i.p., high-index planes, l.i.p., low-index planes. Situation: (a) emerging h.i.p.; (b) partly dissolved h.i.p. site; (c) completely dissolved h.i.p. site.

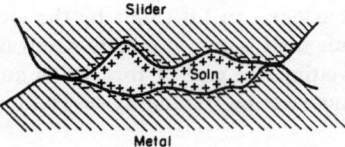

FIGURE 45. Schematic showing a microgap and a microcontact region that exist between the slider and the metal.

time was very skeptical about this.† He was sufficiently skeptical to go to use "Vicker's hardness tester," from the School of Mines, and, according to this, hardness did not vary with potential at all! We published a paper in *Nature*[216] which said that the potential-dependent quantity was *friction*, for the xenophobic Briton had cunningly photographed the bobbin of our pendulum close up with a borrowed movie camera and showed it moved about on the surface as the pendulum swung to and fro.

This was all in 1952, and, when in 1956 I made my first visit to Moscow, I was amazed to meet Professor Rehbender, who, at the age of 80, was careening up and down the corridors of the Institute of Physical Chemistry of the Academy of Sciences at a great rate, leaving me panting behind him. Curiously enough, he seemed to have all sorts of apparatus measuring *friction*. I thought it inappropriate to voice a triumphant comment, but Frumkin helped me understand my stupidity. "It *is* hardness, of course," he said, "but friction *as well*. That is entirely obvious of course." I agreed.

Anyway, later on, Argade, in 1969, worked out a theory, the repulsion of two double layers. The idea was that when the pzc was reached the repulsion between the double layer on the bobbin and that on the metal would be a minimum, and the friction, therefore, maximum.[217] Frumkin, however, on my next visit to Moscow, pointed out some difficulties in the calculation, and I tried again, now down in Australia with Sen, and this time we made a better calculation, on the basis of the model shown in Fig. 45. As a matter of fact, the second calculation was surprisingly quantitative (Fig. 46). I find it an astonishing thought

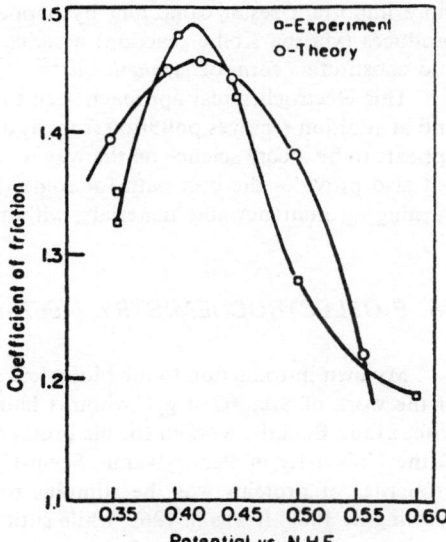

FIGURE 46. Coefficient of friction versus potential for Pt on Pt in $HClO_4$.

† The xenophobic Briton was skeptical of all *foreign* things. He was an Englishman of the older school, complete with blue blazer and well-pressed trousers. He leaned against things a lot and looked on, occasionally giving out a tip as to the mode of work to the ignorant foreigner. He confided in me that he had a "rather negative impression" of many of the students from the outer reaches of the Empire. "My father would just have booted a lot of these chaps off the pavement," he commented.

that one can lift solids by the electrical forces in the double layer. My present opinion of this field is that it greatly needs exploitation. One thinks, for example, of an ionic solution wetting the end of rock drills and a control of the potential there. One might be able to increase the friction by several hundred percent.

17. ORGANO-ELECTROCHEMISTRY (1951–1990, 11 PUBLICATIONS)

Only 12 studies in organo-electrochemistry have so far been carried out. However, much of the work was effected by the powerful critical intellect of Halina Wroblowa and the experience and solidity of J. J. Johnson. Some of it was affected by the mischievous enthusiasm of Anselm Kühn. I shall mention two papers. The first is the study of anodic oxidation of ethylene on noble metals and alloys.[219]

This study is noteworthy because it took ethylene, on which Wroblowa had done several mechanistic studies, and examined the mechanism of the electrochemical oxidation on metals and alloys. The result was a volcano curve.

Thus, the rate is small when the M—C bond is weak because θ_{org} is small. As the bond strength increases, there is an increase in coverage and, therefore, rate.[230] Finally, when the metal–carbon bond strength gets too great, the adsorbed radicals can no longer easily leave the surface so the rate again goes down.

Then it is worth mentioning a 1985 study by Oliver Murphy et al.† about products found in the anodic oxidation of coal.[220] When powdered coal is brought into contact with H_2SO_4 evolution at ~80°C, sufficient material from the porous surfaces dissolves so that a current of ~50 mA/cm^2 can be maintained at the anode while the *cell* potential is merely 1.0–1.1 V (cf. 1.7–1.8 V for the H_2–O_2 cell).

However, although the work allows a cheap method for the formation of hydrogen, a more important result came out, by happenstance—namely, the anodic oxidation of coal produces (via the Kolbe reaction) a series of hydrocarbons which cover C_9, C_{16}, and C_{15} and constitute a form of *synthetic oil*.[221-229]

This electrochemical approach seems to be the cheapest way to produce oil from coal and in addition removes pollution from fly ash. It is interesting to reflect that electrochemistry appears to be a core science on the way to a new age in energy production (solar/hydrogen) but also provides the best path for some (hopefully brief) continuation of the old way of burning up nonrenewable materials, without significant CO_2 injection into the atmosphere.

18. BIOELECTROCHEMISTRY (1957–1990, 22 PUBLICATIONS)

My own introduction to the biological side of electrochemistry relates to my knowledge of the work of Szent-Gyorgyi, whom I had first gotten to know indirectly because my first wife, Linda Bockris, worked for his brother, who was a medical researcher of great renown at the University of Pennsylvania. Szent-Gyorgyi's ideas concerning the alleged electronic properties of proteins was the stimulus to my interest in bioelectrochemistry, which was growing by 1960. It was in 1961, while sitting in the old Midway Airport in Chicago, waiting for a plane, that I realized, as I observed my own breathing, how there might be a similarity between the energy process in the body and that in an electrochemical cell [lungs like porous

---

† The practical work was carried out by Dr. Ferri Saffarazi, who unfortunately resigned suddenly from our group after an explicit verbal disagreement with Lily Bockris concerning the completion of an overdue report. She took her notebooks and the report with her.

electrodes in which the cathodic reduction of oxygen is occurring, and food becoming after 'digestion the anodically oxidized fuel (glucose, etc.)].

However, it was not until I was working with Srinivasan on the book *Fuel Cells: Their Electrochemistry*[204] in 1966, that there was an opportunity to write up what I had realized at Midway Airport. Thus, in discussions with Srinivasan it transpired that the efficiency of metabolism was in the region of 50%, and we knew from our study of energy conversion that this was an unprecedented efficiency, more than twice the efficiency of heat engines, where the hot part of the cycle could be several hundred degrees. We wrote an article for *Nature* in 1967.[231] Shortly afterwards, Rosenberg[232] published his work showing that wet proteins had an *electronic* conductivity much greater than that of dry protons, and this provided the missing link in the concepts forming in my mind—this discovery of electronic conductivity in proteins provided the wire between the cathodes and the anodes which I saw as groups on the proteins, respectively—and I therefore put everything together that I saw at this time to support the idea that it was electron transfer at interfaces, and not the homogeneous electron transfers of redox couples in solution, that constituted the rate-determining steps in metabolism and in many other biochemical reactions. The paper was called "A Basic Biological Step?" and was published in *Nature*.†[233]

Australia was only two years away when this was published, and, shortly after I arrived there, I got a student who was interested in the bioelectrochemical side, Mohammed Shuaib, a Pakistani. He carried out some interesting experiments which gave for the first time a direct experimental proof of the *photoelectrochemical nature* of photosynthesis. Unfortunately, the paper was published in a rather obscure place, and the work is not well known.[234]

The basic idea was that to extract photosystem I and photosystem II from spinach by centrifugation using known methods and to adsorb, successively, but independently, photosystem I and photosystem II on platinized platinum, then irradiating the platinized platinum (no effect) and afterwards the platinized platinum with the photosystem (significant effect). However, the *main* point was that photosystem II gave an anodic reaction, and photosystem I a cathodic reaction; that is, the currents went in opposite directions, depending upon the photosystem. This seemed consistent with the view that the basic mechanism of photosynthesis is the electrochemical dissociation of water and that the rest of the action, the actual formation of the CH_2O and O_2, comes at a later stage. [The separation of the two reactions had already been suggested by Calvin (Fig. 47), but I think what was new was the proof of the *photoelectrochemical* mechanism of the water splitting to hydrogen and oxygen, one of Nature's great processes.]

Up to this time (1979), my contributions to bioelectrochemistry had been largely inspired by the 1961 Midway Airport realization, but I worried that I was always competing with the overwhelmingly well-known work of Hodgkin and Huxley, who received the Nobel Prize in 1963 for giving what was thought by the Swedish committee to be a reasonable electrochemical model for nerve conduction. I thought that it was better to leave this *central massif* alone in this case and to go on to apply my ideas to the mitochondrion, where energy conversion was known to be taking place.

By now we were back in the United States, and I was fortunate enough in 1984 to be visited by my friend, the irrepressible and irreplaceable Felix Gutmann, who then set to work with his usual gusto to build up a fuel cell model in biological energy conversion, in collaboration with M. A. Habib,[235] who had joined me after postdoctoral work in Australia with Dennis Mathews. The basic idea is shown in Fig. 48, and I think that it hardly needs

---

† Some biologists and biochemists have taken up these ideas. Pethig (last of the Szent-Gyorgyi group) is keen on them, and particularly Michael Berry in Adelaide, Australia. However, although they seem obvious and revolutionary to me, most biochemists bridle at the idea of *electrons* moving fast in the solid phase in a living entity.

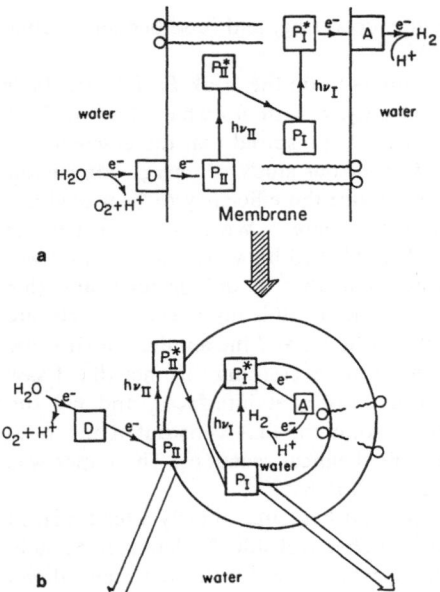

a

b

FIGURE 47. Photoinduced electron transfer schemes for photosynthetic membranes and vesicles.

explanation. The cathodic and anodic sites may be on the same side or different sides of the mitochondrion, perhaps both depending on the type of cell concerned, and the model allows it to be possible (see below) for the potential difference gained by the fuel cell to be applied to other electrochemical reactions which could be seen as "charging the battery."

However, one thing was missing from the model of the mitochondrion as a "fuel cell"—an experimental realization. I wanted to try to make experiments which would give some sharp and clear experimental realization of my ideas, but I was frustrated, of course, by the difficulty of carrying out direct experiments on a mitochondrion itself, around 1 μm in size. I set to work with Flamarion Diniz, a Brazilian research graduate student of extraordinary independence and talent, who tried to achieve the next best thing. We took the apparatus shown in Fig. 49, with typical results shown in Fig. 50. The concepts are quite obvious. The membranes of polypyrrole, prepared in four different varieties according to the doping agent employed, covered a large range of conductivities. On either side of the membrane we put different redox processes and were able to deduce equations which corresponded to the observations. Depending upon the relative magnitudes of the electronic and ionic conductivities of the membrane, and, of course, the i_0s of the electron transfer reactions on either side, I-V shapes of various kinds were realized. We had demonstrated in a model form how redox compounds could make fuel cell happenings occur across a bielectrode.

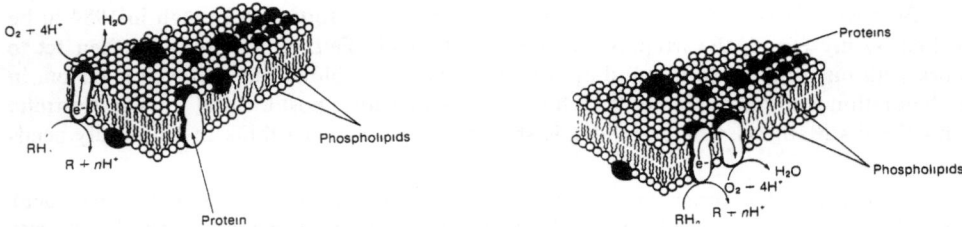

FIGURE 48. Sectional views of the mitochondrial membrane with anodic and cathodic sites on opposite sides and the same side of the membrane.

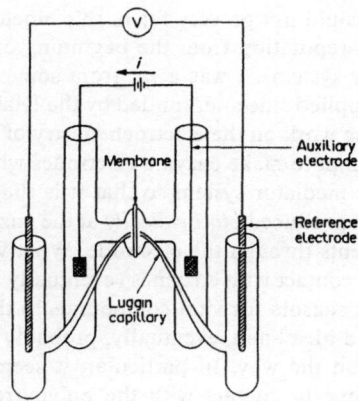

FIGURE 49. Schematic of the bioelectrochemical cell.

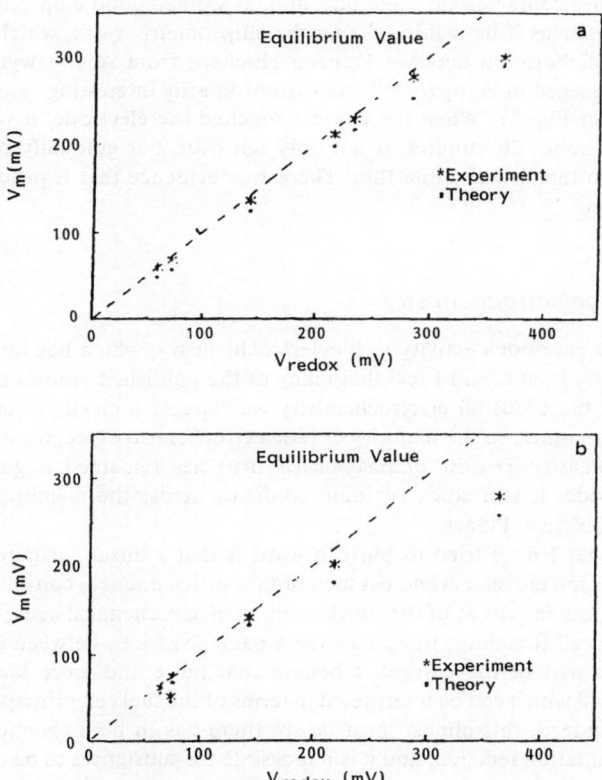

FIGURE 50. (a) Membrane potential as a function of the difference in redox potentials of the solutions on both sides of the membrane for a polypyrrole toluenesulfonate membrane. The dashed line corresponds to the data obtained with a platinum foil replacing the membrane. (b) Membrane potential as a function of the difference in redox potentials of the solutions on both sides of the membrane for a polypyrrole sulfate membrane. The dashed line corresponds to the data obtained with a platinum foil replacing the membrane. [F. Diniz and J. O'M. Bockris, *J. Electrochem. Soc.*, **235**, 1967 (1988).]

I knew, of course, that I could not go very far in this bioelectrochemistry game, where only those who already had a reputation from the beginning of bioelectrochemistry could get federal grants (peer review system). I was a rat from some other pack, and hence not fundable.† So I turned to the applied side and, funded by the Diabetes Foundation, managed to get in some of the underlying work on the electrochemistry of enzymes. The idea Duncan Hitchens and I had was to attempt to make enzyme electrodes which acted *directly*. The ones in use at present all employ a mediator system so that it is the electrodes in the solutions which react with the substances produced *biochemically* at the enzymes; the enzymes themselves do not normally pass currents through the electrode. What we wanted to do was to put biologically active materials in contact with enzymes (eventually a vast number of them) and thereby get innumerable direct sensors for vital compounds in the body.

This, of course, is a "grand idea" and, eventually, probably will be achieved. However, we have hit some difficulties on the way. In particular, it seemed that after the group of enzymes we were studying came in contact with the polypyrrole electrode substrate, the enzyme activity dropped disasterously.

We felt that perhaps we were acting with a system which was too complex to begin with, but instead of leaving it and getting something simpler, we decided to study what did happen when glucose oxidase contacted the electrode, and, as we had good ellipsometric apparatus, I asked Duncan Hitchens if he would take up the ellipsometric work, which was eventually carried out in a collaboration between Duncan Hitchens from Aberystwyth in Wales and Arpad Szucs from Szeged in Hungary.[236] An extraordinarily interesting result was obtained, and this is shown in Fig. 51. When the enzyme touched the electrode, it was standing up, but gradually over about 20 minutes, it not only fell over, but gradually decrepitated and changed its form so that it was quite thin. There was evidence that it produced FAD as a dissociation product.

The Future in Bioelectrochemistry

In spite of the enormous activity in bioelectrochemistry, which has now been going at full tilt for 100 years, I am afraid I feel that many of the published models are on the wrong track. Just as until the 1950s all electrochemistry was largely ionically oriented and bound largely to thermodynamics, so the majority of bioelectrochemistry (see, for example, Volume 10 of the *Comprehensive Treatise of Electrochemistry*) has remained largely in the ionic-thermodynamic mode. It is a study of ionic gradients across the membrane, and the key equation is that of Nernst–Planck.

The concept that I have tried to push forward is that a misunderstanding is going on and that the famous ion movements across membranes which one sees constantly in biological work are the *result, not the cause*, of the functioning of electrochemical cells. Of course, when an electrochemical cell functions, there *must* be a passage of ions between the electrodes to make up the ionic part of the current. I believe that more and more bioelectrochemical systems will be found which can be interpreted in terms of the fuel cell principle and electrodic electrochemistry. Indeed, this almost must be so: there has to be a chemical driving force for some overall oxidation reaction, and it is impossible for substances to be consumed unless there are electron transfer processes. Now, what *I* am saying is that this incontrovertible happening is realized via rate-determining *interfacial* electron transfer reaction, involving

---

† Practicing scientists will know what I mean. It is a lamentable fact that scientists are as emotional as persons in other professions—acting, for example—and when they get to review and grade a paper from someone outside their circle, especially with new ideas, they often get very angry, and begin to wave their arms and even shout.

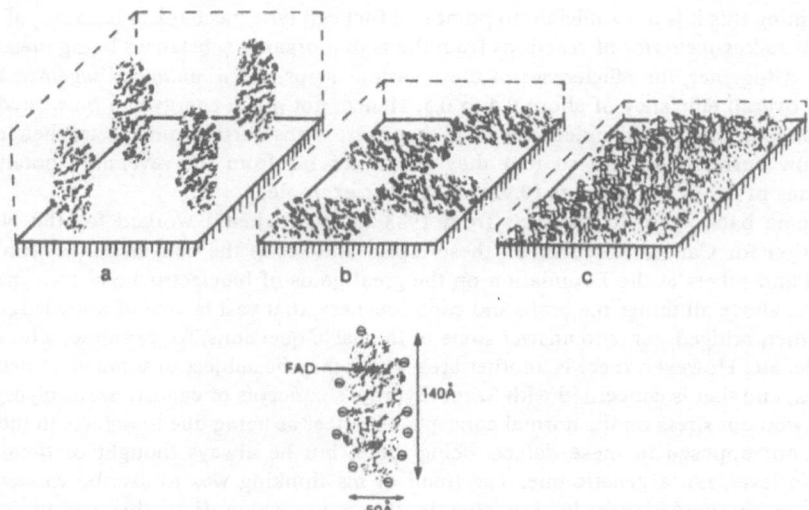

FIGURE 51. Schematic representation of the adsorption processes of glucose oxidase. (a) If the potential is much more positive than the potential of the zero charge. (b) If the potential is close to the PZC. (c) Final stage, when the enzyme unfolded.

many zillions of cathodes and many zillions of anodes, there being an overall chemical reaction in which, of course, no electrons appear (as with the overall reaction in a fuel cell).

Most of the details here have to be filled in, and this may take a decade or two, but there is a need to fill out the rest of the essentials of the story. For it is not enough to have the fuel cell undergo electrode reactions; it must pass a current and produce energy, and this energy must then be stored, though be available to produce electrical work when called upon. How is it stored?

It is known, of course, that the storage of energy in the body takes place by a mechanism of which the overall reaction is[235]

$$ADP + H_2PO_4^- + H^+ \rightarrow ATP + H_2O$$

A suggestion which Gutmann, Habib, and I made was that this chemical reaction was broken down into two parts, namely, the cathodic reaction

$$H_2PO_4^- + H^+ + e^- \rightarrow HPO_3^- + H_2O$$

and the anodic reaction

$$ADP + HPO_3^- \rightarrow ATP + e^-$$

Thus, the electrons needed for these reactions would come from the mitochondrion, and the ATP would then circulate to muscles or wherever energy is needed, the reaction now being reversed and the muscles acquiring the necessary electrical charges for their action by the reverse of the above.

This low-level electrochemical picture has one thing to recommend it. It does what it was designed to do—explains the great efficiency of the metabolic process in biology. Thus, in explaining this it is not sufficient to point to a fuel cell type mechanism because, of course, there is a *successive series* of reactions from the actual organic substances being metabolized to CO_2. Altogether, the efficiencies of these various steps, *when multiplied together*, have to yield an overall efficiency of about 0.4 to 0.5. Hence, not much energy can be wasted in the storage steps. This would indeed be so if they were steps carried out electrochemically at rather low current densities so that they occur not far from the reversible potential, at efficiencies pf 0.9 or thereabouts ($0.92^6 = 0.61$, for example).

I come back again to the days from 1983 to 1986 when I worked for the National Foundation for Cancer Research on these topics and recall the conversations with Szent-Gyorgyi and others at the Foundation on the great goals of bioelectrochemistry. These are, of course, above all things the brain and consciousness, that vast lacuna of knowledge which could, when bridged, serve to answer some of the basic questions, for example, who decides to decide, etc. However, there is another area which may be subject to some electrochemical advances, and that is concerned with Szent-Gyorgyi's concepts of cancer. Szent-Gyorgyi was not one who put stress on the normal concepts of cancer as being due to defects in the DNA. He was not opposed to these defects being there, but he always thought of them on an electronic level, not a genetic one. The trend of his thinking was to ascribe cancer to the absence of an oxidation–reduction step in the redox chain (i.e., this would interrupt metabolism).[237] Thus, in his view, the cause of cancer could be seen as the inhibition of an electron transfer step.

19. CORROSION AND PASSIVITY (1947–1990, 16 PUBLICATIONS)†

My introduction to the field of corrosion was made by T. P. Hoar of Cambridge University, a close colleague of U. R. Evans. Hoar led me at age 26 to Pourbaix, and so I became a founding member of what is now the International Society for Electrochemistry, and, indeed, apart from Pourbaix himself, I am the only still living founder.

During the time with Pourbaix, I was appraised of the virtue of Wagner and Traud and, indeed, I was shortly to know Carl Wagner in visits (1951–52) to MIT hosted by Herb Uhlig.

When I read the Wagner and Traud concepts in German, I became keenly aware of the fact that they lacked a quantitative theoretical basis, and so when I began to write the chapter called "Electrode Kinetics" in the first volume of *Modern Aspects of Electrochemistry*, I gave the polyelectrode section its mathematical garb. I worked out the mixed potential in electrode kinetics terms.[238] The equation I obtained is:

$$\Delta\phi_{corr} = \frac{RT}{F} \ln\left[\frac{I_{0,so} \exp(F\Delta\phi_{e,so}/2RT) + I_{0,M} \exp(F\Delta\phi_{e,M}/2RT)}{I_{0,so} \exp(-F\Delta\phi_{e,so}/2RT) + I_{0,M} \exp(-F\Delta\phi_{e,M}/2RT)} \right]$$

Thereafter, it was easy to substitute in the Butler–Volmer equation the term $V_{mp} - V_{rev}$ for η, where V_{mp} was the mixed potential for the whole electrode, and V_{rev} was the reversible potential of the anodic, or the cathodic, reaction ($V_{mp} \simeq \Delta\phi_{corr}$).

This gave rise to a fair amount of algebra, and I was kept at it for some days but finally came to the corrosion rate in absolute terms as:

$$I_{corr} = I_{0,M}^{\vec{\lambda}_M/(\vec{\lambda}_M + \vec{\lambda}_{so})} I_{0,so}^{\vec{\lambda}_{so}/(\vec{\lambda}_M + \vec{\lambda}_{so})} \exp\left(\frac{\Delta\phi_{e,so} - \Delta\phi_{e,M}}{\vec{\lambda}_M + \vec{\lambda}_{so}} \right)$$

Of course, this assumes metals free of oxide films, etc. It implies that the cathodic and anodic reactions take place to an equal extent on the surface, but later on Herb Uhlig and Winston Revie were able to show that this is indeed the likely case.

---

† Several more works were carried out here but are described in the Section 10.

These equations have been taken up by others and are presented, for example, by Smyrl in the *Comprehensive Treatise of Electrochemistry* as the basis for the treatment of corrosion.[239] I think that they have their place in the history of corrosion science although they do not often get referred to specifically, perhaps because they were published for the first time in a book rather than a journal.

Most of my other works on "corrosion" have been in respect to studies of passivation and of hydrogen embrittlement and the like. Here I mention particularly the work carried out with Bill O'Grady just before I left the University of Pennsylvania for Australia.[241-242] It was, indeed, O'Grady who led this work rather than I, because he suggested the application of Mössbauer spectroscopy. When we could not interpret the frequency which he obtained, he brought up the answer himself by dashing into my room one day, blurting out, "It's amorphous." O'Grady had spotted a paper by Prados and Good,[243] and this paper had shown the characteristics brought into the Mössbauer patterns by the presence of amorphousness, which O'Grady recognized in his own material. O'Grady and I then speculated on the essential reason for this—the trapping of water in the oxide layer, and its forming a polymer with Fe_2O_3. Our guess was helped by the fact that when we dried the material for a prolonged time at 120°C, it reverted to a γ-Fe_2O_3 structure. The first proposal for the amorphous film structure can be seen in Fig. 52.

Essentially the rest of the work that I have directed with my collaborators on passivity consists of a series of spectroscopic investigations and is summarized in my recent paper.[244] Winston Revie, one of Herb Uhlig's former students and my first postdoctoral fellow in Australia,† was lucky to come to a department at the Flinders University where high-vacuum techniques were the order of the day. With the aid of Bruce Baker, we constructed the first ultra-high-vacuum equipment to be used in electrochemistry (Art Hubbard was working at the same time and took over this technique and made it into his own).[245] The equipment is shown in Fig. 53.‡

Winston Revie achieved XPS measurements of the ratio of O to Fe and found it to be 2.0 for the passive film on Fe, showing the composition of the film to be $Fe_2O_3 \cdot H_2O$.[246] When the chloride was added, the ratio was reduced to about 1.6 (cf. the value of 1.5 for Fe_2O_3). It seemed clear that the chloride removed the water, and this removed the amorphousness and depassivated the film.

Oliver Murphy and T. E. Pou, working then back at Texas A&M, went on the rampage to investigate passive layers spectroscopically and, with the aid of Gene Sparrow, carried

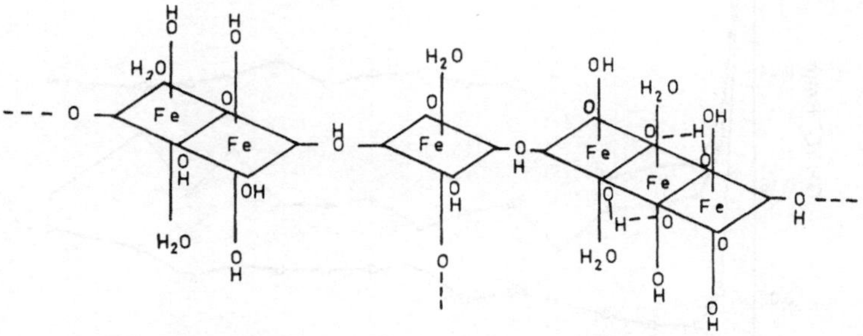

FIGURE 52. The first proposal for the amorphous film structure.

† Note, again, how lucky I was in receiving an outstanding and excellent co-worker just at the beginning of my Australian sojourn. I had had Parsons and Conway to start with in London; Conway, Blomgren, George Myers, and the devoted Linda Bockris in Pennsylvania; dynamic Winston Revie in Australia; and rollicking Oliver Murphy (Khan and Habib, too) back in Texas.

‡ Hubbard's later developments were done in metal vacuum systems, which, on the whole, offer a great improvement; but at this time Baker preferred to work mainly in glass.

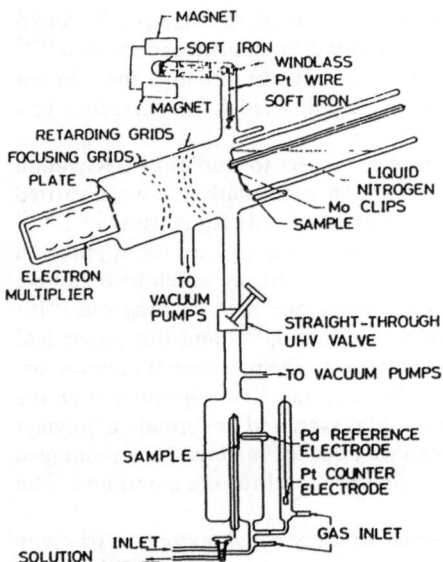

FIGURE 53. Apparatus for combined electro-chemical studies and Auger electron spectroscopy.

out a number of measurements, *inter alia*, using SIMS, which is a direct method for finding out the water to O/Fe ratio, again finding results consistent with the presence of water in the passive film.[247,248] Of course, with SIMS it is possible to find out the depth of the water penetration (Fig. 54).

So, in this brief account it is not possible to give all the involvements of people who did not agree with us about amorphousness, water, and protectivity and people who did—I think

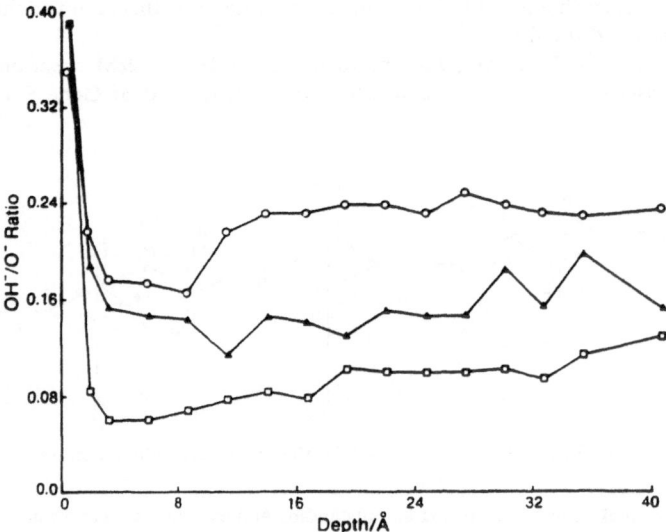

FIGURE 54. Dependence of the OH⁻/O⁻ ratio obtained from SIMS on depth into passive films formed on pure iron electrodes at 0.065 V versus NHE in borate buffer solution and subsequently exposed to a buffered 0.5 M chloride solution at the same potential for various times: ○, 0 s; △, 10 s; □, 30 s.

on the whole we came out well on balance. The idea of amorphousness as related to protectivity has spread from O'Grady's suggestion, and now it is not only a part of the theory of passive layers but also to alloying in metal deposits (cf. Ref. 249).

The question of why the chloride ion and other ions of similar size prove detrimental to passive systems is important in seawater corrosion, and, with the brilliant Jovancicevic from Paris and points further east and T. Mizuno from Hokkaido (and also Carbajal and Zelenay), we tackled the matter by letting chloride ions diffuse into passive layers.[250] Mizuno did the experimental work and obtained a most interesting and unambiguous result. The essence of this is shown in Fig. 55. The chloride versus time relation follows a stable pattern up to a given chloride concentration at a given potential whereupon a sudden change in the chloride versus time relation occurs, the chloride concentration heads on up abnormally, and *just a little later* the current floods forth through the passive layer; that is, the passive film is broken down. Thus, it is chloride diffusing through the film that causes the breakdown, and, of course, we gave models whereby the chloride displaced the water and thus led to dissociation within the amorphous oxide and loss of passivity.

There is also the question of inhibition, and this was something that was worked on in the very early days with Conway,[251] but we wanted to go ahead and see what would happen now with a strong inhibitor, such as octynol, which is used in the protection of stainless steel from boiling HCl. We used ellipsometry in this work with Jovancicevic and the stubborn but knowledgeable Bo Yang from China.[252] The octynol inhibitor has at least three positions, lying down, "getting up," and standing up, and these are shown in the diagrams in Fig. 56.

We found that not only did the octynol assume these three positions, with the surprising standing-up one justified by three-way bonding of carbon to the metal (close-packed vertical octynol presumably being a much better inhibitor than the lying-down version), but also that very thick films of octynol could form. In fact, at high temperatures, somewhat corresponding to those which would be used in practice for boiling HCl, we found that the octynol layers were thicker than 50 Å—it polymerizes *in situ* to what is effectively a paint.

Analysis of the details of the electrode kinetics, however, were somewhat muted in respect to the fact that quantitative interpretation was limited to the submonolayer. As the inhibitor coverage increased, the b value in the Tafel equation increased, too, and this happened both for cathodic and anodic reactions.

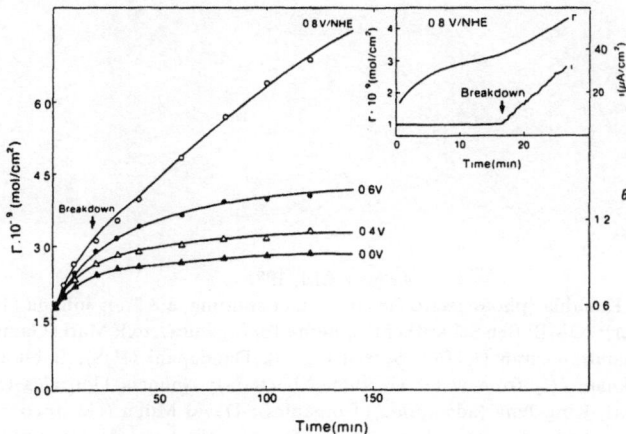

FIGURE 55. Total "surface" concentration of chloride, for $5 \times 10^{-3} M$ Cl$^-$ in borate buffer (pH 8.4) solution and different potentials.

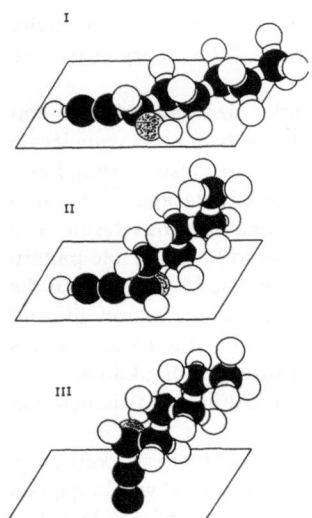

FIGURE 56. Change in orientation of the octyne-3-ol as a function of concentration and time (see text).

Texas A&M, 1985

Front row: Ramesh Kainthla (photoelectrochemical water splitting at 8.2%); Johnna Gauze (secretary); Bill Craven (manager); JOMB; Ben Scharifker (isotherms for organics, etc.); Maria Gamboa (radiotracer, adsorption); K. Chandrasekaran (FTIR). *Second row*: B. Dandapani (H_2S); Babli Kapur (alloys of α-SiH); Nigel Packham (H_2 from water via rumen bacteria); Anuncia Gonzales (STM); Bo Yang (corrosion inhibition); King Jeng (adsorption of organics); David Miller (electroconically conducting polymers). *Third row*: Sasha Gochev (theory of β); Duncan Hitchens (biosensors); Jeff Wass (low-temperature kinetics); Vladimir Jovancicevic (corrosion); Tang (passivity); Ignacio Villegas (photo-electrochemistry); Flamarion Diniz (bioelectrochemical mechanisms).

20. HYDROGEN IN METALS (1966–1990, 20 PUBLICATIONS)

The first person to introduce me to the subject of hydrogen in metals was Dr. Walter Beck, a remarkable gentleman who used to visit me about once a week in the University of Pennsylvania, tapping a stick as though blind. It was difficult for him to conceal his excitement about the topic. He claimed to have worked with the great Walter Nernst ("Ja, I was a student of *Nernst*") and put forward his ideas with great vehemence. He promised to get us financial support from the Navy Air Materials Command Center at Warminster, Pennsylvania, if we were able to do the fundamental work that he thought was particularly valuable.

Dr. Beck was particularly interested in the effects of stress on the solubility of hydrogen. Eventually, my enthusiasm lit up under his tutelage. I persuaded James McBreen, at the time (1962) a raw research student, to come and work on the project.

The result of the net effort of Walter Beck from Berlin, Jimmy McBreen from Ireland, and Leonard Nannis (an American!) working with me in this direction was a quite remarkable one. We had, first of all, to learn how to use the Devanathan-Stachurski cell.[253] This was a device which has been used in a number of laboratories but was originated by M. A. V. Devanathan and myself in a Sunday discussion which took place in the early 1960s. It harked back to the work of Frumkin and Aladjalowa, who had used something similar in their study of hydrogen diffusion through palladium.[254] It is shown in Fig. 57.

We modified the cell to carry membranes which were designed so that when they were put under tensile stress by applying calipers to the bottom and top, the stress was uniform throughout the thin portion through which the hydrogen permeated.†

McBreen's work brought two very pertinent results to the history of the field. One was the first measurement‡ of the partial molar volume of hydrogen in metals. It was possible to observe the variations of the permeation rate with the stress which we were able to apply (see Fig. 57). From the slope of this log J-stress relation, we were able to determine the partial molar volume of H in Fe (Fig. 58).

However, this was only one of two important results obtained by Jimmy McBreen, for the other was a demonstration that the permeation-time relation at constant potential had two radically different types of behavior depending upon the overpotential.[256,257] Thus, if one had a sufficiently low overpotential, one could turn on the current and watch the permeation of hydrogen as a function of time (Fig. 59). It would go on at a constant rate for many hours. The current could be turned off, and the permeation would then rapidly fall to zero; when the current was turned on again, a rapid rise would occur (Fig. 60). The permeation-time line could be repated many times.

On the other hand, directly McBreen reached a hydrogen overpotential of about −0.35 V, the character of the J_H-t line would change radically.

This was at once interpreted by McBreen and myself as being due to the opening of voids inside the (now embrittled) metal. We had reached a critical value of the internal fugacity, so that now the internal pressure of the hydrogen was enough to cause yielding of the iron and spreading of the voids. A simple quantitative treatment was consistent with this model.

---

† We are indebted to F. S. Williams of the Navy Yard at Philadelphia for help in the design of such membranes.

‡ My colleague, Dr. Richard Oriani, may well protest this claim for in parallel work he had used an X-ray method to determine the partial molar volume. He would claim that his method was the first to do this, because, in fact, our own work, which was entirely parallel to his, contained a trivial numerical error, so that the value was 2.303 too high. The correct value for the partial molar volume, therefore, came out a little later, but I am correct in saying that it was the first time the value had been obtained electrochemically and that thermodynamically by a very clean argument.[255]

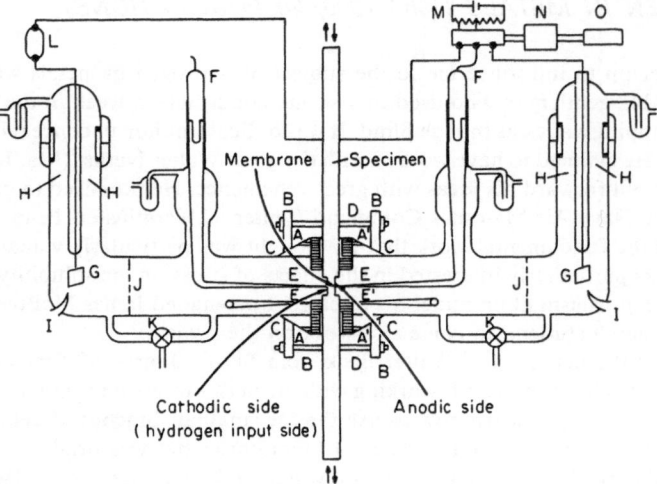

FIGURE 57. Cell and preelectrolysis vessel. A, cathodic compartment; B, anodic compartment; C, SCEs; D, F, auxiliary electrodes; E, test electrode.

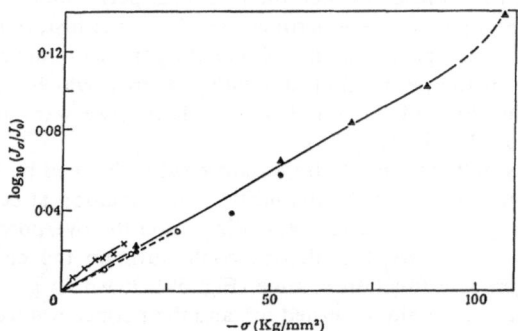

FIGURE 58. Stress dependence of permeation ratio, $T = 25°C$. ×, Armco iron; ●. A.I.S.I. 4340 steel (u.t.s. 260 kLb./in.^{2}); ●, A.I.S.I. 4340 steel (u.t.s. 150 kLb./in.^{2}); ○, de Kazinezy (annealed steel).

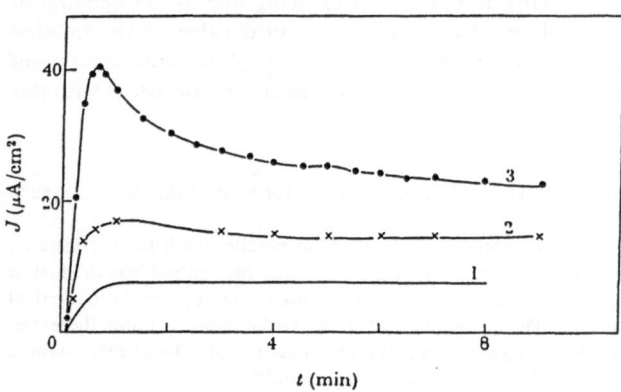

FIGURE 59. Permeation-time curves. $L = 0.77$ mm, $T = 25°C$, $0.1N$ H_2SO_4. Cathodic current: (1) 0.4; (2) 0.8; (3) 4.3 mA/cm^2.

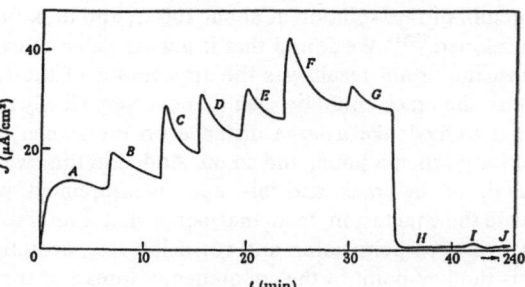

FIGURE 60. Permeation transient. Armco
iron, $L = 0.77$ mm, 0.1 N H_2SO_4.
Cathodic current: A, 0.6; B, 1; C, 4; D,
10; E, 20; F, 40; G, 50; H, 0.6; I, 1; J,
0.6 mA/cm^2.

So now we have a nice electrochemical way of studying traps within the metal, applied later, with Minnevski, to cold fusion.

Another outstanding character who enters the game at this point (1967) is P. K. Subramanian, who was very fertile in his work on hydrogen, following up that of McBreen, P. K., as he rapidly became known in our group, was a man who showed a rising inner anger by putting his hands behind his back and moving up and down vigorously on the balls of his feet, while staring at the ground just in front of him. I have seen him do this particularly in play against Richard Oriani, with whom we had some rather intense debates.

The work with P. K. gave rise to results whereby we were able to calculate the size and pressure in the traps (which the hydrogen was filling after the critical fugacity had started to spread them).[258,259] One of the more interesting results was the graph we obtained when we turned the current off on the cathodic side and allowed hydrogen to dissolve anodically through the outgoing side on the anodic side (through a thin layer of palladium) because now we saw a permeation-time line lasting several hours and containing several bumps. (See Fig. 61.) These we thought were the release of hydrogen from particular impurity centers within the iron.

P. K. was concerned with other interesting forays. One (with which Marvin Genshaw was connected) was concerned with the diffusivity and solubility of hydrogen as a function of composition in iron-nickel alloys.[260] Here, a tantalizing result was obtained. The permeation underwent a rapid decrease as nickel was added to the iron, and so did the corrosion. In fact, the corrosion-composition line was parallel to the hydrogen permeation-composition line, suggesting that the rate of corrosion was determined in this case by the permeation rate of hydrogen. This result merits a mechanistic interpretation which I think is still lacking.

There are few experiments nowadays which can be carried out to give simple, straight-forward, and unambiguous conditions. Therefore, it is with pleasure that I recall one such experiment in modern times, namely, a 1977 experiment carried out with Harvey Flitt at the Flinders University. We took a piece of steel bent into the form of a ring, immersed it in a

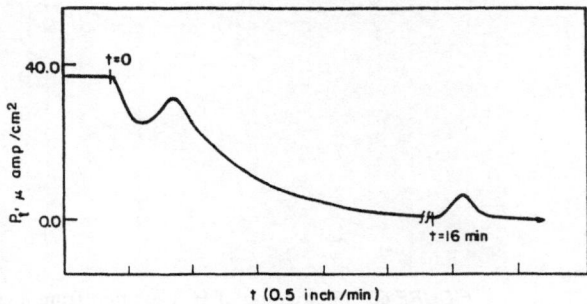

FIGURE 61. Laboratory record of an
anomalous permeation decay transient
showing two humps.

calcium nitrate solution at about 100°C, and took time-lapse photographs of the crack which developed.[261] We found that it always came at exactly 31 minutes after initiation, but the most important result was the appearance of the crack as seen in time-lapse photography. Thus, the crack could be seen to move very slowly indeed, but then, suddenly, it would jump what looked like a large distance on the screen, and then continue its slow meandering pathway, then a jump, and so on. Added to this, we saw a large stream of a gas coming from the tip of the crack, and this, upon measurement, proved to be hydrogen. It was difficult to avoid the conclusion, then, that, somewhat similar to the result we obtained for the parallelism of hydrogen permeation and corrosion rate, evolution of hydrogen near the tip of the crack was the key point in the spontaneous spread of the crack (Fig. 62).

Two more headline points are to be brought out in this work on corrosion related to hydrogen in metals. One is the work of Harvey Flitt, who struggled several years to get his Ph.D. at Flinders, ending up with the writing of the thesis at Texas A&M. Superb high-vacuum techniques were available (with the kind collaboration of Bruce Baker) in the Australian labs. Harvey devised a Nd-YAG laser to make potholes of the order of about one-tenth of a micron in steel samples which had previously been exposed to the electrolytic evolution of hydrogen.[262] The hydrogen which was then released from these tiny invisible indentations was brought into contact with a very high vacuum inside a quadrupole mass spectrometer, and the momentary flash of hydrogen visible on the screen after the laser had struck gave rise eventually to the concentration of hydrogen in the steel specimen. We hoped thereby to find hydrogen in very tiny areas such as at cracks or at joints or welded sections, etc. The technique is one which may well be developed more in the future with better optics.

Everything gets sooner or later to spectroscopy, and we have recently (in the 1990s with Jose Carbajal) been making spectroscopic measurements of hydrogen on the surface of metals to go along with the measurements which Harvey Flitt did with hydrogen inside metals. A typical IR spectrum is shown in Fig. 63. It is possible therefrom to calculate the variation of the surface hydrogen with potential and to get even some estimate of the fraction of the surface covered.[263] The cold fusion experiments of Fleischmann and Pons have not been as yet satisfactorily interpreted. However, their most frustrating characteristic is their irreproducible and sporadic character. This makes some (e.g., P. K. Iyengar) think that this means that the location of the fusion reactions must be on the surface. If this is *not* the case,

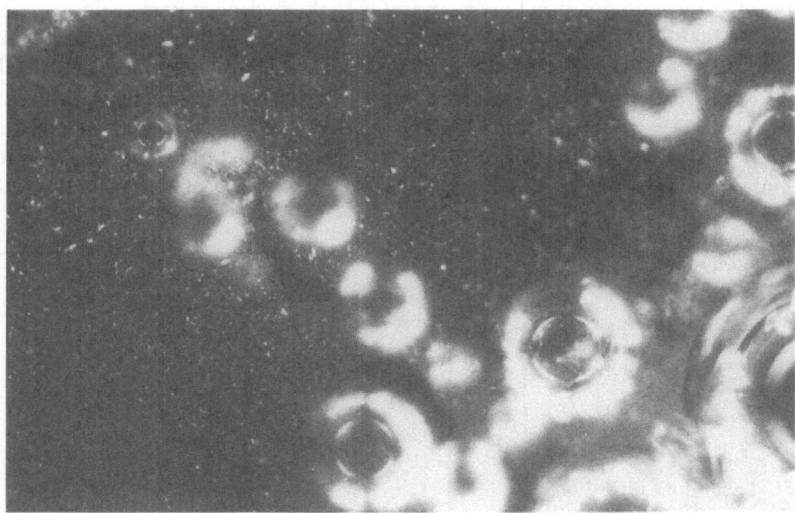

FIGURE 62. Photograph of H_2 evolution from a crack in an iron electrode.

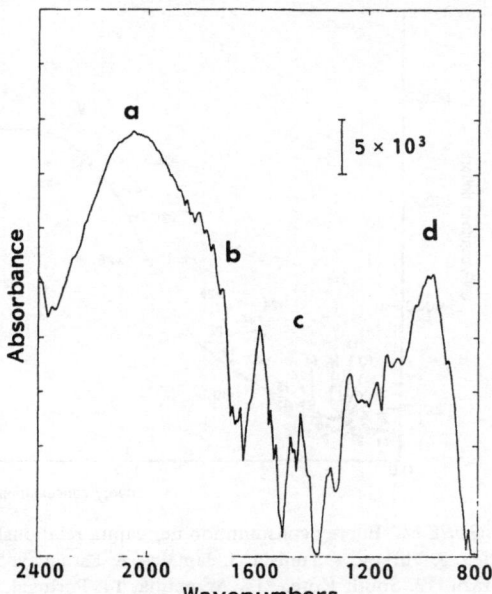

FIGURE 63. Differential IR spectra in absorbance of Fe—H vibration at −0.9 V versus NHE in borate buffer solution. a, Fe—H symmetrical stretch; b, O—H bending; c, B—O vibration; d, Fe—H asymmetrical stretch.

investigations of the type discussed here, and indeed the direct determination of fugacity, must be made for D in Pd.

21. ECONOMICS (1971–1990, 18 PUBLICATIONS)

I have steadily kept up the idea that scientists should not limit themselves to their own field,† and so I have been active in a minor capacity in a number of others. One of these is energy economics, and I would like to give here only one result from this work. The paper was written with B. Dandapani [who prediscovered surface-enhanced Raman spectroscopy (SERS), working with Fleischmann], but I must here recognize that the first plots which woke my mind toward certain conclusions were made by a Chinese graduate student, Ai Jiu, who later decided that discretion was the better part of valor and went to work for another adviser. Before he quit, however, Mr. Ai acceded to my request to go to the library and find out the energy consumption and incomes in some 40 countries. After this he made a log–log plot, and this was later reformulated by Dandapani and is shown in Fig. 64.[264] I believe that the implications of this plot are strange and thought-provoking, because it seems to show that income (which is readily seen to be the *rate* of working of an economy) is related to the energy input into the economy in the same way as an electrochemical current is a function of the bias energy moving the electrode reaction in one direction off equilibrium. The curve not only has an exponential portion and a linear portion at a low "overpotential," but also it has what electrochemists call the limiting current—when the energy per person is sufficiently high, the economy no longer continues to increase with increase in the energy input.

† A good and experienced research scientist (mind bright and fresh, balanced, relaxed, and open to anything) should be able to think and reason in all fields. He will have to ask for information, but he should be able to digest it and turn it into something with less entropy better than those with most other backgrounds.

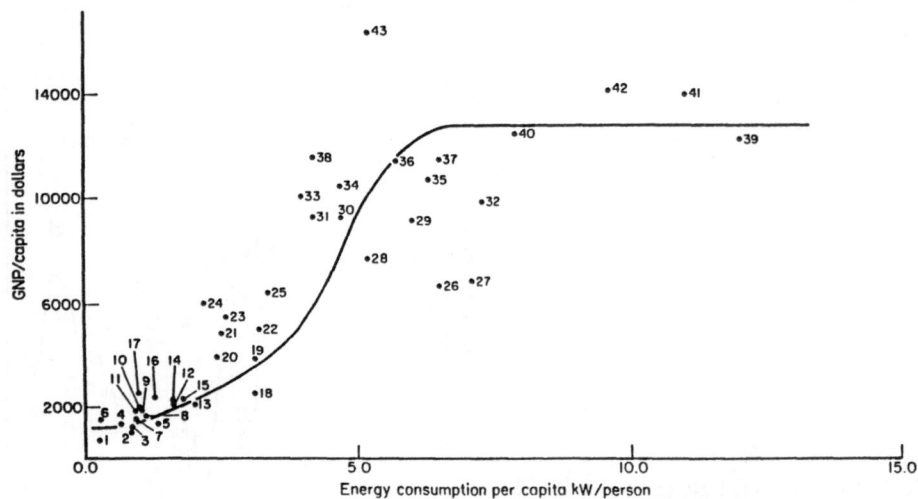

FIGURE 64. Energy consumption per capita relationship with GNP per capita. 1. Ivory Coast, 2. Costa Rica, 3. Turkey, 4. Tunisia, 5. Jamaica, 6. Paraguay, 7. Ecuador, 8. Jordan, 9. Malaysia, 10. Chile, 11. Brazil, 12. South Korea, 13. Argentina, 14. Portugal, 15. Mexico, 16. Algeria, 17. Uruguay, 18. South Africa, 19. Venezuela, 20. Greece, 21. Spain, 22. Ireland, 23. Israel, 24. Hong Kong, 25. Italy, 26. Singapore, 27. Trinidad & Tobago, 28. New Zealand, 29. Belgium, 30. United Kingdom, 31. Austria, 32. Netherlands, 33. Japan, 34. France, 35. Finland, 36. West Germany, 37. Australia, 38. Denmark, 39. Canada, 40. Sweden, 41. Norway, 42. United States, 43. Switzerland, 44. Kuwait. [J. O'M. Bockris and B. Dandapani, *Int. J. Hydrogen Energy* **11**, 101 (1986).]

Thus, countries into which high enough (~5 kW/person) energy flows, such as the United States and Canada, could cut their energy use by about one-third without suffering dire consequences in terms of reduction of living standards. On the other hand, if they cut back too far, they will go over the brink (see Fig. 64), and the reduction in living standards will be more than proportional to the further reduction in energy income.

What relevance has this to electrochemical science? The graph gives the death knell to conservation and makes it essential that a pollution-free economy at high energy per head (6 kW) be engineered for the continuance. The only way of doing this is to go to a solar hydrogen-electric energy scheme, a base which involves water electrolysis.[265-268]

I showed the result to an established economist of repute, one specializing in energy economics (which often means the price of oil). His view was that the result would not be accepted for publication in a journal on economics. "It is not what would be expected," he said. He then employed a student to check the result and found the same. "It simply doesn't fit what we know," he said.†

22. THE HYDROGEN ECONOMY (1971–1990, 19 PUBLICATIONS)

The initiation of the idea of a hydrogen economy occurred during consulting work that I was doing at General Motors in Warren, Michigan. We were already considering the future of fuels for the automobile in 1969. During a discussion session involving half a dozen

---

† Here one has an example of the astronomer royal effect. If your paradigm of the time says it can't be, it isn't.

technical staff members, we had drawn up on the board five columns outlining future energy 'schemes. Three of them ended with hydrogen. I said, "We'll be in a hydrogen-based society." Neil Triner, a member of the group, said, "It will be a 'hydrogen economy'." A few weeks later, an editress wanted me to write an article for an environmental journal, and I thought I would write about a hydrogen economy in the form of a note, which I wrote on Thanksgiving Day, 1970.[269]

I wanted to develop the idea, but in 1971 I had immigrated to Australia. On one of my first trips back, I was in Oxford, England, and called John Appleby who, curiously enough, turned out to be in Paris. We met at the Savoy Hotel in London the same day and, in the bar and over dinner, outlined an article called "The Hydrogen Economy—An Ultimate Economy?".[270]

Then there were the meetings with Veziroglu and his colleagues and the people he drew around him at the University of Miami, in 1974. It was at the Theme Conference of 1974 that we formed the International Society of Hydrogen Energy. The Society has since that time been the coordinating body of hydrogen research and holds its meetings in various parts of the world.

There had been, however, before this some activity on my side in Australia. I took a family vacation at Christmas in 1973 on Dunk Island in the Great Barrier Reef, and there I started to study solar energy. I had undergone a sort of satori earlier at Flinders University when somebody from the Australian CSIRO visited me and pointed out casually as we were walking down the corridor that the result of an extensive and general reliance on solar energy would clearly be that the rich countries (which depend at present on oil) would become the poor ones and the present poor ones (which have sun) would become rich. I suddenly saw the whole lot: linking the solar-rich countries by hydrogen-containing pipelines to the present oil-consuming ones, which obviously had but a few decades more of prosperity—the dream was born (Fig. 65). I wrote it up in a monograph which was published by the Halsted Press in 1975. It was called *Energy: The Solar–Hydrogen Alternative*.[271]

I doubt if I could have written this book had I remained in the United States. Working in Australia has some disadvantages, but it has the advantage of independence from scientific trends and ideas of the establishment, which, as far as energy is concerned, often are independent of any consideration of the half-century future but are controlled by profit this year.

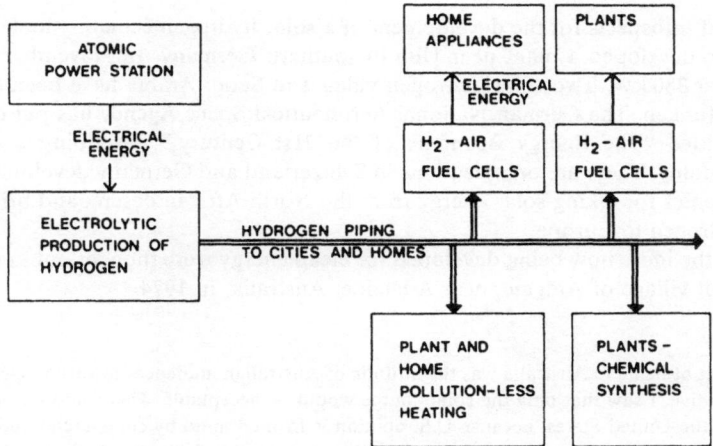

FIGURE 65. A schematic of the hydrogen economy.

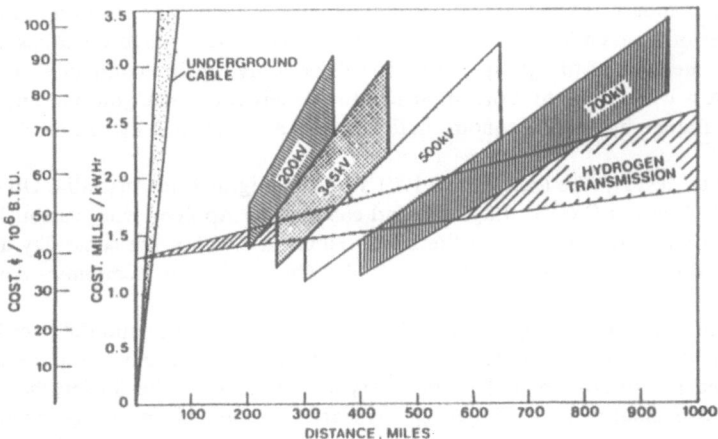

FIGURE 66. The relative costs of transmission of energy.

The scientific view from Australia is egalitarian. One sees Europe, the Soviet Union, Japan, and the United States—they seem to be all around 5,000–10,000 miles away. This tends to make one give to each a roughly equal weight. If, working at an *American* institution in 1973, I had proposed writing a book showing that the entire world could be converted to energy from the sun, distributed as natural gas is now, using hydrogen obtained from water (Fig. 66), I should have been laughed out of court.

The 1975 book certainly was published in England, America, and Europe—the Australian–New Zealand Press people did not know enough to know it was silly. Veziroglu took over the organization of biannual meetings. Soon the Australian–New Zealand Press people were on for a second edition by 1978. However, it was decided that a wider perspective was needed, and *Energy Options* was written in 440 pages dealing with the whole energy picture,[267] including fission and fusion, but with the main driving being that pollutional aspects made the only possibility the solar hydrogen one.†

Energy Options was published in 1980, and I believe it has had an effect on the energy scene.

At present, prospects for the development of a solar hydrogen economy look good. The Germans have developed a plant near Ulm in southern Germany. In November 1989, this was working at 380 kW. Two solar hydrogen villages in Saudi Arabia have been contracted for from Telefunken. The German National Aeronautical Space Agency has put out a large pamphlet entitled "The Energy Economy of the 21st Century," describing a solar-wind hydrogen economy. There are organizations in Switzerland and Germany developing rapidly for major schemes for taking solar energy from the North African deserts and bringing it in a form of hydrogen to Europe.

Many of the ideas now being developed for clean energy were thought out in my garden shed in the hill village of Aldgate, near Adelaide, Australia, in 1974.

---

† Another aspect of being in Australia was the attitude of Australian audiences toward nuclear schemes. It was so negative. I saw that only the solar source would be acceptable. This could never have been acceptable in the United States, because U.S. opinion is formed more by commercial and immediate considerations. American views tend to be those of business men and are switched on in interest only if personal profit seems near.

While outlining the story of how the solar–hydrogen scheme began, I omitted to refer to the work of the Hydrogen Research Center at Texas A&M University, of which I was Director from 1982 to 1987. This was financed at about $1.2 million over a five-year period—half by the National Science Foundation and half by associated industry (some of these latter contributions were made by Hampton Robinson, Jr. and his family and oil companies associated with them). The Hydrogen Research Center worked well because we were free to do what we wanted, and out of it came advances in the calculation of future energy scenarios, relevant fundamental physical chemistry; the splitting of water, electrocatalysis, photoelectrochemistry, the photoreduction of CO_2, and several new methods for hydrogen production.[272]

Two colleagues particularly helped in the Hydrogen Research Center. I refer to its manager, Mr. William Craven, without whom we would not have had the continued support from the Robinson family, and Mrs. Debbi Smith, my secretary during most of the time the Center was active and a person who has contributed enormously to its furtherance, particularly with her lobbying work in Washington.

23. THINKING

The contributions in the field of electrolyte interfaces, with some implications for a pollution-free, high-energy society, that has been discussed in this chapter have been made by about 200 people, each of whom worked with me for between one and ten years (mostly around three to four years). I used to find it puzzling that we were able to produce more than a paper per month during 45 years of continuous research. I thought of myself as lucky in having had such excellent co-workers. However, Halina Wroblowa's paradoxical opinion, "He gets more out of them than they have in them,"† turned me in the 1980s to thinking that I must do research differently. Herewith, a few points which came to my mind when I thought about how I do research.

1. Science, seen in terms of philosophy, does not lead to a permanent kind of truth. For the most part, it leads to working models which for a decade or two bring consistency to bunches of knowledge available at the time. However, it is important not to be much attached to any principles. Ideas are likely to be substantially different in even 10 years and certainly 50 years, as new knowledge demands new paradigms.

2. It is not good to know too much about a field before one tries to make original contributions. David Grahame used to be proud that he attended fewer scientific meetings than his colleagues. "If you go to them, you'll just get the other guy's ideas," he snapped. The time to make yourself familiar with the literature may be after you have formed some ideas of your own (but, of course, before you submit a paper!).

3. In new idea formation, the facts should be carefully shielded from the theories. When the new facts are clean and clear, unassociated with any model, one can confront them with the models qualitatively and mathematize those remaining.

4. It may not be the best to go forward from the last theory. This gets you government grants, but people will not remember your work that way. One should always try the sideways jump out of the groove of one's time.

---

† This view must not overshadow the fact that many who were later (and indeed some were before) luminaries in electrochemistry did work with me, either as graduate students or postdocs. One thinks particularly of Austin Angel, John Appleby, Jimmy McBreen, Srini, Brian Conway, A. R. Despić, Toshiko Emi, Kohei Uosaki, Glen Stoner, Halina Wroblowa, Sasha Damjanovic, Dragutin Dražić, Eleazer Gileadi, Mino Green, M. A. Habib, S. U. M. Khan, Bill O'Grady, Felix Gutmann, Douglas Inmann, Oliver Murphy, Roger Parsons, and Lawrence Young.

5. Seldom stay in a subfield (e.g., electrocatalysis) for more than about 5 years. If you do, you will know all about it: the groove gets deeper.

6. Think models. Physicists have stressed that the behavior of subatomic particles cannot be discussed in terms of models. Whatever may be the applicability of this view, chemical reactions are generally usefully discussable in model terms (but perhaps this may apply less to nuclear reactions).

7. One's discussion partner should have around the same intelligence as you, and be emotionally compatible so silence can reign without embarrassment. A good discussion may last some hours. Your partner does not have to know as much as you to make a discussion useful.

8. Find quiet conditions even if you have to work very early or very late to get them. It is difficult to think in a concentrated fashion for a long time. Low-level sporadic noise (clinking of cups in the kitchen) may break the spell.

9. One seldom gets new ideas without thinking deeply for a long time. However, the ideas do not come then. They plop in at some adventitious moment the next day.

10. University heads decide your salary depending on your research income. If you cease to bring in money, after a pause for decency, the pincers will reach out for you. So, one must remain connected to the human race. Research grants have to have proposals written, but that is not only what gets them funded. Peer review is supposed to bring fairness, but it is the program manager who decides which of your peers he will ask to decide whether you live or die. And he knows your relation to each peer.

24. YOUTH, AGE, AND RESEARCH DIRECTION

As one gets older, one cannot run so fast and so far, and one constantly has to be worried about how much one eats and what (cf. Fig. 67). But the counterpoint is that one knows much more.

Now, research is *thinking new*, primarily, and all the rest is secondary. It is easier to think new if one has all the other things there, ready to draw upon. It is some of that and some of this and this newly juxtaposed with that and there you are: it comes out. Just to give a recent example, a year or so back I was studying the mechanism of dielectric breakdown of liquids with Marek Szklarczyk from Warsaw and conceived of the critical potential of breakdown in water as being that at which the electrons in the metal reach the level of the

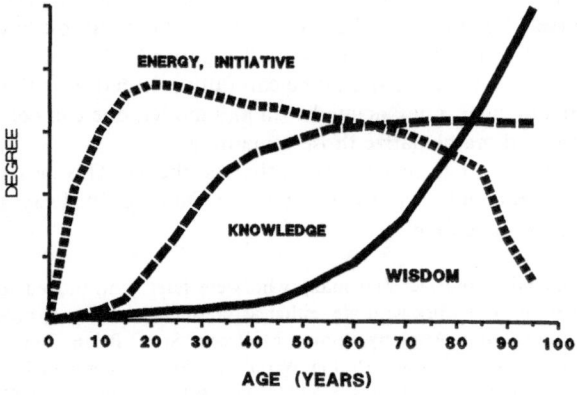

FIGURE 67. A speculative view of aging in respect to research direction.

conductivity band in water and are able to undergo rapid transport across the solution.[273] During this, we noticed that a gas layer was always formed at points of high field strength on the electrode surface (Figs. 68a and 68b), in which sparks could be observed.

Such a concept came to mind when trying to interpret the extraordinarily irreproducible phenomenon of cold fusion and led me to a model in which the key point is the field produced at the dendrite tip causing electron emission and ionization in the gas layer. The resulting D^+ is accelerated in the gas and collides vigorously with D adsorbed on the tip.

One sees the relation to the earlier research and to the obtaining of the new idea.

However, one thing has to be compromised with the advantages of being older. One has been to every interesting place six times. One has gotten the prizes and a good salary. People keep on offering you grants. What keeps you fresh, new, unpaunchy, and *eager*—on Monday morning?

I think a lot has to do with how deeply one can dig into a subject. As suggested elsewhere in this chapter, there is importance in changing the subfield every few years. Interest is the key thing to keep one gunning hard at the new frontier. One must have one's eye upon something new.

Then, of course, there is the body. But there is so much advice on ways to look after the body that anyone who does not apply it and optimize his situation, minimizing his entropy, is a fool. So long as that is looked after and the discussions are exciting, then the research vista can seem fresh and new, and the eagerness grows once more.

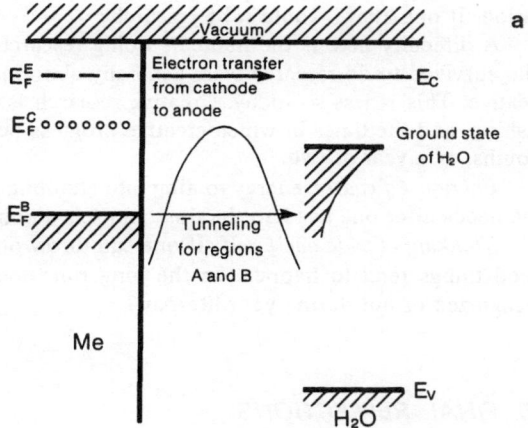

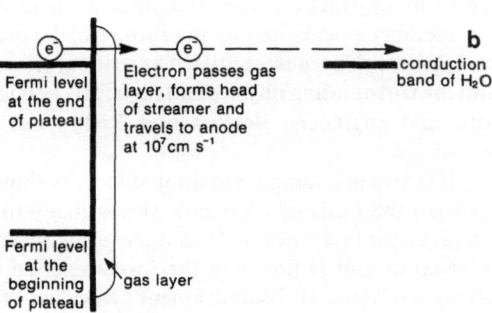

FIGURE 68. (a) Schematic representation of the Fermi level of the electrons in the electrode and different levels in H_2O. E_F^B, E_F^C, and E_F^E correspond to the position of the Fermi level corresponding to the regions B, C, and E, respectively, of the i–V curve. E_v and E_C represent the valence and conduction bands of water. Barrier for the electrons at the metal–solution interface is also shown. (b) Schematic representation of the transfer of electrons from metal to the conduction band of water in the region at D and after the breakdown (region E). [S. U. M. Khan and J. O'M. Bockris, *J. Phys. Chem.* **88**, 1808 (1984).]

25. THE RIGHT ATTITUDES

I think I would like to say just a word about attitudes.

The right attitude toward what one does is paramount: this applies to doing research, too.

There are several of my attitudes which are the common stuff of every successful endeavor.

Positivity: Whatever happens—grant lost, good co-worker resigning, paper proved wrong, etc.—one has to find out what is *good* about the situation and live on that.

Skepticism: For me things are more probable or less probable. But *all* must be doubted.

In particular, I do not believe unreservedly in sense data. I think our knowledge, our physics, our Weltanschauung is limited and doubtful and labile.

Having nothing to do: I am an advocate of sometimes having nothing to do. Colleagues who know me will smile as I am thought to put in 14-hour days etc. However, the most productive periods are when one has nothing to do, when one has isolated oneself from the pressures. Being "frightfully busy" when one is supposed to be a creative scientist is *stupid* because whilst one is dashing about catching planes or hurtling around the university going to committees, one is not going to think of anything new.

New thoughts come in the silence. As the period of calm lengthens, the mind gets brighter, and insights into problems become more likely.

The need to have periods (several hours at a time) of having nothing to do poses a problem for administrators. Directly they have given a grant, they burden us with 101 forms to fill out, reports to make, schedules to fulfill, points of progress to reach. Of course, when one is going into the unknown, it is a bit difficult to know where one will get to in a three-month period. If one knew, would it be research?

A difficulty here is the numbers doing research. Until World War II, the only people who survived to do research were generally abnormally intelligent and perhaps abnormally creative. This is less so today. Creative research is something which does not succumb to rushing, and the times in which creative progress occurs are years (three, five, ten) and not months, or a year or two.

Energy: *Of course*, energy to fling into attaining a goal is helpful to any project. Its what one needs after one has had the time and relaxation to *think what to do*.

Thinking outside one's self: If one can throw one's self outward to give rather than get, good things tend to happen. In the long run, does it matter whether your own work is recognized or not during your lifespan?

26. FINAL REFLECTIONS

Does the authorship of around 650 scientific publications and more than a dozen books bring some philosophical reflections? Of course, it does. One is that such work does give a degree of satisfaction. The development of physical electrochemistry in the West has been considerably accelerated by my former collaborators, and the textbook was and is even now a seminal book because, until it was written, electrochemistry in book form was little connected with the surrounding physics and physical chemistry. For many physicists, biologists, metallurgists, and engineers, Bockris and Reddy has been their sole source of electrochemical knowledge.

It is also encouraging to think that work done by some of my ca. 200 sometime colleagues has been the basis of what now seems likely to be an evolution: fire, heat, and explosions will give over in the next half-century to energy technology which will be based predominantly on electron and H flow and the forces exerted by electric fields—what the official German Aerospace Research Establishment has called "the end-energy economy of the 21st century" (Fig. 69).

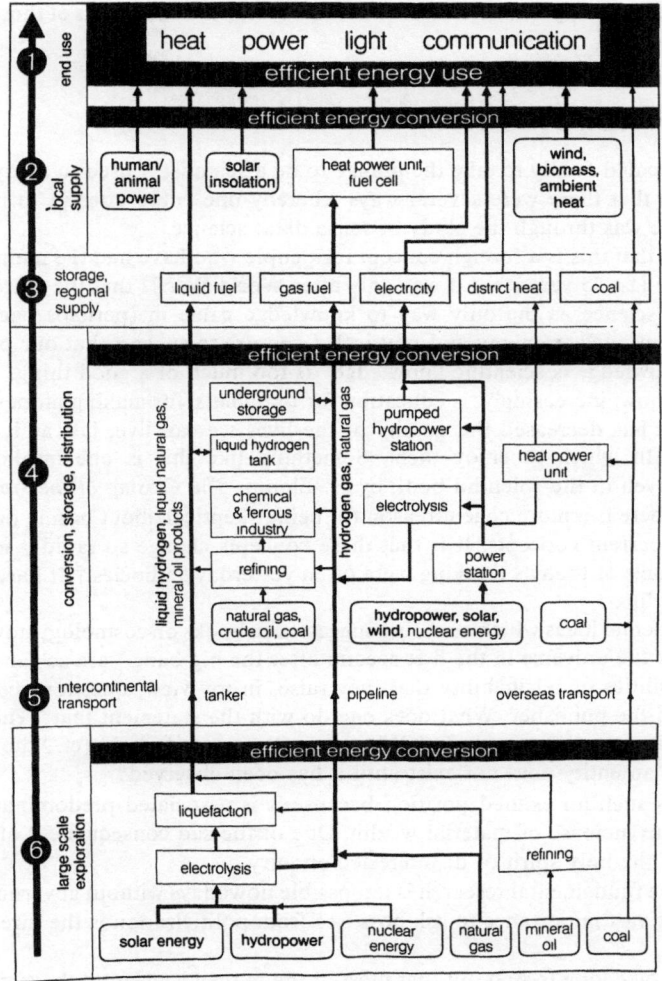

FIGURE 69. The energy economy of the 21st century.

These contributions have been made within science and engineering, economics, and sociology. What about some wider considerations?

This is a tempting question to put to oneself, but one has a feeling at once that a sharp press is needed on the brakes because otherwise a long and separate paper will result. I will limit myself to two—albeit major—points.

A. The Replacement of Morality by Expediency

For a long time now, morality in public dealings has been largely replaced by considerations of expediency. In the 1950s, at least in those areas of the United Kingdom of which I had experience, the consideration, for example, of a colleague for an appointment would actually go along the lines of whether he could do the job well. Now the considerations are so much more how this appointment would affect the fortunes and feelings of those considering his appointment.

B. An Overstress on Science

Before I decided finally to take the plunge to be a scientist, I used to study philosophy. There I learned that there were several ways whereby one could attempt to realize truths, and one of these was through the study of sense data: science.

I am aware that this is a foreign concept for people who have had the main part of their education in the last 25 years, but it was only after World War II that the extreme stress on the primacy of science as the only way to knowledge came in (perhaps because of high expectations from nuclear energy and physics). I venture to suggest that our present public philosophy "knowledge is scientific knowledge" is too much of a good thing.

Education now, increasingly, is education for *usefulness*, for making money. I think that this prostitution has decreased the quality of the lives we can live, just as it has given us more money with which to enjoy them. Something like this is one implication of the consideration given in the splendid book by H. Bloom, *The Closing of the American Mind.*

However, there is a more cogent reason for being skeptical about basing one's life upon science and its current concepts. It is that these concepts change so rapidly with time that the rational kernels of today's thinking were often yesterday's idiocies.† It hardly makes for a sound foundation.

The fundamental ideas which are being suggested in books on cosmology now—a detailed consideration of the universe in the first second after the big bang—are so far disconnected from the possibilities of falsifiability that they raise, in my view, questions concerning the responsibility of the publisher. What does one do with the statement that "The state vector has collapsed" in quantum mechanics? What does it mean to be told (cf. "the Copenhagen principle") that an entity does not exist until it has been observed?

Science has such an exalted position because it is associated predominantly with the possibilities of an increase of material wealth. One of the sad consequences of this attitude is the demise of the holy spirit of disinterested enquiry.

Thus, maybe fundamental research is impossible nowadays without government support. It is therefore of vast importance to ask how free from politicization is the direction of such support?

It does not take long to find out that most of the money is given to the support of areas in which *investments have already been made*! Take, for example, the Department of Energy (DOE) of the United States. Energy and economics are closely coupled (see Section 21). All would agree that there is a delay time of many decades (even as much as ten for some developments, seldom less than two) between the new idea and profitable investment. What does one find? In the DOE 93% of the expenditure is on nonrenewable energy sources, and for the renewables, there is only 7%.

The second aspect one might have associated more with medieval times. It is that if the new report offends the faith, it must be wrong. Further, people who say it *is* right must be either frauds or fools.

It is true that, compared with the situation 100 years ago, there has been a reduction of confidence in science. There seem to be hefty reasons such a reduction in trust from evidence gathered inside the Lager. Gödel's theorem shows that certainty by the use of mathematical logic is a broken reed. The Einstein–Podolski–Rosen effect with Bell's theorem seems to

† There are dozens of amusing examples of this throughout the ages. I like to think of the bewhiskered, portly (perhaps red-faced), wealthy Lord Kelvin being told about the discovery of X rays. "Stuff and nonsense, Sir, fraud" is the beginning of the anecdotal reply.

imply nonlocality; that is, some information transfer must be superluminal. And it is chastening to read Prigogine explaining that it was realized in 1886 that no accurate calculations can be done for the interaction of more than *two* particles.

These points might be enough to produce a shot of humility into those who consider themselves part of a scientific establishment. But such a shot would be mild compared with the *blast* which occurs if one dares to open the *forbidden door* a bit and let in the evidence for the paranormal events.

This is where one sees the rarity of the independent mind, bright, balanced, *calm*, and always skeptical of the present paradigm. Thanks be to the Lord that these structures apply only to the majority. A number of scientists in the 1970s and 1980s *have* risked the ostracism given through the ages to those who worshipped a god other than yours—and have actually dared to make objective, scientific evaluations of the existence of phenomena which do not fit the tenets of faith in the present view. It is chastening to find that the majority of research support in modern times in these areas has come from the Communist Russian government through the Academy of Sciences of the Soviet Union! The effort made in America has had to be funded by private donation. Clearly, no one is going to be convinced in matters of new thinking by being *told* something so I simply ask the readers of this document to take their courage in both hands and purchase the book *Mind Reach* published by Russel Targ and Harold Puthoff from the Stanford Research Institute. There are many others, nearly all scientists in background training, who do dare to look across the border and have the courage to shine their torches into it. Every quarter, two journals (one British, one American) are published where the authors bend over backwards in an extreme degree of scientific caution and an array of experimental methods for elimination of the dangers so well known in the investigation of abnormal things, particularly those involving the rarely found humans called "sensitives."

A panoply of alleged new phenomena are now spread out and waiting further study and evaluation as to the consequences for science if they are verified. They are, by definition, phenomena impossible to conceive in terms of the present physics. Admittedly, the majority of the claimed effects have not been authenticated at a level of scientific research and do remain in a realm where the anecdote and the sworn account (but thousands of the same kinds of anecdotes and accounts for centuries in all countries) are the basis to the assertions. Until the requisite controlled experiments have been done, this large Pandora's box will remain closed.

However, we know of at least three large black files which are in the box, the presence of which should not be ignored. The phenomena to which I refer have been investigated for around 100 years in the main countries of the world (with equipment of the time and always by research scientists) so that it seems unscientific to continue to keep them out of the phenomena with which the scientific paradigm must be consistent.

1. Information can be transferred from some minds to some other minds with an intensity which seems independent of distance and electromagnetic shielding.
2. It is possible for some people to exert forces on distant bodies by concentrating on them.
3. A substantial part of the population (around 16%), when examined under fully scientific conditions, show unconscious knowledge of the near-future choices of a random number generator.

To embrace phenomena of this kind, to make physics consistent with them, needs changes in basic concepts as radical as those wrought by Newton and more radical than any wrought by Planck and Einstein.

Whether bioelectrochemistry—a science—can contribute to the great task of creating a new physics able to embrace precognition and psychokinesis I know not, but it seems one of the sciences likely to lead to knowledge with which to address that great question: what is the relationship between the brain functions studied in electrophysiology, the electrochemistry of the brain, and consciousness. Maybe it will be one of the stronger torches to shine into that darkness that, as always, lies all around us.

27. THE MOST FREQUENT COLLABORATIONS

All comparisons are odious, it is said, and, in deciding to mention six co-workers in particular out of 200, I have tried simply to be guided by the numbers of publications to make my choice an easy one. There are many of my co-workers—and here I think of long, valued, and productive collaborations I have had, and still have, with Asa Despić, Dragutin Dražić, Felix Gutmann, and Oliver Murphy—who have contributed individual things as good as those which I am going to mention below. However, I think it is reasonable to say that with the six names below the amount of work done has been particularly large, and I think much of it quite fruitful.

The first of these is obviously my long-standing collaborator Brian Conway. Our collaboration started when he was my graduate student at Imperial College and has continued sporadically in these last 46 years. (We are at this writing collaborating on the theory of β.) Among the things that I remember particularly with Brian is the first work on poisons, where, during his graduate student days, he established a 10^{-10} moles per liter upper limit on impurities in a solution in dealing with the hydrogen evolution reaction. Much in experimental techniques in electrochemistry sprang from that contribution. The relationship between log i_0 for H_2 evolution and heats of adsorption of H on metals in the 1957 *Journal of Chemical Physics* paper was a big contributor in respect to distinguishing between the two groups of mechanisms for hydrogen evolution. The 1956 paper in the *Journal of Chemical Physics* on the mechanism of proton conduction in solution has been the paper basic to most of the advances in that field which have gone on since the first Bernal and Fowler paper in the 1930s where the idea of the quantum-mechanical tunneling of protons was applied to electrolyte solutions for the first time. Then came the seminal work with Brian on proton conduction in aqueous solutions, published in the *Proceedings of the Royal Society* (London), which I feel is a paper of some merit.

My intellectual relationship with Brian has been as good as anything I could wish for because of the match of personalities and the ability to be together for lengthy times discussing, sometimes for several hours, and I have received from him an immense amount of collaborative inspiration and help.

Sasha Damjanovic must certainly rank next in degree of collaboration for he worked with me for a total of ten years, and I have continued my scientific contacts with him through to the present. I remember in particular his contributions to surface diffusion in metal deposition and dissolution. Without him, I could not have got on to the methods of looking at surfaces that we developed: he was originally a metallurgist and taught me very much. At the time we began our work with spirals and pyramids, they had not been observed before on the surface of electrodes, and it was Damjanovic's help in introducing me to Nomaski interference miscroscopy which allowed us to make those advances. Sasha's encyclopedic contributions to the mechanism of the reduction and the evolution of oxygen are, of course, very well known apart from those done with me, and he is particularly to be thanked and remembered for his work with Marvin Genshaw in establishing the relations between I_{disk}/I_{ring} which are so diagnostic of the types of pathways for oxygen reduction.

Shahed Khan, that brilliant student, as I once called him, has been the person who has kept my quantum-mechanical contributions afloat. He has partnered me in quite a few battles in the literature, and I believe that we can look back to a change in the views of the field in respect to the mechanism of redox reactions. Thus, until the early 1980s, the solvent fluctuation mechanism was dominant, but I think this has changed and part of the reason certainly pertains to Khan's work. The other thing that I associated with Shahed is, of course, writing the book on quantum electrochemistry, which we did while we were still in Australia, and his very considerable and complex paper on the photoelectrochemical kinetics, developed from the paper which was written with Koheii Uosaki in Adelaide.

With those first three people who contributed so much in numbers of publications to my work, there come along another three who rank only slightly lower in terms of intensity of collaboration. The first of these I should mention is my good friend Srinivasan—and when I say that I almost have to say "and Mange," his wife, because of the fact that I have sat at their table so often. The work with Srini started with separation factors, and that is how I got to know something of tritium and how to measure it. Then the work turned to fuel cells and the complex kinetics of porous electrodes in fuel cell design. From this came the beginning of my work on bioelectrochemistry, which was born in a laboratory in the Down State Medical Center (we used to spend weekends there in 1968 writing our book on fuel cells) where we paced up and down together realizing that the only way you could explain 50% efficiency in metabolism was to invent a sort of fuel cell whereby the mitochondrion could convert its energy. More than 30 years of relationship with Srini and his family have cemented the relation, and now he is back on the same campus with me, and, although he is working with John Appleby rather than with me, some of the relationships of earlier times can be continued.

In referring to Halina Wroblowa, I think particularly of the first paper, called "Electrocatalysis," which was written largely over a valued weekend in Philadelphia and presented at the Agard Conference of 1963 in Cannes, France. From this early work it was possible to write three good papers with Halina on organic oxidation reactions and their mechanisms. The best, I think, was the one on ethylene oxidation with a negative reaction order. There was, of course, the recent event of the rechargeable MnO_2 electrode which came from the work which she carried out, entirely independently, at the Ford Motor Company, but with which I have now come into contact because I was a director of the company that is attempting to take the inventions which she made with her colleagues at Ford and commercialize them.

As far as sheer intelligence is concerned, I consider that Halina had the most critical (but that does not mean the brightest) mind of all the 200 persons who worked with me. How tragic it was that our marriage was made unviable basically because of Halina's nobility, by her refusal to give up what she thought of as her responsibilities to her former family. But at least some connection remains through the rechargeable MnO_2 battery.

The last person my paper counting tells me I should cite here is Asham Habib, who is now at the General Motors Technical Center, and probably for the rest of his career. Asham is very different from the rest of the most intense of my collaborators in respect to his attitudes and type of mind. He has not the quick and bright mind which one associates, for example, with Srini, with Khan, and with many people who make many scientific contributions. His is characterized by words like deep and broad thinking, etc.—and herein lies a remote resemblance to that of Conway. This quiet and self-effacing person has been featured in this paper enough so that it is not necessary to mention specific contributions, but I feel that he is a resource for the future of basic electrochemistry in the United States which needs nurturing and developing to bring out the truly very considerable potential which lies below the quiet exterior.

ACKNOWLEDGMENTS

I owe acknowledgments in a paper as broad as this one to several hundred people. These run from Mr. Considine, head of the laboratories at Imperial College, who could "find" pieces of equipment utterly unobtainable anywhere (World War II shortages aftermath); to Allen Saunders, the brilliant glass blower at the University of Pennsylvania, so eager to convert himself to a research worker that he invented experiments to do in the cells he built; to George Myers, 18 years my administrative assistant; to Mrs. Dorothy Hampton, in Adelaide, who (along with remarkable Pam Paddick†), received much of my output in Australia and dealt with the human pressures I put on her in a turbulent time; and now, at Texas A&M, to many who have helped me here, particularly Professor A. E. Martell, who "created" this outstanding department now led forward again by Michael Hall, my present boss; and my present secretary, Jainnie Leighmann. The formation and running of NSF's Hydrogen Research Center would not have been possible without stalwart Bill Craven, its manager, and Deb Smith, who was much much more than an administrative assistant.

This is not an autobiographical document, and hence I have barely mentioned my first wife, Linda, the first mother of the group, who instituted the "at home" for thesis writers and who gave me much devotion. I owe a far greater acknowledgment than the few words of personal regret to Halina Wroblowa, who reigned in my group for the middle period in Philadelphia and was for many a demanding and inspiring supervisor.

Finally, I come to my thanks to Lily, to whom I have been married since 1971. Born into a family of Viennese intellectuals and earlier married to the nationally recognized engineer Manfred Altmann, Lily has been my wife through Adelaide and College Station. Being in her own right a chemist and an environmentalist, Lily has—for the most part let us say—been able to accept a life often made too lonely by my many absences, both those overseas and those in my garden shed. This has been possible because of her remarkable strength of character and her self-discipline and will to do scientific work herself. Having the intellectual means for making suggestions in the work, she is often a direct help to me, to wit, in respect of her outstanding abilities for finding information and for seeing it all *differently*. Thus, she has created for me a home atmosphere in which stimulation to work harder exceeds all urgings to relax. Insofar as I am able to reach *700* publications (its now around 650), much will be due to Lily.

REFERENCES

1. G. Wenglowski, in: *Modern Aspects of Electrochemistry*, No. 4 (J. O'M. Bockris and B. E. Conway, eds.), p. 251, Plenum Press, New York, (1966).
2. E. Justi and A. Winsel, *Kalte Verbrennurg*, Verlag Chemie, Munich (1962).
3. A. J. Allmand and H. J. T. Ellingham, *The Principles of Applied Electrochemistry*, Arnold, London (1924).
4. J. O'M. Bockris, *J. Electroanal. Chem.* **9**, 408 (1965).
5. J. O'M. Bockris, *J. Chem. Ed.* **48**, 352 (1971).
6. J. O'M. Bockris and A. K. N. Reddy, *Modern Electrochemistry*, Plenum Press, New York (1970).
7. J. O'M. Bockris, *Proc. Roy. Aust. Chem. Inst.* **1977**, 19.
8. J. O'M. Bockris, B. E. Conway, and E. Yeager (eds.), *Comprehensive Treatise of Electrochemistry*, Plenum Press, New York (1979–1985).
9. J. O'M. Bockris and D. C. Lowe, *J. Sci. Instrum.* **30**, 403 (1953).
10. E. Blomgren and J. O'M. Bockris, *Rev. Sci. Instrum.* **32**, 11 (1961).
11. E. Blomgren and J. O'M. Bockris, *Nature* **186**, 305 (1960).
12. M. A. Habib and J. O'M. Bockris, *J. Electrochem. Soc.* **132**, 108 (1985).
13. K. Chandreasekaran and J. O'M. Bockris, *Surf. Sci.* **185**, 495 (1987).

† The fastest stenographer of them all and able to take dictation nonstop for three to four hours at uninhibited speed.

14. V. Jovancicevic and J. O'M. Bockris, *Rev. Sci. Instrum.* **58**, 1251 (1987).
15. J. W. Tomlinson, D. C. Lowe, G. W. Mellors, and J. O'M. Bockris, *J. Sci. Instrum.* **31**, 107 (1954).
16. H. G. Effemy, D. F. Parsons, and J. O'M. Bockris, *J. Sci. Instrum.* **32**, 99 (1955).
17. C. A. Angell and J. O'M. Bockris, *J. Sci. Instrum.* **35**, 458 (1958).
18. W. Beck, J. O'M. Bockris, J. McBreen, and L. Nannis, *Proc. Roy. Soc. (London), Ser. A* **290**, 220 (1966).
19. A. Damjanovic, T. H. V. Setty, and J. O'M. Bockris, *J. Electrochem. Soc.* **113**, 429 (1966).
20. J. O'M. Bockris, S. Srinivasan, and M. A. V. Devanathan, *J. Electroanal. Chem.* **6**, 205 (1963).
21. J. O'M. Bockris, J. Bowler-Reed, and J. A. Kitchener, *Trans. Faraday Soc.* **47**, 184 (1951).
22. J. O'M. Bockris and H. Eagan, *Trans. Faraday Soc.* **44**, 151 (1948).
23. B. E. Conway, J. O'M. Bockris, and H. Linton, *J. Chem. Phys.* **24**, 834 (1956).
24. R. S. Bradley, *Trans. Faraday Soc.* **53**, 687 (1957).
25. B. E. Conway and J. O'M. Bockris, *J. Chem. Phys.* **28**, 354 (1958).
26. M. Eigen and L. DeMoyer, *J. Chem. Phys.* **31**, 1134 (1959).
27. B. E. Conway and J. O'M. Bockris, *J. Chem. Phys.* **31**, 1133 (1959).
28. M. Eigen and L. DeMoyer, *Proc. Roy. Soc. (London), Ser. A* **247**, 505 (1958); M. Eigen and L. DeMoyer, in: *The Structure of Electrolytic Solutions* (J. Hamer, ed.), Chapter 5, John Wiley & Sons, New York (1959).
29. A. Borucka, J. O'M. Bockris, and J. A. Kitchener, *Proc. Roy. Soc. (London), Ser. A* **241**, 554 (1957).
30. J. O'M. Bockris and N. E. Richards, *Proc. Roy. Soc. (London), Ser. A* **241**, 44 (1957).
31. R. Furth, *Proc. Cambridge Phil. Soc.* **37**, 252 (1941).
32. L. Nannis and J. O'M. Bockris, *J. Phys. Chem.* **67**, 2865 (1963).
33. T. Emi and J. O'M. Bockris, *J. Phys. Chem.* **74**, 159 (1970).
34. M. Tanaka, K. Balasubramanyan, and J. O'M. Bockris, *Electrochim. Acta* **8**, 621 (1963).
35. D. Inman and J. O'M. Bockris, *Can. J. Chem.* **39**, 1161 (1961).
36. D. Inman, A. K. N. Reddy, S. Srinivasan, and J. O'M. Bockris, *J. Electroanal. Chem.* **5**, 476 (1963).
37. M. K. Nagajaran and J. O'M. Bockris, *J. Phys. Chem.* **70**, 1854 (1966).
38. J. O'M. Bockris and D. C. Lowe, *Proc. Roy. Soc. (London), Ser. A* **226**, 423 (1954).
39. J. O'M. Bockris, J. D. Mackenzie, and J. A. Kitchener, *Trans. Faraday Soc.* **51**, 1734 (1955).
40. W. Eitel, *The Physical Chemistry of Silicates* (German edition 1929 and 1941), American edition, University of Chicago Press, Chicago (1954).
41. W. H. Zachariasen, *J. Amer. Chem. Soc.* **54**, 3841 (1932).
42. J. O'M. Bockris, J. A. Kitchener, S. Ignatowicz, and J. W. Tomlinson, *Discuss. Faraday Soc.* **4**, 265 (1948).
43. J. O'M. Bockris, J. A. Kitchener, S. Ignatowicz, and J. W. Tomlinson, *Trans. Faraday Soc.* **48**, 75 (1952).
44. J. O'M. Bockris, J. A. Kitchener, and A. E. Davies, *J. Chem. Phys.* **19**, 225 (1951).
45. J. O'M. Bockris and D. C. Lowe, *J. Sci. Instrum.* **30**, 403 (1953).
46. J. O'M. Bockris and D. C. Lowe, *Proc. Roy. Soc. (London), Ser. A* **226**, 423 (1954).
47. J. O'M. Bockris, J. D. MacKenzie, and J. A. Kitchener, *Trans. Faraday Soc.* **51**, 17334 (1955).
48. J. W. Tomlinson, J. L. White and J. O'M. Bockris, *Trans. Faraday Soc.* **52**, 299 (1956).
49. J. W. Tomlinson, M. S. R. Heyenes, and J. O'M. Bockris, *Trans. Faraday Soc.* **54**, 1822 (1958).
50. B. E. Conway, J. O'M. Bockris, and I. A. Ammar, *Trans. Faraday Soc.* **47**, 756 (1951).
51. T. J. Webb, *J. Am. Chem. Soc.* **48**, 1589 (1926).
52. F. Booth, *J. Chem. Phys.* **19**, 391 (1951).
53. J. O'M. Bockris, W. Mehl, B. E. Conway, and L. Young, *J. Chem. Phys.* **25**, 776 (1956).
54. E. Blomgren and J. O'M. Bockris, *J. Phys. Chem.* **63**, 1475 (1959).
55. J. O'M. Bockris and D. A. J. Swinkels, *J. Electrochem. Soc.* **111**, 736 (1964).
56. J. O'M. Bockris, M. A. V. Devanathan, and K. Müller, *Proc. Roy. Soc. (London), Ser. A* **274**, 55 (1963).
57. P. Zelenay, M. A. Habib, and J. O'M. Bockris, *Langmuir* **1**, 293 (1986).
58. F. Bitter, *Proc. Roy. Soc. (London), Ser. A* **145**, 629 (1934).
59. M. A. V. Devanathan, *Trans. Faraday Soc.* **50**, 373 (1954).
60. J. R. Macdonald and C. A. Barlow, *J. Chem. Phys.* **36**, 3062 (1962).
61. N. F. Mott and R. J. Watts-Tobin, *Electrochim. Acta* **4**, 79 (1961).
62. J. O'M. Bockris, E. Gileadi, and K. Müller, *Electrochim. Acta* **12**, 1301 (1967).

63. J. O'M. Bockris and S. Argade, *J. Chem. Phys.* **49**, 5133 (1968).

64. S. Trasatti, *J. Electroanal. Chem. Interfacial Electrochem.* **52**, 313 (1974).

65. J. O'M. Bockris and M. A. Habib, *J. Electroanal. Chem.* **65**, 473 (1975).

66. B. E. Conway and L. G. M. Gordon, *J. Phys. Chem.* **73**, 3609 (1969).

67. G. J. Hills and R. Payne, *Trans. Faraday Soc.* **61**, 326 (1965).

68. R. Parsons, *J. Electroanal. Chem.* **59**, 229 (1975).

69. B. B. Damaskin and A. N. Frumkin, *Electrochim. Acta* **19**, 173 (1974).

70. R. Guidelli, *J. Electroanal. Chem.* **110**, 205 (1980).

71. J. O'M. Bockris and M. A. Habib, *J. Electrochem. Soc.* **123**, 24 (1976).

72. S. Levine, G. M. Bell, and D. Calvert, *J. Chem. Phys.* **40**, 518 (1962).

73. J. O'M. Bockris and M. A. Habib, *Z. Phys. Chem.* **98**, 43 (1975).

74. J. O'M. Bockris and M. A. Habib, *J. Res. Inst. Cataly., Hokkaido Univ.* **23**, 47 (1975).

75. J. O'M. Bockris and M. A. Habib, *J. Electroanal. Chem.* **68**, 367 (1976).

76. E. Lange and K. D. Miscenko, *Z. Phys. Chem.* **A149**, 1 (1930).

77. W. Schmickler and D. Henderson, *J. Chem. Phys.* **80**, 3381 (1984).

78. W. Schmickler and D. Henderson, *J. Chem. Phys.* **85**, 1650 (1986).

79. J. O'M. Bockris and A. K. N. Reddy, *Modern Electrochemistry*, Chapter 1, Plenum Press, New York (1973).

80. J. O'M. Bockris and M. A. Habib, *Electrochim. Acta* **22**, 41 (1976).

81. J. T. Law, Ph.D. thesis, Royal College of Science, London (1951).

82. M. Fleischmann, S. Pons, and M. Hawkins, *J. Electroanal. Chem.* **261**, 301 (1989); erratum **263**, 187 (1989).

83. N. J. C. Packham, K. L. Wolf, J. C. Wass, R. C. Kainthla, and J. O'M. Bockris, *J. Electroanal. Chem.* **270**, 451 (1989).

84. R. D. Armstrong, E. A. Charles, I. Fells, L. Molyneaux, and M. Todd, *J. Electroanal. Chem.* **272**, 293 (1989).

85. R. D. Petrasso, X. Chen, K. W. Wenzel, R. R. Parker, C. K. Li, and C. Fiore, *Nature* **339**, 667 (1989).

86. M. Gai, S. L. Rugari, R. H. France, B. J. Lund, Z. Zhae, A. J. Davenport, H. S. Isaacs, and K. G. Lynn, *Nature* **340**, 29 (1989).

87. N. S. Lewis and M. Wrighton, *Nature* **340**, 525 (1989).

88. K. L. Wolf, Proceedings of the EPRI/NSF Workshop on Gold Fusion Phenomena, Washington, D.C., October, 1989.

89. S. Srinivasan and A. J. Appleby, Proceedings of the EPRI/NSF Conference on Cold Fusion Phenomena, Washington, D.C., October, 1989.

90. M. McKubre, private communicatioan to N. J. C. Packham (February 1990).

91. R. A. Oriani, J. C. Nelson, S. K. Lee, and J. H. Broadhurst (in press) (1991).

92. R. Huggins, private communication to N. J. C. Packham (January 1990).

93. M. E. Wadsworth, S. Guruswamy, J. G. Byrne and J. Li, Proceedings of the DOE Workshop on Cold Fusion Phenomena, Santa Fe, New Mexico, May, 1989.

94. R. Champion, private communication to J. O'M. Bockris (September 1989).

95. G. J. Schoessow and J. A. Wethington, private communication to N. J. C. Packham (May 1989).

96. K. L. Wolf, N. J. C. Packham, D. E. Lawson, J. Shoemaker, F. Cheng, and J. C. Wass, Proceedings of the DOE Workshop on Cold Fusion Phenomena, Santa Fe, New Mexico, May, 1989.

97. E. Storms and C. Talcott, *J. Fusion Tech.*, in press (1990).

98. Government of India, Atomic Energy Commission Report BARC-1500 (P. K. Iyengar and M. Srinivasan, eds.), September (1989).

99. R. Adzic, private communication to N. J. C. Packham (February 1990).

100. J. M. Malo, J. Morales, B. Zamora, F. P. Ramirez, and O. Novaro, private communication to N. J. C. Packham, August (1989).

101. S. Guruswamy, private communication to N. J. C. Packham, November (1989).

102. C. D. Scott, J. E. Mrochek, E. Newman, T. C. Scott, G. E. Michaels, and M. Petek, Department of Energy Report ORNL/TM-11322, November (1989).

103. D. R. Rolison and W. E. O'Grady, *Science* (in press) (1991).

104. R. Taniguchi, T. Yamamoto, and S. Irie, *Jpn. J. Appl. Phys.* **28**, 2021 (1989).

105. N. Wada and K. Nishizawa, *Jpn. J. Appl. Phys.* **28**, 2017 (1989).

106. H. O. Menlove, M. M. Fowler, E. Garcia, A. Mayer, M. C. Miller, R. R. Ryan, and S. E. Jones, Department of Energy Report LANL-LAUR:89-1974.

107. R. J. Beuhler, G. Friedlander, and L. Friedman, *Phys. Rev. Lett.* **63**, 1292 (1989).
108. G. H. Lin, R. C. Kainthla, N. J. C. Packham, O. Velev, and J. O'M. Bockris, *J. Electroanal. Chem.* **289**, 451 (1989).
109. P. K. Subramanyan and J. O'M. Bockris, *Electrochim. Acta* **16**, 2169 (1971).
110. K. I. Popov and M. D. Maksimovic, in: *Modern Aspects of Electrochemistry*, No. 19 (B. E. Conway, J. O'M. Bockris, and R. E. White, eds.), p. 193, Plenum Press, New York (1989).
111. R. A. Oppenheimer and M. Phillips, *Phys. Rev.* **48**, 520 (1935).
112. R. A. Marcus, *J. Chem. Phys.* **24**, 966 (1956).
113. J. O'M. Bockris, S. U. M. Khan, and D. B. Mathews, *J. Res. Inst. Catal., Hokkaido Univ.* **22**, 1 (1973).
114. P. Delahay, V. Burg, and A. Dziedzic, *Chem. Phys. Lett.* **79**, 157 (1979).
115. P. Delahay, *Chem. Phys. Lett.* **87**, 607 (1982).
116. S. U. M. Khan and J. O'M. Bockris, *Chem. Phys. Lett.* **99**, 2599 (1983).
117. R. A. Marcus, *J. Chem. Phys.* **43**, 679 (1965).
118. L. I. Krishtalik, *Charge Transfer Reactions in Electrochemical and Chemical Processes*, Consultants Bureau, New York (1986).
119. H. Gerischer, in: *Physical Chemistry: An Advanced Treatise*, Vol. 9A (H. Eyring, ed.), Academic Press, New York (1970).
120. H. Gerischer, *Z. Phys. Chem.* **26**, 223 (1960).
121. J. O'M. Bockris and S. U. M. Khan, *Appl. Phys. Lett.* **48**, 913 (1984).
122. R. Parsons and J. O'M. Bockris, *Trans. Faraday Soc.* **47**, 914 (1951).
123. H. Gerischer and W. Eckardt, *Appl. Phys. Lett.* **43**, 393 (1983).
124. J. O'M. Bockris and S. U. M. Khan, *Quantum Electrochemistry*, Plenum Press, New York (1979).
125. S. U. M. Khan and J. O'M. Bockris, *J. Phys. Chem.* **87**, 2599 (1983).
126. J. O'M. Bockris and A. M. Azzam, *Trans. Faraday Soc.* **48**, 45 (1952).
127. M. J. Weaver, *J. Phys. Chem.* **83**, 1748 (1979).
128. A. J. Appleby, J. O'M. Bockris, R. K. Sen, and B. E. Conway, *Electrochemistry*, Vol. 6 (J. O'M. Bockris, ed.), Butterworths, London (1973).
129. B. E. Conway, in: *Modern Aspects of Electrochemistry*, No. 16 (B. E. Conway, J. O'M. Bockris, and R. E. White, eds.), p. 103, Plenum Press, New York (1986).
130. J. O'M. Bockris and A. Gochev, *J. Phys. Chem.* **90**, 5232 (1986).
131. A. K. N. Reddy, M. A. V. Devanathan, and J. O'M. Bockris, *J. Electroanal. Chem.* **6**, 61 (1963).
132. A. K. N. Reddy, M. A. Genshaw, and J. O'M. Bockris, *J. Chem. Phys.* **48**, 671 (1968).
133. A. K. N. Reddy, M. A. V. Devanathan, and J. O'M. Bockris, *J. Electroanal. Chem.* **8**, 406 (1964).
134. J. O'M. Bockris, M. A. Genshaw, and V. Brusic, *Symp. Faraday Soc.* **4**, 177 (1970).
135. Y. C. Chiu and M. A. Genshaw, *J. Phys. Chem.* **72**, 4225 (1968).
136. Y. C. Chiu and M. A. Genshaw, *J. Phys. Chem.* **73**, 3571 (1969).
137. W. K. Paik, M. A. Genshaw, and J. O'M. Bockris, *J. Phys. Chem.* **74**, 4266 (1970).
138. W. K. Paik and J. O'M. Bockris, *Surf. Sci.* **28**, 61 (1971).
139. W. K. Paik and J. O'M. Bockris, *Surf. Sci.* **27**, 191 (1971).
140. W. K. Paik and J. O'M. Bockris, *Surf. Sci.* **33**, 617 (1972).
141. R. W. Revie, B. G. Baker, and J. O'M. Bockris, *Surf. Sci.* **52**, 664 (1975).
142. B. G. Baker, J. O'M. Bockris, and R. W. Revie, *J. Electrochem. Soc.* **122**, 1460 (1975).
143. O. J. Murphy, J. O'M. Bockris, T. E. Pou, D. L. Cocke, and G. Sparrow, *J. Electrochem. Soc.* **129**, 2149 (1982).
144. T. E. Pou, O. J. Murphy, V. Young, J. O'M. Bockris, and L. L. Tongsen, *J. Electrochem. Soc.* **131**, 1243 (1984).
145. V. Jovancicevic, J. O'M. Bockris, J. L. Carbajal, P. Zelenay, and T. Mizuno, *J. Electrochem. Soc.* **133**, 2219 (1986).
146. V. Jovancicevic, R. C. Kainthla, Z. Tang, B. Yang, and J. O'M. Bockris, *Langmuir* **3**, 388 (1987).
147. H. Neugebauer, G. Nauer, N. Brinda-Konopik, and G. Gidaly, *J. Electroanal. Chem.* **122**, 381 (1981).
148. A. Bewick, K. Kunimatsu, and S. B. Pons, *Electrochim. Acta* **25**, 465 (1980).
149. M. A. Habib and J. O'M. Bockris, *Langmuir* **2**, 388 (1986).
150. K. Chandrasekaran and J. O'M. Bockris, *Surface Sci.* **185**, 495 (1987).
151. J. O'M. Bockris, J. L. Carbajal, B. R. Scharifker, and K. Chandrasekaran, *J. Electrochem. Soc.* **134**, 1957 (1987).
152. A. Fujishima and K. Honda, *Nature* **283**, 37 (1972).
153. A. Fujishima, K. Kohayakama, and K. Honda, *J. Electrochem. Soc.* **122**, 1487 (1975).

154. J. O'M. Bockris and K. Uosaki, *Energy* **1**, 143 (1976).
155. J. O'M. Bockris and K. Uosaki, *J. Electrochem. Soc.* **124**, 98 (1977).
156. J. O'M. Bockris and K. Uosaki, *J. Electrochem. Soc.* **124**, 1346 (1977).
157. K. Ohashi, K. Uosaki, and J. O'M. Bockris, *Energy Res.* **1**, 25 (1977).
158. J. O'M. Bockris and K. Uosaki, *J. Electrochem. Soc.* **125**, 223 (1978).
159. K. Ohashi, J. McCann, and J. O'M. Bockris, *Energy Res.* **1**, 259 (1977).
160. V. Guruswamy, G. Hildreth, O. J. Murphy, and J. O'M. Bockris, *Sol. Energy Mater.* **6**, 43 (1981).
161. M. Szklarczyk and J. O'M. Bockris, *Appl. Phys. Lett.* **42**, 1035 (1983).
162. J. O'M. Bockris and M. Szklarczyk, *Appl. Phys. Commun.* **2**, 295 (1983).
163. J. O'M. Bockris, S. U. M. Khan, O. J. Murphy, and M. Szklarczyk, *Int. J. Hydrogen Energy* **9**, 243 (1984).
164. M. Szklarczyk and J. O'M. Bockris, *J. Phys. Chem.* **88**, 1808 (1984).
165. J. O'M. Bockris, M. Szklarczyk, A. Q. Contractor, and S. U. M. Khan, *Int. J. Hydrogen Energy* **9**, 741 (1984).
166. M. Szklarczyk and J. O'M. Bockris, *Int. J. Hydrogen Energy* **9**, 831 (1984).
167. M. Szklarczyk and J. O'M. Bockris, *J. Phys. Chem.* **88**, 1808 (1984).
168. A. Heller and R. G. Vadimsky, *Phys. Rev. Lett.* **46**, 1153 (1981).
169. A. Heller, E. Aharon-Shalom, W. A. Bonner, and B. Miller, *J. Am. Chem. Soc.* **104**, 6942 (1982).
170. Y. Nakato, T. Ohinshi, and H. Tsubomura, *Chem. Lett.* **1975**, 863.
171. Y. Nakato, S. Tonomura, and J. Tsubomura, *Ber. Bunsenges. Phys. Chem.* **80**, 1002 (1976).
172. W. Kautek, J. H. Gobrecht, and J. Gerischer, *Ber. Bunsenges. Phys. Chem.* **84**, 1034 (1980).
173. M. Szklarczyk, J. O'M. Bockris, V. Brusic, and G. Sparrow, *Int. J. Hydrogen Energy* **9**, 707 (1984).
174. J. O'M. Bockris and R. C. Kainthla, *J. Phys. Chem.* **89**, 2963 (1985).
175. A. Heller, *J. Phys. Chem.* **89**, 2962 (1985).
176. A. Q. Contractor and J. O'M. Bockris, *Electrochim. Acta* **32**, 121 (1987).
177. S. U. M. Khan and J. O'M. Bockris, *J. Phys. Chem.* **88**, 2504 (1984).
178. R. C. Kainthla, S. U. M. Khan, and J. O'M. Bockris, *Int. J. Hydrogen Energy* **12**, 381 (1987).
179. K. L. Chopra, R. C. Kainthla, D. K. Pandya, and A. P. Thakoor, *Phys. Solid Films* **12**, 167 (1982).
180. R. C. Kainthla, B. Zelenay, and J. O'M. Bockris, *J. Electrochem. Soc.* **134**, 841 (1987).
181. O. J. Murphy and J. O'M. Bockris, *Int. J. Hydrogen Energy* **9**, 557 (1984).
182. W. T. Grubb, *Nature* **198**, 883 (1963).
183. F. P. Bowden and E. K. Rideal, *Proc. Roy. Soc.* (*London*), Ser. A **120**, 59 (1928).
184. E. Baers, *Sitzb. Ges. Beford. Ges. Naturw.* **63**, 213 (1928).
185. H. Wroblowa and J. O'M. Bockris, *J. Electroanal. Chem.* **7**, 428 (1964).
186. D. B. Sepa, A. Damjanovic, and J. O'M. Bockris, *Electrochim. Acta* **12**, 764 (1967).
187. J. McHardy and J. O'M. Bockris, *J. Electrochem. Soc.* **120**, 53 (1973).
188. J. McHardy and J. O'M. Bockris, *J. Electrochem. Soc.* **120**, 61 (1973).
189. V. S. Bagotzskii, L. S. Kanevskii, and V. Sh. Palanker, *Electrochim. Acta* **18**, 473 (1973).
190. J. O'M. Bockris and B. E. Conway, *Trans. Faraday Soc.* **45**, 989 (1949).
191. J. O'M. Bockris, R. Parsons, and H. Rosenberg, *Trans. Faraday Soc.* **47**, 766 (1951).
192. J. O'M. Bockris and E. C. Potter, *J. Electrochem. Soc.* **99**, 169 (1952).
193. J. O'M. Bockris and A. K. M. S. Huq, *Proc. Roy. Soc.* (*London*), Ser. A **237**, 277 (1956).
194. B. E. Conway and J. O'M. Bockris, *J. Chem. Phys.* **26**, 532 (1957).
195. J. O'M. Bockris, *Z. Electrochem.* **55**, 105 (1951).
196. A. Damjanovic, M. A. Genshaw, and J. O'M. Bockris, *J. Chem. Phys.* **45**, 4057 (1966).
197. H. S. Wroblowa, Y. C. Pan, and G. Razumney, *J. Electroanal. Chem.* **69**, 195 (1976).
198. V. S. Bagotskii, M. R. Sukhotin, and V. Yu Filinovskii, *Elektrokhimiya* **5**, 1218 (1969); **8**, 84 (1972).
199. J. O'M. Bockris and T. Ottagawa, *J. Phys. Chem.* **87**, 2960 (1983).
200. T. Ottagawa and J. O'M. Bockris, *J. Electrochem. Soc.* **129**, 2391 (1982).
201. V. Jovancicevic and J. O'M. Bockris, *J. Electrochem. Soc.* **133**, 1797 (1986).
202. S. Srinivasan, H. D. Hurwitz, and J. O'M. Bockris, *J. Chem. Phys.* **46**, 3108 (1967).
203. J. O'M. Bockris and B. D. Cahan, *J. Chem. Phys.* **50**, 1307 (1969).
204. J. O'M. Bockris and S. Srinivasan, *Fuel Cells: Their Electrochemistry*, McGraw-Hill, New York (1969).
205. J. O'M. Bockris and A. J. Appleby, *Int. J. Hydrogen Energy* **6**, 1 (1980).
206. J. Ghoroghchian and J. O'M. Bockris, *Int. J. Hydrogen Energy* **10**, 101 (1984).

207. W. Mehl and J. O'M. Bockris, *J. Chem. Phys.* **27**, 818 (1957).
208. B. E. Conway and J. O'M. Bockris, *Proc. Roy. Soc. (London)*, Ser. A **248**, 304 (1958).
209. I. N. Stranski, *Z. Phys. Chem. (Leipzig)* **136**, 259 (1928).
210. J. O'M. Bockris and G. A. Razumney, *Fundamental Aspects of Electrocrystallization*, Plenum Press, New York (1967).
211. J. O'M. Bockris, D. M. Dražić, and A. R. Despić, *Electrochim. Acta* **4**, 325 (1961).
212. D. M. Dražić, in: *Modern Aspects of Electrochemistry*, No. 19 (B. G. Conway, J. O'M. Bockris, and R. E. White, eds.), p. 69, Plenum Press, New York (1989).
213. J. Barton and J. O'M. Bockris, *Proc. Roy. Soc. (London)* **268**, 484 (1962).
214. A. R. Despić, R. G. Raicheff, and J. O'M. Bockris, *J. Chem. Phys.* **49**, 926 (1968).
215. J. O'M. Bockris, Z. Nagy, and D. Dražić, *J. Electrochem. Soc.* **120**, 30 (1973).
216. J. O'M. Bockris and R. Parry-Jones, *Nature* **171**, 930 (1953).
217. J. O'M. Bockris and S. D. Argade, *J. Chem. Phys.* **50**, 1622 (1969).
218. J. O'M. Bockris and R. K. Sen, *Surf. Sci.* **30**, 237 (1972).
219. A. T. Kühn, H. Wroblowa, and J. O'M. Bockris, *Trans. Faraday Soc.* **63**, 1458 (1967).
220. O. J. Murphy, J. O'M. Bockris, and D. W. Later, *Int. J. Hydrogen Energy* **10**, 435 (1985).
221. L. F. Oldfield and J. O'M. Bockris, *J. Phys. Coll. and Chem.* **55**, 1255 (1951).
222. H. Dahms and J. O'M. Bockris, *J. Electrochem. Soc.* **111**, 728 (1964).
223. J. W. Johnson, H. Wroblowa, and J. O'M. Bockris, *Electrochim. Acta* **9**, 639 (1964).
224. J. O'M. Bockris, B. J. Piersma, and E. Gileadi, *Electrochim. Acta* **9**, 1329 (1964).
225. J. O'M. Bockris, H. Wroblowa, E. Gileadi, and B. J. Piersma, *Trans. Faraday Soc.* **61**, 2531 (1965).
226. E. Gileadi, G. Stoner, and J. O'M. Bockris, *J. Electrochem. Soc.* **113**, 585 (1966).
227. J. O'M. Bockris, E. Gileadi, and G. Stoner, *J. Phys. Chem.* **73**, 427 (1969).
228. J. O'M. Bockris and D. Miller, Proceedings of the Workshop Held at Sintra, Portugal, July 1986, pp. 1–36 (1987).
229. D. Miller and J. O'M. Bockris, *J. Electrochem. Soc.* (accepted for publication) (1991).
230. J. Johnson, H. Wroblowa, and J. O'M. Bockris, *J. Electrochem. Soc.* **111**, 863 (1964).
231. J. O'M. Bockris and S. Srinivasan, *Nature* **215**, 197 (1967).
232. B. Rosenerg, *J. Chem. Phys.* **36**, 816 (1969).
233. J. O'M. Bockris, *Nature* **224**, 775 (1969).
234. J. O'M. Bockris and M. Shuaib, *Trans. SAEST* **13**, 4 (1978).
235. J. O'M. Bockris, F. Gutmann, and M. A. Habib, *J. Biol. Phys.* **13**, 3 (1985).
236. A. Szucs, G. D. Hitchens, and J. O'M. Bockris, *J. Electroanal. Chem.* **275**, 133 (1989).
237. P. R. C. Gascoyne, R. Pethig, and A. Szent-Gyorgyi, *Proc. Natl. Acad. Sci. U.S.A.* **78**, 261 (1981).
238. J. O'M. Bockris, in: *Modern Aspects of Electrochemistry*, No. 1 (J. O'M. Bockris and B. E. Conway, eds.), pp. 180–275, Butterworths, London (1954).
239. N. Smyrl, in: *Comprehensive Treatise of Electrochemistry*, Vol. 2 (J. O'M. Bockris, B. E. Conway, E. Yeager, and R. E. White, eds.), pp. 97–147, Plenum Press, New York (1981).
240. W. E. O'Grady and J. O'M. Bockris, *Chem. Phys. Lett.* **5**, 116 (1970).
242. J. O'M. Bockris, A. Damjanovic, and W. E. O'Grady, *J. Colloid Interface Sci.* **34**, 387 (1970).
242. W. E. O'Grady and J. O'M. Bockris, *Surf. Sci.* **38**, 249 (1973).
243. R. Prados and M. L. Good, *J. Inorg. Nucl. Chem.* **33**, 3733 (1971).
244. J. O'M. Bockris, *Corros. Sci.* **29**, 291 (1989).
245. A. T. Hubbard, *Crit. Rev. Anal. Chem.* **3**, 201 (1973).
246. R. W. Revie, B. G. Baker, and J. O'M. Bockris, *J. Electrochem. Soc.* **122**, 1460 (1975).
247. O. J. Murphy, J. O'M. Bockris, T. E. Pou, D. L. Cocke, and G. Sparrow, *J. Electrochem. Soc.* **127**, 2149 (1982).
248. O. J. Murphy, T. E. Pou, J. O'M. Bockris, and L. L. Tongson, *J. Electrochem. Soc.* **131**, 2785 (1984).
249. D. L. Cocke, O. Mendoza, B. A. Horrell, R. E. White, and D. G. Naugle, Proceedings of the Advanced Materials Conference II, Denver, March 6–9, 1989, p. 557.
250. V. Jovancicevic, J. O'M. Bockris, J. Carbajal, P. Zelenay, and T. Mizuno, *J. Electrochem. Soc.* **133**, 2219 (1986).
251. J. O'M. Bockris and B. E. Conway, *Experientia* **3**, 454 (1943).
252. V. Jovancicevic, B. Yang, and J. O'M. Bockris, *Electrochim. Acta* **32**, 557 (1987).
253. M. A. V. Devanathan and Z. O. M. Stachurski, *Proc. Roy. Soc. (London)*, Ser. A **270**, 90 (1962).
254. A. Frumkin and N. Aladjalowa, *Acta Physiochim.* **19**, 1 (1944).

255. J. O'M. Bockris and P. K. Supramaniam, *Scr. Metall.* **6**, 947 (1972).
256. W. Beck, J. McBreen, J. O'M. Bockris, and L. Nannis, *Proc. Roy. Soc.* (*London*), *Ser. A* **290**, 220 (1966).
257. J. O'M. Bockris, J. McBreen, and L. Nannis, *J. Electrochem. Soc.* **112**, 1025 (1965).
258. J. O'M. Bockris and P. K. Subramanian, *J. Electrochem. Soc.* **118**, 1114 (1971).
259. J. O'M. Bockris and P. K. Subramanian, *Electrochim. Acta* **16**, 2169 (1971).
260. W. Beck, J. O'M. Bockris, M. A. Genshaw, and P. K. Subramanian, *Metall. Trans.* **2**, 883 (1971).
261. H. J. Flitt, W. R. Revie, and J. O'M. Bockris, *Corros. Australasia* **1**, 4 (1976).
262. H. J. Flitt, J. Pezy, and J. O'M. Bockris, *Int. J. Hydrogen Energy* **8**, 39 (1983).
263. J. O'M. Bockris, J. L. Carbajal, B. R. Scharifker, and K. Chandrasekaran, *J. Electrochem. Soc.* **134**, 1957 (1987).
264. J. O'M. Bockris and B. Dandapani, *Int. J. Hydrogen Energy* **12**, 439 (1978).
265. J. O'M. Bockris, B. Dandapani, and J. C. Wass, in: *Advances in Solar Energy 5* (K. W. Boer, ed.), p. 171, Plenum Press, New York (1989).
266. J. O'M. Bockris, J. C. Wass, and N. J. C. Packham, in: *Hydrogen Energy Systems VII*, Proc. 7th WHEC Pergamon, Oxford (1988).
267. J. O'M. Bockris, *Energy Options*, Halsted Press, New York (1980).
268. T. N. Veziroglu, *Int. J. Hydrogen Energy* **12**, 99 (1987).
269. J. O'M. Bockris, *Environment* **13**, 51 (1971).
270. J. O'M. Bockris and J. Appleby, *Environment This Month* **1**, 29 (1972).
271. J. O'M. Bockris, *Energy: The Solar–Hydrogen Alternative*, Halsted Press, New York (1975).
272. J. O'M. Bockris, *Int. J. Hydrogen Energy* **13**, 489 (1988).
273. M. Szklarczyk, R. C. Kainthla, and J. O'M. Bockris, *J. Electrochem. Soc.* **136**, 2512 (1989).

Hydrogen Technologies, Organic Electrochemistry

Comparison of Hydrogen with Coal and Synthetic Fossil Fuels

T. N. Veziroglu, H. J. Plass, Jr., and F. Barbir

1. INTRODUCTION

Presently, the earth's population is about 5.3 billion and is growing at the rate of 1.8% per year. However, the demand for energy is growing at a much higher rate (at about 3–4% per year), since the developing countries are trying to increase their energy consumption faster than the industrialized countries. Today, most of the energy demand is met by fossil fuels (i.e., coal, petroleum, and natural gas). On the other hand, it is estimated[1-3] that the world fossil-fuel production, beginning with petroleum and natural gas, will soon start declining. Nonconventional energy sources, such as solar, ocean thermal, wind, ocean currents, tides, waves, thermonuclear, and geothermal, are being considered as possible sources of energy to meet the growing demand.[4-15] However, none of these new energy sources have all the desirable qualities of petroleum and natural gas. For example, some are only intermittently available. Others are only available away from the consumption centers, and none can be used as a fuel for transportation. Therefore, it becomes necessary to find an intermediary or synthetic form of energy which can be produced using the nonconventional primary energy sources being considered.

Many scientists and engineers[16-28] believe that a hydrogen energy system could form the link between the new energy sources and the user. It is the most economical to produce, yields the cleanest fuel, and is recyclable. In the hydrogen energy system it is envisaged that hydrogen will be produced from the new nonfossil energy sources and will be used in every application where fossil fuels are used today, in this system hydrogen is not a primary source of energy. It is an intermediary form of energy, and secondary form of energy, or an energy carrier.

Hydrogen has the most desirable properties for a fuel. It is the lightest and the cleanest fuel. It can be converted to other forms of energy more efficiently than other fuels. It is also the most abundant element in the universe. Many stars and planets are either entirely made up of hydrogen or contain large percentages of it. For example, the most abundant element in the sun is hydrogen. The sun's energy is produced by the fusion of hydrogen atoms or nuclei into helium. The planet Jupiter is made up of liquid and solid hydrogen. Even the interstellar space contains about one hydrogen molecule per cubic centimeter.

T. N. Veziroglu, H. J. Plass, Jr., and F. Barbir ● Clean Energy Research Institute, University of Miami, Coral Gables, Florida 33124.

Electrochemistry in Transition, edited by Oliver J. Murphy *et al.* Plenum Press, New York, 1992.

On the other hand, on Earth hydrogen is not abundant as a free element. It is found in natural gas in small percentages. It forms 0.2% of the atmosphere. These are very small quantitites compared to the fuel needs of the world. Therefore, hydrogen must be produced using some primary energy source if it is to meet our fuel needs.

Those who desire continuation of the present fossil-fuel system claim that synthetic gasoline (SynGas), synthetic jet fuel (SynJet), and synthetic natural gas (SNG) could be manufactured through use of the vast deposits of coal, oil shale, and tar sands, or even of CO_2 from air and from limestone, when we run out of petroleum and natural gas. Consequently, hydrogen (gaseous and liquid) will be compared in this chapter with synthetic natural gas, synthetic jet fuel, and synthetic gasoline from the viewpoint of real economics, that is, by taking into account production costs, environmental damage, and utilization efficiencies. Since coal deposits are expected to last a few centuries, direct use of coal will be included in the comparison as well.

2. PRODUCTION COSTS

The production costs of interest here are for the large-scale production of the synthetic fossil fuels and hydrogen, so that they may meet large-scale demands. As mentioned earlier, hydrogen can be produced by several methods, using various primary energy sources.[29-64] Among the methods are electrolytic, thermal, thermochemical, photolytic, and various "hybrid" methods. Any one of the primary energy sources, including the fossil fuels, can be used as the energy source for the production of hydrogen. As the post-fluid fossil-fuel era is under consideration, the main fossil-fuel resource of interest would, of course, be coal.

Table 1 presents the averages of large-scale production costs taken from recent literature.[65-85] In the case of hydrogen, costs are classified by the primary energy source used in production, namely, coal, hydropower, and solar (direct solar, photovoltaics, wind, waves, ocean currents, tides, ocean thermal, etc.), as the estimated prices also group according to this classification. Nuclear energy has not been included, since it would result in much

TABLE 1
Estimated Average Production Costs of Synthetic
Fuels and Coal

Fuel[a]	Estimated average cost (1990 $/GJ)
Coal GH$_2$	8.58
Hydro GH$_2$	11.51
Solar GH$_2$	15.44
SNG	8.85
Coal LH$_2$	10.73
Hydro LH$_2$	14.39
Solar LH$_2$	19.30
SynGas	17.25
SynJet	13.05
Coal	2.19

[a] Definitions of terms and abbreviations: Hydro, Hydropower; Solar, direct and indirect solar, except hydropower; SNG, synthetic natural gas; SynGas, synthetic gasoline; SynJet, synthetic jet fuel; G, gaseous; L, liquid.

higher hydrogen production cost. In Table 1, all the dollar values have been brought up to 1990 U.S. dollars by taking into consideration the inflation.

Although gaseous hydrogen can be used in most of the applications where gaseous or liquid (and solid) fossil fuels are currently being used, there are some applications where liquid hydrogen must be used, for example, in rocket engines for space travel and in jet engines for air transportation. Consequently, prices of liquid hydrogen must also be considered. If conventional liquefaction methods are used, the price has to be increased by about 50%[86] over that of gaseous hydrogen. However, a revolutionary liquefaction process (magnetic liquefaction) is being developed at the Los Alamos National Laboratory[87] and has a circuit efficiency of 60% as compared with only 30% in conventional systems. Preliminary studies show that the magnetic liquefaction process will need less capital investment and less maintenance than conventional systems. It then becomes reasonable to assume a 25% add-on for liquid hydrogen produced by the new method, which could be available in the 2000s.

As can be seen from Table 1, the estimated production costs of coal gaseous hydrogen (GH_2) and coal liquid hydrogen (LH_2) are on average lower than those for the synthetic fossil fuels, while GH_2 from other energy sources is more expensive than SNG. SynJet is the second least expensive liquid synthetic fuel, followed by hydro LH_2, SynGas, and solar LH_2. Coal has been added to the list, since it could by itself (without converting to a synthetic fuel) be used in some applications, with cost benefits.

3. ENVIRONMENTAL DAMAGE

The combustion products of fossil fuels are damaging the Earth's climate and environment. An important type of pollution, air pollution, is caused mainly by fossil fuels used to obtain energy for transportation, electricity production, heat generation, etc. This year alone fossil fuels will be spewing some 30 billion tons of CO_2, CO, SO_2, NO_x, soot, and ash into the atmosphere.

In major cities around the world, air pollution is getting so bad that at times schools are closed, children sent home, and people asked to refrain from physical activities so that they will not have to breathe excessively. Respiratory diseases are increasing, and the life span is decreasing. The traffic police in Tokyo are regularly administered oxygen so that they can continue to perform their duties in the poisoned atmosphere of the city.

Acid rains produced by fossil fuels are literally killing our lakes. Already hundreds of lakes in Canada, Finland, Norway, Sweden, and the United States have high enough acidity ratios—produced by acid rains—that they are no longer a suitable habitat for fish and aquatic plants. They are literally dead! Every year, hundreds more lakes are dying. This is causing some international tension between the industrial countries in the temperature belt and the countries to the north, since the former produce most of the ingredients for the acid rains, and the latter are on the receiving end, due to the prevailing winds.

Acid rains have also started contaminating sources of some drinking water, as a result of which babies are becoming sick and water pipes are corroding.

It is estimated that by the year 2035 the oceans will have such high acidity that they will no longer be a hospitable environment for fish.

Acid rains are also affecting crops. Studies show that the quality and quantity of farm products, forests, and vegetation in general are suffering. They are also causing irreversible damage to historical buildings and structures.

In addition to pollution and the acid rains, there is a new culprit—CO_2, the main combustion product of fossil fuels. For a long time, CO_2 was believed to be harmless. Now, there is evidence that because of the "greenhouse effect" caused by CO_2, the polar ice caps have started to melt and the oceans have started to rise. At the present time, the rate of rise is estimated to be about 1 to 1.5 cm per year, and this rate is growing with the increase of CO_2 in the atmosphere. It is projected that at this rate, when the melting process is completed the oceans will rise by about 26 feet, or 8 meters, enough to put many coastal cities and plains, where a very large percentage of the Earth's population lives, under water. This will result in a shortage of habitable and arable land. The effects are already observable in the North Sea. The Germans and the Dutch are increasing the height of their dykes to keep the North Sea back. In the United States, the Environmental Protection Agency sent a team of experts to various coastal cities two years ago. They told the local officials that the oceans are rising and that if they are going to put up any structures on the coast or in the water, they should build them higher.

The greenhouse effect is also causing, as we are now observing, climatic changes. The wind patterns are changing. As a result, the cloud movements and the locations where rain falls are changing. This is causing drought in places where historically there was good rainfall, and floods in other places where historically there was less rainfall. The consequences are less agricultural produce (higher produce prices) and less hydroelectric power.

In calculating the cost of fuels to society, their environmental effects and damages must be considered. As described above, investigations are being conducted in many parts of the world to estimate those damages.[84,88-137] Detailed estimates of the fossil-fuel damage on various elements of the biosphere are presented in Table 2. Table 2 indicates the type of damage and the damage per unit of modified fossil-fuel consumption, which is defined as the total of the petroleum and coal consumption plus one-third of the natural gas consumption (i.e., it has been assumed that the environmental damage caused by natural gas is one-third of that caused by the liquid or solid fossil fuels).

As can be seen from Table 2, the total environmental damage of fossil fuels is $10.62/GJ, which is quite a large figure. This is what society pays, in addition to the market prices, for using fossil fuels. On a worldwide basis, it amounts to about U.S. $1,900 billion, or 12.7% of the world's total gross domestic product in 1990. It should be noted that the figure of $10.62/GJ ought actually to be greater, as it does not include the costs of human discomfort and any induced climatic changes.

TABLE 2
Summary of the Estimates of the Fossil Fuel Damage

Type of damage	Damage per unit of fossil fuel energy (1990 $/GJ)
Effect on:	
Humans	3.76
Fresh water sources and resources	0.59
Farm produce, plants, and forests	1.25
Animals	0.40
Buildings	0.73
Coasts and beaches	0.17
Effect of:	
Rising oceans	0.37
Strip mining	0.15
Rising temperatures	3.20
Total	10.62

TABLE 3
Pollution Factor

Fuel[a]	Pollution factor (CO_2 emission, kg/GJ)
Coal GH_2	116.3
Hydro GH_2	0
Solar GH_2	0
SNG	116.3
Coal LH_2	145.4
Hydro LH_2	0
Solar LH_2	0
SynGas	131.6
SynJet	131.6
Coal	85.5
Gasoline	76.5
Natural gas	48.4
Mean fossil (coal + gasoline)	81.0

[a] For definition of terms and abbreviations, see footnote a to Table 1.

The figure \$10.62 refers to the environmental damage caused by 1 GJ of coal or oil. When 1 GJ of a synthetic fuel is produced from coal, more than 1 GJ of coal is consumed. Consequently, the environmental damage due to 1 GJ of a synthetic fuel manufactured from coal is greater than \$10.62. This can be expressed as follows:

$$E_s = E_f \frac{p_s}{p_f} \tag{1}$$

where E_s is the environmental damage due to 1 GJ (or one unit) of a synthetic fuel produced from coal, E_f (= \$10.62) is the fossil-fuel environmental damage, p_s is the synthetic fuel pollution factor, and p_f is the fossil-fuel pollution factor. Because of the dearth of data, total CO_2 generated has been taken as a measure of the pollution factor. Table 3 presents the pollution factors for various synthetic fuels, as well as for coal, gasoline, and natural gas and the average for coal and gasoline.[66]

4. UTILIZATION EFFICIENCY AND EFFECTIVE COST

In comparing the fuels, it is important to compare the utilization efficiencies at the user end. For utilization by the user, fuels are converted to various energy forms, such as mechanical, electrical, and thermal. Studies show that in almost every instance of utilization, hydrogen can be converted to the desired energy form more efficiently than the fossil fuels (or the synthetic fossil fuels). In other words, conversion to hydrogen would result in energy conservation owing to its higher utilization efficiencies.

Investigations[138] show that, for a given number of passengers and a given payload, a subsonic jet passenger airplane would use 19% less energy if it were to use hydrogen (liquid) instead of fossil-based jet fuel. In the case of a supersonic jet plane, the efficiency advantage of hydrogen is even greater[139]; it is 38% better than jet fuel.

Research workers[140-143] have reported a wide range (22-60%) of utilization efficiency advantages for hydrogen use in existing automobile internal combustion (IC) engines. The wide variation in the reported efficiencies originates from the fact that the lower figure applies to the engine alone, while the higher figure applies to the automobile under city driving

TABLE 4

Effective Cost of Synthetic Fuels and Coal for Thermal Energy Generation

Type of fuels	Application	Fuel	Utilization efficiency, η_f or η_s	Effective cost (1990 \$/GJ)
Fossil fuels	Flame combustion	SNG	0.800	24.10
		Syn-fuel oil[a]	0.800	30.30
		Coal	0.800	13.40
GH_2	Flame combustion	Coal CH_2	0.800	23.83
		Hydro GH_2	0.800	11.51
		Solar GH_2	0.800	15.44
GH_2	Catalytic combustion	Coal GH_2	1.000	19.06
		Hydro GH_2	1.000	9.21
		Solar GH_2	1.000	12.35

[a] Considered to be the same as SynJet.

conditions. As hydrogen can burn in lean fuel/air mixtures as well as in rich mixtures, it can cause large improvements in fuel-use efficiencies in the stop–start type city driving as compared with fossil fuels, which can only burn in rich mixtures.

Hydrogen can be converted to electricity in fuel cells with much greater efficiencies than those possible in thermal power plants using fossil fuels. While conversion efficiencies for the latter are in the range of 35–38%, practical efficiencies in hydrogen fuel cells are 50–70%. In the advanced hydrogen fuel cells which are now being developed, it is expected that efficiencies will rise to 80–90%. This is an important, unique property of hydrogen, which can also increase the conversion efficiencies in transport vehicles. Even if the end use required mechanical power (such as in automobiles, buses, or trucks), hydrogen fuel cell/electric motor combinations would yield far greater conversion efficiencies than an internal combus-

TABLE 5

Effective Cost od Synthetic Fuels and Coal for Electric Power Generation

Type of fuels	Application	Fuel	Utilization efficiency, η_f or η_s	Effective cost (1990 \$/GJ)
Fossil fuels	Thermal plant	SNG	0.380	24.10
		Syn-fuel oil	0.380	30.30
		Coal	0.380	13.40
GH_2	Thermal plant	Coal GH_2	0.380	23.83
		Hydro GH_2	0.380	11.51
		Solar GH_2	0.380	15.44
GH_2	Thermal plant with aphodid steam generator[a]	Coal $GH_2 + GO_2$	0.475	21.46
		Hydro $GH_2 + GO_2$	0.475	11.61
		Solar $GH_2 + GO_2$	0.475	14.75
GH_2	Fuel cells	Coal GH_2	0.700	12.94
		Hydro GH_2	0.700	6.25
		Solar GH_2	0.700	8.38

[a] \$3/GJ has been added to fuel cost as oxygen cost.

TABLE 6

Effective Cost of Synthetic Fuels for Surface Transportation

Application	Fuel	Utilization efficiency, η_f or η_s	Effective cost (1990 \$/GJ)
Fossil-fueled transport—IC engines	SNG	0.250	24.10
	SynGas	0.250	34.50
GH$_2$-fueled transport—IC engines	Coal GH$_2$	0.300	19.86
	Hydro GH$_2$	0.300	9.59
	Solar GH$_2$	0.300	12.87
LH$_2$-fueled transport—IC engines	Coal LH$_2$	0.330	22.57
	Hydro LH$_2$	0.330	10.90
	Solar LH$_2$	0.330	14.62
GH$_2$-fueled transport—fuel cells	Coal GH$_2$	0.700	8.51
	Hydro GH$_2$	0.700	4.11
	Solar GH$_2$	0.700	5.51

tion engine running on fossil fuels. Via a fuel cell/electric motor system, hydrogen can be converted to mechanical power more than twice as efficiently as gasoline or diesel fuel.

In some industrial, commercial, and residential applications, such as in heating and cooling, fuels are converted to thermal energy. Experiments[144,145] show that hydrogen can be converted to thermal energy 24% more efficiently than fossil fuels. Gas-turbine electric power plants using liquid hydrogen may have favorable efficiency benefits if cryogenic energy of LH$_2$ is converted to useful work.

As a result of the foregoing, in order to compare the synthetic fuels under consideration, we could define a societal (effective) cost, which takes into account the production cost, environmental cost, and utilization efficiency, as follows:

$$S_s = (C_s + E_s) \frac{\eta_f}{\eta_s} \qquad (2)$$

where S_s is the societal or effective cost of the synthetic fuel under consideration, C_s is the production cost, E_s is the environmental damage, η_s is the utilization efficiency, and η_f is

TABLE 7

Effective Costs of Synthetic Fuels for Air Transportation

Application	Fuel	Utilization efficiency ratio, η_f/η_s	Effective cost (1990 \$/GJ)
Subsonic jet transport	Coal LH$_2$	0.840	25.02
	Hydro LH$_2$	0.840	12.09
	Solar LH$_2$	0.840	16.21
	SynJet	1.000	30.30
Supersonic jet transport	Coal LH$_2$	0.725	21.59
	Hydro LH$_2$	0.725	10.43
	Solar LH$_2$	0.725	13.99
	SynJet	1.000	30.30

the utilization efficiency of the fossil fuels. Substituting Eq. (1) into Eq. (2), one obtains:

$$S_s = \left(C_s + E_f \frac{p_s}{p_f} \right) \frac{\eta_f}{\eta_s} \tag{3}$$

Using the above equation with the data developed earlier, Tables 4–7 have been prepared to give the effective costs for various types of applications for thermal energy generation, electric power generation, surface transportation, and air transportation, respectively. Table 4 shows that the catalytic combustion of GH_2 would produce the most economical thermal energy, while the heat produced from synthetic fossil fuels would be most expensive. The effective cost of coal-generated heat is reasonably low, since there are no conversion (to synthetic fuels) losses. Another point to note is that among the GH_2 fuels, the effective cost of coal GH_2 is the highest, although its production cost is the lowest (Table 1). This is of course due to the addition of the environmental damage.

When electric power generation is considered (see Table 5), hydro GH_2 and solar GH_2 with fuel cells would be the most cost effective, followed by hydro GH_2 with aphodid steam generator and coal electricity. In the case of surface transportation (see Table 6), GH_2 with fuel cells is the most cost effective, followed by GH_2 with IC engines and LH_2 with IC engines. Fossil-fueled surface transport becomes the most expensive. When air transportation is considered (see see Table 7), hydro LH_2 is the most economical, followed by solar LH_2 and coal LH_2. SynJet is the most expensive fuel when societal cost is considered.

5. COMPARISON OF TWO SCENARIOS

In order to compare the real economics of the postpetroleum (and natural gas) era energy systems, two possible scenarios will be assumed:

1. *Coal/synthetic fossil-fuel system.* In this case, it is assumed that the present fossil-fuel system will be continued by the substitution of the synthetic fossil fuels wherever convenient and/or necessary. On a worldwide basis for the first decade of the 21st century, it will be assumed that out of the total "fuel" energy consumption, 40% will be used for thermal energy generation, 30% for electric power generation, 20% for surface transportation, and 10% for air transportation. Of course, there will be additional electricity generation by hydropower and nuclear power, which does not enter into or affect the comparison. It will further be assumed that two-thirds of the thermal energy generation will be achieved using coal and one-third using SNG, electric power generation will be achieved by coal-burning thermal power plants, one-half of the surface transportation will run on SNG and the other half on SynGas, and one-half of air transportation will be subsonic and the other half supersonic—both running on SynJet.

2. *Solar hydrogen energy system.* In this case, it is assumed that the conversion to the hydrogen energy system will take place, and one-third of the hydrogen needed will be produced by hydropower and two-thirds by direct and other (other than hydro-power) indirect solar energies. On a worldwide basis for the first decade of the 21st century, the same percentages of energy demand by sectors as in the first (above) scenario will be assumed. It will further be assumed that two-thirds of the thermal energy generation will be achieved by flame combustion of GH_2 and one-third by catalytic combustion of GH_2, electric power generation will be achieved by fuel cells, one-half of the surface transportation will use GH_2-burning IC engines, and the other half, fuel cells, and one-half of the air transportation will be subsonic and the other half supersonic—both running on LH_2.

TABLE 8
Coal/Synthetic Fossil Fuel System

Application	Energy consumption fraction	Fuel	Effective cost (1990 $/GJ)	Fraction × cost (1990 $)
Thermal energy	0.30	Coal	13.40	4.02
Thermal energy	0.10	SNG	24.10	2.41
Electric power	0.30	Coal	13.40	4.02
Surface transport	0.10	SNG	24.10	2.41
Surface transport	0.10	SynGas	34.50	3.45
Subsonic jet transport	0.05	SynJet	30.30	1.51
Supersonic jet transport	0.05	SynJet	30.30	1.51
Total	1.00			19.33[a]

[a] Overall effective cost per gigajoule.

TABLE 9
Solar Hydrogen Energy System ($\frac{1}{3}$ Hydro H$_2$ + $\frac{2}{3}$ Solar H$_2$)

Application	Energy consumption fraction	Fuel	Effective cost (1990 $/GJ)	Fraction × cost (1990 $)
Thermal energy—flame	0.30	GH$_2$	14.13	4.24
Thermal energy—catalytic	0.10	GH$_2$	11.30	1.13
Electric power—fuel cells	0.30	GH$_2$	7.67	2.30
Surface transport—IC engines	0.10	GH$_2$	11.78	1.18
Surface transport—fuel cells	0.10	GH$_2$	5.04	0.50
Subsonic jet transport	0.05	LH$_2$	14.84	0.74
Supersonic jet transport	0.05	LH$_2$	12.80	0.64
Total	1.00			10.73[a]

[a] Overall effective cost per gigajoule.

TABLE 10
Utilization Efficiency Advantage of Solar Hydrogen System over Synthetic Fossil Fuel System

Application	Energy consumption fraction	Fuel	Utilization efficiency ratio, η_f/η_s	Fraction × η_f/η_s
Thermal energy—flame	0.30	GH$_2$	1.000	0.300
Thermal energy—catalytic	0.10	GH$_2$	0.800	0.080
Electric power—fuel cells	0.30	GH$_2$	0.543	0.163
Surface transport—IC engines	0.10	GH$_2$	0.833	0.083
Surface transport—fuel cells	0.10	GH$_2$	0.357	0.036
Subsonic jet transport	0.05	LH$_2$	0.840	0.042
Supersonic jet transport	0.05	LH$_2$	0.725	0.036
Totals	1.00			0.740[a]

[a] Overall η_f/η_s.

Using the above assumptions and the data generated earlier, Tables 8–10 have been prepared. As can be seen from Table 8 for the coal/synthetic fossil-fuel system, the overall effective cost of energy (by taking the weighted averages for all the sectors considered) will be $19.33/GJ, whereas Table 9 shows that the overall effective cost for the solar hydrogen energy system will be $10.73/GJ. In other words, the solar hydrogen energy system's societal cost will be 45% less than that of the coal/synthetic fossil-fuel system.

Table 10 presents the data for the comparison of the utilization efficiencies by considering worldwide energy consumption by sectors. It can be seen that on a weighted average basis, the solar hydrogen energy system is 26% more efficient than the coal/synthetic fossil-fuel system.

6. CONCLUSIONS

As a result of the foregoing study, the following conclusions are reached: (a) The solar hydrogen energy system is environmentally more compatible than the fossil-fuel system; (b) the utilization efficiencies of the solar hydrogen energy system are greater than those of the fossil-fuel system; and (c) the solar hydrogen energy system is more cost effective than the synthetic fossil-fuel system.

ACKNOWLEDGMENTS

The assistance of M. Ackin, D. Pressley, S. Gursu, N. Lufti, J. Myers, and T. Ozgokmen, all of the Clean Energy Research Institute, is gratefully acknowledged.

REFERENCES

1. M. A. Elliot and N. C. Turner, Symposium on Non-Fossil Chemical Fuels, Division of Fuel Chemistry, 163rd National Meeting of the American Chemical Society, Boston, April 13, 1972.
2. J. D. Parent, A Survey of United States and Total World Production, Proved Reserves, and Remaining Recoverable Resources of Fossil Fuels and Uranium as of December 31, 1977, Institute of Gas Technology, Chicago (Marsh, 1979).
3. W. Fulkerson, R. J. Judkins, and M. K. Sanghui, *Scientific American* **263**(3), 129–135 (1990).
4. T. N. Veziroglu (ed.), *Alternative Energy Sources I* (11 vols.), Hemisphere, Washington, D.C. (1977).
5. T. N. Veziroglu (ed.), *Solar Energy: International Process* (4 Vols.), Pergamon Press, New York (1978).
6. T. N. Veziroglu (ed.), *Solar Energy and Conservation* (3 vols), Pergamon Press, New York (1978).
7. T. N. Veziroglu (ed.), *Alternative Energy Sources II* (9 vols.), Hemisphere, Washington, D.C. (1979).
8. L. B. McGown and J. O'M. Bockris, *How to Obtain Abundant Clean Energy*, Plenum Press, New York (1980).
9. T. N. Veziroglu (ed.), *Alternative Energy Sources III* (9 vols.), Hemisphere, Washington, D.C. (1980).
10. T. N. Veziroglu (ed.), *Alternative Energy Sources IV* (8 vols.), Ann Arbor Science, Ann Arbor, Michigan (1981).
11. T. N. Veziroglu (ed.), *Alternative Energy Sources V* (6 vols.), Elsevier, Amsterdam (1983).
12. T. N. Veziroglu (ed.), *Renewable Energy Sources International Progress* (2 vols.), Elsevier, Amsterdam (1984).
13. T. N. Veziroglu (ed.), *Alternative Energy Sources VI* (4 vols.), Hemisphere, Washington, D.C. (1985).

14. T. N. Veziroglu (ed.), *Alternative Energy Sources VII* (6 vols.), Hemisphere, Washington, D.C. (1987).
15. T. N. Veziroglu (ed.), *Alternative Energy Sources VIII* (2 vols), Hemisphere, Washington, D.C. (1988).
16. T. N. Veziroglu (ed.), *Hydrogen Energy*, Parts A and B, Proc. Hydrogen Economy Miami Energy Conference, Plenum Press, New York (1975).
17. T. N. Veziroglu (ed.), *Proc. 1st WHEC* (3 vols.), Clean Energy Research Institute, University of Miami (1976).
18. T. N. Veziroglu and W. Weifritz (eds.), *Hydrogen Energy System* (5 vols.), Proc. 2nd WHEC, Pergamon Press, Oxford (1979).
19. T. N. Veziroglu, K. Fueki, and T. Ohta (eds.), *Hydrogen Energy Progress* (4 vols.), Proc. 3rd WHEC, Pergamon Press, Oxford (1981).
20. T. N. Veziroglu, W. D. Van Vorst, and J. H. Kelley (eds.), *Hydrogen Energy Progress IV* (4 vols.), Proc. 4th WHEC, Pergamon Press, New York (1982).
21. T. N. Veziroglu and J. B. Taylor (eds.), *Hydrogen Energy Progress V* (4 vols.), Proc. 5th WHEC, Pergamon Press, New York (1984).
22. T. N. Veziroglu, N. Getoff, and P. Weinzierl (eds.), *Hydrogen Energy Progress VI* (3 vols.), Proc. 6th WHEC, Pergamon Press, New York (1986).
23. T. N. Veziroglu and N. Protsenko (eds.), *Hydrogen Energy Progress VII* (3 vols.), Proc. 7th WHEC, Pergamon Press, New York (1988).
24. J. O'M. Bockris, *Energy: The Solar-Hydrogen Alternative*, Australia and New Zealand Book Co., Sydney (1975).
25. T. N. Veziroglu, in: *Heliotechnique and Development*, Development Analysis Associates, Cambridge, Massachusetts (1975).
26. T. Ohta (ed.), *Solar-Hydrogen Energy Systems*, Pergamon Press, Oxford (1978).
27. L. O. Williams, *Hydrogen Power: An Introduction to Hydrogen Energy and Its Aplications*, Pergamon Press, Oxford (1980).
28. L. W. Skelton, *The Solar-Hydrogen Energy Economy: Beyond the Age of Fire*, Van Nostrand Rheinhold, New York (1984).
29. T. N. Veziroglu and O. Basar, Hydrogen Economy Miami Energy (THEME) Conference, Clean Energy Research Institute, University of Miami, p. S15:93 (1974).
30. T. N. Veziroglu, S. Kakac, O. Basar, and N. Forouzanmehr, *Int. J. Hydrogen Energy* **1**, 205 (1976).
31. C. Bilgen and E. Bilgen, *Int. J. Hydrogen Energy* **9**, 197–204 (1984).
32. A. Tofighi and F. Sibieude, *Int. J. Hydrogen Energy* **9**, 293–296 (1984).
33. J. W. Warner and R. S. Berry, *Int. J. Hydrogen Energy*, **11**, 91–100 (1986).
34. S. V. Korobtsev, T. A. Kosinova, B. V. Potapkin, Y. R. Rakhimbabaev, F. D. Rusanov, A. A. Fridman, and E. V. Shulakova, in: *Hydrogen Energy Progress VII*, Vol. 2, Proc. 7th WHEC, (T. N. Veziroglu and N. Protsenko, eds.), pp. 1071–1078, Pergamon Press, New York (1988).
35. J. Schmitz, L. Lucke, F. Herzog, D. Glaubitz, and R. Schulten, in: *Hydrogen Energy Progress VII*, Vol. 2, Proc. 7th WHEC (T. N. Veziroglu and N. Protsenko, eds.), pp. 819–830, Pergamon Press, New York (1988).
36. G. E. Beghi, *Int. J. Hydrogen Energy* **11**, 761–772 (1986).
37. H. Engels, J. E. Funk, K. Hesselmann, and K. F. Knoche, *Int. J. Hydrogen Energy* **12**, 291–296 (1987).
38. E. I. Onstott, in: *Hydrogen Energy Progress VII* Vol. 2, Proc. 7th WHEC, (T. N. Veziroglu and N. Protsenko, eds.), pp. 773–782, Pergamon Press, New York (1988).
39. K. Yoshida, M. Aihara, M. Umita, H. Kameyama, H. Kondo, T. Sato, T. Aochi, Y. Takodoro, M. Nobue, T. Yamaguchi, and N. Saki, in: *Hydrogen Energy Progress VII*, Vol. 2, Proc. 7th WHEC (T. N. Veziroglu and N. Protsenko, eds.), pp. 831–842), Pergamon Press, New York (1988).
40. V. R. Rustamov, V. K. Kerimov, Kh. B. Gezalov, and Kh. Ya. Nasirova, in: *Hydrogen Energy Progress VII*, Vol. 2, Proc. 7th WHEC (T. N. Veziroglu and N. Protsenko, eds.), pp. 943–954, Pergamon Press, New York (1988).
41. H. Kameyama, Y. Tomino, T. Sato, R. Amir, A. Orihara, M. Aihara, and K. Yoshida, *Int. J. Hydrogen Energy* **14**, 323–330 (1989).
42. R. Aureille and J. Pottier, *Int. J. Hydrogen Energy* **9**, 183–186 (1984).
43. O. J. Murphy and J. O'M. Bockris, *Int. J. Hydrogen Energy* **9**, 557–562 (1984).
44. K. W. Quandt and R. Streicher, *Int. J. Hydrogen Energy* **11**, 309–318 (1986).
45. M. A. Liepa and A. Borhan, "High-temperature steam electrolysis: Technical and economic evaluation of alternative process designs," *Int. J. Hydrogen Energy* **11**, 435–442 (1986).

46. W. Donitz, E. Erdle, R. Schamm, and R. Streicher, in: *Hydrogen Energy Progress VII*, Vol. 1, Proc. 7th WHEC, (T. N. Veziroglu and N. Protsenko, eds.), pp. 65–74 Pergamon Press, New York (1988).

47. J. Divisek, J. Mergel, and H. Schmitz, in: *Hydrogen Energy Progress VII*, Vol. 1, Proc. 7th WHEC (T. N. Veziroglu and N. Protsenko, eds.), p. 327, Pergamon Press, New York (1988).

48. A. G. Pshenichnikov and V. E. Kazarinov, in: *Hydrogen Energy Progress VII*, Vol. 1, Proc. 7th WHEC (T. N. Veziroglu and N. Protsenko, eds.), pp. 507–530, Pergamon Press, New York (1988).

49. R. C. Kaithla and J. O'M. Bockris, *Int. J. Hydrogen Energy* 12, 23–26 (1987).

50. T. Chivers and C. Lau, *Int. J. Hydrogen Energy* 12, 561–570 (1987).

51. S. A. Naman and K. Al-Emara, *Int. J. Hydrogen Energy* 12, 629–632 (1987).

52. G. I. Novikov, A. N. Tretyak, B. A. Butylin, A. L. Kuzmenko, I. A. Belov, and V. A. Shnyp, in: *Hydrogen Energy Progress VII*, Vol 2, Proc. 7th WHEC (T. N. Veziroglu and N. Protsenko, eds.), pp. 987–1002 Pergamon Press, New York (1988).

53. L. M. Al-Shamma and S. A. Naman, *Int. J. Hydrogen Energy* 14, 173–180 (1989).

54. S. U. M. Khan and J. O'M. Bockris, *Int. J. Hydrogen Energy* 11, 373–380 (1986).

55. M. M. T. Khan, R. C. Bhardwaj, and C. M. Jadhav, in: *Hydrogen Energy Progress VI*, Vol. 1, Proc. 6th WHEC (T. N. Veziroglu, N. Getoff, and P. Weinzierl, eds.), pp. 312–321, Pergamon Press, New York (1986).

56. Yu. A. Gruzdkov, E. N. Savinov, and V. N. Parmon, *Int. J. Hydrogen Energy* 12, 393–402 (1987).

57. A. J. Abdul-Ghani, S. Abdul-Kareem, and S. N. Maree, *Int. J. Hydrogen Energy* 12, 547–554 (1987).

58. R. M. Quint and N. Getoff, *Int. J. Hydrogen Energy* 13, 269–276 (1988).

59. M. I. Rustamov, N. Z. Muradov, A. D. Guseinova, and Yu. V. Bazhutin, *Int. J. Hydrogen Energy* 13, 533–538 (1988).

60. S. D. Huang, C. K. Secor, R. Ascione, and R. M. Zweig, *Int. J. Hydrogen Energy* 10, 227–232 (1985).

61. J. Miyake and S. Kawamura, *Int. J. Hydrogen Energy* 12, 147–150 (1987).

62. S. Tanisho, Y. Suzuki, and N. Wakao, *Int. J. Hydrogen Energy* 12, 623–628 (1987).

63. S. Roychowdhury, D. Cox, and M. Levandowsky, *Int. J. Hydrogen Energy* 13, 407–410 (1988).

64. G. Spazzafumo and G. Gaggio, in: *Hydrogen Energy Progress VII*, Vol. 1, (T. N. Veziroglu and N. Protsenko, eds.), pp. 315–326, Pergamon Press, New York (1988).

65. T. N. Veziroglu, *Int. J. Hydrogen Energy* 12, 99–129 (1987).

66. J. M. Ogden and R. H. Williams, PU/CEES Report No. 231, Princeton Uniersity, Princeton, New Jersey (1989).

67. A. J. Weiss and C. E. Lumms, in: *Alternative Energy Sources*, Proc. Miami International Conference on Alterntive Energy Sources (T. N. Veziroglu, ed.), Vol. 7, pp. 3021–3055, Hemisphere, Washington, D.C. (1978).

68. P. R. Westmoreland, C. R. Forrester III, and A. P. Sikri, in: *Alternative Energy Sources*, Proc. Miami International Conference on Alternative Energy Sources (T. N. Veziroglu, ed.), Vol. 7, pp. 3113–3132, Hemisphere, Washington, D.C. (1978).

69. L. D. Hadden, in: *Hydrogen Energy System* (T. N. Veziroglu and W. Seifritz, eds.), Vol. 2, pp. 893–1005, Pergamon Press, Oxford (1979).

70. O. H. Krikorian, in: *Hydrogen Energy System* (T. N. Veziroglu and W. Seifritz, eds.), Vol. 2, pp. 791–807, Pergamon Press, Oxford (1979).

71. A. Lavi and L. C. Trimble, in: *Hydrogen Energy System* (T. N. Veziroglu and W. Seifritz, eds.), Vol. 1, pp. 147–167, Pergamon Press, Oxford (1979).

72. R. L. LeRoy and A. K. Stuart, in: *Hydrogen Energy System* (T. N. Veziroglu and W. Seifritz, eds.), Vol. 1, pp. 359–373, Pergamon Press, Oxford (1979).

73. F. Behr, F. Flocke, R. Schulten, H. Sussman, and W. Weirich, in: *Hydrogen Energy Progress*, Proc. 3rd WHEC (T. N. Veziroglu, K. Fueki, and T. Ohta, eds.), Vol. 1, pp. 489–501, Pergamon Press, Oxford (1981).

74. J. O'M. Bockris, *Int. J. Hydrogen Energy* 6, 223 (1981).

75. M. Dokiya, T. Fujishige, K. Kameyama, K. Fukuda, and H. Yokokawa, in: *Hydrogen Energy Progress*, Proc. 3rd WHEC (T. N. Veziroglu, K. Fueki, and T. Ohta, eds.), Vol. 1, pp. 373–387, Pergamon Press, Oxford (1981).

76. M. I. German, in: *Alternative Energy Sources II*, Proc. Miami International Conference on Alternative Energy Sources (T. N. Veziroglu, ed.), Vol. 7, pp. 2749–2762, Hemisphere, Washington, D.C. (1981).

77. M. Sappa, in: *Hydrogen Energy Progress*, Proc. 3rd WHEC (T. N. Veziroglu, K. Fueki, and T. Ohta, eds.), Vol. 3, pp. 1373–1406, Pergamon Press, Oxford (1981).

78. G. H. Schutz, in: *Hydrogen Energy Progress*, Proc. 3rd WHEC (T. N. Veziroglu, K. Fueki, and T. Ohta, eds.), Vol. 1, pp. 463-475, Pergamon Press, Oxford (1981).

79. J. O'M. Bockris, F. Gutmann, and W. Craven, in: *Hydrogen Energy Progress IV*, Proc. 4th WHEC (T. N. Veziroglu, W. D. Van Vorst, and J. H. Kelley, eds.), Vol. 5, pp. 1475-1493, Pergamon Press, New York (1982).

80. D. Z. Chen, I. Gurkan, J. W. Sheffield, and T. N. Veziroglu, in: *Hydrogen Energy Progress IV*, Proc. 4th WHEC (T. N. Veziroglu, W. D. Van Vorst and J. H. Kelley, eds.), Vol. 4, pp. 1523-1537, Pergamon Press, New York (1982).

81. E. Fein, in: *Alternative Energy Sources IV*, Proc. 4th Miami International Conference on Alternative Energy Sources (T. N. Veziroglu,ed.), Vol. 5, pp. 265-281, Ann Arbor Science, Ann Arbor, Michigan (1982).

82. G. N. Krishman and C. W. Marynowski, in: *Hydrogen Energy Progress IV*, Proc. 4th WHEC (T. N. Veziroglu, W. D. Van Vorst, and J. H. Kelley, eds.), Vol. 2, pp. 829-836, Pergamon Press, New York (1982).

83. F. H. Schubert and K. A. Burke, in: *Hydrogen Energy Progress IV*, Proc. 4th WHEC (T. N. Veziroglu, W. D. Van Vorst, and J. H. Kelley, eds.), Vol. 1, pp. 215-224, Pergamon Press, New York (1982).

84. A. H. Awad and T. N. Veziroglu, *Int. J. Hydrogen Energy* **9**(5), 355-366 (1984).

85. J. O'M. Bockris and T. N. Veziroglu, *Environ. Conserv.* **12**(2), 105-117 (1985).

86. C. R. Baker, in: *Hydrogen Energy Progress IV*, Proc. 4th WHEC (T. N. Veziroglu, W. D. Van Vorst, and J. H. Kelley, eds.), Vol. 3, pp. 1317-1333, Pergamon Press, New York (1982).

87. J. A. Barclay, ASME Paper 81-HT-82 (1982).

88. R. C. Vetter (ed.), *Oceanography: The Last Frontier*, p. 320, Basic Books, New York (1973).

89. Environmental Quality, 11th Annual Report of the Council on Environmental Quality, U.S. Government Printing Office, Washington, D.C. (1980).

90. Agricultural Statistics, 1981, U.S. Government Printing Office, Washington, D.C. (1981); National Health Expenditures, by Object: 1960-1980, *Statistical Abstract of the United States* (2nd ed.), p. 100, U.S. Department of Commerce, Bureau of the Census, Washington, D.C. (1981).

91. T. Shabad, "Russia reveals oil spill cost up to $900 million," *New York Times*, February 2, 1982.

92. "Money income of households—Aggregate and mean income, by race and Spanish origin of householder: 1979," in: *Hydrogen Energy Progress IV*, Proc. 4th WHEC (T. N. Veziroglu, W. D. Van Vorst, and J. H. Kelley, eds.), Vol. 4, p. 433, Pergamon Press, New York (1982).

93. R. M. Zweig, in: *Hydrogen Energy Progress IV*, Proc. 4th WHEC (T. N. Veziroglu, W. D. Van Vorst, and J. H. Kelley, eds.), Vol. 4, pp. 1789-1808, Pergamon Press, New York (1982).

94. T. D. Crocker, Paper presented at the Conference on Adjusting to Regulatory, Pricing and Marketing Realities, Williamsburg, Virgina, December 13, 1982.

95. F. A. Record, D. V. Bubenick, and R. J. Kindya, *Acid Rain: Information Book*, Noyes Data Corp., Park Ridge, Illinois (1982).

96. R. A. Taylor, *U.S. News and World Report*, February 28, 1983, pp. 27-28.

97. *Acid Rain: A Review of the Phenomenon in the EEC and Europe*, A report prepared for the Commission of the European Communities, Directorate-General for Environment Consumer Protection and Nuclear Safety by Environmental Resources Ltd., Unipub, New York (1983).

98. *Acid Rain: A Survey of Data and Current Analyses*, A report prepared by the Congressional Research Service, U.S. Government Printing Office, Washington, D.C. (1984).

99. M. J. Gibbs, in: *Greenhouse Effect and Sea Level Rise* (M. C. Barth and J. G. Titus, eds.), Van Nostrand Reinhold, New York (1984).

100. J. M. Fowler, *Energy and the Environment*, 2nd ed., McGraw-Hill, New York (1984).

101. J. S. Hoffman, in: *Greenhouse Effect and Sea Level Rise* (M. C. Barth and J. G. Titus, eds.), Van Nostrand Reinhold, New York (1984).

102. F. H. Borman, *Bioscience* **35**(7), 434 (1985).

103. C. R. Sheppard, *Am. Mining Congress J.* **72**, 9 (1986).

104. T. E. Graedel and R. McGill, *Environ. Sci. Technol.* **20**, 1093 (1986).

105. L. S. Kalkstein *et al.*, in: *Climate Change*, Vol. 3 (J. G. Titus, ed.), pp. 275-293, Environmental Protection Agency/United Nations Environmental Programme (1986).

106. R. N. Palmer and J. R. Lund, *J. Water Resources Planning Manage.* **112**(10), 469 (1986).

107. S. J. Cohen, in: *Climate Change*, Vol. 3 (J. G. Titus, ed.), pp. 163-183, Environmental Protection Agency/United Nations Environmental Programm (1986).

108. D. A. Wilhite, in: *Climate Change*, Vol. 3 (J. G. Tirus, ed.), pp. 73–88, Environmental Protection Agency/United Nations Environmental Programme (1986).

109. J. A. Laurmann, Paper presented at the CEC Symposium on CO_2 and Other Greenhouse Gases: Climatic and Associated Impacts, Brussels, November 1986.

110. "The leaking underground storage tank trust fund," *EPA Journal* 13(1), (1987).

111. J. G. Titus, C. Y. Kuo, M. J. Gibbs, T. B. LaRoche, M. K. Webb, and J. O. Waddell, *J. Water Resources Planning Manage.* 113(3), 216 (1987).

112. R. S. Bradley, H. F. Dizz, J. K. Eischeid, P. D. Kelley, and C. M. Goodess, *Science* 237, 171 (1987).

113. Surface Coal Mining Reclamation: 10 Years of Progress, 1977–1987, U.S. Department of the Interior, Office of Surface Mining Reclamation and Enforcement, Washington, D.C. (1987).

114. S. Nilsson and P. Duinker, *Environment* 29(9), (1987).

115. R. Monastersky, *Science News* 132(November), 326 (1987).

116. K. Mellanby (ed.), Air Pollution, Acid Rain and the Environment, Watt Committee on Energy, Report No. 18 (1988).

117. D. W. Schindler, *Science* 239, 149 (1988).

118. Interim Assessment: The Causes and Effects of Acidic Deposition in: *Effects of Acidic Deposition*, Vol. IV, The National Acid Precipitation Assessment Program, Washington, DC (1988).

119. *Statistical Abstract of the U.S.* 108th ed., United States Department of Commerce, Washington, D.C. (1988).

120. *Climate Alert* 1(4), 4 (1988).

121. *Climate Alert* 1(4), 6 (1988).

122. J. R. Luoma, *Audubon* 90(July), 16 (1988).

123. T. Vogel, *Business Week*, August 8, 1988, p. 36.

124. "Bangladesh: The water this time," *Newsweek*, September 19, 1988.

125. F. Pearce, *New Scientist* 119(September 8), 31 (1988).

126. L. J. Wilson and E. Britton, *Chemical Week* 143(August 17), 9 (1988).

127. B. J. Smith and D. A. Tirpak (eds.), The Potential Effects of Global Climate Change on the United States (draft), U.S. Environmental Protection Agency, Office of Policy, Planning and Evaluation and Office of Research and Development, Washington, D.C. (1988).

128. K. E. Trenberth, G. W. Branstator, and P. A. Arkin, *Science* 242, 1640 (1988).

129. T. A. Sanction, *Time*, January 2, 1989, p. 24.

130. K. E. Kunkel, *Natural History* 98(1), 48 (1989).

131. S. H. Schneider, *Science* 243, 771 (1989).

132. V. Cahan and B. Bremner, *Business Week*, February 13, 1989, p. 95.

133. "Living in the greenhouse," *The Economist*, March 11, 1989, p. 87.

134. E. Marshall, *Science* 244, 20 (1989).

135. R. A. Houghton and G. M. Woodwell, *Scientific American* 260(4), 36 (1989).

136. T. N. Veziroglu, I. Gurkan, and M. M. Padki, *Int. J. Hydrogen Energy* 14, 257–266 (1989).

137. F. Barbir, *MEN 692 Special Problem Report*, Clean Energy Research Institute, University of Miami (1989).

138. G. D. Brewer, *Int. J. Hydrogen Energy* 4, (1979).

139. G. D. Brewer, *Int. J. Hydrogen Energy* 7, (1982).

140. W. Peschka, in: *Hydrogen Energy Progress*, Proc. 3rd WHEC (T. N. Veziroglu, K. Fueki, and T. Ohta, eds.), Vol. 3, pp. 1561–1576, Pergamon Press, Oxford (1981).

141. M. R. Swain, J. M. Pappas, R. R. Adt, and W. J. D. Escher, Society of Automotive Engineers Technical Paper 810350 (1981).

142. F. E. Lynch, in: *Hydrogen Energy Progress IV*, Proc. 4th WHEC (T. N. Veziroglu, W. D. Van Vorst, and J. H. Kelley, eds.), Vol. 3, pp. 1033–1051, Pergamon Press, New York (1982).

143. W. F. Stewart, in: *Hydrogen Energy Progress IV*, Proc. 4th WHEC (T. N. Veziroglu, W. D. Van Vorst, and J. H. Kelley, eds.), Vol. 3, pp. 1071–1093, Pergamon Press, New York (1982).

144. R. E. Billings, in: *Hydrogen Energy System* (T. N. Veziroglu and W. Seifritz, eds.), Vol. 4, pp. 1709–1730, Pergamon Press, Oxford (1979).

145. *Hydrogen Prog. Mag.*, Billings Corp., 6, 7 (1981).

19

Adsorption and Absorption of Hydrogen in Metals

FTIR and RT Approach

Jose L. Carbajal and Ralph E. White

1. INTRODUCTION

Because of catastrophic failures in high-strength-steel structures (oil rigs, oil pipes, ships, and aircraft) due to hydrogen embrittlement, renewed interest has been shown in the study of hydrogen in ferrous metals. An investigation of the relation of surface hydrogen to that absorbed in the metal was thought to be necessary.

For the majority of this century, studies in surface chemistry have involved the mercury/solution interface.[1-3] During the last two decades, studies have been increasingly made upon solids, and research material on the solid/liquid interface has become now quite substantial.[4-7] On the other hand, this material has the limitation that this research has been exclusively on noble metals. The reason for this is ease of measurement, largely with the potentiodynamic approach.[8-10] This preoccupation with noble metals is not dissimilar to the earlier one with mercury. It leaves not only an absence of knowledge of the facts on metals that are far more used in practice than the noble metals, but also restricts knowledge of theory because of the limitation in breadth of the experimental data.[11-13]

When one leaves both mercury and the noble metals, an obvious surface which demands study is that of iron. Conversely, such a surface is more difficult to study than the earlier ones mentioned because of potential simultaneous dissolution of the iron electrode along with processes of adsorption.[14-16] It is therefore necessary to use methods in which current passage across the surface is not part of the measurement. Among these is the radiotracer method, but this gives only information concerning adsorption of the marked ion, and the actual structure on the surface is left to be elucidated by other methods. One of these is the Fourier transform version of infrared spectroscopy (FTIR). These two methods have been combined in the present study along with an electrochemical permeation technique.

The FTIR spectroscopic technique is a novel *in situ* technique that allows the adsorbed species and intermediates to be identified and the variation of their concentration on the

Jose L. Carbajal • Inland Steel Company, Research Laboratory, East Chicago, Indiana 46312.   *Ralph E. White* • Chemical Engineering Department, Texas A&M University, College Station, Texas 77843-3122.
Electrochemistry in Transition, edited by Oliver J. Murphy *et al.* Plenum Press, New York, 1992.

surface with electrode potential and solution concentration to be studied. This technique has several advantages over other *in situ* techniques in respect to molecular specificity, elucidation of adsorption sites and nature of adsorbate–substrate interactions, and detection of adsorption-induced chemical changes.[17] This molecular specificity makes this technique attractive for the study of the iron–hydrogen interaction as a function of potential.

2. FTIR EXPERIMENT

2.1. Principle of the Method

An IR spectrum from the electrode/electrolyte interface is obtained when an IR beam is externally reflected on this interface while the potential of the electrode is held at some potential of interest. Analysis of this reflected spectrum leads to information on the structure of species at or near the electrode surface. Since this technique is performed *in situ* to the electrochemical experiment, dynamic processes under controlled potential or current may be studied, as well as steady-state structures as a function of potential.

2.2. Principle of the FTIR Spectrometer

The heart of the FTIR instrument is a Michelson interferometer, which consists of a beamsplitter and a fixed and a movable mirror. Collimated infrared radiation from a broadband source enters the interferometer and is split into two beams of equal intensity. The transmitted beams are passed into the two arms of the interferometer and are reflected from the mirrors back to the beamsplitter. These beams are recombined constructively and destructively depending on the relative mirror displacement. The interferogram would be the constructive recombination of the beams and would be a function of the mirror displacement (retardation). This signal intensity is recorded at regular retardation intervals to provide a digital representation of the interferogram and is stored in the computer memory. This interferogram can then be Fourier transformed by the computer into the frequency spectrum.

The method used here was surface normalized interfacial Fourier transform infrared spectroscopy (SNIFTIRS). Parallel-polarized infrared radiation at an angle of incidence of 28° on a plane window was used. The spectra reported represent the difference between the IR intensities obtained at a given absorbing potential and under nonadsorbing conditions, normalized with respect to the reference spectrum; that is,

$$\Delta R / R_{ref} = (R - R_{ref}) / R_{ref} \tag{1}$$

where R is the reflected intensity at a potential of interest, and R_{ref} is that at a reference potential (this reference was chosen to correspond to the least negative potential at which the radiotracer measurement indicated negligible adsorption). Since

$$\Delta R / R_{ref} \ll 1 \tag{2}$$

then

$$A \equiv -\log(R / R_{ref}) \cong -\Delta R / R_{ref} \tag{3}$$

and thus the spectra reported correspond to absorbances due to adsorbed species at the electrode/solution interface.

A second technique, infrared reflection–adsorption spectroscopy (IRRAS), can also be used. In this technique the polarization state of the radiation is modulated, thereby perturbing the interaction of the radiation with spatially oriented molecules (adsorbed molecules). This can only happen due to the interaction of parallel-polarized IR radiation with molecular dipole oscillators perpendicular to the interface. The intensity of the parallel and perpendicularly polarized components of the IR radiation, I_p and I_s, are detected. From Greenler's theory, a polarization modulation gives an ac voltage at the detector which is proportional to $I_p - I_s$; since only I_p interacts with adsorbed species, this signal contains the IR spectrum of the surface species. For more detail on these techniques, the reader is referred to an article by Habib and Bockris.[18]

2.3. Electrochemical Cell

The cell used in FTIR experiments has been described elsewhere.[18] A ZnSe disk was used as the IR-transparent window. A Luggin capillary, connecting the reference electrode to the cell, was mounted close to the edge of the working electrode surface. A platinum ring was used as a secondary electrode.

The working electrode was built from a 3-cm-long, 1-cm-diameter iron rod (99.9985% pure; Alfa Ventron), connected to a 1-cm-diameter copper rod and clad with a heat-shrinkable Teflon tube to expose 0.79 cm^2 of flat surface to the solution. This surface was polished with successively finer diamond paste, to 0.25 μm, to a mirror finish.

2.4. FTIR Instrument

For this experiment a Digilab FTS-20E FTIR spectrometer system equipped with a Data General Nova 4 computer was used. A versatile reflection attachment (Harrick model VRA S5D) with retromirror accessory (RMA-4DG) was used to guide the incident IR beam to the electrode surface and the reflected beam back to the detector. The optical configuration has been described elsewhere.[18] A Molectron IGPZ28 (Cambridge Physical Sciences) gold grid polarizer was used to plane polarize the beam. The FTIR detector was a midrange IR fast detector (mercury cadmium telluride 77 K solid-state units).

2.5. Experimental Procedure

The solutions were prepared from pyrolytically distilled water. The solution was a borate buffer, pH 8.4. Deuterium oxide, 100% isotopic purity, was used in the experiment. A potentiostat driven by a waveform generator and an X-Y recorder were used for controlling and monitoring the electrode potential. Synchronization between application of the potential step and the beginning of the data collection was achieved by tapping a TTL (logic gate) signal accompanying the beginning of the current data collection at the Digilab A/D converter and applying the signal to trigger the appropriate potential step at the potentiostat.

The cell was aligned to obtain the maximum reflectance from the working electrode. Then, ca. 30 min was allowed to purge the external chamber with nitrogen. The electrode was held at −0.5 V (versus NHE) during this purging time. Then, the electrode was polarized at successive potentials in the adsorption range. A total of 500–1000 scans were collected at each potential. All spectra at each potential were added and signal-averaged between the spectra at a given cathodic potential and that at the reference potential (−0.5 V versus NHE). All spectra were measured with parallel-polarized IR radiation. The resulting spectra are thus termed p-polarized differential spectra and give information about the adsorbed layer.

The real surface of the iron electrode cannot be determined by coulometric hydrogen desorption due to codissolution of the substrate. An estimation of the roughness factor was made by comparison with electrodes of known surface area. The roughness factor of the iron electrode prepared by chemical vapor deposition was determined to be 8.5 using various methods (see below). The polished iron electrode (0.25-μm diamond paste) of the present experiment should have a roughness factor less than that of the chemical-vapor-deposited electrode. Metal electrodes prepared by the same procedure, that is, polished with 0.25-μm diamond paste, showed a roughness factor of 3, as determined by capacitance measurement. Hence, it seems reasonable to accept a roughness factor of 3 for the oxide-free iron electrode.

2.6. Results

The adsorption spectrum of the iron/electrolyte interface during the hydrogen evolution reaction in borate buffer solution, pH 8.4, is shown in Fig. 1. Three peaks at 2060, 1600, and 980 cm^{-1} are predominant. Two small peaks at 1450 and 1260 cm^{-1} are also observed. The absorbance of the various peaks changes with electrode potential. These changes are shown in Fig. 2. The area under a peak is proportional to the concentration of the species adsorbed at the surface. The relative areas of the peaks at 2060 and 980 cm^{-1} are shown in Figs. 3a and 3b as a function of electrode potential. The peak area A varies logarithmically with electrode potential, and the slope is $\Delta V/\Delta \log A = -0.33$.

In order to detect the effect of isotopes on the adsorption spectrum, a 50:50 mixture of D_2O and H_2O was used. The peak observed at 2060 cm^{-1} in pure aqueous solution was accompanied by a peak at 1485 cm^{-1}. The peak at 1600 cm^{-1} was accompanied by a peak at 1200 cm^{-1}. The peak at 980 cm^{-1} was not accompanied by any other peak, but the additional peak would have been expected at 680 cm^{-1}, which was outside of the spectrometer range. The relative area of the peak at 1485 cm^{-1} is shown in Fig. 4 as a function of electrode potential. The peak area in the presence of deuterium varies logarithmically with electrode potential, and the slope of the plot of log area (A) versus overpotential was -0.34. A Tafel plot of the hydrogen evolution reaction was measured inside the FTIR cell, and the slope is presented in Fig. 5. This slope is the same before and after the experiment. There is the need

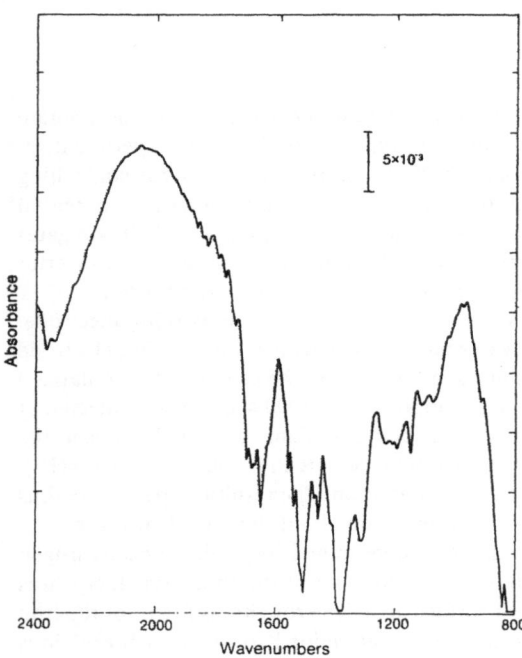

FIGURE 1. Differential IR spectrum in absorbance of Fe—H vibration at -0.9 V versus NHE in borate buffer solution.

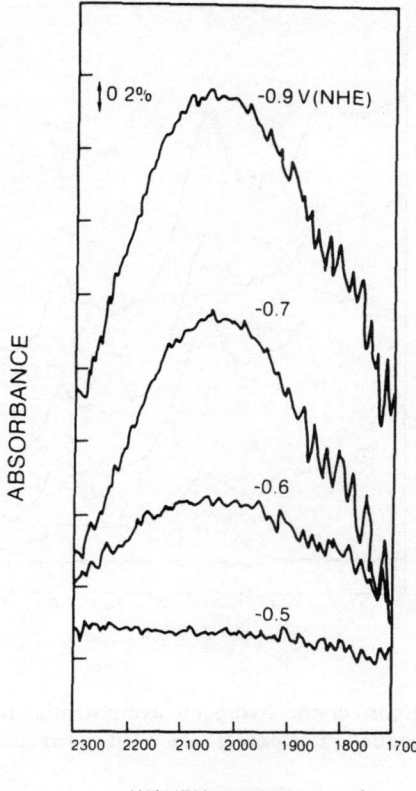

FIGURE 2. Differential IR spectra in absorbance of Fe—H vibration at different potentials in borate buffer solution.

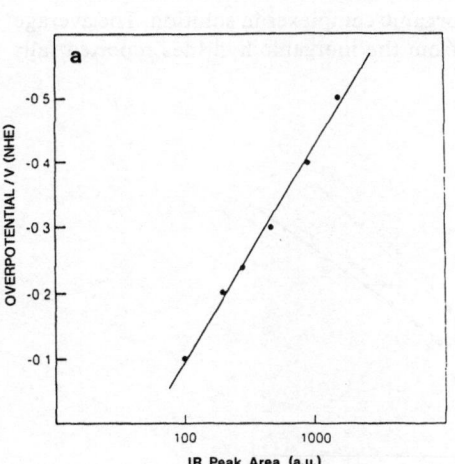

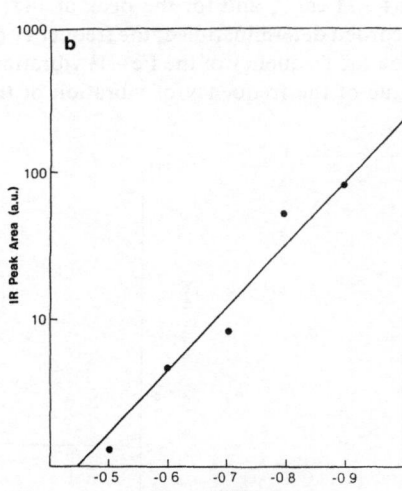

FIGURE 3. Coverage–potential plots for hydrogen on iron electrode in borate buffer solution; coverage is represented by area of IR peaks at 2060 (a) and 980 cm^{-1} (b).

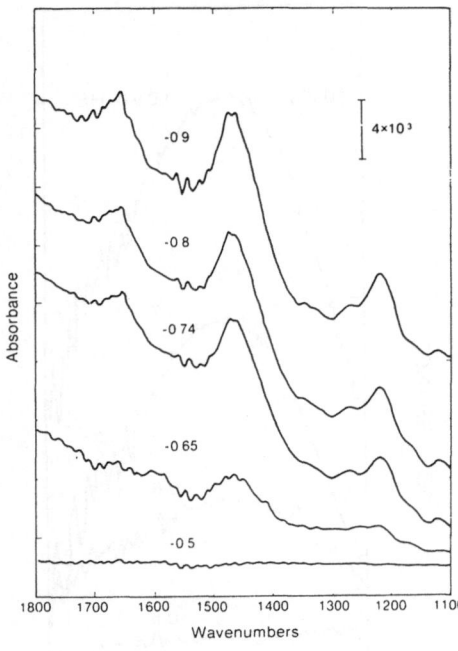

FIGURE 4. Differential IR spectra of Fe—D vibration at different potentials in H_2O/D_2O (1:1) borate buffer solution. Potentials indicated on the figure are in volts versus NHE.

to correct the hydrogen overpotential for the change in pH at the interface due to the production of excess OH^-. This correction is shown in Fig. 5.

2.7. Discussion

The reproducibility of the measurement of the peak maximum for the peak at 2060 cm^{-1} was ± 11 cm^{-1}, and for the peak at 1485 cm^{-1} the reproducibility was ± 5 cm^{-1}. There is no recorded determination of the frequency of vibration of hydrogen adsorbed on iron. Barclay[19] took the frequency of the Fe—H vibration from inorganic complexes in solution. The average value of the frequency of vibration of this bond from the inorganic hydrides reported falls

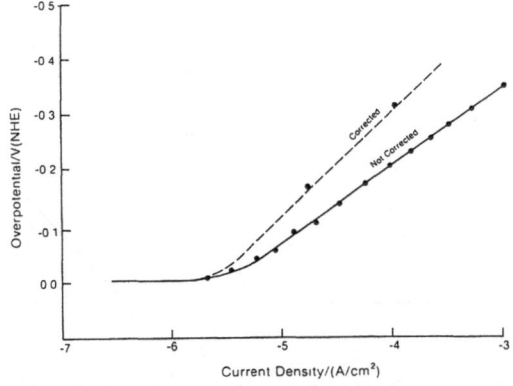

FIGURE 5. Tafel correction.

within a range of $1885 \pm 25 \text{ cm}^{-1}$. However, it is known that as a result of adsorption, the absorbance maximum may exhibit a shift in frequency or a change in intensity.[20] The shifts in frequencies may be to higher or lower frequencies, and their magnitude may be from tens to hundreds of wave numbers.[21] Thus, for example, SCN^- in solution has a peak maximum at 2150 cm^{-1}, but in the adsorbed state on platinum, the peak shifts to lower frequencies, i.e., 2075 cm^{-1}.[22] Similarly, Pons *et al.* showed[23] that acetonitrile strongly adsorbs on platinum at all potentials positive to -0.3 V (versus NHE), as was evidenced by the shifting of the peak of $C\equiv N$ from 2150 cm^{-1} in the spectrum of the bulk solution to 2350 cm^{-1} for the adsorbed state.

Kaeszing and Saillant[24] reported the frequency of vibration for several hundred metal hydrides. The M—H stretching frequency lies around $1900 \pm 300 \text{ cm}^{-1}$ for all the transition-metal hydride complexes. In view of the above evidence, the absorbance at 2060 cm^{-1} appears to be consistent with adsorbed hydrogen bonded to iron.

The frequency of a vibrational transition can be calculated from force constants and the reduced mass; that is,

$$\bar{\nu}_1 = \frac{1}{2\pi c} \sqrt{\frac{k}{\mu_1}} \tag{4}$$

where $\bar{\nu}_1$ is the wave number of the absorbance maximum, c is the velocity of light, and k and μ are the force constant and the reduced mass, respectively. When one of the atoms is isotopically labeled, the force constant remains unchanged. Thus, for an isotopically labeled compound,

$$\bar{\nu}_2 = \frac{1}{2\pi c} \sqrt{\frac{k}{\mu_2}} \tag{5}$$

Then,

$$\frac{\bar{\nu}_1}{\bar{\nu}_2} = \sqrt{\frac{\mu_2}{\mu_1}} \tag{6}$$

When D_2O is used instead of water, an absorbance maximum at 1485 cm^{-1} is observed. The ratio of frequencies ($2060/1485 = 1.39$) agrees with the square root of the reduced mass ratio of Fe—D/Fe—H (1.4). A further confirmation that the wave numbers observed in the present experiment (2060 and 980 cm^{-1}) are associated with adsorbed hydrogen can be made by comparison with the wave numbers observed for hydrogen adsorbed on other transition metals from the gas phase. A broad peak centered at 2105 cm^{-1} has been attributed to the stretching vibration of the hydrogen adsorbed from the gas phase on polycrystalline platinum, Pt—H.[25] When hydrogen is replaced by deuterium, the peak maximum shifts to 1512 cm^{-1}. The shift in frequency agrees with the frequency calculated based on the relevant reduced masses and can be compared with the similar change noted in the present work.

The full width at half-maximum for the Fe—H absorption at 2060 cm^{-1} is 350 cm^{-1}. This large width may be compared with those of other metal hydrides.[25] Hydrogen bonding broadens the absorption spectrum. Since deuterium forms hydrogen bonds less readily than hydrogen, the full width at half-maximum of the Fe—D absorption is 80 cm^{-1}, which is smaller than that of the Fe—H absorption. The relatively strong absorption at 980 cm^{-1} may be assigned to an asymmetric stretching vibration of Fe—H. This vibrational mode has been observed at 1060 cm^{-1} for hydrogen adsorbed on iron from the gas phase.[26] Coadsorbed water shifts this absorption to lower frequencies. Hence, it is reasonable to assume that the asymmetric stretching vibration of hydrogen adsorbed on iron in solution occurs at 980 cm^{-1}. This peak disappears when D_2O is used instead of H_2O. With D_2O, the expected absorbance maximum based on reduced mass calculations is 700 cm^{-1}, which is beyond the range of the instrument used.

2.8. Estimate of the Surface Coverage

Calculation of an absolute hydrogen coverage on the iron electrodes was attempted from FTIR results obtained during cathodic polarization. The formalism used is currently applied to predict the vibrational frequency of species adsorbed on the surface.[29] The present approach does not predict a frequency; rather, it calculates the extinction coefficient and consequently the surface coverage from the IR peak intensity. It is known that

$$\frac{8\pi^2}{3}\frac{m_e\bar{\nu}C}{he^2}|\mu|^2 = \frac{4m_eC_{\varepsilon_0}}{N}A \tag{7}$$

where m_e is the mass of the electron, $\bar{\nu}$ is the wave number of the adsorption peak, C is the velocity of light, μ is the transition dipole moment, h is Planck's constant, e is the electronic charge, ε_0 is the dielectric constant of the medium, and A is the absorptivity. The absorptivity can be related to the absorption coefficient α at a given frequency by

$$A = \int_{\nu_1}^{\nu_2}\alpha_1\,d\nu_1 \tag{8}$$

where ν_1 and ν_2 represent the breadth of the adsorption band. The value of the transition moment μ is not known for Fe—H. However, a value of 1.33 debye/Å was obtained for Si—H, which vibrates at 2100 cm^{-1} as compared to 2060 cm^{-1} for Fe—H. Thus, considering an Fe—H bond length of 1.5 Å, one obtains a transition dipole moment of 1.99 debye. Using this value in Eq. (7), one obtains A, which is the integrated absorption coefficient. From the expression in Eq. (8), one obtains the absorption coefficient $\alpha = 8 \times 10^6$ cm^2/mol.

The concentration of adsorbed species at the interface was obtained by consideration of the extinction coefficient $\varepsilon = \alpha/2.303$ and the optical path through the adsorbate. This experimentally measured absorbance, a, at -0.9 V was 1.8×10^{-2}, at an angle of incidence ϕ of 40. Thus, from $a = \varepsilon Cd$, where C is the concentration of the adsorbing molecules, C was found to be 0.15 mol/cm^3. The value of Γ can be correspondingly obtained from the value of the Fe—H distance. Thus, this comes to be $\Gamma = 2.7 \times 10^{-9}$ mol/cm^2, and if the roughness factor of the surface is 3, then $\theta = 0.28$ for the highest cathodic overpotential (-0.5 V).

A plot of the coverage-overpotential relation for hydrogen adsorbed on iron corrected for reference state is shown in Fig. 6. The reference potential used in the present investigation was -0.5 V (versus NHE). Bockris and Kita found that at this potential there is a fractional coverage of hydrogen of 0.03, which has been taken into account for the coverage correction.

2.9. Ratio of H/D on the Surface

Measuring peak areas for hydrogen and deuterium adsorption from Figs. 2 and 3, one finds a ratio of $A_H/A_D = 6$. If the mode of evolution of hydrogen is a rate-determining proton discharge,[28] followed by a surface recombination reaction, it can be shown that

$$\frac{\theta_H}{\theta_D} = \left[\left(\frac{k_H}{k_D}\right)_{Dis}\left(\frac{k_D}{k_H}\right)_{Com}\left(\frac{a_{H^+}}{a_{D^+}}\right)\right]^{1/2} \tag{9}$$

where Dis and Com represent proton discharge and recombination of H and D, respectively,

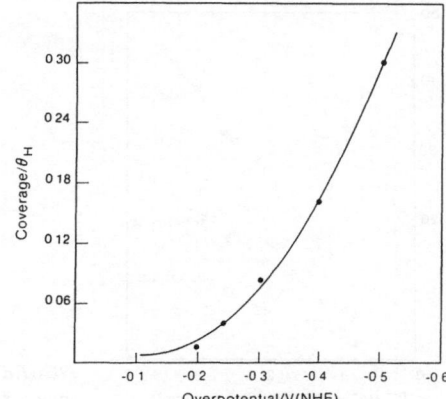

FIGURE 6. Coverage–overpotential relation for hydrogen adsorbed on iron corrected for reference potential.

and a_{H^+}/a_{D^+} is the ratio of the sum of the concentration of ions discharging H over the sum of the concentrations of those discharging D. In order to obtain this quantity, one can use the equation derived by Gold,[29] which relates the sum of the concentrations of the ions needed for the H discharge and D discharge, respectively, to the equilibrium constant of the reaction

$$2D_3O^+ + 3H_2O \rightleftharpoons 2H_3O^+ + 3D_2O \tag{10}$$

Thus, from Eq. (9) and using the value calculated for a_{H^+}/a_{D^+}, one gets $\theta_H/\theta_D = 1.5$. This compares with the experimental value of 6. This apparent discrepancy involves the implicit assumption that the relative spectral intensities per Fe—H bond and Fe—D bond are the same. It is known that O—H bonds give IR intensities which are four times greater than those of O—D bonds. Thus, $1.5 \times 4 = 6$, in agreement with the observed ratio of peak areas.

3. HYDROGEN EVOLUTION MECHANISM

Views in the literature concerning the mechanism of the hydrogen evolution reaction on iron at pH 8.4 are not unified, though suggestive of a slow discharge reaction followed by a recombination of hydrogen.[30,31] This mechanism can be distinguished from the alternate mechanism (fast proton discharge followed by rate-determining electrochemical desorption) by the θ value. In the case of rate-determining proton discharge, this should be low ($\theta \ll 1$), whereas in the case of the rate-determining electrochemical desorption mechanism, θ approaches unity. The spectroscopic results support the first of these two mechanisms.

The Tafel slope obtained (0.18) is higher than the predicted one for a slow discharge mechanism followed by fast recombination. This can be explained in terms of an activated couple discharge mechanism under Temkin conditions.[32] Consequently, the coverage-overpotential relation can also be explained using a Temkin approximation, giving $\delta \ln \theta/\delta V = 1/(4RT/F + 3r/F)$, where r is the Temkin coefficient. Gileadi and Conway[33] reported values of r of 2–5 kJ/mol. Taking $r = 3$ and substituting in the equation given provides 3.07 while the experimental value is 3.08.

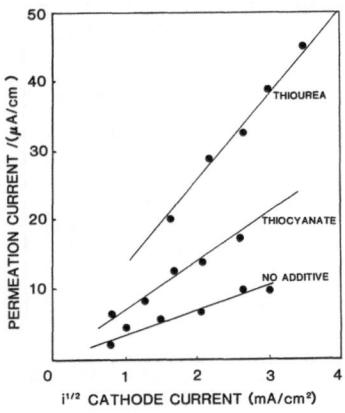

FIGURE 7. Permeation current at steady state as a function of square root of the cathodic current.

4. PERMEATION STUDIES

In permeation studies, the metal whose permeability is to be studied is made to function as a bipolar membrane in the permeation cell. Hydrogen is generated on one side of the metal membrane. Most of the discharged atomic hydrogen combines to form molecular hydrogen and escapes as bubbles. At the same time, part of it dissolves in the metal membrane and diffuses to the opposite side. At this side the potential is maintained sufficiently anodic so that all arriving hydrogen atoms are instantly oxidized to generate an equivalent electrical current. By analyzing such current transients obtained from well-defined initial and boundary conditions, the diffusion coefficient and the solubility of the hydrogen at a given cathodic current density are obtained. The technique was first introduced by Devanathan and Stachurski[34] and later improved and used extensively by Bockris and co-workers.[35]

The permeation of hydrogen into iron specimens was examined with two objectives. It was desired to determine the concentration of hydrogen just "inside" the surface of the electrode, that is, C_0, and to determine the diffusion coefficient of hydrogen. The C_0 value would be a function of surface composition. Thus, the permeation of hydrogen into pure ion membranes was studied as a function of solution composition and as a function of over-potential.

Results from studies of hydrogen permeation in pure borate buffer solution, pH 8.4, and in pure borate buffer with addition of thiourea or thiocyanate as a function of cathodic current are presented in Figs. 7 and 8, showing the plots obtained for the permeation as a

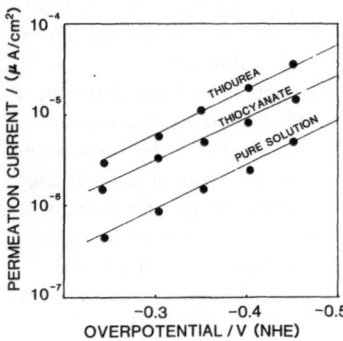

FIGURE 8. Steady-state permeation current as a function of overpotential.

function of cathodic current and the permeation current as a function of overpotential, respectively.

5. RADIOTRACER TECHNIQUE

5.1. Principle of the Method

In the radiotracer technique, the adsorption at the solid/solution interface is monitored by measuring the changes in radiation as a function of potential using labeled compounds (adsorbate) in solution. This adsorption is carried out *in situ* without disturbing the primary experimental conditions. The working electrode (adsorbent) is a very thin film (3000 Å thick) which acts as a window for the counter. This counter is a glass scintillator which emits a photon when struck by radiation reaching it. This light is converted into an electrical signal, preamplified and further amplified in a photomultiplier, and finally converted into counts per minute (cpm). Thus, the amount of radiation being detected is a direct measure of the surface concentration when provisions are made to subtract for the solution background radiation.

5.2. Results

Radiotracer cell and electrode preparation for this experiment has been described elsewhere.[36] The roughness of the electrode was determined using three separate methods. The first consisted in the measurement of the charge required to form a monolayer of passive film on iron, in borate buffer solution, pH 8.4, taken as the charge required to pass over the hump in the passivation curve (Fig. 9a) at 2 mV/s. This charge was then compared with that

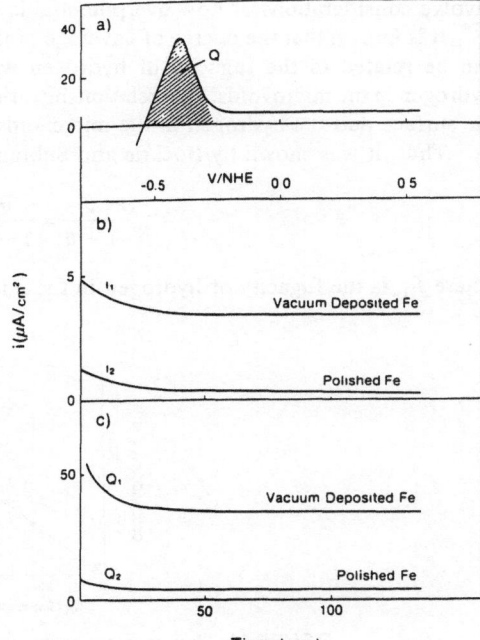

FIGURE 9. Evaluation of the roughness factor of the vacuum-deposited iron electrode on glass scintillator. (a) Measurement of the charge required to form a monolayer of passive film on iron (borate buffer pH 8.4, $v = 2 \, mV/sec$, $Q/Q_m = 6.5$); (b) measurement of the steady-state current developed at 0.3 V/NHE ($i_1/i_2 = 8.7$); (c) measurement of the charge over the transient current arising from introducing the electrode into the solution at a potential of -0.4 V/NHE ($Q_1/Q_2 = 11.4$).

required to form a monolayer of FeO on a smooth surface,[37] 580 $\mu C/cm^2$. The roughness factor obtained by this method was 6.5. A second method consisted in the measurement of the steady-state current developed at 0.3 V (versus NHE) and its comparison with that of a polished iron surface (Fig. 9b). The ratio between the two currents was 8.7, but the roughness factor determined for similarly treated platinum surfaces was 1.15. Thus, the true roughness factor determined by this approach would be 8.7/1.15 = 7.6. Finally, the charge over the transient current arising from introducing the electrode into the solution at a potential of −0.4 V (versus NHE) was determined (Fig. 9c) and compared to that of a polished iron electrode. Taking the roughness factor of the latter as 1.15, the roughness factor obtained for the evaporated iron electrode was 11.4. The value of the roughness factor used in this work was thus 8.5 ± 2.6, which is the mean of the result obtained by the three methods.

The adsorption of thiourea and thiocyanate on an active iron electrode was studied using the radiotracer technique. These results were compared with results obtained from the FTIR technique. Good agreement was observed between the results from the two techniques.

The adsorption of thiourea fits a Bockris–Swinkels isotherm which represents a water displacement model. Thiourea was found to displace three water molecules.[36] The coverage–overpotential relation at constant concentration fits the Bockris–Gileadi–Müller (BGM) theory, which explains a parabolic type of adsorption.[36] Thiocyanate adsorption was found to fit a Bockris–Devanathan–Müller (BDM) isotherm.

6. ADSORBED AND ABSORBED HYDROGEN RELATIONSHIP

6.1. Without Additives

The results of the relation of adsorbed hydrogen to absorbed hydrogen in a borate buffer solution, pH 8.4, are presented in Fig. 10. Each of the points at which θ is related to C occurs at a different overpotential, and hence a rationalization of the relationship observed must involve considerations of how overpotential is related to both θ and C.

It is known that the degree of coverage of the surface of the metal with atomic hydrogen can be related to the fugacity of hydrogen within the metal, that is, the fugacity of the hydrogen in the microvoids. This relationship arises assuming that there is equilibrium between the surface and the hydrogen in the microvoids.

Thus, it was shown by Bockris and Subramanyan[37] that

$$\frac{\theta}{1-\theta} = \frac{\theta_{rev}}{1-\theta_{rev}} f_{H_2}^{1/2} \tag{11}$$

where f_{H_2} is the fugacity of hydrogen in the microvoids.

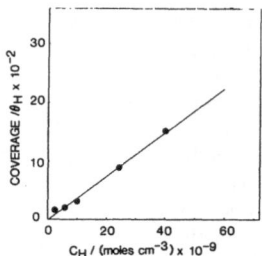

FIGURE 10. Coverage–concentration relation for hydrogen on iron.

Finally, a relationship can be seen between the fugacity of the hydrogen in the microvoid and the overpotential which is associated with coverage θ on the surface. This relationship is dependent upon the mechanism for the hydrogen evolution reaction, and these relationships of fugacity to overpotential for all the common mechanisms of hydrogen evolution have been discussed by Bockris and Subramanyan.

Assuming that the mechanism which we have here is given by the so-called coupled discharge–recombination mechanism, then the relation between fugacity and overpotential can be deduced as follows:

$$H^+ + e^- M \xrightarrow{k_1} M-H \tag{12}$$

$$M-H + M-H \xrightarrow{k_2} 2M + H_2 \tag{13}$$

At any other potential than V_{rev}, the following relationship holds for the Langmuir case:

$$2k_1 \cdots C_{H^+}(1 - \theta) \cdots \exp\left(-\frac{\alpha VF}{RT}\right) = k_2\theta^2 \tag{14}$$

or

$$\frac{\theta^2}{1 - \theta} = \frac{2k_1 C_{H^+}}{k_2} \exp\left(-\frac{\alpha VF}{RT}\right) \tag{15}$$

When $\theta \ll 1$,

$$\frac{\theta^2}{1 - \theta} = \frac{\theta^2}{(1 - \theta)^2} = \frac{2k_1 C_{H^+}}{k_2} \exp\left(-\frac{VF}{RT}\right) \tag{16}$$

The fugacity of H_2 in the cavity is given by Eq. (11) as

$$\left(\frac{\theta}{1 - \theta}\right)^2 = \left(\frac{\theta_{rev}}{1 - \theta_{rev}}\right)^2 f_{H_2} \tag{17}$$

At the equilibrium potential, since the forward reaction rates are no longer coupled, one cannot obtain θ_{rev} from Eq. (15) by letting $V \to V_{rev}$. However, $\theta_{rev}/(1 - \theta_{rev})$ can be obtained by considering the equilibrium state of either Eq. (12) or (13).

From Eq. (13),

$$k_2\theta_{rev}^2 = k_{-2}(1 - \theta_{rev})^2 P_{H_2} \tag{18}$$

Hence,

$$\left(\frac{\theta_{rev}}{1 - \theta_{rev}}\right)^2 = \frac{k_{-2}}{k_2} P_{H_2} \tag{19}$$

From Eqs. (11), (16), and (19),

$$f_{H_2} = \frac{2k_1 C_{H^+}}{k_{-2}P_{H_2}} - \exp\left(-\frac{VF}{RT}\right) \tag{20}$$

or

$$f_{H_2} = \frac{2k_1 C_{H^+} \exp\left(-\dfrac{\alpha V_{rev}F}{RT}\right) \exp\left(-\alpha \dfrac{\eta F}{RT}\right)}{k_{-2}P_{H_2}} \tag{21}$$

The quantity P_{H_2} represents the atmospheric pressure.

Thus, for the coupled discharge-recombination mechanism, the internal hydrogen fugacity, f_{H_2}, cannot be obtained without a knowledge of various constants. By making a reasonable approximation, it is possible to calculate the value of the internal fugacity in terms of overpotential. For example, one assumes that the discharge step is in equilibrium up to an overpotential of $\eta = RT/F$, thus assuming the discharge step at equilibrium (Eq. 12).

$$k_1 C_{H^+}(1 - \theta) \exp\left(-\frac{V_{rev}F}{RT}\right) = k_{-1}\theta \exp(1 - \alpha)\frac{V_{rev}F}{RT} \tag{22}$$

$$\frac{\theta}{1 - \theta} = \frac{k_1}{k_{-1}} C_{H^+} \exp-\left(\frac{V_{rev}F}{RT}\right) \tag{23}$$

Adding an overpotential of up to RT/F, Eq. (23) becomes

$$\frac{\theta_{lim}}{1 - \theta_{lim}} = \frac{k_1 C_{H^+}}{k_{-1}} \exp\left(-\frac{V_{rev}F}{RT}\right) \exp\left(-\frac{\eta F}{RT}\right) \tag{24}$$

Dividing Eq. (24) by Eq. (23) and squaring, one has

$$\left(\frac{\theta_{lim}}{1 - \theta_{lim}}\right)^2 = \left(\frac{\theta_{rev}}{1 - \theta_{rev}}\right)^2 \exp\left(-\frac{2\eta F}{RT}\right) \tag{25}$$

where

$$\theta_{lim} = \theta \qquad \text{for } \eta < \frac{RT}{F} \tag{26}$$

Reexpressing Eq. (16) as a function of η,

$$\frac{\theta^2}{(1 - \theta)^2} = \frac{2k_1 C_{H^+} \exp\left(-\dfrac{\alpha V_{rev}F}{RT}\right) \exp\left(-\dfrac{\alpha\eta F}{RT}\right)}{k_2} \tag{27}$$

$$= K_0 \exp\left(-\frac{\alpha\eta F}{RT}\right) \tag{28}$$

where

$$K_0 = \frac{2k_1 C_{H^+}}{k_2} \exp\left(-\frac{V_{rev}F}{RT}\right) \tag{29}$$

Retaking Eq. (28),

$$\left(\frac{\theta}{1 - \theta}\right)^2 = K_0 \exp\left(-\frac{\alpha\eta F}{RT}\right) \tag{30}$$

Solving for K_0,

$$K_0 = \frac{\left(\dfrac{\theta}{1 - \theta}\right)^2}{\exp\left(-\dfrac{\alpha\eta F}{RT}\right)} \tag{31}$$

From Eq. (25), one has

$$K_0 = \frac{\left(\dfrac{\theta}{1 - \theta_{\text{rev}}}\right)^2 \exp\left(-\dfrac{2\eta F}{RT}\right)}{\exp\left(-\dfrac{\alpha\eta F}{RT}\right)} \tag{32}$$

Rearranging and taking $\alpha = \frac{1}{2}$,

$$K_0 = \left(\frac{\theta_{\text{rev}}}{1 - \theta_{\text{rev}}}\right)^2 \exp\left(-\frac{3}{2}\frac{\eta F}{RT}\right) \tag{33}$$

At $\eta = -RT/F$, Eq. (33) becomes

$$K_0 = \left(\frac{\theta_{\text{rev}}}{1 - \theta_{\text{rev}}}\right)^2 \exp\left(-\frac{3}{2}\right) \tag{34}$$

Finally, substituting Eq. (34) into Eq. (28),

$$\left(\frac{\theta}{1 - \theta}\right)^2 = \left(\frac{\theta_{\text{rev}}}{1 - \theta_{\text{rev}}}\right)^2 \exp\left(-\frac{3}{2}\right) \exp\left(-\frac{\alpha\eta F}{RT}\right) \tag{35}$$

Hence, when $\eta > RT/F$, f_{H_2} is given by, from Eqs. (17) and (35),

$$f_{H_2} = \exp\left(-\frac{3}{2}\right) \exp\left(-\frac{\eta F}{2RT}\right) \tag{36}$$

Knowing that Sievert's law is[37] $C_H = K_S f_{H_2}^{1/2}$, Eq. (36) becomes

$$C_H = K_S \exp\left(-\frac{3}{4}\right) \exp\left(-\frac{\eta F}{4RT}\right) \tag{37}$$

This gives the relation between the concentration of hydrogen inside the metal, C_H, and the overpotential. Further, for a coupled discharge-recombination mechanism it is known that

$$\theta = \left[\frac{k_1}{k_2}\frac{C_{H^+}}{(k^1)^2}\exp\left(-\frac{\alpha\Delta\phi_e F}{RT}\right)\right]^{1/2} \exp\left(-\frac{\eta F}{RT}\right) \tag{38}$$

Substituting Eq. (38) into Eq. (37),

$$C_H = \frac{K_S e^{-3/4}}{\left[\dfrac{k_1}{k_2}\dfrac{C_{H_2}O}{(k^1)^2}\exp\left(-\dfrac{\alpha\Delta\phi_e F}{RT}\right)\right]^{1/2}}\theta \tag{39}$$

or

$$C_H = K\theta_H \tag{40}$$

where

$$K = \frac{K_S e^{-3/4}}{\left[\dfrac{k_1}{k_2}\dfrac{C_{H_2}O}{(k^1)^2}\exp\left(-\dfrac{\alpha\Delta\phi_e F}{RT}\right)\right]^{1/2}} \tag{41}$$

Thus, if $K_s = 3 \times 10^{-6}$ mol/cm^3, $k_1/k_2 = 0.01$, $C_{H_2} = 5.5 \times 10^{-10}$ mol/cm^2, $k^1 = 10^{-9}$ mol/cm^2, and $+\Delta\phi_e = 0.6$, one obtains $K = 2 \times 10^{-7}$ mol/cm^3, which is the experimental value.

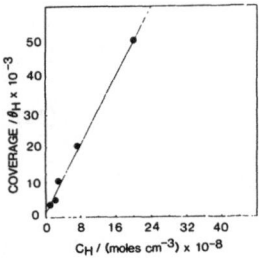

FIGURE 11. Coverage–concentration relation for hydrogen adsorbed on iron in the presence of thiocyanate.

6.2. Thiocyanate

A θ_H–C_H relationship is plotted in Fig. 11 for the solution containing thiocyanate. It can be seen that this relationship follows a straight line. As a result of a great promotional effect on hydrogen permeation, the concentration of hydrogen inside the metal is twice as much compared with that in the case of the pure electrolyte solution alone, despite the fact that the hydrogen coverage is one-third of that obtained for the pure electrolyte case. Thus, the presence of thiocyanate catalyzes the permeation.

Here again, Eq. (41) was applied to find the theoretical value for the constant. Thus, for a value of $k_1/k_2 = 10^{-4}$, one obtains $K = 2 \times 10^{-6}$ mol/cm^3, compared to the experimental value of 3.5×10^{-6} mol/cm^3. The value of $k_1/k_2 = 10^{-4}$ indicates that the discharge step is the rate-determining step in the hydrogen evolution reaction.

From the relation $K = C_H/\theta_H$, an increasing θ_H will give an increase in C_H. The above criterion is a pure chemical effect of concentration acting on the equilibrium between the surface and the bulk metal. With the electrochemical condition of fixed hydrogen evolution current density, there is a steady state at the interface. Hence, a chemical condition is present at the interface, which is not dependent on the overpotential at the fixed current density. The coverage of hydrogen produced on the electrode surface is increased with the overpotential to enhance the rate of the hydrogen evolution reaction. Thus, an ideal promoter is a substance which lowers the bond energy of hydrogen to the electrode surface. In this study thiocyanate and thiourea had a large promotional effect on the adsorption of hydrogen.

Adsorption in a "specific" way on the iron surface by these additives probably causes active sites to be heavily covered, thus giving a lowering of bond energy of adsorbed hydrogen so that the Langmuir adsorption model applies in this case.

7. EQUILIBRIUM CONSTANT

The rates of hydrogen adsorption and desorption for the steady-state condition

$$H \underset{k_{des}}{\overset{k_{ads}}{\rightleftharpoons}} H_{ads} \tag{42}$$

will give the equilibrium constant.

The rate of passage of adsorbed atomic hydrogen from the surface into the metal is given by

$$R_{diff} = k_{ads} 10^{-9} \theta \tag{43}$$

where k_{ds} is the rate constant (mol/cm^2 s) for the process, θ is the coverage, and 10^{-9} is a factor equal to the number of moles per square centimeter on a fully covered surface. The

TABLE 1.
Equilibrium Constant and Free Energy of Adsorption for
the Adsorption/Absorption Process

Overpotential (V)	K	G° (cal/mol)
−0.5	2.26×10^{-6}	7.8
−0.4	2.2×10^{-6}	7.82
−0.3	1.9×10^{-6}	7.88
−0.24	2.3×10^{-6}	7.77
−0.2	3.96×10^{-6}	7.46

rate of transfer of absorbed H to the surface from the layer just inside the metal is

$$k_{abs}C_H(1 - \theta)10^{-8} \tag{44}$$

where C_H is the concentration of absorbed hydrogen in moles per cubic centimeter of iron just inside the metal. At equilibrium the rates of the forward and the back reaction are equal; that is,

$$k_{ads}\theta 10^{-9} = k_{abs}C_H(1 - \theta)10^{-8} \tag{45}$$

$$K = \frac{k_{ads}}{k_{abs}} = \frac{C_H(1 - \theta)10^{-8}}{\theta 10^{-9}} \tag{46}$$

$$= 10\frac{C_H(1 - \theta)}{\theta} \tag{47}$$

Substituting values for the overpotential −0.5 V, one obtains an equilibrium constant of 2.26×10^{-6}, corresponding to a free energy for the process of $G = 7.8$ kcal/mol (32.6 kJ/mol). Table 1 presents the results for other overpotentials.

8. FUGACITY

Using Eq. (48) one can calculate the fugacity in a microvoid, for example, at −0.5 V overpotential:

$$f_{H_2} = \exp\left(\frac{3}{2}\frac{r\theta_{rev}}{RT}\right)\exp\left(-\frac{nF}{2RT}\right)\exp\left(-\frac{3}{2}\frac{r\theta}{RT}\right)$$

$$= \exp\left(\frac{3}{2}\frac{1.36 \times 10^3 \times 0.03}{8.31 \times 300}\right)\exp\left(+\frac{0.5 \times 96{,}484}{2 \times 8.31 \times 300}\right)\exp\left(-\frac{3}{2} \times \frac{1.36 \times 10^3 \times 0.15}{8.31 \times 300}\right)$$

$$= 1.433 \times 10^4 \text{ atm} \tag{48}$$

Table 2 presents the fugacities at other overpotentials. It is obvious that at an overpotential of −0.5 V, the hydrogen fugacity inside the metal can cause the spreading of the microcrack and consequently the embrittlement of the metal, despite the apparent low concentration inside the metal [at this overpotential (−0.5 V) the concentration of hydrogen is 4×10^{-8} mol/cm^3].

TABLE 2
Fugacity–Overpotential   Relationship
in Microvoids in the Metal

Overpotential (V)	f_{H_2} (atm)
−0.5	1.43×20^4
−0.4	1.97×10^3
−0.3	2.88×10^2
−0.24	9.1×10^1
−0.2	4.22×10^1

This low concentration inside the metal is compared with the hydrogen coverage outside the metal, where we found that there are $\sim 10^6$ atoms of hydrogen adsorbed on the surface per one atom of hydrogen absorbed in the metal. However, even at this low concentration the hydrogen inside the metal can cause catastrophic consequences such as the spreading of cracks and eventually the embrittlement of the metal.

The great increase of hydrogen absorbed in the metal in the presence of the two additives studied here will greatly increase the possibility of embrittlement. This may be considered a serious threat to the indiscriminate use of known "corrosion inhibitors" (such as thiourea) because certainly in the present studies thiourea and thiocyanate act as corrosion inhibitors, but at the same time they promote hydrogen permeation a great deal.

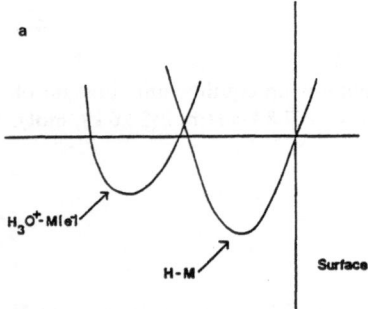

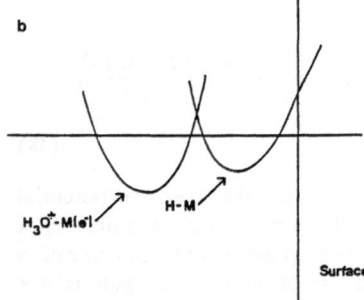

FIGURE 12. Energy sites in hydrogen adsorption: (a) high bond energy site; criteria: discharge favored, combination not favored, absorption not favored; (b) low bond energy site; criteria: discharge, combination favored, absorption favored.

A suitable inhibitor of the permeation of hydrogen is one that adsorbs on the low-energy sites of the surface and blocks them, leaving the high-energy sites for the hydrogen evolution reaction, stopping effectively the hydrogen permeation.

The mechanism of promotion of hydrogen permeation into the metal can be understood in terms of the additive being adsorbed on the high-energy sites of the metal surface, leaving the low-energy sites for the hydrogen evolution. This follows from the lowering of the exchange current density during hydrogen evolution (Fig. 12).

REFERENCES

1. B. Damaskin, A. O. Petrii, and V. Batrakar, *Adsorption of Organic Compounds on Electrodes*, Plenum Press, New York (1973).
2. D. C. Grahame, *Chem. Rev.* **41**, 441 (1947).
3. D. C. Grahame, *J. Am. Chem. Soc.* **63**, 1207 (1941).
4. R. Parsons, *Croat. Chem. Acta* **53**, 133 (1980).
5. G. Valette, A. Hamelin, and R. Parsons, *Z. Phys. Chem.* **113**, 71 (1978).
6. G. Valette and A. Hamelin, *J. Electroanal. Chem.* **45**, 301 (1973).
7. E. Gileadi, G. Stoner, and J. Bockris, *J. Electrochem. Soc.* **113**, 585 (1966).
8. A. Bard, and L. Fulkener, *Electrochemical Methods*, John Wiley & Sons, New York (1980).
9. E. Blomgreen and J. Bockris, *J. Phys. Chem.* **63**, 1975 (1959).
10. M. Breiter, in: *Modern Aspects of Electrochemistry*, No. 10 (B. Conway and J. O'M. Bockris, eds.), Plenum Press, New York (1975).
11. J. O'M. Bockris, *Mater. Sci. Eng.* **53**, 47 (1982).
12. R. Parson, *J. Electrochem. Soc.* **127**, 1766 (1980).
13. A. Wieckowski, E. Ghali, M. Szlarczyk, and J. Sobkowski, *Electrochem. Acta* **28**, 1627 (1983).
14. M. Devanathan, J. O'M. Bockris, and W. Mehl, *J. Electroanal. Chem.* **1**, 143 (1960).
15. H. Flitt and J. O'M. Bockris, *Int. J. Hydrogen Energy* **7**, 411 (1982).
16. A. Wieckowski, *J. Electrochem. Soc.* **122**, 252 (1975).
17. M. Habib and J. O'M. Bockris, *J. Electrochem. Soc.* **132**, 108 (1985).
18. M. Habib and J. O'M. Bockris, *J. Electroanal. Chem.* **180**, 287 (1984).
19. D. Barclay, *J. Electroanal. Chem.* **44**, 310 (1966).
20. J. de Boer, *Z. Phys. Chem.* **B18**, 49 (1932).
21. H. Leftin and M. Hobson, Jr., in: *Advances in Catalysis*, Vol. 14 (E. Eley, H. Pines, and P. Weisz, eds.), Academic Press, New York (1963).
22. S. Pons, *J. Electroanal. Chem.* **150**, 495 (1983).
23. S. Pons, T. Davison, A. Bewick, and P. Schmidt, *J. Electroanal. Chem.* **125**, 237 (1982).
24. H. Kaeszing and R. Saillant, *Chem. Rev.* **72**, 231 (1932).
25. W. A. Pliskin and R. P. Eischens, *Z. Phys. Chem. N.F.* **24**, 11 (1960).
26. A. M. Baro and W. Erley, *Surf. Sci.* **112**, L759 (1981).
27. S. Korseniewski, R. B. Shirts, and S. Pons, *J. Phys. Chem.* **89**, 2297 (1985).
28. J. O'M. Bockris, in: *Modern Aspects of Electrochemistry*, No. 1 (J. O'M. Bockris and B. Conway, eds.), Plenum Press, New York (1964).
29. V. Gold, *Trans. Faraday Soc.* **64**, 2270 (1967).
30. M. Enyo, in: *Comprehensive Treatise of Electrochemistry*, Vol. 7 (B. Conway, J. O'M. Bockris, E. Yeager, S. Khan, and R. White, eds.), Plenum Press, New York (1983).
31. M. Enyo, *Electrochim. Acta* **18**, 155 (1973).
32. H. Flitt and J. O'M. Bockris, *Int. J. Hydrogen Energy* **7**, 411 (1982).
33. E. Gileadi and B. Conway, in: *Modern Aspects of Electrochemistry*, No. 3 (J. O'M. Bockris and B. Conway, eds.), Butterworths, London (1964).
34. M. Devanathan and Stachurski, *Proc. Roy. Soc.* **A270**, 90 (1962).
35. W. Beck, J. O'M. Bockris, J. McBreen, and L. Nanis, *Proc. Roy. Soc.* **A290**, 220 (1966).
36. J. O'M. Bockris, B. Scharifker, and J. Carbajal, *Electrochim. Acta* **32**, 799 (1987).
37. J. O'M. Bockris and P. Subramanyan, *Electrochim. Acta* **16**, 2169 (1971).

20

Investigations on the Fabrication of Active Electrocatalysts for Methanol Electrooxidation

Michio Enyo, Ken-ichi Machida, Atsushi Fukuoka, and Masaru Ichikawa

1. INTRODUCTION

The fabrication of direct-type methanol fuel cells relies upon the development of high-performance electrodes such that the rest (open-circuit) potential should be as negative and the current as high as possible. The theoretical potential of methanol oxidation in acidic solution is 0.05 V (versus RHE), but the rest potential actually observed on a Pt electrode may be 0.4–0.5 V at room temperature. Due also to additional difficulties involved in the oxygen counter electrode, the working output voltage of methanol fuel cells at the present stage is only about 0.5 V, as compared with the theoretical value of 1.18 V.

Chemical modification of the electrode surface has been the subject of recent investigations directed toward the development of active electrocatalysts for methanol oxidation. In this work, we first attempted the attachment on C or ion-exchange membrane (M) substrates of Pt or Pt-based bimetallic clusters which were derived from organoplatinum complexes as precursors. It will be shown below that Pt clusters as small as Pt_9 or Pt_{15} are more active on a specific basis than dispersed Pt catalysts prepared by ordinary techniques. A second approach involved the utilization of gas-phase CO oxidation catalysts for surface modification of Pt electrodes, which might be effective in the electrooxidation of CO-like self-poisoning reaction intermediates on the electrode. Au-based Fe or Ru oxide catalysts were found to be effective in shifting the rest potential of Pt electrodes to more negative values, that is, closer to the theoretical value, as well as in maintaining the oxidation currents for a prolonged time.

2. Pt CLUSTER-SUPPORTED ELECTRODES

Platinum is the most active electrocatalyst for methanol electrooxidation. In practical cases, Pt is dispersed on electrically conductive substrates such as graphite to provide high

Michio Enyo, Ken-ichi Machida, Atsushi Fukuoka, and Masaru Ichikawa • Catalysis Research Center, Hokkaido University, Sapporo 060, Japan.

Electrochemistry in Transition, edited by Oliver J. Murphy *et al.* Plenum Press, New York, 1992.

surface area. Recently, organometallic compounds, such as $(NEt_4)_2[Pt_3(CO)_6]_n$ ($n = 2$–5), have been used as precursors to fabricate metal aggregates (clusters) smaller than 1 nm, which are dispersed homogeneously on supports such as SiO_2, MgO, and Al_2O_3.[1-10] The supported metal cluster catalysts exhibited differences and/or advantages with respect to conventional ones in terms of activity or selectivity for some catalytic reactions such as isomerization of hydrocarbons[9] and methanol synthesis from $CO + H_2$.[11]

2.1. Experimental

Platinum carbonyl cluster complexes, $Na_2[Pt_3(CO)_6]_n$ ($n = 3, 5$), and bimetallic ones, $[NMe_3(CH_2Ph)]_2[Pt_3Fe_3(CO)_{15}]$, $(NMe_4)_2[PtCl_2(SnCl_3)_2]$, and $(NMe_4)_4[Pt_3Sn_8Cl_{20}]$, were used as surface modifiers of graphite disks [4 mm (diameter) × 2 mm (thickness); Union Carbide] and anion-exchange resin membranes (5×5 mm^2; Toyo Soda). The surface modification of graphite[12,13] by attachment of platinum carbonyl cluster complexes was carried out with the use of a silanizer complex, dimethyl(octadecyl)[3-(trimethoxysilyl)propyl]ammonium chloride.

Graphite disks were thermally treated at 200°C in air for 36 h and refluxed with $LiAlH_4$ in anhydrous ether for 4 h, followed by successive rinses with ether, $1M$ HNO_3, and Millipore water. The disks were then refluxed in a toluene solution containing the silanizer for 50 h and rinsed thoroughly with dry toluene. The attachment of Pt clusters, Pt_{3n}, through their carbonyl complexes, on C disks was performed by anion exchange in a 1.5–3.0 mol % methanol solution. The attachment on the anion-exchange membranes was performed by dipping them into a methanol–THF solution containing the complex. No direct observation of the size of the clusters was made in this work; it has been reported earlier from transmission electron microscopic (TEM) observations[14,15] of Pt_{3n}/SiO_2 specimens that the platinum aggregates prepared are indeed of the size anticipated from the source complex. Auger electron spectra (AES) were recorded during the preparation of the electrodes, using a Nichiden-Anelva AES instrument.[16]

Electrical contact to the electrodes was made by holding them on Ta hairpin lead wires in the case of Pt_{3n}/C or placing them between gold meshes in the case of Pt_{3n}/M. The amount of platinum attached on the electrode surface was evaluated from the amount of electrochemical hydrogen adsorption. Most of the electrochemical measurements were carried out potentiodynamically (sweep rate, 10 mV/s) in $0.5M$ H_2SO_4 with or without addition of $1.0M$ CH_3OH at 30°C under a stream of Ar in a conventional three-compartment glass cell. The current density values per unit number of surface platinum atoms on the Pt_{3n}/C electrodes, observed after polarization at 0.60 V for 1 h, are summarized in Table 1, together with those for ordinary platinized Pt electrodes (Pt_∞) and carbon-supported Pt electrodes, Pt_x/C, prepared by impregnation with chloroplatinic acid.

2.2. Results and Discussion

2.2.1. Pt Cluster-Attached Electrodes

AES patterns recorded during the preparation of Pt_{15}-attached electrodes are shown in Fig. 1. For a graphite disk oxidized at 200°C in air for 36 h (Fig. 1a), a strong band attributed to C (273 eV) was observed but that of O (503 eV) was very weak. This indicates that the amount of hydroxyl or carboxyl groups introduced by the oxidation treatment must be relatively small. After modification with the organosilane reagent, two bands due to Si (92 eV) and Cl (180 eV) appeared in the spectrum although no signal from N was observed (Fig. 1b). The graphite disk after ion exchange with $[Pt_{15}(CO)_{30}]^{2-}$ anions gave an additional

TABLE 1

Electrocatalytic Activity of Pt Cluster-Attached and Conventional Pt Electrodes for Anodic Methanol Oxidation[a] $0.5M$ $H_2SO_4 + 1.0M$ CH_3OH, 30°C

Pt unit	Substrate[b]	Treatment	i^c $[\mu A/(10^{15}\ Pt)]$
Pt_9	C	None	0.32
		H_2, 150°C, 10h	4.50
		Air, 150°C, 10h	14.0
Pt_{15}	C	None	1.70
		H_2, 150°C, 10h	11.0
		Air, 150°C, 10h	28.8
Pt_{15}	M	None	29.2
		H_2, 200°C, 10h	38.5
$Pt_\infty{}^d$	—	None	1.5–4.0
$Pt_x{}^e$	C	H_2, 300°C, 10h	10.1

[a] Electrodes immersed in $0.5M$ $H_2SO_4 + 1.0M$ CH_3OH at 30°C.
[b] C, Graphite disks; M, anion-exchange membranes.
[c] Oxidation current at 0.60 V (versus RHE).
[d] Platinized Pt (roughness factor, 120).
[e] Prepared by impregnating hydrated graphite disk substrates in chloroplatinic acid. The size of the Pt particles is deduced to be 3–10 nm.

Auger signal from Pt (66 eV) but no signal from Na (Fig. 1c). After use for electrolysis, a signal from S (152 eV) was observed and that of Cl was lost (not shown): It appeared that Cl was dissolved out during the electrolysis, and SO_4^{2-} was adsorbed instead.

Typical cyclic voltammograms in $0.5M$ H_2SO_4 observed on Pt_9/C and Pt_{15}/C electrodes before and after thermal treatment at 150°C in an H_2 stream for 10 h are shown in Figs. 2A and 2B, respectively. All the samples (including Pt_{3n}/M electrodes, not shown) exhibited very similar broad bands over the range 0.05–0.40 V, which must be caused by the $H^+/H(ads)$ redox reaction. The shape of the band resembles that for evaporated amorphous-like Pt film electrodes,[17] but not that for crystalline Pt. The electrocatalytic activity toward anodic methanol oxidation of Pt_9/C and Pt_{15}/C electrodes before and after the thermal treatment is shown in Figs. 2C and 2D. These electrodes do not have a particularly high activity in their as-obtained form as compared with ordinary Pt, but often attained a higher activity after thermal treatment as described below.

The dependences of the rate of anodic methanol oxidation on Pt_{15}/C and Pt_{15}/M electrodes at 0.60 V upon the temperature of the thermal treatment are shown in Fig. 3. The electrocatalytic activity of Pt_{15}/C electrodes before the thermal treatment was about one order

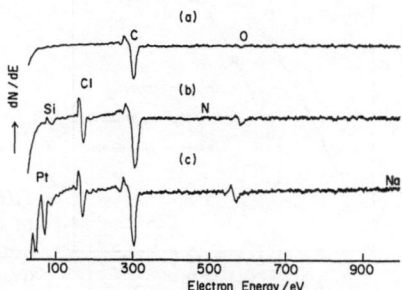

FIGURE 1. AES spectra of a graphite disk thermally treated in air at 200°C for 36 h (a), followed by treatment with $SiC_{26}H_{58}ClNO_3$ (b), and after modification with $[Pt_{15}(CO)_{30}]^{2-}$ anions (c).

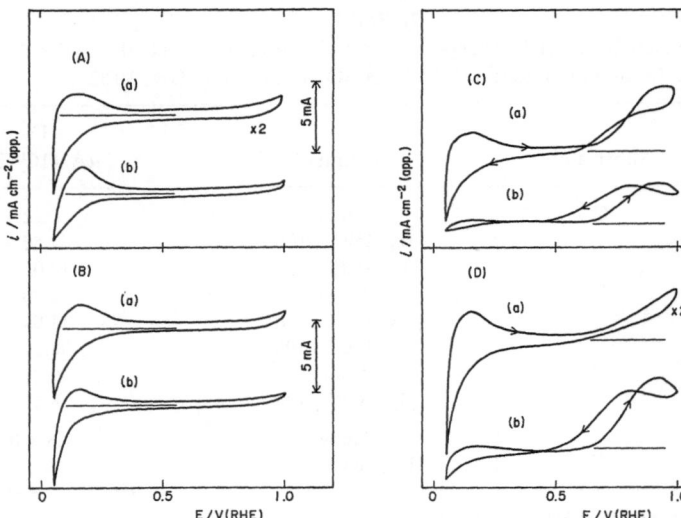

FIGURE 2. Cyclic voltammograms on Pt_9/C (A and C) and Pt_{15}/C (B and D) before (a) and after (b) thermal treatment in H_2 at 150°C for 10 h. (A) and (B) In $0.5M$ H_2SO_4, 30°C; (C) and (D) in $0.5M$ $H_2SO_4 + 1.0M$ CH_3OH, 30°C. Sweep rate, 10 mV/s.

of magnitude lower than that of Pt_{15}/M, but it was considerably enhanced with increasing temperature, eventually approaching that of the Pt_{15}/M electrode, whose activity was changed little by the thermal treatment. It is also seen that the thermal treatment was more effective in an oxidizing atmosphere (air) than in H_2. The oxidation currents observed on the Pt_{3n}/C and Pt_{15}/M electrodes are summarized in Table 1, together with those on an ordinary Pt electrode (Pt_∞) and on Pt_x/C electrodes. The Pt_9/C and Pt_{15}/C electrode after the thermal treatment, as well as the Pt_{15}/M electrodes, exhibited considerably higher activity as compared with that of Pt_∞ or Pt_x/C.

2.2.2. Pt-Based Bimetallic Cluster-Attached Electrodes and the Cluster Size Effect

The specific activity of Pt_{15}/C was higher than that of Pt_9/C (Table 1). To investigate the size effect in more detail, Pt-based bimetallic systems have been employed; these systems

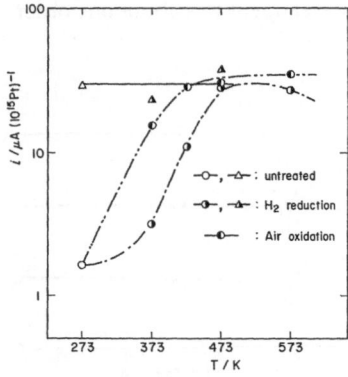

FIGURE 3. Specific electrocatalytic activity vesus temperature of thermal treatment of Pt_{15}/C and Pt_{15}/M electrodes. The ordinate shows specific rate of anodic methanol oxidation at 0.60 V versus RHE. In $0.5M$ $H_2SO_4 + 1.0M$ CH_3OH, 30°C.

have the advantage that metal aggregates containing smaller Pt cluster units than Pt_9 are obtainable. In the left-hand panel of Fig. 4, the AES spectrum of the Pt_3Fe_3/C electrode after thermal treatment at 150°C in air for 10 h (curve a) is shown, together with that of the same sample after use for electrochemical measurements in $0.5M$ H_2SO_4 (curve b). In (a), bands due to C (273), O (503), and Fe (593, 653, 705 eV) are observed, but the signal from Pt (66 eV) is not. In (b), after the electrochemical measurements, the signals of Pt and S appeared while those of Fe disappeared, thus suggesting the loss of Fe by leaching, leaving small-size Pt ensembles. It may, therefore, be concluded that the Pt_3Fe_3 cluster changed to Pt_3. The cyclic voltammograms of Pt_3Fe_3/C electrodes are shown on the right-hand side of Fig. 4. The redox peaks indicate that both the electrodes before (panel A) and after the thermal treatment (panel B) can adsorb H atoms. The methanol oxidation characteristics (broken curves) resembled those on conventional Pt electrodes, but with much smaller current values.

The left-hand panel of Fig. 5 shows AES signals from the Pt_3Sn_8/C electrode at various stages: (a) The electrode with $(Pt_3Sn_8Cl_{20})^{4-}$ attached showed the signals due to Pt, Cl, C, Sn, and O; (b) after thermal treatment at 150°C in air for 2 h followed by 10 h at 300°C in H_2, the loss of Cl and the surface enrichment in Sn relative to Pt are seen; and (c) use in the electrochemical system leads to the disappearance of Sn and appearance of S. Cyclic voltammograms on Pt_3Sn_8/C electrodes are shown on the right-hand side of Fig. 5, before (panel A) and after the thermal treatments (panel B). Before the treatments, the sample was somewhat active for the $H^+/H(a)$ redox reaction (curve a in panel A), but not for methanol oxidation (curve b in panel A). The activity for the hydrogen evolution was decreased after the thermal treatment (curve a in panel B), and the methanol oxidation current was also small (curve b in panel B). This probably means that Pt is covered with Sn, as suggested by an increased level of activity for methanol oxidation after an anodic polarization up to 1.50 V (curve C in panel B).

Figure 6 shows cyclic voltammograms on $PtSn_2/C$ electrodes before (panel A) and after similar thermal treatments (panel B). Analogously to the Pt_3Sn_8/C electrodes described above, the electrode before the thermal treatments was active for the H^+ reduction reaction (curve a in panel A) but not for the methanol oxidation (curve b in panel A). After the thermal treatments, the hydrogen current was almost lost (curve a in panel B), and the methanol

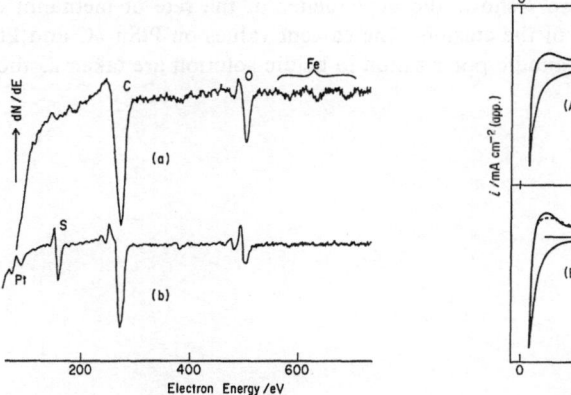
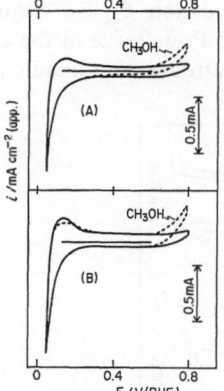

FIGURE 4. *Left*: AES signals from Pt_3Sn_3/C after thermal treatment in air at 150°C for 10 h (a) and after, in addition, several cycles of polarization in the range 0.05–1.50 V versus RHE (b). *Right*: Cyclic voltammograms before (A) and after the thermal treatment (B). In $0.5M$ H_2SO_4 or $0.5M$ $H_2SO_4 + 1.0M$ CH_3OH, 30°C.

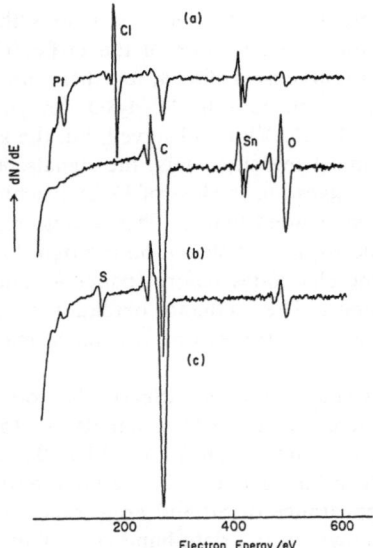

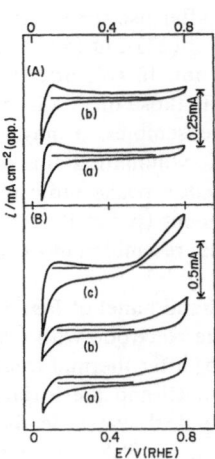

FIGURE 5. *Left*: AES signals from Pt$_3$Sn$_8$/C: (a) untreated; (b) thermally treated in air at 150°C for 2 h and then in H$_2$ at 300°C for 10 h; (c) after several cycles of polarization in the range 0.05–1.50 V versus RHE. *Right*: Cyclic voltammograms on Pt$_3$Sn$_8$/C before (A) and after the thermal treatment (B). In 0.5M H$_2$SO$_4$ (curve a) or 0.5*M* H$_2$SO$_4$ + 1.0*M* CH$_3$OH (curves b and c), 30°C. Curve c was obtained after polarization up to 1.50 V.

oxidation current also remained small (curve b in panel B). The anodic oxidation of methanol was not enhanced even after anodic polarization (curve c in panel B).

Summarizing the above, the anodic methanol oxidation currents on PtSn$_2$/C and Pt$_3$Sn$_8$ electrodes were negligible, but, after the leaching of Sn by the anodic polarization treatment, the Pt$_3$Sn$_8$ electrode showed a relatively high oxidation current of 0.51 μA for 10^{15} surface Pt atoms. In contrast, the current was still negligible on PtSn$_2$/C even after removal of Sn. These results suggest that there exists an optimum cluster size between Pt$_{15}$ and Pt$_\infty$ for the methanol oxidation. Figure 7 shows the dependence of the rate of methanol oxidation at 0.60 V on the Pt unit size of the clusters. The current values on PtSn$_2$/C and Pt$_3$Sn$_8$/C and Pt$_3$Fe$_3$/C electrodes after anodic polarization in acidic solution are taken as those for Pt/C

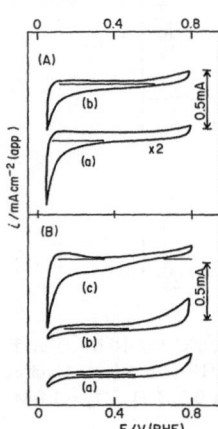

FIGURE 6. Cyclic voltammograms on PtSn$_2$/C before (A) and after thermal treatment in air at 150°C for 2 h and then in H$_2$ at 300°C for 10 h (B). In 0.5*M* H$_2$SO$_4$ (curve a) or 0.5*M* H$_2$SO$_4$ + 1.0*M* CH$_3$OH (curves b and c), 30°C. Curve c was obtained after polarization up to 1.50 V.

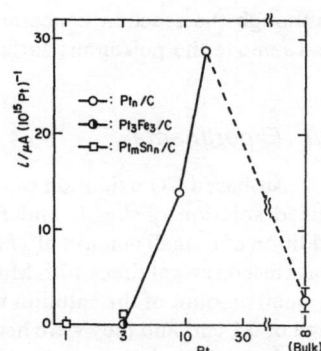

FIGURE 7. Platinum cluster size dependence of the specific electrocatalytic activity for the anodic oxidation of methanol on Pt-based cluster-attached electrodes. In $0.5M$ $H_2SO_4 + 1.0M$ CH_3OH, at 0.60 V, 30°C.

and Pt_3/C electrodes, respectively. As seen, the specific activity for the methanol oxidation greatly depends on the size of the clusters, and there seems to exist a maximum at or above Pt_{15}. It may be postulated that small Pt clusters such as Pt_3 do not provide a sufficient number of sites to accommodate a methanol molecule.

Takasu et al.[18] investigated dispersed platinum aggregates with particle sizes in the range 1.5–3.5 nm, prepared by vacuum evaporation, and reported that the specific electrocatalytic activity decreased monotonically with decreasing platinum particle size. The present investigation, on the other hand, has indicated that the platinum aggregates (Pt_9 and Pt_{15}) derived from platinum carbonyl cluster anions below 1 nm in size[19,20] are still active, or exhibit even higher than ordinary platinum as seen in Fig. 7, for the anodic methanol oxidation. Nevertheless, it may not be safe to compare the results of these experiments directly, as the methods of preparation are widely different; more investigations are probably needed before discussing details of the relation between the electrocatalytic activity and the particle size.

2.3. Conclusion

Platinum-based metal cluster-attached electrodes prepared by the use of organoplatinum cluster compounds provide broad $H^+/H(a)$ bands on cyclic voltammograms in the range 0.05–0.40 V versus RHE. The Pt_9- or Pt_{15}-attached electrodes show a high electrocatalytic activity for the anodic methanol oxidation, considerably greater than that of ordinary Pt electrodes. The optimum cluster size was suggested to be at or above Pt_{15}. This result may open the possibility of preparing highly active Pt electrodes having a uniform particle size of the order of 1 nm.

3. Pt ELECTRODES MODIFIED WITH CO OXIDATION CATALYSTS

It is widely recognized that the electrooxidation of methanol is retarded by the formation of CO-like reaction intermediates which cover the electrode surface, and their removal is difficult as electrooxidation requires high positive potentials such as 0.6 V versus RHE. On the other hand, a catalyst with use of ultra-fine Au particles (~50 Å), which exhibits very high activity toward CO oxidation in air, even in humid atmosphere and at a very low temperature, for example, −60°C, has been reported recently.[21] It is thus of interest to investigate the use of this catalyst as a surface modifier of the electrode so as to oxidize (although the reaction concerned with now is *electro*oxidation and not *oxygen* oxidation) and remove the poisonous surface CO-like species.

3.1. Experimental

Au-based CO oxidation catalysts were prepared by a coprecipitation method.[21] A dilute mixed solution of $AuCl_3$ and $FeCl_3$ (mixing ratio Au : Fe = 1 : 9) was precipitated by the addition of a small amount of $1M$ NaOH. After decantation of the supernatant, the precipitate was rinsed several times with Millipore pure water and was then suspended in propyl alcohol. A small amount of the solution was placed on 50-μm-thick Pt foils, with an apparent surface area of 0.1 cm^2 and they were heated in air at various temperatures for about 0.5 h. An Au–Ru mixed system (mixing ratio 1 : 1) was also prepared similarly.

Both CO and methanol oxidation reactions were tested mainly by a triangular voltammogram technique. A stress was placed on the rest (open-circuit) potential values, as such data would be important in developing electrodes which operate at potentials closer to the theoretical oxidation potential of methanol, and hence with improved energy conversion efficiency.

3.2. Results and Discussion

Figure 8 shows voltammograms observed for CO oxidation in $1M$ H_2SO_4 on the Pt electrode modified with the Au–Ru oxide catalyst, thermally treated in air at 400 or 900°C for 0.5 h. The electrode heated at 400°C (curve b) behaved not much differently from ordinary Pt electrodes (curve a). The rest potential, indicated by the arrow, was also practically unchanged. Rather disappointingly, the sharp peak current at 0.9 V, which is characteristic of CO oxidation, was also unchanged. On the other hand, the behavior of the modified electrode after thermal treatment at 900°C (curve c) was significantly different: (i) the rest potential (the solid arrow) was more negative by about 0.1 V as compared with that for Pt, (ii) the current peak at 0.9 V became less sharp, (iii) a broad wave of oxidation current at 0.3–0.9 V was significant, and (iv) the broad oxidation current tended to grow significantly if the electrode potential was maintained at 0.05 V for 30 s (broken curve). In this case, therefore, the electrode appeared to have an improved electrocatalytic activity toward CO

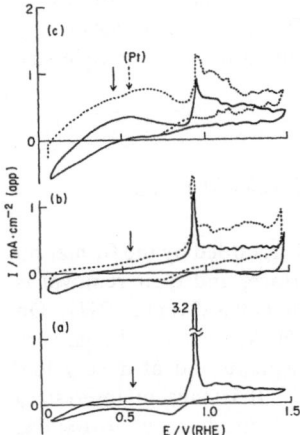

FIGURE 8. Electrooxidation of CO on a Pt electrode (a) and Pt modified with an Au–Ru oxide catalyst followed by thermal treatment at 400 (b) and 900°C (c). In $1M$ H_2SO_4 saturated with CO, 30°C.

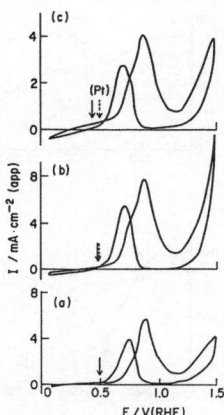

FIGURE 9. Electrooxidation of CH_3OH on a Pt electrode (a) Pt modified with an Au-Ru oxide catalyst followed by thermal treatment at 400 (b) and 900°C (c). In $1M$ $H_2SO_4 + 0.6M$ CH_3OH, 30°C.

oxidation. Analogous observations for methanol oxidation are shown in Fig. 9. Similarly as above, the electrode treated at 900°C revealed improved characteristics in terms of the rest potential, shifted by about 55 mV (the solid arrow on curve c), as compared with the other electrodes. The oxidation current value also was higher near the rest potential, but lower at high potentials.

Figure 10 shows the case of methanol oxidation, in alkaline solution, on the electrodes modified with Au-Fe and Au-Ru catalysts which were treated at 400 or 900°C. The plots in panel A were obtained with a potential scan range of 0.05–1.50 V, and those in panel B with a potential scan range of 0.05–0.60 V. Among the electrodes, the electrode modified with the Au-Fe catalyst and treated at 400°C showed the highest activity in terms of both the rest potential and the oxidation current. The results are in harmony with those reported in the

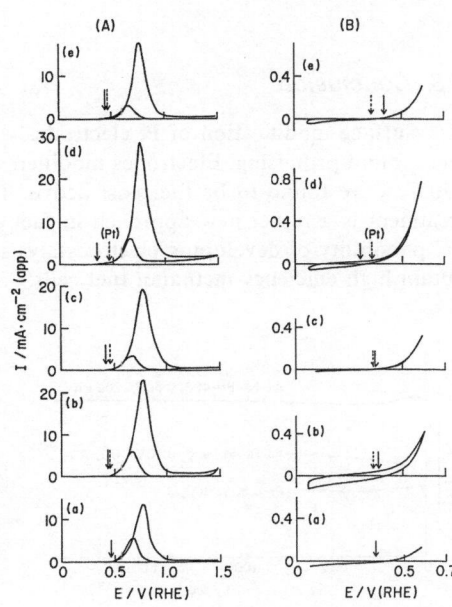

FIGURE 10. Electrooxidation of CH_3OH on Pt (a), on Pt modified with an Au-Ru oxide catalyst followed by thermal treatment at 400 (b) and 900°C (c), and on Pt modified with au Au-Fe oxide catalyst followed by thermal treatment at 400 (d) and 900°C (e). Potential sweep range: (A) 0.05–1.50 V, (B) 0.05–0.60 V. The solid arrows on the abscissa indicate the rest potential value in each system which may be compared with that for Pt (broken arrows). In $1M$ NaOH + $0.6M$ CH_3OH, 31°C.

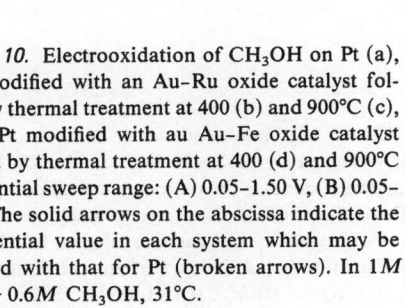

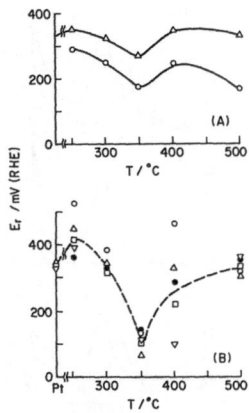

FIGURE 11. Effect of temperature of the thermal treatment of Au–Fe oxide-modified electrodes on the rest potential values in $1M$ NaOH + $0.6M$ CH_3OH (A) and $2M$ H_3PO_4 + $0.6M$ CH_3OH (B).

literature[21] in that the optimum treatment temperature of Au–Fe/SiO_2 catalysts for gas-phase CO oxidation was about 300°C.

Further experiments were carried out to examine the effect of the temperature of the thermal treatment of the Au–Fe catalyst system. In alkaline solution (Fig. 11A), the most negative rest potential value was observed on the electrode treated at 350°C and the value was, although time dependent as indicated by two sets of observations, sometimes as low as 0.20 V, which was significantly more negative than the rest potential of ordinary Pt electrodes, usually at >0.4 V. In acid solution (Fig 11B), an analogous tendency was seen, although the data were rather erratic. The most negative potential, sometimes below 0.1 V, was observed on the electrode treated at 350°C.

Typical cases of quasi-steady-state methanol oxidation currents are shown in Fig. 12. The oxidation current on the electrode treated at 400°C was already higher at 0.40 V than that on an ordinary Pt electrode at 0.50 V, or at the same potential of 0.50 V, the current was roughly five times higher.

3.3. Conclusion

Surface modification of Pt electrodes with Au-based gas-phase CO oxidation catalysts were found promising. Electrodes modified with Au–Fe oxide, thermally treated typically at 350°C, were found to be the most active. This way of fabricating electrodes with thermal treatment is a rather new approach in fuel cell studies. The observation above might open the possibility of developing electrocatalysts more active than electrodes available so far to obtain high-efficiency methanol fuel cells.

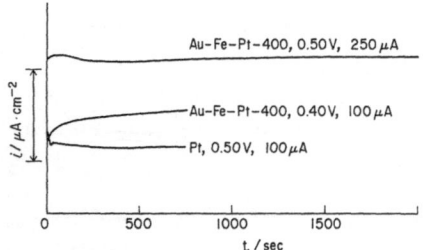

FIGURE 12. Time dependence of methanol oxidation currents. Au–Fe–Pt-400, 0.50 V, 250 μA, for example, denotes the oxidation current on the Au-Fe-oxide-modified Pt electrode which is thermally treated at 400°C, and data are taken at 0.50 V versus RHE with 250 μA for the current scale indicated. In $1M$ NaOH + $0.6M$ CH_3OH, 60°C.

REFERENCES

1. M. Ichikawa, in: *Tailored Metal Catalysts* (Y. Iwasawa, ed.), pp. 183–263, D. Reidel, Dordrecht (1986), and references cited therein.
2. B. C. Gates, in: *Metal Clusters in Catalysis* (B. C. Gates, L. Guczi, and H. Krözinger, eds.), pp. 415–423, Elsevier, Amsterdam (1986).
3. R. Psaro and R. Ugo, in *Metal Clusters in Catalysis* (B. C. Gates, L. Guczi, and H. Krözinger, eds.), pp. 427–496, Elsevier, Amsterdam (1986).
4. M. Ichikawa, in: *Homogeneous and Heterogeneous Catalysis* (Yu. I. Yermakov and V. Likholoba, eds.), pp. 819–835, VNU Science, Utrecht (1986).
5. M. Kaminsky, K. J. Yoon, G. L. Geoffroy, and M. A. Vannice, *J. Catal.* **91**, 338 (1985).
6. A. Choplin, M. Leconte, J.-M. Basset, S. G. Shore, and W.-L. Hsu, *J. Mol. Catal.* **21**, 389 (1983).
7. M. Ichikawa, *Chemtech* **1982** (12), 674.
8. L. Guczi, *Stud. Surf. Sci. Catal.* **29**, 547 (1986).
9. M. Ichikawa, *J. Chem. Soc., Chem. Commun.* **1976**, 11.
10. J. C. Clabrese, L. F. Dahl, P. Chini, G. Longoni, and S. Martinengo, *J. Am. Chem. Soc.* **96**, 2614 (1974).
11. M. Ichikawa and K. Shikakura, *Stud. Surf. Sci. Catal.* **7**, 925 (1981).
12. M. Fujihira, T. Matsue, and T. Osa, *Chem. Lett.* **1976**, 875.
13. M. Yamana, R. Darby, and R. E. White, *Electrochim. Acta* **29**, 329 (1984).
14. S. Iijima, K. Moriyama, A. Fukuoka, and M. Ichikawa, Proceedings of the Annual Meeting of the Chemical Society of Japan 1987, Abstract 1J09 (in Japanese).
15. S. Iijima and M. Ickikawa, *J. Catal.* **94**, 313 (1985).
16. AES Catalogue, Surface Analysis Division of Nichiden Anelva, Studio NOW, Tokyo, pp. 4–12 (1979).
17. K. Machida and M. Enyo, unpublished results.
18. Y. Takasu, Y. Fujii, and Y. Matsuda, *Bull. Chem. Soc. Jpn.* **59**, 3973 (1986).
19. V. S. Bagotzky and Yu. B. Vassilyev, *Electrochim. Acta* **12**, 1323 (1967).
20. B. D. McNicol, *J. Electroanal. Chem.* **118**, 71 (1981).
21. M. Haruta, F. Delannay, S. Iijima, and T. Kobayashi, *Shokubai* (*Catalysts*) **29**, 162 (1987) (in Japanese).

Some Perspectives on Electroorganic Chemistry

B. J. Piersma

1. ETHYLENE OXIDATION ON PLATINUM ALLOYS

A kinetic study of the anodic oxidation of ethylene was my introduction to modern electrochemistry in Philadelphia in the early 1960s. The electrochemical oxidation of hydrocarbons had just been demonstrated, and there was great enthusiasm for the development of fuel cells using organic fuels. Steady-state potentiostatic measurements of current as a function of ethylene partial pressure led us to conclude that on platinum the rate of oxidation decreased as the ethylene concentration was increased. We understood the negative pressure dependence in terms of competition for adsorption sites on the electrode surface, and the water discharge mechanism was proposed.[1] The ethylene work was followed by a study of several alkenes, and the electrode kinetics was well explained by the water discharge mechanism.[2] Interest in the development of fuel cells with hydrocarbon fuels lasted a relatively short time, primarily because of the poisoning of the electrode surface by species involved in the oxidation process. Recent publications[3-5] indicate that some interest still exists in the anodic oxidation of alkenes. Arvia and co-workers have suggested from non-steady-state measurements using cyclic voltammetry that water discharge is not the rate-limiting step in ethylene oxidation at platinum.[6,7] We have recently reexamined our early work and extended the study to include several titanium–platinum alloys (electrodes that were developed for our ongoing study of electrochemical processes in physiological systems).[8] We were not surprised to find that the negative ethylene partial pressure dependence was observed provided measurements were made at potentials negative to diffusion-limiting currents (see Fig. 1). Steady-state Tafel behavior was observed in $1 N$ H_2SO_4 at 80°C with Pt, Ti–5% Pt, Ti–10% Pt, Ti–20% Pt, and Ti–30% Pt anodes with ethylene partial pressures from 0.01 to 1 atm. An interesting correlation of electrocatalytic activity is presented in Fig. 2. We have also observed that Ti–Pt alloys have an increased resistance to corrosion and an increased electrocatalytic activity for ethylene oxidation with a maximum between 20% and 30% Pt.

2. ADSORBED INTERMEDIATES IN ORGANIC ELECTROOXIDATION

Several hundred papers have been published in the last 15 years on the adsorption and oxidation of simple organic molecules (CO, HCO, HCOOH, and CH_3OH). We have read

---

B. J. Piersma • Houghton College, Houghton, New York 14744.

Electrochemistry in Transition, edited by Oliver J. Murphy *et al.* Plenum Press, New York, 1992.

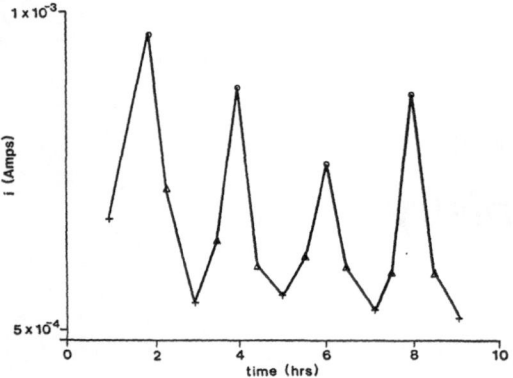

FIGURE 1. Dependence of anodic current on partial pressure for ethylene oxidation on Ti–30% Pt in $1N$ H_2SO_4 at 80°C. Measurements were made at 0.45 V (versus RHE). +, 1 atm ethylene; $\triangle$, 0.1 atm ethylene in nitrogen; $\bigcirc$, 0.01 atm ethylene in nitrogen.

most of these papers in the last few months and have formed some definite opinions. Before about 1980, most of the work done, using electrochemical techniques, particularly cyclic voltammetry, and the conclusions reached were largely qualitative; for example, see the review by Breiter in *Modern Aspects of Electrochemistry.*[9] The recent developments in spectroscopic techniques coupled with electrochemical measurements and the use of single-crystal electrodes have provided detail that we could only think about, even ten years ago. While the reaction mechanisms and identification of adsorbed intermediates have been subjected to excellent analyses by several groups, there are still enough questions remaining to make this a very interesting subject.

It is encouraging to note that many of the conclusions reached in earlier studies without the benefit of the spectroscopic techniques are consistent with the more detailed results. For example, in a study of formic acid and formate ion using high-current-density galvanostatic transient methods coupled with potentiostatic techniques we concluded in 1970 that two electrons per site are required to oxidize the species formed by adsorption of formic acid from $1M$ $HCOOH/1M$ H_2SO_4 at steady state, that the adsorbed residue formed under steady-state conditions is different from the one adsorbed after only short times, and that the irreversible dehydrogenation that occurs on adsorption is dependent on potential and on the presence of hydrogen.[10] We also concluded that the adsorbed species resulting from a

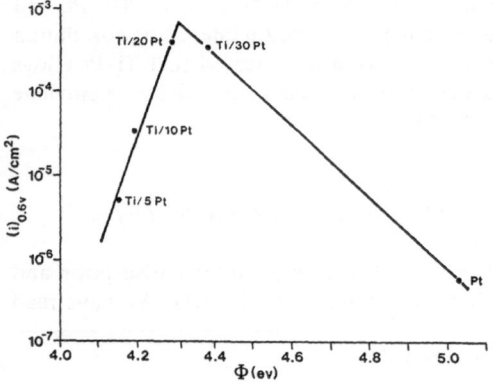

FIGURE 2. Correlation of current densities measured at 0.6 V for ethylene oxidation with electronic work functions for Pt and Ti–Pt alloys.

solution of $1M$ HCOONa could best be explained by the interaction of the free radicals resulting from dehydrogenation to form a complex multilayer polymeric-type structure.[10] Kita *et al.* have very recently suggested that of two different adsorbed intermediates formed in the anodic oxidation of HCOONa, one of them is an unreactive "polymer-like substance."[11]

An initial reading of the literature suggests that significant disagreement remains among results reported by different groups. Thus, the number of electrons per Pt atom site (eps) required for anodic oxidation of the adsorbed intermediate(s) formed from CO, HCOOH, or CH_3OH adsorption on platinum in acidic solution is reported to be 0.66, 1.0, 1.5, 2.0, 2.5, 3.0, and several other values between 1.0 and 2.0. Each of the (eps) values reported obviously requires a different adsorbed species or some combination of two or more adsorbed species. Thus, $(Pt)CO$, $(Pt)_2CO$, $(Pt)COOH$, $(Pt)_2CHO$, $(Pt)_3COH$, $(Pt)_3CO$, and $(Pt)_3(CO)_2$ have all been proposed in the recent literature. Since the work reported by Gilman on CO adsorption on platinum in 1963,[12] it has been generally assumed that adsorbed CO is present in both linear (one site) and bridged (two site) structures. Sobkowski and Czerwinski found that the bridged structure is formed preferentially at lower potentials (negative to 0.4 V versus RHE) and is more easily oxidized and at faster rates.[13] Positive to 0.4 V, the linear structure is preferred, and at about 0.6 V, only the linear structure remains on the surface. For adsorption of CO in the potential region where adsorbed H is present (negative to 0.2 V) the formate radical, $COOH_{ad}$, is thought to be present; for example, at 0.1 V the ratio of $COOH_{ad}$ to CO_{ad} is stated to be $3:2$.[13] The presence of a double peak for the potentiodynamic oxidation of CO in perchloric acid solution has been explained by the presence of the linear and bridged structures.[14] Lamy and co-workers found that on low-index platinum single crystals, both the linear and bridged structures are present when the adsorption time is short, but that the linear structure dominates at longer adsorption times.[15] They also observed that CO adsorption is only slightly dependent on crystal structure and that a third CO species may occur on the Pt(100) surface.

Kunimatsu *et al.* agreed that the linear structure is the dominant surface species for adsorbed CO and that the bridged structure is more easily oxidized.[16] They further suggested that oxidation of CO adsorbed at 0.05 V occurs randomly, but CO adsorbed at 0.4 V is oxidized only at the edges of islands of CO_{ad}. It was also found that CO adsorbed in the presence of adsorbed hydrogen is different from CO adsorbed in the double-layer region.[17] A somewhat different view has been very recently expressed by Toda and co-workers.[18] They suggested that unstable adsorbed CO, constituting about one-third of the adsorbed CO, has the linear structure and that about two-thirds of the surface is covered by a stable, IR-inactive structure, "adsorbed on two Pt atoms with the C=O axis parallel to the electrode surface and one of the Pt atoms bound to two CO molecules."

3. METHANOL ADSORPTION

The study of methanol adsorption and oxidation has received considerable attention so the discussion here is necessarily both selective and limited. Using cyclic voltammetry, Levia and Giordano concluded that the bridged structure of adsorbed CO is the main stable intermediate but that the linear structure and also $(Pt)_2COH$ are present, depending on electrode pretreatment and electrolyte composition.[19] They used perchloric acid and sulfuric acid as electrolytes. Léger and co-workers have demonstrated that methanol oxidation is structure dependent, with the Pt(100) plane being the most electroactive at steady state since it is not autopoisoned by an adsorbed intermediate.[20] The Pt(111) plane, which initially had the highest activity, was rapidly poisoned, and the Pt(110) surface was least active due to strong poisoning by an adsorbed intermediate, suggested to be (Pt)$_3$COH. A more recent study using electrochemically modulated infrared reflectance spectroscopy (EMIRS) led Léger and co-workers to conclude that the adsorbed intermediate species are also dependent on the methanol concentration.[21] For methanol concentrations less than $5 \times 10^{-3} M$, a species with a carbonyl stretch, consistent with (Pt)CHO, is observed. At higher methanol concentrations linearly bonded CO was the primary adsorbed species but for concentrations below $5 \times 10^{-3} M$, almost equal amounts of the linear and bridged CO species were observed. From a careful analysis of charge and coverage data with polycrystalline platinum, Léger and co-workers concluded that two electrons per site are required to oxidize the adsorbed intermediate species over a wide range of potentials (0.1 to 0.7 V) and that this species most likely has the linear CO structure.[22] They further suggested that the rate-limiting step for methanol oxidation is oxidation of this strongly adsorbed species. In contrast, for the Pt(100) surface three adsorbed species, linearly bonded CO, bridge-bonded CO, and (Pt)CHO, coexist at all methanol concentrations, and the relative amounts are potential dependent.[23] The Pt(111) plane also exhibits three different adsorbed intermediate species, and it is suggested that blocking and poisoning of the Pt surface is the result of lateral interactions between the adsorbed CO species.[24] According to Léger and co-workers,[24] "The more strongly blocked surface is the one which exhibits the strongest lateral attraction interactions between the two kinds of CO species." The Pt(100) surface, which has all three adsorbed species present, has attractive surface interactions between linearly adsorbed and bridge-adsorbed CO, resulting in surface blocking. Pt(110) is essentially covered with linearly adsorbed CO, and the repulsive interactions lead to poisoning and not blocking of the surface. Pt(111) is considered intermediate between these extremes.

From studies with differential electrochemical mass spectroscopy (DEMS), Heitbaum and co-workers have concluded that the strongly adsorbed intermediate during methanol electrooxidation is (Pt)$_3$COH but that this may be oxidized to CO in the presence of oxygen.[25] A determination of 3 ± 0.15 electrons per site required for oxidation of the adsorbed intermediate species and studies with deuterium-labeled methanol and D$_2$O give further "proof" that the strongly bound intermediate contains hydrogen.[26] Electrochemical desorption mass spectroscopy (ECTDMS) also has presented evidence that the adsorbate from methanol adsorption contains hydrogen.[27] (Pt)$_3$COH was considered to be the adsorbed intermediate from both methanol and formic acid adsorption. Additional measurements using surface normalized Fourier transform infrared spectroscopy (SNIFTIRS) by Vielstich *et al.* led them to conclude that (Pt)$_3$COH is a true intermediate in the anodic oxidation of methanol.[28] For methanol oxidation in sulfuric acid, Vielstich and co-workers have determined that the surface composition is dependent on the surface coverage and that the surface coverage is time and methanol concentration dependent.[29] They concluded that not more than 10% of the Pt surface is covered by CO species but that the (Pt)$_3$COH species may be geometrically prevented at very high coverage.

In contrast, Christov and co-workers suggested that the structure of the adsorbed species is strongly dependent on electrode pretreatment and that freshly platinized electrodes and electrodes subjected to activation procedures accumulate adsorbed CO during methanol oxidation.[30] The (Pt)$_3$C—OH species is postulated as the adsorbed species at aged platinized electrodes. Other primary alcohols undergo similar destructive chemisorption to yield adsorbed intermediates of the type (Pt)$_2$CROH, with some evidence for the (Pt)$_3$COH species as well.[31] Using EMIRS, Sun and Clavilier have studied methanol adsorption on 11 different platinum single-crystal planes.[32] They confidently stated that the number of electrons per site required to oxidize and desorb the poisoning intermediate is never greater than two and that values close to two are obtained for Pt(100) and Pt(111) surfaces. They found that the number of eps is influenced by the specific adsorption of anions. Finally, they concluded that linearly adsorbed CO is the primary species resulting from adsorption of both methanol and formic acid. A mixture of linear and bridged CO structures leads to eps values of less than two.

4. FORMIC ACID

Rach and Heitbaum have determined that the adsorbate formed from formic acid on platinum in the hydrogen region depends on electrode pretreatment.[33] Adsorbed CO was observed for roughened polycrystalline surfaces while a different species, probably HCO, was obtained on annealed platinum electrodes. Clavilier and Sun argued that three characteristic factors should be considered—the percentage of H sites blocked by the adsorbed species, the H site recovery factor after the adsorbed species is removed, and the number of electrons per site required for the desorption–oxidation of the adsorbed species.[34] From their analyses, they concluded that no species other than linear and bridge-adsorbed CO are present from HCOOH adsorption. For Pt(100) and Pt(111) surfaces, linearly adsorbed CO is the chemisorbed poison. On other Pt surfaces, a mixture of the linear and bridged-adsorbed species is found.

5. OXIDATION MECHANISMS

One of the first reaction mechanisms proposed for CO and methanol oxidation was the "reactant-pair" mechanism suggested by Gilman in 1964.[35] Adsorbed CO, either linear or bridged, reacted with adsorbed H_2O, forming an activated complex. The data obtained for methanol oxidation were considered consistent with this "reactant-pair" mechanism, and it was suggested that other simple organic molecules may react via the proposed activated complex. The presence of oxygen species (e.g., OH radicals) on the electrode surface at potentials negative to 0.9 V was not a consideration.

A study of formic acid oxidation at platinum single-crystal anodes clearly demonstrated the influence of surface structure on oxidation rate.[36] Reaction rates were found to decrease in the order Pt(100) > Pt(110) > Pt(111). Comparisons of cyclic voltammograms for CO, HCOOH, and CH_3OH oxidations led Léger and co-workers to conclude that CO and CH_3OH oxidation proceed via the adsorbed CO intermediate with the linear structure (and perhaps ten percent of the CO as the bridged structure) while HCOOH oxidation involves the (Pt)COOH intermediate.[37] Within the hydrogen region, a strong poisoning intermediate was produced. More recent work has shown that the nature of the adsorbed intermediate for methanol oxidation is concentration dependent, and low methanol concentrations (below $5 \times 10^{-3} M$) result in adsorption of (Pt)CHO.[21] The adsorbed species from methanol are also surface-structure and potential dependent as previously discussed.[23] A distinction has been made between poisoning of the surface, as in the case of Pt(110), for which the surface is essentially covered by linearly adsorbed CO, and blocking of the surface, as in the case of Pt(100), where attractive lateral surface interactions between linearly and bridged-adsorbed CO are operative.[24]

Using single potential alteration infrared spectroscopy (SPAIRS) with voltammetric techniques, Corrigan and Weaver have shown that from formic acid or methanol solution an essentially complete monolayer of adsorbed CO is present on a polycrystalline platinum surface over the potential range of about 0.05 to 0.55 V (versus NHE).[38] Kinetic data suggest that adsorbed CO is an intermediate for methanol oxidation with reaction occurring at the edges of islands of CO_{ad}, but that the adsorption kinetics are not fast enough to account for the oxidation rates of formic acid.[38] Thus, Corrigan and Weaver agree that adsorbed CO is an intermediate for methanol oxidation but acts as a poison for formic acid oxidation. A dual pathway then becomes necessary to explain formic acid oxidation on polycrystalline platinum.[38] Leung and Weaver have found that high coverages of adsorbed CO result from a wide variety of organic molecules when an α-H is present.[39] A lack of facile electrooxidation pathways for molecules in solution is thus related to the sluggish chemisorption kinetics of CO adsorption.[39] Removal of the adsorbed CO which covers the electrode surface is required before more weakly adsorbed, more reactive intermediate species can be formed.

Arvia and co-workers have reached similar conclusions, namely, "that the same kind of adsorbed residue is formed at the platinum surface from either reduced CO_2, formic acid, methanol or ehylene glycol."[40] Bilmes and Arvia have extended their discussions to suggest that linearly adsorbed CO reacts with weakly bonded adsorbed OH and bridged-adsorbed CO reacts with strongly bonded adsorbed OH and found no evidence for the conversion of bridged-adsorbed CO to linearly adsorbed CO.[41] From infrared adsorption studies, Kunimatsu also agreed that linearly adsorbed CO is the predominant surface species adsorbed from methanol and from formic acid.[42] Kunimatsu and Kita have presented a detailed discussion of the potential dependence of methanol and formic acid oxidations and of the poisoning effect of linearly adsorbed CO.[43]

There also exists a considerable amount of literature on the development of electrocatalysts, particularly bimetallic materials, which was presented in the preceding discussion and is therefore not included here.

In summary, while several new techniques such as EMIRS, SNIFTIRS, SPAIRS, DEMS, and ECTDMS have provided very interesting and important information, old questions remain. The nature of the species adsorbed on platinum from solutions of small organic molecules such as CH_3OH and HCOOH appears to depend on several factors, including at least the following:

1. The potential at which adsorption occurs, specifically whether it occurs in the hydrogen region (0 to 0.2 V versus NHE) or the double-layer region (0.2 to 0.8 V). This is probably related to the presence of H atoms on the platinum surface.
2. The solution concentration of the adsorbing species. This certainly accounts for some of the differences reported in the literature using different instrumental techniques.
3. The time allowed for adsorption. It is clear that the surface composition resulting from short adsorption times (on the order of a second or less) is different from that for longer adsorption times.
4. The structure of the platinum surface. One thing the newer techniques combined with studies at platinum single crystals have clearly demonstrated is that differences exist in the structure of species adsorbed at different crystal planes.
5. The influence of adsorbed anions. Competition for adsorption sites by anions has been observed to cause differences in the surface composition.
6. Surface pretreatment and aging. This can be a factor when extensive pretreatments have resulted in disruption of the platinum surface structure; for example, alternative cycling through hydrogen and oxygen evolution results in significant roughening of the surface. I suspect that in some cases, particularly with aging effects, the results are more likely from lack of system control (presence of impurities).

A majority opinion at this time (but not without a strong minority opinion) would be that (Pt)CO (linearly adsorbed CO) is at least a major component, if not the major component, on the surface when small organic molecules are adsorbed on platinum, and this is the species responsible for the deactivation of Pt anodes in organic oxidations. (Pt)CO appears to be an intermediate in methanol oxidation but may be a poison and not a true intermediate in formic acid oxidation.

6. ELECTROORGANIC OXIDATION IN ROOM TEMPERATURE MOLTEN SALTS

The Friedel–Crafts reaction using anhydrous aluminum chloride as catalyst has been the most important method for attaching alkyl side chains to aromatic rings.[44] The role of $AlCl_3$ is to abstract the halogen from the alkyl halide, generating a carbocation, which then acts as an electrophile attacking the aromatic ring. Alternatively, a "sigma complex" may form as an intermediate with the alkyl group being transferred in a single step from the halogen to the aromatic ring. In a recent publication,[45] a mechanism involving the formation of a "sigma complex" as the rate-limiting step was proposed for the acylation of benzene in acidic (0.60 and 0.67 mole fraction $AlCl_3$) MeEtImCl–$AlCl_3$ melts. That work found no acylation reactions in neutral (0.50 mole fraction $AlCl_3$) melt, and the initial reaction rates for substitution increased as the melt acidity increased.

The Lewis acid–base properties of the melt are described by

$$2AlCl_4^- \rightleftharpoons Al_2Cl_7^- + Cl^-$$

with an equilibrium constant on the order of 10^{-17} in the MeEtImCl–$AlCl_3$ melt. In neutral melt (0.50 mole fraction $AlCl_3$) the concentration of $Al_2Cl_7^-$ is negligible, but it increases with addition of $AlCl_3$. This led to the suggestion that $Al_2Cl_7^-$ was the catalyst responsible for promoting the acylation reaction.[45]

In earlier work, Koch et al.[46] were successful in electroinitiating Friedel–Crafts transalkylations with hexamethylbenzene in acidic (0.67 mole fraction $AlCl_3$) ethylpyridinium bromide–$AlCl_3$ melt. Luer and Bartak[47] studied triphenylchloromethane with acidic n-butylpyridinium chloride–$AlCl_3$ melts. They proposed the following for the formation and cathodic reduction of the triphenylmethyl carbocation:

$$Ph_3CCl + Al_2Cl_7^- \rightleftharpoons Ph_3C^+ + 2AlCl_4^-$$

$$Ph_3C^+ + e^- \rightleftharpoons Ph_3C$$

$$Ph_3C \xrightarrow{slow} products$$

The radical dimerizes to produce an oxidizable form of 1-(diphenylmethylene)-4-(triphenylmethyl)-2,5-cyclohexadiene which also isomerizes to [4-(diphenylmethyl)phenyl]triphenylmethane.

We have attempted to determine whether carbocations could be formed and Friedel–Crafts type reactions could be carried out in neutral melts. Several simple chloroalkanes were selected to look for differences in stability and formation rates of carbocations that could be detected electrochemically. We wanted to explore the alkylation and acylation mechanisms, particularly in neutral melt, and attempt to detect "sigma complexes" or other intermediates electrochemically.

The 1-methyl-3-ethylimidazolium chloroaluminate melts were prepared in a helium-filled glove box maintained at <2 ppm combined water and oxygen as previously described.[48] Exactly neutral (0.50 mole fraction of MeEtImCl and of $AlCl_3$) melts were prepared by adjusting the acidity (by addition of solid MeEtImCl or solid $AlCl_3$) until a maximum electrochemical window of about 4.6 V was obtained. (The electrochemical window, defined as the potential range in which essentially no oxidation or reduction of the melt occurs, was determined at a glassy carbon electrode at a sweep rate of 100 mV/s.) The chloroalkanes studied included four chlorobutane isomers and seven chloropentane isomers.

Representative cyclic voltammetric (cv) curves are illustrated in Fig. 3. In general, no reducible species were observed until the melts were made acidic, that is, until $Al_2Cl_7^-$ ions were present. Depending on melt acidity, these reduction peaks were observed with all the

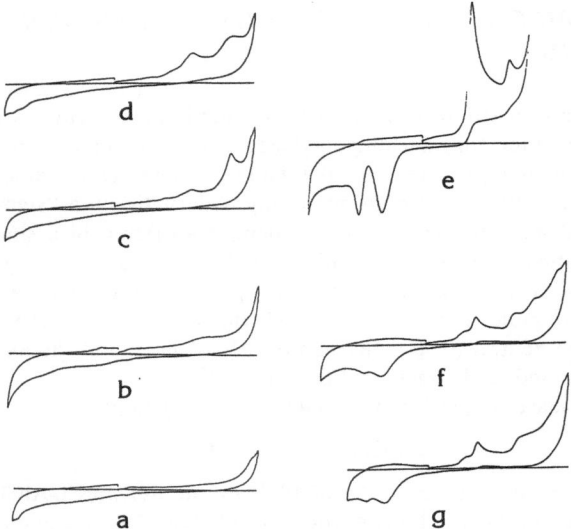

FIGURE 3. Cyclic voltammetric curves for 1-chloro-2-methylpropane and 2-chloro-2-methylpropane. The starting potential is +0.50 V versus an Al wire in 0.60 melt. Cathodic sweeps are to the right, anodic to the left; sweep rate is 100 mV/s. (a) 0.50 MeEtImCl–AlCl$_3$ melt; sweep limits are +2.40 to −1.90 V. (b) 1-Chloro-2-methylpropane in 0.5 melt; +2.4 to −1.9 V. (c) 1-chloro-2-methylpropane in slightly acidic melt; +2.4 to −1.9 V. (d) 1-chloro-2-methylpropane, melt acidity slightly greater than in (c); +2.4 to −1.9 V. (e) 2-chloro-2-methylpropane, slightly acidic melt; +2.3 to −1.9 V. (f) 2-chloro-2-methylpropane, 0.50 melt; +2.45 to −1.30 V. (g) Benzene added to solution in (e).

chlorobutane and chloropentane isomers. The chlorobutane isomers gave reduction peaks typically at −0.3, −0.9, and −1.5 V (versus an Al wire in 0.60 melt) while the chloropentane isomers yielded species with reduction peaks at −0.16, −0.45, and −0.91 V. Addition of an excess of benzene in each case resulted in disappearance of the reduction peaks from the CV curves. The area under the third (most negative) reduction peak increases with time and with melt acidity. Oxidation peaks were observed following the reduction cycle but were not seen without prior reduction. The cyclic voltammetric behavior observed with butyryl/chloride/benzene mixtures over a range of melt acidities shows the complexity and surprising sensitivity of electrochemical behavior for changes of 0.01 mole fraction in melt composition. While drawing conclusions is premature, some interesting considerations are relevant. The electrochemical behavior observed suggests that carbocations are formed and rapidly rearrange to give the more stable species within a given series of isomers. The rate and degree of rearrangement is dependent on melt acidity. The three reduction peaks may represent different carbocations or alternatively may indicate one species which interacts with the melt-forming complexes, for example, with AlCl$_4^-$ and Al$_2$Cl$_7^-$ ions. We believe that room temperature melts represents a new and very interesting class of nonaqueous solvents for electroorganic chemistry.

REFERENCES

1. H. Wroblowa, B. J. Piersma, and J. O'M. Bockris, *J. Electroanal. Chem.* **6,** 401 (1963).
2. J. O'M. Bockris, H. Wroblowa, E. Gileadi, and B. J. Piersma, *Trans. Faraday Soc.* **61,** 2531 (1965).
3. F. Maran, S. Roffia, and E. Vianello, *Electrochim. Acta* **30,** 613 (1985).

4. G. R. Stafford, *Electrochim. Acta* **32**, 1137 (1987).
5. S. M. Piovano, A. C. Chialvo, W. E. Triaca, and A. J. Arvia, *J. Appl. Electrochem.* **17**, 147 (1987).
6. W. E. Triaca, T. Rabockai, and A. J. Arvia, *J. Electrochem. Soc.* **126**, 218 (1979).
7. W. E. Triaca, A. M. Castro Luna, and A. J. Arvia, *J. Electrochem. Soc.* **127**, 827 (1980).
8. B. J. Piersma and W. Greatbatch, *Platinum Metals Rev.* **30**, 120 (1986).
9. M. W. Breiter, in: *Modern Aspects of Electrochemistry*, No. 10 (J. O'M. Bockris and B. E. Conway, eds.), Plenum Press, New York (1975).
10. S. Schuldiner and B. J. Piersma, *J. Phys. Chem.* **74**, 2823 (1970).
11. H. Kita, T. Katagiri, and K. Kunimatso, *J. Electroanal. Chem.* **220**, 125 (1987).
12. S. Gilman, *J. Phys. Chem.* **67**, 78 (1963).
13. J. Sobkowski and A. Czerwinski, *J. Phys. Chem.* **89**, 365 (1985).
14. S. A. Bilmes, N. R. DeTacconi, and A. J. Arvia, *J. Electroanal. Chem.* **164**, 129 (1984).
15. J. M. Léger, B. Beden, and C. Lamy, *J. Electroanal. Chem.* **170**, 305 (1984).
16. K. Kunimatsu, H. Seki, W. G. Golden, J. G. Gordon II, and M. R. Philpott, *Langmuir* **2**, 464 (1986).
17. K. Kunimatsu, W. G. Golden, H. Seki, and M. R. Philpott, *Langmuir* **1**, 245 (1985).
18. Y. Ikezawa, H. Sairo, H. Fugisawa, S. Tsuji, and G. Toda, *J. Electroanal. Chem.* **240**, 281 (1988).
19. E. P. M. Levia and M. C. Giordano, *J. Electroanal. Chem.* **158**, 115 (1983).
20. J. Clavilier, C. Lamy, and J.-M. Léger, *J. Electroanal. Chem.* **125**, 255 (1981).
21. B. Beden, F. Hahn, S. Juanto, C. Lamy, and J.-M. Léger, *J. Electroanal. Chem.* **225**, 215 (1987).
22. A. Papoutsis, J.-M. Léger, and C. Lamy, *J. Electroanal. Chem.* **234**, 315 (1987).
23. S. Juanto, B. Beden, F. Hahn, J.-M. Léger, and C. Lamy, *J. Electroanal. Chem.* **237**, 119 (1987).
24. B. Beden, S. Juanto, J.-M. Léger, and C. Lamy, *J. Electroanal. Chem.* **238**, 323 (1987).
25. J. Willsau, O. Wolter, and J. Heitbaum, *J. Electroanal. Chem.* **185**, 163 (1985).
26. J. Willsau and J. Heitbaum, *J. Electroanal. Chem.* **185**, 181 (1985).
27. S. Wilhelm, W. Vielstich, H. W. Buschmann, and T. Iwasita, *J. Electroanal. Chem.* **229**, 377 (1987).
28. W. Vielstich, P. A. Christensen, S. A. Weeks, and A. Hammett, *J. Electroanal. Chem.* **242**, 327 (1988).
29. S. Wilhelm, T. Iwasira, and W. Vielstich, *J. Electroanal. Chem.* **238**, 383 (1987).
30. S. N. Raicheva, E. I. Sokolova, and M. V. Christov, *J. Electroanal. Chem.* **175**, 167 (1984).
31. M. V. Christov and E. I. Sokolova, *J. Electroanal. Chem.* **175**, 183 (1984).
32. S. G. Sun and J. Clavilier, *J. Electroanal. Chem.* **236**, 95 (1987).
33. E. Rach and J. Heitbaum, *J. Electroanal. Chem.* **205**, 151 (1986).
34. J. Clavilier and S. G. Sun, *J. Electroanal. Chem.* **199**, 471 (1986).
35. S. Gilman, *J. Phys. Chem.* **68**, 70 (1964).
36. J. Clavilier, R. Parsons, R. Durand, C. Lamy, and J.-M. Léger, *J. Electroanal. Chem.* **124**, 321 (1981).
37. C. Lamy, J.-M. Léger, J. Clavilier, and R. Parsons, *J. Electroanal. Chem.* **150**, 71 (1983).
38. D. S. Corrigan and M. J. Weaver, *J. Electroanal. Chem.* **241**, 143 (1988).
39. L.-W. H. Leung and M. J. Weaver, *J. Electroanal. Chem.* **240**, 341 (1988).
40. E. P. M. Levia, E. Santos, R. M. Cervino, M. C. Giordano, and A. J. Arvia, *Electrochim. Acta* **30**, 1111 (1985).
41. S. A. Bilmes and A. J. Arvia, *J. Electroanal. Chem.* **198**, 137 (1986).
42. K. Kunimatsu, *J. Electroanal. Chem.* **213**, 149 (1986).
43. K. Kunimatsu and H. Kita, *J. Electroanal. Chem.* **218**, 155 (1987).
44. R. T. Morrison and R. N. Boyd, *Organic Chemistry*, 3rd ed., p. 378, Allyn and Bacon, Boston (1974).
45. J. A. Boon, J. A. Levisky, J. L. Pflug, and J. S. Wilkes, *J. Org. Chem.* **51**, 480 (1986).
46. V. R. Koch, L. L. Miller, and R. A. Osteryoung, *J. Am. Chem. Soc.* **98**, 5277 (1976).
47. G. D. Luer and D. E. Bartak, *J. Org. Chem.* **47**, 1238 (1982).
48. J. S. Wilkes, J. A. Levisky, R. A. Wilson, and C. L. Hussey, *Inorg. Chem.* **21**, 1263 (1982).

22

Electrochemical Reduction of Carbon Dioxide

Benedict Aurian-Blajeni

1. INTRODUCTION

The present work is intended as both an introduction to and an update of the subject of electrochemical reduction of carbon dioxide. By electrochemical reduction I mean processes in which electrically charged species take part; this type of processes includes photogenerated charge carriers, such as electron–hole pairs photogenerated in semiconductors or 19-electron reducing species.[1] It is an introduction in the sense that it tries to give a general idea of the problems associated with the process and indicates sources for further documentation. It is an update because many of the examples used were published after 1987. Material published prior to this year was summarized in excellent recent monographs on carbon dioxide utilization,[2] the catalytic activation of carbon dioxide in general,[3] or specifically the activation of carbon dioxide by metal complexes.[4]

Although carbon makes up only 0.087% of the weight of the crust of the Earth, it is the main component of the organic life, which includes the human species. We live in a world based on carbon, and the product of the decay of organic matter is ultimately carbon dioxide. Carbon dioxide is ubiquitous in nature, and the distribution of C in the various "-spheres" is shown in Fig. 1. Carbon dioxide is an intrinsic part of both inorganic and organic cycles, bridging the two. Carbon dioxide itself is difficult to assign to either inorganic or organic chemistry since it plays an important role in both. Life itself is based on reactions of carbon dioxide. Besides the biological role, carbon dioxide has an important social role. Prometheus was the first to teach man how to produce carbon dioxide artificially, and we have not stopped since. Carbon dioxide is put to good use but also has potentially devastating effects for human society. Starting with light and ending with carbonated water, many processes produce or use carbon dioxide. Carbon dioxide is used in "inert" form in the food and beverage industries (for carbonated beverages, deep freezing, or as a protective gas for sensitive foods), in agriculture (feeding in greenhouses), in enhanced oil recovery and in extractions with supercritical carbon dioxide. Carbon dioxide is also used as reagent in a number of syntheses:

This work is based on a presentation honoring the activity of John O'Mara Bockris. He taught me the difference between science and science fiction: in the former you dream first, while in the latter you base your dreams on science.

Benedict Aurian-Blajeni ● ChemLogic, Bellingham, Massachusetts 02019.

Electrochemistry in Transition, edited by Oliver J. Murphy *et al.* Plenum Press, New York, 1992.

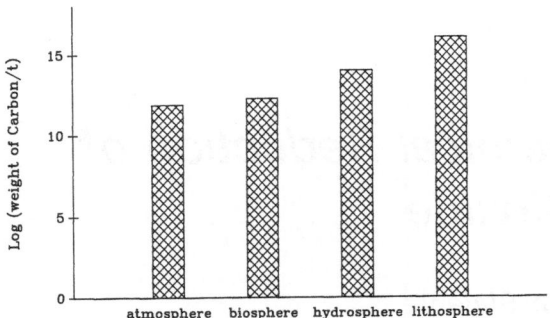

FIGURE 1. Distribution of carbon in nature.

Na_2CO_3, $BaCO_3$, urea, salicylic acid, methanol, propylene carbonate, and ethylene carbonate, to name but a few.[5,6] The removal of carbon dioxide by humans is however, so small that it has no room in a description of the natural cycles.[7]

Why would anybody want to reduce carbon dioxide? One good reason is the greenhouse effect. Humans produce carbon dioxide in low concentrations, not usable for industrial purposes; for example, the emissions of CO_2 from fossil fuel increased more than 25% between 1971 and 1986, from ca. 4.2 to ca. 5.4 tons of carbon/year.[8] The concentration of carbon dioxide in the atmosphere is steadily increasing, and the causes and the possible climatic consequences have been recently discussed.[9] The world, and by this I mean with vanity the human race, might be imperiled by the accumulation of carbon dioxide in the atmosphere. "The details of the natural carbon cycle and the future disposition of injected CO_2 are still unclear . . . Changes in atmospheric CO_2 concentration have the potential to influence the heat balance and an altered climate would influence how humanity secures its continuing welfare."[10] The predicted effects of increasing carbon dioxide concentration are strongly dependent on the model and the values of the parameters used.[11,12] The deleterious effects of CO_2 accumulation are controversial, but, if they are true . . .

Since the oil crisis a decade ago, society also became more aware of the continuous depletion of organic raw materials and of the energy sources based on fossil fuels. Reduction of carbon dioxide was considered both as a means for storing solar energy and as a process for producing chemicals necessary to the chemical industry, such as formaldehyde.

An immediate application for CO_2 reduction is the removal of CO_2 produced in closed spaces such as submarines or spaceships by biological activity or fuel cells based on fuels containing carbon.

The main fixation process of carbon dioxide on Earth is carried out by plants. The general reaction for natural photosynthesis is:

$$CO_2 + 2H_2A \xrightarrow[\text{biosystem}]{h\nu} CH_2O + A_2 + H_2O$$

where A is oxygen in the case of chlorophyll, and sulfur or organic acids in the case of bacteria. Under optimal conditions, the conversion of the electromagnetic energy to chemical energy stored as carbohydrates occurs with 36% efficiency and requires eight quanta/molecule of reduced CO_2. It must be noted, however, that in photosynthesis the reduction of CO_2 is a *dark* process: carbon dioxide is reduced by a photogenerated reductant (see, for example, Ref. 13). Much of the input of energy and matter is spent for the "maintenance" of the system, and the fixation itself is only a small part in a complex process. The carbon dioxide fixation itself is a multistep, multielectron, complex process. The point of this short digression into natural fixation of carbon dioxide is that much effort and ingenuity are required for taming the process into occurring under terrestrial conditions. Nature is a good teacher, and many catalytic systems try to minic natural processes.[4,14] A symbiosis exists between technological and biological processes, the former helping to understand and stimulate the latter, and the latter being used as guidelines for designing new catalysts.

2. ASPECTS OF CO_2 REDUCTION

There are different ways to reduce carbon dioxide, several of which are listed below. Examples are given of the various sources of energy used alone or in combination.

- Radiochemical reduction in aqueous media[15]

$$CO_2 \xrightarrow{\gamma\text{-radiation}} HCOOH, HCHO$$

- Chemical reduction by metals, which occurs at relatively high temperatures[16]

$$2Mg + CO_2 \rightarrow 2MgO + C$$

$$Sn + 2CO_2 \rightarrow SnO_2 + 2CO$$

$$2Na + 2CO_2 \rightarrow Na_2C_2O_4$$

- Thermochemically[17]

$$CO_2 \xrightarrow[T>900°C]{Ce^{4+}} CO + \tfrac{1}{2}O_2$$

- Photochemically[18,19]

$$CO_2 \xrightarrow{h\nu} CO, HCHO, HCOOH$$

- Electrochemically[20]

$$CO_2 + xe^- + xH^+ \xrightarrow{eV} CO, HCOOH, (COOH)_2$$

- Biochemically[21,22]

$$CO_2 + 4H_2 \xrightarrow{bacteria} CH_4 + 2H_2O$$

The bacteria *Methanobacterium thermoautotrophicum* can be immobilized in a fixed bed or on hollow fibers, and 80% of the theoretical yield is attained by feeding stoichiometric ratios for the reaction.
- Biophotochemically.[23,24] The "bio" part of the energy consists in catalysis and information content of an enzyme.

$$CO_2 + oxoglutaric\ acid \xrightarrow{h\nu} isocitric\ acid$$

In studies of this reaction,[23,24] the enzyme was isocitrate dehydrogenase, $Ru(bpy)_3^{2+}$ was used as photosensitizer, $d,-l$-dithiothreitol was the electron donor, and ferredoxin-$NADP^+$ reductase was included for recycling NADPH.
- Photoelectrochemically[25]

$$CO_2 + 2e^- + 2H^+ \xrightarrow[eV,\ semicond.]{h\nu} CO + H_2O$$

Other products than CO are possible, and their distribution and yields depend on many factors (see, for example, Ref. 26).

- Bioelectrochemically[27]

$$CO_2 + \text{oxoglutaric acid} \xrightarrow[\text{eV, methyl viologen}]{\text{enzyme}} \text{isocitric acid}$$

- Biophotoelectrochemically[28]

$$CO_2 \xrightarrow[\text{eV, methyl viologen}]{h\nu, \text{enzyme}, p\text{-InP}} HCOOH$$

when the enzyme is formate dehydrogenase.

Ironically, most of the above reactions require some form of activation, and most of the energy is taken nowadays from fossil fuels, that is, from reduced forms of carbon dioxide.

In the sequel the discussion will be confined to systems in which CO_2 is the only source of carbon. The various factors affecting the reduction of carbon dioxide will be examined. The strategies used to enhance the reduction of carbon dioxide are based on varying one or more of these factors; in much of the published work the advantage of using a particular combination of, for example, solvent, electrolyte, and electrodes is not clear, and the product analysis is not always complete (see Ref. 29). By "enhancement" I mean either improved yields, be they for energy or chemical conversion, or steering the reaction on a desired path resulting in a certain distribution of products. For this purpose, however, one must be acquainted with some of the properties of carbon dioxide relevant for its reduction.

2.1. Reaction Thermodynamics

Ideally, the equilibrium composition of a system is governed by thermodynamics. Carbon dioxide is very stable, as illustrated by its standard formation energy ($\Delta G° = -394.359$ kJ/mol[30]). Carbon dioxide is the most oxidized form of carbon, and therefore the only chemical transformation at "normal" energies would be to reduce it, that is, to transfer electrons *to* it. The electron affinity of CO_2 is -0.6 ± 0.2 eV.[31] Upon transfer of one electron, the structure changes from linear to bent,[32] which results in irreversible reduction. The products of the reduction of carbon dioxide are theoretically just two: carbon monoxide and carbon. However, even elemental carbon can appear in various forms, with different reactivity: graphite, amorphous, or vitreous, and diamond. The situation becomes very complicated when hydrogen is involved, and the number of possibilities is virtually unlimited if other elements, such as phosphorus, nitrogen, or sulfur, are part of the system. In Table 1, the free-energy changes and the standard redox potentials for several reactions of reduction of carbon dioxide at pH 0 are listed. The reduction potentials for various radicals of potential interest for the reduction of CO_2 obtained by thallium titration[33] are shown in Table 2.

On the basis of tabulated thermodynamic data, one may calculate the changes in redox potential with temperature in the range 0–200°C for C(graphite), CO, HCOOH, HCHO, CH_3OH, and CH_4 for the reaction:

$$CO_2(\text{aq un}) + me^- + mH^+ \to PROD + nH_2O(l)$$

where PROD is one of the substances listed. The thermodynamic values for the aqueous unionized (aq un) carbon dioxide were chosen because it is carbon dioxide, and not the carbonate ion(s), that is reduced.[34] There are also indications that the reaction is a reduction-protonation, not a hydrogenation.[35] The variation of the redox potential E with temperature was calculated according to the equation[36]:

$$E(T) = -\Delta G(T)/nF = [\Delta H(T) - T\Delta S(T)]/nF$$

TABLE 1
Free Energy of Reaction, at 25°C, and Standard Potentials for
$$a\mathrm{CO_2(aq\,un)} + ne^- + n\mathrm{H^+(aq\,un)} = \mathrm{PROD} + q\mathrm{H_2O(l)}^{a,b}$$

PROD	State	ΔG (kJ/mol)	$-\Delta G/Fn^c$ (V)
CO	g	11.68	−0.061
	aq un	28.95	−0.150
HCOOH	l	24.63	−0.128
	aq	13.68	−0.071
	aq ion	34.98	−0.181
C	Graphite	−88.28	0.229
HCHO	g	46.32	−0.120
	aq un	148.85	−0.386
CH₂OH	l	−17.42	0.03
	g	−13.11	0.023
	aq un	−26.46	0.046
CH₄	g	−139.00	0.180
	aq un	−122.61	0.159
(COOH)₂	aq ion 2	98.06	−0.508
	aq ion 1	73.62	−0.382
CH₃COOH	l	−92.20	0.119
	g	−76.30	0.099
	aq un	−98.76	0.128
	aq ion	−71.61	01.093
CH₃CHO	l	−67.55	0.07
	g	−68.29	0.071
C₂H₄	g	−108.41	0.094
	aq un	−95.20	0.082
C₂H₅OH	l	−114.21	0.099
	g	−107.92	0.093
	aq un	−121.07	0.105
(CH₃)₂O	g	−52.02	0.045
(CH₂OH)₂	l	−262.51	0.227
C₂H₆	g	−209.38	0.155
	aq un	−193.57	0.143
CO₂	aq		−1.84[d]
			−1.9[d-f]

[a] aq un, Aqueous not ionized.
[b] Refs. 124 and 125.
[c] F = Faraday's constant = 96,487 C.
[d] Ref. 126.
[e] This value refers to CO₂(g) as starting substance.
[f] Ref. 127.

TABLE 2

Thermodynamic Data for Some Radicals of Interest for CO_2 Reduction

$RO^{\cdot-}$	$E^0(RO/RO^{\cdot-})$	pK^a	$E^0(RO,H^+/ROH)^b$
$CH_2O^{\cdot-}$	-1.81	10.71	-1.18
$CH_3CHO^{\cdot-}$	-1.93	11.51	$-1.25-1.25$
$(CH_3)_2CO^{\cdot-}$	-2.1	12.03	-1.39
$CO_2^{\cdot-}$	-1.9	1.4	-1.82^c

^{a} Ref. 128.
^{b} $-E^0(RP,H^+/\dot{}ROH) = E^0(RO/RO^{\cdot-}) - 0.0592\ pK_{\cdot ROH}$.
^{c} Ref. 129.

where G is the Gibbs free energy, and the variations of the enthalpy (H) and entropy (S) with temperature (T) are approximated by the expressions[30]:

$$\Delta H(T) \approx \Delta H^0 + \Delta C_p^0 \Delta T$$

$$\Delta S(T) \approx \Delta S^0 + \Delta C_p^0 \Delta(\ln T)$$

where C_p is the specific heat at constant pressure, and the superscript ^{0} means standard values. Whenever the values for the aqueous form were not available, the values for the gas phase were chosen. The results of this approximate calculation of the redox potential are shown in Fig. 2. The point that this is intended to illustrate is that the temperature is one of the factors influencing the outcome of the reaction. For example, at low temperature the production of carbon monoxide is less favored electrochemically than that of formaldehyde, and the reverse is true at high temperature.

From Table 1, one can see that none of the reactions of reduction of carbon dioxide requires large amounts of energy, except the formation of the CO_2^- radical. The reduction reactions require transfer of an even number of electrons, and in a step-by-step reaction the CO_2^- radical would be formed. There are, however, reports suggesting multiple electron transfer to carbon dioxide.[18]

The reduction of carbon dioxide at high temperature was recently attempted.[37] The electrochemical reduction of CO_2 was performed in molten salts (chloride or carbonate eutectics). Advantage is taken of both the increased temperature and the large potential window of the eutectics [e.g., 3.6 V for (Li,K)Cl eutectic]. At 600°C, the reduction of CO_2 starts at -0.6 V versus Ag and reaches 12% efficiency at 800°C. Corrosion is, however, a severe problem and was considered responsible for observed faradaic yields in excess of 100%.

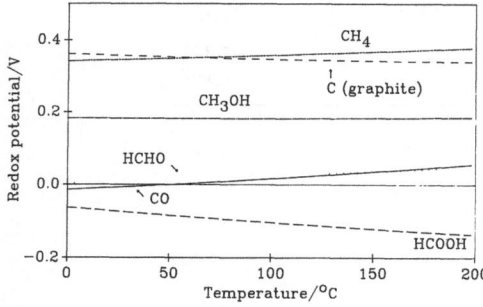

FIGURE 2. Change with temperature of the redox potential for reduction of CO_2 to C_1 species.

2.2. Solubility

Another equilibrium property that must be taken into consideration is the solubility of carbon dioxide in various solvents. For practical applications the high solubility in and diffusivity through plastics must also be considered; for example, the solubility of CO_2 in rubber is quite remarkable, ca. 70 ml CO_2/100 ml rubber (the example values depend on the quality of the rubber).[38] The solubility of CO_2 at 1 atm, expressed as the Bunsen coefficient, α, in water and ethanol obeys the equation:

$$\alpha = A_1 - A_2 t + A_3 t^2$$

where t is the temperature in °C, and A_1, A_2, and A_3 are coefficients dependent on the solvent: for water, $A_1 = 1.7326$, $A_2 = 0.066724$, and $A_3 = 0.0012394$; for ethanol, $A_1 = 4.3294$, $A_2 = 0.094261$, and $A_3 = 0.0012394$.

In Fig. 3, the solubility of carbon dioxide is plotted as a function of temperature for several solvents important for electrochemistry. In all cases the solubility decreases with increasing temperature, but for most organic solvents the solubility is much higher than in water, and unrelated to their dielectric constant. The reported solubility of CO_2 in dimethyl sulfoxide (DMSO) and acetonitrile is 4 times that in water, in propylene carbonate 8 times, and in dimethylformamide 20 times (the exact numbers differ; cf. reports of $0.04M$[39] with $0.14M$ for CO_2 in acetonitrile[40]).[41]

The chemical character of the solvent is important for determining the course and yield of the reaction. The work of Taniguchi et al.[25] showed that protophilic solvents, such as dimethylformamide or dimethyl sulfoxide, decrease the hydrogen evolution in the course of the competition between CO_2 and H^+ for electrons at illuminated semiconductor electrodes. Protophobic solvents, such as propylene carbonate or acetonitrile, favor the hydrogen evolution during the reaction. This is illustrated by the fact that the catalytic reduction of CO_2 to formic acid occurs on modified electrodes in dimethylformamide, but not in acetonitrile.[42]

The solubility is strongly dependent on pressure, and it can be approximated by Henry's law, as seen in Fig. 4. It can be seen from the figure that the proportionality constant depends on temperature and is different for different solvents. That means that in some instances a reversal in the ordering of solvents according to CO_2 solubility can occur.

The effect of pressure on solubility is felt strongly by the electrochemical potential. Attempts to use this effect were made on semiconductor electrodes[43]; the faradaic efficiency was low and decreased with current density. Recent results demonstrate the use of gas-diffusion electrodes for high-rate electrochemical reduction of carbon dioxide[44] in an attempt to increase current density by increasing local pressure, and the current densities for formate production reached 50 mA/cm^2 on Pb impregnated electrodes.[45]

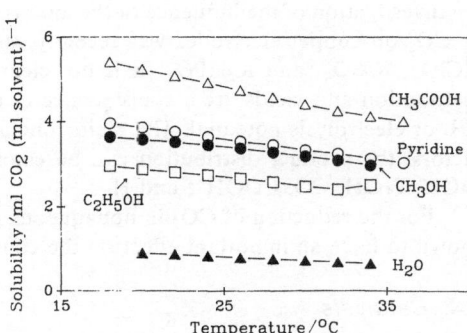

FIGURE 3. Variation of the solubility of CO_2 with temperature for several solvents used in electrochemistry.

† Milliliters of CO_2 at normal temperature and pressure dissolved at 760 mm Hg partial gas pressure.

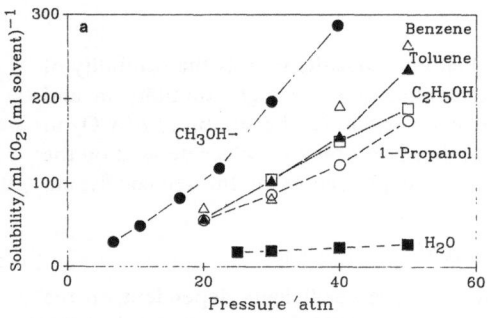 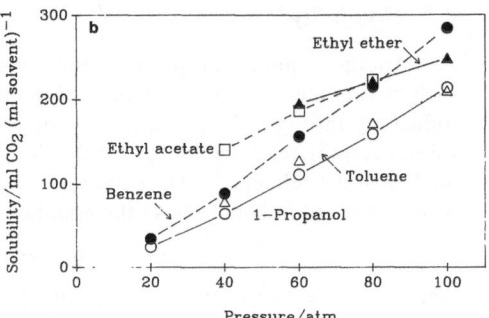

FIGURE 4. Variation of solubility of CO_2 with pressure for several solvents at $T = 20$ (a) and 60°C (b).

For electrochemical applications, moreover, the stability domain of the solvent is important and must be correlated with the properties of the particular electrodes and supporting electrolyte used (see, for example, Ref. 46).

2.3. Electrolytes

The supporting electrolytes influence both the solubility *and* the mechanism of reaction in the case of reduction of carbon dioxide. The exact mechanism of dissolution of CO_2 is not fully understood. For example, the solubility of CO_2 in H_2SO_4 solutions goes through a minimum at a concentration of ca. $38m$, while in $HClO_4$ it goes through a maximum at a concentration of ca. $10m$.[47] The solubility of CO_2 in $Fe(OH)_2$ and Prussian Blue solutions increases over that in pure water.[48] This might be the explanation of increased current efficiencies for CO_2 conversion at illuminated *p*-silicon electrodes modified with Prussian Blue[49] or in more complex systems using the approach of homogeneous/heterogeneous catalysis.[50,51] Among common salts, solutions of $NH_4HB_2O_4$, $Na_4B_4O_7$, $NaBO_2$, $Na_3PO_4 \cdot 12H_2O$, and $Na_4P_2O_7 \cdot 10H_2O$ dissolve more CO_2 than pure water.[48]

Although the effect of solutes on the solubility of carbon dioxide is known and moderately used, their effect as "catalysts" has been less used. Taniguchi *et al.* showed that ammonium ions have a catalytic effect for the reduction of carbon dioxide on semiconductor electrodes in nonaqueous media.[25] In spite of electrode corrosion, the effect of electrolytes on current efficiency for the CO_2 reduction at semiconductor electrodes and product distribution was studied also in aqueous media.[52] It was shown that the alkali carbonates favored formation of formate. Tetraethylammonium perchlorate suppressed H_2 evolution better than other tetraalkylammonium perchlorates and better than alkali carbonates, phosphates, or sulfates. An investigation of the influence of the anions and electrolyte concentration on the reduction of CO_2 on copper electrodes was recently carried out.[53] The salts investigated were KCl, $KClO_4$, K_2SO_4, and K_2HPO_4. It is not clear whether the differences observed in product distribution and yields are a consequence of changes in the composition of the electrolyte, pH, or electrolysis potential. The main conclusion is that by changing one or more of these factors, the product distribution can be changed. The products reported are CH_4, C_2H_4, EtOH, PrOH, CO, $COOH^-$, and H_2,

For the reduction of CO_2 in nonaqueous media, water itself is an electrolyte, and it was shown to have an important effect on the course of the reaction.[25,54]

2.4. Catalysis

The reduction of carbon dioxide necessitates a source of energy. This energy can be thermal, luminous, nuclear, chemical, or electrical or combinations thereof. Some forms of energy require at least one carrier or mediator that will convey the energy to carbon dioxide. For example, in the case of visible and near-UV light, mediators are photosensitizers or semiconductors; in the case of chemical energy, there are biosystems that result in methanation.

As seen in Table 1, none of the reactions above requires a huge amount of energy, nor do any of these reactions have a high standard potential. Some are even favored over hydrogen evolution.The problem consists in the Manichaean character of the reduction of carbon dioxide. On the one hand, carbon dioxide competes with hydrogen for electrons, and, on the other hand, most of the reduction reactions require protons. Therefore, the problem is twofold: to transfer the electrons to carbon dioxide instead of protons, and, after convincing them not to go to the protons, to labilize carbon dioxide sufficiently to receive them. In other words, the reduction of carbon dioxide can be catalyzed either positively, in the sense that the overpotential is reduced, and/or the currents increased, or it can be catalyzed negatively, in the sense that competing reactions are discouraged.

As is often the case in electrochemistry, the nature of the electrodes plays an important role in the system. This is particularly so when semiconductor electrodes are used in photoelectrochemical systems.

2.4.1. Metals

The reaction of electroreduction of carbon dioxide implies both protons *and* electrons, and therefore hydrogen evolution must be discouraged. One of the ways of doing this is to use electrodes with high hydrogen overpotential. An alternative approach was recently proposed in which the two processes are separated by using a Pd membrane that acts as a hydrogen reservoir.[59]

The nature of the metal affects not only the yield, but also the distribution of products. A recent example is the selective formation of formic acid, oxalic acid, and carbon monoxide.[56] The electrocatalysis of carbon dioxide reduction at metallic electrodes in aqueous and nonaqueous solutions was investigated, and the results are summarized in Table 3. Table 3 illustrates also the importance of the solvent on the product distribution.

Ruthenium is a known catalyst for reduction of CO_2 to methane.[57] Ruthenium electrodes are also active in CO_2 reduction.[58,59] The only product reported in a recent study is methane, at a faradaic efficiency of up to 27% at 80°C and 0.289 mA/cm^2, compared to a previous report from the same laboratory of CO and CH_3OH produced in a similar system.[60]

A breakthrough in the use of metallic electrodes occurred recently with the introduction of Cu or Cu-coated electrodes[61,62] for reduction of CO_2 to hydrocarbons in aqueous media

TABLE 3
Influence of the Solvent and Electrode on the Reaction Mechanism

Reaction	Cathode	Solution
$CO_2 + e^- \rightarrow CO_2^-$	All	All
$CO_2^- + H^+ + e^- \rightarrow HCOO^-$	In, Pb, Hg	H_2O
$CO_2^- \rightarrow CO^\cdot + O^{\cdot-}$ $CO^\cdot + O^{\cdot-} + H^+ + e^- \rightarrow CO + OH^-$	Zn, Au, Ag	H_2O
$CO_2^- + CO_2^- \rightarrow (COO)_2^{-2}$	Pb, Tl, Hg	Nonaqueous
$CO_2^- + CO_2 + e^- \rightarrow CO + CO_3^{2-}$	In, Zn, Sn, Au	Nonaqueous

at low temperature and high current density. It was previously known that the reduction of gaseous CO_2 by hydrogen to methanol is catalyzed by copper supported on zinc oxide and aluminium oxide (see Ref. 5 and references cited in Ref. 63). The products reported for the reduction of CO_2 in gaseous phase include superior alcohols. In solution, though, hydrocarbons were produced at high efficiency. Enhanced formation of ethylene and alcohols was observed at ambient temperature and pressure in electrochemical reduction of carbon dioxide at copper electrodes.[64] Faradaic efficiencies of the order of 100% in almost neutral solutions were reported. Products observed were CH_4 (max. 17%), C_2H_4 (max. 48%), C_2H_5OH (max. 22%), n-C_3H_7OH (max. 4%), CO (max. 3%), and $HCOO^-$ (max. 32%). The selectivity was dependent on the electrolyte, and the current density was $5 \, mA/cm^2$. The electrochemical reduction of carbon dioxide to methane at high current densities was shown to occur at Cu electrodes less pure than those initially used.[65,66] It was shown that Cu deposited electrolytically on glassy C is electrocatalytically active as long as the Cu deposit is fresh, the products containing a high percent of CH_4 and C_2H_4.[67] The high methane yields previously reported were reconfirmed, and the importance of the pretreatment of the electrode was demonstrated: cleaning the electrode with HCl results in better conversion than cleaning in HNO_3.[68] A mechanism specific for reduction of carbon dioxide on copper electrodes was recently published, suggesting that the reduction of CO_2 to $=CH_2$ occurs in the adsorbed state.[69] As previously observed, poisoning of the electrode occurs and is associated with the appearance of a black deposit, assumed to be carbon, on the surface. The copper-catalyzed system was also applied for the reduction of gaseous CO_2 and CO at ambient temperature to yield hydrocarbons: CH_4, C_2H_4, and C_2H_6.[70]

Equally important is the observation that the reduction of CO_2 to CO occurs at Au electrodes in $0.5 M$ $KHCO_3$ at 18°C.[71] The reaction starts at -0.8 versus NHE, and the faradaic efficiency is 91% at $-1.1 \, V$ at high current density ($3.7 \, mA/cm^2$ partial current). Other products detected were H_2 and $HCOO^-$. The redox potential reported was $E°(CO_2/CO) = -0.52 \, V$ versus NHE at 25°C and pH 7.

Metals are important in other roles than as electrodes. Reduction of carbon dioxide was studied at p-GaP photocathodes with metal deposits in solutions of tetralklylammonium salts in propylene carbonate.[72] Au, In, and Pb increased the photocurrents and stability, while Zn had little effect. The faradaic efficiency reported for reduction of CO_2 to CO was 50% on bare p-GaP, the balance being distributed between $(COOH)_2$, HCOOH, and H_2. The product distribution depends on the metal deposited: Pb resulted in more oxalic acid, while on the rest of the metals CO was almost exclusively produced.

Illuminated SiC and ZnSe with metals on the surface reduce CO_2 to methanol and ethanol.[73,74] The salts used for surface treatment were $Pb(NO_3)_2$, $Cu(NO_3)_2$, $PdCl_2$, $FeCl_2$, and $PtCl_4$. As in all systems using suspended powders, the conversion efficiency is very small, even though all the metals but Pb improved it. The products were HCOOH, HCHO, MeOH, MeCHO, and EtOH. Platinized TiO_2, with or without other metal additives, illuminated in a solution of Na_2CO_3 resulted in elemental carbon and HCOOH.[75]

Copper proved to have catalytic properties for the reduction of carbon dioxide in photoelectrochemical systems as well.[76] The reaction products resulting from the photoassisted reduction of CO_2 on p-SiC were CH_4, C_2H_4, and C_2H_6. The reaction has an optimum pH of 5.

Reduction of CO_2 to hydrocarbons (methane, ethane, and ethylene) is reported to be catalyzed by Ru and Os colloids in illuminated aqueous solutions.[35] N,N'-Dialkyl-2,2'-bipyridinium electron relays and Ru(II) tris(bipyridine) or Ru(II) tris(bipyrazine) were used as photosensitizers. An important conclusion is that catalyst-relay mixtures can be engineered so as to attain the desired selectivity of reaction. It is also suggested that the reduction is a protonation-electron transfer reaction, not hydrogenation, which implies the necessity of an electron relay.

2.4.2. Semiconductors

Semiconductors occupy a special place in the electrochemical reduction of carbon dioxide, due to the possibility of using them without an external source of electricity, either as photoelectrodes with an external circuit or as microbatteries suspended in solution.[77-79] The charges necessary for the reduction process are generated by illumination. Illumination of semiconductors results in creation of electron-hole pairs. If the charges can be effectively separated, they can be involved in a number of electrochemical reactions. The separation is based on superficial bending of the bands of the semiconductor electrodes. The electron transfer from semiconductor electrodes toward carbon dioxide depends on the nature of the semiconductor. One possible criterion feature favoring electron transfer would be that the position of the conduction band of the semiconductor allows for easy isoenergetic transfer. For example, CdTe has a very conveniently situated conduction band, as well as a reasonable band gap. An important factor is the strength of the bonding of the adsorbed species to the surface. The fact that Cd metal is a good electrode for the reduction of carbon dioxide might bear on the efficiency of CdTe as a photocathode for carbon dioxide reduction. For efficient reduction of carbon dioxide at semiconductor surfaces, there exists an optimal doping level of the semiconductor material.[80] The chemistry of semiconductors is also involved in the creation and stabilization of radicals at the surface of the semiconductors.[81-83] Fourier transform infrared (FTIR) studies performed on p-CdTe and p-GaP indicate the existence of two different species on the surface of the two electrodes. The position of the peaks in the difference spectra indicates a stronger bond of carbon dioxide with the surface of p-GaP than with the surface of p-CdTe, which results in less catalytic activity.

Semiconductors are not very efficient when used in suspensions as "microbatteries."[26] Possible causes are poor charge separation, fast charge carrier recombination on surface states, and reoxidation of the products.

2.4.3. Other Electrode or Solution Modifiers

A most interesting approach employing electrode or solution modifiers is that of homogeneous/heterogeneous catalysis. This approach is schematically illustrated in Scheme 1.

CO_2 is caputred by a metal complex in solution (cat), and electrons are transferred to the carbon dioxide complex from an electrode modified by a suitable catalyst (catfix). A series of papers illustrating this approach for the electrocatalytic reduction of carbon dioxide to methanol was published by a group headed by K. Ogura. The immobilized catalysts were Prussian Blue, indigo, quinone derivatives, and metal porphyrins.[50,84-86] The energy used was electrical, photoelectrical, or photoelectrochemical. The homogeneous catalysts were metal complexes in primary alcohol solutions.[50,87] The catalytic activity of tetraphenyl-porphyrin-metal complexes decreases in the order Co(II) > Ni(II) > Fe(III)Cl > Fe(II) > Cr(III)Cl, approximately following the order for phthalocyanine complexes.

Both homogeneous and heterogeneous catalysis by nonmetallic substances have been included under the same heading here because in many instances similar types of substances

SCHEME 1

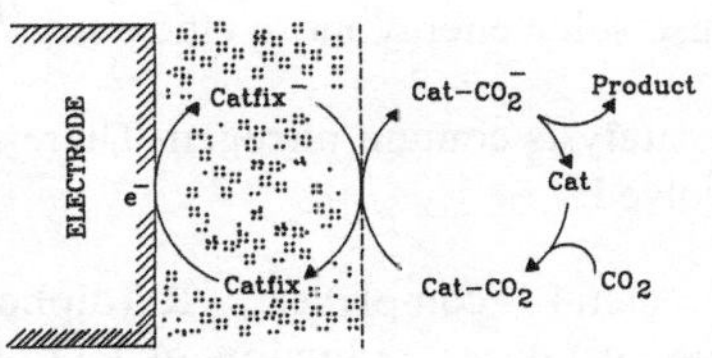

catalyze the reduction of carbon dioxide, be they immobilized on the electrode or in solution. A good example is that of phthalocyanines that catalyze CO_2 reduction in both homogeneous[88-90] and heterogeneous systems.[91] It also appears that the method of preparation affects the catalytic properties of the phthalocyanines[92] and that, in general, not only the elemental composition of the catalysts is important, but also their stereochemistry.[93]

A less obvious example is that of the ammonium type ions. The ammonium ion was found to mediate the reduction of carbon dioxide in nonaqueous solutions and as polyaniline immobilized on silicon in aqueous solutions.[94,95] The influence of NH_3 can be better realized when one notes that the first step in the urea synthesis, formation of ammonium carbamate, is highly favored energetically and kinetically:

$$CO_2 + 2NH_3 \leftrightarrows H_2N\text{-}CO\text{-}ONH_4 \qquad \Delta H = -117 \, kJ/mol$$

Adsorption at the surface of semiconductor electrodes and one-species mediation were also suggested for this process.[83,94] It has been suggested recently that other nitrogen-containing ligands are responsible for increased activity toward the reduction of carbon dioxide in DMSO solutions ($0.1 M$ tetrabutylammonium perchlorate) of Co(II), Ni(II), Cu(II), and Fe(II) tetracoordinated to 1,10-o-phenanthroline.[96] On the other hand, it was found that in an isoelectronic series of complexes of the type $\{MC_6H_5P[CH_2CH_2P(C_6H_5)_2]_2\}L^{2-}$ with M=Ni, Pd, or Pt and L = $P(OCH_3)_3$, CH_3CN, $P(CH_3CH_2)_3$, or $P(CH_2OH)_3$, only the Pd complexes catalyze the reduction of CO_2 to CO in CH_3CN.[97]

The field of electrocatalysis of the carbon dioxide reduction was definitely influenced by the extensive research carried out on oxygen electrodes for fuel cells. An impressive amount of work was invested in catalysts based on metals bound in nitrogen-containing complexes, such as:

- Phthalocyanines.[89,90,98] It is reported that the product distribution changes for Co-phthalocyanine impregnated electrodes (CO) compared to Mn-, Cu-, or Zn-phthalocyanines ($HCOO^-$).[91] The difference might also be due, however, to the change in solvent from water to tetrahydrofuran.[99]
- Porphyrins.[100] Ag and Pd porphyrin complexes, although unstable, are catalytically active for CO_2 reduction in dichloromethane.[101] It appears also that Fe^0-porphyrin is a catalytically active species.[102]
- Polypyridyl metal complexes of, for example, Ni, Co, Ru, Rh, Ir, Re, or Os in homogeneous or heterogeneous solutions.[40,103-109] It was suggested that Ru and Os complexes capture carbon dioxide in their coordination sphere, resulting in a reactive intermediate.[110] In recent work, catalysis by electropolymerized films of bis(vinylter-pyridine)cobalt(2+) was reported in dimethylformamide, but not in acetonitrile. The catalytic effect consists in a displacement of the reduction potential by almost 1 V.[42]
- Tetraazamacrocycles. Complexes of this type have been used as photoelectrocatalysts for reduction of CO_2 on p-Si, p-GaP, and p-GaAs.[54,111-115] New tetraazamacrocycles containing a pyridine group helped reduce CO_2 to CO and other products in MeCN with NBu_4BF_4 as electrolyte.[116] Mono- and bidentate cyclam complexes were used with various degrees of success.[117,118] The Ni cyclams were also tried as "molecular relays" in tandem with a low-band-gap semiconductor (p-GaP or p-GaAs) in an attempt to use solar energy more efficiently.[119,120]

All the above catalysts contain nitrogen. There are several examples, however, in which nitrogen is not involved:

- Phosphino metal complexes. Rh(diphos)$_2$Cl [diphos=2,1-bis(diphenylphos-phino)ethane] catalyzes reduction of CO_2 to formate with acetonitrile as proton source[121] according to the overall reaction:

$$CO_2 \xrightarrow[\text{W}_2(\text{CO})_{10}^{2-}/\text{PPh}_3]{h\nu\,(\lambda>420\,\text{nm})} HCO_2^- (20\%) + CO_3^{2-} (39\%) + HCO_3^- (18\%) + CO$$

$$+ \text{W(CO)}_5\text{PPh}_3 (70\text{-}90\%)$$

- Iron–Sulfur clusters with cubane structure of the type $[Fe_4Se_2(SR)_4]^{2-}$ (R = $CH_2C_6H_5$, C_6H_5). These catalysts result in formation of formate using the tetraalkylammonium electrolyte as hydrogen donor in nonaqueous media.[122]

An interesting class of catalysts is that of photosensitizers, which translate electromagnetic energy (light) into chemical energy. The excited species of some substances are powerful redox agents. A well studied example is that of $Ru(bpy)_3^{3+}$.[123] A recent example is a photosensitizer based on carbonyl. The 19-electron reducing species photogenerated from $W_2(CO)_{10}^{2-}$ in aceto-nitrile react with CO_2 in the presence of PPh_3. The reaction products reported are CO, HCOONa, $NaHCO_3$, $Na_2CO_3^1$.

3. CONCLUSIONS

The large volume of work carried out in the field of carbon dioxide activation is difficult to summarize systematically, because of the variety of approaches and experimental conditions. They were dictated both by the aim of the research (e.g., understanding photosynthesis, producing useful chemicals, or energy storage) and by the training of the investigator.

In the preceding pages I tried to give a succinct account of various factors that influence the reduction of carbon dioxide, and I hope that my attempts to avoid redundancy did not lead to confusion. I apologize to those whose work I omitted because of space limitations. In spite of the huge amount of work already carried out, there are many more avenues to explore in order to find an optimal process. Moreover, there might be more than one set of optimal conditions for a certain process.

REFERENCES

1. N. D. Silavwe, A. S. Goldman, R. Ritter, and D. R. Tyler, *Inorg. Chem.* **28**, 1231 (1989).
2. M. Aresta and G. Forti (eds.), *Carbon Dioxide as a Source of Carbon*, NATO ASI Series, D. Reidel, Dordrecht (1987).
3. W. M. Ayers (ed.), *Catalytic Activation of Carbon Dioxide*, ACS Symposium Series 363, American Chemical Society, Washington, D.C. (1988).
4. A. Behr, *Carbon Dioxide Activation by Metal Complexes*, VCH, New York (1988).
5. G. C. Chinchen, P. J. Denny, J. R. Jennings, M. S. Spencer, and K. C. Waugh, *Appl. Catal.*, **36**, 1 (1988).
6. S. Inoue and N. Yamazaki (eds.), *Organic and Bio-Organic Chemistry of Carbon Dioxide*, Kodansha/John Wiley & Sons, Tokyo/New York (1982).
7. G. E. Hutchinson, *The Biosphere*, Scientific American, New York (1970).
8. D. Dumanowski, *The Boston Globe*, July 18, 1988; data from World Resources Institute.
9. D. Elsom, *Atmospheric Pollution*, Basil Blackwell, Oxford (1987), Chapter 6.
10. *Carbon Dioxide and Climate: A Second Assessment*, National Academy Press, Washington, D.C. (1982).
11. D. Rind, *J. Geophys. Res.* **93**, 5385 (1988).
12. H. G. Marshall, J. C. G. Walker, and W. R. Kuhn, *J. Geophys. Res.* **93**, 791 (1988).
13. K. Asada, in: *Organic and Bio-organic Chemistry of Carbon Dioxide* (S. Inoue and N. Yamazaki, eds.), Chapter 5, Kodansha/John Wiley & Sons, Tokyo/New York (1982).

14. L. C. Allen, in: *Catalytic Activation of Carbon Dioxide* (W. M. Ayers, ed.), ACS Symposium Series 363, American Chemical Society, Washington, D.C. (1988).
15. N. Getoff, G. Scholes, and J. Weiss, *Tetrahedron Lett.*, **1960**, 17.
16. E. L. Quinn and C. L. Jones, *Carbon Dioxide*, Reinhold, New York (1936).
17. C. E. Bamberger and P. R. Robinson, *Inorg. Chim. Acta* **42**, 133 (1980).
18. J. M. Lehn and R. Ziessel, *Proc. Natl. Acad. Sci. U.S.A.*, **79**, 701 (1982).
19. N. Getoff, *Z. Naturforsch.*, *B* **17**, 87 (1962).
20. C. Amatore and J. M. Saveant, *J. Am. Chem. Soc.* **102**, 5021 (1981).
21. H. S. Jee, N. Nishio, and S. Nagai, *J. Ferment. Technol.* **66**, 235 (1988).
22. H. S. Jee, N. Nishio, and S. Nagai, *Biotechnol. Lett.* **10**, 243 (1988).
23. I. Willner, D. Mandler, and J. Riklin, *J. Chem. Soc., Chem. Commun.* **1986**, 1022.
24. D. Mandler and I. Willner, *J. Chem. Soc., Perkin Trans. 2* **1988**, 997.
25. I. Taniguchi, B. Aurian-Blajeni, and J. O'M. Bockris, *Electrochim. Acta* **29**, 923 (1984).
26. M. Ulman, B. Aurian-Blajeni, and M. Halmann, *CHEMTECH* **1984**(April), 235.
27. K. Sugimura, S. Kuwabata, and H. Yoneyama, *J. Am. Chem. Soc.* **111**, 2361 (1989).
28. B. A. Parkinson and P. F. Weaver, *Nature* **309**, 149 (1984).
29. C. O'Connell, S. I. Hommeltoft, and R. Eisenberg, in: *Carbon Dioxide as a Source of Carbon* (M. Aresta and G. Forti, eds.), NATO ASI Series, D. Reidel, Dordrecht (1987).
30. D. D. Wagman, W. H. Evans, V. B. Parker, R. H. Schumm, I. Halow, S. M. Bailey, K. L. Churney, and R. L. Nuttall, *J. Phys. Chem. Ref. Data* **11**, Suppl. 2 (1982).
31. R. N. Compton, P. W. Reinhardt, and C. D. Cooper, *J. Chem. Phys.* **63**, 3821 (1975).
32. K. D. Jordan, *J. Phys. Chem.* **88**, 2459 (1984).
33. H. A. Schwarz and R. W. Dodson, *J. Phys. Chem.* **93**, 409 (1989).
34. W. Paik, T. N. Anderson, and H. Eyring, *J. Phys. Chem.* **46**, 3278 (1972).
35. I. Willner, R. Maidan, D. Mandler, H. Dürr, G. Dörr, and K. Zengerle, *J. Am. Chem. Soc.* **109**, 6080 (1987).
36. W. M. Latimer, *Aqueous Solutions*, Prentice-Hall, New York (1937), p. 7.
37. M. Halmann and K. Zuckerman, *J. Electroanal. Chem. Interfacial Electrochem.* **235**, 369 (1987).
38. T. Yammamoto, *Bull. Inst. Phys. Res. (Tokyo)* **7**, 999 (1922).
39. B. R. Eggins and J. McNeill, *J. Electroanal. Chem.* **148**, 17 (1983).
40. C. M. Bolinger, N. Story, B. P. Sullivan, and T. J. Meyer, *Inorg. Chem.* **27**, 4582 (1988).
41. H. Stephen and T. Stephen, *Solubilities of Inorganic and Organic Compounds*, Macmillan, New York (1963).
42. H. C. Hurrell, A.-L. Mogstad, D. A. Usifer, K. T. Potts, and H. D. Abruña, *Inorg. Chem.* **28**, 1080 (1989).
43. B. Aurian-Blajeni, M. Halmann, and J. Masnassen, *Solar Energy Mater.* **8**, 425 (1983).
44. M. N. Mahmood, D. Masheder, and C. J. Harty, *J. Appl. Electrochem.* **17**, 1159 (1987).
45. D. Masheder and K. P. J. Williams, *J. Raman Spectrosc.* **18**, 387 (1987).
46. H. Lund and P. Iversen, in: *Organic Electrochemistry* (M. M. Baizer, ed.), Marcel Dekker, New York (1973).
47. A. Seidell and W. F. Linke, *Solubilities*, Van Nostrand, New York (1949), supplement to 3rd edition.
48. A. Seidell, *Solubilities of Inorganic and Metal organic Substances*, Vol. 1, Van Nostrand, New York (1940).
49. B. Aurian-Blajeni, I. Taniguchi, and J. O'M. Bockris, unpublished results.
50. K. Ogura and H. Uchida, *J. Chem. Soc., Dalton Trans.* **1987**, 1377.
51. K. Ogura and I. Yoshida, *J. Mol. Catal.* **34**, 676 (1986).
52. H. Yoneyama, K. Sugimara, and S. Kuwabata, *J. Electroanal. Chem. Interfacial Electrochem.* **249**, 143 (1988).
53. Y. Hori, A. Murata, R. Takahashi, and S. Suzuki, *J. Chem. Soc., Chem. Commun.* **1988**, 17.
54. D. J. Pearce and D. Pletcher, *J. Electroanal. Chem.* **197**, 317 (1986).
55. W. M. Ayers and M. Farley, in: *Catalytic Activation of Carbon Dioxide* (William M. Ayers, ed.), ACS Symposium Series 363, American Chemical Society, Washington, D.C. (1988).
56. S. Ikeda, T. Takagi, and K. Ito, *Bull Chem. Soc. Jpn.* **60**, 2517 (1987).
57. F. Solymosi, A. Erdohelyi, and M. Kocsis, *J. Catal.* **77**, 1003 (1981).
58. D. P. Summers and K. W. Frese, Jr. *Langmuir* **4**, 51 (1988).

59. D. P. Summers and K. W. Frese, Jr. in: *Catalytic Activation of Carbon Dioxide* (William M. Ayers, ed.), ACS Symposium Series 363, American Chemical Society, Washington, D.C. (1988).

60. K. W. Frese, Jr. and S. Leach, *J. Electrochem. Soc.* **132**, 259 (1985).

61. Y. Hori, K. Kikuchi, and S. Suzuki, *Chem. Lett.* **1985**, 1695.

62. Y. Hori, K. Kikuchi, A. Murata, and S. Suzuki, *Chem. Lett.* **1985**, 897.

63. O. A. Malinovskaya, A. Ya. Rozovskii, I. A. Zolotarskii, Yu. V. Lender, Yu. Sh. Matros, G. I. Lin, G. V. Dubovich, N. A. Popova, and N. V. Savostina, *React. Kinet. Catal. Lett.* **34**, 87 (1987).

64. Y. Hori, A. Murata, R. Takahashi, and S. Suzuki, *J. Chem. Soc., Chem. Commun.* **1987**, 17.

65. R. L. Cook, R. C. MacDuff, and A. F. Sammels, *J. Electrochem. Soc.* **134**, 1873 (1987).

66. R. L. Cook, R. C. MacDuff, and A. Sammels, *J. Electrochem. Soc.* **135**, 1320 (1988).

67. R. L. Cook, R. C. MacDuff, and A. F. Sammels, *J. Electrochem. Soc.* **134**, 2375 (1987).

68. J. J. Kim, D. P. Summers, and K. W. Frese, Jr., *J. Electroanal. Chem. Interfacial Electrochem.* **245**, 223 (1988).

69. D. W. DeWulf, T. Jin, and A. J. Bard, *J. Electrochem. Soc.* **136**, 1686 (1989).

70. R. L. Cook, R. C. MacDuff, and A. Sammells, *J. Electrochem. Soc.* **135**, 1470 (1988).

71. Y. Hori, A. Murata, K. Kikuchi, and S. Suzuki, *J. Chem. Soc., Chem. Commun.* **1987**, 728.

72. S. Ikeda, Y. Saito, M. Yoshida, H. Noda, M. Maeda, and K. Ito, *J. Electroanal. Chem., Interfacial Electrochem.* **260**, 335 (1989).

73. S. Yamamura, H. Kojima, J. Iyoda, and W. Kawai, *J. Electroanal. Chem. Interfacial Electrochem.* **247**, 333 (1988).

74. S. Yamamura, H. Kojima, J. Iyoda, and W. Kawai, *J. Electroanal. Chem. Interfacial Electrochem.* **225**, 287 (1987).

75. M. W. Rophael and M. A. Malati, *J. Chem. Soc., Chem. Commun.* **1987**, 1419.

76. R. L. Cook, R. C. MacDuff, and A. F. Sammels, *J. Electrochem. Soc.* **135**, 3069 (1988).

77. M. Halmann, *Nature* **275**, 115 (1978).

78. T. Inoue, A. Fujishima, S. Konishi, and K. Honda, *Nature* **277**, 637 (1979).

79. B. Aurian-Blajeni, M. Halmann, and J. Manassen, *Sol. Energy* **1980**, 371.

80. B. Aurian-Blajeni, *J. Electrochem. Soc.* **133**, 2058 (1986).

81. B. Aurian-Blajeni, M. Halmann, and J. Manassen, *Photochem. Photobiol.* **35**, 157 (1982).

82. N. Aurian-Blajeni, M. A. Habib, I. Taniguchi, and J. O'M. Bockris, *J. Electroanal. Chem.* **157**, 399 (1983).

83. K. Chandrasekaran and J. O'M. Bockris, in: *Catalytic Activation of Carbon Dioxide* (W. M. Ayers, ed.), ACS Symposium Series 363, American Chemical Society, Washington, D.C. (1988).

84. K. Ogura and M. Fujita, *J. Mol. Catal.* **41**, 303 (1987).

85. K. Ogura and I. Yoshida, *Electrochimi. Acta* **32**, 1191 (1987).

86. K. Ogura and I. Yoshida, *J. Mol. Catal.* **47**, 51 (1988).

87. K. Ogura and M. Takagi, *J. Electroanal. Chem.* **201**, 359 (1986).

88. K. Hiratsuka, K. Takahashi, H. Sasaki, and S. Toshima, *Chem. Lett.* **1977**, 137.

89. S. K. Kapusta and N. Hackerman, *J. Electrochem. Soc.* **131**, 1511 (1984).

90. C. M. Lieber and S. N. Lewis, *J. Am. Chem. Soc.* **106**, 5033 (1984).

91. M. N. Mahmood, D. Masheder, and C. J. Harty, *J. Appl. Electrochem.* **17**, 1223 (1987).

92. H. Tanabe and K. Ohno, *Electrochim. Acta* **32**, 1121 (1987).

93. M. R. M. Bruce, E. Megehee, B. P. Sullivan, H. Thorp, T. R. O'Toole, A. Downward, and T. J. Meyer, *Organometalics* **7**, 238 (1988).

94. I. Taniguchi, B. Aurian-Blajeni, and J. O'M. Bockris, *J. Electroanal. Chem.* **161**, 385 (1985).

95. B. Aurian-Blajeni, I. Taniguchi, and J. O'M. Bockris, *J. Electroanal. Chem.* **149**, 291 (1983).

96. T. C. Simpson and R. R. Durand, Jr., *Electrochim. Acta* **33**, 581 (1988).

97. D. L. DuBois and A. Miedaner, *J. Am. Chem. Soc.* **109**, 113 (1987).

98. K. Hiratsuka, K. Takahashi, H. Sasaki, and S. Toshima, *Chem. Lett.* **1977**, 1137.

99. D. Masheder and K. P. J. Williams, *J. Raman Spectrosc.* **18**, 391 (1987).

100. K. Hiratsuka, H. Sasaki, K. Takahashi, and S. Toshima, *Chem. Lett.* **1979**, 305.

101. J. Y. Becker, B. Vainas, R. Eger, and L. Kaufman, *Chem. Lett.* **1985**, 1471.

102. M. Hammouche, D. Lexa, J. M. Savéant, and M. Momenteau, *J. Electroanal. Chem. Interfacial Electrochem.* **249**, 347 (1988).

103. J. Hawecker, J.-M. Lehn, and R. Ziessel, *Helv. Chim. Acta* **69**, 1990 (1986).

104. J. Hawecker, J.-M. Lehn, and R. Ziessel, *J. Chem. Soc., Chem. Commun.* **1984,** 328.
105. C. M. Bolinger, B. P. Sullivan, D. Conrad, J. A. Gilbert, and N. Story, *J. Chem. Soc., Chem. Commun.* **1985,** 796.
106. B. P. Sullivan, C. M. Bolinger, D. Conrad, W. J. Vining, and T. J. Meyer, *J. Chem. Soc., Chem. Comm.* **1985,** 1417.
107. H. Ishida, K. Tanaka, and T. Tanaka, *J. Chem. Soc., Chem. Commun.* **1987,** 131.
108. S. Daniele, P. Ugo, G. Bontempelli, and M. Fiorani, *J. Electroanal. Chem.* **219,** 259 (1987).
109. S. Cosnier, A. Deronzier, and J.-C. Moutet, *J. Mol. Catal.* **45,** 381 (1988).
110. M. R. M. Bruce, E. Megehee, B. P. Sullivan, H. Thorp, T. R. O'Toole, A. Downard, and T. J. Meyer, *Organometallics* **7,** 238 (1988).
111. M. G. Bradley and T. Tysak, *J. Electroanal. Chem.* **135,** 153 (1982).
112. B. Fischer and R. Eisenberg, *J. Am. Chem. Soc.* **102,** 7361 (1980).
113. A. H. A. Tinnemans, J. P. M. Koster, D. H. M. W. Thiewissen, and A. Mackor, *Rech. Trav. Chim. Pays Bas* **103,** 288 (1984).
114. D. A. Gangi and R. R. Durand, Jr., *J. Chem. Soc., Chem. Commun.* **1986,** 697.
115. M. Beley, J. P. Collin, and J.-P. Sauvage, *J. Electroanal. Chem.* **206,** 333 (1986).
116. C.-M. Che, S.-T. Mak, W.-O. Lee, K.-W. Fung, and T. C. W. Mak, *J. Chem. Soc., Dalton Trans.* **1988,** 2153.
117. J.-P. Collin, A. Jouaiti, and J.-P. Sauvage, *Inorg. Chem.* **27,** 1986 (1988).
118. A. Mackor, T. P. M. Koster, J.-P. Collin, A. Jouaiti, and J.-P. Sauvage, in: *Energy from Biomass* (D. O. Hall and G. Grassi, eds.), Elsevier Applied Science, New York (1990).
119. J. P. Petit, P. Chartier, M. Beley, and J. P. Sauvage, *New J. Chem.* **11,** 751 (197).
120. M. Beley, J. P. Collin, J. P. Sauvage, and P. Chartier, *J. Electroanal. Chem.* **206,** 333 (1986).
121. S. Slater and J. H. Wagenknecht, *J. Am. Chem. Soc.* **106,** 5367 (1984).
122. M. Tezuka, T. Yajima, A. Tsuchiya, Y. Matsumoto, Y. Uchida, and M. Hidai, *J. Am. Chem. Soc.* **104,** 6834 (1982).
123. R. Ziessel, in: *Carbon Dioxide as a Source of Carbon* (M. Aresta and G. Forti, eds.), p. 113, NATO ASI Series, D. Reidel, Dordrecht (1987).
124. *Tables of Chemical Thermodynamic Properties*, National Bureau of Standards, Washington, D.C. (1982).
125. D. D. Wagman, W. H. Evans, V. B. Parker, R. H. Schumm, I. Halow, S. M. Bailey, K. L. Churney, and R. L. Nuttall, *J. Phys. Chem. Ref. Data* **11** (Suppl. 2) (1982).
126. W. H. Koppenol and J. D. Rush, *J. Phys. Chem.* **91,** 4427 (1987).
127. H. A. Schwarz and R. W. Dodson, *J. Phys. Chem.* **93,** 409 (1989).
128. G. P. Laroff and R. W. Fessenden, *J. Phys. Chem.* **77,** 1283 (1973).
129. G. V. Buxton and R. M. Sellers, *J. Chem. Soc., Faraday Trans.* **69,** 555 (1973).

23

Electrochemical Treatments of Wastes

Lamine Kaba and G. Duncan Hitchens

1. INTRODUCTION

The present chapter is an attempt to give a fundamental basis to the early stages of a potentially valuable electrochemical waste treatment technology for the future. The disposal of domestic waste is a matter of increasing public concern. Earlier, it was regarded as permissible to reject wastes into the apparently infinite sink of the sea, but during the last 20 years, it has become clear that this is environmentally unacceptable.[1, 2] On the other hand, sewage farms and drainage systems for cities and for new housing developments are cumbersome and expensive to build and operate.[3, 4] New technology whereby waste is converted to acceptable chemicals and pollution-free gases at site is desirable. The problems posed by wastes are particularly demanding in space vehicles, where it is desirable to utilize treatments that will convert wastes into chemicals that can be recycled. In this situation, the combustion of waste is undesirable due to the difficulties of dissipating heat in a space environment and to the inevitable presence of oxides of nitrogen and carbon monoxide in the effluent gases.[5, 6] Here, in particular, electrochemical techniques offer several advantages including the low temperatures which may be used and the absence of *any* NO and CO in the evolved gases.

2. THE ELECTROCHEMISTRY OF BIOMASS OXIDATION TO CO$_2$

2.1. "Direct" Oxidation

Aqueous solutions have an electrode potential window in which substances can be oxidized without competition from oxygen evolution up to about 1.6 V on the normal hydrogen scale in acid solutions at 25°C. It should be possible, therefore, to carry out a complete oxidation of all the components in fecal wastes, including urea, to CO$_2$ in aqueous solution. Correspondingly, a number of organic solvents exist that offer an extended potential range for waste oxidation without competition from the oxygen evolution reaction although the reduced conductivity available in such systems (even in the presence of suitable salts) would have to be taken into account[7]; the most anodic potential used in the present work was 1.9 V versus the normal hydrogen electrode (NHE).

Lamine Kaba • Center for Electrochemical Systems and Hydrogen Research, Texas A&M University, College Station, Texas 77843.   *G. Duncan Hitchens* • Lynntech, Incorporated, Bryan, Texas 77803.

Electrochemistry in Transition, edited by Oliver J. Murphy *et al.* Plenum Press, New York, 1992.

On the other hand, if the oxidation of wastes is incomplete in aqueous solution, there is a possibility of introducing them into molten carbonate systems at 650°C and using anodic oxidation under these or similar conditions[8,9]; it may be that a considerable degree of thermal oxidation of the wastes would occur at 650°C with bubbled oxygen in the presence of the typical lithium carbonate/potassium carbonate mixture without the application of potential.

The temperature coefficients of waste oxidation reactions are important in view of the work of Hori et al.,[10] who found that upon lowering the temperature to 0°C, during the reduction of CO_2 on Cu, a 90% yield of methane was obtained. At present, little is known about the temperature coefficient of organic oxidation reactions, but in Kolbe-like oxidation reactions of material formed from the dissolution of coal,[11] it was found that $(dE/dT) \simeq 4 \times 10^{-3}$ V/°C and $(di/dT) \simeq 0.1$ mA/°C. From considerations of the oxidation of coal products, it seemed that an expected range for the oxidation of wastes would be around 1.0–1.5 V (versus NHE), thus not enough to decompose water significantly on platinum.

The electrode materials for waste oxidation in aqueous solutions should be inexpensive, remain unoxidized under the highly anodic conditions (i.e., be an oxide), and be a poor electrocatalyst for oxygen evolution. Lead dioxide is often used for organic oxidations since it is highly conducting (it is used as a battery material) and has a high overvoltage for oxygen evolution.[12] This field is open to new oxide electrodes that have come to be used in recent years.[13] Among these are the perovskites,[14] where the addition of 20–30% barium oxide to materials such as lanthanum nickelate allows a conducting oxide to be used as an electrode which is stable to oxygen attack. In addition, Ebonex (Ti_4O_7), which is an extremely stable anodic or cathodic material in acid environments, may be used.[15]

Although electrochemical oxidations are an important means of obtaining high-value chemical feedstocks from waste biomass,[16] little information is available on the complete oxidation of carbonaceous material to CO_2 as a means of waste disposal. Bockris et al.[17] showed that carbohydrates are completely oxidized to CO_2 during electrolysis. Initially, the electrolysis was performed on simple carbohydrates, such as sucrose, cellobiose (β-D-glucose dimer), and glucose, in 40% H_3PO_4 or $5N$ NaOH at temperatures of 80–100°C. A platinized platinum gauze (52 cm^2) was used as the anode. The reactivity decreased for molecules of increasing complexity. Nevertheless, it was found that cellulose could be broken down to CO_2 with a current efficiency of around 100% and only 2 faradays were involved in the evolution of 1 mol of CO_2, perhaps because of preliminary hydrolysis reactions.

2.2. "Indirect" Oxidation

The electrolysis of waste mixtures can be achieved indirectly by the anodic generation of oxidizing agents which themselves bring about the oxidation of carbonaceous material in the solution. This approach has been utilized for urine purification in regenerative life support systems that were based on the generation of activated chlorine (Cl_2, HOCl, and OCl^-) as the oxidants from Cl^- in urine.[18–20] These studies showed that close to 100% of the organic material could be removed by electrolysis using planar electrodes. Additional features include the production of OCl^-, a strong disinfectant and bleaching agent. Also, the generation of activated chlorine species, such as OCl^-, has been used as an effective means of treating mixtures of urine and fecal waste from humans.[21,22] Through the electrochemical generation of activated chlorine species on platinum electrodes, the waste mixture could be decolorized and clarified using a current of 50 mA/cm^2; furthermore, the odor was quickly lost. No quantitative estimates of the products were available, although the evolution of CO_2, O_2, and Cl_2 was detected. An "indirect" electrochemical method of waste treatment has been developed recently by Delphi Research Inc.[23] In this approach, electrodes were used to regenerate redox couples which were circulated to a waste holding tank. Waste material such as cattle manure and wood chips was converted to an effluent that did not contain carbon monoxide or oxides of nitrogen.

3. METHODOLOGY

3.1. Electrochemical Cells and Apparatus

The current versus potential experiments and constant-potential electrolysis experiments were carried out using a three-compartment glass cell described previously.[11] For experiments involving urine–waste biomass mixtures, a U-tube cell was used (Fig. 1). The cell had an internal diameter of 3 cm and was 25 cm high. The volume of electrolyte was 240 ml. The working electrode was either a platinum foil (100 cm²), designed so that it occupied most of the volume of the cell on both sides, or a lead dioxide rod. A platinum counter electrode was used in both cases.

For each cell type, argon was used to sparge evolved gases from the electrolyte. The outlet gas was carried through a tube packed with glass wool, through a water trap and finally to a barium hydroxide solution. The cell rested on a hot plate, and the temperature was monitored using a thermometer placed in the electrolyte. The electrolyte was stirred with a magnetic follower. After each experiment, the glass cell was cleaned by treatment in an oven at 500°C in air to eliminate organic waste products and washed with sulfuric acid and finally with distilled water.

A Pine Instrument RDE4 potentiostat was used for the experiments. Current versus potential curves were recorded on a Hewlett-Packard X-Y recorder (Model 7044B). All potentials are given versus the normal hydrogen electrode.

3.2. Waste Materials and Solutions

A synthetic mixture of material that represented many of the constituents of sewage was made up in a supply of 4.6 kg that was sufficient for all the experiments and was kept frozen between experiments. Although the nature of fecal waste can vary considerably, generally one-third of its content is made up of microorganisms from the intestinal flora and approximately one-third is made from undigested fiber. The remainder is made up of lipids and

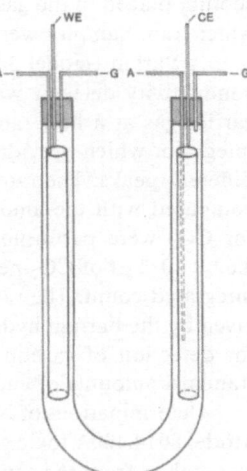

FIGURE 1. The U-tube cell used for electrolysis of waste biomass mixtures. WE, Working electrode; CE, counter electrode; A, gas inlet; G, gas outlet.

TABLE 1
Contents of Artificial Fecal Waste

Waste component	Weight (kg)	Percent of total dry weight
Cellulose	0.60	33
Torpulina	0.43	25
E. coli	0.12	7
Casein	0.17	10
Oleic acid	0.37	20
KCl	0.04	2
NaCl	0.04	2
CaCl$_2$	0.03	1
Water	2.8	—

inorganic materials; the waste mixture shown in Table 1 reflects these proportions. The microorganisms *Torpulina* and *E. coli* were obtained in powdered form (Sigma Chemical Co.), and the remaining contents were obtained from standard chemical suppliers. The mixture had a pastelike consistency and gave off a strong odor. The conversion factor wet waste:dry waste was determined by leaving known samples of waste in the oven at 150°C, the weight being monitored for a number of days until a constant value was obtained. One gram wet weight of waste was equivalent to 0.39 g dry weight. All amounts of waste are given in terms of dry weight. In the experiments involving waste biomass in H_2SO_4, the biomass was weighed, then added to a small amount of $12M$ H_2SO_4, homogenized with a pestle and mortar, and then treated for 20 min with a 150-W sonicator (Fisher). In the experiments involving urine–waste biomass mixtures, urine was collected from volunteers, and 7 g of waste biomass was dissolved in 10 ml of urine and treated ultrasonically as described above. A total of 230 ml of urine was then added to the homogenized mixture.

3.3. Product Sampling and Chemical Analysis

Gases were continuously removed from the working electrode compartment as described above. The rate of CO_2 production was determined through backtitration of the $Ba(OH)_2$ with HCl solution. Gas samples were removed for gas chromatographic analysis through a septum placed in the gas flow immediately downstream from the reaction cell but before the water trap. Samples were taken up in a syringe and transferred to the gas chromatograph.

A Varian (model 3400) gas chromatograph with a carbosieve II column and a thermal conductivity detector was used for the gas analysis. Helium or argon was used as the GC carrier gas at a flow rate of 30 ml/min. Chromatograms were recorded using an HP3390A integrator which provided digital readouts of the retention times and integrated areas of the different peaks. The rate of CO_2 evolution was also monitored by gas chromatography and compared with the amount measured from the barium hydroxide trap. Calibration curves for CO_2 were performed on a regular basis. The CO_2 calibration graph gave a slope of 4.62×10^{-4} μl of CO_2 per integrated count; this is equivalent to 2.06×10^{-11} mol of CO_2 per integrated count. The rate of CO_2 production measured by the GC was within 1–2% of that given by the barium hydroxide method. Gas chromatography was used to assay the off gases for detection of carbon monoxide and methane. The procedure involved the addition of standard amounts of each of these gases to determine their retention time.

Determinations of NO and NO_2 in the gas stream were made using a Matheson–Kitagawa Model 8014.400A toxic gas detector with Kitagawa Precision gas detection tubes; gas samples were taken from the sample port in the gas stream as described above.

Molecular nitrogen analysis of the electrolyte was performed using a Perkin-Elmer Elemental Analyzer, which involves heating 1–3 mg of the sample at a temperature of 900°C, causing the sample to be volatilized. A helium gas stream transfers the vapor to separate detectors for H_2, N_2, and CO_2.

Total organic carbon (TOC) analysis of the electrolyte was determined by means of an O.I.C. model 700 TOC Analyzer in conjunction with a persulfate oxidation CO_2 trapping technique (EPS Method 415). These measurements were made by Intermountain Laboratories, College Station, Texas.

Chlorine evolution was measured by collecting the off gases in the water trap. Chlorine is easily dissolved under these conditions. The water was transferred to the trace analysis unit at Texas A&M, and chlorine was determined by neutron activation analysis. The sample was placed in a 1-MW TRIGA research reactor at a neutron flux of 2×10^{13} n cm/s for 10 min. After a delay of about 5 min from the end of the irradiation, the sample was placed in a high-resolution lithium-drifted germanium [Ge(Li)] γ-ray detector. Characteristic γ emission from the reaction product ^{38}Cl of 1642 and 2167 keV was detected and used for quantification. Samples were compared to standard materials to compute chlorine content. The lower detection limit was less than 0.1 ppm.

4. RESULTS AND DISCUSSION

4.1. Waste Biomass–H_2SO_4 Mixtures

An investigation was made of the general electrochemical properties of the waste biomass mixture shown in Table 1. Firstly, basic information was collected on the electrochemical behavior of each of the principal components of the waste biomass mixture. The results shown in Fig. 2 indicate that the cellular components of the yeast were the most readily oxidizable material. Taking the waste mixture itself, potential sweep measurements on platinum and lead dioxide were performed in a $12M$ H_2SO_4 electrolyte. Figure 3 shows that on platinum, at a temperature of 150°C, a waste oxidation plateau was observed from 1.3 up to 1.7 V; in this potential range vigorous oxygen evolution occurred. Figure 4 shows a current versus potential curve for waste in $12M$ H_2SO_4 at three temperatures on lead dioxide. An oxidation plateau was observed between 0.6 and 1.9 V. Constant-potential electrolysis was performed at 1.8 V on lead dioxide, and TOC measurements were made during the electrolysis. As shown in Fig. 5, there was a 95% reduction in the concentration of organic carbon. Gas chromatographic analysis for carbon monoxide, ammonia, and methane and test kit analysis

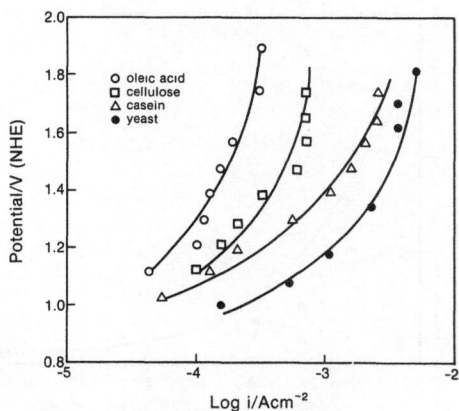

FIGURE 2. Log i versus V for components of the waste biomass mixture. The electrolyte was $12M$ H_2SO_4 at 150°C and contined each material at a final concentration of 14 g/liter.

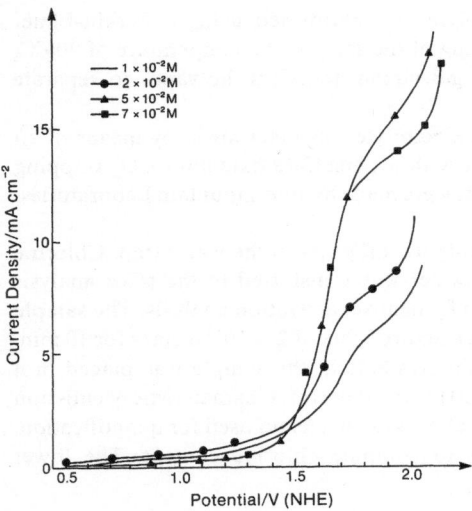

FIGURE 3. Potential sweep measurements with different concentrations of waste biomass using platinum electrodes. The sweep rate was 1 mV/s, and the electrolyte was $12M$ H_2SO_4 at 150°C.

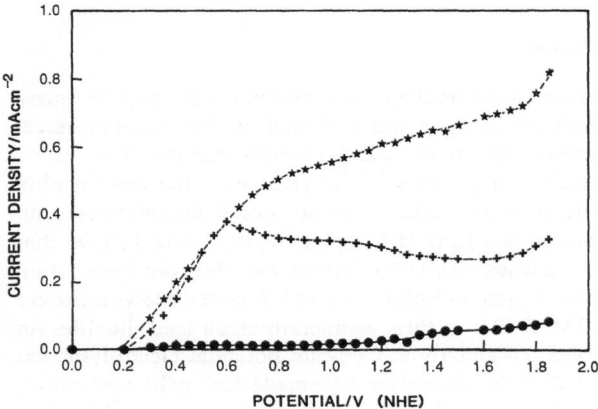

FIGURE 4. Potential sweep measurements for waste biomass on lead dioxide. The electrolyte was $12M$ H_2SO_4 and contained waste biomass at a final concentration of 24 g/liter, and the scan rate was 1 mV/s. ●, 25°C; +, 80°C; ★, 130°C.

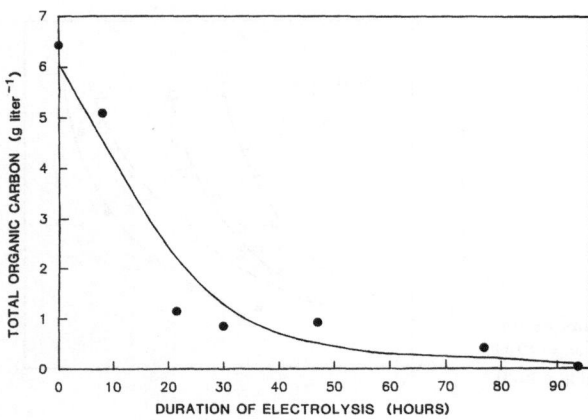

FIGURE 5. Total organic carbon versus electrolysis time using lead dioxide. The electrolyte was $5M$ H_2SO_4 at 80°C, and the final concentration of waste was 14 g/liter.

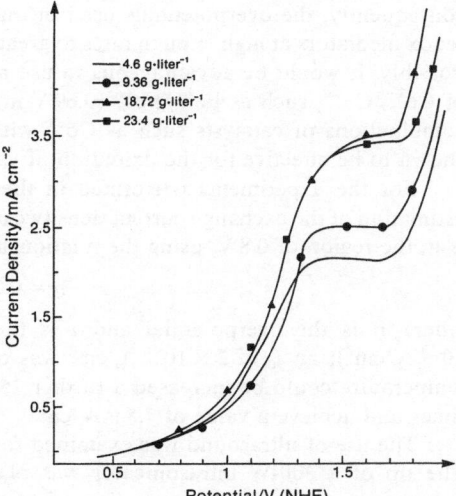

FIGURE 6. Effect of Ce^{4+} on potential sweep measurements on platinum. Concentrations of Ce^{4+} in the electrolyte are indicated on the Figure. The electrolyte was $12M$ H_2SO_4 at 150°C and contained waste at a final concentration of 14 g/liter. The sweep rate was 1 mV/s.

for nitrous oxide and nitrogen dioxide was made on the effluent gases, but none of the above chemicals were detected (lower detection limit was 10 ppm for nitrous oxide and 1 ppm for nitrogen dioxide, carbon monoxide, ammonia, and methane).

Similar electrolysis experiments were performed in which cerium sulfate was added to the electrolyte to a final concentration of $5 \times 10^{-2}M$. The potential sweep measurements shown in Fig. 6 indicate that the mediator causes an increase in current density. The electrolysis experiments performed over extended periods of time shown in Fig. 7 indicate that the fivefold current density increase was, however, not matched by an increase in the rate of waste decomposition, which increased by only a factor of 1.5, as judged from the CO_2 production during electrolysis. Thus, the chemical oxidation of the waste by Ce^{4+} in solution is clearly inadequate to keep up with the more rapid oxidation of the cerous ion at the electrode.

Mediators have been shown to play an important role in the breakdown and upgrading of coal slurries.[24-26] Redox mediators were used based on the notion that they would have a lower i_0 than that of the organic materials and carry out homogeneous oxidation of the materials in solution. The standard redox potential of the Ce^{3+}/Ce^{4+} reaction is 1.45 V;

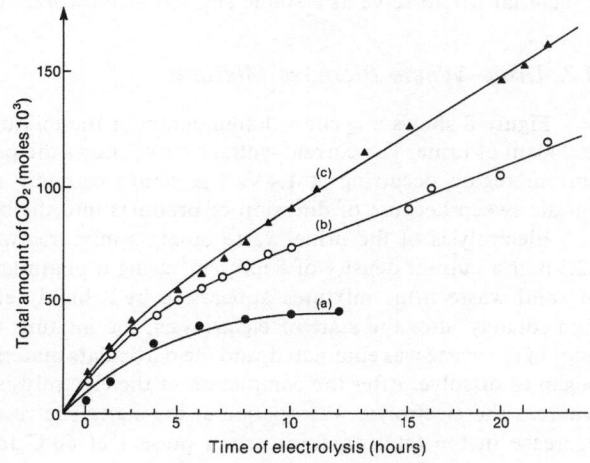

FIGURE 7. Production of CO_2 during waste electrolysis on platinum. The electrolyte was $12M$ H_2SO_4 at 150°C and contained waste at a final concentration of 14 g/liter. (a) Thermal decomposition of biomass. (b) Constant potential electrolysis of biomass; $E = 1.74$ (NHE). (c) Constant potential electrolysis of biomass, in presence of 5×10^{-2} M Ce^{4+}.

consequently, the overpotentials used in the experiments were insufficient to oxidize these redox mediators at high enough rates to greatly enhance the oxidation of material in solution. Possibly, it would be advantageous to use mediators with redox potentials lower than that of Ce^{3+}/Ce^{4+}, such as Fe^{2+}/Fe^{3+} (0.69 V in $1M$ H_2SO_4) or Br_2/Br^- (1.08 V). Furthermore, combinations of catalysts such as Co^{2+} with Fe^{2+}/Fe^{3+} or V^{5+} with Fe^{2+}/Fe^{3+} have been shown to be effective for the oxidation of various types of waste biomass.[23]

For the experiments performed in the absence of redox mediators, an approximate estimation of the exchange current density can be made, assuming that the reversible potential is in the region of 0.8 V, using the relationship given below:

$$\eta = 0.25 \log i/i_0$$

where η is the overpotential and i is the average current in the limiting region (5×10^{-3} A/cm^2); an i_0 of 3×10^{-6} A/cm^2 was calculated at 150°C. It is noteworthy that if the temperature could be increased a further 25°C, the limiting current would be increased 1.5 times and achieve a value of 7.5 mA/cm^2.

The use of ultrasound was examined for the purpose of increasing the limiting current. The tip of a 300-W ultrasonicator was placed into an electrolyte of $0.5M$ sulfuric acid containing waste at a concentration of 14 g/liter. The tip was positioned 1.5 cm from the surface of a 25-cm^2 platinum foil. In the absence of ultrasound but with stirring at 6000 rpm, the limiting current density achieved was 3.8 mA/cm^2. Upon the application of ultrasound at 7 W/cm^2, the current density of waste oxidation increased 2.5 times over the maximum value achieved with stirring alone. The increase in current density was linear with power output, and thus it is reasonable to expect that a greater increase in current density could be attained at greater ultrasound intensities. Huck showed that ultrasound at 2 W/cm^2 increased the limiting current by one order of magnitude for the electrochemical reactions of Fe^{2+}/Fe^{3+} and $[Fe(CN)_6)]^{3-}/[Fe(CN)_6]^{4-}$ due to cavitation.[27] A significantly lower enhancement was obtained in the present case. The power density of 7 W/cm^2 incident on the electrode was determined from the known power output at the tip of the ultrasonic probe and does not take into account losses in ultrasound intensity caused by the presence of electrolyte. Thus, power density at the electrode surface is likely to be lower than the value given above. A further investigation is therefore required to determine if increased power output per unit area of electrode can substantially increase the current density for waste oxidation.

The present approach would benefit from the use of advanced materials such as sub-stoichiometric oxides of titanium with Ti_4O_7 that exhibit high chemical stability in both acidic and alkaline environments and can function as anode and cathode without degradation. Such a material would serve as a stable support electrocatalytic material.

4.2. Urine–Waste Biomass Mixtures

Figure 8 shows a cyclic voltammogram of the mixture containing 7 g of waste biomass in 240 ml of urine. The current–voltage curve shows the peak that corresponds to the limiting current region occurring at 1.4 V. The return cathodic sweep does not correspond to the anodic sweep because of diffusion of products into the bulk solution.

Electrolysis of the urine–waste biomass mixtures was performed galvanostatically for 72.5 h at a current density of 8 mA/cm^2 using a platinum electrode of 100 cm^2. Electrolysis of solid waste–urine mixtures appears to be a highly effective means of waste treatment. Immediately after the start of electrolysis, the mixture was decolorized, the characteristic odor of the waste was eliminated, and the particulate material that was present in the electrolyte began to dissolve. After the completion of the electrolysis, the solution was practically clear whereas the electrolyte was opaque at the start. The results are summarized in Fig. 9. The decrease in the total organic carbon present at 60°C in 72 h is about 78%. The original

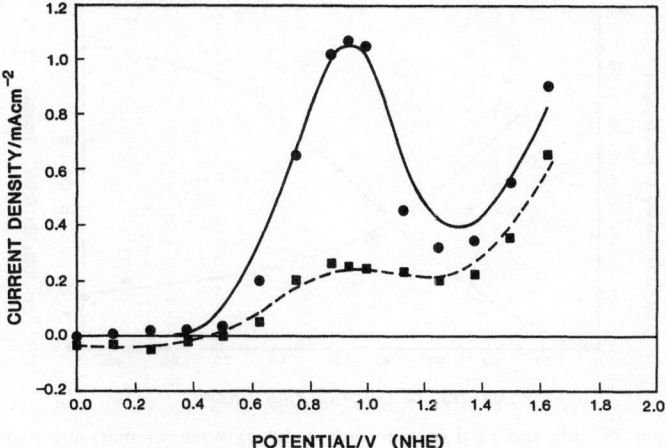

FIGURE 8. Potential sweep measurements of a waste biomass–urine mixture on platinum at a scan rate of 10 mV/s.

nitrogen content of the urine–waste biomass mixture was much higher than the TOC, which is consistent with the relatively high nitrogen content of urine.[13] The nitrogen content decreased by a smaller proportion than the TOC. The loss of nitrogen from the electrolyte as nitrogen gas would account for 57.6% of the evolved gas, with CO_2 making up 42.3%. Using the methods described above, carbon monoxide, nitrogen dioxide, nitrous oxide,

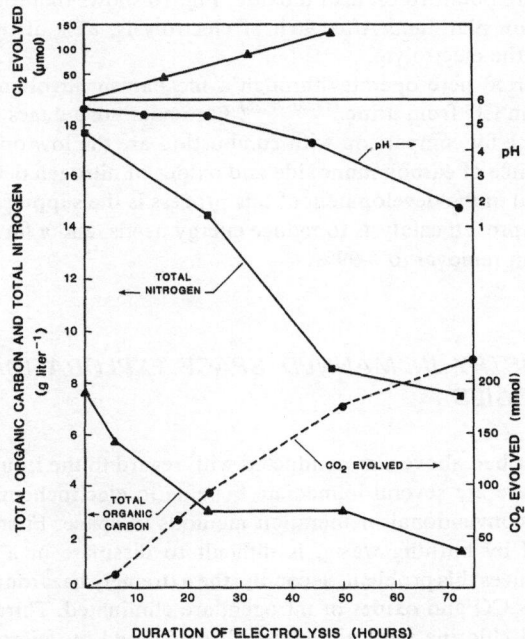

FIGURE 9. Carbon dioxide, chlorine, total organic carbon, total nitrogen, and pH during waste biomass–urine electrolysis. The mixture was at 60°C and contained waste biomass at a final concentration of 13 g/liter. The platinum electrode area was 100 cm², and the applied current was 800 mA.

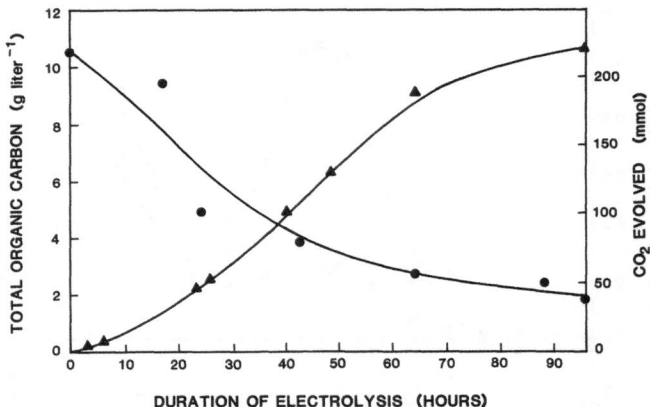

FIGURE 10. Carbon dioxide and total organic carbon during waste biomass–urine electrolysis using lead dioxide. The mixture was at 60°C and contained waste biomass at 13 g/liter. The applied current was 800 mA.

ammonia, and methane were not detected in the gaseous effluent. A total volume of 3 cm³ of chlorine was produced in 48 h, indicating that it would constitute only 0.1% of the evolved gases. It was difficult to detect the odor of chlorine of the off gas. During electrolysis, the pH of the mixture decreased from 6 to 2. The pH decline of the mixture began after about 48 h of electrolysis. Since the pH buffering comes from ammonium salts, the loss of these species leads to reduced buffering in the presence of organic acids.

Similar results were obtained on lead dioxide; Fig. 10 shows that there was a more even decline in TOC than on platinum. After 96 h of electrolysis, 82% of the TOC carbon had been eliminated from the electrolyte.

The process reported here operates through a mechanism involving the generation of OCl^- at the anode from Cl^- from urine.[18,20-22,28] Particular advantages of waste electrolysis using activated chlorine by comparison with combustion are the low operating temperature and the complete absence of carbon monoxide and oxides of nitrogen detected in the effluent gas. An immediate goal in the development of this process is the suppression of residual Cl_2, the development of improved catalysts to reduce energy needs, and a flow system to increase the efficiency of carbon removal to >99%.

5. ELECTROCHEMISTRY IN MANNED SPACE EXPLORATION AND CLOSED ECOLOGIES

The research described above was conducted with regard to the future needs of manned space exploration. There are several immediate benefits to electrochemical waste "incineration" as opposed to conventional incineration methods in space. Firstly, excess heat in a space capsule, caused by burning waste, is difficult to dissipate in a vacuum. The electrochemical option reduces this problem. Secondly, the extremely hazardous toxic by-products of combustion, such as CO and oxides of nitrogen, are eliminated. Thirdly, the disinfection properties of activated chlorine reduces the health risk posed by microorganisms. Finally, the potential exists to generate hydrogen fuel from the cathodic reaction. On deep-space missions, it will be essential to eliminate the need to resupply expendable items. The method of treating solid waste and urine mixtures described above fulfills this requirement in that the only consumable in the waste treatment process is electrical energy.

Keeping humans alive in space for extended periods of time is a challenge that ultimately will require the provision of food, potable water, and a breathable atmosphere from waste products. Fuel cells, batteries, and photovoltaic systems are familiar components of power systems used in space exploration. Electrochemical techniques also offer the possibility to effect the chemical transformations necessary for upgrading and recycling of waste materials by utilizing electrical energy. A number of options for recycling are given below and point to some unique uses for electrochemistry in the future.

5.1. Food Generation.

The efficient conversion of waste material into CO_2, using electrochemical incineration, is a convenient starting point for food production since CO_2 can be used for photosynthesis by plants in a space greenhouse, for instance. Alternatively, electrochemical reduction of CO_2 to low-molecular-weight hydrocarbons offers another route to food regeneration through the cultivation of methylotrophic yeast as a source of protein.[29] Another potential source of single-cell protein in a closed ecology is the cultivation of the bacterium *Hydrogenomonas*.[30] In this case water electrolysis provides O_2 and H_2 for cultivating the microorganisms; the waste products CO_2 and urea provide the nitrogen and carbon sources.

5.2. Cabin Atmosphere Regeneration

The use of water electrolysis for the provision of oxygen is well established. In addition, electrochemical methods can serve the important purpose of CO_2 extraction from the crew cabin atmosphere by electrochemical CO_2 concentration. The use of solid oxide electrolytes offers the possibility to recover breathable oxygen from CO_2-rich atmospheres such as that found on Mars.[31]

5.3. Water Recovery

Electrolysis is a candidate system for removing urea and other organic solutes from urine to facilitate the regeneration of potable water.[18] The recovery of potable water from other types of waste water, such as spacecraft heat exchange condensate and wash water, using electroorganic oxidations and electrochemical techniques for disinfection is an important area for future research. The use of solid polymer electrolytes is attractive for this purpose since they can allow water to be treated directly without the need for a liquid electrolyte.

ACKNOWLEDGMENT

Financial support was provided by Regional Universities Research Grants from NASA Johnson Space Center (JSC), Houston, Texas, and, in particular, we would like to thank A. L. Behrend, D. Price, and C. E. Verostko of JSC for their assistance. Also we would like to thank N. L. Weinberg and O. J. Murphy for their advice during the performance of this work.

REFERENCES

1. A. F. M. Barton, *Resource Recovery and Recycling*, John Wiley & Sons, New York (1979).
2. J. O'M. Bockris (ed.), *Electrochemistry of Cleaner Environments*, Plenum Press, New York (1972).
3. L. Lessing, *Fortune* **1973** (July), 133.

4. R. M. E. Diamant, in: *Environmental Chemistry* (J. O'M. Bockris, ed.), pp. 95–119, Plenum Press, New York (1977).

5. W. Strauss, in: *Environmental Chemistry* (J. O'M. Bockris, ed.), pp. 179–212, Plenum Press, New

6. D. J. Spedding, in: *Environmental Chemistry* (J. O'M. Bockris, ed.), pp. 213–241, Plenum Press, New York (1977).

7. F. Goodridge and C. J. H. King, in: *Techniques of Electroorganic Synthesis, Part 1* (N. L. Weinberg, ed.), p. 7 John Wiley & Sons, New York (1974).

8. J. Greenburg, U.S. Patent 3,642,583 (1972).

9. D. H. Kerridge, *Pure Appl. Chem.* **41**(3), 355 (1975).

10. Y. Hori, N. Kamida, and S. Suzuki, *J. Fac. Eng. Chiba Univ.* **32**, 37 (1981).

11. O. J. Murphy, J. O'M. Bockris, and D. W. Later, *Int. J. Hydrogen Energy* **10**, 453 (1985).

12. H. Sharifian and D. W. Kirk, *J. Electrochem. Soc.* **133**, 921 (1986).

13. S. Trasatti (ed.), *Electrodes of Conductive Metal Oxides*, Elsevier, Amsterdam (1981).

14. J. O'M. Bockris and T. Otagawa, *J. Phys. Chem.* **87**, 2960 (1983).

15. R. R. Miller-Folk, R. E. Noftle, and D. Pletcher, *J. Electroanal. Chem.* **274**, 257 (1989).

16. H. L. Chum and M. M. Baizer, *The Electrochemistry of Biomass and Derived Materials*, ACS Monograph 183, American Chemical Society, Washington, D. C. (1985).

17. J. O'M. Bockris, B. J. Piersma, and E. Gileadi, *Electrochim. Acta* **9**, 1329 (1964).

18. D. F. Putnam and R. L. Vaughan, SAE Life Support and Environmental Control Conference, San Francisco, July 1971.

19. Lockheed Missiles and Space Co., Electrolytic Pretreatment of Urine, NASA Report CR-151566 (1977).

20. J. C. Wright, A. S. Michaels, and A. J. Appleby, *AIChE J.* **32**(9), 1450 (1986).

21. R. G. Tischer, B. P. Tischer, L. R. Brown, and J. C. Mickelson, *Dev. Ind. Microbiol.* **4**, 253 (1963).

22. R. G. Tischer, L. R. Brown, and M. V. Kennedy, *Dev. Ind. Microbiol.* **6**, 238 (1965).

23. P. M. Dhooge, *Proceedings of the 8th Princeton/AIAA/SSI Conference*, American Institute of Aeronautics and Astronautics, Washington, D.C. (1987).

24. P. M. Dhooge, D. E. Stillwell, and S.-M. Park, *J. Electrochem. Soc.* **129**, 1719 (1982).

25. N. C. Taylor, C. Gibson, K. D. Bartle, D. G. Mills, and G. Richardson, *Fuel* **64**, 415 (1985).

26. R. L. Clark, P. C. Foller, and A. R. Wasson, *J. Appl. Electrochem.* **18**, 546 (1988).

27. H. Huck, *Ber. Bunsenges. Phys. Chem.* **91**, 648 (1987).

28. M. Fels, *Med. Biol. Comput.* **16**, 25 (1978).

29. G. R. Peterson, B. O. Stokes, W. W. Schubert, and A. M. Rodriquez, *Enzyme Microb. Technol.* **5**, 337 (1983).

30. H. G. Schlegel, in: *Fermentation Advances* (D. Perlmann, ed.), Academic Press, New York (1969).

31. A. O. Isenberg and R. J. Cusick, SAE Technical Paper 881040, 18th ICES Conference, San Francisco, July 11–13, 1988.

Bioelectrochemistry

Coherent Dynamics and Solitary Waves in Ordered Monolayers

G. C. Huth, F. Gutmann, and G. Vitiello

1. INTRODUCTION

In a recent communication[1] we have studied the energy transfer in J-aggregates, and we have shown that the dynamical formation of coherent solitary waves may explain and fit the observations. In this chapter we review and further discuss this topic. In our presentation we closely follow Ref. 1.

Scheibe aggregates are highly ordered molecular monolayers produced, for example, by the Langmuir–Blodgett technique.[2,3] Such monolayers of, for example, oxycyanine dyes as donors doped with acceptor-type thiacyanine dyes exhibit highly efficient, virtually loss-free, energy transfer from excited host molecules to the acceptor guests.[2-4] The doping level may be as low as a donor-to-acceptor ratio of several tens of thousands. Less efficient energy transfer has been observed in multilayers where the acceptor molecules are situated in a layer adjacent to the monolayer containing the donor molecules; the whole forming a sandwich-type structure. The efficiency of energy transfer has been found to decrease with temperature.[4,5] These experimental results have been rationalized in terms of a model[4,5] based on the concept of a coherent exciton: the classical analog of a domain containing a certain number of molecules all oscillating in phase, and at the frequency of the exciting radiation. This coherent exciton is then thought to move as a unit, in a random walk process, throughout the layer until it possibly reaches an acceptor site. This model indeed agrees well with most of the experimentally observed facts,[2,3] though a difficulty arises if the random walk process is considered quantitatively: the distance x traveled is related to the transfer distance y covered in any one single step by

$$\overline{x^2} = ny^2 \tag{1}$$

where n is the number of random transfer steps. If each step involves energy transfer between adjacent molecules—a distance of, say, 9 Å—x being, say, 500 Å (though much longer transfer

G. C. Huth • Xsirius Scientific, Incorporated, Arlington, Virginia 22201, and Institute of Physics, School of Medicine, University of Southern California, Marina del Rey, California 90291.   *F. Gutmann* • Xsirius Scientific, Incorporated, Arlington, Virginia 22201, and School of Chemistry, Macquarie University, North Ryde, New South Wales 2109, Australia.   *G. Vitiello* • Dipartimento di Fisica, Università di Salerno, 84100, Salerno, Italy.

Electrochemistry in Transition, edited by Oliver J. Murphy *et al.* Plenum Press, New York, 1992.

distances, up to nearly 1000 Å, have been reported), n is calculated to be 3100 steps. This model assumes actual energy transfer via a dipole–dipole interaction, the Förster mechanism. This requires that the energy must reside on any one molecular site for at least a few molecular vibrational periods. These should be of the order of $(10^{13}\,\text{Hz})^{-1}$. Neglecting the transit time for the energy transfer, that is, assuming that energy transfer takes place with the velocity of light in the medium, a total time of at least $10^{-10}\,\text{s}$ is seen to be required in order to accommodate the residence times in 3100 steps. However, at 20 times the velocity of sound, which is of the order of $1000\,\text{m s}^{-1}$, only about $2.5 \times 10^{-12}\,\text{s}$ are available for transfer over 500 Å, which is about two orders of magnitude too short. Even if one assumes that the single step involves transfer over an entire domain diameter—say, 18 Å—there would still be about 800 steps involved, requiring a minimum of about $8 \times 10^{-11}\,\text{s}$, again allowing for only one single vibrational period, a very severe restriction. Even then, the time available is only about one-third of that required.

· One of the main features of this model is that the number of molecules making up the domain, ν, remains constant during the process. It appeared to us that the essential first step in this model, that is, the formation of a domain, deserves further study. We wish to propose that such coherent exciton domains arise from nonlinear dynamical effects in the layer. For our analysis to be valid, the following experimental observations should be accommodated within its framework:

1. The number of molecules ν comprising a domain ranges between 10 at room temperature and 150 at 20 K. Writing Q_∞ rayleighs for the number of fluorescent quanta emitted per unit of time per unit surface area, by the donors only, that is, in the absence of any acceptor molecules, and Q for the corresponding number of quanta emitted by the doped layer, one obtains the relation

$$(Q_\infty/Q)^{-1} = \text{const. } xN^{-1} \qquad (2)$$

where N is the concentration of donor molecules, that is, the number of donors per acceptor molecule. The constant is a function of the chemical and physical properties of the layer.[4] The domain occupancy number ν is determined by the competition, or interplay, between the energy gained by domain extension and thermal fluctuations or thermal energy, which is of the order of kT and tends to counteract the cooperative interaction leading to domain formation. This reasoning leads to

$$\nu \propto T^{-1} \qquad (3)$$

which is in agreement with experiment.

2. Assuming a mean lifetime of $10^{-10}\,\text{s}$ for the coherent exciton, that is, for the domain, its velocity of propagation is determined to be about 10 to 20 times the sound velocity V_0. The above lifetime being a reasonable estimate, it follows that the domain migrates with a velocity of the order of $10^4\,\text{m/s}$.

2. ANALYSIS

2.1. One Dimension

We shall first treat the problem reduced to one dimension; extension to the two-dimensional case will follow.

Consider the molecular excitation wave function $\,^{\bullet}a(r, t)$ which is a solution of the Schrödinger equation:

$$\left[i\hbar \frac{\partial}{\partial t} - \Lambda + J\frac{\partial^2}{\partial r^2} + V(r, t) \right] a(r, t) = 0 \qquad (4)$$

where Λ is the energy of the bottom of the exciton band, $-J$ stands for the resonant interaction between neighboring molecules, and $V(r, t)$ represents the potential arising from the displacement of the excited molecules from their equilibrium positions. The distance r refers to a radial coordinate centered on an arbitrary donor site, which, once selected, is held fast. Let l denote the molecular spacing of the aggregate in the absence of excitation: The displacement β_n of a molecule situated at a site where $r = n$ produces a deformation so that $l \to l - \rho_n$ with $\rho_n = \beta_{n-1} - \beta_n$. Generally, we may assume that this deformation is associated with an anharmonic potential

$$U = \sum_n \phi(\rho_n) \tag{5}$$

with

$$\phi(\rho_n) = \tfrac{1}{2}\omega\rho_n^2 + \tfrac{1}{6}\alpha\omega\rho_n^3 + \cdots = \omega\alpha^{-1}\left\{-\rho_n + \alpha^{-1}\left[\exp(\alpha\rho_n)^{-1}\right]\right\} \tag{6}$$

assuming that the deformation is small, that is, that $\alpha\rho_n \ll 1$. In Eq. (5) we have described the anharmonic potential U in terms of a Toda lattice potential[6,7]; ω is elasticity coefficient of the layer, and α its anharmonicity parameter. In other words, we employ the most general assumption of a deformation controlled by a Toda potential.

In the limiting case of $\alpha = 0$, that is, for a harmonic interaction potential, the longitudinal sound velocity $V_0 = l[\omega/M]^{1/2}$, where M represents the molecular mass. The propagation equation for the deformation ρ_n is

$$M\frac{d^2}{dt^2}\rho_n = \omega\alpha^{-1}[\exp(\alpha\rho_{n+1}) + \exp(\alpha\rho_{n-1}) - 2\exp(\alpha\rho_n)] \tag{7}$$

which has the well-known soliton solution[6-8]

$$\exp(\alpha\rho_n) - 1 = \sinh^2(ql)\,\mathrm{sech}^2[q(nl - Vt)] \tag{8}$$

that is,

$$\rho_n(t) = \alpha^{-1}\sinh^2(ql)\,\mathrm{sech}^2[q(nl - Vt)] \tag{9}$$

which goes to zero as $n \to \infty$.

$$V_s = V_0(ql)^{-1}\sinh(ql) \tag{10}$$

where q is the soliton parameter [see also Eq. (15) below]; q's dimension is a reciprocal length, and thus it may be considered as a propagation constant.

Assuming that

$$ql \ll 2\pi \tag{11}$$

we can apply the continuous approximation so that Eq. (9) may be written

$$\rho(r, t) \cong \alpha^{-1}\sinh^2(ql)\,\mathrm{sech}^2[q(r - r_0 - Vt)] \tag{12}$$

where r_0 stands for the coordinate of the center of the soliton at the initial time, $t = 0$. In this approximation, the deformation $\rho(r, t)$ then results as

$$\rho(r, t) = -l\frac{\partial\beta(r, t)}{\partial r} \tag{13}$$

Insert now for $V(r, t)$ in Eq. (4) the expression

$$V(r, t) = \sigma^2\rho(r, t) \tag{14}$$

where the coupling σ^2 is defined in Eq. (17) below, and $\rho(r, t)$ is given by Eq. (12). The molecular excitation $a(r, t)$ is now obtained by solving Eq. (4) for the lowest energy E:

$$a(r, t) = \alpha^{-1/2} \sinh(ql) \frac{\exp[i\kappa(r - r_0) - i\omega t]}{\cosh[q(r - r_0 - Vt)]} \tag{15}$$

where

$$K = \frac{\hbar V}{2J} \quad \text{and} \quad E = \hbar\omega - \Lambda - \frac{\hbar^2 V^2}{4J} = -Jq^2 \tag{16}$$

The soliton velocity V is given by Eq. (10).

The coupling parameter σ^2 introduced in Eq. (14) is given by

$$\sigma^2 = J\alpha \frac{2q^2}{\sinh^2(ql)} = \frac{2J\alpha}{l^2} \frac{V_0^2}{V^2} \tag{17}$$

In contradistinction to the case of a conventional "free" exciton, which is described by a wave packet, the soliton solution of Eq. (15) does not spread out in time.

From Eqs. (12) and (15) it is seen that

$$\rho(r, t) = |\alpha(r, t)|^2 \tag{18}$$

so that Eq. (4) becomes identical with the nonlinear Schrödinger equation (NSE):

$$\left(i\hbar \frac{\partial}{\partial t} - \Lambda + J \frac{\partial^2}{\partial r^2} \right) a(r, t) = -\sigma^2 |a(r, t)|^2 a(r, t) \tag{19}$$

Due to Eq. (10), the velocity of the soliton (Eq. 15), which describes the aggregate molecular excitation, is allowed to exceed the velocity of sound in the system: thus, $V > V_0$. Also, from Eq. (10) it is seen that as $q \to 0$, $V \to V_0$ and $\rho(r, t) \to 0$ so that the soliton disappears, meaning that the molecular excitation then reverts to behave as a conventional exciton. In consequence, the energy E carried by the solitonic molecular excitation then tends to zero [see Eq. (16)]. It thus follows that the efficiency of energy transfer drops with a decrease in the value of q. The size of the coherent excitation domain may then be approximated by

$$\Delta r \cong \frac{2\pi}{q} \tag{20}$$

[see Eqs. (15) and (12)]. It is evident that the domain size is inversely proportional to q and that thus the parameter q is proportional to temperature. These results agree with experiment. A more detailed analysis of these temperature dependences as well as quantitative computations will be presented in a subsequent communication. On the lines of the present discussion, we note that one may introduce as an experimental datum the ratio V/V_0 and thence, using Eq. (10), compute values for the soliton parameter q for given values of the molecular spacing l and thus determine the size of the coherent excitation domain using Eq. (20). The reverse procedural sequence is of course also feasible. For example, using 2 Å as molecular spacing and the molecular size reported in Ref. 5, we have an *effective* molecular area of 105 Å^2. Using[4] $V/V_0 \simeq 20$, we find that a circular domain of radius given by Eq. (20) includes about 12 molecules coherently excited at room temperature. This is compatible with findings of Refs. 4 and 5. For a domain corresponding to 130 molecules (at $T = 20$ K)[2-4] we compute $V/V_0 \simeq 1.53$.

2.2. Extension to Two Dimensions

We shall now extend the discussion to the two-dimensional case. With r being, as before, the radial distance from a given acceptor site, so that $r^2 = x^2 + y^2$, Eq. (19) is modified thus:

$$\left(i\hbar \frac{\partial}{\partial t} - \Lambda + J\frac{\partial^2}{\partial r^2} + J\frac{1}{r}\frac{\partial}{\partial r}\right) a(r, t) = -\sigma^2 |a(r, t)|^2 a(r, t) \tag{21}$$

In other words, we assume that a nonlinear Schrödinger equation such as Eq. (19) also holds in two dimensions. A solution $a(r, t)$ of Eq. (21) which approaches zero as $r \to \infty$ for any value of t and which remains analytic for $r = 0$ must have for small values of r the form

$$a(r, t) = \sum_{n=0}^{\infty} a_n(t) r^n \tag{22}$$

with

$$a_n(t) = 0 \qquad \text{for odd values of } n \text{ and any value of } t$$
$$a(\infty, t) = 0 \qquad \text{for any value of } t \tag{23}$$
$$a(0, t) = 0 \qquad \text{for any value of } t$$

A choice of general validity, at $t = 0$ and assuming r to be small, is

$$a(r, 0) = \alpha^{-1/2} \sinh(ql)\{\operatorname{sech}[q(r - r_0)] \exp(i\kappa_0 r) + \operatorname{sech}[q(r + r_0)] \exp(-i\kappa_0 r)\} \tag{24}$$

where r_0 denotes, within the accuracy of the method, the mean position at which $|a(r, 0)|^2$ has its maximum; r_0 thus defines the initial radial coordinate position. Solutions of the form of Eq. (22) are associated with ring wave solitons.†[9]

Numerical computations based on computer simulations indicate a peculiar behavior of ring waves, depending on the specific relations between the wave parameters. In particular, three regimes arise which are of special interest:

(a) The radius $r(t)$ increases monotonically with t: a diverging ring wave.
(b) The radius $r(t)$ attains a maximum R and thence monotonically decreases to zero: the converging ring wave collapses.
(c) The radius $r(t)$ decreases monotonically to zero; the ring wave then collapses.

In the Japanese literature, ring waves are also frequently termed "explode-decay solitons."[11]
In the limit

$$qr_0 \ll 1 \tag{25}$$

the second term in Eq. (24) can be neglected, this equation thus reducing to the solution (Eq. 15) of a one-dimensional (i.e., in the radial direction) NSE of the type in Eq. (19), thus justifying the assumption in Eq. (21) for the two-dimensional case. Moreover, it is also seen that, for small values of r

$$\frac{\partial^2}{\partial r^2} a(r, t) \cong \frac{1}{r}\frac{\partial}{\partial r} a(r, t) \tag{26}$$

which again reduces to the one-dimensional NSE as $r \to 0$.

In Eq. (21) the wave function $a(r, t)$ represents the probability amplitude for the molecular excitation produced by the incoming photon. Due to cylindrical symmetry centered at the acceptor site, the probability amplitude $a(r, t)$ appears as a ring-shaped wave function.

---

† Ring wave solitons have been studied and numerical solutions presented in Ref. 10.

For small r, as observed above, Eq. (21) may be approximated by the one-dimensional nonlinear Schrödinger equation [cf. Eq. (19)]. In other words, the acceptor molecule can be considered as an "observer" which, when the excitation is very near to it, sees such an excitation traveling toward it as a one-dimensional soliton coming from a specific (radial) direction.

It is intended in our future plans to perform numerical computations of ring wave behavior as related to the energy transport in Scheibe aggregates. Of the three regimes mentioned above, (b) and (c) appear to be the most interesting ones because they involve a maximum radius from which a converging ring wave collapses toward a center. If such features were indeed confirmed for these aggregates, this would suggest the existence of some sort of a "natural" domain size around an acceptor molecule. This, in a way, would resemble the Debye critical distance in solid-state theory.

3. DISCUSSION

The observed efficiency of lateral energy transfer in Scheibe–Kuhn J-aggregates implies that excitations induced on single molecules, donors, of the aggregate do not interfere destructively while moving in the layer plane. The excitations must move coherently in order to be transport-efficient. This requirement is indeed achieved by the solitonic, or quasi-solitonic, ring wave discussed here. Once the ring wave meets an acceptor, coherence breaks down and the energy contained in the wave is released to the acceptor. The possibility of coherent energy transfer can be intuitively considered as due to the periodicity of a close-packed and highly ordered molecular lattice. The excitation then gives rise to a series of superimposed wave functions which coalesce into a localized wave packet, described by a wave function involving many molecules and traveling with a group velocity governed by the velocity of propagation of an adiabatic lattice deformation.

The extent of this macroscopic quantum system depends on the phase coherence length L. Consider, as a gross simplification, that there are only two possible paths linking a donor to an acceptor, each path associated with a wave function $A_n e^{i\rho}$, where $\rho(t)$ is the phase. Superposing the two wave functions yields

$$|A_1 + A_2|^2 = A_1^2 + A_2^2 + A_1 A_2 \cos(\rho_1 - \rho_2) \tag{27}$$

The cos term is a cross term, an interference term. If the transfer distance $l > L$, then this term can be neglected and all is linear, but the cos term shows that lossless transfer, coherent transfer, is only possible for $l < L$.

An applied magnetic field H will alter the phases of the wave functions and thus affect the secondary fluorescence emitted by the acceptor; that is, H will result in a different maximum donor–acceptor ratio allowing loss-free transfer.

Nonsolitonic exciton transfer is a series of scattering events, resulting in a strongly damped wave. However, in the case of a Scheibe–Kuhn organizate, a collective state may be formed in which nearly all the energy is concentrated near or at the exciton band edges. In a monolayer, the film thickness is comparable to the de Broglie wavelength of an electron or, more accurately, to the Fermi wavelength being the Fermi wave vector. Then, quantum effects become important, further modifying the energy level structure and leading to a "quasi-soliton" as discussed in this chapter. The mean free path of the exciton solitary wave then greatly exceeds the intermolecular (i.e., the lattice) spacing, resulting in energy being transferred coherently. The amplitude of the exciton component must then be interpreted in terms of the coherent, macroscopic wave function. The entire system, consisting of a majority of energy donors plus a small minority of acceptors, all closely and anharmonically linked, is thus associated with one macromolecular wave function. The virtually loss-free and extremely rapid energy transfer to the acceptor(s) observed in J-aggregates therefore represents a macroscopic quantum effect operating even at and above room temperatures. Although each chromophore exhibits, *per se*, linear absorption, the whole donor–acceptor system behaves as a one-quantum nonlinear system. In other words, it becomes energetically advantageous for the phases of the molecular wave functions to lock together. The present model and discussion so far does not take into account any electron transfer effects. However, the solitary wave may trap electrons and thus provide also a highly efficient electron transfer mechanism.

It is likely that the nonlinear, anharmonic coupling which is stipulated involves interactions between lateral hydrogen bonds between the filler and the fatty acid chains. Acoustic Toda lattice solitons do arise in hydrogen-bonded assemblies in which solitons are stable at the above room temperatures.

ACKNOWLEDGMENT

We wish to thank Professor Hans Kuhn and Dr. Lei-Ming Xie of the Shanghai Institute of Metallurgy for many illuminating discussions.

REFERENCES

1. G. C. Huth, F. Gutmann, and G. Vitiello,
2. L. M. Blinov, *Russ. Chem. Rev.* **52**(8), 713 (1983).
3. M. Sugi, *J. Mol. Electron.* **1**, 3 (1985).
4. D. Moebius and H. Kuhn, *Isr. J. Chem.* **18**, 382 (1979), and references cited therein; see also E. A. Bartnik and K. J. Blinovska, *Phys. Lett. A* **134**(7), 448 (1989).
5. D. Moebius and H. Kuhn, *J. Appl. Phys.* **64**, 5138 (1972).
6. M. Toda, *Theory of Non-Linear Lattices*, Springer, Berlin (1981).
7. M. Toda, *Springer Ser. Synerget.* **30**, 6 (1985); cf. also L. E. Reichl, *Phys. Rev. A* **28**(5), 30 (1983).
8. A. S. Davydov, *Solitons in Molecular Systems*, D. Reidel, Dordrecht (1987).
9. P. S. Lomdahl, O. H. Olsen, and P. C. Christiansen, *Phys. Lett. A* **78**(2), (1980).
10. P. L. Christiansen and P. S. Lomdahl *Physica D* **2**, 482 (1981).
11. A. Nakamura and F. Toda, *Macromol. Chem. Suppl.* **14**, 201 (1985).
12. C. B. Harris and D. A. Zwemer, *Annu. Rev. Phys. Chem.* **29**, 473 (1978).
13. S. Yomosa, *J. Phys. Soc. Jpn.* **53**, 3692 (1984).
14. S. Yomosa, *J. Phys. Soc. Jpn.* **52**, 1866 (1983).
15. V. Muto, A. C. Scott, and P. L. Christiansen, *Phys. Lett. A* **136**(1/2), 33 (1989).
16. P. Perez, *Phys. Lett. A* **136**(1/2), 37 (1989).
17. S. Yomosa, *Springer Ser. Synerget.* **30**, 242 (1985).

25

Electrochemical Aspects of Bone Remodeling

Nejat Guzelsu and Alvin J. Salkind

1. INTRODUCTION

The remodeling and the biofeedback properties of bones have been of substantial interest to the scientist. The goals of these studies have been mainly (1) to determine the real feedback mechanism of bones and (2) to utilize the nature of the feedback system of the bone tissue for accelerating the healing process of fractures and nonunions by artificially introducing a similar environment at the fracture site of the bones.

External forces which cause an internal stress field in the bone tissue play an important role in the structural adaptation of bones.[1] Since electrical signals can be detected when bone tissue is mechanically stressed,[2,3] it has been proposed that endogenous bioelectrical signals may be involved in bone homeostasis by a biofeedback system stemming from the electromechanical properties of bone.

Bone is a porous tissue containing a fluid phase and a calcified matrix composed of inorganic bone mineral (hydroxyapatite)[4] and organic components (mainly type I collagen).[5] The porosity of bone includes membrane-lined capillary blood vessels which function to transport nutrients and ions in bone, canaliculi and the lacunae occupied *in vivo* by bone cells (osteoblasts),[6] and micropores present in the matrix.[7] Bone fluid spaces support a compartmental model whereby the extracellular fluid in the calcified matrix is separated from the membrane-lined vascular channel system.[8]

The surface of the bone matrix is negatively charged under physiological conditions,[9-11] and in contact with the fluid phase it attracts opposite charges and repels the particles carrying the same electrical charge. This leads to a potential difference between the bone surface and the bulk of the fluid, the magnitude of the potential depending upon many parameters including chemical composition, pH, and temperature.[9,12]

Under an electric field the charged bone particles move through the suspending fluid due to electrostatic forces. As a unit, part of the attracted opposite-charge-carrying entities close to the surface also move with the bone particle. The surface of this unit is defined as

Nejat Guzelsu • University of Medicine and Dentistry of New Jersey—School of Osteopathic Medicine, Piscataway, New Jersey 08854, and Biomedical Engineering Program, Rutgers University, New Brunswick, New Jersey 08903.   *Alvin J. Salkind* • University of Medicine and Dentistry of New Jersey—Robert Wood Johnson Medical School, Piscataway, New Jersey 08854, and Biomedical Engineering Program, Rutgers University, New Brunswick, New Jersey 08903.

Electrochemistry in Transition, edited by Oliver J. Murphy *et al.* Plenum Press, New York, 1992.

the surface of shear,[13,14] and the potential at this surface with respect to the solution bulk is termed the electrokinetic or zeta (ζ) potential. From particle electrophoresis experiments, it is the ratio of the particle's velocity to the applied electric field, defined as the electrophoretic mobility, which is used in calculating the ζ potential. The direction of particle motion relative to electrode polarity dictates the sign of the potential.

When fluid is forced through a bone plug, the ionic charges in the fluid phase near the bone matrix–fluid interface are carried toward the low-pressure end. This constitutes a streaming current, and the accumulation of charge sets up an electric field. The field causes a conduction current flow in the opposite direction through the bulk of the liquid in the porous structure of bone tissue. At steady state, the conduction current is equal to the streaming current. The resulting electrostatic potential difference between both sides of the bone plug is the streaming potential.[14,15]

Zeta potentials can be calculated from streaming potential experiments on porous plugs by knowing the applied pressure difference across the sample and the generated streaming potential.[14,15] The zeta potential is the common link among the different electrokinetic phenomena (i.e., streaming potentials, particle electrophoresis) and allows comparison of the different measuring techniques.

Here we present the results of particle microelectrophoresis measurements which are performed over a pH range while maintaining ionic strength and temperature constant.[16] In doing so, the isoelectric point (IEP) is identified as the pH where no particle motion can be detected under an applied electric field, hence where the ζ potential is zero. Particles suspended in solutions of pH below and above their IEP display positive and negative ζ potentials, respectively. The overall electrokinetic behavior of different particle groups (intact, partially deorganified, decalcified, and bone mineral) is therefore obtained from their ζ potentials as functions of pH. In addition to particle electrophoresis, the results of streaming potentials of intact cortical wet bone samples parallel to the long bone (z) axis at room temperature are presented.[17] Fresh bovine femur samples were tested with and without chemical treatments, to examine the role of some bone matrix constituents on streaming potentials and zeta potentials.

Zeta potentials of particle electrophoresis and streaming potential results will be compared and the differences will be discussed.

2. MATERIALS AND METHODS

In these experiments bovine femurs were used. A diamond disk cutter was used to obtain cortical bone particles for particle electrophoresis. Bone plug samples were cut perpendicular to the long bone (z) axis to a thickness of approximately 0.50 ± 0.05 mm to measure streaming potentials parallel to the long bone (z) axis.

Two types of buffers were used at room temperature, with an ionic strength of 0.145 with different pH values.[18]

The first type of buffers were prepared without phosphate ion, which is a potential-determining ion for the bone mineral. (Phosphate ions are known to adsorb to bone mineral and alter its surface chemistry and electrokinetic behavior.) The second type of buffers were used to examine the role of phosphate ions. Figure 1 schematically shows the particle electrophonesis measuring system (Mark II, Rank Brothers, U.K.). The flat cell configuration has the advantage of allowing heavy particles falling under gravity (perpendicular to the electric field) to remain in the field of observation during electrophoresis. Electrophoretic mobilities were calculated as:

$$\mu = \frac{\text{Velocity}}{\text{Electric field}} = \frac{(d/t)}{(V/l)} \tag{1}$$

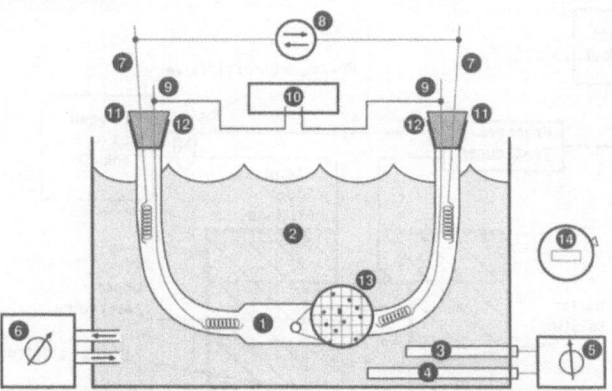

FIGURE 1. Schematic of electrophoretic rectangular cell (1) in a water bath (2) whose temperature is maintained uniformly at 25°C via a temperature probe (3), a heater (4), a thermal control unit (5), and a circulating water pump (6). The cell is equipped with a pair of outer Ag/AgCl electrodes (7) supplying a current of 4.5 mA from a reversible-polarity constant-current source (8) and a pair of inner Ag/AgCl electrodes (9) for voltage sensing to read on a digital multimeter (10). Electrodes enter the cell (1) via sealed ground glass stoppers (11) adapted with standard taper Teflon sleeves (12) to form a proper junction seal. The microscope objective expanded view (13) shows the grid spacing ($d = 33.33 \ \mu$m) over which the suspended particle transit times are noted with a stopwatch (14) during electrophoresis.[16]

where d is the fixed grid distance (μm), t is the calculated transit time (s), V is the potential difference between sensing electrodes (V), and l is the sensing interelectrode distance (cm) (Fig. 1). Zeta potentials were then calculated from mobilities[14,19] as per Smoluchowski's equation:

$$\zeta = \frac{\eta}{\varepsilon_0 D} \mu \tag{2}$$

where η, ε_0, and D at 25°C are, respectively, the viscosity of the water (8.904×10^{-4} N s/m^2), the permittivity of free space (8.8542×10^{-12} F/m), and the dielectric constant of water (78.85), which measures its relative permittivity. The following bone particle groups were used:

(a) Untreated bone particles (control).
(b) Partially deorganified bone. This group is obtained by treating bone particles in $4M$ guanidine hydrochloride (GuCl). This treatment removes some organic constituents of bone by dissociative extraction.[20]
(c) Decalcified bone particles. This group is obtained by treating bone particles with $0.25M$ disodium EDTA.
(d) Extracted bone minerals. This group is obtained by anhydrous ethylenediamine extraction on a soxhlet setup.

Bone plug streaming potentials were measured in a specially designed chamber (Fig. 2). Bone plugs were placed in the chamber between the upper and lower reservoirs and sealed with a single O-ring. Pressure gradients across the sample were supplied by nitrogen gas and detected by a pressure transducer attached to the upper reservoir. The following sample groups were used:

(a) Control samples (intact).
(b) Partially deorganified bone (GuCl treated).
(c) Decalcified bone samples (EDTA treated).
(d) Bone mineral (deorganified by NaOCl).

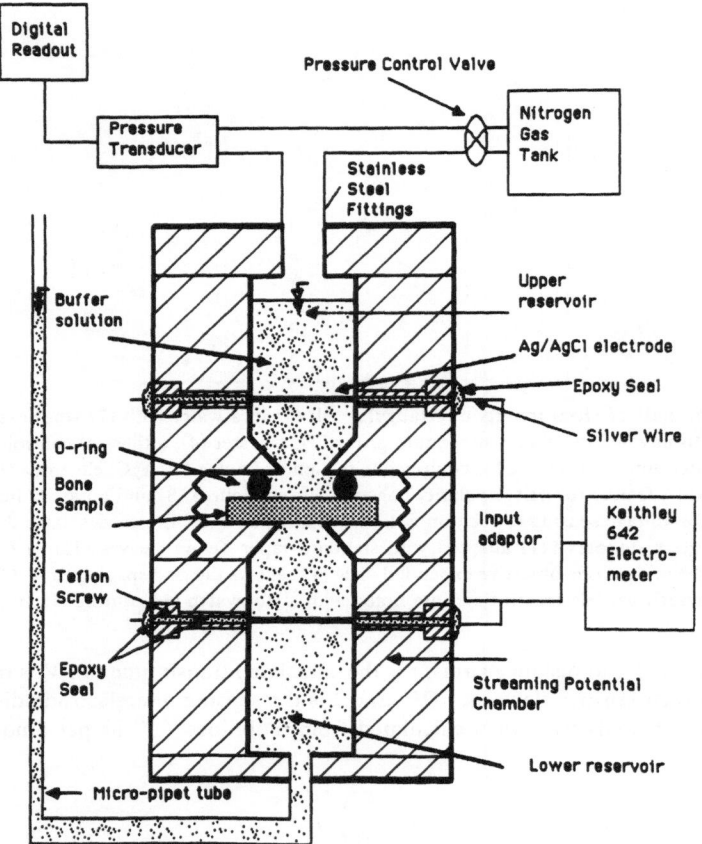

FIGURE 2. A schematic drawing of the streaming potential chamber used in our experiments. This design utilizes a high-impedance electrometer (Keithley 642) for recording streaming potentials and a pressure transducer to record the applied pressure gradient.[17]

3. RESULTS

Results of particle electrophoresis are summarized in Fig. 3 for all the sample groups.[16] The following points are revealed in Fig. 3:

(a) Bone minerals show a high isoelectric point (IEP = 8.6) as compared to the intact bone particle group (IEP = 5.1). Therefore, ζ potentials of opposite signs are noted for hydroxyapatites ($+6 < \zeta < +9$ mV; open squares) with respect to intact bone particle ζ potentials ($\zeta \cong -2.8$ mV) within the physiological pH range.

(b) A ζ-potential sign reversal for hydroxyapatites in phosphate buffers (solid points) is observed.

(c) Within the physiological pH range, there is no significant difference between zeta potentials of intact and EDTA-treated bone particle groups. These two groups, however, have significantly less respective negative zeta potentials as compared to partially organified bone particles.

The streaming potential experimental technique measures the streaming potential and applied pressure gradient across the sample and allows use of the following formula developed

FIGURE 3. Zeta potentials as function of pH at 25°C for intact cow bone particles (O——O), partially deproteinated (GuCl treated) bone particles (△ - - - △), EDTA-treated bone particles (▽), and ethylenediamine-extracted cow bone hydroxyapatites (□). Solid data points indicate results obtained with the use of phosphate-containing buffers. Curves were computer fit with high-order polynomials (not including phosphate data) for $r^2 \geq 0.95$, and individual mean ζ potential standard errors (SE) are only shown by error bars for those values with SE > ±0.5.[16]

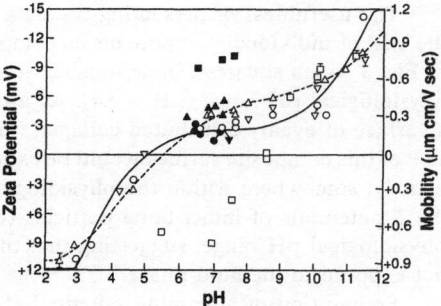

for single capillary and porous plugs to calculate zeta potentials[15]:

$$\frac{E_s}{\Delta P} = \frac{\varepsilon_0 D}{\eta \lambda} \zeta$$

where E_s, ζ, ΔP, ε_0, D, η, and λ are the streaming potential (mV), zeta potential (mV), pressure (Pa), permittivity of free space (8.8542×10^{-12} V/m), relative dielectric constant (78.35), viscosity (8.904×10^{-4} N s/m²), and conductivity of $0.145M$ NaCl, measured with an ac conductivity bridge ($1.452 \, \Omega^{-1} \, m^{-1}$), respectively, at 25°C.

The results of streaming potential experiments are summarized in Table 1. The table shows that control and EDTA-treated samples have similar streaming potential slopes. This indicates that the inorganic phase plays little or no role, while the organic phase of bone is the main contributor to streaming potentials in aphosphate buffer at pH 7.3. The slopes for GuCl-treated samples are higher than those for controls. This indicates that GuCl treatment exposes a more highly charged organic layer to participate in the streaming potential behavior. Totally organified samples (NaOCl-treated) have a negative streaming potential in aphosphate buffer at pH 7.3 and a positive streaming potential in phosphate buffer at pH 7.3 (Table 1). This again supports the idea that the role of the mineral phase in control samples in aphosphate buffer, if any, is minimal. The sign reversal observed for NaOCl-treated samples in phosphate buffer reflects adsorption of negatively charged phosphate to the positively charged bone mineral at pH 7.3 (Table 1). In Table 1, positive slope corresponds to negative ζ potential.[17]

4. DISCUSSION

The organic phase of the bone matrix dominates the electrokinetic behavior of bone particles by screening the bone mineral from the fluid phase.

TABLE 1
Effects of Various Chemical Treatments on Streaming Potentials

Treatment	Equilibration time and buffer[a]	Average thickness (cm)	n^b	Slope[c] (× 10⁻³)	S.D.[d] (× 10⁻⁴)
Controls	24 h, AP	0.49	6	3.258	3.397
GuCl	24 h, AP	0.49	6	5.710	9.730
EDTA	24 h, AP	0.47	7	3.470	9.68
NaOCl	24 h, AP	0.45	9	−4.290	7.220
NaOCl	24 h, P	0.47	4	4.370	0.816

[a]AP, phosphate buffer. [b]n = number of samples. [c]Slope units are mV/kPa. [d]S.D. = Standard deviation.

The usefulness of presenting the data as pH versus ζ potential is found in considering the IEP of individual components as compared to that of intact bone particles. As indicated in Fig. 3 (open squares), bone minerals (IEP = 8.6) have positive potentials throughout the physiological pH range (pH = 7.4). As a thought experiment, if it were possible to prepare a surface of evenly distributed collagen (IEP = 5.1) and bone mineral (IEP = 8.6), then the IEP of this composite surface would be expected to lie between the IEPs of these components, namely, somewhere within the physiological pH range. However, as can be seen in Fig. 3, the ζ potentials of intact bone particles (open circles) are negative in sign throughout the physiological pH range, suggesting that the component of higher IEP (e.g., the mineral) is not exposed to the fluid phase.

Preparation of streaming potential (SP) samples maintains the *in vivo* geometry of a calcified matrix and a lined vascular channel system. Particle electrophoresis (EP) requires bone to be ground to 5-10-μm particles. This removes the *in vivo* geometry and exposes portions of the calcified matrix to the fluid phase previously buried.

The different zeta potentials for comparable treatments between SP and EP can be attributed to the different electrokinetic properties of the lined channel system, where fluid flow occurs in SP, and the calcified matrix, exposed and examined in EP experiments (Table 2). Zeta potentials of SP and EP converge only following total deorganification of the samples (SP-NaOCl and EP-EDY in Table 2). This reflects the removal of the organic layers and compartments that separate the calcified matrix from the channel system in SP and EP.

Deorganification produces a more homogeneous sample of bone mineral alone that is very similar for both SP and EP, yielding closer zeta potentials. Small variations between organified samples (SP versus EP) may be due to the effect of different chemical treatments in obtaining the mineral phase (Table 2).

Deorganified samples tested in phosphate buffer solution have a charge reversal from positive to negative, compared to control samples. This is due to the adsorption of negatively charged phosphate ions to the bone mineral. Phosphate, calcium, and hydroxyl ions are

TABLE 2
Zeta Potential Calculations from Streaming Potential and Particle Electrophoresis[a]

Authors	Tissue	Sample	Technique[c]	Buffer[d,e]	Treatment[f]	Zeta potential (mV)
Guzelsu and Walsh[a]	Bovine femurs	Intact	SP	AP	Control	−6.07
				AP	EDTA	−6.46
				AP	GuCl	−10.64
				AP	NaOCl	+7.99
				P	NaOCl	−8.14
Guzelsu and Regimbal[b]	Bovine femurs	Particles (5–10 μm)	EP	AP	Control	−2.80
				AP	EDTA	−3.1
				AP	GuCl	−5.0
				AP	EDY	+7.1
				P	EDY	−9.7

[a] Ref. 17.
[b] Ref. 16.
[c] SP, Streaming potential; EP, electrophoresis.
[d] pH = 7.3; ionic strength = 0.145.
[e] AP, Aphosphate buffer; P, phosphate buffer.
[f] EDTA, Ethylenediaminetetraacetic acid (decalcified samples); GuCl, guanidine hydrochloride (partially deproteinized samples); NaOCl, sodium hypochlorite (deorganified samples); EDY, ethylenediamine (deorganified samples).

known as potential-determining ions (PDI) for bone mineral. A PDI can adsorb to the mineral surface and alter the surface charge and the electrokinetic potential.

The electrokinetic behavior of the channel system versus the calcified matrix is evident from different values for control and GuCl-treated samples between SP and EP. The electrokinetic nature of the channel system is revealed further by GuCl treatment. GuCl-treated SP samples in aphosphate (AP) buffer have significantly greater negative zeta potentials compared to control SP samples. This is attributed to the dissociation of loosely bound proteins with different rates, exposing a more highly charged organic layer (similar to EP samples).

The zeta potential of the mineral alone at pH 7.3 in aphosphate buffer is positive. Zeta potentials of control, GuCl-treated, and detergent treated (nonidet) samples remain negative at pH 7.3, suggesting that the mineral phase has not been exposed. Data indicate that organic barriers are present in the channel system that either limit or prevent phosphate ion adsorption to the bone mineral in the calcified matrix. The diffusion barriers that limit phosphate penetration are present in control and GuCl-treated samples. Therefore, during the time span of equilibration (24 h for both experiments) phosphate ions could not penetrate and alter the zeta potentials.

Data from the study of streaming potentials suggest that the bone is a compartmental tissue with unique fluid spaces corresponding to the channel system and the calcified matrix. This may account for the differences between intact (SP) and particle (EP) measurements of bone in the literature. In addition, the channel system seems to be composed of organic layers that can limit the penetration of phosphate ions to the mineral phase in intact bone and its contribution to the electrokinetic properties.

Streaming potential experiments reflect the electrokinetic behavior of the lined vascular channel system. Particle electrophoresis experiments reveal information on the electrokinetic properties of the calcified matrix. Both techniques complement each other, yielding information on the effects of chemical treatments and ions on the electrokinetic behavior of bone and its constituents.

REFERENCES

1. J. Wolff, *The Law of Bone Remodelling* (translated by P. Maquet and R. Furlong), Springer, Berlin (1986).
2. C. A. L. Bassett, in: *The Biochemistry and Physiology of Bone* (G. H. Bourne, ed.), Vol. III, Second ed., pp. 1–76, Academic Press, New York (1971).
3. C. Eriksson, in: *The Biochemistry and Physiology of Bone* (G. H. Bourne, ed.), Vol. IV, Second ed., pp. 329–384, Academic Press, New York (1976).
4. A. S. Posner and F. Betts, *Acc. Chem. Res.* **8**, 273 (1975).
5. H. C. W. Skinner, in: *The Scientific Basis of Orthopaedics* (J. A. Albright and R. A. Brand, eds.), pp. 105–134, Appleton-Century Crafts, New York (1979).
6. M. W. Johnson, *Calcified Tissue Int.* **36**, S72 (1984).
7. S. Hughes, R. Davies, R. Khan, and P. Kelly, *Clin. Orthop. Rel. Res.* **134**, 332 (1978).
8. C. J. Vander Weil, S. A. Grubb, and R. V. Talmage, *Clin. Orthop. Rel. Res.* **134**, 350 (1978).
9. C. Eriksson, *Clin. Orthop. Rel. Res.* **121**, 295 (1976).
10. N. Guzelsu and J. Donofrio, *J. Bioelectricity* **2**, 187 (1983).
11. D. A. Baretta and S. R. Pollack, *J. Orthopaed. Res.* **4**, 337 (1986).
12. D. N. Misra, in: *Methods of Calcified Tissue Preparation* (G. R. Dickson, ed.), pp. 435–465, Elsevier, New York (1984).
13. J. O'M. Bockris and A. K. N. Reddy, *Modern Electrochemistry*, Vols. 1 and 2, Plenum, New York (1970).
14. R. J. Hunter, *Zeta Potential in Colloid Science*, Academic Press, London (1981).

15. J. Th. G. Overbeek, in: *Colloid Science* (H. R. Kruyt, ed.), pp. 174–244, Elsevier, Amsterdam (1952).
16. N. Guzelsu and R. L. Regimbal, *J. Biomech.* **23,** 661 (1990).
17. N. Guzelsu and W. R. Walsh, *J. Biomech.* **23,** 673 (1990).
18. G. L. Miller and R. H. Golder, *Arch. Biochem.* **29,** 420 (1950).
19. D. C. Henry, *Proc. Roy. Soc.* **133,** 106 (1931).
20. J. D. Termine, S. Wientroub, and L. W. Fisher, in: *Methods of Calcified Tissue Preparation* (G. R. Dickson, ed.), pp. 547–563, Elsevier, New York (1984).

26

Electrochemistry Applied to Medical Diagnostics

Marvin A. Genshaw

1. INTRODUCTION

The present discussion is limited to direct applications of electrochemical sensors for measuring analytes. Measurement of electroactive species produced by reactions outside of the integrated sensor is not considered.

For the general discussion of the reasons for performing tests and general techniques, the books edited by Tietz[1] and Henry *et al.*[2] have been used as references. For general references on the electrochemical methods, the book by Eisenman[3] and those edited by Koryta[4] and Pungor[5] are recommended.

Potentiometric and amperometric measurements are the electrochemical methods usually applied for measurement of clinical samples. Other methods are rarely useful. Diffusional barriers and enzymes are often utilized in the sensors.

The basic equation used in potentiometric measurements is

$$V = V_0 + (RT/nF) \log a \tag{1}$$

The measured potential is logarithmically dependent on the activity of the ion. For a monovalent ion at 25°C, the slope of a plot of log a versus activity is 60 mV per decade of activity.

To illustrate such a sensor, consider the potassium electrode. It consists of an inner reference element, such as a silver electrode, coated with the insoluble salt silver chloride. Surrounded by a solution of potassium chloride, this provides an inner reference electrode with constant potential. A membrane selective for potassium ions separates the inner reference electrode from the test sample. Selectivity for potassium is achieved by dispersing a selective compound in the membrane. Valinomycin is an antibiotic which has a high specific affinity for potassium. When valinomycin is present in a membrane, the membrane becomes a cation-exchange membrane and has a very high specificity for potassium. The combination of an inner reference electrode, inner solution, and selective membrane forms an ion-selective electrode (ISE).

Marvin A. Genshaw • Diagnostics Division, Miles Incorporated, Elkhart, Indiana 46515.

Electrochemistry in Transition, edited by Oliver J. Murphy *et al.* Plenum Press, New York, 1992.

The potential of the ion selective electrode must be measured with respect to a second reference electrode. A saturated calomel electrode or silver/silver chloride reference electrode is usually used. The reference electrode is connected to the test sample by means of a salt bridge. The bridge is often saturated potassium chloride. The salt bridge provides a nonselective, electrolytic connection between the reference element and the test sample.

For amperometric measurements, a catalytic metal electrode is used to carry out an oxidation or reduction reaction. For example, consider the oxidation of hydrogen peroxide. At a platinum electrode, hydrogen peroxide may be oxidized to oxygen:

$$H_2O_2 = O_2 + 2H^+ + 2e^- \tag{2}$$

or reduced to water:

$$H_2O_2 + 2H^+ + 2e^- = 2H_2O \tag{3}$$

The reaction of Eq. (2) readily occurs at the platinum electrode in neutral solutions at +0.7 V versus a silver/silver chloride electrode; the reaction of Eq. (3) occurs at −0.7 V versus a silver/silver chloride electrode.

For amperometric measurements the catalytically active electrode is usually covered by a membrane which limits the rate of transport of the analyte of interest to the electrode surface. The diffusion-limited rate of transport provides a linear relation between concentration and current:

$$i = kc \tag{4}$$

The amperometric method of measurement requires that the sensed species participate in an electrochemical reaction at the electrode surface. Only a few species of medical diagnostic interest are electroactive. Other species must be transformed into an electroactive species. Enzymes are often used to produce the electroactive species. Enzymes are biochemical catalysts which often have high selectivity for a single species. In addition, the products of the reaction often include an electroactive species such as hydrogen peroxide or nicotinamide adenine dinucleotide (NAD). These reaction products can be oxidized or reduced to provide a current.

The diffusion membrane is often chosen to provide some selectivity for the desired species over other potentially interfering species such as ascorbic acid (vitamin C) or uric acid.

As an example of an amperometric sensor, consider the glucose electrode. The glucose electrode consists of a platinum sensing electrode and a silver electrode acting as a combined reference and counter electrode. A counter electrode is necessary to complete the electrical circuit. An electrochemical reaction occurs at it to convert the ionic current in the test solution to the electronic current in wires connecting to the ammeter.

The platinum electrode is coated with immobilized glucose oxidase. Glucose and oxygen react to form gluconolactone and hydrogen peroxide in the presence of glucose oxidase. The enzyme glucose oxidase provides a very high specificity for glucose. The hydrogen peroxide produced by the reaction can either be oxidized back to oxygen or reduced to water at the platinum surface. Since two electrons are involved in either reaction, the sensitivities are similar. Other factors need to be considered. The oxidation restores the oxygen consumed in the reaction of glucose with oxygen and thus aids the complete reaction of glucose. If the hydrogen peroxide is reduced, the potential is cathodic enough so that the oxygen is also reduced. Thus, the sensor senses the quantity of oxygen consumed by the reaction of the glucose. Measuring oxygen consumption reduces the interference from blood components that react with hydrogen peroxide but results in an undesirable sensitivity to the sample oxygen concentration.

To linearize the response, a diffusion membrane is placed over the electrode and immobilized enzyme layer. With faster enzymatic and electrode reactions than membrane diffusion, the response is diffusion limited and is linear in analyte concentration. In addition, convective transport of the analyte to the sensor surface is not limiting. In some commercial sensors, an additional membrane separates the platinum from the glucose oxidase layer to improve the rejection of electroactive interferences with peroxide measurement such as ascorbic acid and uric acid.

Potentiometric measurements allow determining concentrations over a wider concentration range than amperometric measurements. However, to achieve accurate concentration measurements by potentiometry, variations in temperature and liquid junction potential must be held very small. A one-millivolt potential error corresponds to a concentration error of 4% for a univalent ion and 8% for a bivalent ion. The dependence of concentration error on millivolt error is a direct consequence of the logarithmic relation between activity and potential. Amperometric determinations require less stringent conditions than potentiometric measurements to achieve similar accuracy.

2. BLOOD GASES

The measurement of blood gases includes the measurement of pH, oxygen, and carbon dioxide. In some cases carbon monoxide may also be measured, but this is not done electrochemically.

The blood gases are of primary importance in determining the acid–base balance of the body. The reactions of oxygen produce acidic by-products. The exhalation of carbon dioxide removes acid from the body. In addition to the pulmonary role in acid–base balance, the renal function is also very important.

The measurement of blood gases is important in the treatment of respiratory and metabolic disorders and cardiopulmonary impairment. Premature infants often have respiratory impairment and may require monitoring of blood gases. For most of these applications, an immediate result is of critical importance. During surgery, continuous monitoring of blood gases may be done to ensure that life is sustained.

The precise measurement of blood pH is critical in determining the acid–base balance of the patient. Although arterial blood samples are preferred, venous samples may be used. The blood pH may change in either metabolic or respiratory imbalance. A change of blood pH outside of the range of 6.8 to 7.8 leads to a significant risk of mortality. The normal range for blood pH is 7.31 to 7.45. It may be measured to an accuracy of 0.01 units.

Glass pH electrodes are usually used with a thermostated cell to ensure constant temperature. The blood sample must either be measured immediately after it is taken or kept in a hermetically sealed container. Otherwise, loss of carbon dioxide changes the pH of the sample. Efforts to develop *in vivo* monitoring based on pH-sensitive field effect transistors have also been reported.[6]

A number of critical factors are involved in measuring blood pH. The measurements should be made at 37°C since the blood pH varies with temperature by about 0.015 pH units/°C, and the electrode response is also temperature sensitive. The readings may have to be corrected for patient body temperature if the patient's temperature is more than 2°C from normal. The sample should be measured quickly since loss of carbon dioxide results in a change in pH. Metabolism of glucose in the sample reduces the pH by about 0.00035 units per minute.[7]

The liquid junction potential is also significant since it may amount to several millivolts. Erythrocytes are reported to give shifts in the junction potential by about a millivolt from the value in plasma. The calibrators must be very close to the composition of blood to minimize the difference in liquid junction potential between the calibrators and blood samples.

Since carbon dioxide is a significant factor in determining the blood pH and is also a critical parameter in determining the source of pH deviations from normality, carbon dioxide is usually measured with pH.

Carbon dioxide is measured by covering a pH electrode with a thin layer of bicarbonate buffer. A carbon dioxide-permeable membrane, such as Teflon, separates the sample from the buffer.[8, 9] The membrane is impermeable to carbonate and other ions. Carbon dioxide permeating the membrane comes to equilibrium with the bicarbonate buffer and determines the pH of the buffer. The pH change is sensed by the pH electrode.

The pH of the blood and the bicarbonate buffer system are related through the Henderson-Hasselbalch equation[10]:

$$pH = pK' + \log([HCO_3^-]/[H_2CO_3]) \tag{5}$$

From the known bicarbonate concentration and the measured pH, the carbon dioxide partial pressure can readily be calculated. The normal levels for carbon dioxide are 33–48 mm Hg in arterial blood and 38–52 mm Hg in venous blood.

Carbon dioxide has more recently been measured by the use of a carbonate ISE in place of the pH electrode.[11, 12]

Oxygen is normally measured by oxygen reduction at a platinum electrode. A thin layer of a buffered potassium chloride solution covers the surface of the platinum electrode and an adjacent silver/silver chloride reference–counter electrode. The platinum electrode is held 0.7 V negative to the silver electrode. A membrane of polypropylene (or other polymer) is used as a diffusional barrier for the oxygen to provide a current, which is linear in the sample's oxygen concentration.[9, 13]

The partial pressure of arterial oxygen is normally 80–104 mm Hg while that of venous oxygen is 20–49 mm Hg. The accuracy of the determination is 2–5%.

3. SERUM AND BLOOD ELECTROLYTES

Electrolytes play an important role in nerve and muscle function. Sodium is important in blood pressure regulation, and potassium in heart function. Relatively small imbalances in these electrolytes can lead to very serious risks to health. Lithium is not an essential electrolyte but is frequently used in therapy for treating depression. An excess is toxic. The measurement of ionized calcium has proven valuable in the diagnosis of primary hyperparathyroidism, and a deficient level of serum calcium leads to convulsions.

The development of ion-selective electrodes has led to their wide use in the measurement of electrolytes in clinical samples.[14, 15] ISE methods have significantly displaced the older methods of flame photometry, colorimetric methods, and titrations.[16]

The change in technology from measuring cation concentration to cation activity has caused changes in medical decision levels.

The development of electrodes which are selective for potassium and sodium made the measurement of cations in serum practical by electrochemical methods. The initial electrodes developed were selective glasses. The sodium-selective glass electrode is sufficiently selective for sodium over potassium and the other ions present in serum to make a direct potentiometric measurement practical. The glass electrodes for potassium and other cations are not sufficiently selective to make their direct determination practical.

Highly selective chelating agents for potassium, such as valinomycin with a selectivity of about 5000 to one for potassium over sodium, made an ion-selective electrode for potassium in serum possible.[17, 18] Selectivity is still a critical factor for analyzing clinical samples.

The first clinical analyzers required dilution of the sample. The dilution into a buffer resulted in a constant activity for the ions. Thus, ion activity and concentration were proportional. These measurements gave good correlations of measured concentrations to flame photometer results.

More recently, analyzers for undiluted samples have been introduced. These analyzers provide a direct measurement of the activity of the ions rather than the concentration.

Comparison of potassium and sodium measurements with the flame photometer to those obtained by undiluted ISE has shown differences. A major difference is that the ISE measures the activity in the aqueous portion of the serum sample. The flame measures the total concentration. Since a normal serum contains about 7% lipids, the flame values are about 7% lower than those from the ISE. One can compensate for this bias by calibrating with solutions containing lipids. However, many pathological samples are high or low in lipids or proteins. The ISE provides a correct measurement of the ionic activity (concentration), which is the physiologically meaningful value. The value measured with the flame may be misleading.[19-23] The normal ranges of electrolytes used for making medical decisions will be different for flame photometry (concentration) and ISE (activity).

Other differences arise due to liquid junction effects. The liquid junction potential is somewhat dependent on the composition of the sample. Liquid junction potential effects appear as a change in the concentration of the ion. For measurements on whole blood, the erythrocytes have been shown to cause changes in the liquid junction potential of about a millivolt. Also, the activity is dependent on the concentration of other ions. Since the ISE is not absolutely specific, interferences are also sensed. For some time lithium could not be measured because the lithium ISE is insufficiently selective since lithium may be present at $1/1400$ of the concentration of sodium.[18] This has now been partially overcome, and clinical systems for lithium in the therapeutic range are available.[24] A simultaneous measurement of sodium allows correction of the lithium response for interference by sodium.

The measurement of serum calcium is somewhat more complex. Much of the calcium is specifically bound by proteins, mainly albumin. For this purpose, the ISE is the preferred method. One can estimate the total calcium from the ionized calcium, particularly if the albumin is measured.[25, 26]

Chloride is measured by a silver/silver chloride electrode with an anion-selective membrane.

In Table 1 the normal concentration ranges and changes in ISE potential for the common electrolytes are shown. The ranges are based on published data.[27, 28] The normal range is the range of concentrations observed in a healthy population. The dV is the millivolts change in electrode potential for a change in concentration from one extreme of the normal range to the other. The S.D. MS is the maximum allowable standard deviation in millivolts to obtain medically significant results.

For sodium the potential range for normal variability is very small. The electrode system must be controlled very well to permit useful sodium determinations. For other ions such as

TABLE 1
Potential Ranges for Electrolytes

Electrolyte	Normal range (mM)	dV (mV)	S.D. MS[a] (mV)
Lithium	0.7–1.5	19.1	1.0
Sodium	130–150	3.7	0.4
Potassium	3–6	18.0	1.2
Calcium	8–11	8.3	0.6
Chloride	90–110	5.2	0.5

[a] Maximum allowable standard deviation to obtain medically significant results.

potassium, the demands on potential resolution are not as great. As is evident from the table, the measurement of sodium, calcium, and chloride places great demands on the electrochemical determination. The requirements for measuring potassium and lithium are not as stringent in potential resolution requirements. However, high selectivity over sodium is required to ensure that variations in sodium are not sensed rather than the desired potassium or lithium. For a sodium concentration of 150mM to appear as a lithium concentration of less than 0.05mM, a selectivity of 3000:1 (150/0.05) is needed. To give an error of less than 0.05mM in lithium for sodium variations in the normal range of ±10mM, a selectivity of 200:1 (10/0.05) is sufficient. Calibration with lithium standards in a 140mM sodium calibrator is then necessary.

Disposable electrodes for the measurement of electrolytes are also commercially available. Kodak's system uses adjacent pairs of electrodes with the serum sample placed on one and a calibration solution on the other. The differential measurement then provide a good measurement of the ion activity even though electrodes are changing in absolute potential as they become hydrated.[29]

4. SERUM AND BLOOD METABOLITES

Determining metabolite concentrations is very important in diagnosing many diseases. The term metabolite encompasses a diverse range of compounds. For carbohydrates, a number of genetic defects or acquired conditions are associated with defects in metabolism. A common condition is diabetes mellitus, in which the metabolism of glucose is impaired, resulting in levels of blood glucose higher than normal. Cholesterol and triglyceride levels have received considerable attention recently due to their relation to coronary heart disease. Renal disease will result in increases in urea and many other products usually excreted by the kidneys.

Electrodes responding to many metabolites and enzymes have been reported.[30-37] However, only a few have become commercial devices at present. Several factors have limited their application. Human fluids are broths with many components. Usually, some of these interfere with measurement of the component of interest. For example, many electrode systems sense by the oxidation of a species. Naturally found reducing agents such as ascorbic acid (vitamin C) and uric acid are easily oxidized at electrodes and produce an interfering response. Also, species such as proteins and lipids coat electrodes and foul them. This coating progressively degrades the response. The metabolites of interest are also found in more than one form. Cholesterol is found both free and as esters and bound in complex protein globules restricting access to electrodes. Other species may be quite similar to a metabolite so that high specificity is required in the sensor. The problems have made in vivo measurements very difficult.[38]

Electrochemical glucose sensors based on glucose oxidase were described in the introduction. The glucose sensor has application both in analytical determination of blood glucose and in the closed-loop application of the artificial pancreas.[39] The pancreas is an organ which secretes insulin. Insulin is the hormone regulating the blood glucose concentration. The artificial pancreas is a combination of an electrochemical glucose sensor with a device to infuse insulin. With a closed-loop system, control of glucose equivalent to that of a person with a normally functioning pancreas is possible. Much research is devoted to producing an implantable glucose sensor for an artificial pancreas. For implantation, the enzyme-based glucose electrode is not totally satisfactory because the enzyme slowly degrades. Alternate approaches involving the direct electrooxidation of glucose have been investigated.[40] It is possible to partially or completely oxidize glucose electrochemically at a suitable catalytic electrode. The ultimate products are carbon dioxide and water.

Disposable glucose electrodes are also being commercialized.[41] They are used with a drop of blood obtained from a finger prick.

Not all metabolite sensors are based on electrooxidation or reduction reactions. A number have also been described which rely on pH changes produced by enzymatic reactions. A simple case is the urea electrode.[42] Urea is decomposed to carbonate and ammonium by the enzyme urease. The resulting increase in pH is measured by a pH electrode. To compensate for the variable sample pH, a second pH electrode is used to measure the sample pH. Alternately, a measurement of the rate of change of pH can be made. In addition to the urea electrode, a number of similar sensors are made which rely on deaminase enzymes to produce ammonia from poteins.[42]

5. DRUG AND HORMONE SENSORS

The level of a number of drugs in serum is very important in long-term therapy for a wide variety of diseases. At insufficient concentrations the drug is ineffective. At an excess concentration, the drug may be toxic. The absorption, metabolism, and secretion of drugs is variable from patient to patient. Thus, monitoring of drug levels is often important. The concentration of drugs is often very low (1 to 100 μM) and thus is most difficult to measure electrochemically.[43]

Hormone levels are also of diagnostic importance since pathological conditions result from abnormal levels. The concentrations of hormones are extremely low (10 to 100 nM).

One approach to the determination of drugs is to prepare selective electrodes directly responding to the drug of interest. This can be done for drugs which are ionic. A membrane containing the drug is used to make an electrode similar to other ion-selective electrodes. This electrode will respond to the drug concentration just as an ISE. A major problem in the use of this type of sensor with clinical samples is the lack of selectivity for the drug ion over other organic ions. Serum samples contain a virtual soup of organic ions. Thus, the sensors work well only inaqueous environments containing only the drug of interest and inorganic buffers, and not serum samples.[44]

An alternate approach to the analysis of low levels of species has proven more practical. For example, an electrode responding to insulin will be describd. An electrode is prepared with antibody to insulin attached to its surface or incorporated into a membrane covering the electrode surface.[45] Antibodies are proteins synthesized biologically which have high specific affinity for a single molecular species. Antibody is utilized in the biological recognition of foreign species such as bacterial or viral invaders of the body. The attachment of an antibody to a foreign species is an initial step in the neutralization and removal of the species.

In addition, a preparation is made of the species of interest (i.e., insulin) chemically linked to an enzyme which can participate in a reaction producing an electroactive species.[46, 47] An example is glucose oxidase, which will produce hydrogen peroxide by catalyzing the reaction of glucose with oxygen. An aliquot of a solution of enzyme-labeled insulin is mixed with the clinical sample. The electrode is exposed to this mixture. The antibody on the electrode surface reacts with the insulin and the glucose oxidase-labeled insulin. The high affinity of the antibody results in all antibody molecules reacting with insulin, whether labeled or unlabeled. The fraction of the antibody molecules on the electrode surface with an attached glucose oxidase molecule is inversely proportional to the concentration of insulin present in the clinical sample.[47]

After washing, the electrode is placed in a standard glucose solution. The response of the electrode will be dependent on the relative amounts of the insulin present in the sample (unlabeled) and the glucose oxidase-labeled insulin added. The antibody electrode approach can produe a very sensitive electrode sensor. This approach has the disadvantage that it requires several steps. Direct antibody-based potentiometric electrodes responding to the drug digoxin have also been report.[48, 49]

6. INTEGRATED SENSORS WITH ELECTRONICS

An array of sensors with an integrated array of electronic components such as pre-amplifiers, multiplexers, and analog-to-digital converts offers some advantages.[50] The integration of sensor elements with electronics makes the system less subject to noise. In addition, the simultaneous measurement of a number of parameters makes compensation for the interactions between parameters practical. For example, many sensing elements are sensitive to temperature. Measurement of sample temperature permits compensating for the temperature response. Other parameters which might be compensated are sample pH, ionic strength, and interfering components. The beginnings of such systems already exist in analyzers which provide many results on a single sample. Diamond Sensor Systems sells an on-line system for use in open-heart surgery. The system measures oxygen, carbon dioxide, pH, potassium, calcium, and hematocrit.[51-53]

A major factor limiting the commercial production of these devices is the difficulty in producing arrays of complex sensors with all elements functional. Such devices must either be calibrated or made very reproducibly. The cost benefit of such integrated devices is also uncertain.

REFERENCES

1. N. W. Tietz (ed.), *Textbook of Clinical Chemistry*, W. B. Saunders, Philadelphia (1986).
2. R. J. Henry, D. C. Cannon, and J. W. Winkelman (eds.), *Clinical Chemistry, Principles and Technics*, 2nd ed., Harper & Row, Hagerstown, Maryland (1974).
3. G. Eisenman, *Glass Electrodes for Hydrogen and Other Cations, Principles and Practice*, Marcel Dekker, New York (1967).
4. J. Koryta (ed.), *Medical and Biological Applications of Electro-chemical Devices*, Wiley-Interscience, New York (1980).
5. E. Pungor (ed.), *Ion-Selective Electrodes*, Elsevier, Amsterdam (1978).
6. S. J. Schelpel, G. Koning, B. Oeseburg, A. J. M. Langbrock, and W. G. Zijtstra, *Med. Biol. Eng. Comp.* **1987** (Jan.), 63.
7. J. A. R. Kater, J. E. Leonard, and G. Matsuyama, *Ann. N. Y. Acad. Sci.* **148,** 54 (1968).
8. J. W. Severinghouse and A. F. Bradley, *J. Appl. Physiol.* **13,** 515 (1958).
9. T. Matsuo, M. Esashi, and K. Shibatani, *Symposium on Biosensors*, IEEE, Los Angeles, Sept. 15–17, 1984, pp. 33–34.
10. K. A. Hasselbalch, *Biochem. Z.* **78,** 112 (1917).
11. W. J. Scott, E. Chapoteau, and A. Kumar, *Clin. Chem.* **32,** 137 (1986).
12. M. E. Meyerhoff, *Symposium on Biosensors*, IEEE, Los Angeles, Sept. 15–17, 1984, pp. 51–53.
13. J. Y. Lucisano, J. C. Armour, and D. A. Gough, *Anal. Chem.* **59,** 736 (1987).
14. P. Sekelj and R. B. Goldbloom, in: *Glass Electrodes for Hydrogen and Other Cations, Principles and Practice* (G. Eisenman, ed.), pp. 520–555, Marcel Dekker, New York (1967).
15. L. J. Russell and K. M. Rawson, *Biosensors* **2,** 301 (1986).
16. P. R. Demko, K. M. Pasko, and D. Curran, *Am. Clin. Prod. Rev.* **1987** (May), 32.
17. U. Oesch, D. Ammann, and W. Simon, *Clin. Chem.* **32,** 1448 (1986).
18. V. P. Y. Gadzekpo, G. J. Moody, and J. D. R. Thomas, *Analyst* **111,** 567 (1986).
19. N. Fogh-Andersen, P. D. Wimberley, J. Thode, and O. Siggaard-Andersen, *Clin. Chem.* **30,** 433 (1984).
20. J. D. Czaban, A. D. Cormier, and K. D. Legg, *Clin. Chem.* **28,** 1703 (1982).

21. J. D. Czaban, A. D. Cormier, and K. D. Legg, *Clin. Chem.* **28**, 1936 (1982).
22. R. L. Coleman and C. C. Young, *Clin. Chem.* **28**, 1705 (1982).
23. F. S. Apple, D. D. Kock, S. Graves, and J. H. Ladenson, *Clin. Chem.* **28**, 1931 (1982).
24. L. Balulescu, A. Cormier, and M. Hough, *Am. Clin. Prod. Rev.* **1988** (Jan.), 30.
25. R. Freaney, T. Egan, M. J. McKenna, M. C. Doolin, and F. P. Muldowney, *Clin. Chim. Acta* **158**, 129 (1986).
26. T. F. White, J. R. Farndon, S. C. Conceicao, M. F. Laker, M. K. Ward, and D. N. S. Kerr, *Clin.Chim. Acta* **157**, 199 (1986).
27. R. N. Barnett, *Am. J. Clin. Pathol.* **50**, 671 (1968).
28. F. R. Elevitch (ed.), *Proceedings of* 1976 *Aspen Conference on Analytical Goals in Clinical Chemistry,* College of American Pathologists, Northfield, Illinois (1977).
29. T. L. Shirey, *Clin. Biochem.* **16**, 147 (1983).
30. M. Mascini and G. G. Guilbault, *Biosensors* **2**, 147 (1986).
31. F. W. Scheller, F. Schubert, R. Renneberg, H.-G. Muller, M. Janchen, and H. Weise, *Biosensors* **1**, 135 (1985).
32. G. G. Guilbault, in: *Solid Phase Biochemistry* (W. H. Scouten, ed.), pp. 479–505, Wiley, New York (1983).
33. C. R. Lowe, *Trends Biotechnol.* **2**, 50 (1984).
34. W. H. Mullen, S. J. Churchouse, F. H. Keedy, and P. M. Vadgama, *Clin. Chim. Acta* **157**, 191 (1986).
35. M. R. Weaver and P. M. Vadgama, *Clin. Chim. Acta* **155**, 295 (1986).
36. M. Mascini, F. Mazzei, D. Moscone, G. Calabrese, and M. M. Benedetti, *Clin. Chem.* **33**, 591 (1987).
37. M. A. N. Rahni, G. G. Guilbault, and N. G. de Olivera, *Anal. Chim. Acta* **181**, 219 (1986).
38. M. Thompson and E. T. Vanderberg, *Clin. Biochem.* **19**, 256 (1986).
39. A. P. F. Turner and J. C. Pickup, *Biosensors* **1**, 85 (1985).
40. H. Lerner, J. Giner, J. S. Soeldner, and L. K. Coltur, *Ann. N. Y. Acad. Sci.* **428**, 263 (1984).
41. D. R. Matthews, R. R. Holman, E. Brown, J. Steemson, A. Watson, S. Hughes, and D. Scott, *Lancet* **1987** (April 4), 778.
42. P. Vadgama, *Analyst* **111**, 875 (1986).
43. T. C. Pinkerton and B. L. Lawson, *Clin. Chem.* **28**, 1946 (1982).
44. R. P. Buck, and V. V. Cosofret, *Symposium on Biosensors*, IEEE, Los Angeles, Sept. 15–17, 1984, pp. 54–55.
45. N. Yamamoto, Y. Nagasawa, S. Shuto, H. Tsusomura, M. Sawai, and H. Okumura, *Clin. Chem.* **26**, 1569 (1980).
46. J.-L. Boitieux, G. Vesmet, and D. Thomas, *Clin. Chem.* **25**, 318 (1979).
47. B. Mattiasson and H. Nilsson, *FEBS Lett.* **78**, 251 (1977).
48. M. Y. Keating and G. A. Rechnitz, *Anal. Chem.* **56**, 801 (1984).
49. J. Briggs, *Nature* **329**, 565 (1987).
50. H. G. DeYoung, *High Technol.* **3**, 41 (1983).
51. W. Schramm, T. Yang, and A. R. Midgley, *MD&Di* **1987** (November) 52.
52. B. Burgess, P. Burleigh, and H. Diamond, *Symposium on Biosensors*, IEEE, Los Angeles, Sept. 15–17, 1984, pp. 48–50.
53. *Lab. Med.* **18**, 732 (1987).

27

Utilization of Dissolved Oxygen in Water to Prevent Microbial Fouling on Metals

Hari P. Dhar

1. INTRODUCTION

When I was invited to make a presentation on the occasion of Professor Bockris's receiving the ACS award for his contributions in contemporary chemistry and on the occasion of his 65th birthday, I chose to speak on the topic I researched in his laboratory. The achievement from my work was a modest few publications and a patent. Perhaps more appropriately, it was my first experience in working with microorganisms and applying principles of electrochemistry to them. An amount of elation followed among the members of the research group after my observations that bacteria lost motility under an applied electric field. The following is a review of the work I carried out on the development of an electrochemical method for prevention of microbial fouling. In an earlier article,[1] I summarized various electrochemical methods available for such fouling prevention.

1.1. The Problem of Biofouling

Vast quantities of water are removed daily from rivers, lakes, streams, and oceans for use in industrial processes. Industries utilizing natural waters frequently encounter a serious problem of fouling of metal surfaces. This fouling is initiated through colonization of the surface by bacteria and microalgae in a brief period of time. The bacterial fouling is believed[2,3] to be the initial event in some of the more serious biofouling processes, whereby the surface is colonized by all water borne organisms, for example, diatoms, protozoa, fungi, barnacles, bryozoa, oysters, mollusks, etc. In addition, the process of algal fouling becomes important in the sunlight-exposed parts of the metallic structure. The general problem of biofouling is particularly significant in that the heat transfer efficiency of the equipment surface is substantially reduced and the frictional resistance of water flow across the surface is increased[4,5]; both result in inefficiency and energy losses. Also, the presence of a fouling layer on a metal surface in many instances initiates corrosion and pitting of the surface.[6,7]

Industries affected by biofouling include power plants producing electricity through the steam cycle. The condensers through which the cold water circulates to extract the heat from

---

Hari P. Dhar ● BCS Technology, Incorporated, Bryan, Texas 77803.

Electrochemistry in Transition, edited by Oliver J. Murphy *et al.* Plenum Press, New York, 1992.

the steam get easily fouled. Many industries are located along the seacoast because of the immediate availability of seawater that is needed to meet a large demand, for example, for a single-pass cooling water. Marine fouling in these industries invariably occurs. In an ocean thermal energy conversion (OTEC) power plant, the heat exchanger has to work without any loss of efficiency for the process to be economical. Other possible coastal industries could be electrolysis plants to produce hydrogen and recover heavy water from seawater.

1.2. Existing Antibiofouling Methods

The formation of microbial deposits on metallic surfaces is a well-recognized and long-standing problem. A monograph[8] was published on the procedures followed in controlling biofouling under various situations. Thermal, chemical, and mechanical methods have been used in attempts to prevent or retard biofouling. Mechanical cleaning requires that the fouled equipment be shut down and subjected to ultrasonic vibrations and washed or mechanically scraped, for example, through insertion of rubber balls or by other mechanical devices. Such a cleaning procedure results in significant lost operating time for the equipment. Chemical treatments have used toxic materials, such as formalin, chlorine and chlorine-based compounds, oxone, and organosulfur compounds. These toxic materials are generally added to the bulk water supply. The value of a biocidal chemical needed to control fouling in an open system would be prohibitive, and many of these biocides are of environmental concern.

In situ electrochemical techniques have been employed to generate toxic materials. Such techniques have included the anodic generation of chlorine from chloride ions in saltwater environments.[9] The cathodic evolution of hydrogen[10] at relatively high current densities, for example, $1-10 \text{ mA/cm}^2$, was partially successful in retarding microbial fouling but resulted in increased calcareous deposits because of the highly alkaline condition that was created near the surface. The anodic electrochemical generation of oxygen and the production of anodic conditions near the surface have been suggested[10,11] as a means to biofouling and scale control. A disadvantage of this process is that the anode must be coated with a special corrosion-resistant and electrocatalytic coating.

2. BACKGROUND

2.1. Origin of the Present Idea

At the start of work on electrochemical prevention of biofouling in the laboratory of Professor Bockris at Texas A&M University, the working hypothesis was to electrocute bacteria with application of potentials on electrodes in saline solutions. The expectation was that the settling bacteria on the electrode surface would suffer electrocution and eventually fall off the surface.

As bacteria are easily visible under a microscope with dark- or bright-field illumination, in the design of the necessary experiment, the bacteria had to be observed visually at the time a potential was applied at the electrode. A transparent electrode, for example, tin oxide-coated glass, was ideal for that purpose. Finally, when the experiment was set up, the bacteria were noted to lose motility on application of a potential that was below the range in which the saline electrolyte breaks down. Both the current density and cell potential were very small. Usually, an electrocution requires[10,11] a high potential (100 V or higher), and the smaller the organism, the more difficult it is to electrocute. Therefore, the observations must be associated with a cause other than electrocution. Later, the effect was proved to be arising from the generation of hydrogen peroxide, a common bactericide, on the electrode surface. The oxygen present in the electrolyte (3×10^{-4} mol/liter) containing the bacteria was consumed to generate the hydrogen peroxide.

2.2. Principle of H_2O_2 Generation

Electrochemically, hydrogen peroxide is produced as one of the reduction products of oxygen; the other product is water or OH^-. In an alkaline medium, the following reaction occurs for the generation of hydrogen peroxide:

$$O_2 + 2H_2O + 2e^- = H_2O_2 + 2OH^- \qquad (E_0 = -0.146 \text{ V}) \tag{1}$$

In the acid medium, the reaction proceeds as follows:

$$O_2 + 2H^+ + 2e^- = H_2O_2 \qquad (E_0 = 0.68 \text{ V}) \tag{2}$$

With assumptions of the activity of hydrogen peroxide ($a_{hp} = 10^{-6}$ mol liter) and oxygen pressure ($P_{ox} = 0.02$ atm or 2 kPa), the reversible electrochemical potential, E_{rev}, for the generation of hydrogen peroxide can be calculated from the Nernst equation. For an electrolyte of pH 7, the value is 0.45 V (versus NHE). All potentials are reported here versus a normal hydrogen electrode.

It is possible to construct i–V curves for the production of hydrogen peroxide and water by making assumptions for the exchange current densities appropriate for these reactions. At a cathodic potential at any current at least ten times below the limiting current,

$$E_i = E_{rev} - \eta_i \tag{3}$$

where η_i is the overpotential at current density i.

For a cathodic reaction, the overpotential is given by

$$\eta_i = (RT/\alpha F) \ln i_0/i \tag{4}$$

The following values for a noble metal were used in constructing the i–V plots shown in Fig. 1: $i_0 = 1 \times 10^{-11}$ A/cm^2 for reduction of oxygen to hydrogen peroxide, $i_0 = 1 \times 10^{-12}$ A/cm^2 for the reduction of oxygen to OH^-, and $i_0 = 1 \times 10^{-8}$ for the reduction of H^+ to hydrogen. According to the calculated situation with the parameters used, the efficiency for the hydrogen peroxide reaction is under 10%, the competing reaction being the four-electron reduction of oxygen to water or OH^-. For a base metal, the proportion of H_2O_2 among the reduction products of oxygen would be greater. It is to be noted from Fig. 1 that in the potential region where H_2O_2 is produced, evolution of hydrogen does not yet occur.

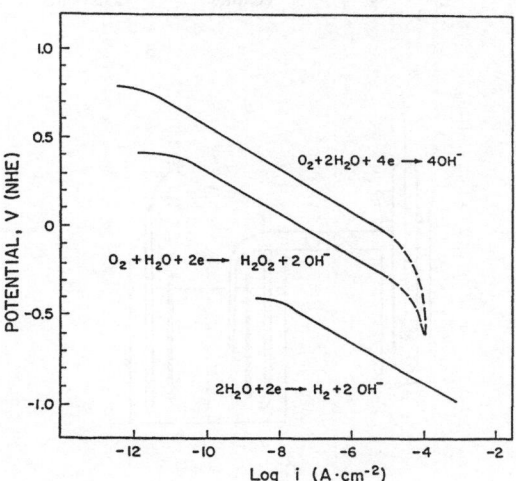

FIGURE 1. A hypothetical construction of potential versus log i relationships for the cathodic processes $O_2 \rightarrow OH^-$, $O_2 \rightarrow H_2O_2$, and $H_2O \rightarrow H_2$ for an electrolyte of pH 7.

3. INITIAL EXPERIMENTS: THIN-CELL STUDIES

The aim of the initial experiments[12,13] was to make direct observations of bacteria during the application of electrochemical potentials. A thin divided cell as shown in Figs. 2A and 2B was used. The dimensions of each compartment of the cell ws 3 cm × 2 cm × 0.02 cm. A Nafion cationic membrane no. 425 was used to separate the cathode and anode compartments. The membrane was kept immersed in water for 2–3 days to allow leaching of any acid organic groups before its use in the electrochemical cell.

The bacteria used in various tests were obtained from the culture collection of the Department of Veterinary Microbiology, Texas A&M University. Two kinds of marine bacteria, *Vibrio anguillarum* and *Pseudomonas atlantica*, were subjected to experiments. The bacteria were cultured batchwise in marine broth (Difco) at 21°C for 24 h. A portion of the culture medium with the bacteria ($10^9 \, ml^{-1}$) was diluted in saline solutions and sterilized seawater to 10^7 bacteria/ml before experimental studies. A fresh culture was used in each experiment.

The thin cell containing the bacterial species was placed on the platform of a microscope fitted with the dark-field and phase-contrast viewing optics. As one of the signs that a bactericidal chemical was produced on application of a potential at the cathode compartment, the bacteria were observed to lose motility with application of potentials, the range of which was in a region where hydrogen peroxide generation was expected. The current density in the cell was approximately $5 \, \mu A/cm^2$. The presence of hydrogen peroxide in the cathode compartment and chlorine in the anode compartment was detected by chemical tests as described below. Trivial calculations show that at 10% current efficiency of the experimentally obtained value of $5 \, \mu A/cm^2$, the region within 0.02 cm of the surface has a hydrogen peroxide concentration of around 5×10^{-6} mol/liter. The latter value was determined in separate viability experiments for bacteria to be in the lethal range.

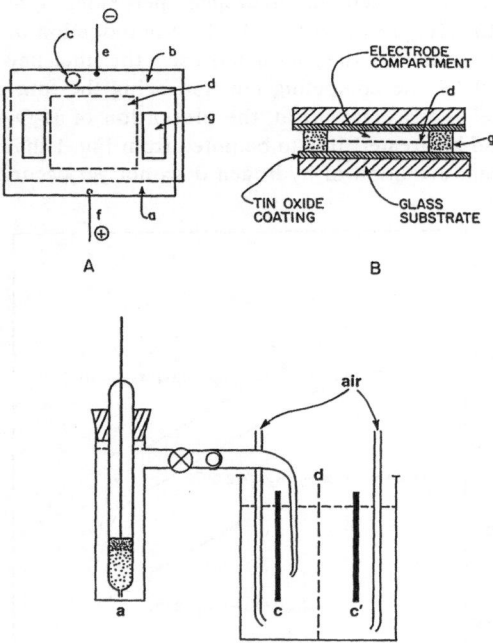

FIGURE 2. Experimental cell. Top view (A) and horizontal sectional view (B) of the thin cell. a, Anode; b, cathode; c, reference electrode position; d, membrane; e and f, electrical connections; g, glass spacer. (C) Vertical sectional view of the beaker type cell. a, Reference electrode; b, cell; c and c′, working and counter electrodes; d, membrane.

3.1. Test for Hydrogen Peroxide and Chlorine

Chemical spot tests[14] for hydrogen peroxide and chlorine were carried out in the cathode and anode compartments, respectively. A cell of 10-ml volume and with a cathode surface area of 30 cm^2 was prepared for this purpose. Transparent tin oxide glass was used for cell fabrication. Hydrogen peroxide reduces potassium ferricyanide to potassium ferrocyanide, which with acidic ferric chloride produces a Prussian Blue coloration. A test for hydrogen peroxide was positive in the electrolyte from the cathode compartment at the end of 15 min of an experiment at a constant electrode potential of −0.3 V. A light blue coloration, as opposed to a yellow coloration for a blank test, was obtained. A spot test for chlorine in the anodic electrolyte was carried out with a fluorescein–KBr solution mixture. To 2 ml of solution from the anode chamber one drop of the reagent was added. A faint red coloration indicated the presence of chlorine, which was also detected by the smell. Spot tests for hydrogen peroxide in the anode chamber and chlorine in the cathode chamber were negative. Thus, there was no migration of hydrogen peroxide from catholyte to anolyte and chlorine from anolyte to catholyte, due to the membrane that separated them.

3.2. Further Tests in the Thin Cell

Besides being used for *in situ* observations, the thin cell was also used in assessing bacterial concentrations in the solution. The assessment of bacterial viability in the electrolyte during the application of potentials was made by collecting samples from each compartment of the cell. For this assessment, the electrochemical potential was applied to the cell for a period of time ranging from 5 to 30 min. For each experiment, a fresh sample of bacterial solution containing approximately 10^4 cells/ml was introduced into the electrochemical cell. At the end of an experiment, 0.1 ml of solution was collected for counting of bacteria. The method of serial dilution was used for such countings. The data presented in Fig. 3 represent bacterial viability as a function of time during the production of hydrogen peroxide and chlorine in the cathode and anode compartments, respectively. These data show that there is a decrease in bacterial concentration in the *solution* with increasing time of application of the electrochemical potential and also that, as is widely known, chlorine is a stronger bactericide than hydrogen peroxide. Figure 4 shows data similar to those in Fig. 3, but for the cathode compartment only and for cells of different thicknesses varying from 0.02 to 0.15 cm. At a constant time, the bacterial concentration in the solution decreases with decreasing cell thickness. The reason for such observations appears to be related to a higher concentration of hydrogen peroxide in a thinner cell. A higher concentration in a thinner

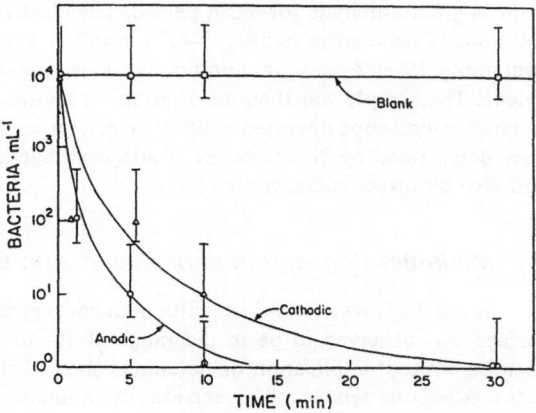

FIGURE 3. Bacterial concentration versus time of experiment in the thin cell showing blank, cathodic, and anodic experimental results. Cathode potential, −0.36 V; cell potential, 1.95 V; electrolyte, 3% NaCl; current density, 0.05 A/m^2.

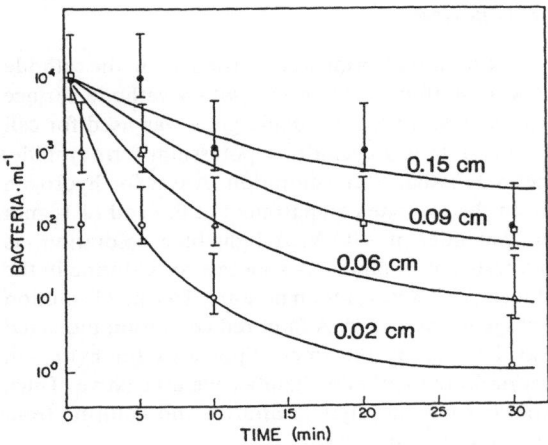

FIGURE 4. Bacterial concentration versus time of experiment in the cathode compartment of thin cells of various thicknesses as indicated. Cathode potential, -0.36 V; electrolyte, 3% NaCl; current density, 0.05 A/m².

cell is possible because of the low volume of electrolyte and because the rate at which the dissolved oxygen in the thicker cell reaches the cathode surface is apparently diffusion limited.

4. EXPERIMENTS IN A LARGER CYLINDRICAL CELL

Experiments[13,15,16] in a larger cell (Fig. 2C) were aimed at keeping the electrode surface free of bacteria. The objective was to polarize the electrode in the saline electrolyte containing bacteria and to examine the efficacy of prevention of bacterial settlement. The cell had a volume of 100 ml. It was divided into two compartments by a Nafion cationic membrane no. 425. The electrodes were cut in pieces of 5-cm² area. They were degreased in acetone, followed by immersion in alcohol, and dried in an oven. The electrical connections were established with silver epoxy over which a coating of epoxy resin was applied. The experiments were carried out at room temperature (22°C).

The electrode samples were kept immersed in the cell containing 10^7 bacteria/ml of electrolyte for 3-h periods, while at the same time a steady potential in the range of 0.0 to -0.8 V was applied to the working electrode. The potential was also applied in the form of pulses and sine waves. The pulse length was varied from 0.01 to 100 s and the pulse height from 0.24 to -1.0 V. The sinusoidal peak potential was -0.8 V.

The absorbed bacteria on an electrode surface were fixed in 3% saline solution containing 4 vol % glutaraldehyde for a 2-h period. The electrode was then placed for 10 min in saline solutions of decreasing salinity, 3–0.2% NaCl, in distilled water, in aqueous acetone solution containing 25–30% acetone, and finally in acetone–xylene solution containing 50 and 100% xylene. The sample was then air-dried in a vacuum evaporator and examined in a scanning electron microscope operated at 25-kV accelerating voltage. The number of attached bacteria were determined by direct counts of adsorbed bacteria from scanning electron micrographs and also by direct microscopic counts.

4.1. Minimum Concentration of Bacteria on Electrodes

As will be shown below, the minimum concentration of adsorbed bacteria on a polarized surface was observed to be in the range of 10^8 to 10^9 m⁻². Experiments on adsorption of bacteria without application of potentials showed that adsorption of these amounts is very rapid. After one second of immersion, the number of adsorbed bacteria on an unpolarized surface was 5×10^8 m^{-2}. An electrolyte containing 2×10^7 bacteria/ml will have approximately 2×10^9 bacteria/m^2 in a 0.01-cm-thick layer of solution.

Thus, one explanation for the high minimum is that bacteria coming in contact with a surface instantly attach themselves. Another explanation is that as no precaution was taken to prevent samples withdrawn from the electrolyte from passing through the enriched air/water interface, a large number of bacteria can be dragged out with a sample, and this can account for the large minimum concentration adsorbed on the surface.

The consequence of these observations is that the true number of living bacteria on the surface in the electrolyte is less than recorded here.

4.2. Constant-Potential Biofouling Experiments

4.2.1. Three Percent Saline Solution

Figure 5 shows adsorbed bacteria as a function of steady anodic and cathodic potentials. Potentials negative to 0.0 V are effective in reducing the number of bacteria on the cathode surface. This potential region is consistent with the values speculatively calculated in Fig. 1. Up to 500-fold diminution in the number of adsorbed bacteria was achieved in these experiments. Initially, the anode surface seems to adsorb more bacteria than the cathode surface. At higher anodic potentials, the bacterial concentration is less at the anode than that at the cathode.

4.2.2. Seawater

Adsorption patterns of two kinds of bacteria as a function of constant potential are shown in Figs. 6 and 7. No significant difference was observed in adsorption results in synthetic and natural seawaters. A diminution in bacterial concentration similar to that in

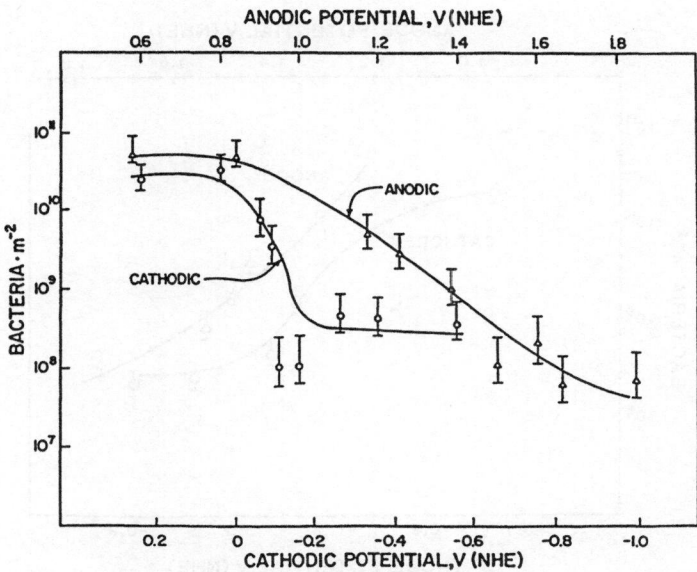

FIGURE 5. Adsorbed *V. anguillarum* on tin oxide cathode and anode versus steady electrode potentials. Time of each experiment, 3 h; electrolyte, 3% NaCl.

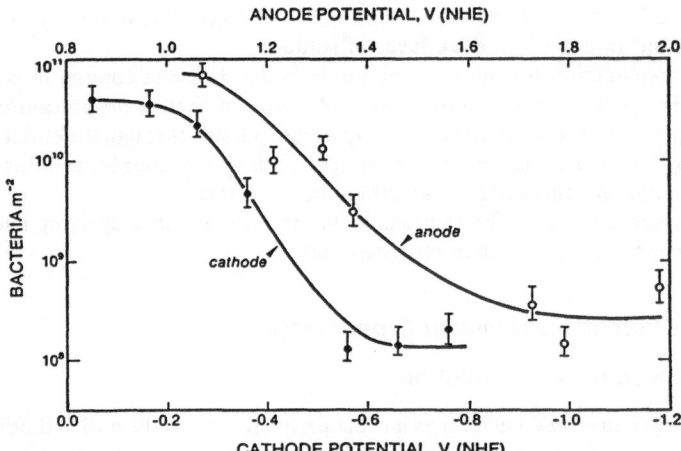

FIGURE 6. Adsorbed *V. anguillarum* on tin oxide cathode and anode versus steady electrode potentials. Time of each experiment, 3 h; electrolyte, natural seawater.

3% saline solution was obtained for the anode and cathode. However, for the cathode in seawater, the maximum polarization for constant adsorption extends to -0.6 V instead of -0.3 V, observed for the saline solution. Of this -0.3-V shift of potential, -0.12 V is associated with the pH difference of two units between saline solution (pH 6.5) and seawater (pH 8.5). The additional shift of -0.18 V can be interpreted qualitatively as arising from the fact that seawater has organic impurities that can partly be oxidized by hydrogen peroxide, and also from the possibility that the lethal effect of pH generated during hydrogen peroxide production would be minimal because of a considerable buffering capacity of seawater.

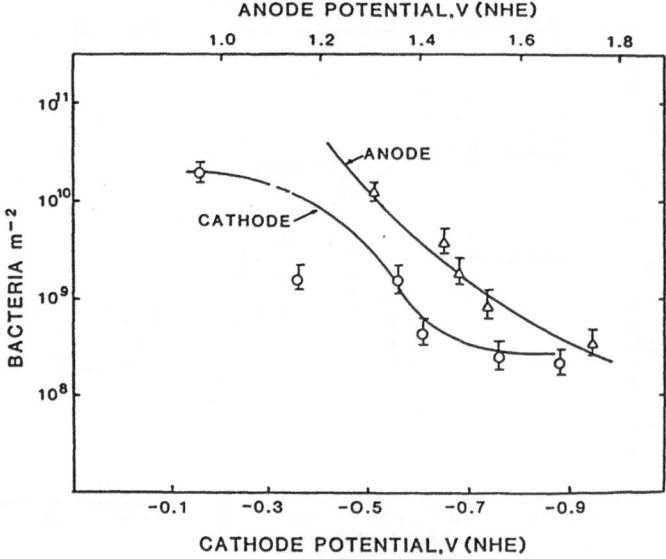

FIGURE 7. Adsorbed *P. atlantica* on tin oxide cathode and anode versus steady electrode potentials. Time of each experiment, 3 h; electrolyte, synthetic seawater.

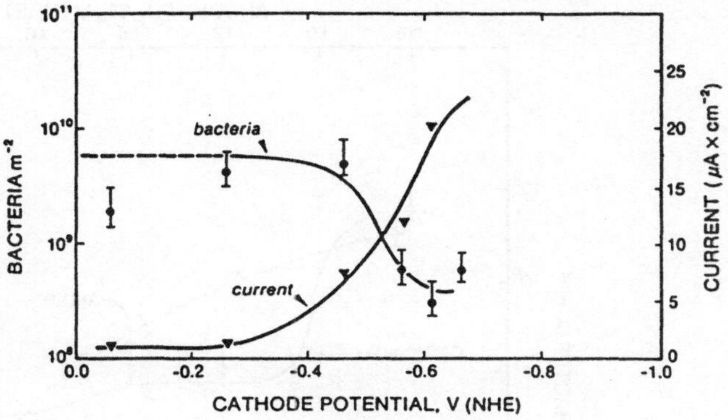

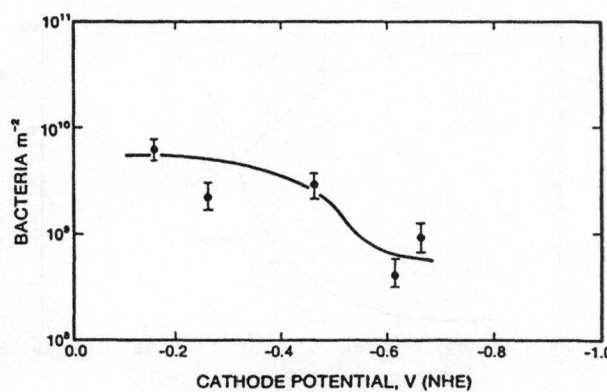

FIGURE 8. Adsorbed *V. anguillarum* on Ti cathode versus steady electrode potentials. Electrolyte, natural seawater; time of each experiment, 3 h. Current-potential plot is also shown.

A comparison of bacterial adsorption on the anode in Figs. 5 and 6 shows that no anodic shift in potential is evident on changing to seawater. This may be due to the fact that evolution of chlorine is not dependent on the change of pH of the electrolyte.

Bacterial adsorption on a titanium cathode as a function of cathode potential is shown in Fig. 8. The corresponding cathodic current, ranging from 1–20 μA cm^2, is also shown. The cell potential ranged from 0.8 to 1.8 V. These data, in general, are comparable to those for the tin oxide cathode. Results of another series of studies with a titanium cathode carried out in a beaker having two liters of synthetic seawater and no membrane are shown in Fig. 9. The bacterial concentration decreases with potential, indicating that the electrochemical method is applicable to situations in which the cathode compartment is not restricted.

4.3. Pulsing-Potential Biofouling Experiments

4.3.1. Three Percent Saline Electrolyte

The effect of potential pulses, their heights and lengths, on the fouling prevention were examined. In Fig. 10, the effect of pulse height for an *off* and *on* period of 10 s is shown. During the *off* period, the electrode rested at a potential of 0.24 V, and during the *on* period the electrode reverted to the chosen cathode potential. The data show that pulses are as effective as constant potentials in decreasing bacteria on a surface.

FIGURE 9. Adsorbed *V. anguillarum* on Ti cathode at steady electrode potentials. Experiments were carried out in 2 liters of synthetic seawater. Time of each experiment, 3 h.

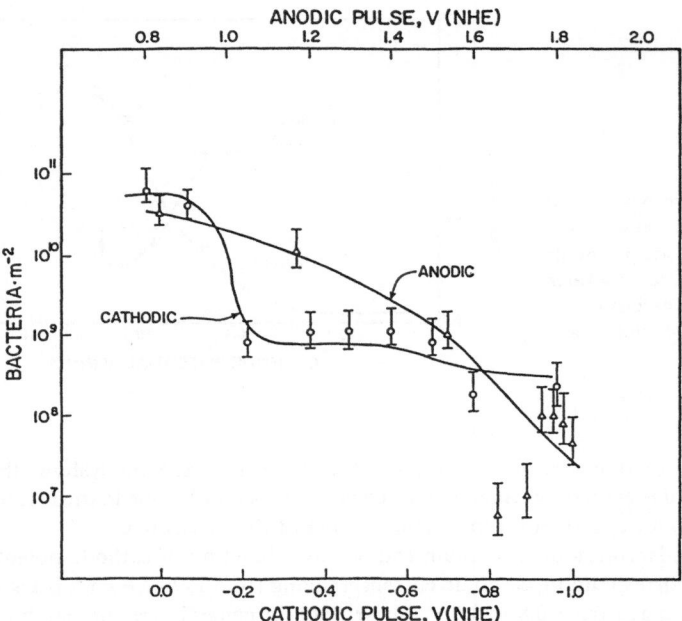

FIGURE 10. Adsorbed *V. anguillarum* on tin oxide cathode and anode versus pulse height potentials. Pulse length, 10 s on, 10 s off; rest potential for off pulse, 0.24 V; electrolyte, 3% NaCl; time of each experiment, 6 h.

4.3.2. Seawater

A pulse pattern of 10 s *on* and 5 s *off* was chosen for a set of experiments in seawater. Figure 11 presents results of adsorbed bacteria versus pulse height potentials. The magnitude of current at the height of the pulse is also shown. The decrease in bacterial content on the surface correlates with the increase in the oxygen reduction current.

To examine the effect of the variation of *off* and *on* pulses on the adsorption of bacteria, experiments were carried out at a fixed potential of -0.6 V. Figure 12 shows results of adsorption versus time of *off* pulse, the *on* pulse remaining constant at 0.1 s. Increasing the *off* pulse decreased the efficacy of pulsing. Similar results were obtained from experiments carried out

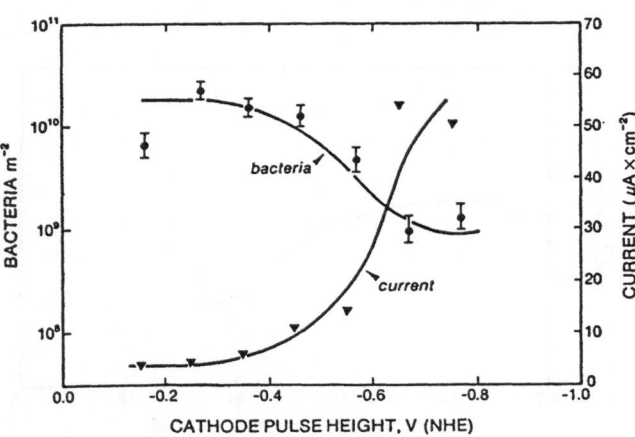

FIGURE 11. Adsorbed *V. anguillarum* on tin oxide cathode versus pulse height potentials. Pulse length, 10 s on, 5 s off; rest potential for off pulse, -0.06 V; electrolyte, natural seawater; time of each experiment, 3 h. Current-potential plot is also shown.

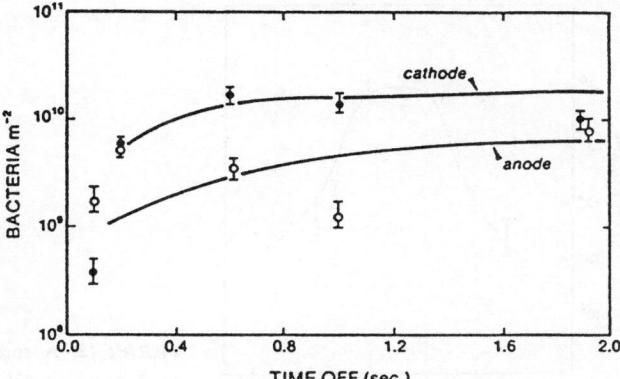

FIGURE 12. Adsorbed *V. anguillarum* on tin oxide electrodes versus time of off pulse. On pulse, 0.1 s; pulse height, −0.6 V; electrolyte, synthetic seawater; rest potential for off pulse, 0.24 V; time of each experiment, 3 h.

with *on* pulses varying from 0.005 to 0.1 s and *off* pulses varying from 0.005 to 19 s and from experiments where *on* : *off* ratios were varied at constant coulombs.

5. INTERPRETATION OF OBSERVED EFFECT

Four models were considered[12] to explain the observed decrease in adsorbed bacteria on the electrode surface: (1) changes in surface energy of the electrode, (2) depletion of oxygen near the polarized surface, (3) change of pH near the surface, and (4) production of hydrogen peroxide and chlorine on the electrodes.

The contributions from the first two effects were shown[12] to be negligible. The pH effect might arise from the effect of pH on the double layer surrounding the bacteria and thus upon their interaction with the electric field provided by the electrode. The reduction of oxygen to H_2O_2 ot to OH^- is accompanied by a change in pH in the vicinity of the electrode surface, causing an increase in bulk pH. Correspondingly, at the anode, the likely reactions are the evolution of oxygen, which would consume OH^-, or the evolution of Cl_2, which would react with water to produce HCl and HOCl. Hence, the cathode compartment will tend to become alkaline and the anode compartment, acidic. At the anode, therefore, the negative charge that exists on bacteria in solution will be decreased, and attraction to the positive electrode lessened. On the other side, the pH is increased, and it may be argued that increasing OH^- adsorption will increase the bacterial charges, thus decreasing adsorption of charged bacteria on the cathode. The pH effect will also arise if the conditions at the surface deviate substantially from those at the isoelectric point of bacteria, which is mostly in the acidic pH range. The adsorption of bacteria on tin oxide glass from electrolytes of known pH without application of potentials show that at extreme pHs the bacterial concentration on the surface decreases (Fig. 13).

With respect to hydrogen peroxide, the reversible potential at pH 6.5 (3% saline) is 0.45 V, and at pH 8.5 (seawater) it is 0.33 V. The theoretically calculated potential range for H_2O_2 production as given in Fig. 1 is 0.45 to −0.5 V. Bacterial life on an electrode surface seems to begin to be affected at 0.0 V. Thus, if the H_2O_2 view is applicable, a cathodic overpotential of approximately 1.0 V is required. The experimental potential region where the effect of H_2O_2 is evident spans approximately 0.6 to 0.8 V. From a consideration of the limiting current obtained in the electrochemical experiments, at 10% current efficiency for hydrogen peroxide production, the concentration of H_2O_2 is approximately 5×10^{-6} mol/liters at approximately 0.02 cm away from the interface, and this concentration is in the lethal range. Thus, reasonable assumptions are consistent with the H_2O_2 model. Also, the fact that

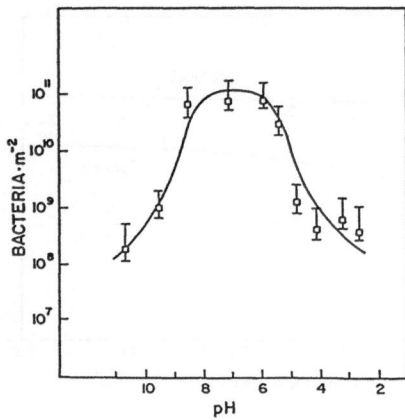

FIGURE 13. Adsorbed *V. anguillarum* on tin oxide glass samples versus pH of 3% NaCl solutions. Time of each experiment, 3 h.

the bacterial concentration starts decreasing at 0.0 V at a low current density when the change of pH at the interface is negligible reaffirms the bactericidal action of hydrogen peroxide.

With respect to the anodic evolution of chlorine, if an assumption is made that the current efficiency for chlorine evolution is 10%, the rest being due to the evolution of oxygen, the reversible potential for chlorine generation can be calculated, with certain assumptions,[12] to be 1.22 V. This potential is approximately 0.2 V more positive than that at which bacterial adsorption on the anode begins to be affected.

In addition to the bactericidal actions of H_2O_2 and the effects of pH changes at the cathode on bacterial adsorption, long-range intermolecular forces between the electrode and bacteria should be considered.[15] The total free energy of such an interaction comprises two terms, G_A, due to van der Waals forces, and G_E, due to overlap of electrical double layers associated with charged groups present on the particle and the electrode. The term G_A is always attractive, irrespective of charges on the particle and electrode. At high electrolyte concentrations corresponding to the salinity of seawater and for particle and surface of the same charge, G_E becomes a small repulsive term, leading to the net attractive $G_A + G_E$. In cases where the surface and the particle are of opposite charges, the net effect of the two terms, G_A and G_E, is a strong attraction, decreasing only with increasing bulk electrolyte concentration, Thus, according to these arguments, attractive interactions between bacteria and electrodes are expected irrespective of the nature of charges on them. The data presented here suggest the dominant nature of the bactericidal action of the generated electrochemical species on the surface.

Calcareous Deposits

The electrodes were examined for any deposits such as $CaCO_3$ and $Mg(OH)_2$. No such deposits were evident on electrodes polarized for periods of up to 24 h. Theoretical considerations[15] suggest that in seawater $Mg(OH)_2$ should precipitate at a pH of 9.2 and $CaCO_3$ at a slightly lower pH. The calculated pH at the interface, assuming a diffusion-controlled process and 100% four-electron reduction of O_2 to OH^-, is 9.5 at a current density of $20 \, \mu A/cm^2$. The diffusion length has been assumed to be 0.01 cm in agitated solution. These considerations suggest the possibility of precipitation of both Ca and Mg under the high current density of the experimental conditions. However, seawater electrochemistry is complex,[17] and in mineral accretion from seawater, current densities in excess of $1 \, mA/cm^2$ are used[18]; in addition, flow conditions and distances between electrodes affect precipitation.

REFERENCES

1. H. P. Dhar, in: *Modern Bioelectrochemistry* (F. Gurmann and H. Keyzer, eds.), pp. 593–605, Plenum Press, New York (1986).
2. K. C. Marshall and G. Bitton, in: *Adhesion of Microorganisms on Surfaces* (G. Bitton and K. C. Marshall, eds.), pp. 1–5, John Wiley & Sons, New York (1980).
3. W. A. Corpe, *Proceedings of the 4th International Congress on Marine Corrosion and Fouling,* Antibes, France, pp. 97–100, Centre de Recherches et d'Etudes Oceanographiques, Boulogne, France (1977).
4. W. A. Corpe, *Proceedings of the Symposium on Microbiology of Power Plant Thermal Effluents,* Iowa City, Iowa, pp. 57–65 University of Iowa, Iowa City, Iowa (1978).
5. B. F. Picologlou, N. Zelver, and W. G. Characklis, *J. Hydraul. Div. Am. Soc. Civ. Eng.* **106,** 733 (1980).
6. W. P. Iversion, *Adv. Appl. Microbiol.* **32,** 1 (1987).
7. B. Little and P. Wagner, Electrochemical Society Meeting Abstract No. 55, *J. Electrochem. Soc.* **136,** 113C (1989).
8. L. D. Jensen (ed.), *Biofouling Control Procedures,* Marcel Dekker, New York (1977).
9. J. E. Bennett and J. E. Elliott, U.S. Patent 4,256,556 (1978).
10. E. S. Castle, *Ind. Eng. Chem.* **43,** 901 (1951).
11. E. Littauer and D. M. Jennings, *Proceedings of the 2nd International Congress on Marine Corrosion and Fouling,* Technical Chamber of Greece, Athens, pp. 527–536 (1968).
12. H. P. Dhar, D. H. Lewis, and J. O'M. Bockris, *Can. J. Microbiol.* **27,** 998 (1981).
13. H. P. Dhar, D. H. Lewis, and J. O'M. Bockris, *J. Electrochem. Soc.* **128,** 229 (1981).
14. F. Feigl, *Spot Tests in Inorganic Analysis,* pp. 353–371, Elsevier, New York (1958).
15. H. P. Dhar, D. W. Howell, and J. O'M. Bockris, *J. Electrochem. Soc.* **129,** 2178 (1982).
16. H. P. Dhar, J. O'M. Bockris, and D. H. Lewis, U.S. Patent 4,440,611 (1984).
17. W. Stumm and J. J. Morgan, *Aquatic Chemistry,* Wiley Interscience, New York (1981).
18. W. H. Hilbertz, *IEEE J. Ocean. Eng.* **4**(3), 94 (1979).

VII

*Electrodeposition and Electrodissolution,
High-Temperature Electrochemistry*

28

Identification of Phase Structure of Alloys by Anodic Linear Sweep Voltammetry

Aleksandar R. Despić

1. INTRODUCTION

Linear sweep voltammetry (LSV) represents such an old and widely accepted electrochemical technique[1] that one is reluctant to talk about its use to an audience composed largely of electrochemists. However, its exceptional qualities are hardly known to nonelectrochemists, and, hence, it is rarely used in other fields. These qualities are especially (a) very simple instrumentation, consisting of a triangular voltage generator, an X–Y recorder, and a simple electrochemical cell (addition of a potentiostat being a welcome but not essential improvement); (b) high selectivity with respect to the chemical nature and physicochemical stability of substances in a mixture, reflected in the reduction or oxidation peak potentials; and (c) extreme sensitivity to the very presence of reducible or oxidizable substances, since quantities of the order of a percent of a monomolecular (atomic) layer resting at a surface (10^{-11} mol/cm^2) can easily be detected.

A good example is the linear sweep voltammogram (LSV) obtained in a system where deposition and dissolution of a monolayer of a metal on a foreign substrate takes place at potentials positive to the thermodynamic reversible potential of the metal with respect to its ions in solution ("underpotential deposition"), which is shown in Fig. 1. The integrals of the cathodic or anodic current, that is, the areas under the corresponding parts of the LSV give a quantitative measure of the amount of the metal deposited during the cathodic sweep. Furthermore, the appearance of several peaks at different electrode potentials provides an insight into the presence of several different surface structures, exhibiting different bond energies between the metal and the substrate as well as between the metal atoms themselves.

A method which can give such useful information about the structure of a metal would clearly be of interest in metallurgy and metallography. This fact encouraged attempts to apply the technique to metallurgical analysis of phase structure appearing at a surface of a metallographic sample using only the anodic branch of the LSV (ALSV), that is, recording the current response to a positive shift of potential from a negative limit which is determined

Aleksandar R. Despić ● Faculty of Technology and Metallurgy, University of Belgrade, Belgrade, Yugoslavia.

Electrochemistry in Transition, edited by Oliver J. Murphy *et al.* Plenum Press, New York, 1992.

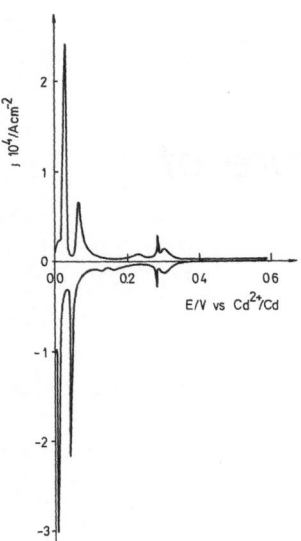

FIGURE 1. Linear sweep voltammogram for underpotential deposition of cadmium on a silver (111) single-crystal surface from a solution containing $0.1M$ $CdSO_4 + 0.02M$ $H_2SO_4 + 0.5M$ Na_2SO_4. Sweep rate: 10 mV/s.

by active hydrogen evolution to potentials at which the most noble components undergo dissolution.

On the occasion of a review of the potential of electrochemistry on the eve of the 21st century, it was considered worthwhile to demonstrate another useful direction of expansion of its use.

For a long time, the ALSV method and other electrochemical techniques were used in metallography of steels and ferrous alloys,[2] where some phases could be detected with the help of special electrolytes, in which passivity inhibited dissolution of other phases. However, difficulties in establishing such conditions in such a manner as to obtain clear meaning of the data limited the value of such methods as compared to the conventional methods of surface- and bulk-phase analysis.

Zejnilović and Jović drew attention to the wealth of information contained in the ALSV of some electrodeposited nonferrous alloys not undergoing passivation.[3] Hence, a detailed analysis of the potentialities of the method in phase-structure determinations was carried out in the last few years, whose results[4-8] encourage us to recommend the method both for research on properties of new alloys obtained under different conditions as well as for routine analysis of industrial products. A detailed consideration is given below.

2. THEORETICAL BACKGROUND AND PRACTICAL ACCOMPLISHMENT OF PHASE-STRUCTURE ANALYSIS

An alloy consists usually of a number of phases of different chemical composition, dispersed very often into fine grains intimately mixed with grains of other phases. When the surface of such an alloy is brought into contact with an electrolytic solution, each phase reaches a certain electrical potential difference with respect to the solution, reflecting the thermodynamics of the electrochemical reaction by which the chemical constituents tend to pass into the solution. On the positive side of this potential, the phase would dissolve by electrochemical reaction, while on the negative side, if no ions corresponding to the metal constituents of the phase are present in the solution, no process should take place.

The thermodynamic treatment of the potential of a single phase deserves attention. In earlier work on electrochemistry of alloys, a thermodynamic formalism based on the Nernst equation was used, taking into account only the tendency of one constituent, that having the most negative dissolution potential ("negative constituent"), to dissolve while considering other components as forming an electrochemically inert matrix.[9-11]

This model of the electrochemical behavior of an alloy has a basic shortcoming. The dissolution potential depends on the concentration of corresponding ions in solution. If, for example, of the two constituents of a binary alloy, the one with the more positive standard potential ("positive constituent") has no corresponding ions in solution, its dissolution potential would be more negative than that of the other constituent with the more negative standard potential in a solution containing a finite concentration of its ions. Hence, if, for example, one succeeded in efficiently removing copper ions from the surface of a copper–zinc alloy immersed in a solution containing only zinc ions as soon as they formed, the copper should continuously dissolve on account of deposition of zinc; that is, the alloy would corrode.†

Moreover, activities of the negative constituent calculated using the Nernst equation in alloys exhibiting relatively positive dissolution potentials turn out to be so small as to lose any physical meaning.

For these reasons, another approach is suggested as being closer to physical reality.

Assume that a binary alloy phase has a composition $A_m B_n$. It can be considered as a single chemical entity obtained by mixing the two metals in stoichiometric amounts. The standard free energy of formation will be

$$\Delta_f G^\ominus = G^\ominus(A_m B_n) - mG^\ominus(A) - nG^\ominus(B) \tag{1}$$

the last two terms being by convention equal to zero.

The entity can also be formed in an electrochemical cell from ions of both metals, A^{p+} and B^{q+}, present in solution, on an electrode made of the alloy phase as cathode and a standard hydrogen electrode as anode. The cell reaction is then

$$mA^{p+} + nB^{q+} + \left(\frac{mp + nq}{2}\right)H_2 = A_m B_n + (mp + nq)H^+ \tag{2}$$

The standard free energy change in this reaction should be

$$\Delta G^\ominus_{cell} = G^\ominus(A_m B_n) + (mp + nq)G^\ominus(H^+) - mG^\ominus(A^{p+}) - nG^\ominus(B^{q+}) - \left(\frac{mp + nq}{2}\right)G^\ominus(H_2) \tag{3}$$

Since, by convention,

$$(mp + nq)G^\ominus(H^+) - \left(\frac{mp + nq}{2}\right)G^\ominus(H_2) = 0 \tag{4}$$

the standard free energy change in the cell reaction is

$$\Delta G^\ominus_{cell} = G^\ominus(A_m B_n) - mG^\ominus(A^{p+}) - nG^\ominus(B^{q+}) \tag{5}$$

---

† In reality, immeasurably small concentrations of copper make copper attain the same potential as zinc, and thus the dissolution virtually stops and the alloy comes to a steady-state potential at which no process is recorded. Nevertheless, copper cannot be considered as an inert matrix since it undergoes active exchange with ions in solution, and its potential can be categorized as a "floating potential."[12]

The electromotive force of this cell, which in this case is identical to the electrode potential of the alloy phase on the standard hydrogen scale, is given by

$$E(A_m B_n) = E^{\ominus}(A_m B_n) + \frac{RT}{(mp + nq)F} \ln a(A^{p+})^m a(B^{q+})^n \tag{6}$$

where $a(A^{p+})$ and $a(B^{q+})$ are the activities of the corresponding ions in solution. The standard electrode potential of the alloy phase, $E^{\ominus}(A_m B_n)$, is related to the thermodynamic quantities by

$$E^{\ominus}(A_m B_n) = \frac{-\Delta G^{\ominus}_{cell}}{(mp + nq)F} = \frac{-1}{(mp + nq)F} [G^{\ominus}(A_m B_n) - mG^{\ominus}(A^{p+}) - nG^{\ominus}(B^{q+})] \tag{7}$$

The standard free energies of formation of the ions relative to that of the hydrogen ion (taken as zero) are related to the standard potentials of the corresponding metals as

$$\Delta_f G^{\ominus}(A^{p+}) = pFE^{\ominus}(A^{p+}/A) = G^{\ominus}(A^{p+}) \tag{8}$$

and

$$\Delta_f G^{\ominus}(B^{q+}) = qFE^{\ominus}(B^{q+}/B) = G^{\ominus}(B^{q+}) \tag{9}$$

Substituting Eqs. (8) and (9) into Eq. (7) and then substituting the resulting $E^{\ominus}(A_m B_n)$ into Eq. (6), one can relate the standard free energy of the alloy to the thermodynamic reversible potential of the alloy phase as follows:

$$E(A_m B_n) = \frac{1}{(mp + nq)} \left[mpE^{\ominus}(A^{p+}/A) + \frac{mRT}{F} \ln a(A^{p+}) + nqE^{\ominus}(B^{q+}/B) + \frac{nRT}{F} \ln a(B^{q+}) \right]$$
$$- \frac{G^{\ominus}(A_m B_n)}{(mp + nq)F} \tag{10}$$

where the terms in the brackets reflect the potentials of the pure metal constituents in solutions of their ions.

A very important condition for the stability of the alloy phase in the solution is that $E(A_m B_n)$ is equal to or more positive than the more positive of the two potentials of the pure metals. If this is not the case, a replacement reaction will take place, with the alloy dissolving and the pure metal depositing in a separate phase. For the above condition to be fulfilled, one could manipulate the activities of the ions in solution. The best way to do this is to immerse the alloy in a solution not containing ions of the more positive metal constituent. In such a case, some of the alloy would dissolve so as to establish the concentration of that ionic species required in order for the potential of the alloy to equal the potential that would be obtained for the more positive pure metal phase.

This sets the framework for ALSV analysis of the phase structure of an alloy. For any real (negative) value of $\Delta G^{\ominus}(A_m B_n)$, no alloy phase could have a more negative electrode potential than the pure negative constituent. Hence, the potential at which to start the ALSV, in a solution containing only ions of that constituent at some concentration, is its reversible potential. The best way to ensure this is to use this pure metal immersed in the same solution as the reference electrode.

As the potential is swept from that potential in the positive direction, the reversible potentials of the different alloy phases are reached one by one. Once such a potential is passed, the corresponding alloy phase dissolves. The current density of dissolution increases as the potential is swept further in the positive direction. However, two phenomena should be noted:

(a) Dissolution creates locally an increased concentration of the positive constituent, corresponding to a more positive reversible potential of the pure metal phase than that attained in the course of the anodic sweep. Hence, that potential turns out to be cathodic with respect to the reversible potential, so that this metal constituent must deposit either in the form of another alloy phase with a more positive reversible potential or in the form of pure metal crystallites. Hence, the net recorded current reflects the dissolution of the negative constituent only.

(b) Dissolution of one alloy phase from a mixture with other phases, or through redeposited positive constituent, must involve transport difficulties. Hence, the dissolution current comes under diffusion control, which makes the current–potential (time) relation typical of such a situation for a cathodic metal deposition process, that is exhibiting a maximum and subsequent decay. The LSV theory should be applicable for relating the peak potential to the reversible potential of the dissolving phase.

In the important practical case of a limited thickness of the investigated alloy layer, if complete dissolution of the phase is achieved, the dissolution current should fall sharply to zero, unless another phase starts dissolving.

The integral of an ALSV peak can be used (if deconvolution, to eliminate effects of dissolution of other alloy phases, can be properly done) as a quantitative measure of the amount of the negative constituent dissolved from the corresponding alloy phase. Moreover, if the composition of the phase was known, the amount of that phase in the alloy can be calculated.

For this to be possible, one must know the depth to which dissolution penetrated.

Although some theoretical estimates can be made when a bulk alloy sample is used, a more accurate result can be obtained if a thin layer of the alloy is used and the entire phase is dissolved.

This leads to a recommended procedure for quantitative analysis, whereby a thin slice of the alloy is cut out using a microtome, an electrical contact is made by sealing it onto an inert (more noble) metal electrode, and the ALSV is performed at a sweep rate sufficiently slow to dissolve the entire sample.

3. TESTING THE METHOD ON DIFFERENT NONFERROUS BINARY ALLOY SYSTEMS

The ALSV method was used for analysis of binary alloys obtained in thin layers by electrochemical deposition onto inert substrates.

Three different types of alloys were tested. Phase diagrams were consulted[13] to obtain:

(a) an alloy with no miscibility (solubility) between the components in the solid phase (eutectic type);

(b) an alloy with complete miscibility of the components in the solid phase (ideal solid solution); and

(c) an alloy of components which form intermediate phases and/or intermetallic compounds.

Examples of these types were found in the binary alloys Cu–Pb, Cu–Ni, and Cu–Cd, respectively. Typical complete (cathodic and anodic) LSVs recorded from solutions containing both ions are shown in Figs. 2–4, together with the corresponding phase diagrams.

A clear difference in the anodic part of the LSV (ALSV) exists among the three types. The eutectic type exhibits an ALSV with a clear separation between the lead- and copper-dissolution peaks. In fact, the ALSV can be obtained by superposition of the ALSVs recorded from two solutions containing each of the depositing species at the same concentration. The

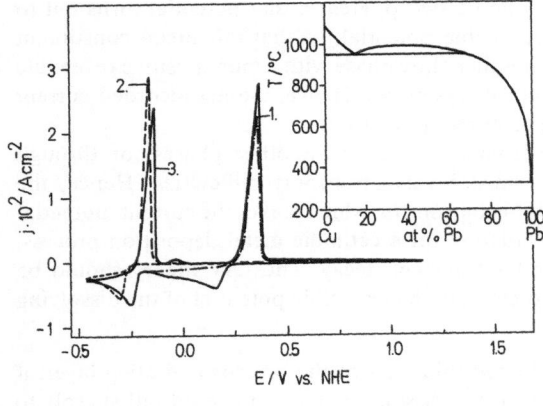

FIGURE 2. Phase diagram of the copper-lead binary alloy and linear sweep voltammograms of a fixed glassy carbon electrode in solutions containing $0.01M$ $Cu^{2+} + 1M$ HBF_4 (1), $0.01M$ $Pb^{2+} + 1M$ HBF_4 (2), and $0.01M$ $Cu^{2+} + 0.01M$ $Pb^{2+} + 1M$ HBF_4 (3). Sweep rate, $v = 50$ mV/s.

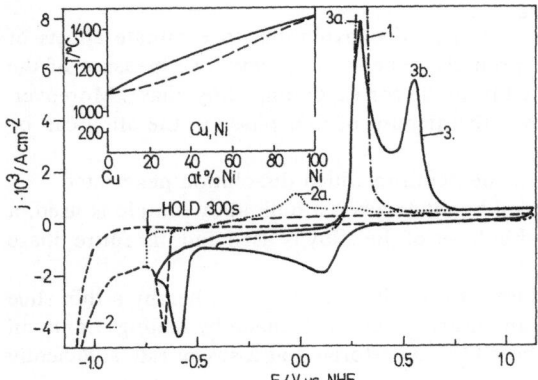

FIGURE 3. Phase diagram of the copper-nickel binary alloy and linear sweep voltammograms of a fixed glassy carbon electrode in solutions containing $0.01M$ $Cu^{2+} + 0.5M$ Na_2SO_4 (1), $0.01M$ $Ni^{2+} + 0.5M$ Na_2SO_4 (2), and $0.01M$ $Cu^{2+} + 0.01M$ $Ni^{2+} + 0.5M$ Na_2SO_4 (3). Sweep rate, $v = 10$ mV/s.

somewhat slower dissolution of lead (as indicated by the shift of the peak) reflects the difficulties that lead ions encounter in getting out through a network of copper crystallites in the eutectic. A minor interaction of lead with copper, not envisaged by the phase diagram, is reflected in a small hump at potentials somewhat more positive than the lead peak potential. Indeed, some solubility of lead in copper (up to 10%) is reported in the literature.[14] It is seen that the beginning of dissolution (reversible potential) of both lead and copper coincides with that of the pure metal phases.

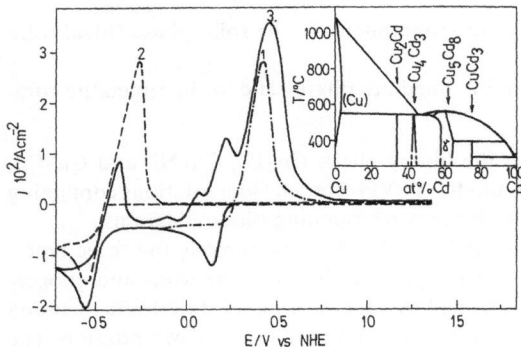

FIGURE 4. Phase diagram of the copper-cadmium binary alloy and linear sweep voltammograms of a fixed glassy carbon electrode in solutions containing $0.1M$ $Cu^{2+} + 0.5M$ $Na_2SO_4 + 1 \times 10^{-3}M$ H_2SO_4 (1), $0.1M$ $Cd^{2+} + 0.5M$ $Na_2SO_4 + 1 \times 10^{-3}M$ H_2SO_4 (2), and $0.1M$ $Cu^{2+} + 0.1M$ $Cd^{2+} + 0.5M$ $Na_2SO_4 + 1 \times 10^{-3}M$ H_2SO_4 (3). Sweep rate, $v = 50$ mV/s.

In contrast to this behavior, the Cu–Ni alloy exhibits the LSV shown in Fig. 3. Although copper and nickel deposit cathodically at potentials which follow the electromotive series, the order of dissolution exhibited in the ALSV is inverted. The first dissolution peak potential and the cathodic portion of the LSV coincide with those of deposition and dissolution of pure copper. However, the amount of dissolved copper, as represented by the area under the peak, is significantly smaller for the alloy than for pure copper, although the amounts deposited cathodically should be even larger, since deposition lasted longer (until the cathodic potential limit was reached). The obvious conclusion is that the second peak reflects the simultaneous dissolution of both nickel and the remaining copper. This is in accordance with earlier findings concerning this alloy.[15, 16] Thus, the alloy behaves as a more noble metal than pure copper, which is attributed to surface phenomena (passivation) rather than to thermodynamics. (Although the activities of the two components are reduced due to dilution, this can hardly account for such a major shift of the peak potential (250 mV), and the standard free energy of mixing should be zero.)

The most complex LSV is that of the Cu–Cd alloy. The cathodic part is simple, reflecting deposition of copper and a subsequent separate deposition of cadmium, at the same cathodic peak potential as that of deposition from a solution containing only cadmium ions. However, the anodic part exhibits four distinct peaks before the dissolution peak of copper. They could be assumed to reflect dissolution of the four intermetallic compounds shown in the phase diagram.

Other systems give ALSVs which can be considered as combinations of the above types. The LSV obtained in a solution containing cadmium and zinc ions is shown in Fig. 5, together with the corresponding phase diagram. In the presence of the positive constituent in the solution, the part of the LSV in the potential region more negative than the reversible potential of the latter is, of course, shifted downward with respect to that obtained in a solution containing ions of the negative constituent only, for the value of the diffusion limiting current of the more positive constituent, which is trivial. However, it is seen that, for an amount of deposit (area under the cathodic part of the LSV) which is larger when deposition is carried out from the solution containing both ions than that obtained by separate deposition of the two metals from the corresponding solutions, the first dissolution peak, pertaining to zinc, is significantly smaller and the second peak larger in the former than in the latter case, indicating that an amount of zinc is codeposited with cadmium into a solid solution. The latter is dissolved as a single entity, much in the same way as in the case of the Cu–Ni alloy.

The Cu–Zn alloy, whose LSV is shown in Fig. 6, is illustrative of the same phenomena, reflecting the presence of at least three intermediate phases, since the first dissolution peak is shifted significantly in the positive direction compared to the peak of pure zinc, there is a small wave before the dissolution peak of copper, and, finally, the latter is larger than that for pure copper.

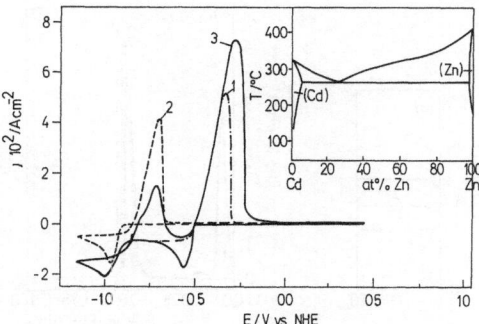

FIGURE 5. Phase diagram of the cadmium–zinc binary alloy and linear sweep voltammograms of a fixed glassy carbon electrode in solutions containing $0.1M$ $Cd^{2+} + 0.5M$ Na_2SO_4 (1), $0.1M$ $Zn^{2+} + 0.5M$ Na_2SO_4 (2), and $0.1M$ $Cd^{2+} + 0.1M$ $Zn^{2+} + 0.5M$ Na_2SO_4 (3). Sweep rate, $v = 10$ mV/s.

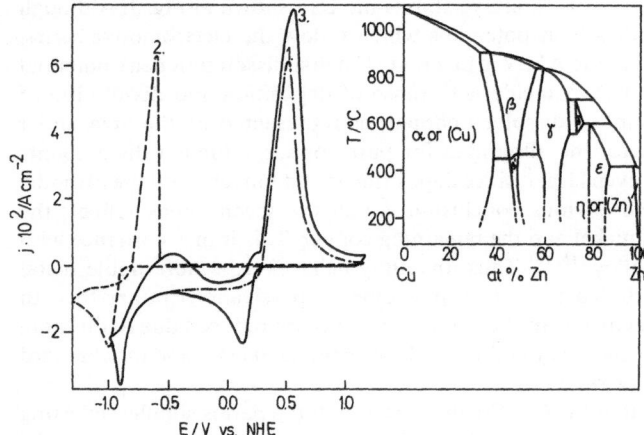

FIGURE 6. Phase diagram of the copper–zinc binary alloy and linear sweep voltammograms of a fixed glassy carbon electrode in solutions containing $0.1 M$ $Cu^{2+} + 0.5 M$ Na_2SO_4 (1), $0.1 M$ $Zn^{2+} + 0.5 M$ Na_2SO_4 (2), an $0.1 M$ $Cu^{2+} + 0.1 M$ $Zn^{2+} + 0.5 M$ Na_2SO_4 (3). Sweep rate, $v = 50\ mV/s$.

4. QUANTITATIVE ANALYSIS OF PHASE STRUCTURE

The same example of the Cu–Zn alloy can be used for some quantitative estimates. Thus, the ratio of the amounts of the three phases, most likely the γ (or ε), β, and α phases, could be found by integrating the corresponding areas as approximately $1:4.16:0.65$.

However, a better approach to quantitative analysis involves a method in which only an ALSV, without the interference of the cathodic processes, is used. This is obtained when a layer of an alloy on an inert substrate is immersed in a solution containing only ions of the negative constituent and the ALSV is performed at a sweep rate sufficiently slow that the entire layer is dissolved. The areas under the peaks should be directly proportional to the amount of the corresponding phase in the alloy.

A good illustration is seen in Fig. 7. Layers of the Cu–Pb eutectic were deposited from a bath containing ionic species of both components, using potentiostatic pulses of constant duration. The current response is shown in Fig. 7a. When the potential was kept sufficiently positive so that no deposition of lead takes place, a pure copper layer was obtained. The amount of deposited copper could be calculated from the integral of the response using the Faraday law and assuming a two-electron process. The ALSV exhibited a single peak (Fig. 7b), and integration of this peak gave the amount of copper dissolved. The ratio of the two areas gives an insight into the relative efficiency of the two processes. It was found to be ~99%.

At increasingly negative potentials of cathodic deposition, the current exceeds the diffusion limiting current of copper, and lead is codeposited. Since the first deposition was

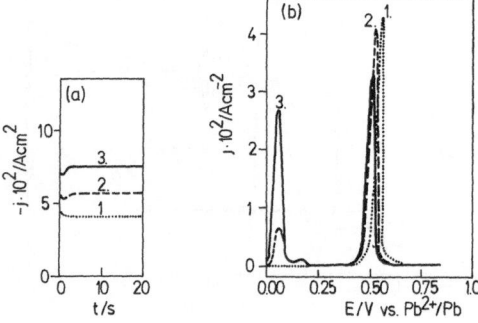

FIGURE 7. Potentiostatic current–time transients for deposition (a) and corresponding linear sweep voltammograms for dissolution (b) of copper–lead alloy. (a) Solution: $0.05 M$ $Cu^{2+} + 0.1 M$ $Pb^{2+} + 1 M$ HBF_4; Rpm = 1000; (1) $\eta = -0.40\ V$ versus Cu^{2+}/Cu, (2) $\eta = -0.50\ V$ versus Cu^{2+}/Cu, (3) $\eta = -0.52\ V$ versus Cu^{2+}/Cu. (b) Solution: $0.01 M$ $Pb^{2+} + 1 M$ HBF_4; Rpm = 1000; sweep rate, $v = 2\ mV/s$.

carried out already at the copper diffusion limiting current, and excess quantity of electricity under the other two current–time response curves reflects directly the amounts of deposited lead. Lead peaks appear in the ALSVs. The efficiency of deposition of lead (the integral ratios) was found to be 90% and 95% for the two different alloys.

5. ALLOYS CONTAINING A MULTITUDE OF DIFFERENT PHASES

The most interesting case is that of an alloy forming a number of intermetallic compounds, such as the Cu–Cd alloy, whose phase diagram was shown in Fig. 4. Using the method described above of recording the ALSV in a solution containing only ions of the negative constituent, a very clear picture of the situation in an alloy layer can be obtained as seen in Fig. 8. It is seen that different phases start dissolving at distinctly different potentials, functioning as the reversible potentials in the sense discussed in Section 2.

The peaks could be assigned to the different intermetallic compounds by assuming that the order should be that of increasing copper content in the compound. Hence, peaks A, B, C, and D should correspond to $CuCd_3$, Cu_5Cd_8, Cu_4Cd_3, and Cu_2Cd, respectively.

An attempt was made to detect these phases by X-ray analysis. However, the grain size of the deposit proved to be too small for the X-ray peaks to be discerned. Hence, a set of special samples was made by potentiostatic electrochemical deposition, adjusting the deposition potentials in such a way as to obtain different excesses of deposition current over the diffusion limiting current of copper, which provide contents of cadmium and copper corresponding to the stoichiometric requirements of the different compounds. It is seen in Fig. 9 that in alloys with an excess of cadmium (3:1 and 8:5) all peaks appear but at different intensities. Such alloys, as well as some made with a large excess of cadmium, were submitted to thermal treatment in an inert atmosphere, in order to increase grain size. The X-ray analysis revealed indeed the presence of all the four compounds. The ALSV of samples treated in such a way exhibited the expected peaks.[17]

The reversible (floating) potentials could be more precisely determined if such a layer was submitted to a stepwise potentiostatic dissolution, one by one of the phases being dissolved at its peak potential. In this way, the potentials listed in Table 1 were found.

Knowing the activity of the cadmium ions (likely to be about $4 \times 10^{-4}M$) and assuming that at the interface the concentration of cuprous (or cupric) ions at a given potential is equal to that for this same value of the reversible (floating) potential of copper, one could calculate the standard potentials as well as the standard free energies of the intermetallic compounds using Eqs. (6) and (10), respectively. These values are also listed in Table 1. It is seen that the standard free energy per atom in the compound (last column in Table 1) increases with increasing copper content, reflecting increasing bond strength and stability of the compound.

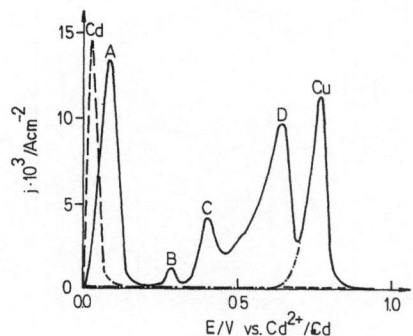

FIGURE 8. Anodic linear sweep voltammogram (ALSV) for dissolution of electrodeposited Cu–Cd alloy, together with the ALSVs for dissolution of pure copper and pure cadmium. Solution: $0.01M$ $Cd^{2+} + 0.5M$ $Na_2SO_4 + 1 \times 10^{-3}M$ H_2SO_4; Rpm = 1000; sweep rate, $v = 0.5$ mV/s.

TABLE 1

Standard Potentials and Standard Free Energy Changes of Intermetallic Compounds

Intermetallic compound	E_{exp} (V vs. Cd^{2+}/Cd)	$\dfrac{m}{m+2n}$	$\dfrac{n}{m+2n}$	E_{exp} (V vs. SHE)	$E^{\ominus}$ (V vs. SHE)	$-G^{\ominus}$ (kJ/mol)	$\dfrac{-G^{\ominus}}{m+n}$ (kJ/mol)
$CuCd_3$	0.015	0.143	0.429	−0.488	−0.258	8.8	2.2
Cu_5Cd_8	0.210	0.238	0.381	−0.293	−0.023	324.2	24.9
Cu_4Cd_3	0.340	0.400	0.300	−0.163	0.171	197.2	28.2
Cu_2Cd	0.500	0.500	0.250	−0.003	0.309	96.5	32.2

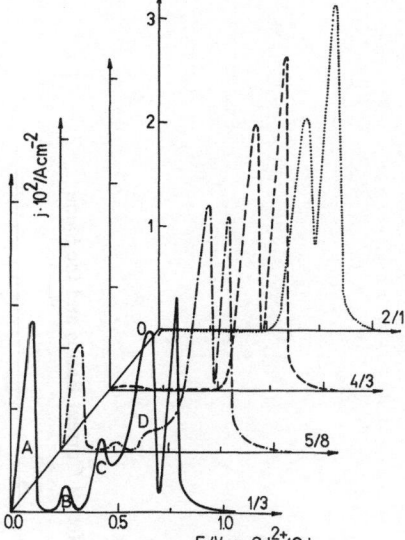

FIGURE 9. The ALSVs for dissolution of 3.2-μm-thick Cu–Cd alloys electrodeposited on a silver electrode at different current ratios i_{Cu}/i_{Cd} (marked on the figure) from the same solution containing $0.025M$ $Cu^{2+} + 0.2M$ $Cd^{2+} + 0.005M$ $H_2SO_4 + 0.5M$ Na_2SO_4. The alloys were kept for 60 min under inert atmosphere before dissolution. Sweep rate, $v = 0.5$ mV/s, Rpm = 1000.

Insofar as the quantitative relationship between contents of different phases is concerned, as seen in Fig. 9, it varies with the ratio of deposited amounts of cadmium and copper. Table 2 lists the approximate charge corresponding to each of the peaks for different deposition current ratios. It is seen that the ratios of the charge under the first four peaks to the total charge are systematically less than that calculated assuming (a) that the peaks reflect dissolution of cadmium only and (b) that dissolution of copper proceeds by a single-electron process forming cuprous ions. This could be due to some of the copper using two electrons for oxidation.

It is interesting to note, however, that, whatever the deposition current ratio, the fourth peak, corresponding to Cu_2Cd, takes up most of the charge and a fairly constant amount, signifying that this compound, having the largest free energy of formation per atom, is favored to the extent that there is copper available to satisfy its stoichiometric requirements. The fact that the charge under the copper peaks for the alloys with high cadmium content is lower than that under the alloy dissolution peaks may be due to some loss of copper, which falls off from the rotating electrode because of being too loosely bound, or at least loses electrical contact with the substrate. Indeed, after the anodic dissolution ended, in some cases some copper powder could still be found on the electrode.

6. POTENTIAL-STEP DISSOLUTION OF INDIVIDUAL PHASES

Once the ALSV is recorded and the maximum dissolution rates of the different alloy phases have been determined from the dissolution peaks, the quantitative estimation of the amounts of metals in each phase can be done also by stepwise dissolution. Thus, in the example of the Cu–Cd alloy, the potential was stepped successively by means of a potentiostat to potentials corresponding to the dissolution peaks of the four intermetallic compounds described above, and the current–time transients recorded. The result is shown in Fig. 10 for a cadmium-rich alloy.

The transients exhibit different characteristics. Thus, the phase richest in cadmium, which is present in a relatively large amount, is seen to dissolve smoothly, with two shoulders

TABLE 2

Charge under the ALSV Peaks of Alloys Obtained by Deposition at Different Current Ratios and the Same Thickness (3.2 μm)

i_{Cu}/i_{Cd}	$\sum_1^5 Q$ (C/cm)	Q_1 (C/cm)	Q_2 (C/cm)	Q_3 (C/cm)	Q_4 (C/cm)	Q_5 (C/cm)	$\sum_1^4 Q / \sum_1^5 Q$ 100%	
							Exptl.	Calc.
2/1	10.38	—	—	—	4.06	6.32	39	50
4/3	9.81	0.05	0.04	—	4.75	4.97	49	60
5/8	9.43	1.36	0.11	0.55	4.63	2.78	71	76
1/3	10.59	2.52	0.29	1.30	4.03	2.45	77	86

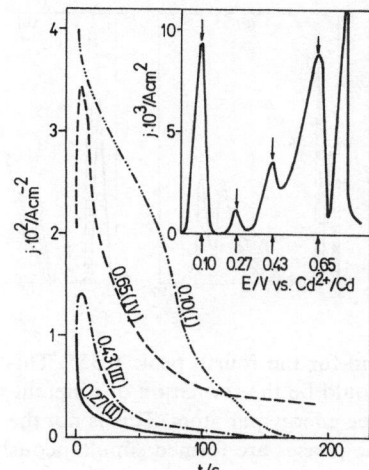

FIGURE 10. Potentiostatic current–time transients for dissolution of individual intermetallic compounds on a 4.4-μm-thick Cu–Cd alloy electrodeposited on a silver electrode at a current ratio $i_{Cu}/i_{Cd} = 1/3$ from the same solution as in Fig. 9.

indicating the presence of two different substructures. The transient decayed to zero, indicating that the entire amount of cadmium present in that phase was dissolved. The total amount of electricity, obtained as the integral of this transient, agrees well with that obtained by the integration of the ALSV peak. This could be expected because the first ASLV peak is well separated from the other peaks, the current falling to zero between them.

For the other peaks, the results from the potential-step dissolution method should be preferred, since the overlap of the peaks makes peak integration less precise as it depends on the deconvolution method. Indeed, for the third peak, the potentiostatic dissolution method gave a somewhat higher result than was obtained from peak integration.

It is interesting to note that the current–time responses for the dissolution of the third and the fourth intermetallic compound exhibited maxima. Since it is likely that these potential steps were applied when the structure had already been made porous by the dissolution of the first two phases, such a shape of the transients may be attributed to an initial increase of the surface of the grains uncovered by dissolution and a subsequent decrease because of the reduction of the grain size in the dissolution process.

Applying the same method to a copper-rich alloy yielded seemingly similar transients. However, the first transient exhibits a maximum although it is the only alloy phase present. This could be ascribed to (negative) nucleation of the dissolution process, and the transient indeed resembles those characteristic of nucleation-controlled metal deposition.[18]

7. RECORDING SOLID-STATE REACTIONS BY ALSV

The alloys deposited electrochemically are hardly ever in an equilibrium state. When an ALSV of the type shown in Fig. 8 is recorded, the question arises of whether all the phases reflected in the dissolution peaks are present from the very beginning or whether some phases are transforming into others during the process of dissolution, as cadmium is dissolved, for example, from a cadmium-rich alloy of the composition $CuCd_3$.

It is not difficult to prove that the first situation is the more likely one. If at the beginning only $CuCd_3$ existed and transformed successively into the phases richer in copper, the areas under the peaks should follow a rational sequence defined by stoichiometry. One could thus calculate that the area under the second peak should be equal to 0.607 times the area under the first peak; for the third peaks, the ratio to the area under the first peak should be 0.178,

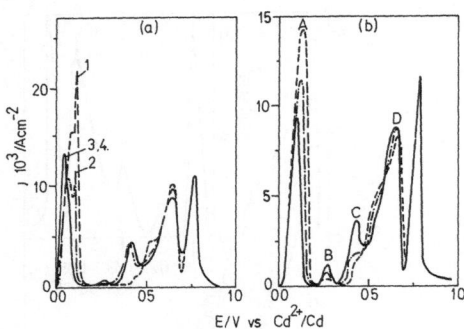

FIGURE 11. Change of ALSV with the amount of time that the electrodeposited alloy was kept under inert atmosphere before dissolution. Alloys were deposited on glassy carbon (a) and silver (b) electrodes from the same solution as in Fig. 9 at a current ratio 1/3. Amount of time under inert atmosphere: (1) 0 min; (2) 60 min; (3) 120 min; (4) 900 min. Thickness of the deposit: (a) $\delta = 1.6 \, \mu$m; (b) $\delta = 3.2 \, \mu$m. Sweep rate of ALSV, $v = 0.5$ mV/s; rpm = 1000.

and for the fourth peak, 0.357. This is obviously not the case. Indeed, a more likely event would be the formation of different phases in accordance with the ordering of the standard free energy per atom. This is not the case either. Hence, the conclusion can be made that all the phases are formed simultaneously.

Some solid-state reactions, however, could take place and change the phase structure with time. This can be seen if the ALSVs of one and the same overall alloy composition are taken at different time intervals after the alloy has been formed by deposition. Such transformations are reflected in the series of ALSVs shown in Fig 11, with the time dependences of the amounts of different phases (expressed as the quantities of electricity obtained from cadmium dissolution at different peaks) shown in Fig. 12. It is seen that the amount of the phase richest in cadmium changes on account of the increase in the amount of the intermetallic compounds with intermediate cadmium and copper contents. The Cu_2Cd phase (peak D) does not seem to undergo any change. Of course, for reactions of the type

$$8CuCd_3 + 7Cu \rightarrow 3Cu_5Cd_8$$

or

$$CuCd_3 + 3Cu \rightarrow Cu_4Cd_3$$

to be possible, there must exist a considerable excess of pure copper to start with, which is not unlikely to occur in an electrochemical deposition process.

The above example demonstrates that the ALSV method can be conveniently used not only for qualitative and quantitative phase-structure determination, but also for investigation of phase transformation kinetics.

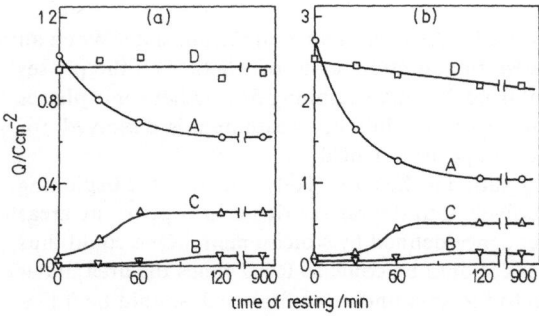

FIGURE 12. Change of charge under the peaks of the ALSVs shown in Fig. 11 with amount of time that the electrodeposited alloy was kept under inert atmosphere before dissolution.

ACKNOWLEDGMENT

The author is indebted to Dr. V. Jović for his assistance in the preparation of this chapter.

REFERENCES

1. P. Delahay, *New Instrumental Methods in Electrochemistry*, Interscience Publishers, New York (1954).
2. E. T. Shapovalov, L. I. Baranova, and G. O. Zektser, *Electrokhimicheskie Metody v Metallovendenii i Fazovom Analize*, Izd. Metallurgiya, Moscow (1988).
3. R. M. Zejnilović and V. D. Jović, Book of Abstracts, 4th Yugoslav Symposium of Analytical Chemistry, Split, Yugoslavia, 1985, p. 172
4. R. M. Zejnilović and V. D. Jović, Book of Abstracts, 29th Symposium of Serbian Chemical Society, Belgrade, Yugoslavia, 1987, p. 124.
5. V. D. Jović, R. M. Zejnilović, A. R. Despić, and J. S. Stevanović, Extended Abstracts, 10th Yugoslav Symposium of Electrochemistry, Bečići, Yugoslavia, 1987, pp. 358-360.
6. V. D. Jović, R. M. Zejnilović, A. R. Despić, and J. S. Stevanović, Extended Abstracts, 38th ISE Meeting, Maastricht, The Netherlands, 1987, Vol. I, pp. 383-385.
7. V. D. Jović, R. M. Zejnilović, A. R. Despić, and J. S. Stevanović, *J. Appl. Electrochem.* **18**, 511 (1988).
8. V. D. Jović, A. R. Despić, J. S. Stevanović and S. Spajić, *Electrochim. Acta* **34**, 1093 (1989).
9. W. Reinders, *Z. Physik. Chem.* **42**, 225 (1902).
10. R. Kremann and R. Muller, in: *Ostwald-Drucker Handbuch der Algem. Chem.* (P. Walden and C. Drucker, eds.), Band VIII, *Electromotorische Kräfte, Elektrolyse und Polarisation*, Vol. 1, pp. 626-628, and 644-648, Akad. Verlagsgeseischaft, Leipzig (1930).
11. F. A. Kroger, *J. Electrochem. Soc.* **125**, 2028 (1978).
12. A. R.Despić, in: *Comprehensive Treatise of Electrochemistry*, Vol. 9 (B. E. Conway, J. O'M. Bockris, E. Yeager, S. U. M. Khan, and R. E. White, eds.), Chapter 7, Plenum Press, New York (1983).
13. M. Hansen and K. Andrenko, *Constitution of Binary Alloys*, McGraw-Hill, New York (1958).
14. E. Raub and A. Engel, *Z. Metallkd.* **41**, 485 (1950).
15. J. O'M. Bockris, B. T. Rubin, A. Despić, and B. Lovreček, *Electrochim. Acta* **17**, 973 (1972).
16. K. M. Gorbunova and Y. M. Polukarov, in: *Advances in Electrochemistry and Electrochemical Engineering*, Vol. 5, (C. W. Tobias, ed.) John Wiley & Sons, New York (1976).
17. V. D. Jović, S. Spaić, A. R. Despić, J. S. Stevanović, and M. Pristavec, *Materials Sci. and Technology*, **7**, 1021 (1991).
18. M. Fleischmann, J. A. Harrison, and H. R. Thirsk, *Trans. Faraday Soc.* **61**, 2472 (1965).

29

The Effects of the Slow Adsorption of Anions and Some Organics on Iron Dissolution Kinetics

D. M. Dražić, V. J. Dražić, and V. Jevtić

1. INTRODUCTION

The mechanism of anodic dissolution of iron in acid solutions has been the subject of experimental investigations and theoretical discussions by many authors during the last 30 years. Many of the proposed reaction mechanisms can be divided into three groups: (i) those which, following Heusler's "catalytic mechanism,"[1] postulate a single, two-electron charge transfer rate-determining step; (ii) mechanisms based on that originally proposed by Bockris, Dražić, and Despić[2] (BDD mechanism), with two consecutive, single-electron charge transfer steps, including adsorbed OH species in the reaction mechanism, and (iii) the anion participation mechanism, a modification of the BDD mechanism, advocated by Kolotyrkin and co-workers,[3] Bech-Nielsen and co-workers,[4,5] and others. A detailed analysis of the relevant literature was presented in a recent review.[6] The purpose of this chapter is to show that most common inorganic anions (SO_4^{2-}, ClO_4^-, Cl^-, Br^-, I^-), similarly to many organic molecules, adsorb on the iron surface and inhibit the anodic reaction. Therefore, there is no justification for incorporating the anions into the formation of the activated complex of the main anodic dissolution reaction, except perhaps in very high anion concentration solutions, when the main dissolution reaction is totally inhibited.

2. EXPERIMENTAL FACTS

2.1. S-Shaped Polarization Curves

As shown by Kuznetsov and Iofa,[7] Schwabe and Voigt,[8] Heusler and Cartledge,[9] Moegensen et al.,[5] Dražić et al.,[10,11] Mikheeva and Florianovich,[18] and others, the anodic polarization curves observed after allowing sufficient time for the adsorption equilibrium to be established in halide-ion-containing H_2SO_4 solutions exhibit a characteristic S shape when the iron electrode is polarized in the positive potential direction.

D. M. Dražić, V. J. Dražić, and V. Jevtić • Faculty of Technology and Metallurgy and Institute of Electrochemistry ICTM, University of Belgrade, Belgrade, Yugoslavia.

Electrochemistry in Transition, edited by Oliver J. Murphy *et al.* Plenum Press, New York, 1992.

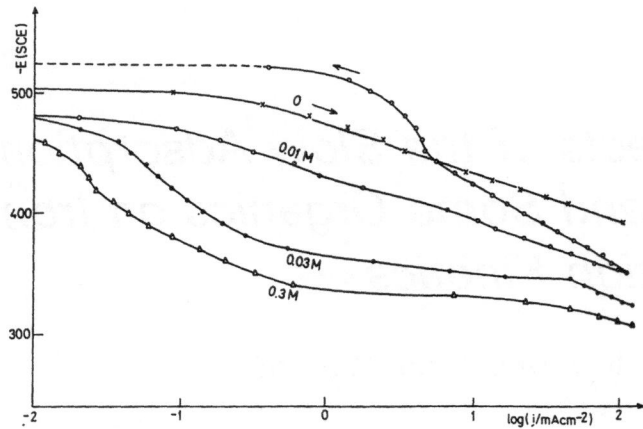

FIGURE 1. Anodic polarization curves for Armco iron electrodes after 24 h in $0.5M$ H_2SO_4 solution containing NaCl at the concentrations indicated on the figure.

This can be seen in Fig. 1 for the case of the Armco iron electrode in $0.5M$ H_2SO_4 with different amounts of NaCl. The anodic polarization curves shown were obtained 24 h after the introduction of the electrode into the deaerated solution. The experimental details are given elsewhere.[10] As seen, with increasing concentration of Cl^- ions, the inhibition of the anodic reaction, compared to the dissolution rate in the Cl^--free H_2SO_4 solution, is more pronounced. Similar behavior was observed[13] for solutions containing Br^- and I^- ions, and also some organics. Figure 2 presents the anodic polarization curve for the iron electrode in $0.5M$ H_2SO_4 containing $3 \times 10^{-4}M$ thiourea.[12] As seen, the characteristic S-shaped polarization curve similar to those in halide-containing solutions, was obtained, indicating that the nature and mechanism of the influence of halide ions on the inhibition of the anodic reaction is the same as for organic inhibitors.

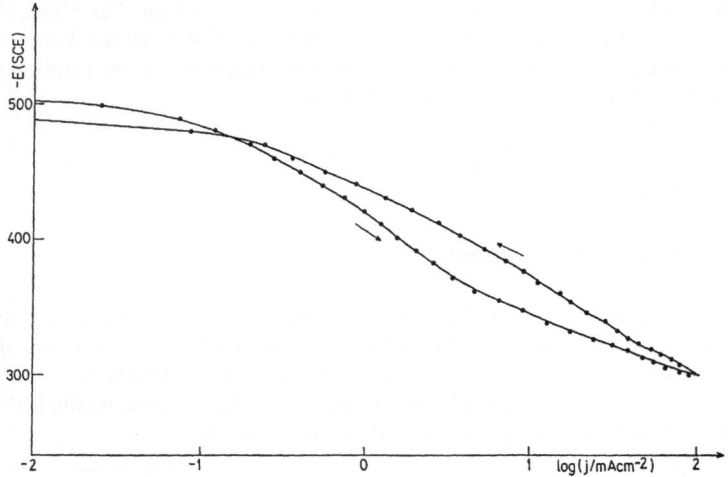

FIGURE 2. Anodic polarization curve for Armco iron electrode after 24 h in $0.5M$ H_2SO_4 containing $3 \times 10^{-4}M$ thiourea [$CS(NH_2)_2$].

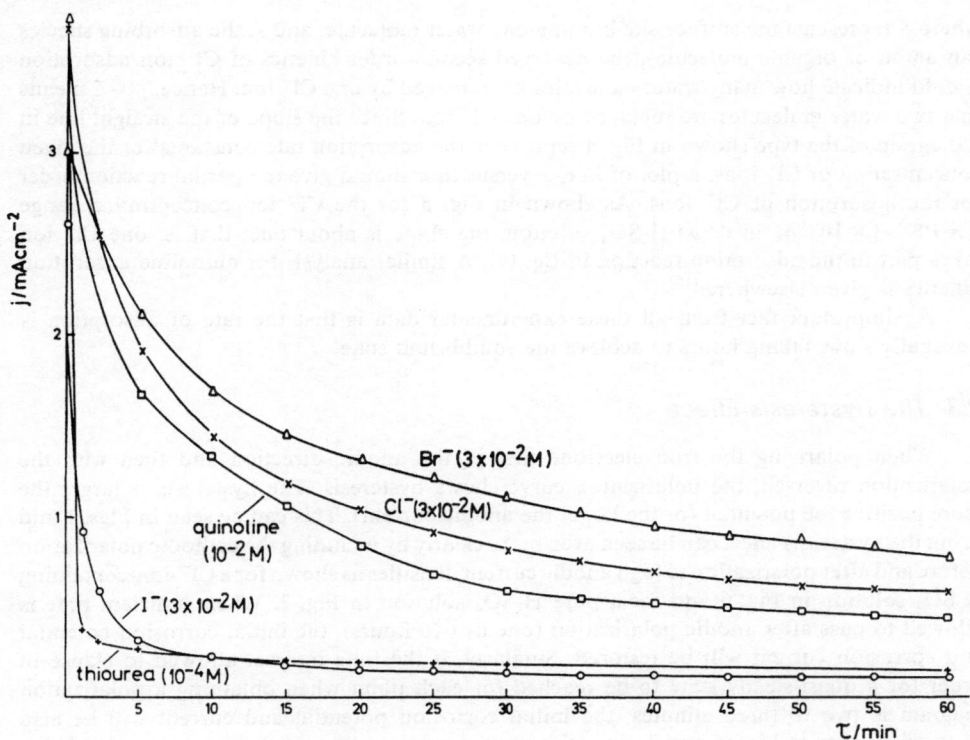

FIGURE 3. Current decays of the anodically polarized iron electrode in $0.5\,M$ H_2SO_4 solution at -470 mV (versus SCE) after the addition of various species at the concentrations indicated on the figure.

2.2. Adsorption Rate Measurements

If one adds halide ions or organic inhibitors to an anodically prepolarized iron electrode in pure H_2SO_4 solution, a slow decay of the current is observed until a new steady-state current is achieved. This is shown in Fig. 3 for the addition of different species to $0.5\,M$ H_2SO_4 after a steady-state current at -470 mV (versus SCE) had been established. If the current decay reflects the decrease of the current due to the blocking of the surface by the adsorbing species, then the decay curves reflect the inhibitor adsorption kinetics. For example, as shown in Fig. 4, in a $1/i$–t diagram, the data for the current decay after the addition of NaCl follow a straight line, indicating a second-order relationship.

If the adsorption is considered as a water replacement reaction,

$$p(\text{S}-\text{H}_2\text{O}) + \text{A} \rightarrow (p\text{S})-\text{A} + p\text{H}_2\text{O} \tag{1}$$

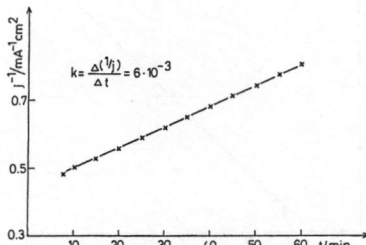

FIGURE 4. Plot of $1/i$ versus t for Cl^- ion adsorption, with data from Fig. 3.

where S represents the surface site binding one water molecule, and A the adsorbing species (an anion or organic molecule), the observed second-order kinetics of Cl^- ion adsorption should indicate how many water molecules are replaced by one Cl^- ion. Hence, $p = 2$ means that two water molecules are replaced by one Cl^- ion. Since the slope of the straight line in a diagram of the type shown in Fig. 4 represents the adsorption rate constants for the given concentration of Cl^- ions, a plot of $\ln c_{Cl^-}$ versus $\ln k$ should give the partial reaction order for the adsorption of Cl^- ions. As shown in Fig. 5 for the Cl^- ion concentration range 5×10^{-3}–$3 \times 10^{-1} M$ in $0.5 M$ H_2SO_4 solution, the slope is about one; that is, one Cl^- ion takes part in the adsorption reaction in Eq. (1). A similar analysis for quinoline adsorption kinetics is given elsewhere.[10]

An important fact from all these experimental data is that the rate of adsorption is unusually slow, taking hours to achieve the equilibrium state.

2.3. The Hysteresis Effect

When polarizing the iron electrode first in the anodic direction and then with the polarization reversed, the polarization curve shows hysteresis. The hysteresis is larger the more positive the potential (or the larger the anodic current). This can be seen in Figs. 1 and 2, but the hysteresis effect can be seen even more clearly by including the cathodic polarization before and after polarization to high anodic current densities as shown for a Cl^--ion-containing H_2SO_4 solution in Fig. 6 and for a pure H_2SO_4 solution in Fig. 7. When sufficient time is allowed to pass after anodic polarization (one or two hours), the initial corrosion potential and corrosion current will be restored. Similarly, if the time interval allowed to elapse in order for a quasi-steady state to be reached for each point when obtaining a polarization diagram is two to three minutes, the initial corrosion potential and current will be also restored, as seen in Figs. 6 and 7, since this procedure also allows sufficient time to pass after the anodic polarization. Similarly, Dražić and Vaščić[14] observed a slow decay of the current in Cl^--ion-containing solution when the potential was fixed potentiostatically at -470 mV (versus SCE) after the reversal of the anodic polarization.

It has been shown elsewhere[15,16] that the polarization curves in the vicinity of the reversible or corrosion potential (for $\eta < 40$ mV) show an inflection point, which, contrary to common belief, is not always at $\eta = 0$, but only in the special case when the Tafel coefficients are equal, that is, $b_c = b_a$, or $\alpha_c = \alpha_a$ if $b = RT/\alpha F$. Otherwise, if $b_c \neq b_a$ (or $\alpha_c \neq \alpha_a$) the potential of the inflection point depends on the ratio b_c/b_a. In the case of iron corroding in

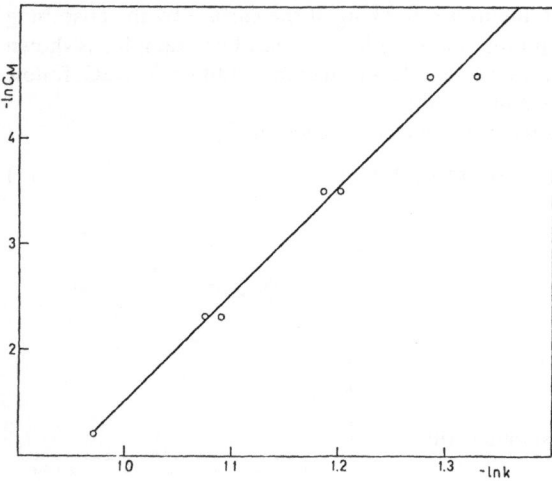

FIGURE 5. Correlation between the Cl^- ion concentration and the relative pseudo rate constants in a log–log plot. The slope is 0.85.

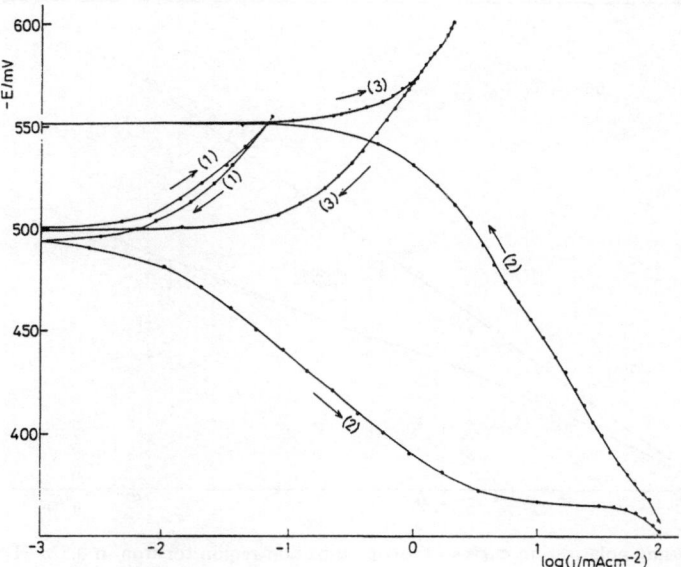

FIGURE 6. The hysteresis effects on the polarization curves for Armco iron in $0.5M$ H_2SO_4 containing $3 \times 10^{-2}M$ NaCl. Sequence of polarizations: (1) cathodic increase and decrease; (2) anodic increase and decrease; (3) repeated cathodic increase and decrease.

H_2SO_4, $b_c \approx 120$ mV/decade and $b_a \approx 40$ mV/decade, which should give an inflection point at $\eta_{ip} = -28$ mV. For determination of the corrosion rate, one can use the Stern–Geary method,[17] taking the slope of a tangent at $\eta = 0$ (linear polarization, faradaic resistance, faradaic impedance, and similar methods based on the Stern–Geary approximation) or by taking the slope at the real inflection point.[16] Hence, one would expect[15] that the analysis of the inflection point potential can provide knowledge of b_a (or α_a), which is very important when the anodic dissolution mechanism is discussed. However, when one plots the low cathodic polarization data for iron in $0.5M$ H_2SO_4 taken from the experiment shown in Fig. 7, as is done in Fig. 8, one observes the inflection point at $\eta = -27$ mV for iron in contact with the solution for 24 h, but also a considerable increase of the slope (i.e., increase of the

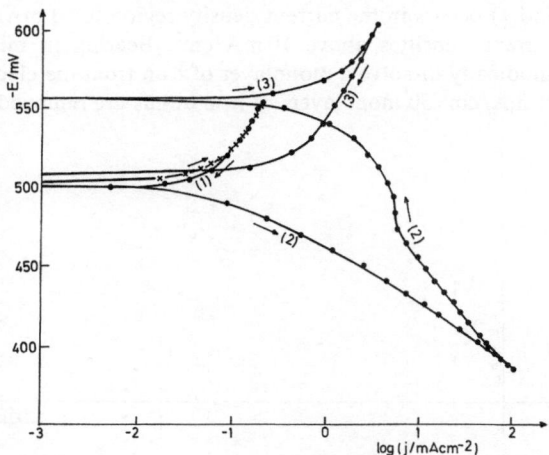

FIGURE 7. The hysteresis effects on the polarization curves for Armco iron as in Fig. 6 but for pure $0.5M$ H_2SO_4 solution.

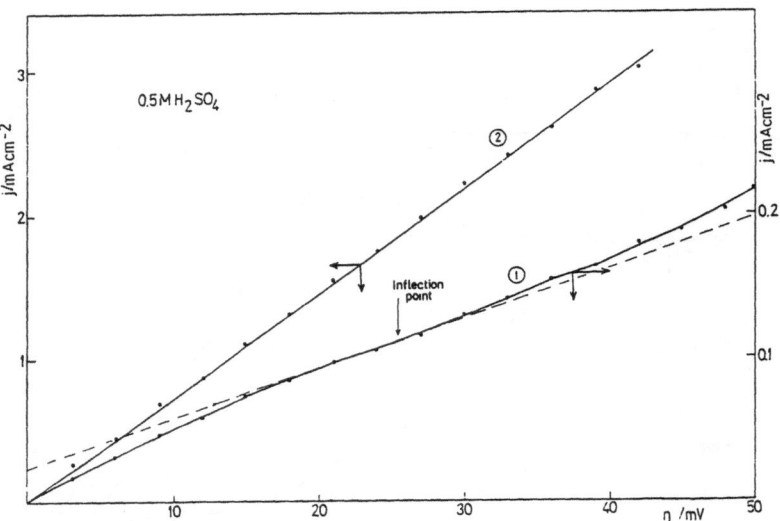

FIGURE 8. Cathodic polarization curves in low-polarization region for iron in $0.5M$ H_2SO_4. Curve 1: data from curve 1 in Fig. 7; curve 2: data from curve 3 in Fig. 7.

corrosion rate) and disappearance of the inflection point when the cathodic polarization curve (curve 2) was recorded after the anodic one.

Faradaic impedance data presented in a complex plane diagram give the charge transfer resistance at the intersection of a half-circle with the abscissa. It is important to draw attention to the fact that, as shown in Fig. 9, the diameter of the semicircle and the value of the charge transfer resistance (i.e., the corrosion rate) depend on the amplitude of the ac signal.

It is obvious that the hysteresis effect appearing after anodic polarization as shown in Figs. 1, 2, 6, and 7 also affects the low-polarization or faradaic impedance measurements, producing a range of corrosion rates, depending on how positive the anodic polarization was during the corrosion measurements or how large the amplitude of the ac signal.

2.4. The Electromechanical Desorption Model

In most cases the first bend of the S-shaped anodic polarization curve (cf. Figs. 1, 6, and 7) occurs in the current density region 0.1–1 mA/cm^2 while the second bend appears at current densities above 10 mA/cm^2. Bearing in mind that ca. 0.5 mC/cm^2 is needed to anodically dissolve a monolayer of iron from the electrode surface, one can estimate that at 25 mA/cm^2 50 monolayers of iron atoms are removed from the surface each second. During

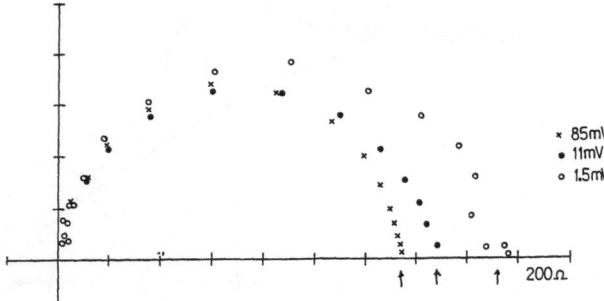

FIGURE 9. Complex impedance diagram for Armco iron electrode in $0.5M$ H_2SO_4 for three different ac signal amplitudes (85, 11, and 1.5 mV).

this process the metal atoms, including those *under* the adsorbed species (ions or molecules), are removed from the surface, which simultaneously also removes (i.e., desorbs) the adsorbed particles. Hence, when considering the adsorption–desorption equilibrium on an anodically dissolving surface (which is the case for all corroding metals), the additional electromechanical desorption process (i.e., electrochemical undermining of the adsorbed particles) should be taken into account, and the larger the anodic current density, the more significant is the electromechanical desorption. A detailed analysis of adsorption on an anodically dissolving surface based on this model has been given elsewhere.[13]

It can be considered, hence, that the last, high-current-density part of the S-shaped anodic polarization curve represents the metal surface free of all adsorbed species, which, at lower current densities, inhibit the anodic reaction. The only species which can come in contact with the iron atoms at a very dynamic metal surface at such large current densities are the solvent molecules, that is, water molecules. With almost all the inhibiting species removed from the surface, the kinetics of the anodic reaction is the fastest in the water electrolyte, which is similar to the situation reported by Mikheeva and Florianovich,[18] who observed that permanent *in situ* mechanical grinding during anodic polarization increased the anodic dissolution rate by more than an order of magnitude.

The observed unusually slow adsorption of inhibiting ions and molecules is a natural consequence of the rather complex dynamic process occurring when inhibitor species are added into a solution with a metal electrode anodically dissolving at a considerable rate (i.e., over $1 \, mA/cm^2$). Instead of following the rather simple kinetics of the adsorption reaction in Eq. (1), the decay curves shown in Fig. 3 represent the final result of the balance between a continuous process of adsorption according to Eq. (1) and predominantly electromechanical desorption, which removes the adsorbed particles from the surface immediately after their adsorption. If the electromechanical desorption rate is higher (high current densities) than the adsorption rate, the surface will become free of adsorbed species. If the anodic current is somewhat lower than that needed to keep the surface clean, some of the ions or molecules adsorbed by the reaction in Eq. (1) will be electromechanically desorbed, but the remaining adsorbed species will somewhat decrease the anodic current, and hence, the rate of electromechanical desorption. By a kind of iterative process, the particles adsorb and desorb with a continuous decrease in the free surface, and thus in the anodic current. Obviously, for this rather complex adsorption process much more time is needed than for simple chemisorption of the type represented by the reaction in Eq. (1). Hence, the slow adsorption of species on the continuously dissolving surface is the result of the additional electromechanical desorption process and should not be compared to adsorption on nondissolving surfaces (e.g., Pt, Au, etc.) or when the electromechanical desorption rates are negligible ($i_a < 10 \, \mu A/cm^2$).

The observed hysteresis effects manifesting themselves in various ways, as in Figs. 1, 2, 6 and 7, 8, and 9, are the consequences of the slow adsorption of anions (including HSO_4^- and SO_4^{2-} ions) when the anodic current and electromechanical desorption disturb the adsorption equilibrium.

3. THE MECHANISM OF THE ANODIC DISSOLUTION OF IRON

As mentioned in Section 1, some authors propose an anodic dissolution mechanism which involves the participation of anions in the reaction mechanism. In view of the observed slow adsorption of anions, including SO_4^{2-} or HSO_4^-, and their inhibiting action, great care must be taken in the experimental collection of diagnostic criteria for reaction mechanism evaluation. For example, the often observed "superpolarization peaks" on the anodic galvanostatic transients and the "transient Tafel line" with a slope of RT/F, which have been

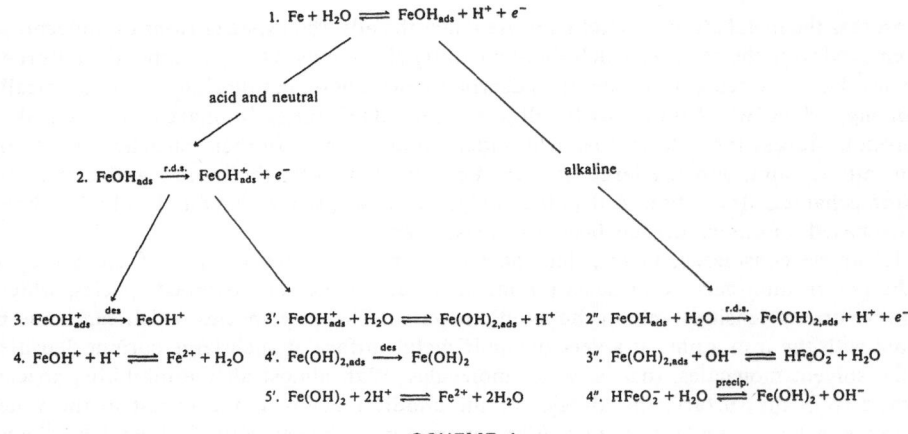

1. $Fe + H_2O \rightleftharpoons FeOH_{ads} + H^+ + e^-$

acid and neutral

2. $FeOH_{ads} \xrightarrow{r.d.s.} FeOH^+_{ads} + e^-$ alkaline

3. $FeOH^+_{ads} \xrightarrow{des} FeOH^+$ 3′. $FeOH^+_{ads} + H_2O \rightleftharpoons Fe(OH)_{2,ads} + H^+$ 2″. $FeOH_{ads} + H_2O \xrightarrow{r.d.s.} Fe(OH)_{2,ads} + H^+ + e^-$

4. $FeOH^+ + H^+ \rightleftharpoons Fe^{2+} + H_2O$ 4′. $Fe(OH)_{2,ads} \xrightarrow{des} Fe(OH)_2$ 3″. $Fe(OH)_{2,ads} + OH^- \rightleftharpoons HFeO_2^- + H_2O$

5′. $Fe(OH)_2 + 2H^+ \rightleftharpoons Fe^{2+} + 2H_2O$ 4″. $HFeO_2^- + H_2O \xrightarrow{precip.} Fe(OH)_2 + OH^-$

SCHEME 1

used[1,19] as a crucial argument for the "catalytic mechanism," were shown[20] to be artefacts, and it was demonstrated that "transient Tafel slopes" ranging from 40 to 150 mV/decade could be obtained, depending on how long the surface of the electrode was in contact with the solution. In terms of the electromechanical desorption model, the "superpolarization peaks" and the subsequent decays reflect the establishment of an adsorption–desorption rate balance. Similarly, many experimentally obtained "Tafel lines" and Tafel slopes should be carefully reexamined in terms of the changing coverage by adsorbed anions and their adsorption–desorption equilibrium.

The fact that halide ions at all concentrations examined (up to $0.3 M$) inhibit the dissolution reaction in H_2SO_4 solution and also that hysteresis effects are seen in pure H_2SO_4 solution (and also in $HClO_4$ solutions) shows that anions do not participate in the main dissolution reaction, at least at the concentrations examined.

It is justified, therefore, to assume that mechanisms including only solvent molecules in the reaction scheme are the most probable. One such mechanism is the BDD mechanism,[2] originally proposed for acid solutions. Later on, its general validity in the appropriately modified form (allowing for the necessary pH equilibria) was extended to the neutral[21] and alkaline[22] pH range. The general reaction mechanism[23] is shown in Scheme 1. The reaction steps 1–4 are for acid solutions, 1–5′ for neutral solutions, and 1–4″ for alkaline solutions. The main common features of the mechanisms are the two consecutive single-electron charge transfer steps, the formation of $FeOH_{ads}$ as the adsorbed intermediate even in acid solutions [adsorbed $Fe(OH)_{2,ads}$ was proven experimentally[24] to be stable in acid solutions], and the feature that the change in the anodic slope from 40 to 30 mV/decade that is sometimes observed can be explained with the same mechanism by a simple change of the position of the rate-determining step from step 2 to step 3, as a result of, for example, a change in the binding energy for the adsorbed intermediate due to internal stresses mechanically introduced in the lattice[25] or by absorbed hydrogen.[26] The detailed analysis of the mechanism is given elsewhere.[6]

REFERENCES

1. K. E. Heusler, *Z. Elektrochem.* **62**, 582 (1958).
2. J. O'M. Bockris, D. M. Dražić, and A. R. Despić, *Electrochim. Acta* **4**, 284 (1961).
3. G. M. Florianovich, L. A. Sokolova, and Ya. M. Kolotyrkin, *Elektrokhimiya*, **3**, 1027, 1359 (1967).

4. G. Bech-Nielsen, *Electrochim. Acta* **21,** 627 (1976).

5. M. Moegensen, G. Bech-Nielsen, and E. Maahn, *Electrochim. Acta* **25,** 919 (1980).

6. D. M. Dražić, in: *Modern Aspects of Electrochemistry,* No. 19 (B. E. Conway, J. O'M. Bockris, and R. E. White, eds.), p. 69, Plenum Press, New York (1989).

7. V. A. Kuznetsov and Z. A. Iofa, *Zh. Fiz. Khim.* **31,** 259 (1947).

8. K. Schwabe and C. Voigt, *Electrochim. Acta* **14,** 853 (1969).

9. K. E. Heusler and G. H. Cartledge, *J. Electrochem. Soc.* **108,** 732 (1961).

10. V. J. Dražić, D. M. Dražić, and V. Jevtić, *J. Serb. Chem. Soc.* **52,** 711 (1987).

11. D. M. Dražić, V. J. Dražić, and V. Jevtić, *Electrochim. Acta* **34,** 1251 (1989).

12. D. M. Dražić, V. J. Dražić, and V. Jevtić, Glas de l'Académie Serbe des Sciences et des Arts, Classe des Sciences techniques, 1988 **28,** 133 (1990).

13. V. J. Dražić and D. M. Dražić, *Proceedings of the 7th European Symposium on Corrosion Inhibitors,* No. 9, p. 99, Ann. Univ. Ferrara, N. S., Sez. V. (1990).

14. D. M. Dražić and V. Vaščić, IX Yugoslav Symposium on Electrochemistry, Dubrovnik, June 1985, Proceedings, Serbian Chemical Society, Belgrade (1985), p. 153.

15. D. M. Dražić and V. Vaščić, *J. Electroanal. Chem.* **185,** 229 (1985).

16. D. M. Dražić and V. Vaščić, *Corros. Sci.* **25,** 483 (1985).

17. M. Stern and A. L. Geary, *J. Electrochem. Soc.* **104,** 56 (1957).

18. F. M. Mikheeva and G. M. Florianovich, *Zashch. Met.* **23,** 33, 41 (1987).

19. K. E. Heusler, in: *Encyclopedia of the Elements,* Vol. 9 (A. Bard, ed.), p. 229, Marcel Dekker, New York (1982).

20. D. M. Dražić and S. K. Zečević, *Corros. Sci.* **25,** 209 (1985).

21. D. M. Dražić and Chen Shen Hao, *Glasnik Hem. društva Beograd* **47,** 649 (1982).

22. D. M. Dražić and Chen Shen Hao, *Electrochim. Acta* **27,** 149 (1982).

23. D. M. Dražić and Chen Shen Hao, Extended Abstracts, 33rd ISE Meeting, Lyon, 1982, p. 221.

24. D. M. Dražić, V. J. Dražić, and M. Ž. Atanacković, *Glasnik Hem. društva Beograd* **47,** 661 (1982).

25. G. Eichkorn, W. J. Lorenz, L. Albert, and H. Fischer, *Electrochim. Acta* **13,** 183 (1968).

26. L. Ž. Vorkapić and D. M. Dražić, *Glasnik Hem. društva Beograd* **42,** 545 (1977).

30

Electroless Deposition of Metals and Alloys

Milan Paunovic

1. ELECTROCHEMICAL ASPECTS

1.1. The Overall Reaction

The process of electrodeposition of metal M

$$M^{z+}_{solution} + ze^- \xrightarrow{\text{electrode}} M_{lattice} \qquad (1)$$

is a process in which z electrons, supplied by an external power supply, are injected into the electrode and captured by metal ions M^{z+} in the solution, resulting in M in the lattice.

In electroless deposition of metals there is no external power supply as a source of electrons. In this process a reducing agent, Red, in the solution is the electron source; the electron-donating species Red gives electrons to the catalytic surface, and metal ions M^{z+} in the solution capture electrons from the catalytic surface.

The overall reaction of electroless metal deposition is

$$M^{z+}_{solution} + Red_{solution} \xrightarrow{\text{catalytic surface}} M_{lattice} + Ox_{solution} \qquad (2)$$

where Ox is the oxidation product of the reducing agent Red. The catalytic surface can be the substrate S or catalytic nuclei of metal M' dispersed on a noncatalytic substrate. The oxidation–reduction reaction (Eq. 2) proceeds only on a catalytic surface. Thus, the reaction of electroless metal deposition is a heterogeneous catalytic electron-transfer reaction in which electrons are transferred across the interface, from a reducing agent Red to metal ions M^{z+}. The reaction given in Eq. (2) must be conducted in such a way that a homogeneous reaction between M^{z+} and Red, in the bulk of the solution, is suppressed.

Metals that can be electrolessly deposited are Ag, Au, Co, Cu, Ni, Pd, Pt, Ru, and Sn. Commonly used reducing agents are $HCHO$, NaH_2PO_2, KBH_4, $(CH_3)_2NH \cdot BH_3$, and NH_2NH_2.

Milan Paunovic ● IBM T. J. Watson Research Center, Yorktown Heights, New York 10598.

Electrochemistry in Transition, edited by Oliver J. Murphy *et al.* Plenum Press, New York, 1992.

1.2. Total and Component Current–Potential Curves

The overall reaction given by Eq. (2) can be electrochemically described in terms of three current-potential (i-V) curves, shown schematically in Fig. 1. The dashed curve in Fig. 1 shows the current-potential curve for the overall reaction, i_{total}-V. This curve intersects the potential axis; at the intersection the current is zero. The solid curves show the current-potential curves for the partial reactions: (i) i_c-V, the current-potential curve for the reduction of M^{z+} ions, recorded from the rest potential $E_{eq,M}$, in the absence of the reducing agent Red, and (ii) i_a-V, the current-potential curve for the oxidation of the reducing agent Red, recorded from the rest potential $E_{eq,Red}$, in the absence of M^{z+} ions.[1]

Two major characteristics of this system of curves are: (i) at the potential where i_{total} intersects the potential axis.

$$i_a = i_c, \qquad \text{for } i_{total} = 0 \tag{3}$$

(ii) at any point of the i_{total}-V curve

$$i_{total} = i_a + i_c \tag{4}$$

Thus, the total current density, i_{total}, is the result of the addition of the current densities of the two partial processes. In other words, i_{total} can be decomposed into i_a and i_c.

1.3. Electrochemical Model

1.3.1. The Mixed-Potential Theory

On the basis of considerations of the current-potential curves, Paunovic[1] and Saito[2] used the mixed-potential theory of corrosion processes, developed by Wagner and Traud,[3] to interpret the process of electroless deposition of metals.

According to the mixed-potential theory, the overall reaction given by Eq. (2) can be decomposed into one reduction reaction, the cathodic partial reaction

$$M^{z+}_{solution} + ze^- \xrightarrow{\text{catalytic surface}} M_{crystal\ lattice} \tag{5}$$

and one oxidation reaction, the anodic partial reaction

$$Red_{solution} \xrightarrow{\text{catalytic surface}} Ox_{solution} + me^- \tag{6}$$

The partial reaction given by Eq. (6) is the source of electrons used for the reduction of M^{z+} ions, Eq. (5).

Thus, the overall reaction (Eq. 2) is the result of the combination of two different partial reactions, Eqs. (5) and (6). The equilibrium (rest) potential of the reducing agent, $E_{eq,Red}$ (Eq. 6), must be more negative than that of the metal electrode, $E_{eq,M}$ (Eq. 5), in order that Red can function as an electron donor and M^{z+} as an electron acceptor.

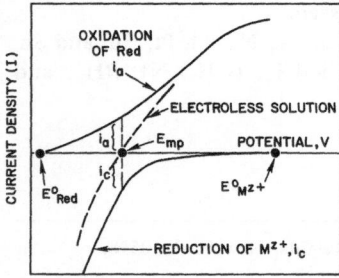

FIGURE 1. Total and component current-potential curves ($i_a + i_c = i_{total}$) for the overall reaction. (From Ref. 4.)

The partial reactions given by Eqs. (5) and (6) occur simultaneously at the catalytic surface, and each of these reactions strives to establish its own equilibrium potential, E_{eq}. The result of this process is establishment of a steady state with the compromised potential called the steady-state mixed potential, E_{mp} in Fig. 1.

According to the mixed-potential theory, the partial electrochemical processes of oxidation and reduction occur simultaneously, and spatially separated on the substrate. Thus, there is a statistical division of the catalytic sites on the substrate into the anodic and cathodic sites. Since these catalytic sites are part of the same piece of metal, they are "short-circuited" and there is a flow of electrons between these sites.

Two basic characteristics of the steady-state mixed potential E_{mp} are[1,4]:

(i) A net electrochemical reaction occurs in each redox system since both reactions, Eqs. (5) and (6), are removed from their equilibrium by establishment of the mixed potential. The result of these partial net reactions is the overall reaction of electroless deposition of metal M, Eq. (2).

(ii) The condition for the steady state is that the rate of the reduction of M^{z+}, deposition of M, is equal to the rate of oxidation of the reducing agent Red. In terms of current densities this condition is

$$(i_c)_{E_{mp}} = (i_a)_{E_{mp}} = i_{\text{electroless M deposition}} \qquad (7)$$

since a net current cannot flow in the isolated system.

1.3.2. Test of the Mixed-Potential Theory

1.3.2.1. Absence of Interfering Reactions. The basic concepts of the mixed-potential theory were tested by Paunovic[1] for the case of the electroless copper deposition from a cupric sulfate solution containing ethylenediaminetetraacetic acid (EDTA) as a complexing agent and formaldehyde (HCHO) as the reducing agent (Red). The test involves comparison between the direct experimental and the "theoretical" values, derived from the current-potential curves for partial reactions, for E_{mp} and the rate of deposition, i_{dep}. Figure 2 shows this comparison. This figure illustrates an alternative method of presentation of current-potential curves for electroless metal deposition. In this method of presentation the sign of the current density is neglected. Figure 2 shows that in this type of presentation the current-potential curves for the two partial reactions, i_a-V and i_c-V, intersect. According to the mixed-potential theory, the coordinates of this intersection have the following meaning: (i) the abscissa, the current density of the intersection, is the deposition current, i_{dep}, that is, the rate of electroless deposition in terms of mA/cm^2; (ii) the ordinate, the potential of the intersection, is the mixed potential E_{mp}.

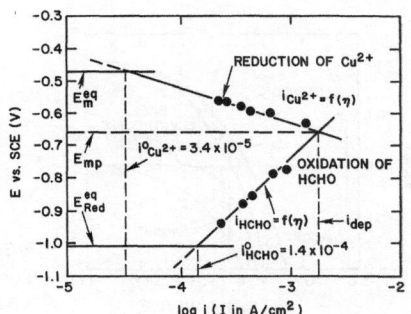

FIGURE 2. Current–potential curves for the reduction of Cu^{2+} ions and for the oxidation of reducing agent Red, formaldehyde, combined into one graph (Evans diagram). Solution for the Tafel line for the reduction of Cu^{2+} ions: $0.1M$ $CuSO_4$, $0.175M$ EDTA, pH 12.50, $E_{eq}(Cu/Cu^{2+}) = -0.47$ V versus SCE; for the oxidation of formaldehyde: $0.05M$ HCHO and $0.075M$ EDTA, pH 12.50, $E_{eq}(HCHO) = -1.0$ V versus SCE; temperature, 24°C (±0.5°C). (From Ref. 1.)

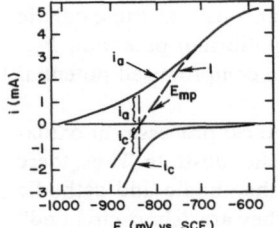

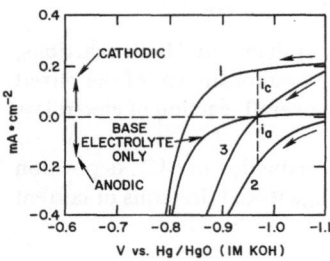

FIGURE 3. Wagner–Traud diagram for the electroless Ni(B) deposition. $E_{mp} = -840$ mV versus SCE. Electrode area, 0.68 cm^2. (From Refs. 8 and 9.)

FIGURE 4. Current–potential curves at gold electrode at 75°C. Base electrolyte, KOH and KCN. Curve 1: $2 \times 10^{-4} M$ KAu(CN)$_2$ without KBH$_4$; curve 2: 0.1M KBH$_4$ without KAu(CN)$_2$; curve 3: $2 \times 10^{-4} M$ KAu(CN)$_2$ and 0.1M KBH$_4$. Potential scanned at 5.56 mV/s. (From Ref. 10.)

Examination of Fig. 2 and the results of direct experimental measurements shows that there is a relatively good agreement between the direct experimental and the "theoretical" values.[1] Thus, one can conclude that the mixed-potential theory is essentially verified for this case of electroless copper deposition. These conclusions were later confirmed by Donahue,[5] Molenaar et al.,[6] and El-Raghy and Abo-Salama.[7] The mixed-potential theory was tested and verified for the case of electroless nickel deposition (Fig. 3).[8,9]

Okinaka verified the mixed-potential theory for the case of electroless gold deposition.[10] Figure 4 shows that the partial cathodic current density of [Au(CN)$_2$]$^-$ reduction measured at the E_{mp} in the absence of BH$_3$OH$^-$ is equal to the partial anodic current density of oxidation of BH$_3$OH$^-$ measured at the E_{mp} in the absence of [Au(CN)$_2$]$^-$.

1.3.2.2. Presence of Interfering Reactions. In some cases the total current density, i_{total}, is the result of a simple addition of the current densities of the two partial reactions, i_a and i_c. In the presence of interfering reactions the verification of the mixed-potential theory involves superposition of current–potential curves for the investigated electroless process with those of an interfering reaction in order to interpret the total i-V curve. Two important examples will be discussed.

The first example is electroless deposition of gold from solutions where the concentration of K[Au(CN)$_2$] is greater than $2 \times 10^{-4} M$.[10] This example shows two types of interference. The first interference is oxidation of the substrate. Figure 5 shows that the total anodic partial

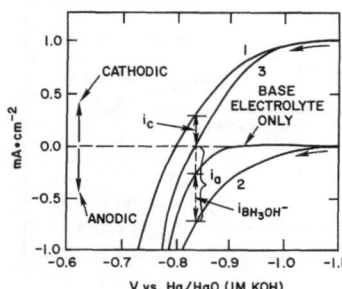

FIGURE 5. Current–potential curves at gold electrode at 75°C. Conditions were the same as for Fig. 4 except that the KAu(CN)$_2$ concentration was $10^{-3} M$. (From Ref. 10.)

current density i_a is the sum of two components: $i_a(BH_3OH^-)$, the anodic partial current of the oxidation of $BH_3OH_3^-$ and $i_a(Au)$, the anodic partial current of oxidation of Au, measured in the base electrolyte alone. The second interference is adsorption of complexed Au ions. Adsorption of $[Au(CN)_2]^-$ interferes with the anodic oxidation of BH_3OH^-. As a result of this interference, as found by Okinaka,[10] the gravimetrically determined deposition rates in this case are equivalent to i_c rather then $i_a(BH_3OH^-)$.

The second example is electroless deposition of copper from solutions containing dissolved oxygen.[11,12] In this case the interfering reaction is reduction of oxygen. As a result of this side reaction, the total cathodic partial current density i_c is the sum of two components: $i_c(M^{z+})$, the cathodic partial current of reduction of metal ions M^{z+}, and $i_c(O_2)$, the cathodic partial current of reduction of oxygen.

1.3.2.3. Interaction between Partial Reactions. The classic mixed-potential theory assumes that the two partial reactions are independent of each other.[3] In some cases (e.g., Section 1.3.2.1), this is a valid assumption. However, it was shown later that the partial reactions are not always independent of each other.[10,13-17] For example, Schoenberg has shown that the reducing agent in the electroless copper deposition, the methylene glycol anion (formaldehyde in an alkaline solution), enters the first coordination sphere of the copper tartrate complex and thus influences the rate of the cathodic partial reaction.[13] In another example it was shown that, in the electroless gold deposition, adsorption of the cathodic electroactive species, $[Au(CN)_2]^-$ influences the anodic oxidation of the reducing agent BH_3OH^-.[10]

The examples of interactions between partial reactions, and the examples of interfering reactions, discussed in section 1.3.2.2, illustrate the variety of factors that should be taken into account when applying the mixed-potential theory to electroless processes.

1.4. Kinetics and Mechanism of Partial Reactions

1.4.1. The Cathodic Partial Reaction

The mechanism of the partial cathodic reaction involves at least two basic elementary steps[18]:

(i) the formation of the electroactive species, and
(ii) the charge transfer from the catalytic surface to the electroactive species (electron capture).

The electroactive species M^{z+} are formed by dissociation of the complex $[ML_x]^{z+xp}$ since metal ions in a solution for electroless metal depsotion are, in general, complexed with a ligand:

$$[ML_x]^{z+xp} \to M^{z+} + L^p \tag{8}$$

where p is the charge of the ligand L, z the charge of the noncomplexed metal ion, and $(z + xp)$ the charge of the complexed metal ion.

The transfer of z electrons from the catalytic surface to the electroactive species M^{z+}

$$M^{z+} + ze^- \to M \tag{9}$$

proceeds in steps,[19-21] with the first charge transfer (one electron transfer) as the rate-determining step (RDS):

$$M^{z+} + e^- \xrightarrow{\text{RDS}} M^{(z-1)+} \tag{10}$$

Thus, from the kinetic aspects, the cathodic partial reaction is an electrochemical reaction, Eq. (9), which is preceded by a chemical reaction, Eq. (8). This kinetic scheme can be studied electrochemically by chronopotentiometry and potential sweep methods.[18,22,23]

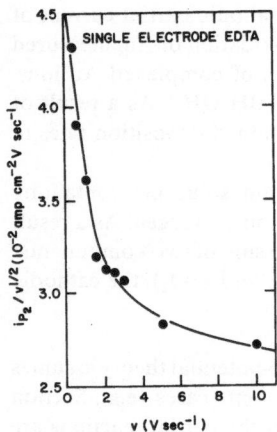

FIGURE 6. Potential sweep function for the dissociation of Cu(II) EDTA complex. (From Ref. 18.)

Paunovic[18] used chronopotentiometric $[\tau^{1/2} = f(i)]$ and potentiodynamic $[i_p/v^{1/2} = f(v)]$ functions to determine the relative rates of dissociation of Cu^{2+} complexes with various ligands (τ, the transition time; v, the rate of potential scan; i_p, the peak current). Figure 6 shows the change of $i_p/v^{1/2}$ versus the scan rate for the reduction of the Cu(II)EDTA complex.

A comparison of the relative rates of dissociation of complexes of Cu^{2+} with various ligands and the rate of electroless copper deposition from these complexes shows that there is a definite correlation between the rate of dissociation and the rate of electroless copper deposition.[18]

The major factors determining the rate of the partial cathodic reaction are concentrations of metal ions and ligands, pH of the solution, and the type and the concentration of additives.

Concentrations of metal ions and ligands and pH determine kinetics of the partial cathodic reaction in a general way given by the fundamental electrochemical kinetic equations.[20,24-26]

Additives can have two opposing effects: acceleration and inhibition.[27-30] The accelerating effect of guanine and adenine on the cathodic reduction of Cu^{2+} ions in electroless copper solution is shown in Fig. 7. The inhibiting effect of NaCN, for the same reaction, is shown in Fig. 8.

1.4.2. The Anodic Partial Reaction

The anodic partial reaction, like the cathodic partial reaction, proceeds in at least two elementary steps[4]:

(i) the formation of the electroactive species, and

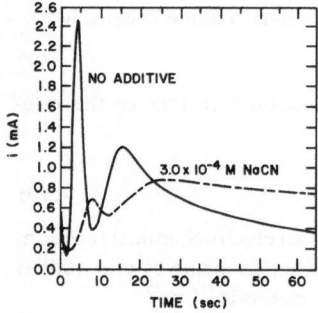

FIGURE 7. Cyclic voltammograms of a copper electrode in the electroless copper solution, showing effects of additives on the reduction of cupric ions. (From Ref. 28.)

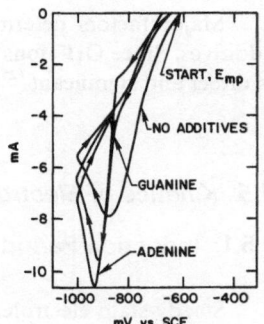

FIGURE 8. Potentiostatic current-time transient of a Pt electrode in electroless copper solution in the absence and in the presence of additive; E = −900 mV versus SCE. (From Ref. 30.)

(ii) the charge transfer from the electroactive species to the catalytic surface (electron injection).

A general mechanism for formation of electroactive species of the reducing agent Red

$$R-H \xrightarrow{\text{breaking RH bond}} R_{ads} + H_{ads} \tag{11}$$

where R—H is the reducing agent Red, and R_{ads} is the electroactive species originating from Red, was proposed by van den Meerakker.[31]

The adsorbed hydrogen, $H_{adsorbed}$, can be desorbed (i) in the chemical reaction

$$H_{adsorbed} \to 1/2 H_2 \tag{11a}$$

or (ii) in the electrochemical reaction

$$H_{adsorbed} \to H^+ + e^- \tag{11b}$$

For example, in electroless deposition of copper, when Red is formaldehyde and the substrate is Cu, $H_{adsorbed}$ desorbes in the chemical reaction (11a). However, if the substrate is Pt or Pd hydrogen desorbes in the electrochemical reaction (11b).[4]

According to this mechanism, the electroactive species R_{ads} (or Red_{ads}) is formed in the process of dissociative adsorption (dehydrogenation) of the reducing agent Red, represented here as R—H, on the catalytic surface. This process usually proceeds through an intermediate, Red'. For example, when the reducing agent is formaldehyde, that is, Red = HCHO, the intermediate R' is $H_2C(OH)O^-$ and the electroactive species $R_{ads} = Red_{ads}$ is $[HC(OH)^-]_{ads}$.[4] All other reducing agents so far examined follow the above kinetic scheme.[4]

The charge transfer from the electroactive species $R_{ads}(Red'')$ to the catalytic surface (electron injection) in an alkaline medium is

$$R_{ads} + OH^- \to ROH + e^- \tag{12}$$

or

$$R_{ads} + yOH^- \to Ox + xH_2O + me^- \tag{13}$$

In some cases of oxidation of reducing agents, there are parallel reactions, for example, in the case of oxidation of BH_4^- and $H_2PO_2^-$; parallel reactions are probably cathodic reactions resulting in the incorporation of B and P into the metal deposit, respectively.[31,32-34]

A comparison of mechanisms for the cathodic and the anodic partial reactions show that both mechanisms involve a chemical reaction that precedes the charge transfer step. Haruyama and Ohno have shown that the catalytic activity of metals for electroless deposition is mostly determined by the catalytic processes in the anodic partial reaction.[35]

Major factors determining the rate of oxidation of the reducing agent are pH and additives. Since OH^- ions are reactants in the charge transfer step, Eq. (13), the effect of pH is direct and significant.[25,30] Additives can have an inhibiting or an accelerating effect.[28,36]

1.5. Kinetics of Electroless Metal Deposition

1.5.1. Induction Period

Steady-state electroless metal deposition at the mixed potential E_{mp} is preceded by a non-steady-state period, called the induction period.[1,18,37]

Paunovic[18] decomposed the induction period for the overall process into the time dependence of the open-circuit potential (OCP) of the oxidation and reduction partial reactions, that is, the individual induction periods for each partial process. In the case of HCHO and dimethylamine borane (DMAB), the rate of the establishment of the OCP of the reducing agent is the rate-determining process in the establishment of the steady-state mixed potential.

1.5.2. Steady-State Kinetics

The rate of electroless metal deposition in terms of electrochemical kinetic parameters of partial reactions is given by[25]

$$i_{dep} = (i_M^0)^p (i_{Red}^0)^q \exp(V) \tag{14}$$

where

$$V = (E_{RM} - E_{RRed})/(b_M + b_{Red})$$

$$b_M = RT/\alpha_M n_M F$$

$$b_{Red} = RT/\alpha_{Red} n_{Red} F$$

$$p = b_M/(b_M + b_{Red})$$

$$q = b_{Red}/(b_M + b_{Red})$$

$$i^0 = nFk^0 (c_{Ox}^0)^{1-\alpha} (c_{Red}^0)$$

and E_R, i^0, α, k^0, F, T, and c are the rest (equilibrium) potential, the exchange current density, the transfer coefficient, the rate constant, the Faraday constant, temperature, and concentration, respectively. Subscripts M and Red designate that a given parameter is related to the partial cathodic and anodic reaction, respectively.

Equation (14) shows major factors determining the rate of electroless deposition. It can be used for computer simulation of electroless processes.[25]

Various empirical rate equations were derived for electroless deposition of copper and nickel.[5,7,38]

Paunovic and Vitkavage used polarization data in the vicinity of the mixed potential to determine the rate of deposition.[39,40] The following equations were used for the *in situ* computerized determination of the electroless deposition rate[39,41]:

$$i_{dep} = \frac{\sum\limits_{j=1}^{n} i_j F_j}{\sum\limits_{j=1}^{n} E_j^2} \tag{15}$$

$$E_j = 10^{\eta_j/b_a} - 10^{-\eta_j/b_c} \tag{16}$$

in which i_j and η_j are current density and overpotential, respectively, at the jth point on the i-V curve; b_a and b_c are the anodic and the cathodic Tafel slope, respectively.

Ohno[42] used ac polarization data and Ricco and Martin[43] used an acoustic wave device for *in situ* determination and monitoring of the rate of deposition.

1.6. Activation of Noncatalytic Surfaces

Noncatalytic surfaces (noncatalytic metals, noncatalytic semiconductors, and nonconductors) have to be activated, that is, made catalytic, prior to the electroless deposition. This activation is performed by generating catalytic metallic nuclei on the surface of a noncatalytic material. Two types of processes have been used to produce catalytic metallic nuclei: electrochemical and photochemical.

1.6.1. Electrochemical Activation

The catalytic metallic nuclei of metal Me on the noncatalytic surface S can be generated in an electrochemical oxidation–reduction reaction:

$$Me^{n+} + Red \xrightarrow[\text{heterogeneous reaction}]{\text{homogeneous}} Me + Ox \tag{17}$$

where Me^{n+} is the metallic ion, and Me the metal catalyst. The preferred reducing agent Red is Sn^{2+}. The preferred nucleating agent Me^{n+} is Pd^{2+}. The palladium catalytic sites on the activated surface are dispersed on the surface of a substrate in an island network.[44-46] The activation process has been studied by electrochemical[47,48] and surface analytical techniques.[48-52]

In some cases the noncatalytic metallic substrate S can play the role of Red in Eq. (17) in a displacement reaction.[53-55]

Many variations of the basic process in Eq. (17) are known.[56-59] One important variant is the example when a polymer is used as a carrier of a catalytic salt to implant the salt on the surface of the polymer.[57] Another example is the use of surfactant to enhance the surface reaction producing Pd catalytic sites.[59]

1.6.2. Photochemical Activation

Catalytic metallic nuclei of Pd, Pt, Au, and Cu can be generated in an intramolecular electron transfer resulting from photon absorption. For example, catalytic copper nuclei can be formed in the photochemical reaction

$$CuAc \xrightarrow[\lambda < 350 \text{ nm}]{h\nu} Cu + Ox \tag{18}$$

where Ox is the oxidation product of acetate ion, Ac.[59] Other photochemical methods have been reviewed by Paunovic.[60]

2. STRUCTURAL ASPECTS

2.1. Mechanism of Electroless Crystallization

Mechanistically, electroless crystallization proceeds in two basic states[40]: (i) the thin-film stage (up to 3 μm) and (ii) the bulk stage.

2.1.1. Thin-Film Stage

The mechanism of the thin-film formation is characterized by three simultaneous crystal-building processes[40,61-63]: nucleation (formation), growth, and coalescence of three-dimensional crystallites (TDC).

In the initial stages of electroless deposition the average density of TDC increases with the time of deposition; in this stage the nucleation is the predominant process. Later, the average density of TDC reaches a maximum and then decreases with time. In the stage of decreasing density of TDC, the coalescence is the predominant crystal-building process.[61] A continuous electroless film is formed by lateral growth and coalescence of TDC.

The rate of nucleation J (N/cm^2 s) can be determined by the curve-fitting method using the equation

$$N(t) = N_s[1 - \exp(-Jt/N_s)] \tag{19}$$

where $N(t)$ and N_s are TDC density at time t and the saturation value, respectively.[62]

2.1.2. Bulk Stage

After the formation of the continuous thin film, the deposition of a thick (1–25 μm) copper or nickel film proceeds, in most cases, by the following processes[40,64-67]: (i) the preferential growth of favorably oriented grains, (ii) restriction (inhibition) of vertical growth of nonfavorably oriented grains, (iii) lateral joining of preferentially growing grains, (iv) cessation of growth of initial grains, and (v) nucleation and growth of a new layer of grains.

Three-dimensional crystallites (TDC) formed in the thin-film stage grow vertically and laterally. In this process of vertical and lateral growth, a preferentially growing, favorably oriented grain (TDC) increases its width and subsequently joins laterally with other preferentially growing grains (TDC). After this lateral joining of growing grains, the width of preferentially oriented grains becomes constant. The result of these processes of electroless crystallization is the columnar structure of the deposit.[64-66]

There is no adequate theory of lamellar growth of Ni(P). Periodic fluctuations in the content of phosphorous in electroless Ni are possible causes of the lamellar structure.

2.2. Typical Structures

Three typical structures of electrolessly deposited metals are columnar, lamellar, and columnar with laminations.

Metals having the columnar structure are Cu[64-66] and Ni.[34,68] The microstructure of electroless Cu shows columnar grains with 0.7-μm diameter in a plane parallel to the substrate, the length perpendicular to the substrate being about 6 μm. One columnar grain usually consists of many subgrains.[65,66]

Lamellar structure is observed in the following cases: Ni(P), Ni(B)–Mo, and Ni(P)–Mo.[69-71] Under conditions reported by Graham et al.,[70] the thickness of lamellae is 5 μm for the Ni(P) deposit from alkaline–chloride solution and 0.5 μm for the Ni(P) deposit from acid–sulfate solution. The average grain size of Ni(P) is about 150 Å, and the orientation of grains is random.[70]

Columnar structure with laminations is observed in Ni(P) deposits.[68,73]

3. TECHNOLOGICAL APPLICATIONS

3.1. Unique Properties of Some Electroless Deposits

Electroless deposition of nickel and cobalt was "accidentally encountered"[73-76] by Brenner and Riddell in 1946 during electrodeposition of nickel–tungsten and cobalt–tungsten alloys (in the presence of sodium hypophosphite) on steel tubes in order to produce material with better hardness than steel. Since then, hardness, wear resistance, and corrosion resistance have been major properties determining technological applications of electroless Ni(P) in the electronic, aerospace (stators for jet engines), automotive, machinery, oil and gas production, power generation, printing, and textile industries.[34,77-81] Electroless Ni(P) is harder and has better corrosion resistance than electrodeposited Ni-P.[82] Nonmagnetic electroless Ni(P), or NiCu(P), is used as underlayer in high-density metallic memory disk fabrication to improve the mechanical finish of the surface.

Magnetic properties determine technological applications of electroless Co(P). Electroless Ni(P), Co(P), Co(B), and their alloys show unique characteristics and can be used for production of thin films for high-density magnetic recording media.[34,83-85] Electrolessly deposited CoNi(P) films were deposited for production of longitudinal recording media while Co(P) alloys with Ni, Re, and Mn were developed for production of perpendicular magnetic recording media[86-89] (Fig. 9).

Magnetic properties of thin films can be altered by varying additives in the deposition solution, deposition parameters, nucleation conditions, structure of deposit, texture, extent of incorporation of other elements (e.g. P, B, Ni, Re, Mn, etc.), and magnetic annealing.[83]

The density of complexity of the problems and the sophistication of interpretation of the phenomena involved in deposition of magnetic thin films can be illustrated by the example of electroless Co(P) thin films. For example, the coercivity (H_c) of the Co(P) film (with 9.6% P) deposited at pH 9.5 increases with increasing film thickness and reaches a maximum at a thickness of ~800 Å.[90] The Co(P) film becomes continuous at a thickness of 300 Å when deposited on the polished Al-based Ni(P) substrate.[91]

As the thickness of the film increases from 200 to 800 Å, the grain size increases to 200–700 Å.[92] At 800 Å the average grain size is comparable to the critical size of a single-domain particle. As the film thickness and the grain size increase further, the multidomain grains form[90,93,94] (Fig. 10). The domain structure (the thickness of the transition region between domains) and the mobility of domain walls affect magnetic properties of materials.[93,94]

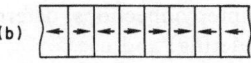

FIGURE 9. Schematic representation of two magnetization modes in magnetic recording: (a) perpendicular magnetization; (b) longitudinal magnetization.

FIGURE 10. Schematic representation of an unmagnetized ferromagnetic material with multidomain grains.

Thus, description of the structure of deposited magnetic material includes, besides the shape, the size, and the orientation of grains, also the size of domains and the structure of domain walls. Other factors to be considered in this example are: (i) the pH effect on transition from hexagonal-close-packed (hcp) to face-centered cubic (fcc) structure. (ii) state of P (e.g., Co_2P versus a pure P), not only %P, and (iii) method of the Ni(P) substrate preparation.[95] All these factors combine to determine magnetic properties of the electroless deposit.

The above example demonstrates the high degree of interdisciplinary character of the background essential to understand the control of electrolessly deposited magnetic media. Also, it shows that a large number of factors should be considered when selecting the deposition process.

3.2. Unique Characteristics of the Electroless Deposition Process

The technologically most important characteristics of the electroless deposition process are: (i) ability to deposit metal on nonconductors (after prior activation), (ii) ability to deposit metal on isolated, electrically not connected areas, (iii) ability to deposit metal with uniform thickness independent of the shape of the substrate (no current distribution problem), and (iv) low cost.

The most important use of these unique characteristics is in the electronics industry in the fabrication of chip carriers and printed circuit boards.[96] In chip carriers and printed circuit boards, individual circuit components and individual circuits are interconnected by metallic film wires (interconnections) on an insulating material. The technology of production of these conductors by electroless copper deposition, or by initial electroless deposition and subsequent electrodeposition, is dependent on the progress in the electronics industry. The progress in the electronics industry since the invention of the first semiconductor device in 1947 and the invention of integrated circuits (IC) in the early sixties was the major driving force for the electrochemistry and electrochemical technology of electroless metal deposition. Continuous demand for a higher level of integration (narrower conductor lines, smaller spacings between lines, and longer and narrower through-holes) continuously increases demand for electroless metal deposition of higher and higher qualities, especially with respect to physical (ductility, tensile strength, toughness) and electrical properties. Physical properties of electrolessly deposited copper in the through-holes (holes connecting two or more layers of interconnections) determine reliability of the printed circuit boards during processing (e.g., soldering) and use.[97]

Okinaka and Nakahara showed that the formation of small voids and small gas bubbles containing hydrogen are major factors determining ductility of electroless copper.[98] Nakahara and Okinaka showed that ductility promoters, such as cyanide ions, and higher temperatures facilitate desorption of hydrogen gas generated in the reaction given by Eq. (11). Some ductility-promoting additives, for example, 2,2'-dipyridyl and $K_2Ni(CN)_4$, inhibit both the inclusion of hydrogen and the formation of voids.[72]

Impurities in the deposit, lattice defects (e.g., vacancies, dislocations, and grain boundaries), and surface roughness are major factors determining electrical resistivity of thin films.[83,99-100]

The applicability of electroless copper in the electronics industry depends upon physical properties (ductility, tensile strength, toughness), structure (grain size, lattice defects, inclusions, surface roughness), and electrical properties of the deposit.

4. FUTURE DIRECTIONS

4.1. Heterogeneous Catalysis

Three fundamental problems in electroless deposition that need further in-depth studies are (i) the dependence of the catalytic activity on the structure of the catalyst, (ii) catalysis in the dissociation reaction of the complexed ion in the partial cathodic reaction, and (iii) catalysis of adsorption and desorption processes involving electroactive species in the anodic partial reaction.

Osaka et al.[52] studied the dependence of the catalytic activity of an evaporated Pd catalyst on its structure for the case of electroless Ni(P) deposition and concluded that discontinuous clusters of Pd particles (crystallites) are more catalytic than a continuous Pd deposit. This dependence of catalytic activity on structure was interpreted on the basis of the dependence of the specific activity (activity/area of catalyst) on the size-dependent density of selective surface sites (e.g., corners, edges, kinks).

Catalytic effects were detected in the dissociation reaction of the complexed ion in the cathodic partial reaction of Ni deposition using DMAB as the reducing agent.[32,101] It was shown that the electroactive species of the anodic partial reaction catalyzes the cathodic partial reaction in electroless deposition of copper.[102] Catalytic effects in the anodic partial reaction was studied mostly for the formaldehyde and borohydride reducing agents.[4,10,102-103] More data are needed for the NaH_2PO_2 reducing agent.[34]

A review of published results shows that there are not sufficient experimental results to form a basis for the formulation of a comprehensive theory of catalytic phenomena in electroless metal deposition.

4.2. Mixed-Potential Theory

Future studies on the elucidation of electroless deposition processes will result in division of deposition processes into two basic groups: (i) processes that follow the classical mixed-potential theory, where it is considered that electroless metal deposition is the result of two independent partial reactions, and (ii) processes that follow the modified mixed-potential theory, where it is considered that the electroless metal deposition is the result of two partial reactions accompanied by (a) the presence of interfering reactions and/or (b) interactions between partial reactions.

More studies on the mechanisms of interfering reactions and interactions between partial reactions are needed.

4.3. Correlation between Kinetic Parameters, Structure, and Physical Properties of Deposits

There are only a few publications treating correlation between (i) electrochemical kinetic parameters and structure and (ii) structure and physical properties of deposits (e.g., Ref. 65 and references therein).

There are presently many activities in this area, and upcoming publications are expected to contribute to our understanding of the above problems.

4.4. New Materials

Metal silicides are used in the electronics industry (integrated circuits) to form contact to Si. They are formed by evaporating or sputtering a metal (e.g., Pt, Pd, Ni, Co) onto Si followed by thermal processing (sintering) to form silicide.[104]

Chang and Lee have shown that nickel silicide ($NiSi_2$) can be formed by electrolessly depositing a thin film of Ni(P) on Si and subsequent reaction of Ni with the substrate at temperatures between 800 and 900°C.[105]

The use of electrolessly deposited metals to form metal silicides should be expanded to other metals, and new, similar technologies should be developed.

4.5. New Technologies for Integrated-Circuit Fabrication

The feasibility of using selective electroless metal deposition for integrated-circuit (IC) fabrication has been demonstrated.[53,106–117] Ting, Paunovic, and co-workers[53,115] studied electroless deposition of Ni, Co, Pd and Cu for filling contact holes (metal contact to silicon) and via holes (connecting one metal layer to the next) and for formation of conductors. These metals were selectively deposited on Si, Al (Al–1%Si), Ti, $TiSi_2$, and $CoSi_2$ substrates through patterned silicon dioxide, silicon oxynitride, photoresist, and polyimide. Figure 11 shows an example of a copper-filled via hole (5 μm or larger) array formed in 5-μm-thick polyimide. The excellent selectivity (deposition only on the patterned substrate but not on patterning material) was achieved in submicron dimensions.[53,115]

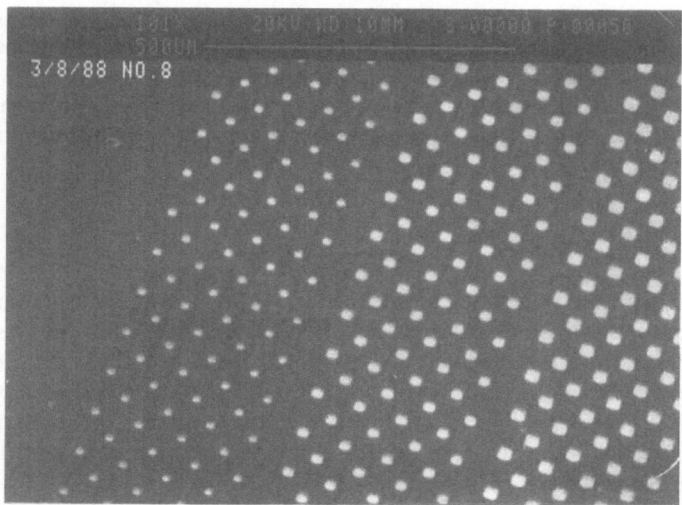

FIGURE 11. Copper-filled via holes in 5-μm polyimide. (From Ref. 57.)

These studies are in the early stage of development. Although deposition processes for different combinations of substrate and patterning material have been well developed, more efforts are required to improve the process yield, demonstrate manufacturability, and address device reliability issues. Since the selective electroless metal deposition process represents a very attractive alternative to the conventional IC fabrication processes,[114] we can expect very intensive activity in this area.

It is important to note that the development of the electroless deposition technology for IC fabrication represents an excellent opportunity to electrochemists. This opportunity stems from the fact that new electroless deposition processes, producing deposits of different structure and properties, are needed to meet requirements of new, submicron computer technologies.

4.6. Computerized Monitoring and Control of the Deposition Process

Increasing demands for electroless metal deposits of the highest possible quality will increase demand for the better control of the deposition process. This can be achieved by the computerized, *in situ* monitoring and control of the process. The initial work on these problems was done by Paunovic and co-workers,[39,41,118] Okinaka, Turner, and co-workers,[119,120] and Anderson *et al.*[121] A greater activity is expected in this area.

5. CONCLUDING REMARKS

A complete review of a scientific subject includes three parts: retrospect (the earlier contributions), present status, and prospects (the future research goals). It is very instructive to consider the early work on electroless metal deposition in this last section of this chapter since the understanding of the basics of electroless deposition is necessary in order to understand differences between the three periods in the development of electroless deposition.

In the first period, from 1835 to 1947, thin metallic films were deposited on nonconductors from solutions very similar to present electroless deposition solutions. For example, in 1835 Leibig deposited films of silver on glass using acetaldehyde as a reducing agent for silver ions in an ammoniacal solution in a glass vessel.[122] Liebig also deposited copper films on glass using copper tartrate and ammonia. Later, researchers used other reducing agents to deposit metallic films on nonconductors: for example, formaldehyde, hydrazine, glucose, and ethyl alcohol.[122] In most cases, deposition of thin metallic films was followed by electrodeposition in order to build thick deposits. Metal deposition on nonconductors in this period was characterized by (i) instability of solutions (simultaneous occurrence of the homogeneous and heterogeneous metal deposition reactions) and (ii) limited thickness of the deposit (thin films only).

In the second period, from 1947–1968, true electroless metal deposition processes were developed. The electroless Ni and Co deposition process developed by Brenner and Riddell[73-76] had basic characteristics of an electroless metal deposition process: (i) metal deposition occurs only on the catalytic surface, (ii) the homogeneous reaction is suppressed, and only the heterogeneous process proceeds and (iii) deposit thickness is not limited by the process and can reach any desired value.

Finally, in the third period, from 1967 up to the present, Paunovic[1] and Saito[2] used the mixed-potential theory of corrosion processes, developed by Wagner and Traud,[3] to interpret electroless deposition of metals and alloys on the basis of the electrochemical principles. After these initial electrochemical studies, and further studies by many other researchers, electroless deposition became a recognized part of electrochemical science and technology. Our present understanding of electroless processes is on the same level of understanding as electrodeposition processes and organic electrochemistry.

ACKNOWLEDGMENT

The author wishes to thank Dr. Lubomyr T. Romankiw for many helpful and clarifying comments.

REFERENCES

1. M. Paunovic, *Plating* **55**, 1161 (1968).
2. M. Saito, *J. Met. Finish. Soc. Jpn.* **17**, 14 (1966).
3. C. Wagner and W. Traud, *Z. Electrochem.* **44**, 391 (1938).
4. M. Paunovic, in: *Electroless Deposition of Metals and Alloys* (M. Paunovic and I. Ohno, eds.), Proceedings Vol. 88-12, pp. 3-19, The Electrochemical Society, Pennington, New Jersey (1988).
5. F. M. Donahue, *Oberfläche-Surface* **13**(12), 301 (1972).
6. A. Molenaar, M. F. E. Holdrinet, and L. K. H. van Beek, *Plating* **61**, 238 (1974).
7. S. M. El-Raghy and A. A. Abo-Salama, *J. Electrochem. Soc.* **126**, 171 (1979).
8. M. Paunovic, AES 1st Electroless Plating Symposium, March, 1982.
9. M. Paunovic, *Plat. Surf. Finish* **70**(2), 62 (1983).
10. Y. Okinaka, *J. Electrochem. Soc.* **120**, 739 (1973).
11. J. W. Jacobs and J. M. G. Rikken, in: *Electroless Deposition of Metals and Alloys* (M. Paunovic and I. Ohno, eds.), Proceedings Vol. 88-12, p. 75, The Electrochemical Society, Pennington, New Jersey (1988).
12. T. Hayashi, *Met. Finish.* **85**(6), 85 (1985).
13. L. N. Schoenberg, *J. Electrochem. Soc.* **118**, 1571 (1971).
14. I. Ohno and S. Haruyama, *Surf. Technol.* **13**, 1 (1981).
15. P. Bindra and J. Tweedie, *J. Electrochem. Soc.* **130**, 1112 (1983).
16. A. Vashkialis and I. Iachiayskene, *Electrochemistry* (Academy of Sciences U.S.S.R.) **17**, 1816 (1981).
17. H. Wiese and K. G. Weil, *Ber. Bunsenges. Phys. Chem.* **91**, 619 (1987).
18. M. Paunovic, *J. Electrochem. Soc.* **124**, 349 (1977).
19. B. E. Conway and J. O'M. Bockris, *Electrochim. Acta* **3**, 340 (1961).
20. J. O'M. Bockris and A. K. N. Reddy, *Modern Electrochemistry*, Plenum Press, New York (1970).
21. J. O'M. Bockris and G. A. Razumney, *Fundamental Aspects of Electrocrystallization*, Plenum Press, New York (1967).
22. M. Paunovic, *J. Electrochem. Soc.* **14**, 447 (1967).
23. R. S. Nicholson and I. Shain, *Anal. Chem.* **36**, 706 (1964).
24. P. Delahay, *Double Layer and Electrode Kinetics*, Interscience, New York (1965).
25. M. Paunovic, *J. Electrochem. Soc.* **125**, 173 (1978).
26. S. R. Morrison, *Electrochemistry at Semiconductor and Oxidized Metal Electrodes*, Plenum Press, New York (1980), Chapter 3.
27. L. N. Schoenberg, *J. Electrochem. Soc.* **119**, 1491 (1972).
28. M. Paunovic and R. Arndt, *J. Electrochem. Soc.* **130**, 794 (1983).
29. D. Vitkavage and M. Paunovic, *Plat. Surf. Finish.* **70**(4), 48 (1983).
30. M. Paunovic, *J. Electrochem. Soc.* **132**, 1155 (1985).
31. J. E. A. M. van de Meerakker, *J. Appl. Electrochem.* **11**, 395 (1981).
32. M. Lelental, *J. Electrochem. Soc.* **120**, 1650 (1973).
33. G. Bech-Nielsen, C. Q. Jessen, and J. C. Reeve, *J. Electrochem. Soc.* **133**, 1521 (1986).
34. P. Cavallotti and G. Salvago, in: *Electrodeposition Technology, Theory and Practice* (L. T. Romankiw and D. R. Turner, eds.), Proceedings Vol. 87-17, p. 327. The Electrochemical Society, Pennington, New Jersey (1987).

35. S. Haruyama and O. Ohno, in: *Electroless Deposition of Metals and Alloys* (M. Paunovic and I. Ohno, eds.), Proceedings Vol. 88-12, p. 20, The Electrochemical Society, Pennington, New Jersey (1988).

36. J. Hoarkans, *J. Electrochem. Soc.* **131,** 1615 (1984).

37. J. Dumesic, J. A. Koutsky, and W. Chapman, *J. Electrochem. Soc.* **121,** 1405 (1974).

38. V. A. Lloyd and G. O. Mallory, AES 1st Electroless Plating Symposium, March, 1982.

39. M. Paunovic and D. Vitkavage, *J. Electrochem. Soc.* **126,** 2282 (1979).

40. M. Paunovic, in: *Electrodeposition Technology, Theory and Practice* (L. T. Romankiw and D. R. Turner, eds.), Proceedings Vol. 87-17, p. 349, The Electrochemical Society, Pennington, New Jersey (1987).

41. J. Duffy, M. Paunovic, S. Christian, and J. McCormack, U.S. Patent 4,814,197 (March 21, 1989).

42. I. Ohno, in: *Electroless Deposition of Metals and Alloys* (M. Paunovic and I. Ohno, eds.), Proceedings Vol. 88-12, p. 129, The Electrochemical Society, Pennington, New Jersey (1988).

43. A. J. Ricco and S. J. Martin, in: *Electroless Deposition of Metals and Alloys* (M. Paunovic and I. Ohno, eds.), Proceedings Vol. 88-12, p. 142, The Electrochemical Society, Pennington, New Jersey (1988).

44. R. Sard, *J. Electrochem. Soc.* **117,** 864 (1970).

45. J. P. Marton and M. Schlesinger, *J. Electrochem. Soc.* **115,** 16 (1968).

46. C. H. deMinjer and P. F. J. v.d. Boom, *J. Electrochem. Soc.* **120,** 1644 (1973).

47. J. Hoarkans, *J. Electrochem. Soc.* **131,** 1615 (1984).

48. T. Osaka, H. Takematsu, and K. Nihei, *J. Electrochem. Soc.* **127,** 1021 (1980).

49. R. L. Cohen and K. West, *J. Electrochem. Soc.* **119,** 433 (1972).

50. R. L. Cohen, J. F. D'Amico, and K. W. West, *J. Electrochem. Soc.* **118,** 2042 (1971).

51. R. L. Meek, *J. Electrochem. Soc.* **122,** 1478 (1975).

52. T. Osaka, I. Koiwa, and L. G. Svendsen, *J. Electrochem. Soc.* **132,** 2081 (1985).

53. C. H. Ting and M. Paunovic, *J. Electrochem. Soc.* **136,** 456 (1989).

54. D. C. Zipperian, S. Raghavan, and M. D. Pritzker, in: *Electroless Deposition of Metals and Alloys* (M. Paunovic and I. Ohno, eds.), Proceedings Vol. 88-12, p. 113, The Electrochemical Society, Pennington, New Jersey (1988).

55. H. M. Naguib, C. Jang, T. F. Klemme, K. Wong, A. Rangappan, W. W. Yao, and R. T. Fulks, IEEE V-MIC Conference, June 15–16, 1987.

56. V. A. Rukhlya, T. N. Vorob'eva, V. V. Sviridov, and V. A. Lastochkina, *Zh. Prikl. Khim.* **61**(3), 653 (1988).

57. L. T. Romankiw, *IBM Tech. Discl. Bull.* **13**(5), 1199, 2000 (1970).

58. A. Viehbeck, Proceedings of the International Symposium on Metallization of Polymers, American Chemical Society, Montreal, Canada, Sept. 24–28, 1989.

59. C. J. Sambucetti, J. Varsik, and B. Laboy, in: *Electrochemical Technology in Electronics* (L. T. Romankiw and T. Osaka, eds.), Proceedings Vol. 88-23, p. 59, The Electrochemical Society, Pennington, New Jersey (1988).

60. M. Paunovic, *J. Electrochem. Soc.* **127,** 441c (1980).

61. M., Paunovic and C. Stack, in: *Electrocrystallization* (R. Weil and R. G. Baradas, eds.), Proceedings Vol. 81-6, p. 205, The Electrochemical Society, Pennington, New Jersey (1981).

62. M. Paunovic and C. H. Ting, in: *Electroless Deposition of Metals and Alloys* (M. Paunovic and I. Ohno, eds.), Proceedings Vol. 88-12, p. 170, The Electrochemical Society, Pennington, New Jersey (1988).

63. R. Sard, *J. Electrochem. Soc.* **117,** 864 (1970).

64. S. Nakahara and Y. Okinaka, *Acta Metall.* **31,** 713 (1983).

65. M. Paunovic and R. Zeblisky, *Plat. Surf. Finish.* **72**(2), 52 (1985).

66. J. Kim, S. H. Wess, D. Y. Jung, and R. W. Johnson, *IBM J. Res. Develop* **8,** 697 (1984).

67. H. J. Choi and R. Weil, *Plat. Surf. Finish.* **68**(5), 110 (1981).

68. A. Brenner and G. Riddell, *Proc. Am. Electroplaters Soc.* **34,** 156 (1947).

69. A. W. Goldenstein, W. Rostoker, F. Schlossberger, and G. Gutzeit, *J. Electrochem. Soc.* **104,** 104 (1957).

70. A. Graham, R. W. Lindsay, and H. J. Read, *J. Electrochem. Soc.* **112,** 401 (1965).

71. G. O. Mallory, *Plat. Surf. Finish.* **63**(6), 34 (1976).

72. Y. Okinaka and H. K. Straschil, *J. Electrochem. Soc.* **133,** 2608 (1986).

73. A. Brenner and G. Riddell, *J. Res. Nat. Bur. Standards* **37,** 31 (1946).

74. A. Brenner, *Plat. Surf. Finish.* **71**(7), 24 (1984).

75. A. Brenner and G. Riddell, *Am. Electroplaters Soc. Annu. Proc.* **33**, 23 (1946).

76. A. Brenner and G. Riddell, *Proc. Am. Electroplaters Soc.* **34**, 56 (1947).

77. L. T. Romankiw and T. A. Palumbo, in: *Electrodeposition Technology, Theory and Practice* (L. T. Romankiw and D. R. Turner, eds.), Proceedings Vol. 87-17, p. 13, The Electrochemical Society, Pennington, New Jersey (1987).

78. G. Salvago, G. Fumagalli, and F. Brunella, in: *Electrodeposition Technology, Theory and Practice* (L. T. Romankiw and D. R. Turner, eds.), Proceedings Vol. 87-17, p. 509, The Electrochemical Society, Pennington, New Jersey (1987).

79. D. DiMilia, J. Horkans, C. McGrath, M. Mirzamaani, and G. Scilla, in: *Electrochemical Technology in Electronics* (L. T. Romankiw and T. Osaka, eds.), Proceedings Vol. 88-23, p. 479, The Electrochemical Society, Pennington, New Jersey (1988).

80. G. O. Mallory, *Plating* **61**(11), 1005 (1974).

81. D. W. Baudrand, *Plat. Surf. Finish.* **68**(12), 57 (1981).

82. R. Weil, J. H. Lee, and K. Parker, *Plat. Surf. Finish.* **76**(2), 62 (1989).

83. L. T. Romankiw and D. A. Thompson, in: *Properties of Electrodeposits* (R. Sard, H. Leidheiser, Jr., and F. Ogburn, eds.), Proceedings, p. 389, The Electrochemical Society, Princeton, New Jersey (1975).

84. M. Schwartz and G. O. Mallory, *J. Electrochem. Soc.* **123**, 606 (1976).

85. Y. H. Chang, C. C. Lin, M. P. Hung, and T. S. Chin, *J. Electrochem. Soc.* **133**, 985 (1986).

86. T. Osaka and H. Matsubara, in: *Electroless Deposition of Metals and Alloys* (M. Paunovic and I. Ohno, ed.), Proceedings Vol. 88-12, p. 244, The Electrochemical Society, Pennington, New Jersey (1988).

87. I. Koiwa, H. Matsubara, T. Osaka, Y. Yamazaki, and T. Namikawa, *J. Electrochem. Soc.* **133**, 685 (1986).

88. I. Koiwa, M. Toda, and T. Osaka, *J. Electrochem. Soc.* **133**, 597 (1986).

89. S. A. Armyanov and G. S. Sotirova, *J. Electrochem. Soc.* **136**, 1575 (1989).

90. T. Chen, D. A. Rogawski, and R. M. White, *J. Appl. Phys.* **49**(3), 1816 (1978).

91. M. R. Khan and J. I. Lee, *J. Appl. Phys.* **57**(1), 4028 (1985).

92. J. S. Judge, J. R. Morrison, D. E. Speliotis, and G. Bate, *J. Electrochem. Soc.* **112**, 681 (1965).

93. R. Feynman, *Lectures on Physics*, p. II-37-9, Addison-Wesley, New York (1964).

94. C. Kittel, *Introduction to Solid State Physics*, 5th ed., p. 488, John Wiley & Sons, New York (1976).

95. D. DiMilia, J. Horkans, C. McGrath, M. Mirzamaani, and G. Scilla, *J. Electrochem. Soc.* **135**, 217 (1988).

96. R. R. Tummala and E. J. Rymaszewski (eds.), *Microelectronic Packaging Handbook*, Van Nostrand Reinhold, New York (1989).

97. M. Paunovic, *Plat. Surf. Finish.* **70**(11), 16 (1983).

98. Y. Okinaka and S. Nakahara, *J. Electrochem. Soc.* **123**, 475 (1976).

99. A. Gangulee, A. M. Tuxford, L. T. Romankiw, and A. F. Mayadas, in: *Properties of Electrodeposits* (R. Sard, H. Leidheiser, Jr., and F. Ogburn, eds.), Proceedings, p. 374, The Electrochemical Society, Princeton, New Jersey (1975).

100. C. Kittel, *Introduction to Solid State Physics*, 5th ed., p. 171, John Wiley & Sons, New York (1976).

101. J. F. Hamilton and P. C. Logel, *J. Catal.* **29**, 253 (1973).

102. H. Wiese and K. G. Weil, in: *Electroless Deposition of Metals and Alloys* (M. Paunovic and I. Ohno, eds.), Proceedings Vol. 88-12, p. 53, The Electrochemical Society, Pennington, New Jersey (1988).

103. R. S. Buck and L. R. Griffith, *J. Electrochem. Soc.* **109**, 1005 (1962).

104. K. N. Tu and J. W. Mayer, in: *Thin Films—Interdiffusion and Reactions* (J. M. Poate, K. N. Tu, and J. W. Mayer, eds.), Chapter 10, John Wiley & Sons, New York (1978).

105. Y. S. Y. S. Chang and J. Y. Lee, Proceedings of 1984 International Electronic Devices and Materials Symposium, National Tsing Hua University, Hsinchu, Taiwan, p. 491.

106. Y. Harada, K. Fushimi, S. Madokoro, H. Sawai, and S. Ushio. *J. Electrochem. Soc.* **133**, 2428 (198).

107. C. H. Ting, M. Paunovic, and G. Chiu, Extended Abstracts, Meeting of the Electrochemical Society, Philadelphia, Pennsylvania May 10–15, 1987, Vol. 87-1, Abstract 239.

108. C. S. Wei, D. B. Fraser, A. T. Wu, M. Paunovic, and C. H. Ting, Extended Abstracts, Meeting of the Electrochemical Society, Atlanta, Georgia, May 15–20, 1988, Vol. 88-1, Abstract 156.

109. P. L. Pai, M. Paunovic, and C. H. Ting, Extended Abstracts, Meeting of the Electrochemical Society, Chicago, Illinois Oct. 9–14, 1988, Vol. 88-2, p. 362.

110. P. L. Pai, W. G. Oldham, C. H. Ting, and M. Paunovic, Extended Abstracts, Meeting of the Electrochemical Society, Honolulu, Hawai, October 1987, Vol. 87-2, Abstract 481.

111. C. H. Ting, in: *Electroless Deposition of Metals and Alloys* (M. Paunovic and I. Ohno, eds.), Proceedings Vol. 88-12, p. 223, The Electrochemical Society, Pennington, New Jersey (1988).

112. C. H. Ting, M. Paunovic, and G. Chiu, in: *Electroless Deposition of Metals and Alloys* (M. Paunovic and I. Ohno, eds.), Proceedings Vol. 88-12, p. 252, The Electrochemical Society, Pennington, New Jersey (1988).

113. H. M.Naguib, C. Jang, T. F. Klemme, K. Wong, A. Rangappan, W. W. Yao, and R. T. Fulks, IEEE V-MIC Conference, June 15-16, 1987, p. 93.

114. F. Vratny, U.S. Patent 4,122,215 (1978).

115. C. H. Ting, M. Paunovic, P. L. Pai, and G. Chiu, *J. Electrochem. Soc.* **136,** 462 (1989).

116. A. M. T. P. van der Putten and J. W. G. de Bakker, Extended Abstracts, Meeting of the Electrochemical Society, Hollywood, Florida, October 15-20, 1989, Vol. 89-2, Abstract 467.

117. A. M. T. P. van der Putten, Extended Abstracts, Meeting of the Electrochemical Society, Hollywood, Florida, October 15-20, 1989, Vol. 89-2, Abstract 322.

118. M. Paunovic, *J. Electrochem. Soc.* **127,** 365 (1980).

119. Y. Okinaka, D. R. Turner, C. Wolowodiuk, and D. W. Graham, Extended Abstracts, Meeting of the Electrochemical Society, Las Vegas, Nevada October 17-22, 1976, Abstract 275.

120. D. R. Turner and Y. Okinaka, Extended Abstracts, Meeting of the Electrochemical Society, Pittsburgh, Pennsylvania, October 15-20, 1978, Vol. 78-2, Abstract 164.

121. N. C. Anderson, M. E. Miner, L. T. Romankiw, and S. F. Starcke, U.S. Patent 4,842,886 (1989).

122. S. Wein, *Metallizing Non-Conductors*, Metal Industry Publishing Co., New York (1945).

31

Electrochemical Nucleation on Active Sites

Benjamin R. Scharifker

1. INTRODUCTION

The nucleation of a new phase with growth of nuclei controlled by the rate of mass transport from the bulk of a solution is relevant to many areas of electrochemistry. It is therefore important to develop a theoretical framework able to describe the rate of growth of individual nuclei as well as the overall kinetics of the process, including interactions between nuclei as they grow and the rate of activation—or deactivation—of active sites on the electrode surface.

The description of the growth rate of a single nucleus is particularly complicated by the advancing nucleus–solution interface that accompanies growth. The moving boundary introduces a convective contribution to mass transport,[1] but, fortunately, under the experimental conditions in which most electrochemical phase transformations take place, the growth of nuclei can be accurately described in terms of semi-infinite radial diffusion of depositing species from the bulk of the solution.[2-4] This has been experimentally confirmed by measuring the current of growth of single nuclei of mercury, silver, copper, and lead[5-7] on platinum or carbon fiber microelectrodes.

The deposition kinetics of a new phase through nucleation and growth of multiple nuclei is not the straightforward summation of the growth currents of single nuclei. A hemispherical diffusion field develops around each growing center, inhibiting the growth of neighboring nuclei. Thus, the growth current of n nuclei is less than n times the current of growth of a nucleus in isolation.[8] Likewise, the diminution in concentration of the depositing species around growing nuclei decreases the supersaturation in their vicinity, inhibiting the formation of new nuclei within an exclusion zone around them. The probability of finding a nucleus close to another is thus lower than expected from a uniform distribution of nuclei in space.[9] Also, the inhibition of nucleation within exclusion zones eventually arrests the nucleation process over the entire surface of the electrode, and thus a limiting number of nuclei obtains.

A further complication arises by the existence of active sites for nucleation on the electrode surface. The kinetics of nucleation is then usually expressed as the product of the

Benjamin R. Scharifker • Departamento de Química, Universidad Simón Bolívar, Caracas 1080-A, Venezuela. This work is dedicated to Professor John O'M. Bockris on the occasion of receiving the American Chemical Society award on Chemistry of Contemporary Technological Problems and on his 65th birthday.

Electrochemistry in Transition, edited by Oliver J. Murphy *et al.* Plenum Press, New York, 1992.

nucleation rate constant, A, times the number density of active sites, N_0. Both quantities vary with overpotential, and in order to establish the exact relationship between overpotential and the kinetics of nucleation, it is necessary to determine separately N_0 and A.

In this chapter we review the method of obtaining N_0 and A simultaneously from single-step potentiostatic current transients. Some experimental results are discussed and used to gain insight into the energetics of the nucleation process. Then we analyze the effects of exclusion zones and active sites on the spatial distribution of nuclei and on their limiting number densities.

2. GROWTH OF MULTIPLE NUCLEI ON A PLANAR SURFACE

The diffusion-controlled growth of multiple nuclei on a surface has been described by consideration of the projection of hemispherical diffusional fields onto the plane of the electrode.[10-12] In those terms, the radius of a (planar) diffusion zone is given by

$$r_d = [(8\pi cM/\rho)^{1/2} Dt]^{1/2}, \tag{1}$$

where c and D are, respectively, the bulk concentration and diffusion coefficient of the depositing species in solution, M is the molar mass of the deposit, ρ is its density, and t is time. The fractional area of the surface which is covered by diffusion zones is[11]

$$\theta = 1 - \exp\{-N_0\pi kD[t - (1 - e^{-At})]\} \tag{2}$$

where $k = (8\pi cM/\rho)^{1/2}$. The current density to the electrode surface is

$$I = (zFD^{1/2}c/\pi^{1/2}t^{1/2})(1 - \exp\{-N_0\pi kD[t - (1 - e^{-At})]/A\}) \tag{3}$$

Figure 1 shows the current transient for different values of the dimensionless parameter $\alpha = N_0\pi kD/A$. It follows that the current I_m and the time t_m corresponding to the maximum of the transient can be used to determine simultaneously the values of N_0 and A, from single-step potentiostatic experiments, by solving the following system of transcendental equations[11]:

$$\ln(1 - \pi^{1/2}I_m t_m^{1/2}/zFD^{1/2}c) + N_0\pi kDt_m - (N_0\pi kD/A)[1 - \exp(-At_m)] = 0$$

$$\ln\{1 + 2N_0\pi kDt_m[1 - \exp(-At_m)]\} - N_0\pi kDt_m + (N_0\pi kD/A)[1 - \exp(-At_m)] = 0$$

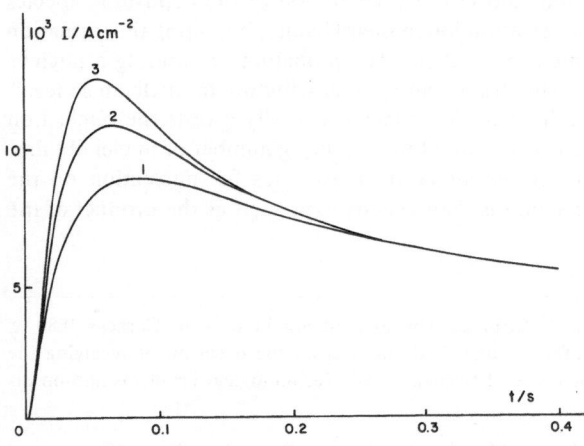

FIGURE 1. Current transients for $D = 1 \times 10^{-5}\,\text{cm}^2/\text{s}$, $c = 10\,\text{m}M$, $M/\rho = 18.3\,\text{cm}^3/\text{mol}$, $z = 2$, and $AN_0 = 10^9\,\text{cm}^{-2}\,\text{s}^{-1}$. (1) $\alpha = 0.16$, (2) $\alpha = 0.50$, (3) $\alpha = 2.0$. (From Scharifker and Mostany.[11])

The accuracy of the method outlined above for the determination of N_0 and A has been recently demonstrated.[13] If N_0 is very small and/or A is very large ($\alpha \to 0$), nucleation is controlled by the rapid exhaustion of active sites, for which case $\pi^{1/2} I_m t_m^{1/2} / zFD^{1/2} c = 0.7153$. Conversely, if $\alpha \to \infty$, then nucleation is controlled by the spreading of exclusion zones, and $\pi^{1/2} I_m t_m^{1/2} / zFD^{1/2} c = 0.9034$. These two limiting cases correspond to what customarily have been called "instantaneous" and "progressive" nucleation, respectively.

3. VARIATION OF NUMBER DENSITIES OF ACTIVE SITES AND NUCLEATION RATES WITH OVERPOTENTIAL

Figure 2 shows the dependence of the number density of active sites for the nucleation of lead onto vitreous carbon with overpotential[14] at different concentrations of lead in solution. Two aspects of these plots deserve special consideration. One of them is that the values of N_0 (10^6 cm^{-2} < N_0 < 10^9 cm^{-2}) are much smaller than the atomic density of the substrate ($\sim 10^{15}$ cm^{-2}). This result indicates that active sites may constitute a serious limitation for nucleus formation. The other remarkable result is the exponential increase of the number of sites with overpotential, which will be discussed below. These two features seem to be common for the nucleation of metals on for example, vitreous carbon, as they also obtain during the nucleation of mercury[13] and silver[15] onto this substrate.

The nucleation rate depends strongly on the overpotential η. The appropriate function that relates A with η, while still being a matter of controversy, provides the means of fully

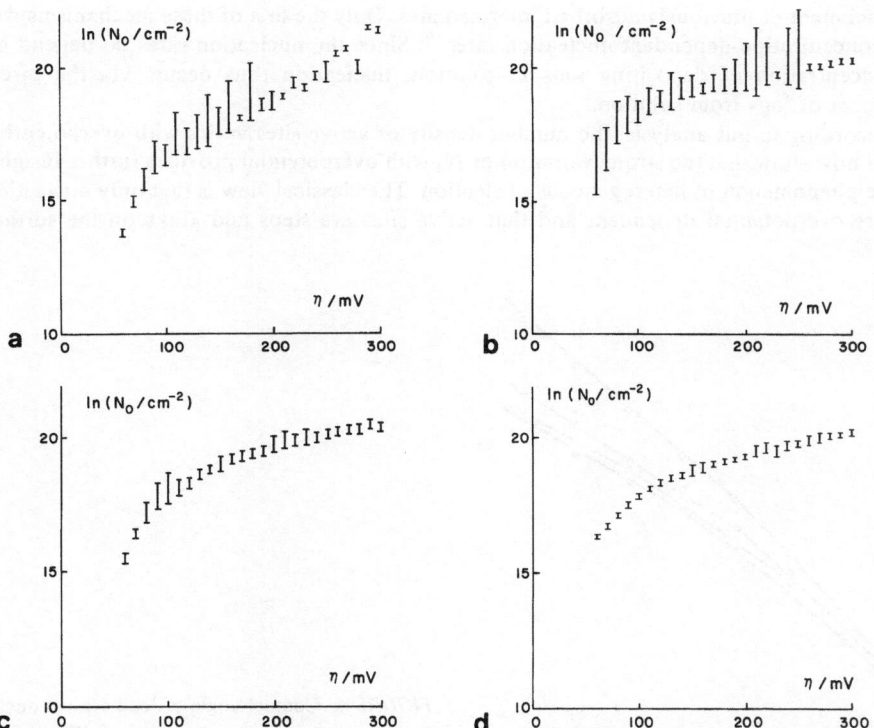

FIGURE 2. Dependence of the number density of active sites on overpotential at different concentrations of Pb^{2+} in solution: (a) 1, (b) 5, (c) 10, (d) 20mM. (From Mostany et al.[14])

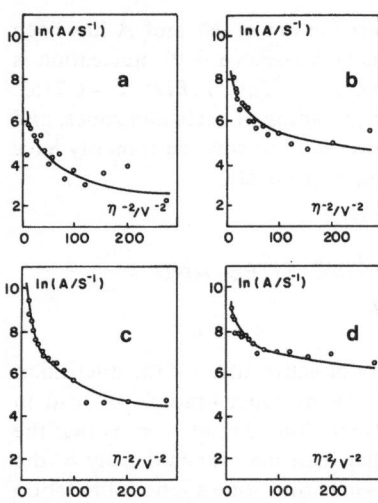

FIGURE 3. Dependence of the nucleation rate on overpotential at different concentrations of Pb^{2+} in solution: (a) 1, (b) 5, (c) 10, (d) 20 mM. (From Mostany et al.[14])

characterizing the energetics of the clustering process.[16,17] Plots of ln A versus $1/\eta^2$ for the nucleation of lead onto vitreous carbon from solutions containing Pb^{2+} ions at various concentrations are shown in Fig. 3. The plots do not yield straight lines, as the classical theory requires,[17] but they can be correctly predicted if the effects of the line tension at the substrate–deposit–solution contact are taken into account.[14] Electrochemical nucleation may occur by the direct attachment of monomers from the bulk to subcritical clusters or through the attachment of previously adsorbed intermediates. Only the first of these mechanisms will show concentration-dependent nucleation rates.[9] Since the nucleation rates do depend on the concentration of depositing ions in solution, nucleation thus occurs via the direct attachment of ions from solution.

According to our analysis, the number density of active sites varies with overpotential. We will now show that the strong variation of N_0 with overpotential provides further insights into the phenomenon of heterogeneous nucleation. The classical view is that only nucleation rates are overpotential dependent and that active sites are steps and kinks on the surface

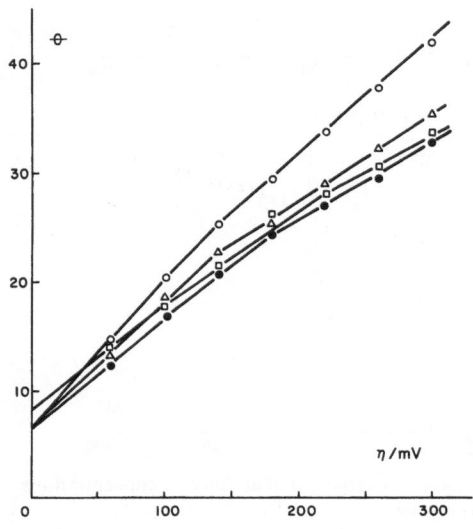

FIGURE 4. Contact angle of lead critical nuclei as a function of overpotential, in 1 (●), 5 (△), 10 (○) and 20 (□) mM Pb^{2+} solutions. (From Mostany et al.[14])

where nucleation of the new phase occurs faster due to more favorable surface energies. Traditionally, it has been thought that the rates of nucleation are single-valued, in the sense that they are uniform throughout the different sites comprising an electrode surface. Recently, the phenomenon of nucleation rate dispersion, which would account for an experimentally observed overpotential-dependent number of active sites during potentiostatic experiments, has been described.[7]

The work of formation of the critical nucleus is a function of the substrate–deposit surface energy, which can be phenomenologically expressed through the contact angle. If there is a distribution of surface energies centered around a nonzero value, and if nucleation takes place at an appreciable rate over a small fraction of sites (those that correspond to the low-energy foot of the distribution) that increases with overpotential, it then follows that the averaged nucleus–substrate contact angle should also increase with the overpotential. Figure 4 shows the contact angle of lead critical nuclei deposited on vitreous carbon as a function of overpotential.[14] The behavior observed thus indicates that there exists a distribution of site energies on the surface, a larger fraction of sites becoming "active" as the overpotential is increased.

4. SATURATION NUMBER DENSITY OF NUCLEI

A growing nucleus develops around itself an exclusion zone for additional nucleation. The experimental study of this effect in electrochemical systems was initiated by Markov et al.,[18] who concluded that it was due to a local deformation of the electric field around the nucleus. This should be considerably reduced in the presence of a supporting electrolyte. Nevertheless, concentration depletion around a nucleus growing under diffusion control remains and decreases the probability of nucleation in its neighborhood, as in the analogous case of surface diffusion in nucleation from vapors.[19,20] Nucleation is therefore confined to the fraction of the surface not included within exclusion zones. As at long times exclusion zones cover the entire area of the electrode, a saturation number of nuclei would be eventually attained.

The problem of the arrest of the nucleation process due to development of exclusion zones around growing nuclei has been recently addressed by several research groups.[7,21-23] The discussion given by Deutscher and Fletcher[7] is illuminating, as it illustrates that exclusion zones are very seldom due to ohmic drops of potential in solution, but are much more likely to arise as a consequence of concentration depletion around growing nuclei. Jacobs,[22] in analyzing the saturation nuclear number densities obtained during the nucleation of Au on GaAs, has pointed out that an exact theoretical treatment of the nucleation process should consider the dependence of nucleation rate on concentration. This has been taken into account by Wijenberg et al.,[23] who, through computer simulations, have shown that a limiting value of nuclei is attained at long times even without the existence of active sites on the surface. The nonlinear concentration field around growing nuclei[3] and the convoluted boundary conditions that account for the overlap and coalescence of diffusion zones make it necessary to introduce a number of approximations in order to keep the problem tractable. One especially simplifying approximation has been that of equating the projection of the diffusional fields generated during nucleus growth onto the plane of the electrode with the exclusion zones that arrest the process of nucleation.[5,10,11,14,21,24]

Identifying thus the exclusion zones with the diffusion zones discussed above, it follows that the probability that a randomly chosen nucleation site shall not be crossed by the perimeter of an exclusion zone is, at any time, $(1 - \theta)$. The effect of exclusion zones on the nucleation process are best understood by examining the condition at the limit of high number

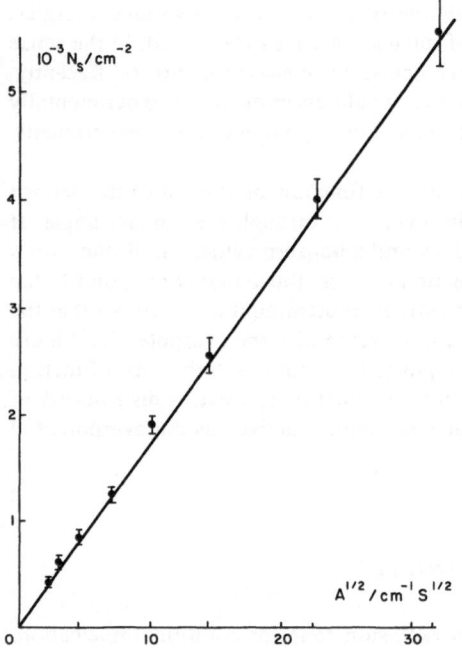

FIGURE 5. Saturation number density of nuclei as a function of $(AN_0)^{1/2}$. Error bars are standard deviations over five simulations for each value of AN_0.

density of active sites on the surface. Thus, when $N_0 \to \infty$ the saturation number density of nuclei is given by[21]:

$$N_s = \lim_{t \to \infty} N(t) = AN_0 \int_0^{\infty} \exp(-AN_0 \pi kDt^2/2) \, dt$$

$$= (AN_0/2kD)^{1/2} \tag{4}$$

Figure 5 shows N_s as a function of $(AN_0)^{1/2}$ as obtained from Monte Carlo simulations of the process on a square surface with periodic (toroidal) boundary conditions. In order to ensure that the $N_0 \to \infty$ condition was fulfilled, the number of sites in the simulations was $N_0 = 10^{12}$ cm^{-2}, arranged in a square lattice. The straight line drawn in the figure is the result expected from Eq. (4), and it can be seen that the agreement obtained between theory and simulation is excellent.

5. SPATIAL DISTRIBUTION OF NUCLEI

The probability density function of the distribution of distances to the nearest neighbor for uniformly distributed particles in the plane is given by[9]:

$$p(r) \, dr = 2\pi rn \exp(-\pi r^2 n) \, dr \tag{5}$$

where n is the number density of uniformly distributed particles. Milchev et al.[9] have shown that the experimentally determined nearest-neighbor distribution functions are shifted to larger distances than expected from uniform distributions of nuclei, attributing this behavior to the development of exclusion zones for nucleation around already established nuclei.

This is seen in Fig. 6, showing the nearest-neighbor distribution of 124 lead nuclei deposited onto a circular electrode with a radius $r_e = 0.1$ cm. In order to account for the

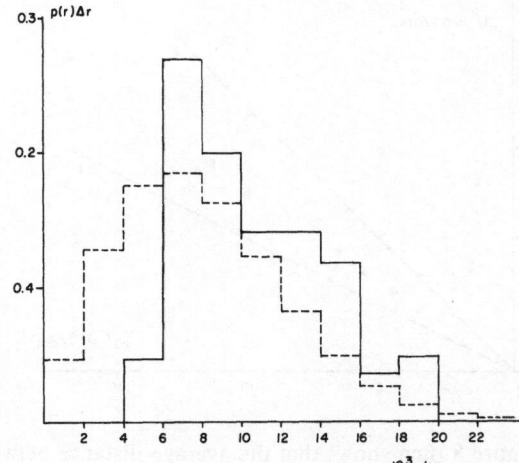

FIGURE 6. Nearest-neighbor distribution of 124 lead nuclei deposited onto a circular electrode of area $3.14 \times 10^{-2}\,\text{cm}^2$ (——), and histogram expected for a uniform distribution of 3.8×10^3 nuclei/cm^2 (- - -).

effects of the edge, only the nearest neighbors to nuclei located within a distance $r \leq r_e(1 - \pi^{1/2}/N_s^{1/2})$ from the center of the electrode were considered,[25] leaving only 85 nuclei for the analysis.

For uniformly distributed nuclei, the maximum of the distribution occurs at a distance

$$r_{\max} = (2\pi n)^{-1/2} \tag{6}$$

Figure 7 shows nearest-neighbor distributions obtained from simulations with different nucleation rates, on square surfaces with periodic boundary conditions and virtually infinite number densities of active sites. It appears that when the distances are normalized with respect to $r_{\max}$, the distribution functions remain invariant to changes in the nucleation rate. The maxima in the distributions obtained, nevertheless, are shifted to longer distances than those predicted on the basis of Eq. (6) for a uniform (random) distribution of nuclei. The spatial distribution of nuclei growing under diffusion control therefore does not correspond to a random distribution of nuclei on the plane of the electrode.

A similar conclusion is drawn from analysis of the average distances to the nearest neighbors, $\langle d \rangle$, shown in Fig. 8. Starting from Eq. (5), it follows that for a random, uniform distribution of particles in the plane,

$$\langle d \rangle = 1/2n^{1/2} \tag{7}$$

For an ordered distribution of particles, for example, in a square array, the following relation holds:

$$\langle d \rangle = 1/n^{1/2} \tag{8}$$

FIGURE 7. Normalized nearest-neighbor distributions of nuclei obtained from simulations with $D = 1 \times 10^{-5}\,\text{cm}^2/\text{s}$, $c = 0.14M$, and $M/\rho = 10.3\,\text{cm}^3/\text{mol}$ over a surface with a virtually infinite number density of active sites. $AN_0 = 6.37$ ($\triangle$), 15.9 ($\blacktriangle$), 31.8 ($\blacksquare$), and 159 cm^{-2} s^{-1} ($\times$). The bold line is the normalized nearest-neighbor distribution for uniformly distributed nuclei.

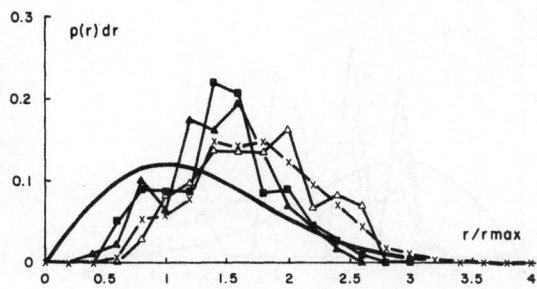

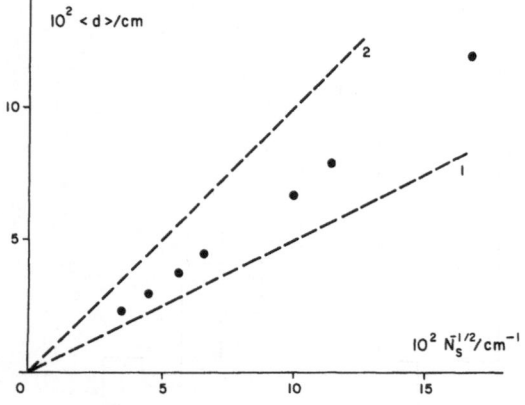

FIGURE 8. Average distance to the nearest neighbor, $\langle d \rangle$, as a function of $N_s^{-1/2}$ for uniform distribution (1) and square lattice (2). Points are results from simulations at different nucleation rates.

Figure 8 then shows that the average distance between nearest neighbors obtained from the simulations is located between the expected values for the uniform distribution and those expected for an ordered, square array. The inhibiting effect of the exclusion zones thus develops a correlation in the location of nuclei.

Nearest-neighbor distributions obtained from simulations on finite number densities of active sites, arranged in square lattices, are shown in Fig. 9. Here fine structures develop at short distances, which are due to spatial correlations, in this case introduced by the location of active sites.

The distribution of nuclei on the electrode surface is thus expected to be nonrandom, on the grounds of both the inhibitory effects of the exclusion zones as well as the possible existence of spatial correlations introduced by the presence of a limited number of active sites on the surface.

6. SOME FUTURE TRENDS IN THE STUDY OF NUCLEATION KINETICS

The kinetics of the nucleation and growth of a new phase on an electrode surface is dominated by several processes, among them the rate of appearance of new nuclei, their growth rate, and the rate of deactivation of the surface for further nucleation. It is now clear that all these processes are intimately related to each other, making it difficult—if not impossible—to isolate one of them from the others in order to facilitate the study of electrochemical phase formation.

Some ways to separate them, however, have been devised. One of them, that of isolating single nuclei using microelectrodes, has been particularly useful in dealing with the growth rate and nucleation kinetics of *single* nucleation events.[26] Attempts to globalize such a

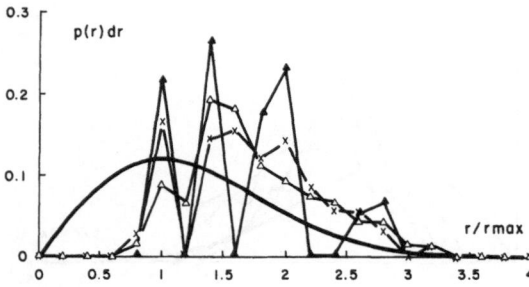

FIGURE 9. Normalized nearest-neighbor distributions of nuclei obtained from simulations with D, c, and M/ρ as in Fig. 7 and $AN_0 = 31.8 \text{ cm}^{-2} \text{ s}^{-1}$, over surfaces with varying number densities of active sites: ▲, 8.15×10^3; ×, 32.6×10^3; △, $130 \times 10^3 \text{ cm}^{-2}$. The bold line is the normalized nearest-neighbor distribution for uniformly distributed nuclei.

procedure for multiple nuclei, for example, by using ensembles of microelectrodes and analyzing the data through a transform procedure in order to recover the nucleation rates from potentiostatic current transients,[7] are inappropriate because they neglect the mutual interactions that arise during the growth of nuclei and that are largely responsible for the deactivation of the surface for further nucleation.

Global methods, as opposed to techniques that intend to separate particular aspects of the overall process, arise thus as a necessity, given the sometimes intractable difficulties that appear in techniques that attempt to separate the different aspects involved in the process.

Thus, it appears that the most appropriate way of studying the kinetics of phase formation on electrodes is by using sufficiently sized electrodes, in order to allow the development of a statistically significant number of nuclei interacting with each other. The kinetics of nucleation should then be extracted from experiment via (i) the analysis of current transients within the framework of theoretical treatments that take into account the different aspects involved in the process, and (ii) by the analysis of direct, microscopic information of the electrode surface, obtained during the phase formation process. From the discussion given above about the saturation nuclear number densities and the spatial distribution of nuclei on electrode surfaces, it seems clear that the location of nuclei on the surface contains information about the nucleation process that remains still to be extracted.

7. CONCLUSION

Global analysis of potentiostatic transients during processes with interaction between growing nuclei appear to be the most convenient way of obtaining, through current measurements, reliable values of nucleation rates and number densities of active sites on the surface. It has been shown that both quantities are strong functions of the overpotential, the latter due to a distribution of the deposit–substrate surface energy over the electrode surface. The saturation number density of nuclei is in general less than the number density of active sites, due to inhibition of nucleation in the vicinity of growing nuclei. This is also manifest in analysis of the nearest-neighbor distribution of nuclei. Studies of the location of nuclei on electrode surfaces may then provide new information on the kinetics and energetics of electrochemical nucleation.

REFERENCES

1. J. Crank, *The Mathematics of Diffusion*, pp. 286–325, Clarendon Press, Oxford (1975).
2. G. J. Hills, D. J. Schiffrin, and J. Thompson, *Electrochim. Acta* **19**, 657 (1974).
3. S. Fletcher, *J. Chem. Soc. Faraday Trans. 1* **79**, 467 (1983).
4. S. K. Rangarajan, *Faraday Symp. Chem. Soc.* **12**, 101 (1977).
5. B. R. Scharifker and G. J. Hills, *J. Electroanal. Chem.* **130**, 81 (1981).
6. G. J. Hills, A. Kaveh-Pour, and B. R. Scharifker, *Electrochim. Acta* **28**, 891 (1983).
7. R. L. Deutscher and S. Fletcher, *J. Electroanal. Chem.* **239**, 17 (1988).
8. A. Milchev, B. R. Scharifker, and G. J. Hills, *J. Electroanal. Chem.* **132**, 277 (1982).
9. A. Milchev, E. Vassileva, and V. Kertov, *J. Electroanal. Chem.* **107**, 323 (1980).
10. G. A. Gunawardena, G. J. Hills, I. Montenegro, and B. R. Scharifker, *J. Electroanal. Chem.* **138**, 225 (1982).
11. B. R. Scharifker and J. Mostany, *J. Electroanal. Chem.* **177**, 13 (1984).
12. B. R. Scharifker, *J. Electroanal. Chem.* **240**, 61 (1988).
13. V. Tsakova and A. Milchev, *J. Electroanal. Chem.* **235**, 237 (1987).
14. J. Mostany, J. Mozota, and B. R. Scharifker, *J. Electroanal. Chem.* **177**, 25 (1984).

15. J. Mostany, J. Parra, and B. R. Scharifker, *J. Appl. Electrochem.* **16**, 333 (1986).
16. A. Milchev, S. Stoyanov, and R. Kaischew, *Thin Solid Films* **22**, 255 (1974).
17. A. Milchev and J. Malinowski, *Surf. Sci.* **156**, 36 (1985).
18. I. Markov, A. Boynov, and S. Toschev, *Electrochim. Acta* **18**, 377 (1973).
19. M. J. Stowell, *Phil. Mag.* **21**, 125 (1970).
20. I. Markov, *Thin Solid Films* **8**, 281 (1971).
21. B. R. Scharifker, *Acta Cient. Venez.* **35**, 211 (1984).
22. J. W. M. Jacobs, *J. Electroanal. Chem.* **247**, 135 (1988).
23. J. H. O. J. Wijenberg, W. H. Mulder, M. Sluyters-Rehbach, and J. H. Sluyters, *J. Electroanal. Chem.* **256**, 1 (1988).
24. M. Sluyters-Rehbach, J. H. O. J. Wijenberg, E. Bosco and J. H. Sluyters, *J. Electroanal. Chem.* **236**, 1 (1987).
25. B. D. Ripley, *Spatial Statistics*, pp. 144–190, John Wiley & Sons, New York (1981).
26. R. De Levie, in: *Advances in Electrochemistry and Electrochemical Engineering* (H. Gerischer and C. W. Tobias, eds.), Vol. 13, pp. 1–43, John Wiley & Sons, New York (1984).

32

Cathodic Reduction of Oxygen on a Partially Immersed Electrode in Chloride and Carbonate Melts

Hiroo Numata, Haruyuki Takagi, and Shiro Haruyama

1. INTRODUCTION

In recent years, fuel cells have been developed as clean energy sources to help solve the problems of the exhaustion of fossil fuels and the progression of environmental pollution. Molten carbonate fuel cells (MCFC) are regarded as a promising energy supply because their advantages include high energy efficiency, low hazardous pollution, utilization of coal, gas, etc. Since the slow kinetics of the reduction reaction of oxygen is a major factor contributing to efficiency loss in fuel cells, a sintered NiO electrode is used to decrease the polarization. In the cells, the electrolyte creeps into pores of the porous electrode and a three-phase boundary, that is, liquid–solid–gas, is formed at the liquid meniscus. Although the reduction of oxygen on the three-phase boundary is a very important reaction in an electrochemical process, very few studies of it have been conducted.[1-3]

In this study, the kinetic parameters of the partially immersed electrodes are measured using an electrochemical method to elucidate the mechanism of the cathodic reduction of oxygen on the three-phase boundary. Furthermore, the rate of the cathodic reduction of oxygen in an alkali chloride melt is compared with that in an alkali carbonate melt to examine the influence of the electrolytic bath, especially anions. Thus, new information concerning fuel cell reactions is obtained by comparing the reaction parameters in the two electrolytes.

2. EXPERIMENTAL

2.1. Electrolyte

An equimolar mixture of reagent grade NaCl–KCl was dried in an oven at 453 K for more than 20 h. The mixture of NaCl and KCl (0.09 kg) was contained in an alumina (99.5%)

Hiroo Numata, Haruyuki Takagi, and Shiro Haruyama ● Department of Metallurgical Engineering, Tokyo Institute of Technology, Tokyo, Japan. H. Takagi's present address is Japan Air Line, Ltd., Tokyo, Japan. S. Haruyama's present address is Tokyo National College of Technology, Tokyo, Japan.

Electrochemistry in Transition, edited by Oliver J. Murphy *et al.* Plenum Press, New York, 1992.

crucible with an inner diameter of 3.7×10^{-2} m and deareated and dehydrated by the bubbling of dry nitrogen gas.

The concentration of the oxide ion was changed by the addition of reagent grade Na_2CO_3 followed by argon bubbling. It is established that Na_2CO_3 decomposes in molten chloride under an argon atmosphere, yielding oxide ion[4, 5]:

$$Na_2CO_3 \rightarrow 2Na^+ + O^{2-} + CO_2 \tag{1}$$

The partial pressure of oxygen was controlled by the ratio of flow rates of oxygen and argon gases. The argon–oxygen gas mixture was dried by passing it through silica gel, $CaCl_2$, and P_2O_5. The temperature was kept at 1023 ± 3 K except where otherwise noted.

2.2. Electrodes

The experiments used a three-electrode cell as shown in Fig. 1. A working electrode was prepared by partially immersing a plate electrode (Au, 35×20 mm, thickness 0.5 mm) connected to an Au wire (diameter 1.0 mm) into the electrolyte. The gold wire was sealed with a Pyrex tube, which was in turn inserted into an alumina tube. The gap between the glass and alumina tubes was filled with alumina cement. All the electrodes were fixed to a silicone-rubber stopper. The working electrode was lowered so that the electrode surface at the bottom just contacted the electrolyte.

In this study, a preliminary experiment was conducted employing various shapes of electrodes. As a result, the reproducibility of the data using an Au plate as the partially immersed electrode was found to be very fine. The reduction current was measured as the current value (A/m) per unit-length meniscus. The counter electrode was a gold plate, 1×10 m^2 in area. The reference electrode was an Ag/Ag^+ (0.05) electrode.

2.3. Levich Film Model

For a reaction at a porous electrode in an aqueous solution, surface diffusion and film models have been established.[6-9] The measurement of the current on a hydrogen anode in

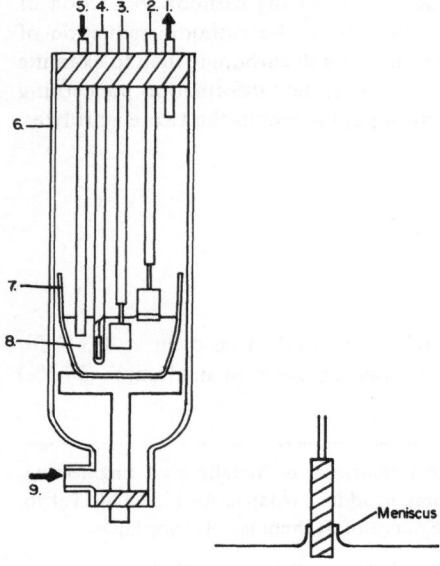

FIGURE 1. Cell arrangement for electrochemical measurements: 1. gas outlet, 2. working electrode 3. counter electrode, 4. reference electrode, 5. bubbling tube, 6. transparent silica tube, 7. alumina crucible, 8. NaCl-KCl melt, 9. gas inlet.

sulfuric acid verified the validity of the film model with the assumption that the diffusion of the hydrogen molecule was the rate-determining step. In this model, the partially immersed electrode formed a three-phase boundary with the meniscus and the thin-film regions; the electrode surface above the melt level was covered by a thin film of electrolyte as schematically shown in Fig. 2. Dissolved gas diffused through the thin film (thickness Δ, length l) and charge transfer took place on the reaction site. The reaction product was transported back to the bulk electrolyte by electromigration. Alternatively, Levich[10] developed the film model, where the electrode process was characterized as delayed discharge control when the rate of diffusion was rapid. The diffusional current (i_{dif}) in terms of the film thickness and the discharge current (i_{dis}) are expressed by:

$$i_{dif} = \frac{2FDHP}{\Delta}(1 - C_0) \tag{2}$$

$$i_{dis} = Ai_0^{(P)}\{C_0 \exp(\alpha nF\psi/RT) - \exp[-(1 - \alpha)nF\psi/RT]\} \tag{3}$$

C_0 is defined as C/HP, and H, P, D, and A are the Henry constant, partial pressure of oxygen, diffusion coefficient, and roughness factor, respectively. ψ_0 expresses the dimensionless overpotential at the entering edge of the film, which is defined as $\psi/2RT$, and $i_0^{(P)}$ is the exchange current density at the partial pressure P. By introducing $\nu = i_{dif}/i_{dis}$, both the currents were equalized and obeyed a steady-state condition. The relationship of total current (I) and overpotential (ψ_0) was developed for two cases:

Case 1: $\nu \geqslant 5, 1 > 1_\Delta$

$$I = 8\sqrt{\frac{\lambda kT\Delta Ai_0^{(P)}}{e}}\left(\frac{\tanh(\psi_0/4)}{1 - \tanh^2(\psi_0/4)}\right) \tag{4}$$

where l_Δ is defined as $(\lambda\Delta kT/Aei_0^{(P)})^{1/2}$, and λ is the specific conductivity of the electrolyte.

Case 2: $\nu < 5, \psi_0 < 1$

$$I = 2\sqrt{\frac{\lambda kT\Delta Ai_0^{(P)}2\nu}{e(1 + 2\nu)}}\psi_0 \tag{5}$$

Thus, in case 1, the cathodic reduction is controlled by the slow discharge reaction throughout the film; that is, this is the case when the rate of diffusion is relatively rapid. The current (I) varied exponentially with potential during the cathodic polarization as shown in Fig. 3; $\nu = 97.5$ and 9.75. In Fig. 3, the cathodic current increased steeply with increasing ψ_0; that is, it was characterized by a Tafel-like behavior. In case 2, the cathodic reaction occurs under diffusion control, where the current (I)-overpotential curves were at first of mixed kinetics and then diffusion limited. Although the overall polarization curve appeared to be an inverse parabola, it exhibited a linear increase close to the immersion potentials.

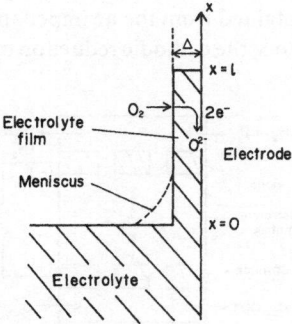

FIGURE 2. Schematic diagram of an electrolyte film at the three-phase boundary of a partially immersed electrode. ($\frac{1}{2}O_2 + 2e^- \rightarrow O^{2-}$).

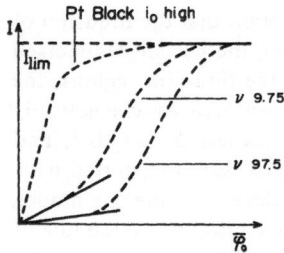

FIGURE 3. Schematic diagram of polarization curves for oxygen reduction at the three-phase boundary for different exchange current densities; $\nu = i_{dif}/i_{dis}$.

The anodic oxidation of hydrogen on a platinized platinum electrode was exemplified in Fig. 3 (i_0 high). Because of the limited numerical value of the thickness and the length of the electrolyte film, the discussion will be confined to the qualitative description of current (I)-overpotential curves.

3. RESULTS AND DISCUSSION

3.1. Fully Immersed $O^{2-}/\frac{1}{2}O_2$ Electrode (Pt, Au, and Pd) in NaCl–KCl and (Au) in Na_2CO_3–K_2CO_3 Melts

In Fig. 4, the structure of a lithiated NiO porous electrode in a MCFC and the cell reaction are shown. The electrolyte, K_2CO_3-Li_2CO_3, seeped within the pores, equilibrated with the pressure on the gas side, and formed the three-phase boundary on the meniscus. The electrochemical studies of the meniscus electrode in molten salts were complex because three aspects of the reaction had to be considered, namely, the processes in the thin-film electrolyte, the processes in the meniscus region, and the electromigration process. The kinetic parameters of the fully immersed electrode were studied, and the nature of the rate-determining step was investigated by considering the thin-film diffusion and electromigration coupled to the charge transfer reaction.

The oxygen reduction reaction in molten chloride was studied by means of conventional current–potential experiments, cyclic voltammetry, and chronopotentiometry. It is widely accepted that the oxygen reduction reaction involves a rapid charge transfer reaction, followed by recombination or an additional charge transfer reaction. Several workers[11-13] have shown that oxygen reduction proceeds via two steps involving MO (adsorbed oxygen) and O_2^{2-} reaction intermediate species. Since the charge transfer process in molten salts is very rapid[14, 15] and is usually inhibited by the diffusion process, the determination of a mechanism required an elaborate technique. The galvanostatic double-pulse studies by Numata et al.[16, 17] have shown that the exchange current density was of the order of 10^4 A/m^2, whereas values obtained from the ac impedance method were two orders of magnitude lower than this value. Thus, the cathodic reduction of oxygen in molten chloride proceeded via a two-step mechanism

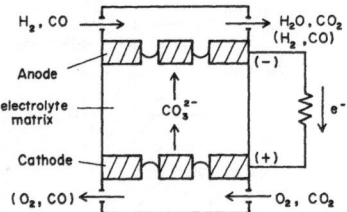

FIGURE 4. Schematic diagram of a single cell and the cell reaction of a molten carbonate fuel cell. Negative electrode: $H_2 + CO_3^{2-} \rightarrow H_2O + CO_2 + 2e^-$; $CO + CO_3^{2-} \rightarrow 2CO_2 + 2e^-$. Positive electrode: $\frac{1}{2}O_2 + CO_2 + 2e^- \rightarrow CO_3^{2-}$.

strongly inhibited by diffusion where the rapid and slow charge transfer processes corresponded to Eqs. (7) and (8), respectively.

$$O_2(\text{bulk}) \rightarrow O_2(\text{electrode}) \qquad \text{r.d.s.} \qquad (6)$$

$$O_2 + 2e^- \rightarrow O_2^{2-} \qquad \text{slow} \qquad (7)$$

$$O_2^{2-} + 2e^- \rightarrow 2O^{2-} \qquad \text{rapid} \qquad (8)$$

On the other hand, the cathodic reduction of oxygen in molten carbonate exhibited a different behavior from that in molten chloride. Thermodynamic considerations[18] showed that the stable chemical entities in molten carbonate are O_2^- and O_2^{2-}, their formation from an oxygen molecule depending on the electrode potentials and the basicity of the melt ($pO^{2-} = -\log a_{O^{2-}}$), whereas molecular oxygen is quite stable in molten chloride. Figure 5 shows the cathodic polarization curves of oxygen on an Au electrode in a carbonate melt under different partial pressures of oxygen. The cathodic currents in molten carbonate are much lower than those in chloride melts and exhibited neither a Tafel region nor a diffusion-limiting current. With an increase in the sweep rate, however, the voltammograms exhibited two current waves around -0.9 and -1.4 V, respectively. The maximum currents of both the waves were proportional to the square root of the sweep rate. These observations indicated that the two current waves corresponded to different reaction processes, both of which were controlled by diffusion as was suggested by Appleby and Nicholson.[19]

First wave

$$O_2 + 2CO_3^{2-} \leftrightarrow 2O_2^{2-} + 2CO_2 \qquad (9)$$

$$O_2^{2-}(\text{bulk}) \rightarrow O_2^{2-}(\text{electrode}) \qquad (10)$$

$$O_2^{2-} + 2CO_2 + 2e^- \rightarrow 2CO_3^{2-} \qquad (11)$$

Second wave

$$O_2^{2-} + e^- \rightarrow O^{2-} + (O^-) \qquad (12)$$

$$(O^-) + CO_2 + e^- \rightarrow CO_3^{2-} \qquad (13)$$

$$O^{2-} + CO_2 \rightarrow CO_3^{2-} \qquad (14)$$

The reaction step can be split into a charge transfer and a chemical reaction step:

$$O_2^{2-} + e^- \rightarrow O^{2-} \qquad (15)$$

$$O^{2-} + CO_2 \rightarrow CO_3^{2-} \qquad (16)$$

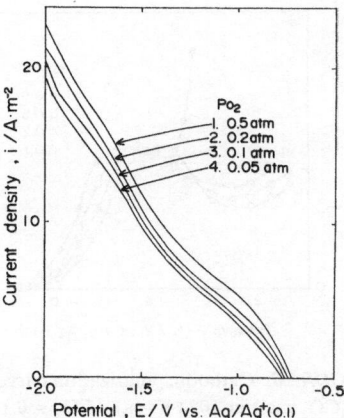

FIGURE 5. Cathodic polarization curves of an $O^{2-}/\frac{1}{2}O_2$ (Au) fully immersed electrode in a Na_2CO_3-K_2CO_3 melt at 1023 K; $pCO_2 = 0.1$ atm; $CO_3^{2-} + 4e^- \rightarrow C + 3O^{2-}$; $2C + O_2 \rightarrow 2CO$.

The cathodic reduction of oxygen in a molten chloride is a fast charge transfer reaction preceded by the diffusion of the dissolved oxygen.[16] This means that the cathodic reaction at the three-phase boundary became large because the diffusion current was increased due to the presence of a thin electrolyte film. It is suggested that the large reduction currents are observable when the electrode processes are free from the mass transport limitation.

As stated above, the cathodic reduction of oxygen in molten carbonate showed two irreversible diffusional peaks, either of which would be controlled by a retarded chemical reaction. When the cathodic reaction occurred on a thin electrolyte film in MCFC, the diffusion became rapid. This resulted in the appearance of the charge transfer reaction which was rate-limited by the chemical reaction.

3.2. Cathodic Polarization Curves of a Partially Immersed $O^{2-}/\frac{1}{2}O_2(Au)$ Electrode in Na_2CO_3–K_2CO_3 Melt

The partially immersed electrode comprised three phases, that is, the gas phase, the liquid phase, and the solid phase. This was useful in simulating the electrode reaction of the porous NiO electrode. By using a gold mesh electrode, Ogura et al.[20] studied the oxygen reduction reaction on the three-phase electrode in Li_2CO_3–K_2CO_3.

Figures 6a and b show the cathodic polarization curves on the partially immersed Au electrode under various pO_2 and pCO_2 conditions. With respect to the individual curves, the current had a linear relation to the potential in the vicinity of the immersion potential. At potentials below −1.5 V, a limiting current was obtained. It has been established that the cathodic reduction at potentials below −1.5 V is attributable to the reduction of CO_3^{2-} accompanied by the formation of CO gas[21]:

$$CO_3^{2-} + 4e^- \rightarrow C + 3O^{2-} \tag{17}$$

$$C + CO_2 \Leftrightarrow 2CO \tag{18}$$

The current increased with potential in an exponential manner. This polarization behavior coincided with the current–potential curve introduced from the film model of Levich that was referred to earlier (case 1, $\nu \geq 5$; see Fig. 3). In other words, the rate of diffusion of oxygen was very rapid, and the charge transfer reaction was in a rate-determining step in a reaction where the oxygen molecules diffused through the electrolyte film and were reduced on the electrode surface. As stated earlier,[19,21] the reduction reaction of the oxygen on the fully immersed electrode was a discharge reaction accompanying a retarded chemical reaction. However, when the reaction took place on a thin electrolyte film, the diffusion became more rapid. This resulted in a fast charge transfer and a rate-determining chemical reaction.

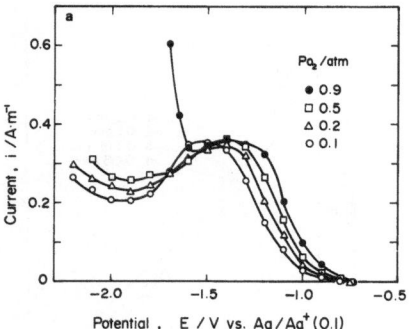 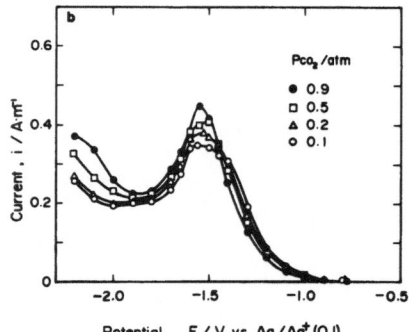

FIGURE 6. Cathodic polarization curves of an $O^{2-}/\frac{1}{2}O_2$ (Au) partially immersed electrode in a Na_2CO_3–K_2CO_3 melt at 1023 K: (a) $pCO_2 = 0.1$ atm; (b) $pO_2 = 0.1$ atm.

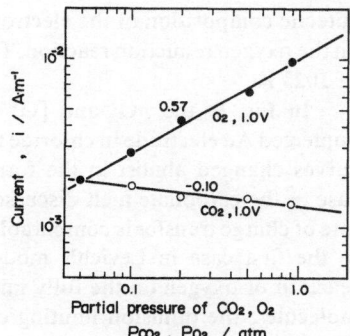

FIGURE 7. Dependence of the cathodic current of a partially immersed Au electrode on pO_2 and pCO_2 at -1.0 V in Na_2CO_3–K_2CO_3 melt at 1023 K.

3.3. Effect of pO_2 and pCO_2 on i_0

As Levich's model explains, the oxygen reduction reaction is governed by the charge transfer reaction. The pO_2 and pCO_2 dependence of the reduction current at a fixed potential (-1.0 V) is shown in Fig. 7. The figure shows an increase of the reduction current which parallels the increase of pO_2 and the decrease of pCO_2. The slopes of these lines give reaction orders of 0.57 and -0.10, respectively. In Fig. 8, logarithmic plots of the cathodic polarization curves are shown for several values of pO_2. In this figure, the reduction current is increased on moving away from the immersion potential and an explicit Tafel line is depicted. Furthermore, the current increases showing oscillation at the less noble potential of -1.5 V as stated earlier. The Tafel line was extrapolated to the equilibrium potential, yielding an i_0 value on the order of 2×10^{-3}–10×10^{-3} A/m, and the reaction orders for pO_2 and pCO_2 were 0.40 and -0.13, respectively. The values are very close to the above results.

3.4. Cathodic Polarization Curves of a Partially Immersed $O^{2-}/\frac{1}{2}O_2$ (Au) Electrode in a NaCl–KCl Melt

As with the Au electrode in molten carbonate discussed above, the oxygen reduction current of the partially immersed Au electrode in molten chloride has been measured. The

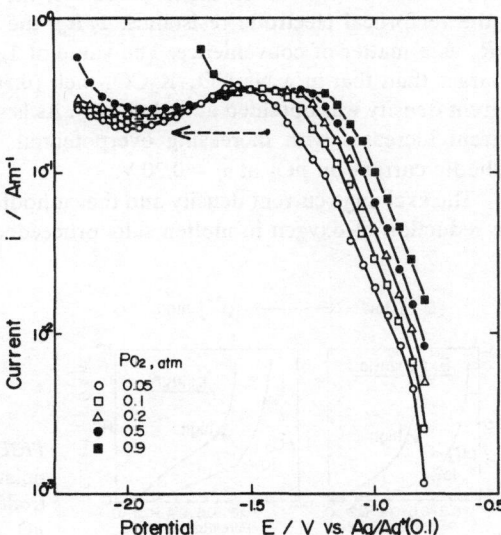

FIGURE 8. Logarithmic plot of cathodic polarization curves of an $O^{2-}/\frac{1}{2}O_2$ (Au) partially immersed electrode in a Na_2CO_3–K_2CO_3 melt at 1023 K; $pCO_2 = 0.1$ atm.

eutectic composition of the electrolyte was selected in both cases to study the effect of anions on the oxygen reduction reaction. Thus, the experiment was performed at the same temperature of 1023 K.

In Fig. 9, the pO_2 and $[O^{2-}]$ dependences of the reduction current of the partially immersed Au electrode in chloride melt are schematically illustrated. The cathodic polarization curves changed almost in the form of an exponential function. This is the same as in the case of the carbonate melt discussed above. At a low concentration of $[O^{2-}]$, 26 mol/m^3, the rate of charge transfer is comparable to or less than the diffusion current,[22] which corresponds to the first case in Levich's model (curves of an exponential form). Since the reduction reaction of oxygen on the fully immersed Au electrode is rate-limited by the diffusion of O_2 molecules, the diffusion limiting current is very high with the partially immersed electrode in chloride melt because of the thinness of the electrolyte film.

At constant $[O^{2-}]$, with increasing pO_2, the rate of charge transfer and the diffusion current simultaneously increased. However, the polarization curves close to the immersion potential exhibited a linear portion whose slopes increased to a large extent under conditions of high pO_2 and high $[O^{2-}]$. This phenomenon was attributable to the discharge reaction being rate-limited by diffusion at $\psi_0 < 1$ (see Eq. 5) or of mixed kinetics.[10] On the other hand, when the concentration of oxide ion increased at constant pO_2, the shape of the curves exhibited a clear transition from an exponential relation at low $[O^{2-}]$ to a linear one at high $[O^{2-}]$, and the slopes increased greatly with increasing $[O^{2-}]$. Thus, the reduction of oxygen in a chloride melt was rate limited by diffusion or under mixed control as long as the potential was close to the immersion potential and the O^{2-} concentration was relatively high.

3.5. Comparison of Cathodic Reduction on Partially Immersed Au in NaCl–KCl and Na₂CO₃–K₂CO₃ Melts

The cathodic polarization curves on Au for $pO_2 = 0.2 \times 10^5$ Pa in a NaCl–KCl melt are shown in Fig. 10 together with those in a Na₂CO₃–K₂CO₃ melt. In Fig. 10, the cathodic overpotential (η) is plotted as the abscissa, and the magnitude of the overpotential was determined as a potential deviation from an equilibrium potential, irrespective of the different melts. The linear relation between η and current was obtained for both melts in the vicinity of the immersion potential. The electrode resistance, R_p, was calculated from the slope of the η versus current curves. Since the rate of the charge transfer reaction was proportional to the reciprocal electrode resistance, $1/R_p$, the exchange current density is expressed as $1/R_p$ as a matter of convenience. The value of $1/R_p$ in a NaCl–KCl melt (2.3–6.7 mho/m) is larger than that in a Na₂CO₃–K₂CO₃ melt (0.50–1.4 mho/m). Similarly, a high exchange current density was obtained at higher pO_2. As is shown in Fig. 10, the difference in cathodic current increases with increasing overpotential. Figure 11 shows the dependence of the cathodic current on pO_2 at $\eta = 0.20$ V.

The exchange current density and the cathodic current at an overpotential indicated that the reduction of oxygen in molten salts proceeded under charge transfer control within the

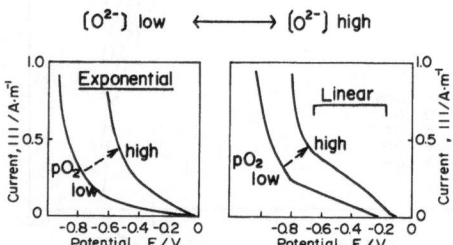

FIGURE 9. Schematic representation of cathodic polarization curves of a partially immersed Au electrode in a NaCl–KCl melt at 1023 K with changing pO_2 and $[O^{2-}]$.

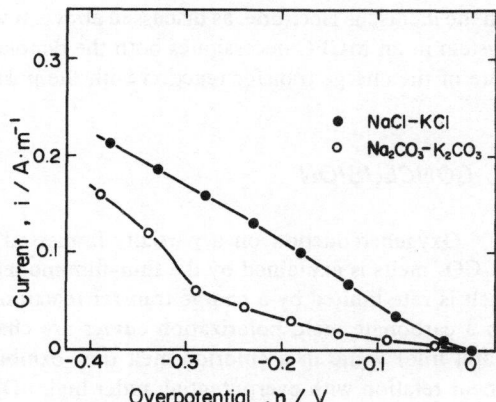

FIGURE 10. Cathodic polarization curve of oxygen reduction on a partially immersed Au electrode in Na_2CO_3-K_2CO_3 and NaCl-KCl melts at 1023 K; pO_2 = 0.2 atm.

potential range investigated. The rate constant was higher in the NaCl-KCl melt than in the Na_2CO_3-K_2CO_3 melt. If the cathodic reduction was controlled by the charge transfer reaction associated with the diffusion process, the dependence of the current on pO_2 was expected to be more complex. This clearly was not the case because R_d (diffusion resistance) contributed little to R_p due to the increased diffusion limiting current through the thin electrolyte film.

Although there is difficulty in defining the current per unit area of the partially immersed electrode from the values of A per unit length of meniscus, it is worthwhile considering the reaction mechanism in connection with the different anions in these systems.

In a NaCl-KCl melt, the rate of cathodic reduction was determined by the charge transfer reaction (case 1) as the diffusion limiting current of oxygen was very high with the partially immersed electrode. As stated earlier, this polarization behavior coincided with that expected from a consideration of the kinetics of the fully immersed electrode. However, the increase of $[O^{2-}]$ caused a reduction in the diffusion rate of oxygen, whereas the rate of the charge transfer reaction increased, resulting in a linear increase of the current (case 2).

In a Na_2CO_3-K_2CO_3 melt, the cathodic reduction was controlled by the diffusion of O_2^-, which is associated with a neutralization reaction of O^{2-} (reaction product) with CO_2. With partially immersed electrodes, the charge transfer reaction was not limited by the diffusion of the reactant, and the overall reaction was controlled by the chemical reaction instead of being under mixed control as was the case at the fully immersed electrode. Qualitatively, the existence of the neutralization reaction in a Na_2CO_3-K_2CO_3 melt resulted in a small improvement in the reduction current on a porous electrode system. On the other hand, in an MCFC, the CO_2 molecule plays the role of a depolarizer, which, in turn, inhibits the oxygen reduction

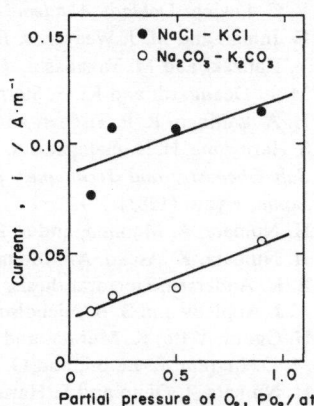

FIGURE 11. Plots of cathodic currents at η = 0.2 V versus pO_2 in Na_2CO_3-K_2CO_3 and NaCl-KCl melts at 1023 K.

on the meniscus electrode, as discussed above. It was concluded that the design of an electrode system in an MCFC necessitates both the depolarization action and the improvement of the rate of the charge transfer reaction with the positive electrode.

4. CONCLUSION

Oxygen reduction on a partially immersed Au electrode in NaCl–KCl and Na_2CO_3–K_2CO_3 melts is explained by the thin–film model, where the cathodic current in a carbonate melt is rate-limited by a charge transfer reaction accompanied by a slow chemical reaction. In a carbonate melt, polarization curves are characterized by an exponential function (the Tafel line) while in a chloride melt they exhibit not only an exponential form but also a linear relation with overpotential under high pO_2 and O^{2-} concentrations. It was found that the cathodic current at a constant overpotential (0.20 V) in a chloride melt exhibits much higher values than that in a carbonate melt. The difference in the cathodic currents can be attributed to the different reaction mechanisms in these melts.

ACKNOWLEDGMENT

The authors wish to thank the NKK Corporation in Japan for its financial support.

REFERENCES

1. H. Numata, K. Asako, T. Kawasaki, A. Momma, and S. Haruyama, *Proceedings of the Joint International Symposium on Molten Salts*, Proceedings Vol. 87-7, pp. 557, The Electrochemical Society, Pennington, New Jersey (1987).
2. S. H. Lu and J. R. Selman, *J. Electrochem. Soc.* **131**, 2062 (1984).
3. H. Numata, T. Shimada, M. Tamura, I. Ohno, and S. Haruyama, Extended Abstracts, Meeting of the International Society of Electrochemistry, p. 343, Kyoto, September, 1989.
4. A. Momma and S. Haruyama, Special Article on Molten Salt Chem., *J. Chem. Soc. Jpn.* **6**, 029 (1982).
5. R. Combes, R. Feys, and B. Tremillon, *J. Electroanal. Chem.* **83**, 383 (1977).
6. F. G. Will, *J. Electrochem. Soc.* **110**, 152 (1963).
7. D. N. Bennion and C. W. Tobias, *J. Electrochem. Soc.* **113**, 593 (1966).
8. S. Srinivasan and H. D. Hurwitz, *Electrochim. Acta* **12**, 495 (1967).
9. S. Haruyama and M. Mukai, *Denki Kagaku* **35**, 310 (1967).
10. V. G. Levich, *Doklady Akademii Nauk SSSR* **157**, 404 (1964).
11. D. Inman and M. J. Weaver, *J. Electroanal. Chem.* **51**, 45 (1974).
12. Y. Kanzaki and M. Takahashi, *J. Electroanal. Chem.* **58**, 339 (1975).
13. M. L. Deanhardt and K. H. Stern, *J. Electrochem. Soc.* **127**, 2600 (1980).
14. H. A. Laitinen, R. P. Tischer, and D. K. Roe, *J. Electrochem. Soc.* **107**, 546 (1960).
15. S. Haruyama, H. Numata, and A. Nishikata, *Proceedings of the 1st International Symposium on Molten Salt Chemistry and Techniques*, pp. 153–156, Molten Salt Committee, Electrochemical Society of Japan, Kyoto (1983).
16. H. Numata, A. Momma, and S. Haruyama, *J. Electrochem. Soc.* **135**, 72 (1988).
17. H. Numata, K. Asako, A. Momma, and S. Haruyama, *J. Jpn. Inst. Met.* **652**, 295 (1988).
18. B. K. Andersen, Doctoral thesis, Technical University of Denmark, Lyngby, Denmark (1975).
19. A. J. Appleby and S. B. Nicholson, *J. Electroanal. Chem.* **53**, 105 (1974).
20. H. Ogura, Y. Ito, K. Murata, and T. Shirogami, *Denki Kagaku* **54**, 886 (1986).
21. M. D. Ingram, B. Baron, and G. J. Janz, *Electrochim. Acta* **11**, 1629 (1966).
22. H. Numata, I. Ohno and S. Haruyama, submitted to *Denki Kagaku*.

VIII

Passivity, Corrosion, Steels, and Minerals

33

Spectroscopic Characterization of the Passive Film on Iron before and after Exposure to Chloride Ion

Oliver J. Murphy

1. INTRODUCTION

The ability to use active metals, for example, iron and its alloys, as structural materials in aqueous environments is a result of the growth of passive films on their surfaces.[1] Localized corrosion processes, such as pitting corrosion, crevice corrosion, or stress corrosion cracking, give rise, in many cases, to catastrophic failures in metallic structural components. These corrosion processes are identified with a number of well-defined stages in their development; in the case of pitting corrosion, these include (i) passive film breakdown, (ii) pit nucleation, and (iii) pit termination or propagation.[2] Common to each form of localized corrosion is the presence of an aggressive anion (usually chloride ion) in the environment contacting the metal. The nature of passive films and the breakdown of passivity are central to, and of paramount importance for, the prevention of localized corrosion processes from taking place. Since breakdown is associated with the passive film only, its structure and chemical composition are of great importance. Confidence in the use of existing metals, and in the formulation of new alloys for more aggressive environments, can only be attained by gaining a complete understanding of the interfacial and bulk physical and chemical aspects of passive films.

Since the initial discussions on the passivity of iron over 150 years ago by Faraday,[3] a very large number of experimental and theoretical investigations have been carried out to elucidate the nature of the phenomenon. However, there is, even at present, very little agreement on the precise physical and chemical characteristics of the passive film and on the essential elements that render the underlying iron metal surface protected or passive.[4] Investigations on the passive film grown on iron have been carried out in aqueous solutions, spanning the whole pH range, that include a wide variety of electrolyte anions. Although much of the earlier work was carried out in aqueous acid solutions, in particular, sulfuric acid solutions, the more recent investigations over the past few decades have primarily used near-neutral borate buffers. The attractiveness of the latter solutions is associated with the high current efficiency for passive film growth, compared to other anionic electrolytes (e.g., sulfate, perchlorate, or phosphate) with similar solution pH values.[5,6] Inhibition of iron

Oliver J. Murphy • Lynntech, Incorporated, Bryan, Texas 77803.

Electrochemistry in Transition, edited by Oliver J. Murphy *et al.* Plenum Press, New York, 1992.

and passive film dissolution during passive film growth in borate solutions minimizes roughening of the electrode surface and, thus, facilitates subsequent spectroscopic investigations, in particular, ellipsometric studies.

Passive film growth on iron has been accomplished both electrochemically by the application of a constant potential or constant current[7] and chemically on using oxidizing anions in solution, for example, chromate or nitrite.[8] The former method involving electrode potential control is preferred, since the extent of the oxidizing driving force can be accurately controlled, and *in situ* removal of air-formed oxide films can be carried out by electrochemical reduction prior to passive film growth. The nature, thickness, and chemical composition of the passive film is controlled by the applied potential and by the time (or aging) for film growth.[9] Other experimental variables, such as solution composition, pH, and temperature, also influence the structure and composition of the passive film on iron.

2. EXPERIMENTAL TECHNIQUES FOR INVESTIGATING THE STRUCTURE, COMPOSITION, AND ELECTRONIC PROPERTIES OF PASSIVE FILMS ON IRON

Techniques used for determining the nature of passive films can be classified into three categories: (i) classical electrochemical methods, for example, galvanostatic, potentiostatic, and potentiodynamic methods[9-11]; (ii) *in situ* spectroscopic techniques, in particular, optical-based techniques, in conjunction with potentiostatic electrochemical control, for example, ellipsometry,[7,8,12,13] Raman,[14-16] Fourier transform infrared (FTIR),[17,18] and photoacoustic spectroscopy,[19] Mössbauer spectroscopy,[20-24] X-ray absorption spectroscopy,[25-27] electrochemical impedance spectroscopy,[28-30] potential-modulated reflectance spectroscopy,[31,32] and photoelectrochemical techniques,[33,34] and (iii) *ex situ* spectroscopic methods, for example, Auger spectroscopy,[31,35-37] X-ray photoelectron spectroscopy (XPS),[38,39] secondary ion mass spectroscopy (SIMS),[38,39,40-42] ion scattering spectroscopy (ISS),[39] and electron diffraction.[43-45]

Almost all of the early work and some of the more recent investigations relied on the use of either galvanostatic or potentiostatic passive film reduction profiles in the absence or presence of simultaneous solution analysis for dissolved ionic iron species.[10,11,46] In the case of galvanostatic reduction, the existence of more than one arrest in electrode potential-time curves was attributed[10,11,46] to the presence in the passive film of well-known oxides of iron containing Fe^{3+} and Fe^{2+} cations. Based on the amount of dissolved iron species in solution, it was assumed[10,46] that all of the Fe^{3+}-containing oxide layer was reduced to a soluble Fe^{2+} species and that the second arrest involved the reduction of an inner Fe^{2+}-containing oxide layer to metallic iron. However, depending on the conditions of passive film growth, the Fe^{3+}-containing oxide layer can also be reduced to the metallic state,[23] probably via an intermediate Fe^{2+} reduction step involving a solid oxide phase. Deposition of metallic iron from at least part of the dissolved Fe^{2+} in solution is likely to occur during reduction events corresponding to the second potential arrest,[12] leading to further complications in interpretations based solely on electrochemical oxidation-reduction profiles.

Thicknesses, d, of passive films grown under various anodic conditions were calculated[10,11] on the basis of the anodic charge expended on film growth or the cathodic charge consumed on reducing the films by means of the following expression:

$$d = QM/zF\rho r \tag{1}$$

where Q is the charge involved in the anodic or cathodic process, M is the molecular weight of the oxide film, z is the number of electrons transferred, F is the Faraday constant, ρ is the density of the oxide, and r is the roughness factor of the electrode. A considerable uncertainty in the roughness factor associated with passive films and possible complications in the passive film reduction processes, together with the arbitrary assumption of a passive oxide film composition corresponding to well-known bulk stoichiometric oxides of iron with established densities, for example, Fe_3O_4 and γ-Fe_2O_3, makes the use of Eq. (1) far from satisfactory.

Early structural information on the passive film on iron was obtained from electron diffraction patterns derived in an *ex situ* manner in electron microscopes.[43-45] Although in some cases identifiable patterns were obtained (Fe_3O_4 and γ-Fe_2O_3),[43,44] the technique has undesirable features: (i) removal of the passivated iron specimen from electrochemical control in its natural aqueous environment; (ii) exposure of the passive film to high vacuum outgassing; and (iii) subjection of the thin passive layer to heating and reduction effects arising from irradiation with an electron beam. Further complications in determining the passive film structure using this technique are associated with the almost-identical electron diffraction patterns obtained from Fe_3O_4 and γ-Fe_2O_3[12,45]; these (or derivatives of them) are considered to be the principal components of the passive film on iron.

The development of *in situ* optical spectroscopic techniques for investigating the nature of solid electrode/solution interfaces over the past three decades has given rise to a considerable advancement in the understanding of the nature of passivity. The earliest, and most significant, of these developments was the introduction of automatic ellipsometric spectroscopy,[8,47] which has been used to obtain the optical spectra, the thickness, and the kinetics of the passive film formation and reduction processes on iron under various electrochemically controlled conditions.[7,8,12,13] Recent improvements in electronic detection equipment, along with improved electrochemical cell designs and novel experimental techniques, have begun to allow the use of vibrational spectroscopies, for example, *in situ* Raman[14,16,48] and FTIR spectroscopy,[17,18] to probe the structure and interfacial aspects of passive films. A simple fingerprinting approach, where spectra derived from the passive film are compared to spectra from known oxides, oxyhydroxides, and hydroxides of iron, can lead to chemical and crystallographic information with regard to the composition of the passive film. *In situ* photoacoustic spectroscopy (PAS) has been developed recently[19] and employed for determining the components of the passive oxide film grown electrochemically on iron in near-neutral phosphate/borate solutions.

Mössbauer spectroscopy as an analytical tool for characterizing passive films on iron *in situ* under potential control in an electrochemical cell was developed in 1973[20,21] and has since been used for this purpose by other investigators.[22-24] Mössbauer spectroscopy is particularly sensitive to the chemical environment and oxidation state of the Mössbauer-active iron atoms in the passive film. Comparison with the Mössbauer parameters derived from standard anhydrous or hydrated iron oxides can lead to the elimination of some, or all, of these compounds as constituents of the passive film in its natural aqueous environment.[21,23] The development of X-ray absorption spectroscopic techniques [extended X-ray absorption fine structure spectroscopy (EXAFS) and X-ray absorption near-edge structure spectroscopy (XANES)] for probing the structure of solid electrode/solution interfaces[49,50] will have the greatest impact in the long term for unraveling the structure and chemical composition of passive films. These techniques can provide information with regard to Fe—O bond lengths (and other metal–ligand bond lengths, e.g., Fe—Cl), the coordination number about the central iron metal atom, electronic transitions involving Fe and ligand atoms, and the valence state(s) of iron atoms in the passive film. Owing to the significant penetration depths of X-rays in condensed matter, coupled with the high-intensity X-ray beam outputs from synchrotron sources, signal attenuation due to the presence of an aqueous electrolyte layer is not a major difficulty, as is the case with *in situ* FTIR spectroscopy.[51] A number of *in situ* and *ex situ* investigations of the structure, electronic properties, and chemical composition of passive films grown on iron have been reported.[25-27]

The nature of the electronic properties of the passive film on iron, in particular, its electrical conductivity, is important with regard to the mechanisms of (i) passive film growth, (ii) electrochemical reactions taking place at passive film surfaces, and (iii) breakdown of passive films.[1,2] Electrochemical impedance spectroscopy provides an *in situ* probe for evaluating the electrical properties of passive films.[28-30] Initial impedance analyses, involving the use of Mott–Schottky plots, have indicated that the passive film has semiconducting properties of n-type in character.[28,30] However, a more recent impedance study over a wide frequency and potential range demonstrated that the film is neither a pure semiconductor nor a pure insulator, but has variable electronic characteristics, depending on the applied electrode potential.[29] The passive film was labeled a "chemiconductor", which was defined as a material whose stoichiometry can be varied by oxidative and/or reductive valence state changes (Fe^{2+} at low potentials and Fe^{4+} at high potentials with the associated incorporation and expulsion, respectively, of H^+ species in the film).

Additional information on the electronic structure, in particular, the precise electronic transitions taking place within passive films, has been obtained from *in situ* potential-modulated reflectance spectroscopy[31,32] and photoelectrochemical techniques.[33,34] Electronic transitions observed in reflectance spectra obtained from passive films on iron grown in solutions over a wide pH range indicated that the films are composed of an Fe^{2+}-containing compound, together with a layer of a ferric oxyhydroxide that is increasingly dehydrated with increasing electrode potential.[32] Photocurrent measurements performed on passive films as a function of incident photon energy and electrode potential gave results indicating that the passive film exhibits electronic properties similar to those of amorphous semiconductors.[33]

The development of *ex situ*, high-vacuum, surface analytical techniques (e.g., Auger, XPS, SIMS, and ISS) in the 1960s and 1970s has greatly facilitated structural and chemical composition determinations of thin films and chemisorbed species on solid surfaces. Although subject to the drawbacks of other *ex situ* methods, since the mid-1970's these surface-sensitive techniques have provided valuable information on the nature of the passive film on iron. Auger and XPS can provide quantitative information with regard to elemental concentrations in thin films, and, in addition, XPS can yield the oxidation state of various elements in these films. For instance, by monitoring the O $1s$ binding energies, the various chemical environments of oxygen atoms present, that is, whether an oxygen atom exists in the form of a lattice oxide ion (O^{2-}) or as a hydroxyl ion (OH^-) or is incorporated as lattice water (H_2O) in the passive film, can be ascertained. On carrying out angle-resolved XPS analyses or Auger and XPS analyses, in conjunction with ion sputtering, it is possible to determine relative changes in the concentration and chemical environments of any particular element as a function of depth into passive films. ISS, again in association with ion sputtering, can provide quantitative information with regard to elemental atomic ratios as a function of depth into the passive layer on iron.

In contrast to Auger, XPS, and ISS, SIMS is uniquely capable of detecting all elements, including hydrogen and its isotopes, down to low concentrations, if present in the passive film. In view of the controversy that has existed concerning the presence of hydrogen-containing species in the passive film on iron and the possible role they play in passivity,[4] the availability of a spectroscopic technique (SIMS), even if it is restricted to the *ex situ* mode of operation, that can determine the hydrogen content as a function of depth into the passive film is essential.

3. STRUCTURAL MODELS FOR THE PASSIVE FILM ON IRON

Spectroscopic studies, over the past two decades in particular, have given a better understanding of the structure and composition of the passive film on iron. Based on a broad range of results obtained using these spectroscopic techniques, a number of proposed models with respect to the nature of the passive film can be summarized in terms of two generalized models.

3.1. The Crystalline Oxide Model

In the crystalline oxide model,[19,43,44,46] the protection rendered to the iron metal surface by the passive layer that grows upon it is understood to be related to the crystal structure(s) of the polycrystalline oxide component(s) in the layer. The passive film is considered to be either a duplex layer, consisting of an inner layer of Fe_3O_4 and an outer layer of γ-Fe_2O_3, or almost exclusively γ-Fe_2O_3 with a concentration gradient of Fe^{2+} species at the metal/passive film interface, sufficient to fulfill the thermodynamic requirement for an Fe^{3+} oxide phase being in contact with an Fe^0 metal phase, without the formation of a distinct intermediate Fe^{2+}-containing phase. The model has for its foundation data derived from relatively thick oxide films grown on iron as a result of thermal oxidation in air or oxygen at elevated temperatures.[2] Because of the relatively thick films formed in the latter cases, it was easily established that the oxide consisted of an inner layer of Fe_3O_4 and an outer layer of α-Fe_2O_3. However, for thermal oxidations carried out at temperatures below about 200°C, oxide films were found to be composed of Fe_3O_4 and γ-Fe_2O_3 layers with cubic spinel-type structure, while, at temperatures above 200°C, α-Fe_2O_3 forms as an outer layer over the inner complex layer.

The essential element of the crystalline oxide model of passivity is the near-perfect crystalline oxide structures that form, not only in two dimensions parallel to the metal surface, but also in the third dimension perpendicular to the metal surface. It is these well-formed crystalline oxide materials (e.g., Fe_3O_4/γ-Fe_2O_3 or γ-Fe_2O_3), which grow by field-assisted oxide growth mechanisms involving primarily ionic conduction by means of oxide ion transport through the growing passive film,[42] that shield the underlying metal surface from coming in contact with the aqueous solution, hence preventing metal dissolution from taking place. The fact that water is involved in passive film growth in aqueous solutions is accounted for in terms of either hydrogen stabilization of cation-deficient γ-Fe_2O_3, particularly in the outer layers,[52,53] or adsorbed water on the surface of the film.[38]

3.2. The Hydrated Amorphous Polymeric Oxide Model

Since there is general agreement that the thickness of the passive film on iron is of the order of 5 nm, it is unlikely that there is any significant long-range order in the film,[54] at least in the dimension perpendicular to the metal surface. It is difficult to visualize how a microcrystalline oxide material (or even a nanocrystalline material) with grain boundaries in such a layer could render protection to the underlying metal surface, given the thickness constraint on the passive film. Experimental data derived, in particular, from in situ Mössbauer,[20,21,23,24] spectroscopy, tritium radiotracer,[55,56] photoelectrochemical,[33,34] and X-ray absorption spectroscopy measurements,[25-27] have led to a view of the passive film as being glassy or amorphous in nature and containing hydrogen species. Inherent to this model is a significant reduction in long-range order effects in any dimension, and the essential component of the film is the bound water molecules, which transform the properties of "bulk iron oxide" to those of the protective, hydrated amorphous passive layer.[21,39,57] The essential concept is that the bound water is supposed to keep the thin film amorphous, and the word polymer is used with the meaning that the incorporated water molecules hold together the "iron oxide chains" in such a fashion that it is difficult for the Fe^{2+} ions from the metal base beneath the film to diffuse to hydration sites at the passive film/solution interface. The incorporated water molecules act, in essence, as a binder, holding the polymer oxide chains together in an amorphous structure, precluding the formation of grain boundaries that would enhance the transport of ions. This view of passivity for iron is supported by the more protective passive films present on the surface of chromium-containing ferrous alloys, which are known to be more highly disordered[58] and hydrated[56] than the passive film on iron, due to the higher hydration and glass-forming tendencies of chromium.

4. RECENT SPECTROSCOPIC INVESTIGATIONS OF THE PASSIVE FILM ON IRON

4.1. Thickness of the Passive Film

A representative current–potential profile for a cathodically cleaned, pure iron electrode in borate buffer solution (pH 8.4) is presented in Fig. 1.[39] The active region in which iron undergoes dissolution, as represented by Eq. (2):

$$Fe \rightarrow Fe^{2+} + 2e^- \qquad (2)$$

covers the potential range from -0.47 to -0.35 V (versus NHE). The pre-passive region encompasses the potential interval from -0.35 to -0.23 V (versus NHE), where iron hydroxy species of the form $Fe(OH)_2$ are formed. The passive region, characterized by a low background iron dissolution current (~ 1 $\mu A/cm^2$), stretches from -0.23 to $+1.17$ V (versus NHE). At higher potentials, the current again increases due to transpassive dissolution of the metal, in conjunction with oxygen gas evolution from water.

A controversial area in passivity investigations for many decades has been the thickness of the passive film itself. The passive film had been considered by a number of investigators[59] to consist of a two-dimensional layer (or a monolayer) of chemisorbed O or OH species. Due to the uncertainty in coulometric estimations of the passive film, as determined from Eq. (1), ellipsometric measurements made in the 1960s[8,47] were the first electrochemically independent and accurate measurements of film thickness. Passive film thicknesses, as measured by ellipsometry and by coulometry (assuming that the film consists of γ-Fe_2O_3), for the same samples as a function of electrode potential, are given in Fig. 2.[56] A discrepancy of 15% in the film thickness was noted between the two methods. This discrepancy most likely arises from the incorrect assumption of a dense, stoichiometric Fe_2O_3-containing passive film in the coulometric estimation of thickness.

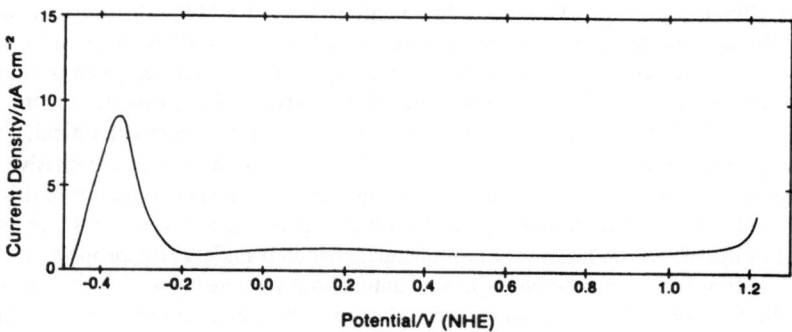

FIGURE 1. Current–potential curve for a clean iron electrode in borate buffer solution, pH 8.4. Scan rate: 0.2 mV/s.

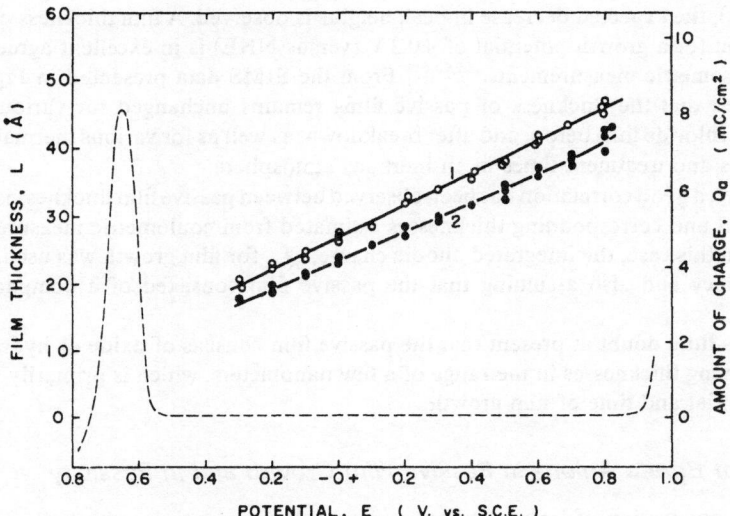

FIGURE 2. Film thicknesses by ellipsometry (curve 1) and by coulometry (curve 2) as a function of passivating potential ($t_a = 60$ min. in borate buffer). Current-potential curve (- - -) shows the passive film thickness at various potentials in the passive potential range for iron.

SIMS has recently been used as an independent method of estimating the passive film thickness on iron.[40] Positive SIMS O^+ peak height signals, as a function of depth into passive films, grown in borate buffer solutions and subsequently exposed to chloride-containing borate solutions for various times before and after breakdown, are shown in Fig. 3a. Similarly, positive SIMS Fe^+ peak heights, as a function of depth into passive films formed in borate buffer solution and subsequently thermally treated in an inert gas atmosphere for various times, are presented in Fig. 3b. After a sharp initial increase in signal intensity over the first 0.2-0.5 nm in the outer surface regions, an approximately constant signal is obtained for all films in both cases (cf. Figs. 3a and 3b), indicative of a single-phase homogeneous material throughout the bulk. At the metal/passive film interface (about

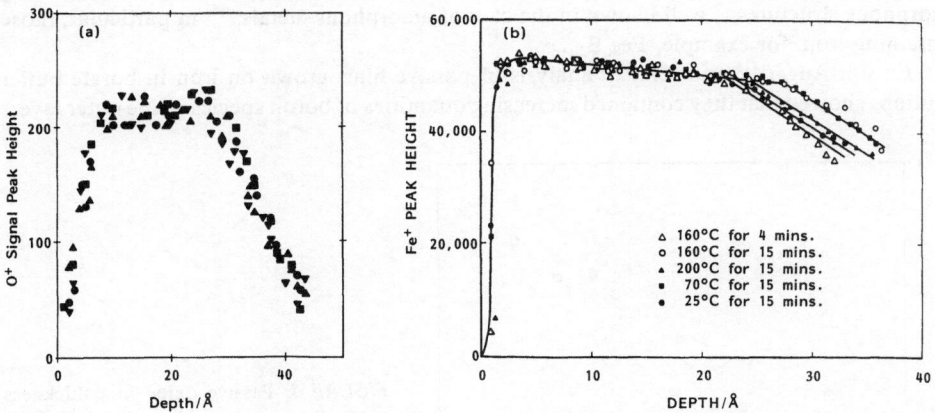

FIGURE 3. (a) Variation of O^+ peak height as a function of depth into passive films exposed to chloride-containing borate buffer solution for various times before and after breakdown. (b) Fe^+ peak height as a function of depth into passive films formed in borate buffer solution at 0.3 V (versus NHE) and then thermally treated in an inert gas atmosphere.

2.7 ± 0.3 nm), the expected decrease in peak heights is observed. A film thickness of the order of 2.7–3.0 nm for a growth potential of +0.3 V (versus NHE) is in excellent agreement with earlier ellipsometric measurements.[56,60,61] From the SIMS data presented in Fig. 3, it can be concluded that the thickness of passive films remains unchanged for various times of exposure to chloride ions before and after breakdown, as well as for various thermal treatment temperatures and treatment times in an inert gas atmosphere.

Recently, a good correlation has been observed between passive film thicknesses measured by ^{18}O SIMS and corresponding thicknesses estimated from coulometric measurements (cf. Fig. 4).[62] In this case, the integrated anodic charge, Q_A, for film growth was used, assuming 100% efficiency and also assuming that the passive film consisted of a compact layer of γ-Fe_2O_3.

There is little doubt at present that the passive film consists of oxide or hydrated oxide material, having thicknesses in the range of a few nanometers, which is primarily dependent on the potential and time of film growth.

4.2. Role of Borate Anions in Passive Film Growth and in Passivity

Anodic passivation of iron in sulfate and perchlorate solutions with bulk solution pH values of 3.0 or 8.4 has been found to be highly inefficient,[5] consuming a very large amount of anodic charge before a background current indicative of passivation was observed. However, small additions of borate buffer of pH 8.4 (<5%) to near-neutral sulfate or perchlorate solutions resulted in a dramatic increase in passivation efficiency, as indicated by over a two orders of magnitude reduction in the anodic charge consumed. The beneficial action of borate (in addition to its pH buffering abilities) toward iron passivation was attributed to the strongly interacting nature of borate anions with iron surfaces, giving rise to inhibition of the latter, together with promotion of surface oxide film formation. The amount of boron-containing species and their precise location, whether incorporated into the passive film during growth or adsorbed on the surface of the passive film, could play an important role in the passivity of iron in borate solutions. Borate and tungstate anion incorporation into barrier anodic alumina films grown on aluminum metal in solutions containing these anions has been well documented.[63,64] Incorporated electrolyte residue can modify considerably not only the electronic properties, but also the chemical composition and structural features of anodic oxide films. The action of the metalloid boron in promoting and stabilizing amorphous structures is well known in the case of amorphous metals,[65] in particular, those containing iron, for example, $Fe_{80}B_{20}$.

Ex situ Auger[31] and XPS[18] analysis of passive films grown on iron in borate buffer solutions showed that they contained increasing quantities of boron species in the outer layers

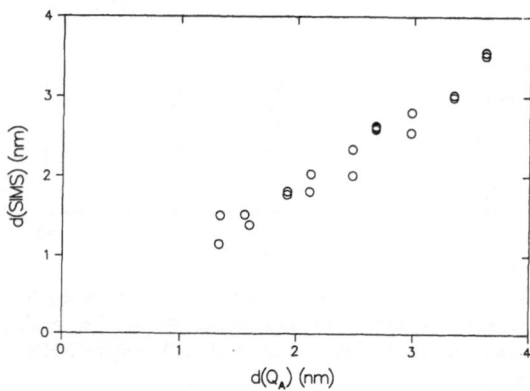

FIGURE 4. Passive oxide film thickness on Fe for a variety of anodization times and potentials. The thickness was calculated from both the integrated anodic charge during the potential step $[d(Q_a)]$ and by using the ^{18}O SIMS technique.

TABLE 1
XPS Analysis (for Boron) in the Passive Film on
Iron Grown in Borate Buffer Solution, pH 8.5

E (V vs. NHE)	Angle (deg)	% Boron
0.0	90	3.5
0.3	90	2.5
0.9	90	2.0
0.9	15	12.0

of the films, as the anodic potential for film growth increased. From XPS analysis (Table 1), the apparent concentration of boron decreased with increasing potential, that is, increasing film thickness. The observation that, for a growth potential of 0.9 V (versus NHE), the concentration of boron obtained at a low angle of detection was significantly higher than that obtained from photoelectrons detected normal to the surface, indicates that boron-containing species are located primarily in the outer regions of the passive films.

Borate adsorption on passive films on iron, investigated with *in situ* FTIR spectroscopy, showed that borate ions adsorb strongly on the surfaces of passive films.[18] Difference spectra pertaining to the passive film/borate solution interface at a number of electrode potentials recorded using FTIR reflection–absorption spectroscopy (FTIRRAS) are presented in Fig. 5a. A strong B—O stretching band at 1400 cm^{-1} can be observed, which displays an intensity that is dependent on electrode potential. The integrated peak area associated with the latter band plotted as a function of electrode potential is presented in Fig. 5b. The exponential

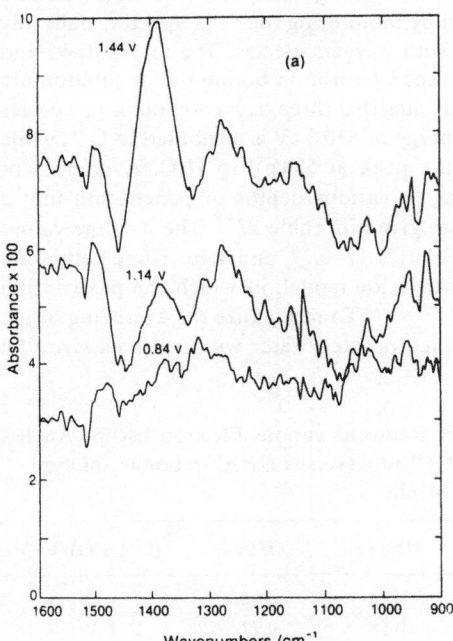
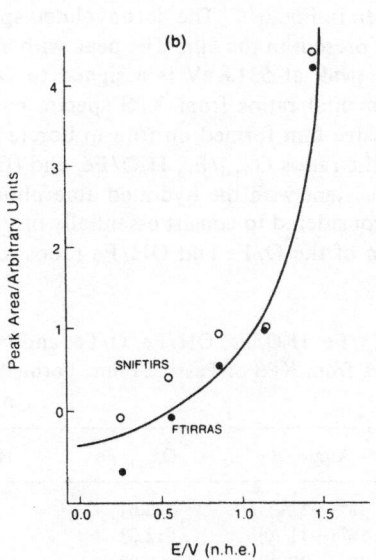

FIGURE 5. (a) Difference FTIRRAS spectra of the passive iron electrode/solution interface at different electrode potentials. (b) Integrated band (1400 cm^{-1}) intensity as a function of electrode potential.

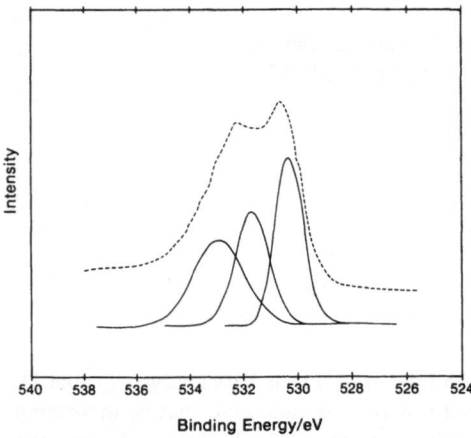

FIGURE 6. XPS spectrum for O 1s obtained for the passive film formed on iron in borate solution at 0.3 V (versus NHE).

dependence of coverage on potential is indicative of a surface species, while increasing coverage with increasing positive potentials indicates that the adsorbed species are negatively charged, that is, negative ions. Weak incorporation, together with strong adsorption, of borate ions is likely to play an important role in stabilizing the passive film in these solutions. Inhibition of passive film dissolution and diffusion of Fe^{2+} cations from the metal surface underneath the film to surface hydration sites are likely to be greatly enhanced by a high coverage of strongly adsorbed borate ions.

4.3. Hydrogen-Containing Species in Passive Films

Although XPS is not capable of directly detecting hydrogen in passive films, an indirect measurement of the latter species can be obtained by monitoring the O 1s spectra, since any hydrogen in the films is likely to be associated with oxygen species. The as-obtained and deconvoluted O 1s spectra for the passive film formed on iron in borate buffer solution are given in Fig. 6.[39] The deconvoluted spectrum indicates that three oxygen-containing species are present in the film. The peak with a binding energy of 530.3 eV is attributed to O^{2-}, while the peak at 531.8 eV is assigned to OH^-, and the peak at 533 eV to H_2O. A number of elemental ratios from XPS spectra corresponding to various depths of penetration into a passive film formed on iron in borate solution are given in Table 2.[39] The average values of the ratios O_{total}/Fe, H_2O/Fe, and $(OH + O)/Fe$ are 2.17, 0.77, and 1.46. These values are consistent with the hydrated amorphous polymeric oxide model, in which the passive film is considered to consist essentially of $Fe_2O_3 \cdot H_2O$.[39,66,67] To rationalize the averaging of the sum of the O/Fe and OH/Fe ratios, the lattice-incorporated water within the passive film

TABLE 2
O_{total}/Fe, H_2O/Fe, OH/Fe, O/Fe, and $(OH + O)/Fe$ Ratios at Various Electron Escape Angles, θ, from XPS of Passive Films Formed on Iron at +0.30 V (versus NHE) in borate solution (pH 8.4) for 50 min

Angle, θ	O_{total}/Fe	H_2O/Fe	OH/Fe	O/Fe	$(OH + O)/Fe$
8° (~5 Å)	2.61	0.91	0.83	0.99	1.82
18° (~11 Å)	2.03	0.84	0.58	0.77	1.35
38° (~22 Å)	2.09	0.69	0.58	0.83	1.41
58° (~30 Å)	1.96	0.65	0.52	0.76	1.28

may be considered to be bound by means of hydrogen bonding to the lattice O^{2-} ions, so that the XPS sees a pseudo-hydroxylated structure.[39] The schematic shown in Fig. 7 represents a possible structural representation of the passive film which is consistent with a value of $(OH + O)/Fe = 1.46$.

An Auger spectrometer with a vacuum-tight transfer system placed between the electrochemical cell and the analysis chamber has been used to obtain spectra from iron specimens passivated electrochemically in borate buffer solutions.[66,67] Thus, any interferences, such as subsequent air oxidation as a result of air exposure, were avoided. Spectral differences were observed between the as-grown passive film, the passive film after heating to 150°C, and an air-formed oxide film. Auger spectra of the passive film yielded an O/Fe ratio of 1.8 ± 0.2 while a similar thermally treated film gave a value of 1.2 ± 0.2. These experiments lend further support to the hydrated amorphous polymeric oxide model of passivity.

A method of estimating the content of hydrogen-containing species in passive films is by measuring OH^- peak heights using negative SIMS.[38,40,68] This technique, in conjunction with depth profiling, on thermally treated passive films grown in borate solutions, can yield information on the relative distribution of hydrogen-containing species as a function of depth and the enthalpy, ΔH, associated with the removal of hydrogen species (or water) from passive films.[40] In Fig. 8, negative SIMS OH^- peak heights (proportional to the concentration of water in the films as a result of OH^- ion formation from water fragmentation on ion bombardment) are presented as a function of annealing temperature for films grown at $+0.3$ V (versus NHE) and subsequently annealed for 15 min at various temperatures in a flowing inert gas atmosphere.[40] For all the films examined, the water contents are higher in the outer passive film regions, and for a given depth an exponential-like decrease in the amount of water remaining in the films after heating to temperatures of up to 200°C is observed.

From plots of the logarithm of the OH^- peak heights against the inverse of the absolute temperature, an average value of the enthalpy, ΔH, for water removal from the passive films of 2.0 kcal/mol was calculated from the slopes of the straight lines obtained.[40] It would be expected that the thermal breaking of the coordinative bonds of H_2O molecules with the Fe_2O_3 molecular entities in the films (cf. Fig. 7) would yield a heat of reaction, ΔH, of the

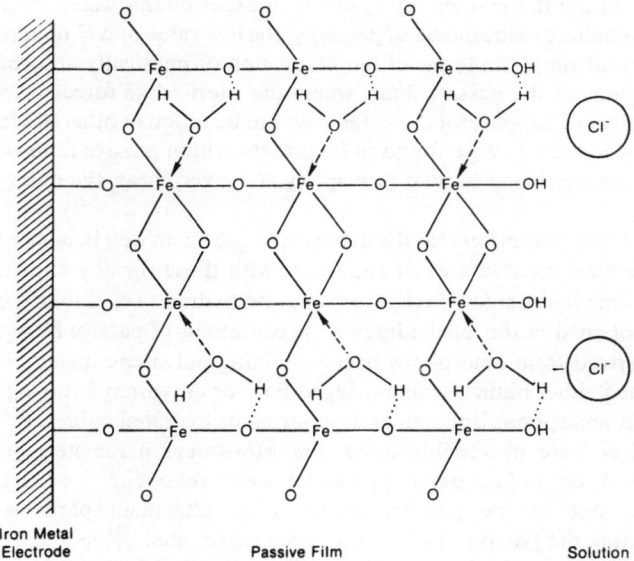

FIGURE 7. Schematic representation of the hydrated passive film on iron.

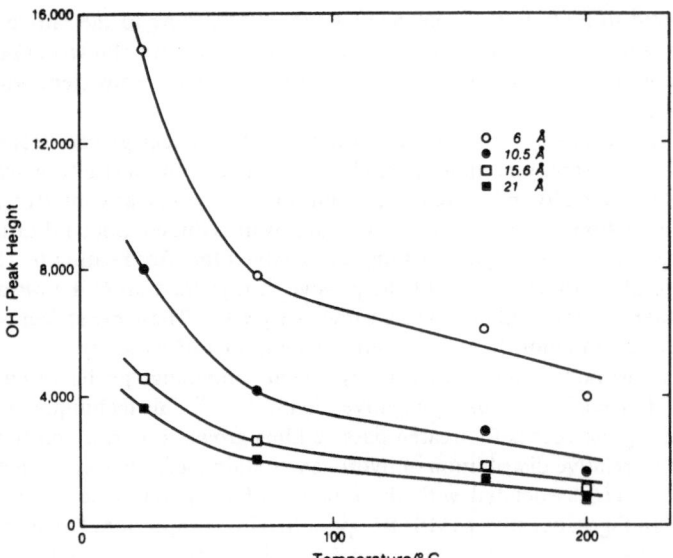

FIGURE 8. Variation of OH⁻ peak height with annealing temperature at various depths into passive films formed in borate buffer solution at 0.3 V (versus NHE) and then thermally treated in an inert gas atmosphere.

order of 20–100 kcal/mol. However, in terms of the hydrated polymeric oxide model of passivity, the low value of ΔH obtained can be accounted for by the overall processes taking place[40]; that is, two compensating reactions are involved—an endothermic bond-breaking reaction involving water removal and an exothermic crystallization reaction associated with the formation of an anhydrous crystalline lattice, probably γ-Fe_2O_3. Thus, it is considered that the heat of crystallization associated with the formation of anhydrous γ-Fe_2O_3 makes available almost all the thermal energy required to expel bound water from within passive films. With the crystalline oxide model of passivity, the low value of ΔH obtained (2 kcal/mol) would be interpreted simply in terms of the desorption of physically adsorbed water on the outer surface regions of the passive films, where the short-range forces involved are of the order of 1–10 kcal/mol. Support for the latter view can be found in other SIMS investigations, where few, if any, hydroxyls were found to be present within passive films on iron,[41,42] the majority of such species being located principally at, or very near, the outer surface regions of the films.[38]

In situ Mössbauer parameters for the passive film grown on iron in borate buffer solutions by the electrochemical method are not consistent with those for any known stoichiometric iron oxide, including hydroxides, oxyhydroxides, and hydrates (cf. Table 3).[21-24] However, the parameters obtained in situ, depending on the conditions of passive film growth, matched similar values derived from amorphous iron-containing polymeric materials (the structures of which contained either binuclear oxo-bridged Fe^{3+} or dihydroxy bridging bonds between the iron atoms) or amorphous iron oxides or amorphous hydrated oxides.[20,21,23,24] It is only when passive films were irreversibly dried that Mössbauer parameters corresponding to an anhydrous bulk crystalline oxide (γ-Fe_2O_3) were observed.[21] Misinterpretations of Mössbauer data, due to the possible existence of ultrasmall particles of crystalline γ-Fe_2O_3 constituting the passive film,[22] have been discounted.[23]

X-ray absorption spectra of passive films on iron, recorded in the region near the K-edge for iron, showed remarkable differences between films investigated in situ[69,70] in an aqueous

TABLE 3

Mössbauer Parameters of Well-Characterized Oxides, Hydroxides, and Oxyhydroxides of Iron

Compound	Isomer shift[a] (mm/s)	Quadrupole split (mm/s)	Internal magnetic field (kOe)
$Fe(OH)_2$	1.44	2.92	0
FeO	1.18	0.8	0
$Fe(OH)_3$	0.59	0.65	480 (5 K)
α-Fe_2O_3	0.61	0.42	517 (300 K)
$Fe_2O_3 \cdot nH_2O$	0.64	0.62	470 (4.2 K)
$Fe_2O_3 \cdot 2H_2O$	0.62	0.64	—
γ-Fe_2O_3			
Td	0.535	0.84	488 (300 K)
O_h	0.675	0.68	499 (300 K)
α-$FeOOH$	0.70	0	515 (77 K)
β-$FeOOH$	0.640 ± 0.006	0.700 ± 0.006	466 (77 K)
γ-$FeOOH$	0.648 ± 0.006	0.594 ± 0.006	463 (72 K)
$\delta FeOOH$	0.64 ± 0.06	0.48 ± 0.06	519 (83 K)
$\delta FeOOH$	0.76 ± 0.2	0 ± 0.01	525^b (80 K)
			505^c (80 K)
Fe_3O_4			500^b (300 K)
Fe^{3+}	0.61 ± 0.10	0 ± 0.1	482^c (300 K)
Fe^{2+}	0.96 ± 0.10	0 ± 0.1	450 (300 K)
Amorphous $Fe(OH)_3 \cdot 0.9H_2O$	—		460 (4 K)
Amorphous iron(III) oxide (thin film)	0.7 ± 0.1	$0.96 - 1.06$	460–490 (4.2 K)
Amorphous iron(III) oxide (thick film)	0.6 ± 0.1	1.01	470 (5 K)
Passive film (*in situ*)	0.7 ± 0.01	1.02 ± 0.07	470 (4 K)

[a] Isomer shift relative to SNP.
[b] T_d.
[c] O_h.

environment and similar passive films studied under *ex situ* conditions.[25,69] These large differences between *in situ* and *ex situ* XANES spectra for passive films can be seen in Fig. 9.[69] In support of earlier Mössbauer investigations,[21] the *ex situ* XANES spectrum of the passive film (cf. Fig. 10) is very similar to that obtained from γ-Fe_2O_3.[70]

Ex situ EXAFS measurements on passive films in a helium atmosphere at atmospheric pressure indicated spinel-type structure in the films, similar to those found for γ-Fe_2O_3 and

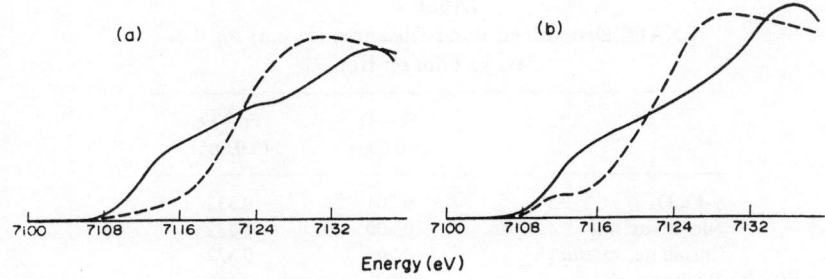

FIGURE 9. K-Edge profiles versus energy for *ex situ* (- - -) and *in situ* (——) films: (a) nitrite passivated; (b) chromate passivated.

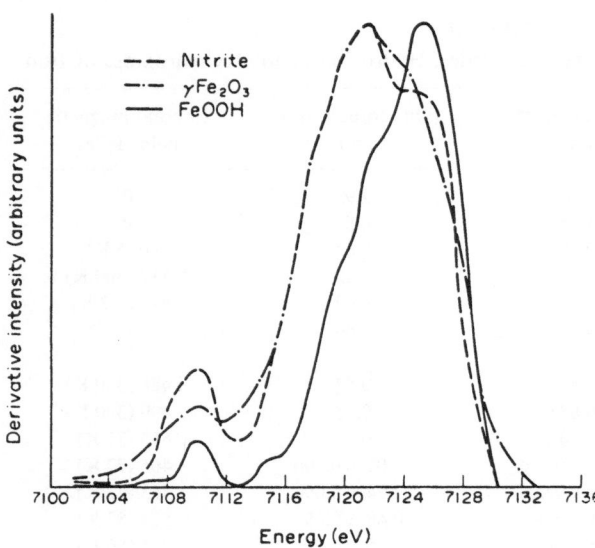

FIGURE 10. Derivative near-edge spectra from passive film formed in nitrite solution, γ-Fe_2O_3, and γ-FeOOH.

Fe_3O_4 bulk oxide standards[26]; however, the passive films displayed Fe—Fe distances that were significantly different from those in the anhydrous crystalline standards. Less sharp profiles for the peaks in the EXAFS spectra for the passive films demonstrate a lesser degree of order in the films. Similarly, a greater degree of covalency was found for the films compared to the crystalline oxide standards, again indicating a less crystalline ordered structure for the passive film on iron.[54] Fe—O and Fe—Fe bond distances derived from *in situ* and *ex situ* EXAFS data for nitrite- and chromate-passivated iron substrates, as well as for stoichiometric crystalline γ-Fe_2O_3, are given in Table 4.[70] Again, like the XANES[69] and Mössbauer[21] data, the *in situ* EXAFS measurements gave results which differed greatly from the *ex situ* derived data, which must reflect the effect of the loss of water (or other hydrogen-containing species) on the structure of the passive film.[54] Also, the *in situ* data (cf. Table 4) display significant differences compared to those obtained from γ-Fe_2O_3, in particular, the Fe—Fe bond distances.[70]

Quasi *in situ* radiometric techniques[53,55,56] involving tritiated water for passive film growth on previously cathodically reduced iron substrates have been used to determine the total content of hydrogen-containing species (e.g., water) in passive films. The amount of water was found to increase with the film thickness (cf. Fig. 11),[56] indicating that the measured water contents were not just due to adsorbed water on the surface. Tritium

TABLE 4
EXAFS Determined Bond Distances (in nm) for the
Passive Film on Iron

	Fe—O (±0.002)	Fe—Fe (±0.005)
γ-Fe_2O_3	0.201	0.332
Nitrite, *ex situ*	0.200	0.332
Chromate, *ex situ*	0.204	0.322
Nitrite, *in situ*	0.201	0.302
Chromate, *in situ*	0.208	0.305

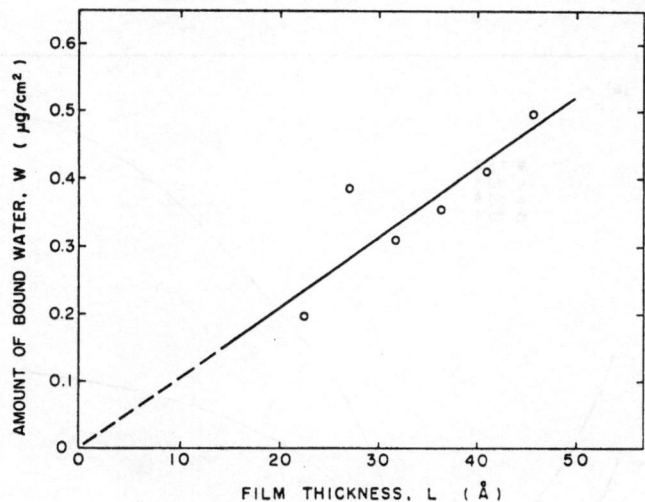

FIGURE 11. Amount of water in the passive film on iron as a function of the film thickness.

radiotracer measurements on passive films grown for one hour by electrochemical oxidation at various potentials in borate buffer are given in Table 5.[56] From the data presented, it can be concluded that up to two water molecules are associated with each Fe_2O_3 molecular unit in the passive layer.

4.4. Exposure of Passive Films on Iron to Chloride Ions

The simultaneous measurement of OH^- and Cl^- peak heights by negative SIMS for passive films grown in borate buffer solutions and subsequently exposed to chloride-containing

TABLE 5
Tritium Measurements and Related Calculations of the Water Content in the Passive Film
on Iron

Potential (vs. SCE), E (V)	Amount of electric charge, Q_t (mC/cm^2)	Amount of bound water, W (μg/cm^2)	n in $Fe_2O_3 \cdot nH_2O$	Film thickness, L (Å)	Ratio Q_a/L (mC/Å cm^2)
−0.2	3.45	0.197	1.83	22.5	0.1533
−0.1	3.81			24.8	0.1536
0	4.21	(0.386)	(2.95)	27.0	0.1559
+0.1	4.57			28.4	0.1609
+0.2	5.00	0.310	1.98	31.7	0.1577
+0.3	5.25			34.0	0.1544
+0.4	5.60	0.355	2.03	36.3	0.1543
+0.5	6.02			38.7	0.1556
+0.6	6.60	0.410	2.00	41.0	0.1610
+0.7	6.80			43.3	0.1570
+0.8	7.61	0.496	2.09	45.7	0.1665
Mean			1.99		0.1573

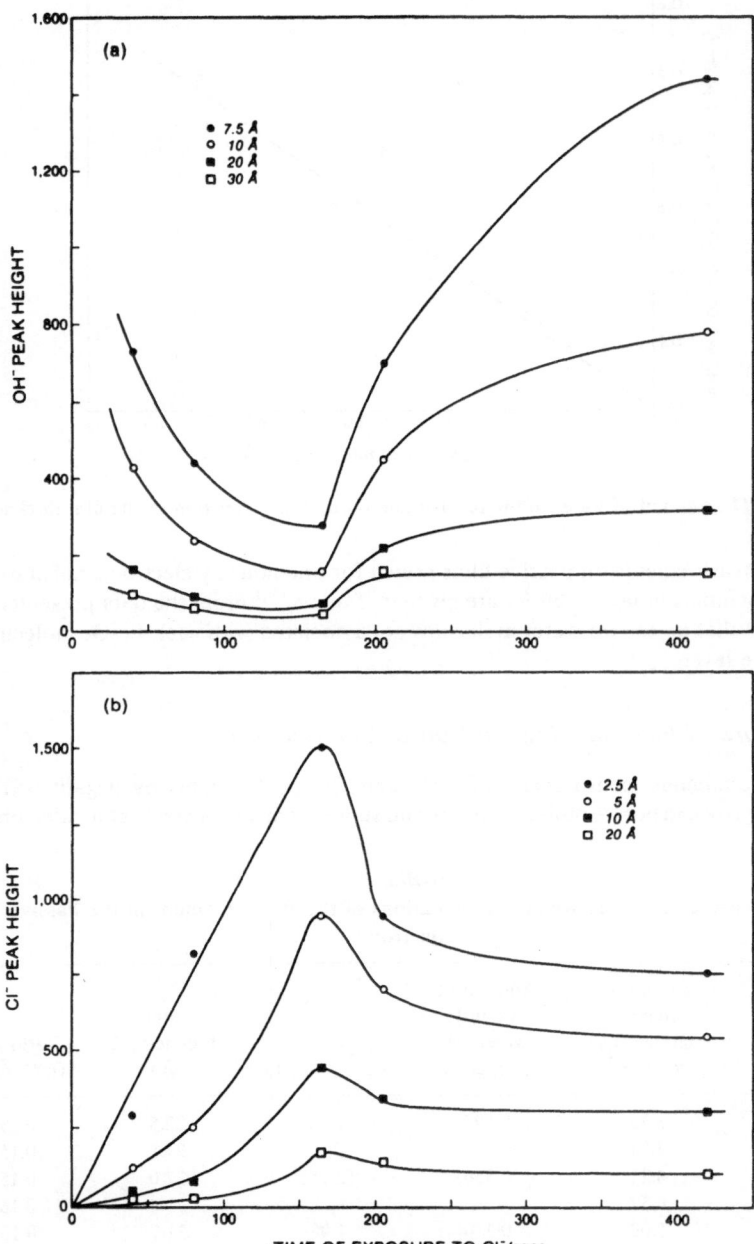

FIGURE 12. (a) OH⁻ peak height as a function of the time of exposure to chloride ions for passive films formed in borate buffer solution at 0.3 V (versus NHE) and then exposed to a chloride-containing borate solution at this potential. (b) Variation of Cl⁻ peak height intensities with time of exposure to chloride ions for passive films formed in borate buffer solution at 0.3 V (versus NHE) and then exposed to a $0.5 M$ chloride-containing borate solution at this potential for various times.

borate solutions at the potential of passive film growth can be useful in obtaining information on the breakdown of the passive film by chloride ions.[40,71] OH^- and Cl^- peak heights for passive films exposed to chloride ions for various times are given in Figs. 12a and 12b, respectively.[40] Opposite trends in the peak height intensities associated with these two anionic species can be seen. The trends reverse direction after an exposure time of 163 s, which was observed to be the breakdown time for the passive film on iron under the experimental conditions used.[71] It is to be noted from Fig. 12b that chloride ions penetrate into the inner regions of the passive films, extending right into the metal/passive film interface, and that the local concentration of chloride ions is highest in the outer regions of the films for all exposure times.

The decrease in OH^- peak heights for exposure times of up to 163 s (Fig. 12a) is associated with events leading to breakdown of the passive films. With the hydrated polymeric oxide model of passivity, diffusion of chloride ions throughout the thickness of the passive film may occur ·along certain paths (cf. Fig. 7), where bound water molecules coordinatively bonded to lattice Fe^{3+} cations are located. The negative SIMS data presented in Fig. 12 support this view, since the buildup of chloride ions throughout the passive film thickness is closely followed by a parallel reduction in the concentration of hydrogen-containing species up to the critical time at which breakdown occurs. The increase in OH^- peak heights after breakdown (about 160 s; cf. Fig. 12a) is understood to be associated with the precipitation of corrosion products [e.g., $Fe(OH)_2$ or FeOOH] at the outer surface regions of the unprotective oxide film, as well as from porous plugs within the chloride ion permeation channels (cf. Fig. 7).

Decreasing Cl^- peak heights observed for times after breakdown (Fig. 12b) could be due to chloride removal of some Fe^{3+} cations from within the films in the form of chloro complexes. This is supported by the variation of positive SIMS Fe^+ peak heights as a function of depth into passive films after various times of exposure to chloride solutions (Fig. 13).[40] On reaching a maximum at about 0.3-0.7 nm, the Fe^+ peak heights decrease with depth, this

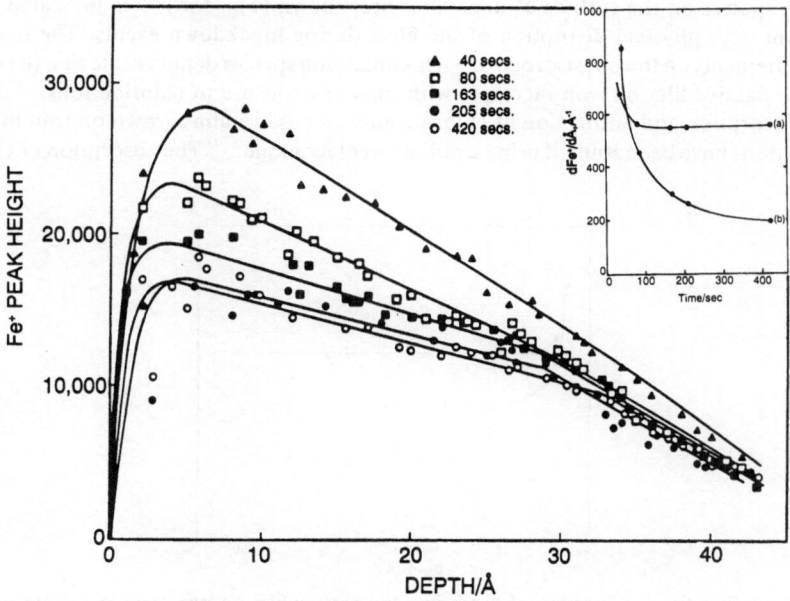

FIGURE 13. Variation of Fe^+ peak height with depth for passive films exposed to 0.5M chloride-containing borate buffer solutions for various times.

being more pronounced for films with the shortest exposure times to chloride ion. This is to be compared with similar data for passive films not exposed to chloride ions (cf. Fig. 3b), where almost constant Fe^+ peak heights are to be noted. Also, a rotating ring-disk investigation of passivated iron disk electrodes exposed to chloride solution indicated the presence of Fe^{3+} species in solution for times shortly after immersing the specimens and prior to pitting.[72]

O/Fe atomic ratios as a function of depth from ISS spectra for a passive film formed in borate solution at 0.3 V (versus NHE) and for a similar film after breakdown in a chloride-containing borate solution at the potential of film growth are presented in Fig. 14.[39] For the former film, after the first 0.5–0.7 nm, the ratio is close to 2, while for the latter film it is 1.5, as required by the hydrated amorphous polymeric oxide model of passivity. The constant O/Fe ratios in the bulk of the films are indicative of a single oxide phase in each case throughout the thickness of the films.

Iron electrodes passivated in borate solutions and subsequently exposed to borate solutions containing chloride ions, followed by cooling to dry ice temperatures under argon until the initiation of XPS measurements, displayed integrated C/O signal ratios that were a function of the degree of depassivation (Fig. 15a).[73] The C/O signal ratio uniformly increased, reaching a maximum just before breakdown, while after breakdown the signal ratio abruptly decreased and thereafter remained at a value of 50% below the maximum. This was observed under various experimental conditions of analysis in the XPS instrument. Since little CO_3^{2-} was detected in these samples, it is likely that chloride ion exposure in solution increased the surface energy, which was lowered by subsequent reaction with hydrocarbon impurities in the solution. Halide treatment of metal oxide surfaces is known to cause increased cracking of hydrocarbons, since hydroxyl removal by Cl^- ions leads to an increase in the acidity of remaining OH^- groups and defect groups on the surface. The trend of increasing C/O signal ratios (Fig. 15a) with increasing time of exposure to chloride ion supports the view that chloride ion causes removal of hydroxyls or water from the passive film.

From integrated O 1s peak areas, a sharp increase in the amount of physisorbed oxygen-containing species on the surface of films after breakdown (Fig. 15b)[73] is indicated, which is consistent with physical disruption of the films during breakdown events. The low-angle XPS measurements on the physisorbed oxygen-containing species demonstrate that the surface area of the passive film on iron increases with time of exposure to chloride ions.

The adsorption and absorption of chloride ions on passive films grown on iron in borate buffer solutions have been studied using a radiotracer technique.[74] The absorption of chloride

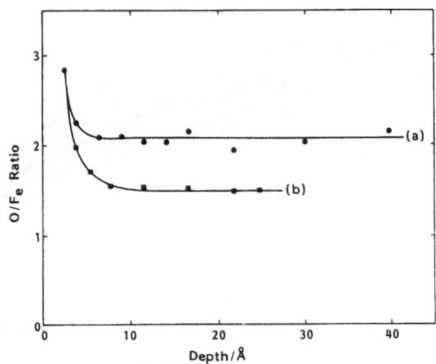

FIGURE 14. O/Fe ratio as a function of depth into the passive film on iron from ISS depth profiling: (a) film formed in borate buffer solution at 0.3 V (versus NHE); (b) passive film after breakdown [0.3 V (versus NHE) in borate buffer solution containing 0.5M NaCl].

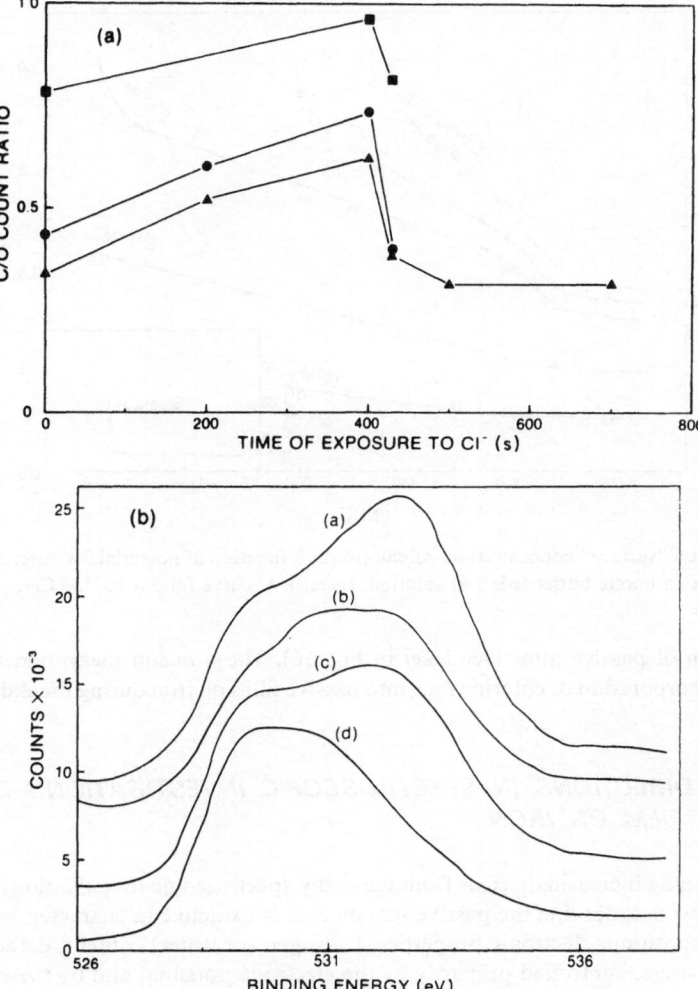

FIGURE 15. The surface concentration of carbonaceous species (C/O ratio) plotted as a function of time of exposure to Cl^-: ▲, at liquid N_2 temperature and 45° measurement angle; ●, at 46°C and 45° measurement angle; ■, at 46°C and 5° measurement angle. Note how the low-angle measurements show that the carbonaceous material is in the outer regions of the layer by the generally higher C/O ratios; O/Fe ratio was approximately constant. (b) XPS spectra for the O 1s region for 30 s after breakdown (a), 400-s Cl^- exposure (before breakdown) (b), 100-s Cl^- exposure (before breakdown) (c), and no Cl^- exposure of the passive layer (d). All spectra taken at liquid N_2 temperature and 5° measurement angle.

ions into the passive film starts at about −0.1 V (versus NHE), followed by a linear increase with increasing electrode potential. The steady-state value of absorption of chloride ions at the breakdown potential varies with concentration of chloride in the borate buffer solution and with potential. The total adsorbed and absorbed concentration of Cl^- ions, as a function of electrode potential, for different Cl^- concentrations in solution are given in Fig. 16.[74] Three regions can be identified which are characterized by different dependences on electrode potential—the region below −0.15 V (versus NHE) is characterized by constant adsorption, the intermediate region of potential by a linear increase in absorption, and the region at more anodic potentials by a stepwise increase in absorption, which appears to be associated with

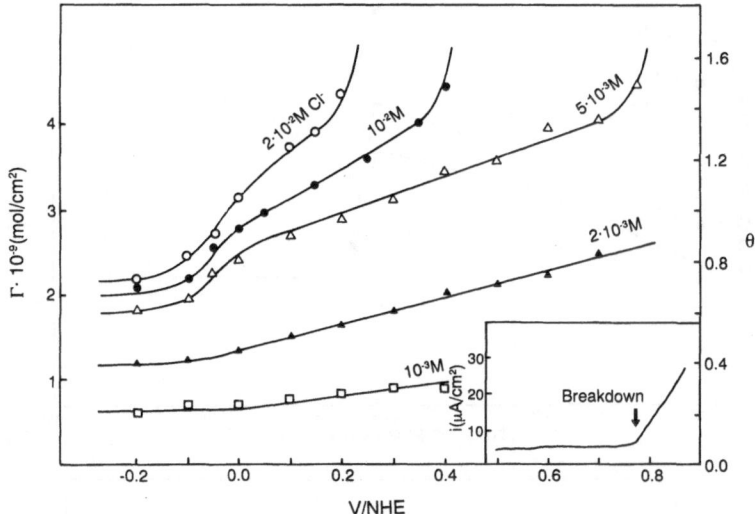

FIGURE 16. Total "surface" concentrations of chloride as a function of potential for different concentrations of chloride in borate buffer (pH 8.4) solution. Inset: i–V curve for $5 \times 10^{-3} M$ Cl$^-$.

the breakdown of passive films (see inset in Fig. 16). These *in situ* measurements strongly support the incorporation of chloride ions into passive films on iron during breakdown events.

5. FUTURE DIRECTIONS IN SPECTROSCOPIC INVESTIGATIONS OF THE PASSIVE FILM ON IRON

It is becoming increasingly clear from the many spectroscopic investigations carried out over the last few decades that the passive film on iron is oxidelike in character, with variable thickness, composition, electronic properties, hydrogen (or water) content, defect structure, disorder, and stress, controlled primarily by the electrode potential and by the composition of the aqueous phase. Sufficient data are now available from both *in situ* and *ex situ* spectroscopic measurements, involving a number of experimental techniques, that it can be unequivocally stated that the properties of the so-called passive film are changed considerably (often irreversibly) by removing passivated specimens from aqueous environments. Further advancements in knowledge of the nature of passivity, in this decade and in the 21st century, will depend very much on the greater use of existing *in situ* spectroscopic techniques, either individually or, preferably, in combination, together with the development and application of the new or emerging techniques outlined below.

5.1. Glancing Angle X-Ray Absorption Spectroscopy

Glancing angle X-ray absorption spectroscopic techniques (XANES and EXAFS) using synchrotron X-ray radiation will be the most important techniques in future studies to elucidate the nature of passive films, since they have the potential to provide information depthwise into the film with regard to structure, bond lengths, disorder, coordination number and oxidation states of iron atoms, covalent character, and the electronic properties of the films. Glancing angle approaches involving these techniques are presently being developed for investigations of anion adsorption and underpotential deposition at solid metal/solution interfaces[49,50] and for the study of inhibiting species (Cr^{6+} and Cr^{3+} oxy species) present in low concentrations in thin anodic oxide films grown on aluminum substrates.[75,76]

5.2. Grazing Incidence Asymmetric Bragg X-Ray Diffraction

A nondestructive structural probe, which can be used *in situ*, is X-ray diffraction using a grazing incidence asymmetric Bragg geometry.[77] As the angle of incidence is varied from angles less than the critical angle for total external reflection to angles greater than the critical angle, the X-ray penetration depth increases from about 20 to several thousand angstroms. The observed diffraction originates from this region of variable depth, and a structural depth profile of a thin film can be obtained by measuring the diffraction pattern as a function of the angle of incidence. The technique has been applied[77] to a 1000-Å-thick iron oxide film, prepared by a combination of reactive sputtering in an $Ar-O_2$ plasma and elevated temperature oxidation at 230°C. The iron oxide film was found to have a 45-Å-thick surface layer of α-Fe_2O_3 on top of a predominantly γ-Fe_2O_3 layer, which also contained about 2.6 at.% α-Fe_2O_3. Conventional X-ray diffraction investigations *in situ* on the underpotential deposition of Tl and Pb on Ag and Au electrodes,[78] as well as on a Prussian Blue modified electrode,[79] have been carried out. *In situ* measurements of the potential-dependent structure of underpotentially deposited Pb on Ag(111) and the initial stages of bulk Pb deposition on the adlayer have been studied using grazing incidence X-ray scattering.[80]

5.3. Grazing Angle Neutron Diffraction

The principle of the grazing angle neutron diffraction technique is the excitation of a diffracted beam under conditions of total external reflection of an incident neutron beam.[81] The fact that the total reflection occurs simultaneously with the diffraction process ensures that the diffracted intensity originates in a layer of the sample close to the surface. The approach is ideally suited for investigations of materials to which neutrons are sensitive, for example, hydrogen-rich films. The technique has been developed recently and used in investigations involving a perfect silicon crystal[82] and an imperfect crystal, the heterostructure $Cr/Nb/Al_2O_3$.[83] There are no major barriers for its adaptation to *in situ* investigations of passive films in the electrochemical environment. As a structural probe, the technique has much in common with glancing angle X-ray absorption spectroscopy.

5.4. Rutherford Backscattering Spectroscopy

Rutherford backscattering spectroscopy (RBS) is a well-established quantitative technique for thin-film and solid/solid interface analysis in vacuum environments.[63,64] The technique is element specific, can be used in the depth profiling mode, is nondestructive, and can yield quantitative information without the need for standards.[84] Although RBS facilitates the determination of the average composition and stoichiometry within a thin surface layer, it suffers from the lack of depth resolution (of the order of 100 Å at present, depending on the detected element and instrumental parameters). Recently, RBS has been adapted for *in situ* investigations of the solid electrode/solution interface (e.g., metal deposition on various substrates and oxide formation on titanium).[85,86] Further improvements in the technique could make it extremely useful for determining Fe and O contents in the passive film on iron, as well as providing quantitative measurements on the extent of adsorption and incorporation of anionic species, such as I^- and Br^- anions.

5.5. Nuclear Magnetic Resonance Spectroscopy

Proton and ^{13}C nuclear magnetic resonance (NMR) spectroscopy have been used widely in organic chemistry for structural determinations of synthesized compounds. Solid-state NMR, spectroscopy, in conjunction with magic angle spinning techniques, is being increasingly used to obtain structural information on inorganic solids (e.g., ceramics).[87] NMR, in principle, could also provide much information on the structure and composition of thin passive oxide films and on electrode/solution interfaces *in situ*. The technique is presently being developed for this environment,[88] and initial results are available on surface species adsorbed on colloidal platinum particles.[89]

5.6. Optical Spectroscopic Methods

Further refinements and improvements for using FTIR, Raman, and UV-visible spectroscopies should be implemented to allow their greater use in *in situ* environments involving aqueous solutions.[48] Vibrational data on Fe—O and Fe—O—H could provide indirect structural and compositional information concerning the passive film on iron. Further Raman and FTIR studies can lead to a determination of the nature and energetics of species adsorbed at the passive film/solution interface (e.g., aggressive anions). Greater use of UV-visible spectroscopy could yield a better understanding of the electronic properties of the passive film grown under well-defined conditions. Photoelectrochemical techniques, in conjunction with this latter spectroscopic method, are highly desirable, not only for unraveling the overall electronic properties of the passive film, but also for elucidating the electronic and ionic conductivity of the film as a function of electrode potential.

6. CONCLUSION

The development of both *in situ* and *ex situ* spectroscopic techniques over the past few decades has led to a significant increase in our understanding of the nature of passivity in respect to iron and other active metals. It is now well established that the passive film has a thickness of the order of a few nanometers and that its structure and composition is most likely that of a single layer of a defective, nonstoichiometric, hydrated, Fe^{3+}-containing amorphous oxide. The wealth of data obtained from electrochemical and spectroscopic investigations can be accounted for by two structural models, the crystalline oxide model and the hydrated amorphous polymeric oxide model. Measurements made, in particular, with a variety of *in situ* techniques, give better agreement with the latter model. Significant differences observed between *in situ* and *ex situ* spectroscopic data indicate that removal from aqueous environments (and potential control) changes the properties of the passive film to those of well-known crystalline oxides. To further unravel the complexities of the passive film will require the widespread use of *in situ* spectroscopic techniques, in particular, the use of two or more of them simultaneously. Future directions in spectroscopic investigations into the 21st century will involve new and emerging *in situ* techniques, such as glancing angle X-ray absorption spectroscopy, grazing incidence asymmetric Bragg X-ray diffraction, grazing angle neutron diffraction, Rutherford backscattering, nuclear magnetic resonance, and optical methods.

REFERENCES

1. H. Kaesche, *Metallic Corrosion*, NACE, Houston, Texas (1985).
2. Z. Szklarska-Smialowska, *Pitting Corrosion of Metals*, NACE, Houston, Texas (1986).
3. M. Faraday, in: *Experimental Researches in Electricity*, Vol. 2, London (1844); reprinted by Dover Publishers, New York (1965).
4. Proceedings of the U.S.-German Seminar on the Passivity of Metals, *Corros. Sci.* **29** (1989).
5. B. MacDougall and J. A. Bardwell, *J. Electrochem. Soc.* **135**, 2437 (1988).
6 N. Sato, K. Kudo, and T. Noda, *Z. Phys. Chem., N.F.* **98**, 271 (1975).
7. J. O'M. Bockris, N. A. Genshaw, V. Brusic, and H. Wroblowa, *Electrochim. Acta* **16**, 1859 (1971).
8. J. Kruger, *J. Electrochem. Soc.* **110**, 654 (1963).
9. N. Sato and M. Cohen, *J. Electrochem. Soc.* **111**, 624 (1964).
10. M. Nagayama and M. Cohen, *J. Electrochem. Soc.* **110**, 670 (1963).
11. J. A. Bardwell and B. MacDougall, *J. Electrochem. Soc.* **135**, 2157 (1988).
12. C. T. Chen and B. D. Cahan, *J. Electrochem. Soc.* **129**, 17 (1982).
13. W. Kozlowski and A. Szklarska-Smialowska, *J. Electrochem. Soc.* **131**, 234 (1984); **131**, 499 (1984); **131**, 723 (1984).
14. A. Hugot-Le Goff and C. Pallotta, *J. Electrochem. Soc.* **132**, 2807 (1985).
15. A. Hugot-Le Goff and C. Pallotta, *Mater. Sci. Forum* **8**, 451 (1986).
16. J. C. Rubim and J. Dunnwald, *J. Electroanal. Chem.* **258**, 327 (1989).
17. J. O'M. Bockris, M. A. Habib, and J. L. Carbajal, *J. Electrochem. Soc.* **131**, 3032 (1984).
18. B. R. Scharifker, M. A. Habib, J. L. Carbajal, and J. O'M. Bockris, *Surf. Sci.* **173**, 97 (1986).
19. K. Ogura, A. Fujishima, and K. Honda, *J. Electrochem. Soc.* **131**, 344 (1984).
20. W. E. O'Grady and J. O'M. Bockris, *Surf. Sci.* **38**, 249 (1973).
21. W. E. O'Grady, *J. Electrochem. Soc.* **127**, 555 (1980).
22. M. E. Brett, K. M. Parkin, and M. J. Graham, *J. Electrochem. Soc.* **133**, 2031 (1986).
23. J. Eldridge and R. W. Hoffman, *J. Electrochem. Soc.* **136**, 955 (1989).
24. M. C. Lin, R. G. Barnes, and T. E. Furtak, in: *AIP Conference Proceedings 80—Phys. Steel Ind.*, American Institute of Physics (1982).
25. G. G. Long, J. Kruger, D. R. Black, and M. Kuriyama, *J. Electrochem. Soc.* **130**, 240 (1983).
26. G. G. Long, J. Kruger, D. R. Black, and M. Kuriyama, *J. Electroanal. Chem.* **150**, 603 (1983).
27. R. W. Hoffman, in: *Passivity of Metals and Semiconductors* (M. Froment, ed.), p. 147, Elsevier, Amsterdam (1984).
28. U. Stimming and J. W. Schultze, *Electrochim. Acta* **24**, 859 (1979).
29. B. D. Cahan and C. T. Chen, *J. Electrochem. Soc.* **129**, 474 (1982).
30. K. Azumi, T. Ohtsuka, and N. Sato, *J. Electrochem. Soc.* **134**, 1352 (1987).
31. W. Hopfner and W. J. Plieth, *Werkst. Korros.* **36**, 373 (1985).
32. G. Larramona and C. Gutierrez, *J. Electrochem. Soc.* **136**, 2171 (1989).
33. U. Stimming, *Electrochim. Acta* **31**, 415 (1986).
34. P. C. Searson, R. M. Latanision, and U. Stimming, *J. Electrochem. Soc.* **135**, 1358 (1988).
35. M. Seo, M. Sato, J. B. Lunsden, and R. W. Staehle, *Corros. Sci.* **17**, 209 (1977).
36. R. W. Revie, J. O'M. Bockris, and B. G. Baker, *Surf. Sci.* **52**, 664 (1975).
37. R. W. Revie, B. G. Baker, and J. O'M. Bockris, *J. Electrochem. Soc.* **122**, 1460 (1975).
38. S. C. Tjong and E. Yeager, *J. Electrochem. Soc.* **128**, 2251 (1981).
39. T. E. Pou, O. J. Murphy, V. Young, J. O'M. Bockris, and L. L. Tongson, *J. Electrochem. Soc.* **131**, 1243 (1984).
40. O. J. Murphy, T. E. Pou, J. O'M. Bockris, and L. L. Tongson, *J. Electrochem. Soc.* **131**, 2785 (1984).
41. D. F. Mirchell, G. I. Sproule, and M. J. Graham, *Appl. Surf. Sci.* **21**, 199 (1985).
42. R. Goetz, D. F. Mitchell, B. MacDougall, and M. J. Graham, *J. Electrochem. Soc.* **134**, 535 (1987).
43. M. Nagayama and M. Cohen, *J. Electrochem. Soc.* **109**, 781 (1962).

44. C. L. Foley, J. Kruger, and C. J. Bechtoldt, *J. Electrochem. Soc.* **114**, 994 (1967).
45. K. Kuroda, B. D. Cahan, Gh. Nazri, E. Yeager, and T. E. Mitchell, *J. Electrochem. Soc.* **129**, 2163 (1982).
46. J. A. Bardwell, B. MacDougall, and M. J. Graham, *J. Electrochem. Soc.* **135**, 413 (1988).
47. J. O'M. Bockris, M. A. V. Devanathan, and A. K. N. Reddy, *Proc. Roy. Soc.* (*London*), *Ser. A* **279**, 327 (1964).
48. B. Pettinger, A. Friedrich, and U. Tiedemann, *J. Electroanal. Chem.* **280**, 49 (1990).
49. O. R. Melroy, M. G. Samant, G. Borges, J. G. Gordon, L. Blum, J. H. White, M. J. Albarelli, S. M. McMillan, and H. D. Abruna, *Langmuir* **4**, 728 (1988).
50. J. H. White and H. D. Abruna, *J. Electroanal. Chem.* **274**, 185 (1989).
51. M. A. Habib and J. O'M. Bockris, *J. Electroanal. Chem.* **180**, 287 (1984).
52. M. C. Bloom and L. Goldenberg, *Corros. Sci.* **5**, 623 (1965).
53. H. T. Yolken, J. Kruger and J. P. Calvert, *Corros. Sci.* **8**, 103 (1968).
54. J. Kruger, *Corros. Sci.* **29**, 149 (1989).
55. G. Okamoto and T. Shibata, *Nature* **206**, 1350 (1965).
56. K. Kudo, T. Shibata, G. Okamoto, and N. Sato, *Corros. Sci.* **8**, 809 (1968).
57. G. Okamoto, *Corros. Sci.* **13**, 471 (1973).
58. C. L. McBee and J. Kruger, *Electrochim. Acta* **17**, 1337 (1972).
59. H. H. Uhlig, in: *Passivity of Metals*, (R. P. Frankenthal and J. Kruger, eds.), p. 1, The Electrochemical Society, Pennington, New Jersey (1978).
60. J. Kruger and J. P. Calvert, *J. Electrochem. Soc.* **114**, 43 (1967).
61. N. Sato, K. Kudo, and R. Nishimura, *J. Electrochem. Soc.* **123**, 1419 (1976).
62. J. A. Bardwell, B. MacDougall, and G. L. Sproule, *J. Electrochem. Soc.* **136**, 1331 (1989).
63. P. Skeldon, K. Shimizu, G. E. Thompson, and G. C. Wood, *Surf. Interface Anal.* **5**, 247 (1983).
64. D. L. Cocke, C. A. Polansky, D. E. Halverson, S. M. Kormali, C. V. Barros-Leite, O. J. Murphy, E. A. Schweikert, and F. Filpus-Luyckx, *J. Electrochem. Soc.* **132**, 3065 (1985).
65. M. D. Archer, C. C. Corke, and B. H. Harji, *Electrochim. Acta* **32**, 13 (1987).
66. R. W. Revie, J. O'M. Bockris, and B. G. Baker, *Surf. Sci.* **52**, 664 (1975).
67. R. W. Revie, B. G. Baker, and J. O'M. Bockris, *J. Electrochem. Soc.* **122**, 1460 (1975).
68. O. J. Murphy, J. O'M. Bockris, T. E. Pou, D. L. Cocke, and G. Sparrow, *J. Electrochem. Soc.* **129**, 2149 (1982).
69. J. Kruger, G. G. Long, M. Kuriyama, and A. I. Goldman, in: *Passivity of Metals and Semiconductors* (M. Froment, ed.), p. 163, Elsevier, Amsterdam (1983).
70. J. Kruger and G. G. Long, in: *Surfaces, Inhibition and Passivation* (E. McCafferty and R. J. Brodd, eds.), p. 210, Proceedings Vol. 86-7, The Electrochemical Society, Pennington, New Jersey (1986).
71. O. J. Murphy, J. O'M. Bockris, T. E. Pou, L. L. Tongson, and M. D. Monkowski, *J. Electrochem. Soc.* **130**, 1792 (1983).
72. K. E. Heusler and L. Fischer, *Werkst. Korros.* **27**, 551 (1976); **27**, 788 (1976).
73. D. L. Cocke, P. Nilsson, O. J. Murphy, and J. O'M. Bockris, *Surf. Interface Anal.* **4**, 94 (1982).
74. V. Jovancicevic, J, O'M. Bockris, J. L. Carbajal, P. Zelenay, and T. Mizuno, *J. Electrochem. Soc.* **133**, 2219 (1986).
75. A. J. Davenport, H. S. Isaacs, and M. W. Kendig, *J. Electrochem. Soc.* **136**, 1837 (1989).
76. A. J. Davenport and H. S. Isaacs, *Corros. Sci.* **31**, 105 (1990).
77. M. F. Toney and S. Brennan, *J. Appl. Phys.* **65**, 4763 (1989).
78. M. Fleischmann and B. W. Mao, *J. Electroanal. Chem.* **247**, 297 (1988).
79. T. Ikeshoji and T. Iwasaki, *Inorg. Chem.* **27**, 1123 (1988).
80. O. R. Melroy, M. F. Toney, G. L. Borges, M. G. Samant, J. B. Kortright, P. N. Ross, and L. Blum, *J. Electroanal. Chem.* **258**, 403 (1989).
81. J. F. Ankner, H. Zabel, D. A. Neumann, C. F. Majkrzak, J. A. Dura, and C. P. Flynn, Extended Abstracts, Vol. 90-1, Abstract 792, The Electrochemical Society, Pennington, New Jersey (1990).
82. J. F. Ankner, H. Zabel, D. A. Neumann, C. F. Majkrzak, A. Matheny, J. A. Dura, and C. P. Flynn, *Materials Science Society Symposium Proceedings* **166**, 109 (1990).
83. J. F. Ankner, H. Zabel, D. A. Neumann, C. F. Majkrzak, J. A. Dura, and C. P. Flynn, *J. Phys. J. Phys. Colloq.* **C7**, 189 (1989).
84. C. V. Barros-Leite, G. B. Baptista, B. K. Patnaik, D. L. Cocke, N. Magnussen, O. J. Murphy, E. A. Schweikert, and L. Quinones, *J. Trace Microprobe Techniques* **4**, 37 (1986).

85. R. Kotz, J. Gobrecht, S. Stucki, and R. Pixley, *Electrochim. Acta* **31**, 169 (1986).

86. J. S. Forster, D. Phillips, J. Gulens, D. A. Harrington and R. L. Tapping, *Nucl. Instrum. Methods Phys. Res.* **B28,** 385 (1987).

87. M. McMillan, J. S. Brinen, J. D. Carruthers, and G. L. Haller, *Colloids Surf.* **38,** 133 (1989).

88. K. W. H. Chan and A. Wieckowski, *J. Electrochem. Soc.* **137,** 367 (1990).

89. R. D. Newmark, M. Fleischmann, and B. S. Pons, *J. Electroanal. Chem.* **255,** 325 (1988).

34

Corrosion of Thin-Film Storage Media: A Review

Vlasta Brusic, Jean Horkans, and Donald J. Barclay

1. INTRODUCTION

The computer, one of the most important technological innovations of the 20th century, is rapidly decreasing in size while increasing in power. Small, powerful computers require small, high-capacity devices for permanent storage of data. The particulate disk, which has served in generations of computers, is not capable of meeting the projected density requirements and is being replaced by higher capacity thin-film disks and optical disks. The new kinds of disks, however, have less environmental stability than particulate disks. Thus, a stable disk is being replaced by a more vulnerable disk just as computers are being moved from controlled environments into the uncontrolled environments of homes, offices, and plants.

The fundamental limit to the achievable density is the width of the transition between domains of opposite magnetization. The minimum transition width, a, for longitudinal media equals $M_r \delta / H_c$; for a given remanent moment, M_r, it is decreased by decreasing the thickness of the magnetic layer, δ, and by increasing its coercivity H_c.[1] In a particulate disk, the magnetic layer consists of a suspension of magnetic particles in an organic binder spun onto an Al disk support and cured. The practical lower limit of this layer is $\sim 0.5~\mu m$, limiting the bit density that can be achieved with this technology. Much higher storage density is attainable in a continuous thin magnetic film, which may be on the order of 50–500 nm in thickness. Thin-film media are usually metal alloys (generally alloys of Co) but can also be other materials, such as metal-oxide films. For most thin-film disks, the structure consists of an Al support, a nonmagnetic film of NiP that can be polished to the necessary smoothness, the magnetic layer, a thin layer of hard coating material, and a lubricant. Information is written or read by a magnetic head flying in close proximity (0.1–0.3 μm) to the rotating disk.

Magneto-optic disks are capable of even higher density storage than thin-film disks. The media are rare-earth transition metals (RE-TMs), such as FeTb, normally covered by a dielectric material. The amorphous RE-TM alloys have a remarkable combination of properties which make them suitable for optical recording[2]: a Curie temperature around 250°C,

---

Vlasta Brusic and Jean Horkans ● IBM T. J. Watson Research Center, Yorktown Heights, New York 10598.   *Donald J. Barclay* ● IBM United Kingdom, Winchester, Hampshire SO21 2JN, United Kingdom

Electrochemistry in Transition, edited by Oliver J. Murphy *et al.* Plenum Press, New York, 1992.

perpendicular anisotropy, a high H_c at room temperature, and relatively high Kerr rotation. Writing is accomplished by a laser with a long focal length; reading exploits the magneto-optic Kerr effect.

The improved performance of disks can be viewed as being inimical to their environmental stability. The magnetic materials are chosen primarily for their magnetic properties, not for their invulnerability toward corrosion or for their compatibility with other materials in the disk structure. Thinner films, smaller bits, and lower flying heights greatly magnify the importance of small, corrosion-induced defects. Meanwhile, our global environment is worsening,[3] and ambient levels of corrosive gases such as SO_2 and NO_2 in a typical office easily reach $1-10^2 \mu g/m^3$.[4]

This chapter will focus on the relevance of electrochemical science to the environmental stability of storage media. Other electrochemical aspects of thin-film disk storage will be only briefly described but have been discussed in a longer review by the authors.[5] The reader is also referred to several excellent general reviews of magnetic storage[6-12] and magneto-optic storage.[2]ptfiff

2. ELECTROCHEMICAL DEPOSITION OF THIN-FILM MEDIA

The Co alloys used as thin-film media can be evaporated, sputtered, electrodeposited, or electrolessly deposited. The RE-TM magneto-optic media are produced only by vacuum deposition. In all cases, the magnetic properties of the materials are strongly dependent on their structures, and these in turn are strongly dependent on the deposition conditions. Nonelectrochemical deposition techniques are beyond the scope of this chapter but have been reviewed elsewhere.[7-9] Here we will briefly describe the deposition of Co-alloy media by electrochemical techniques.

It is the high magnetic moment of Co that makes it the metal of choice for magnetic media. Pure metallic Co films have a relatively low coercivity, however, and are not suitable as media for high-density storage. The desired coercivity (which is currently in the range of ~500–1500 Oe) is attained by alloying the Co with one or more other materials, which function to provide a second-phase precipitate at grain boundaries that pins domain wall motion and thus increases the coercivity.[13] In order to control the magnetic characteristics, such film properties as grain size and preferred orientation, alloy composition, and stress must be properly controlled.

Sputtered media commonly are CoCr or CoCrX alloys (X may be a metal such as Ni), in which Cr enrichment at grain boundaries provides the requisite nonmagnetic second phase. The most common electrochemically deposited media are CoP and CoNiP, for which high coercivity is achieved by the incorporation of phosphorus. The latter materials are usually deposited electrolessly but can also be electrodeposited.

2.1. Electrolessly Deposited Media

Although magnetic films of electroless CoP and CoNiP have been studied for years, earlier structural studies mainly dealt with lower H_c films and are not relevant to the current, thinner, high-H_c layers. Mirzamaani and co-workers[14,15] have determined the interrelationships between structure and magnetic properties of CoP thin films and have described the differences from the behaviour of thicker films.

Phosphorus is typically incorporated at 5–10 at.% in these deposits, which are crystalline.[16] Most authors agree that the Co has an hcp structure, but Chow et al.[17] found that the formation of hcp or fcc structure depended on the solution composition. The preferred orientation (PO) of the deposit also depends on the solution chemistry.[18,19] Mirzamaani et al.[14] showed that shape anisotropy and stress anisotropy can dominate crystal anisotropy under certain conditions.

The mechanisms giving CoP with high coercivity and a high squareness of the M-H loop are discussed by Chen et al.[20] The magnetic parameters are determined by the size and nature of the microcrystallites of the film, which in turn are determined by the deposition conditions. Various pH dependences of the coercivity and the grain structure of CoP and CoNiP have been reported.[21-23] The apparently contradictory observations stem, at least in part, from the failure to compare the properties at constant film thickness and perhaps also from different solution chemistries. Judge et al.[22] show that even the chemical used to adjust the pH will affect the magnetic properties.

Alternative alloys for thin-film media, CoSnP and CoZnP, have been suggested because of improved resistance to corrosion.[24,25] Both Zn and Sn increase the coercivity of the deposit. Electroless CoB has been produced in both amorphous[26] and crystalline[27] states. Both films, however, had coercivities less than or equal to 300 Oe, making them unsuitable for thin-film media.

2.2. Electrodeposited Media

Most electrodeposited media have had compositions (mainly CoP and CoNiP) similar to those obtained by electroless deposition. Incorporation of Ni in CoP substantially increases the coercivity.[28,29] Sallo and co-workers[30-32] examined the structures of CoP and CoNiP electrodeposits. Low- and high-coercivity materials were found to have very different structures, the low-H_c deposits being lamellar and the high-H_c deposits having rodlike structures. The structure is influenced by the substrate.[33] Disks have been fabricated with plated CoNiP media layers.[33-35]

Luborsky[36] has studied electroplated Co and CoNi alloys with P and with other metals from groups VIB and VA. The incorporated elements can influence the grain size of the deposits and can segregate at grain boundaries, thus influencing the coercivity. Other potentially interesting electroplated alloys are CoMnP,[37] CoSnP,[38] and CoPt.[39,40]

3. CORROSION OF THIN-FILM DISKS

What is known about disk corrosion stems from laboratory experiments. No observations of corrosion in the actual working environment have as yet been reported. In fact, a recent reference[12] states that such corrosion is not observed. Disk corrosion will be reviewed here with attention to the disk file environment, the experimental approaches used, the effects of alloy composition, the effects of overcoats, and the projected lifetime of the product.

3.1. File Environment

A disk housing is schematically represented in Fig. 1. It contains an aperture to allow "breathing", the flow of air in and out of the file. Volpe[41] has shown that the air exchange is largely independent of the file operating mode (i.e., on/off) and is dominated by diffusion through the small hole. For a hole with a diameter of 7 mm and depth of 2 mm, the leak rate was measured to be on the order of 100 cm^3/h (NTP). A gaseous atmospheric pollutant, if present in the outside air in typical amounts of 1-100 μg/m^3, would diffuse through the hole at a rate of only 1-100 μg per year. Pollutants absorb on the walls of the file, further reducing their supply to the disks. Also, the elevated temperature of the running drive keeps the relative

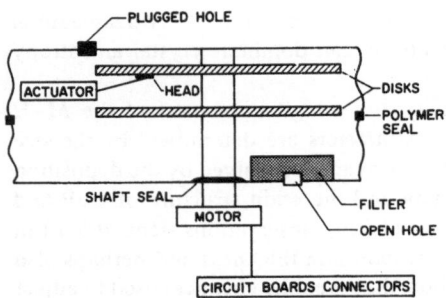

FIGURE 1. Schematic of the disk file with a filter for aerosol particles over the aperture.

humidity of the air inside the enclosure at less than 40%. Thus the housing protects its contents, even when the breathing hole is covered only by a filter to trap aerosol particles.[41] Some designs go further and employ chemical filters to remove pollutants and desiccants to ensure low humidity levels.[42]

3.2. Experimental Approach

Koester and Arnoldussen[8] point out that corrosion of thin-film disks can occur at a low overall rate while causing significant problems because of high corrosion rates at localized corrosion sites. The methods of evaluating disk corrosion must be representative of the working environment and sensitive to minute changes. Typically, accelerated corrosion tests and electrochemical measurements are used, often in conjunction with surface analyses.

In accelerated corrosion tests, the natural disk environment should be modeled by varying the humidity and the concentration of pollutant gases, such as Cl_2, SO_2, NO_2, and H_sS.[43] It is most common, however, to make only temperature/humidity (T/H) variations. This approach is reasonable, in view of the demonstration that even open files provide protection against a corrosive atmosphere.

The corrosion rate is a sensitive function of the amount of water on the surface.[43-45] Tomashov[44] has suggested the dependence shown in Fig. 2A. At very low humidities (region I), where the surface is practically dry, the metal oxidation rate is very low. As the water layer thickens from a noncontinuous film to a visible condensation at 100% relative humidity (region II), the corrosion rate increases to a maximum. Further thickening of the condensed film (region III) will cause some lowering of the corrosion rate, because the thicker moisture film impairs oxygen arrival to the surface. Under total immersion in an electrolyte (region

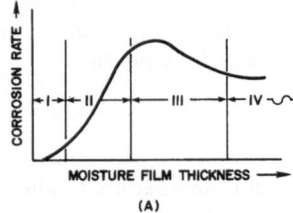

(A)

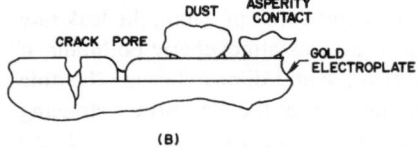

(B)

FIGURE 2. (A) The dependence of the average corrosion rate on the thickness of an adsorbed water layer. (B) The local variation of the adsorbed water with surface geometry and asperities. (Reprinted from Ref. 46, courtesy of the Electrochemical Society.)

IV), the corrosion rate levels off as the effective diffusion layer reaches its maximum thickness. Tomashov estimates the thickness of the water layer at 1–10 nm in region I, 1 μm at the end of region II, and 1 mm at the crossover of regions III and IV. These values are reasonable for the two last regions. Recent data, however, indicate a much thinner water layer in regions I and II. On Co at room temperature, for example, below 30% relative humidity (RH) the water layer is less than one monolayer thick; it reaches about three layers at 50% and eight layers at 70%; a sharp increase in thickness occurs at yet higher humidities.[43]

The strength of T/H tests is the possibility of using them to establish acceleration factors and to estimate product life. The region of interest for corrosion of electronic devices[46] and magnetic media is region II, between about 30 and 90% RH. Sharma et al.[46] determined the amount of adsorbed water on Au, Cu, and Ni. Using an estimate of 0.6 eV for the thermal activation and assuming a single corrosion mechanism throughout the range of RH, they obtain an acceleration factor of 154 at 65°C/80% RH with respect to 25°C/35–40% RH. Despite the usefulness of determining acceleration factors, there are weaknesses in the method. For one, the oxidation mechanism may be electrochemical at the high humidity of the accelerated tests and chemical at the low humidity to which the data are extrapolated; the assumption of a single mechanism will overestimate the rate of corrosion.

The role of adsorbed water in atmospheric corrosion is predicted to be similar to that of bulk water once its thickness reaches about three monolayers[43]; for many materials this thickness is reached at a critical humidity above 50%. Real surfaces, on the other hand, are subject to localized condensation (and a much higher corrosion rate) at crevices, pores in overlayers, points of contacts with asperities, etc. (Fig. 2B). The relationship determining capillary condensation is $P_1 = P_0 e^{-2\sigma v/RTr}$, where P_1 and P_0 are the pressures of saturated vapor above a concave meniscus of radius r and above a flat surface, respectively, of a liquid with a surface tension σ and a molecular volume v. If the surface is hydrophilic and contains small crevices or pores, condensation can occur locally at very low humidities.[44] At 25°C, a RH of 99% is necessary for condensation in a pore with a 69-nm radius, but condensation occurs at ~60% RH in a 2-nm pore and at ~40% RH in a 1-nm pore. Thus, thin-film disks with hydrophilic, porous overcoats and perhaps also with crevices from the purposeful texturing of the surface can easily suffer localized corrosion at relatively low humidities.

Accelerated tests are routinely conducted at humidity levels above 60% and should find the same corrosion mechanism as electrochemical tests.[45] Several studies have shown the correlation between the accelerated tests and electrochemical measurements.[47-48] One example (Fig. 3) is the study by Brusic et al.[47] of corrosion and passivation of amorphous and crystalline Fe(B)Cr alloys. The corrosion current was measured electrochemically in a solution of pH 3, and the weight gain determined after the environmental chamber test. In both cases, significant corrosion resistance was observed only if alloys contained more than about 10 at.% Cr.

Although only a few electrochemical studies of magnetic media have been conducted, electrochemical methods are useful in elucidating the corrosion mechanism. Quantitative measurements can be rapidly obtained in a variety of electrolytes, including distilled water. Walmsley et al.,[49] who employed ac impedance techniques for corrosion testing of CoCr disks, have commented than when the acceleration factors are known crudely or not at all, electrochemical techniques provide an excellent measure of the instantaneous corrosion rate as well as a means of comparison of disks. Brusic et al.[50] designed the miniature cell shown in Fig. 4 for measurement of disk corrosion. The sample (working electrode) is masked with plating tape to expose only a 0.32-cm^2 area. Filter paper disks separate the working electrode from a Pt-mesh counter electrode and a mercurous sulfate reference electrode. The cell uses a 20-μl droplet of electrolyte. The small distance between the electrodes results in a low ohmic resistance, even for water. The thinness of the electrolyte layer results in a large supply of oxygen at the surface; thus, the conditions are close to those of atmospheric corrosion.

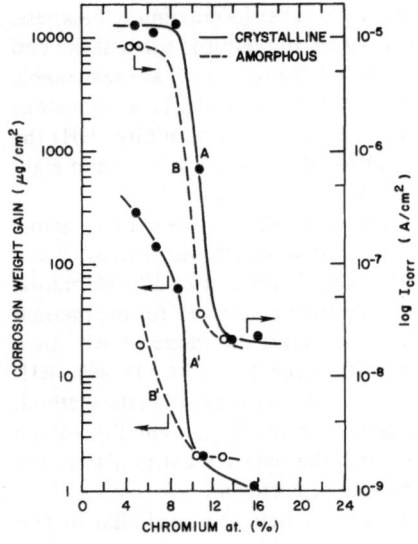

FIGURE 3. A comparison of the corrosion rates of FeCr and FeBCr alloys obtained in electrochemical (curves A and B, respectively) and accelerated corrosion tests (curves A' and B', respectively). A,A' = $Fe_{1-x}Cr_x$; B,B' = $(Fe_{81}B_{19})_{1-x}Cr_x$. (Reprinted from Ref. 47, courtesy of the Electrochemical Society.)

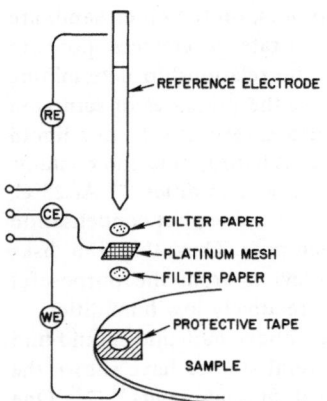

FIGURE 4. Experimental setup for electrochemical corrosion evaluation in a droplet of electrolyte.

3.3. Corrosion of CoP and Co Alloys

Although Co-based films do not corrode in dry contaminated air,[51] they become vulnerable when humidity and oxygen are present. Judge et al.[52] have shown that the magnetic properties of films exposed to 43.5 to 65°C and 80% RH deteriorate with time (Fig. 5), although there is some protection by the corrosion product (identified as Co_3O_4). CoP is more corrosion resistant than Co. Helfand et al.[53] propose a reaction of P in the alloy with water to form adsorbed hypophosphite. Even in the absence of aggressive ions, however, the

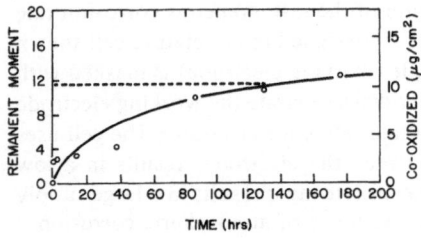

FIGURE 5. Deterioration of the magnetic moment of corroding CoP. (Reprinted from Ref. 52, courtesy of the Electrochemical Society.)

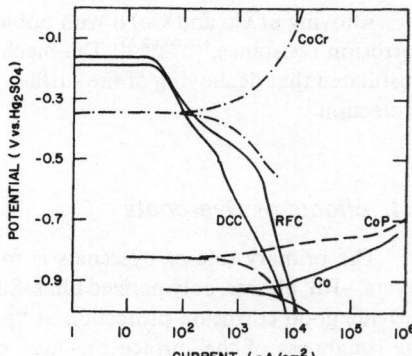

FIGURE 6. Potentiodynamic curves on Co, CoP, CoCr, and rf- (RFC) and dc-sputtered carbon (DCC) in deionized water.

P-containing film provides little protection (Fig. 6). The potentiodynamic polarization curve shows a CoP dissolution rate at open circuit of about 3×10^{-6} A/cm^2, only two times lower than that of sputtered Co films. At higher potentials, the P-containing film completely loses its protectiveness. The anodic Tafel slope is very low, approximately 44 mV/decade[49,54] and extends through several decades. The CoP is subject to galvanic attack by more noble overlayers, since it exhibits a fairly low corrosion potential in water, a low anodic Tafel constant, and no self-passivation.[49]

Alloying with the oxide-forming transition metals decreases the corrosion rate of Co, with the corrosion resistance increasing in the order Co $\simeq$ CoP < CoNi < CoCr $\simeq$ CoCrTa < CoCrTaZr.[49,54–59] Electrochemical measurements of CoCr (Fig. 7A) show that the critical Cr level for self-passivation is about 17%.[60] The protective film is less than 5 nm thick and rich in a passivating metal oxide, such as chromium oxide, which is thought to stabilize the nonpassivating cobalt oxides. Thin CoCr films invariably corrode at higher rates than the bulk alloys (Fig. 7B), even though they show a similar dependence of the dissolution rate on the Cr content.[60] Some CoCr films, however, have endured 60 days[61] and even six months testing at 90% RH at 60°C.[56]

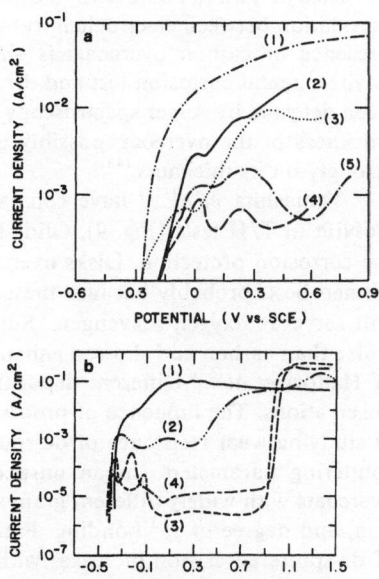

FIGURE 7. (a) Potentiodynamic polarization curves on Co foil (1) and CoCr films with 9% (2), 17% (3), 20.7% (4), and 20.9% Cr (5). (b) Similar data for bulk CoCr with 5% (1), 10% (2), 20% (3), and 15% Cr (4), respectively. (Reprinted from Ref. 60, with the permission of the IEEE.)

Alloying of Co and CoNi with noble metals, notably Pt, results in a material with a high corrosion resistance.[38,40,62,63] The mechanism of Pt protection has yet been published. It is postulated that dealloying of the surface layer results in a Pt-rich film that provides corrosion protection.

3.4. Effects of Overcoats

The primary role of overcoats is to provide wear protection. The most common over-layers—Rh, plasma-polymerized films, SiO_2, and carbon—are all chemically stable and would provide good corrosion protection at "perfect" coverage. The thinness of the overcoats and the roughness of the surface preclude complete coverage and open the path for localized corrosion at the sites where the magnetic layer is exposed to the environment. Coatings with marginal adhesion, such as plasma-polymerized films,[64] can allow corrosion at the sites of adhesion failure. The local corrosion can show a galvanic enhancement, as in the case of Rh overcoats.[64-67] Carbon is also capable of galvanic enhancement, but contradictory results have been reported. Garrison[65] observed that carbon can enhance galvanic corrosion; Smallen et al.[55] report that carbon prevents lateral growth of corrosion products and thereby limits corrosion; Nagao et al.[68] report that carbon protects CoCr; and Black et al.[69] found faster corrosion of CoCrMo when overcoated with carbon.

Electrochemical measurements of carbon films on glass[49] are shown in Fig. 6. Super-posed are the potentiodynamic polarization curves on CoP. The data suggest that the more noble carbon will cause a localized increase in the CoP corrosion rate that, depending on the area ratio of C to CoP, can be many orders of magnitude. The predicted enhancement is observed on C-coated CoP disks exposed to T/H tests with or without corrosive gases: although the average corrosion rate is low, very high rates occur at the sites of galvanic contact.[49] In contrast to carbon, glassy oxides protect CoP and Co and do not enhance the corrosion rate at sites of incomplete coverage.[49]

Other Co alloys, such as CoCr, are more noble than CoP and thus electrochemically more compatible with the carbon. Nevertheless, CoCr (with 20 at.% Cr) will dissolve in the presence of a more noble carbon overcoat rather than form a Cr-rich protective layer.[70] The correlation between electrochemical corrosion measurements and environmental tests in the presence of carbon overcoats is shown in Fig. 8.[48] Cobalt ions produced during the environmental corrosion test and driven by diffusion to the top of the overcoat surface have been detected by Auger spectroscopy. The corrosion shows no systematic dependency on the thickness of the overcoat, possibly because the porosity of the overcoat does not depend strongly on its thickness.[48]

Yamashita et al.[71] have compared SiO_2, Al_2O_3, ZrO_2, and carbon as overcoats for CoNiPt in T/H tests (Fig. 9). Glide-height tests and defect mapping were used to evaluate the corrosion protection. Disks overcoated with alumina- and yttria-stabilized zirconia per-formed best, probably because these materials are pore-free and, being nonstoichiometric, will serve as oxygen scavengers. Surprisingly, in the same study, sputtered SiO_2 behaved worse than carbon and showed enhanced corrosion at pinholes,[71] in contrast to the findings of Hattori et al.[64] Different deposition parameters may partly account for the discrepant observations. The influence of process parameters has been demonstrated by Khan et al.[72] in studying wear resistance of dc-sputtered carbon overcoat on CoCr disks. By varying the sputtering parameters (in an unspecified manner), these authors have prepared carbon overcoats with widely different grain size and grain-size distribution, work-function distribu-tion, and degree of sp^3 bonding. Franco et al.[73] have shown that the corrosion protection of dc-sputtered carbon increases with a decrease in deposition pressure and an increase of power, presumably following an increase in the density of the overcoat.

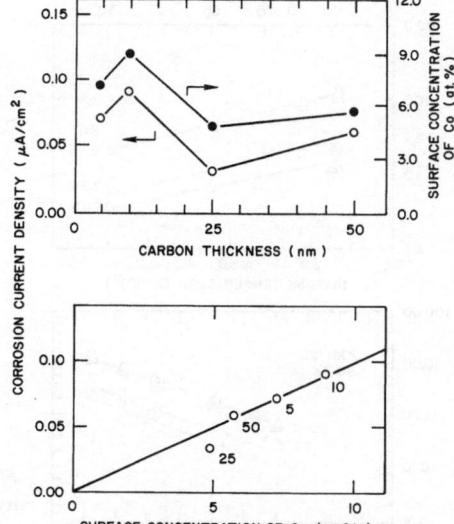

FIGURE 8. Comparison of the corrosion current densities obtained with magnetic disks in solution and surface concentration of Co as a function of carbon overcoat thickness determined on similar disks exposed to the T/H tests. Lower curve shows relationship between the corrosion current and the surface concentration of Co. (Reprinted from Ref. 48, courtesy of the Electrochemical Society.)

3.5. Life Projections

The ultimate goal of corrosion testing is to predict the useful life of the product. Direct experimental evaluation of acceleration factors has been sparse. The estimate of Sharma et al.[46] has also been used by Fisher et al.[58] Novotny et al.[70] attempted to determine experimental acceleration factors by monitoring the accumulation of CoCr corrosion products on the carbon surface. The data follow the relationship $\Delta C = a e^{\Delta RH/22} e^{-\Delta G^*/RT} t^{1/2}$, where ΔC is the amount of Co over carbon increasing with time (t), relative humidity (RH), and temperature (T); and ΔG^* is the thermal activation energy. The reaction rate was parabolic with time. Effects of temperature and humidity are illustrated in Fig. 10. At a given humidity, the Co concentration increases with T with a ΔG^* of about 0.4 eV. At a given temperature, the corrosion rate increases by about an order of magnitude as the humidity increases from 30 to 90%.[70] Using the data of Novotny et al.,[70] the acceleration factor for 90°C/90% RH with respect to 30°C/40% RH is calculated to be 150.

The values of the acceleration factors may vary from film to film. Tagami and Hayashida[74] found that the thermal activation energy, depending on the conditions of sputtering CoCr, varied from 0.07 to 0.3 eV. Such low thermal acceleration factors would invalidate "accelerated tests" performed at low temperature and humidity for a relatively short time (e.g. two weeks).

Long-term exposure to air containing pollutants is yet more difficult to simulate. Volpe[41] has achieved acceleration by forcing contaminated air through the file breathing hole at 100 times the leak rate. If the flow ratio is taken to be the acceleration factor, the test simulated

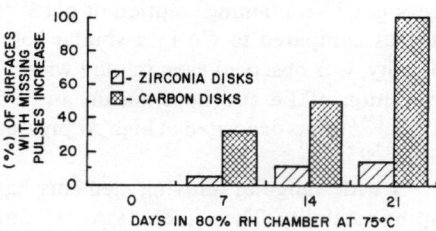

FIGURE 9. Comparison of defect densities on zirconia and carbon disks after corrosion tests. Missing pulse ≥1 at 67% clip level. (Reprinted from Ref. 71, with the permission of the IEEE.)

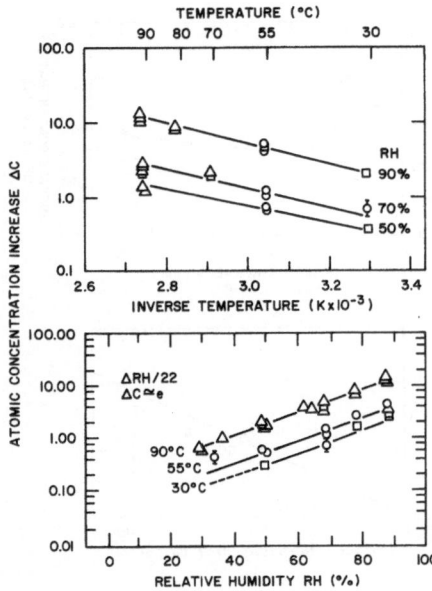

FIGURE 10. Effects of temperature and relative humidity on the amount of Co detected above the carbon layer after 72 hrs. (Reprinted from Ref. 70, with the permission of the IEEE.)

three years of machine life. Similar tests with chemical filters[42] resulted in undetectable corrosion.

4. CORROSION OF MAGNETO-OPTICAL DISKS

4.1. FeTb and Its Alloys

An outstanding deficiency of RE-TM alloys is their lack of thermodynamic stability: the amorphous alloys are metastable and relax to the polycrystalline state, and the presence of an easily oxidized rare-earth element (often terbium) results in instability with respect to oxidation by dielectric materials, such as SiO_2, and by O_2 and H_2 in the environment. In general, structural relaxation and solid-state dielectric interactions are slow processes (with activation energies on the order of 2 eV), and degradation by these processes can be ignored during the expected lifetime of stored data (typically 15 years). The corrosion processes are not so benign: RE-TM alloys react spontaneously with water, undergoing pitting and other corrosion reactions[75] that effectively destroy the media.

The corrodibility of RE-TM alloys can be predicted without recourse to experiment. The potential–pH diagram of a typical rare earth[76] shows a very negative metal dissolution potential (−2.3 V), with insoluble oxyhydroxides forming only above pH 8. Thus, rare-earth-containing alloys do not benefit from the passivation processes that stabilize other reactive elements such as aluminum or chromium. The first electrochemical measurements of FeTb alloys in Cl^--containing solution of pH 3[77] showed about 10 times faster dissolution of this alloy as compared to Co in a similar solution.[54] Pitting, probably caused by a localized impurity, was observed even in pure water. The pitting mechanism was shown to be selective dissolution of Fe and its diffusion away from the pits, which are filled with residual Tb oxides.[75] Films deposited at high Ar pressure and high bias voltage are particularly corrosion-prone.[78]

A wide range of alloying elements has been added to RE-TM alloys in an attempt to improve stability. These include Al, Ti, and Pt,[79,80] Au,[79] Be,[81] In,[82] and Co, Cr, Mn, and Ni.[80] Many of these elements enhance the RE-TM passivation. For example, In increases the thermal activation energy from about 0.3 eV for FeTb alone to 1.3 eV for the ternary alloy. Practically all of the addition elements reduce pitting. Platinum additions up to 12% Pt inhibit pitting and reduce the overall corrosion rate by a factor of 2.5.[83] Platinum increases the nobility of the alloy and most likely enhances Fe passivation. The mechanism responsible for the increased stability, however, is not fully understood.

4.2. Substrates and Overcoats

The corrosion protection of magneto-optic disks starts with the choice of the substrate, which is typically glass or a polymer. Environmental stability is better for glass; economic considerations favor polymers. Much activity has centered on developing materials that have acceptable optical characteristics, such as transmission and birefringence, and that have low solubility and diffusion coefficients for water. Polycarbonates are the best compromise today.

Further protection of the medium layer is provided through judicious selection of a protective dielectric coating and by addition of a metal capping layer. A wide range of materials have been examined: these include SiO_2, Al_2O_3, TiO_2, and Si_3N_4,[84] SiO_2 codeposited with Tb to trap the active O_2 in the oxide,[85] layers of Tb/SiO_2,[86] Al,[87] and amorphous C.[88] Adequate silicon and aluminum nitride coatings are said to be produced by rf magnetron sputtering.[77] An electrochemical test was interpreted as showing that AlN has the higher chemical stability of the two,[77] but the test actually indicates only that AlN provides better coverage than SiO_2. Coverage and other relevant properties are a function of the deposition parameters and are often not optimized.

4.3. Projected Life

There are no substantiated data on the long-term stability of RE-TM-based optical storage media. Accelerated testing at high temperature and humidity yields activation energies of 0.3, 1, and 1.3 eV for FeTb, FeTbCo, and FeTbIn, respectively. Thus, at least for some alloys, acceleration factors lead to predicted lifetimes in excess of ten years.[89,90] The estimated survival of MO films embedded in coevaporated Tb-SiO_2 layers is more than 20 years at 40°C and 90% relative humidity.[85]

5. FINAL REMARKS

The drive to find improved media for recording technology has uncovered new alloys. Electrochemistry has played a seminal role in this process. Fabrication of the first thin-film disks was by electroless deposition, which has opened the door to a much better understanding of the correlation between the magnetic properties and film structure. The new magnetic film media are thermodynamically unstable and continuously challenge electrochemical ingenuity in the fight for long life. The overall corrosion picture for the thin-film media is one of guarded optimism. In the case of thin-film disks, the vulnerability to corrosion is lessened by the file environment, which is protective and can be made totally safe. The durability of the MO media depends entirely on the protectiveness of the substrate and overcoats. Despite optimistic interpretations of the available results, however, there is an understandable drive to develop a corrosion-free disk, perhaps incorporating γ-Fe_2O_3 as a thin film,[91] and to provide impervious coats for the MO layers.

REFERENCES

1. J. K. Howard, *J. Vac. Sci. Technol.* **4**, 1 (1986).
2. M. H. Kryder, *J. Appl. Phys.* **57**, 3913 (1985).
3. D. A. O'Sullivan, *Chem. Eng. News* **67**(13), 7 (1989).
4. D. W. Rice, R. J. Cappell, W. Kinsolving, and J. J. Laskowski, *J. Electrochem. Soc.* **127**, 891 (1980).
5. V. Brusic, J. Horkans, and D. J. Barclay, *Adv. Electrochem. Sci. Eng.* **1**, 249 (1990).
6. J. C. Mallinson, *IEEE Trans. Magn.* **MAG-21**, 1217 (1985).
7. G. Bate, in: *Ferromagnetic Materials*, vol. 2 (E. P. Wohlfarth, ed.), p. 381, North-Holland, New York (1980).
8. E. Koester and T. C. Arnoldussen, in: *Magnetic Recording, Vol. I: Technology* (C. D. Mee and E. D. Daniel, eds.), p. 98, McGraw-Hill, New York (1987).
9. A. H. Eltoukhy, *J. Vac. Sci. Technol.*, *A* **4**, 539 (1986).
10. G. Bate, *J. Appl. Phys.* **37**, 1164 (1966).
11. T. C. Arnoldussen and E.-M. Rossi, *Annu. Rev. Mater. Sci.* **15**, 379 (1985).
12. E. R. C. Johns, paper no. 340 presented at NACE Corrosion Meeting, New Orleans, April 1989.
13. J. D. Livingston, *J. Appl. Phys.* **52**, 2544 (1981).
14. M. Mirzamaani, L. Romankiw, C. McGrath, J. Mahlke, and N. C. Anderson, *J. Electrochem. Soc.* **135**, 2813 (1988).
15. D. DiMilia, J. Horkans, C. McGrath, M. Mirzamaani, and G. Scilla, *J. Electrochem. Soc.* **135**, 2817 (1988).
16. A. W. Simpson and D. R. Brambley, *Phys. Status Solidi B* **43**, 291 (1971).
17. S. L. Chow, N. E. Hedgecock, M. Schlesinger, and J. Rezek, *J. Electrochem. Soc.* **119**, 1614 (1972).
18. P. Cavallotti and G. Salvago, in: *Proceedings of Symposium on Electrodeposition Technology* (L. T. Romankiw and D. R. Turner, eds.), p. 327, The Electrochemical Society, Pennington, New Jersey (1987).
19. R. O. Cortijo and M. Schlesinger, *J. Electrochem. Soc.* **131**, 2800 (1984).
20. Tu Chen, D. A. Rogowski, and R. H. White, *J. Appl. Phys.* **49**, 1816 (1978).
21. E. L. Nicholson and M. R. Khan, *J. Electrochem. Soc.* **133**, 2342 (1986).
22. J. S. Judge, J. R. Morrison, D. E. Speliotis, and J. R. DePew, *Plating* **54**, 533 (1967).
23. G. W. Lawless and R. D. Fisher, *Plating* **54**, 709 (1967).
24. H. Matsuda and O. Takano, *J. Jpn. Inst. Met.* **52**, 414 (1988).
25. M. Soraya, *Plating* **54**, 549 (1967).
26. Y. H. Chang, C. C. Lin, M. P. Hung, and T. S. Chin, *J. Electrochem. Soc.* **133**, 985 (1986).
27. K. Matsui, T. Suzuki, T. Yamade, S. Maruno, and T. Kawaguchi, *Plating* **37**, 36 (1950).
28. J. S. Sallo and J. M. Carr, *J. Electrochem. Soc.* **109**, 1040 (1962).
29. G. W. Reimherr, NBS Technical Note 247 (1964).
30. J. S. Sallo and K. H. Olsen, *J. Appl. Phys. Suppl.* **32**, 203S (1961).
31. J. S. Sallo and J. M. Carr, *J. Appl. Phys.* **33**, 1316 (1962).
32. S. Sallo and J. M. Carr, *J. Appl. Phys.* **34**, 1309 (1963).
33. A. Tago, T. Masuda, and T. Taketa, *Rev. Electr. Commun. Lab. of NTT (Tokyo)* **25**, 1315 (1977).
34. T. H. Bonn and D. C. Wendell, U.S. Patent 2,644,787 (1953).
35. D. Pearce, D. Rice, and G. Tang, *Solid State Technol.* **31**(11), 113 (1988).
36. F. E. Luborsky, *IEEE Trans. Magn.* **MAG-7**, 502 (1970).
37. V. V. Bondar', M. M. Mel'nikova, and Yu. M. Polukarov, *Proc. Met.* **1**, 467 (1965).
38. D. J. Barclay and W. M. Morgan, *IBM Tech. Discl. Bull.* **08-76**, 1098 (1976).
39. D. J. Barclay, D. S. Mansbridge, W. M. Morgan, and C. T. Prowting, *IBM Tech. Discl. Bull.* **04-74**, 3769 (1974).
40. D. J. Barclay and W. M. Morgan, *IBM Tech. Discl. Bull.* **11-74**, 1591 (1974).
41. L. Volpe, Proceedings of the Symposium on Corrosion Effects of Acid Deposition and Corrosion of Electronic Materials, 168th Meeting of the Electrochemical Society, Las Vegas, Nevada, 1985, pp. 379–386.
42. C. K. Day, C. S. Harkins, S. P. Howe, and P. Poorman, *Hewlett Packard J.* **36**, 25 (1985).
43. P. B. P. Phipps and D. W. Rice, in: *Corrosion Chemistry* (G. R. Brubaker and P. B. P. Phipps, eds.), p. 235, American Chemical Society, Washington, D.C. (1979).

44. N. D. Tomashov, *Theory of Corrosion and Protection of Metals* (translated and edited by B. H. Tytell, I. Geld, and H. Preiser), MacMillan, New York (1966).

45. J. M. West, *Basic Corrosion and Oxidation*, John Wiley & Sons, New York (1980).

46. S. P. Sharma, J. H. Thomas III, and F. E. Bader, *J. Electrochem. Soc.* **125**, 2002, 2005 (1978).

47. V. Brusic, J. A. Aboaf, R. D. McInnes, J. W. An, and D. Rice, Paper 15 presented at the Spring Meeting of the Electrochemical Society, St. Louis, Missouri, 1980.

48. V. Novotny and N. Staud, *J. Electrochem. Soc.* **135**, 2931 (1988).

49. R. G. Walmsley, B. R. Natarajan, and D. Wong, *IEEE Trans. Magn.* **MAG-24**, 3000 (1988).

50. V. Brusic, M. Russak, R. Schad, G. Frankel, A. Selius, D. DiMilia, and D. Edmonson, *J. Electrochem. Soc.* **136**, 42 (1989).

51. R. R. Dubin, K. D. Winn, L. P. Davis, and R. A. Cutler, *J. Appl. Phys.* **53**, 2579 (1982).

52. J. S. Judge, J. R. Morrison, D. E. Speliotis, and G. Bate, *J. Electrochem. Soc.* **112**, 681 (1965).

53. M. A. Helfand, C. R. Clayton, and R. B. Diegle, Paper No. 208 presented at the Honolulu Meeting of the Eletrochemical Society, October, 1987.

54. J-B. Ju and W. Smyrl, Proceedings of the ASM 3rd International Conference on Electronic Packing (M. E. Nicholson, ed.), p. 119, Minneapolis, April, 1987.

55. M. Smallen, P. B. Mee, A. Ahmad, W. Freitag, and L. Nanis, *IEEE Trans. Magn.* **MAG-21**, 1530 (1985).

56. R. Sugita, T. Kunieda, and F. Kobayashi, *IEEE Trans. Magn.* **MAG-17**, 3172 (1981).

57. T. Yamada, N. Tani, M. Ishikawa, Y. Ota, K. Nakamura, and A. Itoh, *IEEE Trans. Magn.* **MAG-21**, 1429 (1985).

58. R. D. Fisher, J. C. Allan, and J. I. Pressesky, *IEEE Trans. Magn.* **MAG-22**, 352 (1986).

59. Y. Shiroishi, Y. Sugita, M. Aiharo, H. Fukui, and T. Kobayashi (Hitachi), private communication.

60. T. G. Wang and G. W. Warren, *IEEE Trans. Magn.* **MAG-22**, 340 (1986).

61. G. L. McIntire and C. F. Brucker, *IEEE Trans. Magn.* **MAG-24**, 2221 (1988).

62. M. Yanagisawa, N. Shiota, H. Yamaguchi, and Y. Suganuma, *IEEE Trans. Magn.* **MAG-19**, 1638 (1983).

63. Y. Hoshi, M. Matsuoka, and M. Naoe, *J. Appl. Phys.* **57**, 4022 (1985).

64. S. Hattori, A. Tago, and O. Ishii, *Rev. Electr. Commun. Lab.* (*Japan*) **55**, 2254 (1984).

65. M. Garrison, *IEEE Trans. Magn.* **MAG-19**, 1683 (1983).

66. E. M. Rossi, G. McDonough, A. Tietze, T. Arnoldussen, A. Brunsch, S. Doss, M. Henneberg, F. Lin, R. Lyn, A. Ting, and G. Trippel, *J. Appl. Phys.* **55**, 2254 (1984).

67. S. K. Doss and G. A. Condas, *Metall. Trans.* **18A**, 158 (1987).

68. M. Nagao, K. Sano, M. Kojima, H. Iwasaki, A. Nahara, and T. Kitamoto, Digest of the Intermag. Conference, Tokyo, Abstract BG-14, April 1987.

69. J. Black, P. Oppenheimer, and D. M. Morris, *J. Biomed. Mater. Res.* **21**, 1213 (1987).

70. V. Novotny, G. Itnyre, A. Homola, and L. Franco, *IEEE Trans. Magn.* **MAG-23**, 3645 (1987).

71. T. Yamashita, G. L. Chen, J. Shir, and T. Chen, *IEEE Trans. Magn.* **MAG-24**, 2629 (1988).

72. M. R. Khan, N. Heiman, R. D. Fisher, S. Smith, M. Smallen, G. F. Hughes, K. Viers, B. Marchon, D. F. Ogletree, M. Salmeron, and W. Siekhaus, *IEEE Trans. Magn.* **MAG-24**, 2647 (1988).

73. L. Franco, M. M. Chen, G. Castillo, G. Gorman, L. Viswanathan, and J. Duran, paper presented at AVC, 36th National Symposium and Topical Conference, Boston, Massachusetts, October 23–27, 1989.

74. K. Tagami and H. Hayashida, *IEEE Trans. Magn.* **MAG-23**, 3648 (1987).

75. M. M. Farrow and E. E. Marinero, *J. Electrochem. Soc.* **137**, 808 (1990).

76. M. Pourbaix, *Atalas of Electrochemical Equilibria in Aqueous Solution*, Pergamon Press, Oxford (1974).

77. T. K. Hatwar, S. C. Chin, and D. D. Stinson, *IEEE Trans. Magn.* **MAG-22**, 946 (1986).

78. F. E. Luborsky, *IEEE Trans. Magn.* **MAG-22**, 937 (1986).

79. S. Tanaka, *J. Appl. Phys. Jpn.* **24**, L375 (1985).

80. N. Imamura, S. Tanaka, F. Tanaka, and Y. Nagao, *IEEE Trans. Magn.* **MAG-21**, 1607 (1985).

81. T. Fujii, T. Tokushima, and N. Horiai, Digest of the Intermag. Conference, Tokyo, April 1987, Abstract CG-08.

82. T. Iijima, Digest of the Intermag. Conference, Tokyo, April 1987, Abstract CG-12.

83. Y. Nagao, S. Tanaka, F. Tanaka, and N. Imamura, Digest of the Intermag. Conference, Pheonix, April, 1986, Abstract FC-7.

84. M. Sato, in: *Optical Mass Data Storage, Proc. SPIE* **529**, 33 (1985).

85. M. Miyazaki, I. Shibata, S. Okada, K. Itoh, and S. Ogawa, *J. Appl. Phys.* **61**, 3226 (1987).
86. S. Okada, M. Miyazaki, I. Ishibata, K. Naito, K. Itoh, and S. Ogawa, Digest of the Intermag. Conference, Tokyo, April 1987, Abstract DB-05.
87. C. D. Wright, *IEEE Trans. Magn.* **MAG-23,** 162 (1987).
88. C. J. Robinson, Topical Meeting on Optical Data Storage, Paper ThC4, 1987.
89. T. Iijima, *Appl. Phys. Lett.* **50,** 1835 (1987).
90. R. P. Freese, R. N. Gardner, T. A. Rinehart, D. W. Slitari, and L. H. Johnson, in: *Optical Mass Data Storage, Proc. SPIE* **529,** 6 (1985).
91. S. Futami, K. Kawata, and Y. Okamura, in: *Electrochemical Technology in Electronics* (L. T. Romankiw and T. Osaka, eds.), p. 425, The Electrochemical Society, Pennington, New Jersey (1988).

35

Characterization of Sulfide Mineral Surfaces

R. Woods

1. INTRODUCTION

The metal sulfides constitute the most important group of ore minerals since they are the source of the world's supplies of nonferrous metals. Sulfide minerals are, in general, electronic conductors and can sustain coupled anodic and cathodic reactions at the interface with an electrolyte. The significance of conductivity in determining the properties of sulfides has been recognized for some considerable time. Fox, commenting in 1830 on the properties of metalliferous veins in mines in Cornwall, England, that contained iron, copper, and lead sulfides, stated[1] "many of the phenomena...bear striking analogies to common galvanic combinations and the discovery of electricity in veins seems to complete the resemblance."

Corrosion-type reactions play an important role in a number of processes of practical importance in the winning of metals from their ores. Mixed-potential systems are evident in the alteration of the composition of ore bodies in the ground; in oxidation during mining, storage, and transport; in concentration of the valuable minerals by flotation methods; and in hydrometallurgical extraction of the metal values. Thus, in order to understand these processes, a knowledge of the electrochemical and surface properties of sulfide minerals is required.

A number of techniques have been employed for the characterization of sulfide mineral surfaces. In this chapter, the application of these techniques to the understanding of froth flotation will be reviewed. The flotation process is used to extract the valuable components of an ore with a view to producing essentially monomineral concentrates of a grade suitable for feeding to pyrometallurgical or hydrometallurgical operations. Flotation involves crushing the ore to liberate separate grains of the various valuable minerals and gangue components, pulping the ore particles with water, and then selectively rendering hydrophobic in turn the surfaces of the minerals of interest by interaction with organic collector species. A stream of air bubbles is then passed through the pulp; the bubbles attach to and levitate the hydrophobic particles and collect them in a froth layer that flows over the weir of the flotation cell. It is now generally accepted that the interaction of thiol collectors with sulfide minerals involves anodic oxidation of the collector coupled with the reduction of oxygen. The mixed-potential nature of collector/mineral interaction has been demonstrated experimentally, in particular for the galena/xanthate system.[2-5]

R. Woods ● CSIRO Division of Mineral Products, Port Melbourne, Victoria 3207, Australia.

Electrochemistry in Transition, edited by Oliver J. Murphy *et al.* Plenum Press, New York, 1992.

2. CHARACTERIZATION TECHNIQUES

2.1. Voltammetry

Linear potential sweep voltammetry (LPSV) has proved to be a valuable technique for characterizing the surfaces of sulfide minerals. It has been applied to identifying the products of surface oxidation by a number of authors.[6-11] Surface oxidation is important since it can influence flotation characteristics. Excessive oxidation can inhibit flotation through interfering with the interaction of the mineral with collectors, while the products of mild oxidation can result in flotation being induced in the absence of collectors.[12]

In the LPSV technique, the extent of reaction and the identity of the products can be derived from the characteristics of the voltammogram. The reaction is identified from the potential at which it takes place, the influence of pH on this potential, the reactions of the products on the reverse scan, and the effect of rotating the electrode to disperse soluble products. The products of oxidation of a range of sulfide minerals in alkaline solutions, determined in this way, are presented in Table 1.[13]

Note that the thermodynamically stable product of the oxidation of all metal sulfides is sulfate, but the formation of this species requires a high overpotential, and, in general, metastable species are produced. Correlation of the above data with the efficiency of collector-less flotation[12] indicated that this phenomenon was related to the formation of a surface sulfur species listed as "S" in Table 1. Later, X-ray photoelectron spectroscopic (XPS) studies (see below) demonstrated that the initial products could not be identified as elemental sulfur but rather as metal-deficient sulfides that have sulfur-rich surfaces.

In practical flotation circuits, interaction of collectors with mineral surfaces is responsible for inducing floatability. The coverage of collector species is generally limited to the sub-monolayer level and such quantities are readily amenable to study by LPSV. This technique has been used to investigate the xanthate/galena (PbS),[14-17] and xanthate/chalcocite (Cu$_2$S),[18-20] dithiophosphate/chalcocite,[21] and xanthate/pyrite (FeS$_2$)[11,22] systems. In the former three cases, the charge transfer chemisorption of a monolayer of collector occurs followed by the development of a metal thiol component involving the metal component of the sulfide. In the case of pyrite, anodic oxidation of the collector results in the formation of the disulfide of the collector.

TABLE 1
Products of Anodic Oxidation of Sulfide Minerals in Basic Solutions[a]

Mineral	Products			
	Oxide	Sulfide	Major sulfur	Minor sulfur
Fe$_{1-x}$S	Fe(OH)$_3$	—	S	SO$_4^{2-}$
FeS$_2$	Fe(OH)$_3$	—	SO$_4^{2-}$	S
Cu$_5$FeS$_4$	Fe(OH)$_3$	Cu$_5$S$_4$	—	—
CuFeS$_2$	Fe(OH)$_3$	CuS	S	—
Cu$_2$S	Cu(OH)$_2$	Cu$_{2-x}$S	—	—
CuS	Cu(OH)$_2$	—	S	—
PbS	Pb(OH)$_2$	—	S	S$_2$O$_3^{2-}$
(Fe, Ni)$_9$S$_8$	Fe(OH)$_3$	n.i.[b]	S	SO$_4^{2-}$

[a] From Ref. 13.
[b] n.i. = Not identified.

Modifying reagents are used in the flotation process to achieve selectivity. Depressants stop a mineral from floating while activators induce flotation. The action of such reagents has also been studied by LPSV. Depressant action of alkali and sulfide on pyrite,[22] alkali at arsenopyrite (FeAsS),[11] and cyanide on chalcocite[23] has been shown to involve the introduction of an anodic reaction at the mineral surface more favorable than collector oxidation. This shifts the mixed potential to values below that at which the mineral can interact with collector.

2.2. Hydrophobicity and Floatability

Electrochemical studies can identify the species formed at the mineral surface under a range of conditions. From the flotation point of view, it is important to know how each of these species influences the hydrophilic/hydrophobic nature of the mineral surface, and hence floatability. Techniques have been developed to obtain this information based on measurement of the contact angle at the three-phase boundary between mineral, solution, and gas bubble and of the flotation recovery of mineral particles, while maintaining control of the potential across the mineral/solution interface.

Contact angle measurements at galena surfaces in the presence of ethyl, propyl, and butyl xanthates demonstrated that the development of hydrophobicity coincided with anodic oxidation to chemisorbed xanthate.[15] Investigations with a galena particulate bed electrode demonstrated that flotation occurred in the chemisorption potential region. Analogous studies with pyrite[15] showed that the development of hydrophobicity and of floatability was related to the deposition of dixanthogen on the mineral surface.

More recent studies on the ethyl xanthate/galena system[24,25] have shown that the relationship between flotation and potential depends on the pretreatment of the mineral. Figure 1a shows the flotation response observed for galena particles of conventional flotation size as a function of the potential, which was controlled by the addition of the redox reagents dithionite and hypochlorite. The recovery for mineral ground in a reducing environment corresponds to the potential region in which the anodic peak due to the chemisorption of xanthate is observed. For mineral ground under conditions that allow oxidation to occur, the onset of flotation occurs at the potential at which an anodic current was observed on a preoxidized galena surface. This current is due to the formation of lead ethyl xanthate from

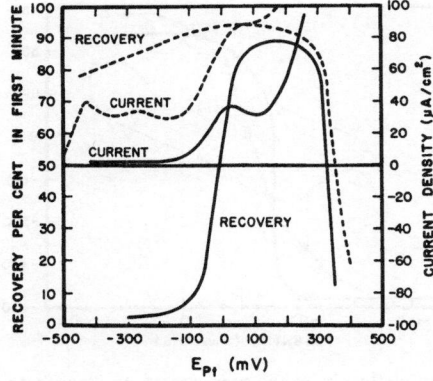

FIGURE 1. Anodic current and flotation response of galena in the presence of ethyl xanthate at pH 8. Solid line indicates preparation in reducing conditions; broken line indicates preparation in oxidizing conditions. (From Ref. 25.)

excess lead in the surface formed by loss of sulfur as thiosulfate under the preoxidation conditions. Studies of the flotation of galena in which the potential was controlled by sodium sulfide additions[26] gave results that were in close agreement with the curve in Fig. 1 corresponding to grinding under reducing conditions.

Figure 2 presents correlations between the flotation response of chalcocite in the presence of ethyl xanthate and voltammetric investigations of xanthate oxidation at a particulate bed electrode.[19] It can be seen that the flotation edge corresponds to the initial peak on the voltammogram that arises from chemisorption of xanthate. The voltammetric peak at higher potentials is due to the development of a copper (I) ethyl xanthate phase; it can be seen that this species also gives rise to a hydrophobic surface. There is good agreement between the results of different authors for the flotation response of chalcocite in the presence of ethyl xanthate for studies in which the potential was controlled by the addition of redox reagents[20,27] (solid curve in Fig. 2) and by electrochemical means[19] (dashed curve in Fig. 2).

Flotation response has also been determined as a function of potential for bornite (Cu_5FeS_4), chalcopyrite $(CuFeS_2)$, and pyrite.[28] The flotation edge for pyrite corresponded to the potential of dixanthogen formation while those for the copper–ion sulfides occurred at lower potentials.

Studies of the influence of depressants on the potential dependence of the flotation of copper minerals have confirmed that depressant action of sulfide and related species[28] involves shifting the potential to values below the flotation edge of the particular mineral/collector system.

2.3. Rotating Ring-Disk Electrodes (RRDE)

Information on the identity of soluble species formed during reactions at mineral surfaces can be obtained by application of the rotating ring-disk electrode (RRDE) technique. This approach has recently been applied to the study of the depressant action of hydrosulfide ions on the flotation of galena, pyrite, and chalcopyrite.[30] As pointed out above, the presence of sulfide decreases the potential of the mineral to values below its flotation edge. To achieve such a shift in aerated solutions, the current due to the anodic oxidation of hydrosulfide ions must balance that arising from the cathodic reduction of oxygen. However, it has been shown[31] that the oxidation of hydrosulfide ions at sulfide mineral surfaces can induce flotation through the deposition of a hydrophobic sulfur layer. The oxidation of hydrosulfide

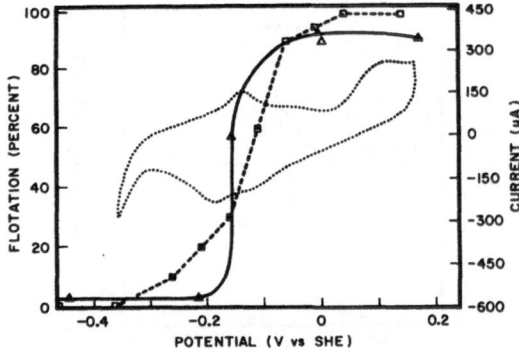

FIGURE 2. Flotation response of chalcocite in the presence of ethyl xanthate at pH 9.2. (From Ref. 19.) The potential was controlled by addition of redox reagents (——) or by electrochemical means (- - -). Voltammogram is for xanthate oxidation at a particulate bed electrode.

ion at a chalcopyrite disk electrode, and the corresponding current recorded at a gold ring held at low potentials, is shown in Fig. 3.

The ring current shows that a soluble intermediate is formed in the oxidation of hydrosulfide ion to elemental sulfur, and in the reverse process. The intermediate has been identified as polysulfide ions, the stoichiometry of which depends on the pH and hydrosulfide ion concentration. Open-circuit potential measurements show that the potential of pyrite or chalcopyrite surfaces lies in the region of polysulfide formation at high hydrosulfide concentrations. Hydrosulfide ion will act as a depressant in the former and as a collector in the latter situation. Nitrogen is being introduced as the carrier gas in some flotation plants to decrease loss of hydrosulfide ions through reaction with oxygen, and this procedure will have the additional effect of extending the concentration range in which depression occurs.

2.4. Photoelectrochemistry

Sulfide minerals are semiconductors, and hence additional information on reactions at their surfaces can be obtained from the influence of irradiation on electrochemical behavior. Studies of phenomena excited by illumination that are relevant to flotation have been confined to galena, although there is a considerable literature relating to the photoelectrochemical properties of a range of metal sulfides and other chalcogenides in relation to solar energy conversion.

It has been considered for some time[2] that the stoichiometry of a galena surface would change with potential from lead-rich, n-type at low potentials to sulfur-rich, p-type at high potentials. Recent photoelectrochemical studies[32] have demonstrated this change experimentally. Figure 4 shows voltammograms for a galena electrode in sodium tetraborate solution (pH 9.2) together with the in-phase component of the photocurrent. The peaks labeled A, B, C, and D have been identified[8] as the deposition of a monolayer of a lead–oxygen species and sulfur, the deposition of these species in multilayer quantities together with some thiosulfate ion formation, the reduction of the surface species to re-form lead sulfide, and the reduction of the excess of the lead–oxygen species to lead metal, respectively. The excess of the lead–oxygen species arises from loss of sulfur as thiosulfate. The lead–oxygen species will be lead borate in this medium,[32] while the initial sulfur product will be a metal-deficient sulfide (see below). In the initial region of the potential scan from the flat-band potential,

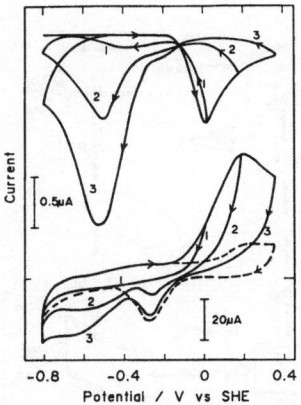

FIGURE 3. Voltammograms for a chalcopyrite disk electrode (lower curve) and corresponding ring current (upper curve) in 10^{-3} M HS$^-$ at pH 9.2. Sweep rate, 10 mV/s. (From Ref. 29.)

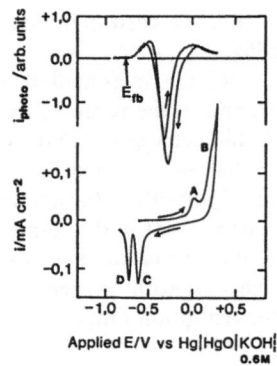

FIGURE 4. Voltammogram and photocurrent for galena. Scan rate, 20 mV/s. (From Ref. 31.) See text for identification of peaks A–D.

E_{fb}, there is only a small current on the voltammogram. However, the photocurrent reveals that the surface changes from its initial n-type condition (positive photocurrent) to p-type (negative photocurrent) as the potential is scanned through this region.

The adsorption of ethyl xanthate has also been studied by the application of photo-electrochemical techniques.[16] In these studies a range of galena samples from highly n- to highly p-type were investigated. Voltammograms for these galena samples in the presence of ethyl xanthate were identical, although the initial open-circuit potential, the flat-band potential, varied by 0.4 V over the range of materials studied. This supports the conclusion that the surface composition is a function of the potential and is the same irrespective of the bulk semiconductor type. Band diagrams are presented in Fig. 5 and show the flat-band potentials (V_{fbp} in this figure) for highly n- and highly p-type lead sulfide in relation to xanthate chemisorption/desorption.

2.5. X-Ray Photoelectron Spectroscopy (XPS)

Electrochemical characterization of surfaces lacks the molecular specificity required to give unequivocal identifiaction of species formed on electrode surfaces. For this reason, electrochemists are augmenting such approaches with a variety of nontraditional techniques

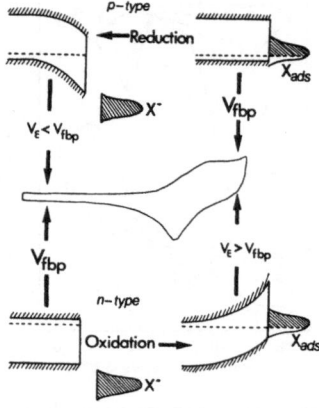

FIGURE 5. Band diagrams for lead sulfide electrodes of different conductivity type. (From Ref. 16.)

that provide information on the elemental and molecular composition, the atomic geometry, and the electronic structure of the interface. These include a number of *in situ* and *ex situ* spectroscopies.[33]

Electron spectroscopies have proved to be particularly powerful for the study of the solid/gas interface. These methods are finding an increasing application for the study of species formed at the solid/electrolyte interface, and there is a growing literature on their application to mineral processing problems. Of the various available techniques, X-ray photoelectron spectroscopy (XPS) is particularly appropriate for the study of mineral surfaces since a knowledge of the chemical environment of atoms is usually required in addition to elemental composition. Such information is important in identifying the products of surface oxidation of sulfide minerals since, for example, different sulfur-containing products need to be distinguished from each other and from the sulfur present in the mineral itself.

Studies of surface oxidation have been carried out by XPS at galena,[34-36] bornite,[35,37,38] chalcopyrite,[34,35,39] pyrrhotite ($Fe_{1-x}S$),[35,40,41] pyrite,[42] arsenopyrite,[10] sphalerite (ZnS),[34,43] and cobaltite (CoAsS).[44] Figure 6 shows some results for galena.[35,36] Curve A presents the $Pb(4f)$ and $S(2p)$ spectra for a surface after only brief exposure to air. Each doublet corresponds to a single environment and can be assigned to galena itself. Curve B shows the spectra after exposure to air for one day. The $S(2p)$ spectrum is unchanged, but

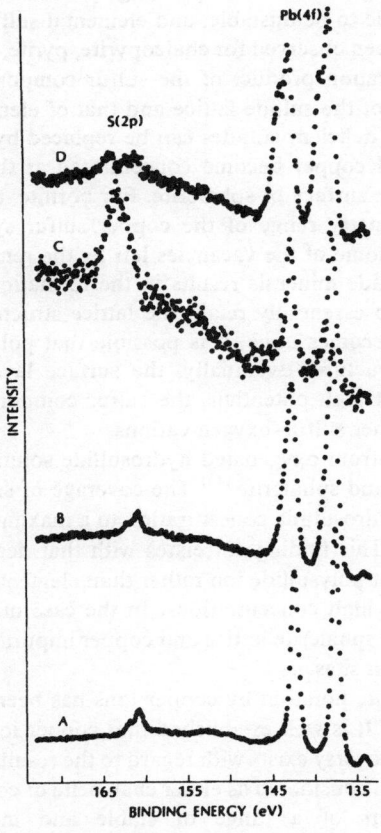

FIGURE 6. $S(2p)$ and $Pb(4f)$ photoelectron spectra for galena after various treatments: (A) after brief exposure to air; (B) after exposure to air for one day; (C) after oxidation in dilute acetic acid; (D) after treatment with hydrogen peroxide and washing with water. (From Ref. 35.)

the Pb($4f$) peaks show additional intensity on the high-binding-energy side. The shifted lead peaks were assigned to a lead-oxygen species and constituted ~40% of the lead intensity after this time. This indicates that the initial oxidation process involves removal of lead atoms from the sulfide lattice to form an overlayer of a lead hydroxyoxide. The quantity of lead removed appears to be much greater than can be accounted for by removal of the excess lead in the n-type material, since the range of nonstoichiometry in galena is only 0.1%.[45] Thus, it was concluded[36] that a metal-deficient sulfide is formed that has a stoichiometry beyond the stable range for the mineral.

Curve C of Fig. 6 shows spectra for galena after oxidation in dilute acetic acid, a medium in which oxidized lead species are soluble. The sulfur concentration is now considerably enhanced, and the S($2p$) binding energy is shifted to higher values. It was found that the shift in binding energy increased as the oxidation proceeded and the sulfur excess in the surface region increased. However, the S($2p$) peak did not appear at the binding energy for elemental sulfur for any of the surfaces treated in acetic acid. The surface became enriched in copper during oxidation in this medium. Indeed, the composition of the surface giving rise to curve C contained as much copper as lead. Curve D is for a surface after treatment with hydrogen peroxide and washing with water. In this case, there are two sulfur doublets, one corresponding to galena and the other to elemental sulfur.

These observations have been interpreted in terms of the development of series of metastable sulfides of increasing metal deficiency. At high potentials, the sulfur excess becomes sufficiently large for the sulfide to be unstable, and elemental sulfur is formed as a separate phase. Similar behavior has been observed for chalcopyrite, pyrite, pyrrhotite, and sphalerite. In each case, the initial oxidation product of the sulfur component of the mineral has a binding energy between that of the sulfide lattice and that of elemental sulfur. Some of the cation vacancies in the metal-deficient sulfides can be replaced by impurity elements in the mineral. Thus, antimony and copper become concentrated at the surface of galena, and copper and lead diffuse to the surface of sphalerite. For bornite, the initial product, Cu_5S_4, is within the stable stoichiometry range of the copper/sulfur system.[45] However, silver diffuses from the bulk to fill some of the vacancies left by the removal of iron.

Further oxidation of sulfide minerals results in the formation of more metal-deficient species which are assumed to essentially retain the lattice structure of the origin mineral. When the metal deficiency becomes large, it is possible that polysulfide linkages between sulfur atoms stabilize the structure. Eventually, the surface layer becomes unstable and elemental sulfur separates. At high potentials, the sulfur component can also be oxidized completely to sulfate or to other sulfur–oxygen cations.

The deposition of sulfur from oxygenated hydrosulfide solutions has been determined at pyrite,[46] chalcopyrite,[46] and sphalerite.[43] The coverage of sulfur on pyrite was found to increase with increase in hydrosulfide concentration to a maximum at ~10^{-2} mol dm^3 and then to fall rapidly (Fig. 7). This finding correlates with that derived from RRDE studies (see above) which showed that polysulfide ion rather than elemental sulfur is the product of hydrosulfide ion oxidation at high concentrations. In the case of sphalerite, the deposited sulfur essentially extended the sphalerite lattice and copper impurities in the mineral diffused to the surface to occupy cation sites.

The activation of sphalerite flotation by copper ions has been the subject of a number of XPS investigations.[43,47-49] It is well established that copper ions displace zinc from the surface of sphalerite, but controversy exists with regard to the resulting copper–sulfur species. The product has generally been considered as either chalcocite or covellite (CuS) even though these are only two members of a range of stable and metastable copper sulfide stoichiometries.[45] Recent studies[43] have shown that cation exchange is associated with surface oxidation, and it is suggested that the surface be considered as a metal-deficient, copper-substituted sphalerite rather than a distinct copper sulfide phase.

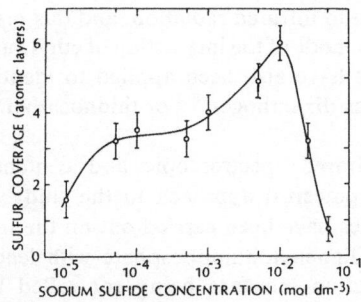

FIGURE 7. Coverage of sulfur on pyrite after exposure to aerated sodium sulfide solutions. (From Ref. 45.)

The XPS technique has also been applied to the study of the attachment of collector species to the mineral surface.[50-55] Although thiol collectors such as the xanthates contain elements that already are present on a sulfide mineral surface, viz., S, C, and O, it has proved possible to separate spectra of the adsorbate from those of the substrate. XPS investigations of the adsorption of ethyl xanthate[51] and diethyldithiophosphate[52] on chalcocite have shown that the initial reaction is the relatively rapid formation of a well-oriented monolayer of adsorbed xanthate. An overlayer of disordered copper(I) ethyl xanthate molecules then grows slowly. This mechanism is consistent with the findings of the electrochemical studies discussed above. The attachment of ethyl xanthate to sphalerite containing lead impurities was shown[54] to arise from diffusion of lead atoms to the mineral surface, where they reacted with the collector.

2.6. Infrared Spectroscopy

The interaction of electromagnetic radiation with mineral surfaces provides another means of characterizing sulfide mineral surfaces in relation to flotation properties.[56] Although such investigative techniques do not have the same elemental specificity as electron spectroscopies, they provide additional structural information and have the advantage that measurements can be performed *in situ*.

Absorption bands in the infrared region of the spectrum are valuable for studying the structure of thiol collector compounds relevant to flotation systems. Thus, it has been shown[57] that the formation of heavy-metal thiol complexes, or the disulfide of the thiol, results in a considerable reduction in the polar character of the head group. Infrared (IR) spectroscopy also has the capability of investigating the nature of collector species adsorbed at mineral surfaces. Clearly, the most valuable approach is to carry out investigations *in situ*. Early studies were carried out in this mode using the attenuated total reflectance (ATR) technique with the metal sulfide films evaporated onto the ATR reflection element. It was shown[58] that the initial monolayer of ethyl xanthate on lead sulfide differed from lead ethyl xanthate. It was suggested that this was due to the initial species having a 1:1 xanthate/lead stoichiometry whereas the bulk compound has a 2:1 stoichiometry.

Modern instrumentation applying Fourier transform (FT) methods has allowed the *in situ* study of species adsorbed on mineral particles placed adjacent to the cell window. Such FTIR-ATR investigations have confirmed the differences between the initial monolayer and the bulk compound for ethyl xanthate adsorbed on natural lead sulfide (galena)[26,59] and identified similar differences for dithiophosphate on chalcocite.[52] Similarly, the species formed when ethyl xanthate interacts with sphalerite containing lead impurities was found[54] to correspond to the monolayer form.

Zinc sulfide is transparent to infrared radiation, and this property has been exploited in an IR study in the transmission mode of the interaction of ethyl xanthate with copper-activated sphalerite.[60] Infrared studies have also been applied to identifying the bonding between minerals and collectors such as dixanthogen[61] or thionocarbamates[62] that adsorb without charge transfer.

The combination of infrared spectroscopic and contemporaneous electrochemical measurements offers a most powerful approach to the study of the electrode/electrolyte interface.[63] FTIR-ATR studies have been carried out on the interaction of ethyl xanthate, diethyldithiophosphate, and diphenyldithiophosphate with lead sulfide particles, with the potential of the mineral/solution interface being controlled by the addition of sodium sulfide.[26,59] It was shown that ethyl xanthate commenced adsorbing from 10^{-3} M xanthate solution at pH 6 at ~ -0.1 V, and a monolayer was formed at ~ -0.5 V. These values agree with those derived from voltammetric investigations[14,15] (see Fig. 1a).

Spectroelectrochemical investigations have recently been carried out on the interaction of ethyl xanthate with chalcocite, chalcopyrite, pyrite, and galena surfaces.[64] In this study, powdered mineral was embedded into a carbon paste electrode for electrochemical control of the solid/solution interface. It was found that the IR signal intensity for adsorbed xanthate as a function of potential correlated with xanthate adsorption determined voltammetrically and with measurements of the potential dependence of flotation recovery (see Fig. 2). Similar good correlations were obtained with pyrite, and it was confirmed that the product of the interaction of xanthate with this mineral is dixanthogen. Figure 8 shows the corresponding correlations obtained for chalcopyrite.[63] The IR data show that diethyl dixanthogen $(EX)_2$ is the initial product on this mineral, with copper(I) ethyl xanthate (CuEX) being also formed at higher potentials.

It can be seen from Fig. 8 that the appearance of the IR signal due to the presence of dixanthogen also occurs at the same potential as that at which an oxidation current is observed on a voltammogram. The figure also shows the flotation recovery from Ref. 28. Since the chalcopyrite particles used in that study did not float in the absence of collector, it was assumed that they were too large to display self-induced floatability. However, flotation commences at potentials below those at which the IR and voltammetric investigations indicate that xanthate becomes attached to the surface. This led to the suggestion[64] that flotation in the low-potential region is due to the removal of oxidation products that inhibit self-induced

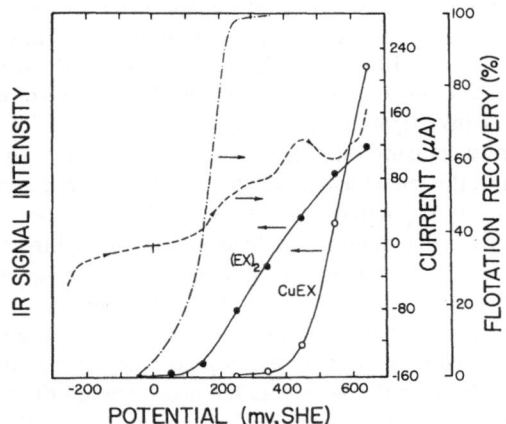

FIGURE 8. Comparison of FTIR (———), voltammetric ($M \cdot M$), and flotation (- - -) studies of the ethyl xanthate/chalcopyrite system. $(EX)_2$ and CuEX represent dixanthogen and copper (I) xanthate, respectively; solution pH 9.2. (From Ref. 63.)

flotation through reaction with xanthate. Such an explanation is consistent with studies of the flotation of sulfide ores in which it was considered[65] that the function of the collector in some situations could be one of overcoming the inhibiting effects of metal hydroxides derived from the ore as much as one of increasing the hydrophobic character of the floating mineral.

3. CONCLUSIONS

The last few decades of the 20th century have seen the establishment of a range of techniques for the characterization of the surfaces of sulfide minerals and the application of these methods to develop a detailed understanding of the individual processes that occur in practical systems for the extraction of metal values from sulfide ores. As we move into the 21st century, we will see this knowledge being applied to improving the efficacy of mineral processing methods and the development of better separation strategies.

REFERENCES

1. R. W. Fox, *Roy. Soc. (London) Phil. Trans.* **1830**, 399.
2. R. Tolun and J. A. Kitchener, *Bull. Inst. Min. Metall.* **73**, 313 (1964).
3. J. R. Gardner and R. Woods, *Aust. J. Chem.* **26**, 1635 (1973).
4. S. M. Ahmed, *Int. J. Miner. Process.* **5**, 175 (1978).
5. K. C. Pillai and J. O'M. Bockris, *J. Electrochem. Soc.* **131**, 568 (1984).
6. P. E. Richardson and E. E. Maust, Jr., in: *Flotation—A. M. Gaudin Volume* (M. C. Fuerstenau, ed.), Vol. 1, pp. 364–392, SME /AIME, New York (1976).
7. R. L. Paul, M. J. Nicol, J. W. Diggle, and A. P. Saunders, *Electrochim. Acta* **23**, 625 (1978).
8. J. R. Gardner and R. Woods, *J. Electroanal. Chem.* **100**, 447 (1979).
9. I. C. Hamilton and R. Woods, in: *Proceedings of the International Symposium on Electrochemistry in Mineral and Metal Processing* (P. E. Richardson, S. Srinivasan, and R. Woods, eds.), pp. 259–285, The Electrochemical Society, Pennington, New Jersey (1984).
10. G. W. Poling and M. J. V. Beattie, in: *Principles of Mineral Flotation* (M. H. Jones and J. T. Woodcock, eds.), pp. 137–146, AusIMM, Melbourne (1984).
11. M. Lamache, D. Bauer, and J. Pegouret, *Electrochim. Acta* **26**, 1845 (1981).
12. W. J. Trahar, in: *Principles of Flotation* (M. H. Jones and J. T. Woodcock, eds.), pp. 117–135, AusIMM, Melbourne (1984).
13. R. Woods, in: *Principles of Flotation* (M. H. Jones and J. T. Woodcock, eds.), pp. 91–115, AusIMM, Melbourne (1984).
14. R. Woods, *J. Phys. Chem.* **75**, 354 (1971).
15. J. R. Gardner and R. Woods, *Aust. J. Chem.* **30**, 981 (1977).
16. P. E. Richardson and C. S. O'Dell, *J. Electrochem. Soc.* **132**, 1350 (1985).
17. M. D. Pritzker and R. H. Yoon, in: *Proceedings of the International Symposium on Electrochemistry in Mineral and Metal Processing* (P. E. Richardson, S. Srinivasan, and R. Woods, eds.), pp. 26–53, The Electrochemical Society, Pennington, New Jersey (1984).
18. A. Kowal and A. Pomianowski, *J. Electroanal. Chem.* **46**, 411 (1973).
19. C. S. O'Dell, R. K. Dooley, G. W. Walker, and P. E. Richardson, in: *Proceedings of the International Symposium on Electrochemistry in Mineral and Metal Processing* (P. E. Richardson, S. Srinivasan, and R. Woods, eds.), pp. 81–95, The Electrochemical Society, Pennington, New Jersey (1984).
20. C. Basilio, M. D. Pritzker, and R. H. Yoon, SME/AIME Annual Meeting, New York, 1985, Preprint No. 85–86.
21. S. Chander and D. W. Fuerstenau, *J. Electroanal. Chem.* **56**, 217 (1974).
22. N. D. Janetski, S. I. Woodburn, and R. Woods, *Int. J. Miner. Process.* **4**, 227 (1974).
23. S. Castro and J. Larrondo, *J. Electroanal. Chem.* **118**, 317 (1981).

24. P. J. Guy and W. J. Trahar, *Int. J. Miner. Process.* **12,** 15 (1985).
25. P. J. Guy and W. J. Trahar, in: *Flotation of Sulphide Minerals* (K. S. E. Forssberg, ed.), pp. 91–110, Elsevier, Amsterdam (1985).
26. J. Leppinen, On the Interaction of Thiol Collector Ions and Lead Sulfide Surface, University of Turku (1986).
27. G. W. Heyes and W. J. Trahar, *Int. J. Miner. Process.* **4,** 317 (1977).
28. P. E. Richardson and G. W. Walker, in: *XVth International Mineral Processing Congress, Cannes, France,* Vol. II, pp. 198–210 GEDIM, St. Etienne, France (1985).
29. D. R. Nagaraj, S. S. Wang, P. V. Avotins, and E. Dowling, *Trans. Inst. Min. Metall.* **95,** C17 (1986).
30. R. Woods, D. C. Constable, and I. C. Hamilton, in: *Proceedings of the International Symposium on Electrochemistry in Mineral and Metal Processing II* (P. E. Richardson and R. Woods, eds.), pp. 113–130, The Electrochemical Society, Pennington, New Jersey (1988).
31. G. W. Heyes and W. J. Trahar, in: *Proceedings of the International Symposium on Electrochemistry in Mineral and Metal Processing* (P. E. Richardson, S. Srinivasan, and R. Woods, eds.), 219–232, The Electrochemical Society, Pennington, New Jersey (1984).
32. S. Fletcher and M. D. Horne, in: *Electrochemistry, Current and Potential Applications* (T. Tran and M. Skyllas-Kazacos, eds.), pp. 152–153, Electrochemistry Division, Royal Australian Chemical Institute, Kensington, New South Wales (1988).
33. E. Yeager, *Surf. Sci.* **101,** 1 (1980).
34. D. Brion, *Appl. Surf. Sci.* **5,** 133 (1980).
35. A. N. Buckley and R. Woods, in: *Proceedings of the International Symposium on Electrochemistry in Mineral and Metal Processing* (P. E. Richardson, S. Srinivasan, and R. Woods, eds.), pp. 286–302, The Electrochemical Society, Pennington, New Jersey (1984).
36. A. N. Buckley and R. Woods, *Appl. Surf. Sci.* **17,** 401 (1984).
37. A. N. Buckley and R. Woods, *Aust. J. Chem.* **36,** 1793 (1983).
38. A. N. Buckley, I. C. Hamilton, and R. Woods, *J. Appl. Electrochem.* **14,** 63 (1984).
39. A. N. Buckley and R. Woods, *Aust. J. Chem.* **37,** 2403 (1984).
40. A. N. Buckley and R. Woods, *Appl. Surf. Sci.* **22/23,** 280 (1985).
41. A. N. Buckley and R. Woods, *Appl. Surf. Sci.* **20,** 472 (1985).
42. A. N. Buckley and R. Woods, *Appl. Surf. Sci.* **27,** 437 (1987).
43. A. N. Buckley, R. Woods, and H. J. Wouterlood, in: *Proceedings of the International Symposium on Electrochemistry in Mineral and Metal Processing II* (P. E. Richardson and R. Woods, eds.), pp. 211–233, The Electrochemical Society, Pennington, New Jersey (1988).
44. A. N. Buckley, *Aust. J. Chem.* **40,** 231 (1987).
45. D. J. Vaughan and J. R. Craig, *Mineral Chemistry of Metal Sulfides,* Cambridge University Press, Cambridge (1978).
46. A. N. Buckley, R. Woods, and H. J. Wouterlood, *Aust. J. Chem.* **41,** 1003 (1988).
47. R. K. Clifford, K. L. Purdy, and J. D. Miller, *AIChE Symp. Ser.* **71,** 138 (1975).
48. D. L. Perry, L. Tsao, and J. A. Taylor, in: *Proceedings of the International Symposium on Electrochemistry in Mineral and Metal Processing* (P. E. Richardson, S. Srinivasan, and R. Woods, eds.), pp. 169–184, The Electrochemical Society, Pennington, New Jersey (1984).
49. V. I. Nefedov, Ya. V. Salyn, P. M. Solozhenkin, and G. Yu. Pulatov, *Surf. Interface Anal.* **2,** 170 (1984).
50. K. C. Pillai, V. Y. Young, and J. O'M. Bockris, *Appl. Surf. Sci.* **16,** 322 (1983).
51. J. Mielczarski and E. Suoninen, *Surf. Interface Anal.* **6,** 34 (1984).
52. J. Mielczarski and E. Minni, *Surf. Interface Sci.* **6,** 221 (1984).
53. K. C. Pillai, V. Y. Young, and J. O'M. Bockris, *J. Colloid Interface Sci.* **103,** 145 (1985).
54. J. Mielczarski, *Int. J. Miner. Process.* **16,** 179 (1986).
55. J. Mielczarski, *J. Colloid Interface Sci.* **120,** 201 (1987).
56. E. W. Giesekke, *Int. J. Miner. Process.* **11,** 19 (1983).
57. G. W. Poling, in: *Flotation—A. M. Gaudin Memorial Volume* (M. C. Fuerstenau, ed.), Vol. 1, pp. 334–363, SME/AIME, New York (1976).
58. J. Leja, L. H. Little, and G. W. Poling, *Trans. Inst. Min. Metall.* **72,** 414 (1963).
59. J. Leppinen and J. Mielczarski, *Int. J. Miner. Process.* **18,** 3 (1986).
60. S. C. Termes and P. E. Richardson, *Int. J. Miner. Process.* **18,** 167 (1986).

61. J. Mielczarski, *Colloids Surf.* **17**, 251 (1986).

62. J. O. Leppinen, C. I. Basilio, and R. H. Yoon, *Colloids Surf.* **32**, 113 (1988).

63. J. K. Foley, C. Korzeniewski, J. L. Daschbach, and B. S. Pons, in: *Electroanalytical Chemistry*, Vol. 14 (A. J. Bard, ed.), pp. 309–340, Marcel Dekker, New York (1986).

64. J. O. Leppinen, C. I. Basilio, and R. H. Yoon, in: *Proceedings of the International Symposium on Electrochemistry in Mineral and Metal Processing II* (P. E. Richardson and R. Woods, eds.), pp. 49–65, The Electrochemical Society, Pennington, New Jersey (1988).

65. L. K. Shannon and W. J. Trahar, in: *Advances in Mineral Processing* (P. Somasundaran, ed.), pp. 408–425, SEM/AIME, Littleton, Colorado (1986).

Electrochemical Energy Conversion

Electrode Kinetic and Electrocatalytic Aspects of Electrochemical Energy Conversion

Supramaniam Srinivasan

1. FUNDAMENTAL ASPECTS

1.1. Types of Fuel Cells

NASA's Space Programs stimulated the initiation of fuel cell research and development in the late 1950s.[1-7] The types of fuel cells that were focused on in the 1960s for space applications were the ones using solid polymer and alkaline electrolytes. The General Electric Company was responsible for developing the solid polymer electrolyte fuel cells, and United Technologies Corporation (Pratt and Whitney Division) for developing the alkaline fuel cells. The initial difficulties regarding the stability of the proton-conducting membrane, polystyrene sulfonic acid, for the solid polymer electrolyte fuel cells used in the Gemini flights and the excellent performance of the alkaline fuel cells are the reasons for the choice of the latter system as a power source for the Apollo, space shuttle, and other space flights. The fuel cell performance is best with the "pristine" reactants hydrogen and oxygen. These reactants, stored cryogenically, are the logical ones for space flights. In the 1970s, with the invention of a highly stable and conducting solid polymer electrolyte, Nafion, by the Du Pont Company, there was a breakthrough in fuel cell technology using such electrolytes. By the substitution of the polystyrene sulfonic acid with Nafion, General Electric Company showed a significant improvement in the performance of solid polymer electrolyte fuel cells. The Dow Chemical Company has developed a membrane which is more promising than Nafion in respect to conductivity and water management characteristics. These advances make the solid polymer electrolyte fuel cell a strong competitor of the alkaline fuel cell, particularly for the lunar- and Mars-based missions of NASA.

Because of the attractive features of fuel cells—high theoretical efficiency, low to medium temperature operation, independence of performance level on rated power output, no loss in performance at part load, and minimum vibrations and noise—there was also a great incentive to explore the potential for developing fuel cells for terrestrial applications.[8-11]

Supramaniam Srinivasan • Center for Electrochemical Systems and Hydrogen Research, Texas A&M University, College Station, Texas 77843-3577.

Electrochemistry in Transition, edited by Oliver J. Murphy *et al.* Plenum Press, New York, 1992.

Hydrogen is not a primary fuel, and air in place of oxygen is the more convenient and economical fuel for fuel cells for terrestrial applications. Thus, programs emerged in the 1960s for the direct oxidation of hydrocarbon fuels (natural gas and higher hydrocarbons derived from petroleum), alcohols, and even coal in fuel cells with air instead of oxygen as the oxidant. The electrolytes used in these fuel cells ranged from low-temperature acids and alkali hydroxides to high-temperature molten carbonate and solid electrolytes. Unfortunately, it was demonstrated within a relatively short time that the organic fuels have a relatively low electroactivity. This, together with the high degree of irreversibility of the oxygen reduction, resulted in the relatively poor performances of such types of fuel cells. Following the energy crisis in 1973, attention was focused on the development of phosphoric acid fuel cells for terrestrial applications (power generation and a power source for electric vehicles) using hydrogen produced by reforming and shift-converting of hydrocarbons or alcohols. For the high-temperature fuel cells with molten carbonate or solid oxide electrolytes, CO is as good a fuel as hydrogen and, thus, these gases produced by the above-mentioned reformer reaction or coal gasification were envisaged as the fuels.

The solid polymer electrolyte fuel cell using reformed methanol appears to be the most attractive candidate as a power source for electric vehicles.[12-14] However, with increasing interest in a "hydrogen economy" and the need to use noble metal (Pt) electrocatalysts in this type of fuel cell, developing alkaline fuel cells using "pristine" hydrogen for the vehicular application is the best alternative. There is renewed interest in developing direct methanol fuel cells, particularly for transportation applications. Methanol is the most electroactive organic fuel, and some progress has been made in enhancing the kinetics of methanol oxidation to carbon dioxide and water.

1.2. Important Role of Electrode Kinetics and Electrocatalysis

A "snapshot version" of the state of fuel cell technology in respect to type, performance, potential application, and problems is presented in Table 1. It is clear from this table that further progress in electrode kinetics and electrocatalysis is vitally important for development and manufacture of advanced fuel cells with the desired performance characteristics (power density, efficiency, and lifetime) for space and terrestrial applications. For the low- and medium-temperature fuel cells, a typical plot of the cell potential versus current density, which solely reflects the performance of a fuel cell in respect to power density and efficiency, is illustrated in Fig. 1. Three distinct regions—activation, ohmic, and mass transport controlled—are indicated in this plot. The predominant cause of the difficulties in attainment of high energy efficiencies and high power densities in these types of fuel cells is the low electrocatalytic activities of most electrode materials for the oxygen reduction reaction. The hydrogen electrode shows a linear relationship of its half-cell potential versus current density (up to the maximum desired current density) in the fuel cells using phosphoric acid (operating temperature, $T = 200°C$), potassium hydroxide ($T = 80°C$), or a proton-conducting membrane [Nafion or the Dow membrane ($T = 85°C$)] as the electrolyte. This is not the case with the oxygen electrode, where a semiexponential relation between its half-cell potential and current density is observed. Thus, at low current densities, the entire loss in the fuel cell potential from the reversible value is due to activation overpotential at the oxygen electrode when using hydrogen and oxygen as reactants. The cell potential (E) versus current density (i) relation (until the end of the linear region—see Fig. 1) can be expressed by an equation of the form

$$E = E_0 - b \log i - Ri \tag{1}$$

where

$$E_0 = E_r + b \log i_0 \tag{2}$$

TABLE 1
Snapshot Version of Fuel Cell Technologies and Applications

Type of fuel cell as classified by electrolyte or fuel	Highest reported performance			Highest rated power output (kW)		Electrochemical challenges	Applications
	Temperature (°C)	Efficiency (%)	Power density (mW/cm^2)	Present	Projected		
Phosphoric acid	200	42	220	4,500	10,000	Oxygen electrocatalysis, cathode corrosion	On-site integrated (electric and heat) power generation, transportation
Alkaline	80	70	4000	20	300	Hydrogen electrocatalysis, cathode corrosion	Space, transportation
Molten carbonate	650	50	150	20	10,000	Oxygen electrode, cathode corrosion	Power generation, cogeneration (electric and heat)
Solid oxide	1000	40	240	5	10,000	Expensive fabrication of component layers, electrolyte for lower temperature operation	Power generation, regenerative fuel cell
Solid polymer	90	45	300	5	10,000	Oxygen electrocatalysis, water management	Space, defense, transportation, standby power
Direct methanol	80	25	20	0.05	100	Methanol electrocatalysis, anode poisoning	Transportation, remote power, standby power

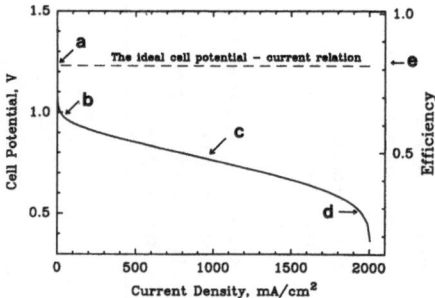

FIGURE 1. Typical plot of cell potential versus current density for fuel cells, illustration regions of control by the three types of overpotential.[7] a. Thermodynamic reversible cell potential (E); b, cell potential losses due to activation overpotential (lack of electrolysis); c. linear drop in cell potential mainly due to ohmic losses in solution between electrodes; d. mass transport losses cause of decrease in cell potential; e. intrinsic maximum efficiency.

and b and i_0 are, respectively, the Tafel slope and the exchange current density for the oxygen reduction reaction, and R accounts for the linear variations of overpotential, predominantly ohmic, with current density, which are observed in the intermediate current density range. The situation is more complex when organic fuels (hydrocarbons and alcohols) are used directly in fuel cells. The exchange current densities for these reactions are as low or even lower than those for the oxygen reduction reaction. For these fuel cells, the exponential decreases of current densities with overpotential for both the anodic and cathodic reactions account for their very poor performances.

It is worthwhile rationalizing the shape of the E versus i plot (Fig. 1) by differentiation of Eq. (1):

$$\frac{dE}{di} = -\frac{b}{i} - R \tag{3}$$

At a low current density, the differential resistance of the cell (dE/di) is high because of the first term on the right-hand side of Eq. (3). Thus, at low current densities, there is a sharp fall of cell potential with increasing current density. At higher current densities, the term (b/i) becomes small relative to R, which has a constant value. Hence, the cell potential varies linearly with current density in the intermediate range. At high current densities, mass transport limitations, due to the low rates of supply of reactants to the electrocatalytic sites or of removal of products away from these sites, may predominate and cause the sudden drop of cell potential (Fig. 1). Advances have been made in the fabrication of fuel cell electrodes with optimized structures; thus, there are hardly any mass transport limitations at current densities as high as 1 A/cm² (in the advanced alkaline and solid polymer electrolyte fuel cells, mass transport limitations have not been observed at current densities as high as 5 A/cm²).

The NASA Space Program in the 1960s and the programs sponsored by the U.S. Department of Energy, Electric Power Research Institute, and Gas Research Institute in the 1970s and 1980s provided a boost to research on the "electrochemistry of fuel cells." With support of about $100,000 per year for five years in the 1960s, pioneering work was carried out on the electrode kinetics and electrocatalysis of fuel cells by Bockris and co-workers at the University of Pennsylvania.[7] The major accomplishments were (i) elucidation of the structure of the double layer and of the effect of adsorption of species on the electrode kinetics; (ii) an understanding of the difficulty of attaining the reversible potential of the oxygen electrode reaction and of finding good electrocatalysts for this reaction; (iii) determination of the complexities of the electrooxidation of organic fuels, providing an explanation for the high overpotential for this reaction and for the poisoning phenomenon; (iv) investigation of the role of geometric and electronic factors on electrocatalysis; and (v) analysis of kinetics at porous gas diffusion electrodes. Several of these studies laid the foundation for the progress in fuel cell technology in respect to attainment of high performance and the lowering of capital and operating costs.

Several other laboratories have contributed to the significant advances in the "electrode kinetics and electrocatalysis of fuel cell reactions." These include the electrochemistry laboratories at (i) General Electric Company, (ii) United Technologies Corporation, (iii) Exxon, (iv) Shell, (v) General Motors, (vi) Siemens, (vii) Case Western Research Institute, (viii) Brookhaven National Laboratory, (ix) Lawrence Berkeley Laboratory, (x) Los Alamos National Laboratory, (xi) Texas A&M University, and (xii) Institute for French Petrole. Since the early 1960s, over 3000 original articles have been published on the contributions in this field. Excellent reviews are found in several books and journal review articles.[1-14] The highlights of the accomplishments in this field are presented in this chapter (Section 2).

1.3. Major Challenges

1.3.1. High Performing Oxygen Electrodes

As illustrated in the previous subsection, the successful development and commercialization of low- to medium-temperature fuel cells, particularly for terrestrial applications, depends on enhancing the kinetics of the electroreduction of oxygen. The major challenges in reaching this goal have been and, to some extent, still are as follows:

(i) The first challenge is to attain the reversible potential. The oxygen electrode has an exchange current density of 10^{-9} A/cm^2 or lower at low to medium temperatures. Due to competing reactions, even on the best metallic electrocatalyst (Pt), it is practically impossible to attain the reversible potential. The competing reactions are metal dissolution, oxide formation, and/or oxidation of carbon supports for electrocatalysts and of organic impurities. The net result is that the observed rest potential is a mixed potential which is lower than the reversible potential of the oxygen electrode reaction by generally as high a value as 0.2 V. This loss in cell potential has a primary effect of reducing the efficiency of fuel cells by as much as 16%. By operating at higher temperatures and pressures, the reaction becomes more reversible and this efficiency loss is somewhat reduced.

(ii) The oxygen reduction reaction is a four-electron transfer reaction, and from the point of view of attaining the highest coulombic efficiency, it is necessary to reduce oxygen to water in acid media and to hydroxyl ions in alkaline electrolytes. Complications arise due to a stable by-product which is sometimes formed in a two-electron reaction, that is, hydrogen peroxide in acids and a peroxide ion in alkaline solutions. Formation of such species reduces the coulombic efficiencies of fuel cells. This problem is not too serious at the present time because electrocatalysts have been found where the four-electron transfer reaction occurs to the extent of 100%.

(iii) The stability of oxygen electrodes during long-term (continuous or intermittent) operation of fuel cells is of concern. The region of potential in which the oxygen electrode operates in fuel cells is one in which even the most noble metals are not 100% stable in strong electrolytes. Even with platinum, which is one of the best electrocatalysts, the extent of corrosion, although slight, is high enough to cause some concern about the stability of platinum electrodes for the projected lifetime of fuel cells for terrestrial applications, which is over 40,000 h for power generation (continuous operation) and for electric vehicles (intermittent operation). In the case of electric vehicles, the operating time may be only 3000 h, and during the balance time, the fuel cell power plant will be on open circuit. The corrosion is more severe when the oxygen electrode potential is close to the open-circuit value (which is the desirable potential from an efficiency point of view). Another complication is that microcrystallite electrocatalysts, necessary for enhancing surface area and minimizing noble metal loading, are supported on conducting substrates—generally high-surface-area carbon.

Quite often the support is more corrodible than the electrocatalyst in the region of potential of interest. This problem has been greatly overcome by prior heat treatment of the carbon support for its graphitization (the graphitized carbon has a lower corrosion rate than the original carbon).

(iv) A final challenge is to enhance the electrode kinetics of oxygen reduction. Progress has been made with alloy electrocatalysts. The best results to date have been obtained with alloys of platinum with chromium, vanadium, cobalt, or nickel. The non-noble metal component is more corrodible than the noble one. Thus, preferential dissolution of the former may occur, which causes a degradation in performance of the oxygen electrode. Worse still is that the transition metal can migrate to the hydrogen electrode and inhibit its electrocatalytic activity. By optimizing the composition and structure of the alloy, these problems are greatly minimized.

1.3.2. Inhibition of Poisoning of Hydrogen Electrodes

The hydrogen electrode generally exhibits an excellent performance in fuel cells. The overpotential at this electrode, at the operating current densities (200 to 400 mA/cm^2) of fuel cells, is only about 20 mV while it is at least 10 to 15 times higher at the oxygen electrode. Even in the high-power-density solid polymer electrolyte fuel cell, which operates at less than 100°C, the overpotential at the hydrogen electrode, with a low platinum loading (<0.4 mg Pt/cm^2), has such a value. The main reason for this is that the hydrogen electrode is "pseudo-reversible" in acid electrolytes. In an alkaline electrolyte, the exchange current density is about two orders of magnitude lower. However, by using an alloy electrocatalyst (80% Pd, 20% Pt), International Fuel Cells-United Technologies Corporation has been successful in attaining high levels of performance even at current densities as high as 8 A/cm^2 (the cell potential at this current density is 0.5 V in the most advanced fuel cell with hydrogen and oxygen as reactants). In the fuel cells using molten carbonate or solid oxide electrolytes, the overpotential at the hydrogen electrode is negligible due to the extremely high operating temperature (650°C for the former and 1000°C for the latter).

The major challenging problem at the hydrogen electrode is inhibition of its poisoning by impurities such as CO, H$_2$S, and SO$_2$ which are generally present in varying quantities when the hydrogen is produced (i) by reforming of natural gas and of the higher hydrocarbons obtained from petroleum, and (ii) by coal gasification. The fuel processor in fuel cells is equipped for CO oxidation (shift converter) and sulfur removal (desulfurizer). However, small amounts are still present in the processed gas. One of the main reasons for increasing the operating temperature of the phosphoric acid fuel cell to 180–190°C is that at such a temperature it can tolerate 1 to 1.5% CO. The CO poisoning problem in the solid polymer electrolyte fuel cell is a most challenging one because its operating temperature is below 100°C—even 100 ppm of CO in the processed fuel is sufficient to "kill" the hydrogen electrode. A solution to this problem, proposed by Ballard Technologies Corporation, is to pass the reformed gases along with small amounts of oxygen over a platinum on alumina catalyst at about 100°C and then to use an alloy electrocatalyst for hydrogen oxidation (the latter method was also carried out by General Electric Company previously). The CO, which is more strongly absorbed than hydrogen on the Pt catalyst, is in turn oxidized by oxygen to carbon dioxide. The combined effect of carbon monoxide and sulfur impurities is even more injurious than carbon monoxide alone. A COS type species is strongly adsorbed on the platinum electrocatalyst. Sulfur is a deadly poison for the hydrogen electrode in molten carbonate and solid oxide fuel cells. A highly stable nickel sulfide is formed which greatly diminishes the activity of the nickel electrocatalyst. In the solid oxide fuel cell, the poisoning is reversible because when pure hydrogen is subsequently used in the fuel cell, the original electrocatalytic activity is regenerated. In the molten carbonate fuel cell, the situation is more complex because the sulfur species is partially oxidized, resulting in the conversion of the carbonate electrolyte to sulfite or sulfate to a small extent, which is more than sufficient to significantly affect the properties of the electrolyte.

1.3.3. Direct Oxidation of Methanol, the Most Electroactive Organic Fuel—The Fuel Cell Researcher's Dream

The enthusiasm has been great since the 1960s to use hydrocarbons, such as natural gas, the higher homologs, and alcohols, such as methanol or ethanol, as the anodic reactants directly in fuel cells. However, until the present time, there has been very limited success in this direction. The rationale for attempting to use these fuels directly, instead of hydrogen derived from them, is that the reformer is one of the three major subsystems in a fuel cell power plant (the other two are the electrochemical cell stack and the power conditioner), and the elimination of this subsystem will have the positive effects of decreasing the weight, volume, and cost of the power plant. The basic reason for the difficulty in developing fuel cells utilizing organic fuels directly is the high degree of irreversibility of the electroorganic reactions. This compounded with the irreversibility of the oxygen reduction reaction results in fuel cells with low efficiencies and power densities. The problem becomes even more complex because, of all the hydrocarbon or alcohol fuels, methanol is the most electroactive, but even its oxidation involves a six-electron transfer reaction. This overall electron reaction may involve three intermediate reactions, each of which is a two-electron transfer reaction, leading to the intermediates and/or products formaldehyde, formic acid, and carbon dioxide. The only desirable product is carbon dioxide. A further complication is that a poisoning species, identified to be COH or CO, is adsorbed on the electrode, and it causes a decrease in the performance of the electrode as a function of time. All these problems result in the rest potential of the electrode (with a platinum electrocatalyst) being higher than the reversible potential by about 300 to 400 mV and a high overpotential for methanol oxidation even at low current densities (thus causing an inherent loss in energy efficiency of a methanol–air fuel cell of about 30%).

In high-temperature fuel cells, the problems have been overcome by incorporating a reformer catalyst behind the fuel cell electrode (internal reforming). However, the enthusiasm is still great to directly oxidize methanol in low-temperature fuel cells for the transportation (electric vehicles) application. The challenge is, thus, to find electrocatalysts which will exhibit rest potentials close to the reversible value and high activities for methanol oxidation in the fuel cell electrolytes, such as proton-conducting membranes and phosphoric acid.

2. THE 1960–1990 TIME PERIOD—THE GOLDEN AGE IN ELECTRODE KINETICS AND ELECTROCATALYSIS OF FUEL CELL REACTIONS

2.1. Supported Platinum Electrocatalysts—The Answer to Lowering of Noble Metal Loadings

The most advanced low- to medium-temperature fuel cells are the ones with phosphoric acid, potassium hydroxide, and perfluorinated sulfonic acid membrane (Du Pont's Nafion or the Dow membrane) electrolytes. During the first decade of active fuel cell research, attention was mostly focused on unsupported platinum electrocatalysts. The platinum particles were relatively large and their BET surface areas relatively small ($<50 \text{ m}^2/\text{g}$). Thus, it was necessary to use a relatively high noble metal loading ($>4 \text{ mg}/\text{cm}^2$) to obtain a reasonable fuel cell performance. Kordesch and his co-workers at Union Carbide Corporation made the first significant advance in reducing the noble metal loading by applying noble metal catalysts

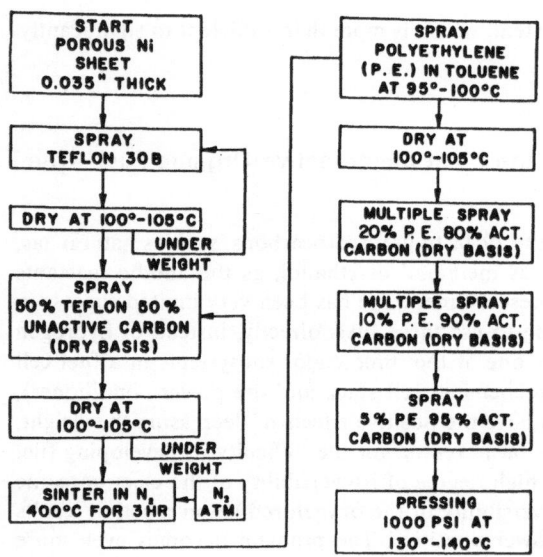

FIGURE 2. Electrode manufacturing procedure used by Kordesch and co-workers at Union Carbide Corporation.[15]

in small quantities to the front surface of the electrode.[15] From a historical point of view, it is worth reviewing the electrode manufacturing procedure used by Kordesch and co-workers as represented in Fig. 2. The catalization of the electrode was carried out by painting (brush or roller) a noble metal salt (say, chloroplatinic acid) solution on the carbon-deposited surface of the electrode. After the electrode was dried, the noble metal salt was decomposed to the oxide by heating in an oven at 100°C. The catalyst was located close to the top layers of the electrode because of the increasing repellency of the polyethylene and/or Teflon treated underlayers. The noble metal loading was reduced by at least a factor of four by this procedure. It is also interesting to note that electrocatalysts other than platinum were also deposited by a similar procedure, and performances somewhat inferior to that of platinum were obtained when these electrodes were tested with air for oxygen reduction (Fig. 3). Figure 3 also shows the behavior of a thick air electrode which was previously prepared at Union Carbide. Mass transport limitations are apparent on this electrode. The electrode with the spinel catalyst showed a higher level of performance at higher current densities.

The next advance in reducing the amount of platinum electrocatalyst in fuel cell electrodes was made by Juda and co-workers[16] at Ionics Incorporated (later Prototech). Just as in the case of gas-phase catalysis, high-surface-area supported electrocatalysts were prepared. For this purpose, a high-surface-area carbon (furnace blacks such as Vulcan XC-72) with a BET surface area of about 200 m^2/g was used as the support on which platinum crystallites were

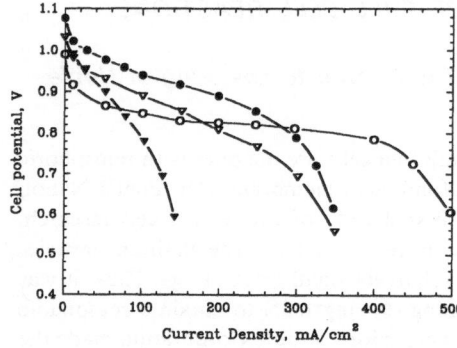

FIGURE 3. Cell potential versus current density plots for thin electrodes with different catalysts and a thick electrode.

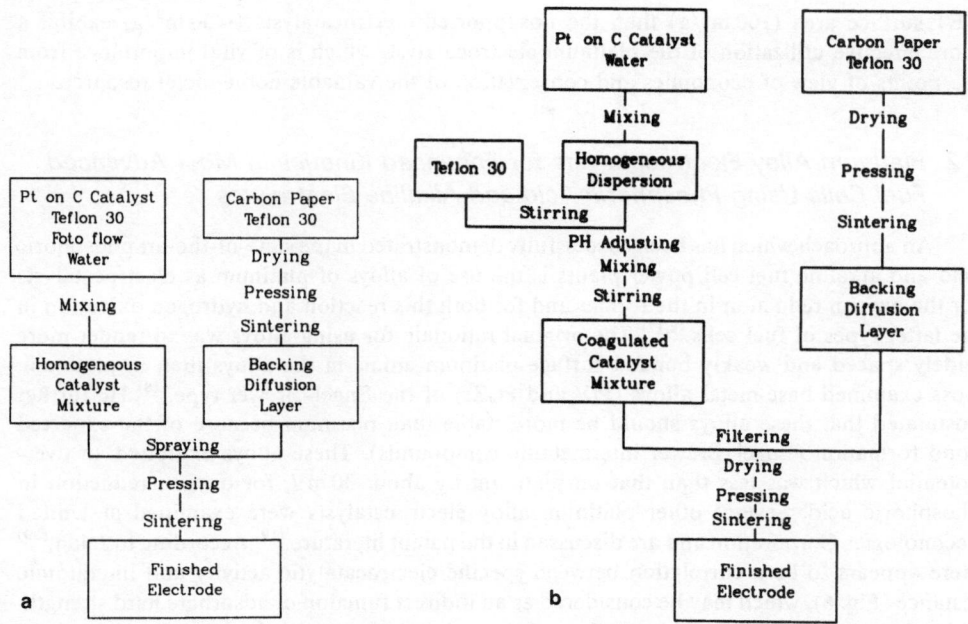

FIGURE 4. Flow charts for fabrication of fuel cell electrodes with carbon-supported electrocatalysis: (a) spraying method; (b) filtration method.

deposited by a colloidal method (platinum hydroxides) or by electroless reduction of chloroplatinic acid. These carbon-supported platinum electrocatalysts were heat treated, and the resulting platinum crystallites had a particle size of about 25 Å. By use of such carbon-supported electrocatalysts in fuel cell electrodes, it was possible to reduce the noble metal loadings for fuel cell electrodes to about 0.35 mg/cm². The state-of-the-art phosphoric acid fuel cell, which is in the most advanced state of development as compared to other types of fuel cells, uses these types of electrodes. The present techniques for electrode fabrication are based on filtration, rolling, or spraying methods. The first is most suitable for research investigations, and the second and the third are the most appropriate for large-scale manufacture. Figure 4 demonstrates flow charts for fabrication of fuel cell electrodes with low noble metal loadings.

The comparable levels of performance of fuel cells with low- and high-platinum-loading electrodes is illustrated in Fig. 5.[13] The electrolyte in this fuel cell is a perfluorinated sulfonic acid polymer membrane (Dow membrane). The supported platinum particles with a higher

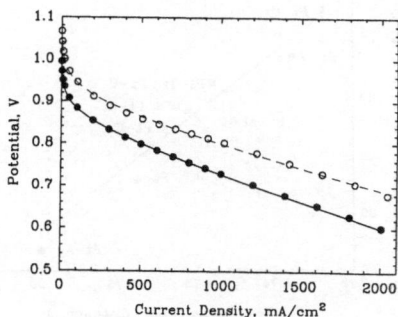

FIGURE 5. Potential versus current density plots for solid polymer electrolyte fuel cells with low (●, 0.45 mg/cm²) and high (○, 10 mg/cm²) platinum-loading electrodes.[13]

BET surface area $(100 \, m^2/g)$ than the unsupported electrocatalysts $(<50 \, m^2/g)$ exhibit a more effective utilization of the platinum electrocatalyst, which is of vital importance from the points of view of economics and conservation of the valuable noble metal resource.

2.2. Platinum Alloy Electrocatalysts for Enhanced Kinetics in Most Advanced Fuel Cells Using Phosphoric Acid and Alkaline Electrolytes

An approach which has been successfully demonstrated in the state-of-the-art phosphoric acid and alkaline fuel cell power plants is the use of alloys of platinum as electrocatalysts for the oxygen reduction in the former and for both this reaction and hydrogen oxidation in the latter types of fuel cells.[8,17] The original rationale for using alloys was to render more widely spaced and weakly bonded surface platinum atoms in the alloys than in platinum. Ross examined base-metal alloys (PtV and Pt_4Zr) of the Engel–Brewer type.[18] He further postulated that these alloys should be more stable than platinum because of the expected bond formation (Engel–Brewer intermetallic compounds). These alloys exhibited an overpotential which was less than that on platinum by about 30 mV, for oxygen reduction in phosphoric acid. Several other platinum alloy electrocatalysts were examined at United Technologies Corporation and are discussed in the patent literature.[19] According to Jalan,[20] there appears to be a correlation between specific electrocatalytic activity and interatomic distance (Fig. 6), which may be considered as an indirect function of adsorbate hard strength. This plot may be correlated with the volcano plot[20,21] representing the dependence of electrocatalytic activity on nearest-neighbor distance (Fig. 7). Figures 6 and 7 show that the Pt-V and Pt-Cr alloys exhibit close to the maximum electrocatalytic activities. Postmortem Rutherford backscattering investigations[22,23] of cutaway sections of fuel cells with these electrocatalysts which had operated for about 3000 h in phosphoric acid fuel cells showed that the less noble component (V or Cr) leached out of the cathode, migrated through the electrolyte, and deposited on the anode. These fuel cells showed a loss in performance with time. More recently, greater success has been achieved in phosphoric acid fuel cell power plants with the alloy electrocatalysts.[24] The electrocatalysts have not been disclosed, but in all probability they are binary or ternary alloys of Pt with Cr, Co, or Ni. Johnson Matthey is the most active organization in the preparation of supported-alloy electrocatalysts (crystallites with sizes of the order of 30 Å). A recent study at International Fuel Cells (Fig. 8) shows that a Pt-Cr-Co alloy has a higher electrocatalytic activity than platinum or other alloys that were tested.[25] Another possible explanation for enhanced electrocatalysis by alloys is that the mechanism of oxygen reduction may be via a redox reaction of the transition metal.[7] It has also been reported that the alloy electrocatalysts have exhibited a more stable performance with time, probably because the alloying component of Pt inhibits its sintering.[26]

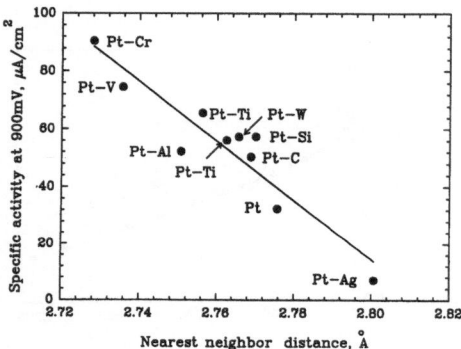

FIGURE 6. Correlation between platinum-alloy electrocatalytic activity for oxygen reduction in phosphoric acid and Pt-Pt nearest-neighbor distance.[21]

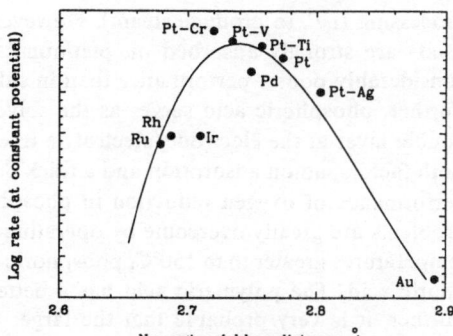

FIGURE 7. Composite volcano plot of rate for oxygen reduction in phosphoric acid and Pt-Pt nearest-neighbor distance.[21]

One of the major discoveries in alkaline fuel cell research and development is that unsupported gold crystallites are excellent electrocatalysts for oxygen reduction. The gold particles have a diameter of approximately 200 Å. An alloy (90% Au, 10% Pt) electrocatalyst (20 mg/cm^2) is used in the alkaline fuel cell power plants developed by International Fuel Cells-United Technologies Corporation for NASA's space shuttle flights.[27] Surprisingly, the platinum is added to prevent the sintering of gold and not to enhance its electrocatalytic activity. The performance of this electrode is excellent, and it exhibits the best reported value in terms of overpotential for oxygen reduction in aqueous electrolytes. It was stated in Section 1.3.2 that the exchange current density for hydrogen oxidation on platinum is about two orders of magnitude lower in alkaline than in acid electrolytes. It is for this reason that the space alkaline fuel cell uses a platinum–palladium alloy (80% Pt, 20% Pd; noble metal loading, 10 mg/cm^2) as the electrocatalyst for the hydrogen electrode.

2.3. Superacid Electrolytes Approach Alkaline Electrolytes in Respect to Fast Oxygen Reduction Kinetics

The main reason for the better performance of oxygen reduction in alkaline than in acid electrolytes is that anion adsorption is minimal with the former and could be significant with the latter. Sulfuric acid is not used in fuel cells because of its instability at the potential of the hydrogen electrode: it gets reduced to sulfite or sulfide. Perchloric acid, even though attractive from the point of view of minimal anion adsorption, has not been considered as an electrolyte because of safety reasons. The other problem is that these acids cannot be used at temperatures much above 100°C, which are necessary to remove product water from the cell. The acid electrolyte of choice was phosphoric acid. The state-of-the-art phosphoric acid (acid concentration close to 100%) fuels cells operate at nearly 200°C. Such a high temperature is beneficial for product water removal and for supply of medium-grade heat for the fuel

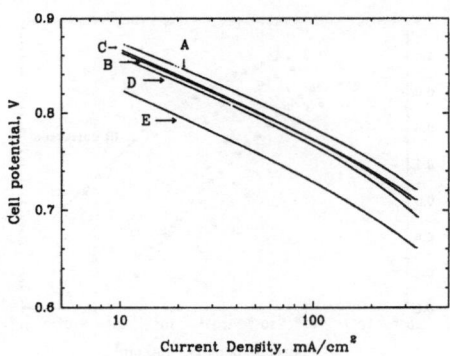

FIGURE 8. Plots of cell potential versus current density in phosphoric acid fuel cells with H$_2$/air as reactants at 191°C and oxygen electrodes containing Pt or Pt-alloy electrocatalysts: (A) Pt-Cr-Co; (B) Pt-Cr; (C) Pt-V-Co; (D) Pt-V; (E) pure Pt.

processing (i.e., to produce steam). However, phosphoric acid anions (and maybe the acid itself) are strongly adsorbed on platinum,[28] and oxygen reduction in this acid shows a considerably poorer performance than in sulfuric acid or perchloric acid at less than 100°C. Further, phosphoric acid serves as the solvent at these high concentrations, and, thus, the double layer at the electrode/electrolyte interface is thicker than in aqueous electrolytes.[28] Both factors, anion adsorption and a thick double layer, account for the considerably poorer performance of oxygen reduction in phosphoric than in sulfuric or perchloric acid. These problems are greatly overcome by operating phosphoric acid fuel cells at close to 200°C. At temperatures greater than 150°C, phosphoric acid is predominantly polymerized to pyrophosphoric acid. The polymeric acid has a better ionic conductivity than the monomeric acid. Further, it is very probable that the larger pyrophosphate anion ($P_2O_7^{4-}$) is less adsorbed than the phosphate (PO_4^{3-}) anion. In addition, there is a decrease of anion adsorption with increasing temperatures.

A revolution in oxygen reduction kinetics occurred when a perfluorinated sulfonic acid [trifluoromethanesulfonic acid (TFMSA)] was first tested as an electrolyte by Appleby and Baker.[29] The fluorine atom is electron-withdrawing, which makes it easier for the protons to dissociate from the molecule. Such an acid is an extremely strong one (i.e., high proton activity). Perfluorinated sulfonic acids belong to the class of superacids. The additional advantage of introducing the fluorine atom in place of hydrogen is that sulfonic or fluorinated compounds exhibit a higher oxygen solubility. The solubility of oxygen in TFMSA is more than ten times higher than in phosphoric acid. Further, because of the fact that the anions are polar, hydrated water molecules will be strongly held and anion adsorption is minimal. These three factors significantly improve the oxygen reduction kinetics. Figure 9 demonstrates the considerably higher activity of oxygen reduction on platinum in this acid than in phosphoric acid. Efforts were made to develop fuel cells using trifluoromethanesulfonic acid at Energy Research Corporation.[30] The main problems which arose were (i) the acid concentration management with the aqueous electrolyte (say about $6M$), due to the difficulty of product water removal at a temperature of less than 100°C; and (ii) with the monohydrate at temperatures above its melting point (135°C), flooding of the electrode occurred due to the hydrophobic behavior of this electrolyte.

These difficulties led to investigations with the higher homologs of trifluoromethanesulfonic acid, but these were not successful because of the wetting of the Teflon-bonded electrodes by these acids and their consequent flooding—increasing the ratio of CF_2/SO_3H in these acids makes their surface tension characteristics similar to those of Teflon.[31] The ionic conductivities of these acids were also not as high as those of the parent acid TFMSA. However, the breakthrough with the perfluorinated sulfonic acids was the discovery of the polymeric acid Nafion by Du Pont, and of a more advanced version by Dow. The attractive feature of the polymeric acid is that it is a proton-conducting membrane, and only water has

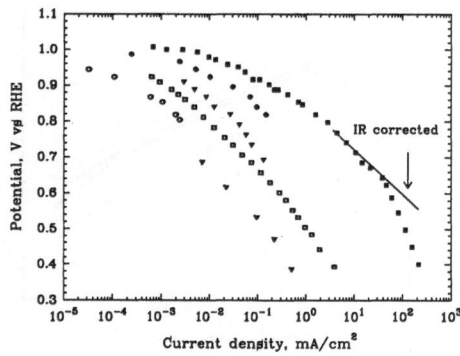

FIGURE 9. Tafel plots for oxygen reduction on platinum in 85% H_3PO_4 and in $1.1M$ CF_3SO_3H at 25°C. [From E. J. Taylor, Ph.D. thesis, University of Virginia (1981).]

FIGURE 10. The skeleton structures of the Du Pont (left) and the Dow (right) perfluorinated sulfonic acid membranes.

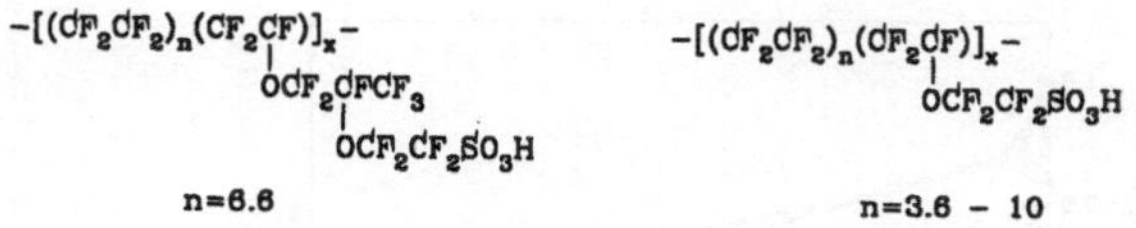

$$-[(CF_2CF_2)_n(CF_2CF)]_x-$$
$$\underset{|}{OCF_2CFCF_3}$$
$$\underset{|}{OCF_2CF_2SO_3H}$$

n=6.6

$$-[(CF_2CF_2)_n(CF_2CF)]_x-$$
$$\underset{|}{OCF_2CF_2SO_3H}$$

n=3.6 – 10

to be introduced into the membrane to make it conducting. The structures of the Du Pont Nafion and Dow membranes are depicted in Fig. 10. In fuel cells, it is only a thin membrane (say 100–175 μm thick) that is used as the electrolyte layer. Thus, in addition to excellent hydrogen and oxygen electrode kinetics, which have been demonstrated in several studies,[12-14,32-35] the additional advantage is that the thin membranes have excellent ionic conductivities, which is necessary to minimize ohmic losses and achieve high power densities. A novel microelectrode technique was recently used at Texas A&M University[36] to determine the electrode kinetic parameters for oxygen reduction at the platinum/solid polymer electrolyte (Nafion) interface and mass transport parameters for oxygen in the Nafion. The exchange current densities obtained were the best to date for oxygen reduction in acid electrolytes at low temperatures.

2.4. Proton Conductor Impregnation into Electrode and Localization of Platinum near Front Surface of Electrode in Solid Polymer Electrolyte Fuel Cells

Until 1986, high levels of performance were demonstrated in solid polymer electrolyte fuel cells (by the General Electric Company, Hamilton Standards Division of United Technologies Corporation, Ballard Technologies Corporation, Ergenics, and Siemens) only by using electrodes with a high platinum loading (4 mg/cm^2). In these fuel cells, the membrane and electrode assemblies were prepared either by electroless deposition of platinum or by hot-pressing electrodes with a high platinum loading (unsupported platinum) onto the membrane.[14] In both these methods, a relatively high electrochemically active surface area is achieved by (i) use of a high loading of platinum with a reasonably high surface area, and (ii) some penetration of the porous platinum network into the solid polymer electrolyte. The use of high platinum loading electrodes in fuel cells is acceptable for space and defense applications. This is not so for terrestrial applications such as fuel cell power sources for electric vehicles. According to a techno-economic assessment at Los Alamos National Laboratory, the leading candidate for a fuel cell power source for electric vehicles (particularly automobiles) is the solid polymer electrolyte fuel cell power plant.[37] However, a practical realization of this assessment is possible only if the noble metal loading can be reduced at least tenfold or if a substitute for the noble metal electrocatalyst can be used.

An ingenious method was developed to utilize electrodes with a low platinum loading by Raistrick at Los Alamos National Laboratory.[38] The main idea in this method is to extend the three-dimensional reaction zone by impregnation of a proton conductor (for example, Nafion) into the active layer of an electrode (for example, Prototech fuel cell electrodes used in phosphoric acid fuel cells) with a low platinum loading (0.40 mg/cm^2). By use of such an electrode which was mechanically pressed onto a membrane, Raistrick demonstrated an enhancement of oxygen reduction kinetics. Srinivasan et al.[12] were, however, the first to demonstrate that by use of this technique it is possible to attain a high level of performance in a fuel cell. This performance is equivalent to that in a fuel cell with high-platinum-loading electrodes (Fig. 11). These workers found it necessary to hot-press the Nafion-impregnated electrodes onto the membrane to achieve the high level of performance. Proof that the three-dimensional reaction zone is extended was confirmed by carrying out cyclic voltammetric studies. Figure 12 illustrates this phenomenon and clearly shows (hydrogen adsorption/desorption, double-layer charging, and oxide formation/reduction) that Nafion

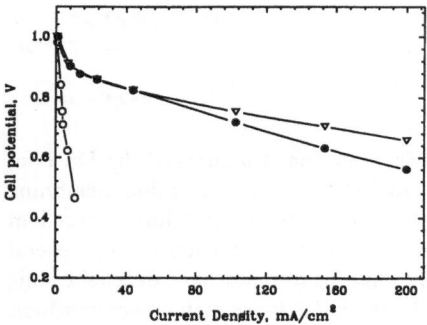

FIGURE 11. Cell potential versus current density plots for H_2/O_2 fuel cells at 50°C and 1 atm pressure (electrode area, 5 cm^2): ∇ and $\bigcirc$, Nafion-impregnated and as-received Prototech electrodes (0.35 mg cm^2 Pt) hot-pressed to Nafion membranes; ●, Ge/HS-UTC membrane and electrode (4 mg cm^2 Pt). Solid lines are computer-generated plots.[12]

impregnation enhances the electrochemically active surface area. Further, this figure also demonstrates that the electrochemically active surface areas are nearly the same with the high- and low-platinum-loading electrodes. It is worthwhile noting that Raistrick used Fig. 11, which was previously published in Ref. 12, in support of his patent.[38]

2.5. Promising Alternatives to Platinum Electrocatalysts for Low-Temperature Fuel Cells

Though at least 3000 papers have appeared in the literature on the electrocatalysis of fuel cell reactions, platinum and its alloys have been the "electrocatalysts of choice" for the hydrogen oxidation and oxygen reduction reactions in low- and medium-temperature fuel cells with acid (phosphoric acid, perfluorinated sulfonic acid polymer) and alkaline (potassium hydroxide) electrolytes. One of the major challenges in electrocatalysis research has been, and still is, to replace platinum. The reasons are twofold. Firstly, platinum is quite an expensive metal—even at the present low level of loading with supported electrocatalysts, the cost of the platinum in the most advanced phosphoric acid fuel cell is $75/kW. One argument which is presented in favor of still utilizing platinum is that the electrochemical cell stack in acid and alkaline electrolyte fuel cells is mainly a carbon structure and the platinum is recoverable and still has its value. This may be the case, but the use of platinum still raises the capital costs. Secondly, the annual production rate of platinum is over 90 million grams, and if platinum will be used in fuel cell power plants for electric vehicles, the present production rate of platinum will be sufficient for only 10% of the automobiles manufactured in the United States. An alternate way of examining this problem is that in the catalytic converters of

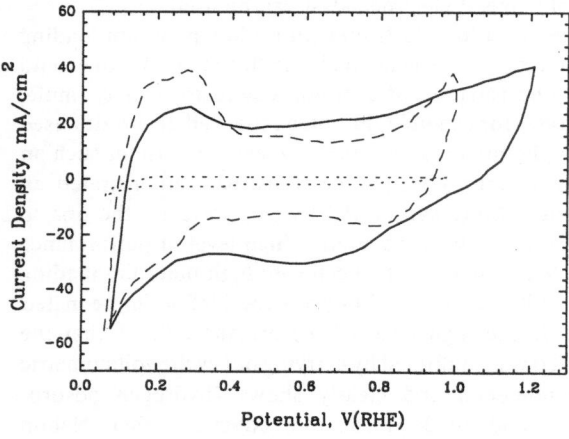

FIGURE 12. Cyclic voltammograms on oxygen electrodes in fuel cells (electrode area, 5 cm^2): —, PEM D-120 cell with Nafion-impregnated Prototech electrodes (0.35 mg cm^2 Pt) hot-pressed to Dow membrane; PEM #3 cell with GE/HS-UTC membrane and electrode assembly (4 mg/cm^2 Pt); ···, PEM #5 cell with as-received Prototech electrodes (0.35 mg/cm^2 Pt) hot-pressed to Nafion membrane.[12]

internal combustion engine-powered vehicles manufactured at the present time, the platinum loading is 1 g/vehicle. About 20% of the platinum mined from its natural resources is used for this application. A fuel cell-powered vehicle will require at least 20 g of platinum!

Thus, the impetus for finding alternate electrocatalysts is high. There has been partial success in utilizing non-noble metal electrocatalysts in alkaline fuel cells—for example, nickel for the hydrogen electrode and lithiated nickel oxide for the oxygen electrode were used in the Bacon–Pratt and Whitney fuel cell, and Raney nickel for the anode and Raney silver for the cathode are the electrocatalysts in the Siemens fuel cell. Success has been achieved in the development of porous gas diffusion electrodes with heat-treated (to 800°C) organometallic macrocycles (for example, cobalt tetraphenylporphyrin) supported on high-surface-area carbon.[39] Yeager and co-workers[40] have carried out detailed, fundamental studies on the structure of the heat-treated macrocycles and on the mechanism of oxygen reduction on these electrocatalysts. Regarding the former, there is the view that even though most of the organic constituents are burned away by the heat treatment, the nitrogen ring still remains with the metal. Further, by use of this method of preparation, it appears that very high surface area supported electrocatalysts can be prepared. In respect to the mechanism of oxygen reduction on these electrocatalysts, since the Tafel slope is only 60 mV/decade, it appears that a chemical step following the first electron transfer step or even a subsequent electron transfer step, which would be expected to have a lower Tafel slope, is rate determining. In a more technologically oriented investigation, it has been shown that fuel cell electrodes prepared with the above-mentioned electrocatalyst exhibited an excellent performance (400 mA/cm^2 at 0.8 V) as an air cathode for a period of 8000 h.[39]

A project to select and evaluate transition metal electrocatalysts for electrodes for phosphoric acid fuel cells was embarked on at International Fuel Cells.[41] The transition-metal (Fe, Co$^-$) organic macrocycles included tetramethoxyphenylporphyrin, phthalocyanines, tetrazoannulenes, and tetraphenylporphyrins. These compounds were not found to be stable in hot, concentrated phosphoric acid. However, when heat-treated carbon-supported electrocatalysts, as discussed in the preceding paragraph, were used for the oxygen electrodes, the phosphoric acid fuel cells showed performances comparable to those obtained with platinum electrocatalysts at temperatures up to about 120°C. Above this temperature, the electrocatalysts corroded in the concentrated phosphoric acid.

2.6. Bifunctional (Metal–Metal Oxide) Electrocatalysts for Methanol Oxidation

One of the most attractive fuel cells, particularly for transportation applications, is the direct methanol–air fuel cell. Though methanol is the most electroactive organic fuel, it is at least three orders of magnitude less active than hydrogen. However, there are positive signs, as a result of several investigations over nearly the last 30 years, that better electrocatalysts can be found. An interesting approach, presented in the 1960s by Shropshire, from Exxon Research and Engineering Laboratories, to accelerate methanol oxidation was to change the pathway for the reaction via redox catalysis.[42] An Mo^{5+}/Mo^{6+} redox couple was used. Thus, the electrochemical reaction was the oxidation of Mo^{5+} to Mo^{6+}; the Mo^{6+} in turn chemically oxidized the methanol and was reduced to Mo^{5+} and the cycle was repeated. The reactions may be represented by :

$$Mo^{5+} \rightarrow Mo^{6+} + e^- \tag{4}$$

$$CH_3OH + H_2O + 6Mo^{6+} \rightarrow CO_2 + 6H^+ + 6Mo^{5+} \tag{5}$$

The difficulty of this scheme was that the Mo^{5+}/Mo^{6+} couple was added to the electrolyte, and these species complicate reactions at the cathode. An alternate scheme will be to incorporate Mo or another metal of a similar type in the electrocatalysts.[43]

Another approach for electrocatalysis of methanol oxidation was initiated by Motoo, Watanabe, and co-workers in Japan.[44,45] The idea, in this method, is to have a bifunctional metallic electrocatalyst of the type Pt–Ru or Pt–Sn. The Ru or Sn metal constituent is in an oxidized state. The methanol intermediate is C—OH or CO type species adsorbed on the platinum. This species can be readily oxidized by the nascent O or OH species on the Ru or Sn. The O or OH species is regenerated electrochemically.

A third route was proposed by Adzic et al.[46] In this scheme, an underpotentially deposited (upd) metal (Pb, Sn, Ri) is used. The upd metal may act in the same manner as above, or alternatively the UPD metal inhibits the adsorption of hydrogen and, in turn, the generation of the poisoning species (COH) which is formed by combination of an adsorbed CO with adsorbed H.

Recently, Kita, and co-workers have examined the electrocatalysis of methanol oxidation at a platinum/solid polymer electrolyte interface.[47] The main difficulties with this approach are that (i) the solid polymer electrolyte fuel cell can operate only at a temperature of less than 100°C, and the poisoning species (CO or COH) are too strongly adsorbed at these temperatures; and (ii) methanol migrates from the anode to the cathode and depolarizes it. Possible solutions to this problem are the use of the dual-function electrocatalyst for the anode and an electrocatalyst inert toward methanol at the cathode.

2.7. A Solution to Electrode Kinetics and Electrocatalysis Problems—High-Temperature Fuel Cells with Molten Carbonate or Solid Oxide Electrolytes

It was realized as early as the 1960s that electrocatalytic problems in fuel cells can be overcome by operation at high temperatures. Firstly, the rates of activation-controlled reactions can be significantly increased at very high temperatures, and, secondly, the strengths of adsorption of poisoning species (CO, S) to the electrocatalytic sites can be greatly reduced. The electrolytes which have proven to be the most successful for high-temperature fuel cells are molten carbonates (e.g., Li_2Co_3/K_2CO_3) and solid oxides (e.g., 88–90% ZrO_2, 12–10% Y_2O_3). The former are carbonate ion conducting, and the latter are oxide ion conducting. A disadvantage of the former type of electrolyte is that CO_2 and O_2 are required as cathodic reactants and CO_2 is formed at the anode. This necessitates a CO_2 recycling scheme. The latter type of electrolyte is ideal because oxygen is reduced at the cathode to the oxide ion, which is transported to the anode and reacts with the hydrogen to form water vapor and electrons.

Another advantage of the high-temperature fuel cells is that carbon monoxide, which is a product in the fuel processing reaction, is as electroactive as hydrogen and is not a poison, as in low-temperature fuel cells. It is possible that the CO oxidation may occur via a shift conversion reaction as represented by

$$CO + H_2O \rightarrow CO_2 + H_2 \tag{6}$$

$$H_2 + O^{2-} \rightarrow H_2O + 2e^- \tag{7}$$

A fourth advantage of the molten carbonate and solid oxide fuel cells is that the fuel processing, which is an endothermic process, occurring at a temperature of 700°C for natural gas reforming and over 1000°C for coal gasification, can utilize high-grade waste heat from the electrochemical cell stacks. Since more waste heat than can be utilized by the fuel processor will still be available, designs of these power plants include bottoming cycles (gas turbines) for additional electricity generation. The overall efficiencies of these high-temperature fuel cell systems for electricity generation are projected to be about 55–60%, whereas the corresponding values for the phosphoric acid, alkaline, and solid polymer electrolyte fuel cell power plants are less than 50%.

As stated in the first paragraph of this section, the electrocatalytic problems are minimal in the molten carbonate and solid oxide fuel cells. The challenges in these technologies are to find electrode materials that are stable at these high temperatures. The instability problems which have arisen in the molten carbonate fuel cell are (a) sintering of the anode nickel electrocatalysts, (b) corrosion of the nickel oxide cathode, which is lithiated, in the highly corrosive 62 mol% Li_2CO_3–38 mol % K_2CO_3 electrolyte, and (c) irreversible poisoning of the anode electrocatalyst by sulfur impurities at even a parts per million level; in the solid oxide fuel cell, the instability problems are (a) low tolerance to sulfur impurities by the anode electrocatalyst (however, if pure hydrogen replaces the fuel-processed hydrogen, which contains the sulfur, the poisoning effect is eliminated), (b) slow diffusion of Mn from the perovskite electrocatalyst (a p-type Sr-doped $LaMnO_3$) into the electrolyte, and (c) difficult matching of the thermal expansion coefficients of the cell components—anode, cathode, electrolyte, and interconnection (for connection of cells in series; the current material for interconnection is magnesium-doped lanthanum chromite). Most of these problems have been overcome to significant extents, and multimillion dollar programs are under way in the United States, Japan, and some European countries (for example, Italy and the Netherlands) to develop multi-megawatt-size molten fuel cell power plants.

3. PROGNOSIS ON ADVANCES IN ELECTRODE KINETICS AND ELECTRODE KINETICS OF FUEL CELL REACTIONS IN THE NEXT DECADE AND IN THE 21ST CENTURY

3.1. Phosphoric Acid Fuel Cells

3.1.1. New Platinum Alloy Electrocatalysts

In order to attain higher power densities (e.g. over 300 mA/cm^2 at 0.7 V) and energy efficiencies (cell potential of over 0.7 V corresponding to an efficiency of about 45%), a "brute force" mechanism is being used at International Fuel Cells-United Technologies Corporation—raising the operating temperature to slightly above 200°C and the pressure to 8 atm. Thus, the challenge is to find stable materials in this corrosive environment. The materials which are susceptible to some degree of corrosion are the cathode electrocatalyst and the cathode support. The promising approaches to minimize corrosion problems are (i) to stabilize the alloy electrocatalysts for the oxygen electrode, perhaps by incorporating a third component such as Zr or Ti—refractory metals are highly stable and are known to stabilize alloys of transition metals, and, further, these refractory metals also form defect oxides, with high electronic conductivity, which can also contribute to the electrocatalytic activity—and (ii) to investigate alternate support materials that can be employed instead of Vulcan XC-72R or Black-Pearls 2000 carbon support. The carbon materials are still good supports, but carbon corrosion occurs to a small extent as the cell potential approaches the open-circuit potential. It was mentioned in Section 1.3.1 that the corrosion is minimized by graphitization of the support but at the sacrifice of reducing the surface area. There is still some interest in alternate supports for carbon. Though some carbides (TiC) are more stable than carbon, the challenge is to greatly enhance their surface area (m^2/g) to nearly the same values as for the furnace blacks.

3.1.2. Mixed Electrolytes in Electrode

Phosphoric acid is used as the electrolyte in the most advanced fuel cell mainly because of (i) its inertness and (ii) the capability to operate at temperatures as high as 200°C, which is necessary to (a) provide at least medium-grade waste heat for the fuel (natural gas, methanol) processor and (b) tolerate 1–2% CO in the fuel cell stream at the anode electrocatalyst. However, the oxygen reduction kinetics in phosphoric acid is considerably slower than in, say, the perfluorinated sulfonic acids. The main reasons for the slower kinetics of oxygen reduction in phosphoric acid are anion adsorption and low solubility of oxygen in the electrolyte (Section 2.3). By operating phosphoric acid fuel cells at high temperatures ($>180°C$), the first problem is mostly overcome because phosphoric acid is mostly polymerized to pyrophosphoric acid, which has a higher ionization constant than phosphoric acid, and the anion $H_3P_4O_7^-$ is less specifically adsorbed than $H_2PO_4^-$. On the other hand, the anions of perfluorinated sulfonic acids are hardly specifically adsorbed, and, in addition, the solubilites of oxygen in these acids are considerably higher than in phosphoric acid. The second factor is an important one in that the electrode kinetics of oxygen reduction is first-order with respect to oxygen concentration, and a tenfold change in it can reduce the overpotential at the operating current density by about 100 mV. Attempts have been made or proposed to use mixed electrolytes within the electrode structure, one for forming an electrolyte film within the pores and the second as the bulk electrolyte in the pores and in the matrix.[48-50] Since oxygen diffuses through the backside of the electrode (via the Teflon binder and/or electrolyte in the pores), it can reach the enhanced saturation value in the electrolyte film adjacent to the electrocatalyst. There is some evidence in the literature that this approach is useful in increasing the rate of oxygen reduction. The challenge is to find suitable proton-conducting polymeric electrolytes that can withstand the high operating temperature in phosphoric acid fuel cells.

3.2. Alkaline Fuel Cells

3.2.1. Modified Macrocyclic Electrocatalysts

Heat-treated cobalt tetraphenylporphyrin, supported on high-surface-area carbon, has been shown to be an excellent electrocatalyst for oxygen reduction. The burning question is: Why does the 800°C heat treatment not completely destroy the organic structure, and, if it does, why cannot the same metallic oxide electrocatalysts be formed by decomposition of the colloidal metal hydroxides? The highly sophisticated work of Yeager and co-workers[40] provides some evidence that the nitrogen ring still remains after the heat treatment. A detailed understanding of the reasons for the high electrocatalytic activity and stability of this heat-treated macrocycle is vitally important. It is very possible that oxygen reduction occurs via a redox mechanism in this electrocatalyst. It is, therefore, worthwhile to examine other macrocycles (other porphyrins and phthalocyanines) in an effort to find alternate electrocatalysts.

There is growing interest in developing low-cost alkaline fuel cells using "pristine" hydrogen as the fuel for terrestrial applications. Appleby strongly feels that the strongest candidate for fuel cell-powered vehicles is the alkaline fuel cell, mainly because it does not need platinum electrocatalysts.[51] Use of heat-treated macrocycles is, thus, a promising approach for the oxygen electrode in alkaline fuel cells. Another approach is to develop mixed oxide electrodes, which is dealt with in the next subsection.

3.2.2. Oxide Electrocatalysts and/or Supports

In the late 1970s and the early 1980s there was great enthusiasm for developing mixed oxides with a spinel or perovskite structure as electrocatalysts for oxygen reduction, as well as oxygen evolution. These are the only promising types of electrocatalysts for bifunctional operation, as required for regenerative fuel cells and rechargeable metal/air batteries. Tseung and Yeung investigated the spinel nickel–cobalt oxide in great detail.[51] It is a very efficient electrocatalyst for oxygen evolution but only a good one for oxygen reduction if the electrode potential is higher than 0.75 V (versus RHE). Below this potential, the oxide is reduced to a lower valence state, and consequently it loses electronic conductivity and electrocatalytic activity. A second problem is the preparation of these electrocatalysts in high-surface-area form, as with supported platinum electrocatalysts. Tseung and Yeung were successful in obtaining reasonably, but not sufficiently high-surface-area $NiCo_2O_4$ particles using the freeze-drying technique. With the advent of technologies for developing high-temperature, superconducting oxides with a perovskite structure, there is promise for successfully developing oxide electrocatalysts with the high surface area required for fuel cells.

3.3. Solid Polymer Electrolyte Fuel Cells

3.3.1. Platinum Alloy Electrocatalysts

Platinum alloy electrocatalysts have hardly been examined for oxygen reduction in solid polymer electrolyte fuel cells. In order to further improve the performance of solid polymer electrolyte fuel cells in terms of efficiency and power density, a promising approach is to use Pt alloy electrocatalysts (e.g., alloys of Pt with Cr, Co, or Ni), which have been successfully developed for phosphoric acid fuel cells. In the phosphoric acid fuel cells, the overpotentials for oxygen reduction have been reduced by 30 to 50 mV at all current densities. This improvement is significant because the power density can increase by a factor of at least two to three, which will have the effect of proportionately reducing the weight, volume, and capital cost of the electrochemical cell stack.

For space applications, it will be necessary to develop methods for the preparation and characterization of unsupported platinum alloy electrocatalysts. For terrestrial applications, supported electrocatalysts will have to be prepared. These electrocatalysts will have to be in the form of fine particles (about 30 Å for supported and 50 Å for unsupported). The best way of elucidating oxygen reduction kinetics will be by (i) preparing porous gas diffusion electrodes; (ii) impregnating a proton conductor into the electrode structure; (iii) hot-pressing this electrode and one with the platinum electrocatalyst (for the hydrogen electrode) to the proton-conducting membrane; (iv) assembling the single-cell test fixture; and (v) determining the potential versus current density behavior. The cyclic voltammetric technique can be used to determine the electrochemically active surface area. Life testing will have to be carried out to determine the performance change as a function of time and, hence, the mechanisms of degradation (say, corrosion) of the less noble metal. It will be ideal if microelectrodes of these alloy electrocatalysts can be prepared and used for investigating electrode kinetics of oxygen reduction at the platinum alloy/solid polymer electrolyte interface. A fundamental study to determine whether the alloy electrocatalysts function via a redox mechanism or exhibit enhanced activities due to changes in geometric or electronic properties of platinum will be most valuable.

3.3.2. New Proton Conductors

Perfluorinated sulfonic acids (Du Pont's Nafion and the Dow membrane) have proven to be very promising proton-conducting membranes for solid polymer electrolyte fuel cells. Some other membranes from Japanese and German companies are beginning to appear. The main problem with these membranes is that their cost of production seems to be extremely high—say, as much as 60 times the cost of Teflon with a similar backbone structure. Further, the conductivity of the membranes depends on their water content and reaches a saturation value when the water content is about 50% by volume. Thus, these membranes cannot be used at temperatures much above 100°C even under acceptable pressurized conditions (for example, up to 10 atm).

Therefore, a twofold technical approach is necessary to find alternate membranes: (i) development of alternate low-cost superacid membranes; and (ii) investigations of proton-conducting solids which can operate at temperatures up to about 400°C. There has been limited success in these directions, which is probably because there has only been a cursory examination with alternate electrolytes such as perfluoroorganosulfonic acids[53] [$HO_3S(CF_2)nSO_3H$], perfluoroorganophosphinic acids [$HO_2P(O)R$, $R = CF_3(CF_2)_n$], perfluoroorganophosphinic acids [$HOP(O)R_2$, $R = CF_3(CF_2)_n$], and perfluorosulfonimides [$RSO_2(NH)$, $R = (CF_2)_n$]. The costs of syntheses of some of these acids are expected to be considerably lower than those of the perfluorinated sulfonic acids. Work in this direction will be fruitful.

Fast proton-conducting solids, capable of operating up to about 400°C, will be ideal electrolytes for fuel cells using natural gas or methanol. For transportation applications, an attractive operating temperature range is 150 to 200°C for fuel cells using methanol. The CO poisoning problem can be greatly minimized. The potential candidates for electrolytes are hydrated salts, where proton conduction can operate by a hopping mechanism ($HUO_2PO_4 \cdot 4H_2O$).[54] In this compound, the layers are connected by hydrogen bonds from water and H_3O^+ ions in intervening layers. Proton vacancies facilitate proton transport. Water is necessary to prevent drying, which can decrease the conductivity. The conductivity of this acid is $4 \times 10^{-3} \, \Omega^{-1} \, cm^{-1}$ at 25°C. Other similar acids are dodecamolybdophosphoric acid ($H_3Mo_{12}PO_{40} \cdot 29H_2O$) and dodecatungstophosphoric acid ($H_3W_{12}PO_{40} \cdot 29H_2O$), both of which have a conductivity of $0.2 \, \Omega^{-1} \, cm^{-1}$ at 25°C.[55]

An alternative approach, which was attempted in the 1960s and 1970s, is to make β-alumina proton conducting.[56] There was only partial success because stability problems were encountered. Such an approach has been reactivated at the University of Pennsylvania by Farrington and co-workers.[57] Enhanced proton conductivity was found by doping with ammonia.

3.3.3. Bifunctional Electrocatalysts for Regenerative Fuel Cells

The solid polymer electrolyte fuel cell system is gaining momentum for space, defense, and terrestrial applications. A promising application in space is the Pathfinder Mission (long-term lunar and Mars missions). For these flights, which could last 15 years, the best power source is the photovoltaic energy conversion system. The only energy storage system which can be coupled with this system—to meet the demands of duty cycle (14 days light, 14 days darkness) and lifetime (15 years)—is the hydrogen energy storage system (i.e. use of a photovoltaic power source to electrolyze water and hydrogen stored as gas and used in a fuel cell to provide power during the dark period). For this energy storage system, it will be most attractive if the electrolyzer and fuel cell can be combined into a single unit (a regenerative fuel cell).[58] The challenge, in this case, is to develop a bifunctional oxygen electrode at which oxygen evolution would occur during water electrolysis, and oxygen reduction during fuel cell operation. Until the present time, the difficulty has been to find an electrocatalyst which will function efficiently in both modes, mainly because oxygen evolution kinetics is best on oxide electrodes (e.g., $Pt_{0.5}Ru_{0.25}Ta_{0.25}O_x$)[59] and oxygen reduction is fastest on the oxide-free metal (e.g., Pt). A second problem is that the optimum structures of the electrode are different for gas evolution (say, high-surface-area gauze) and for gas consumption (Teflon-bonded porous gas diffusion electrodes). A further complication is that if gas evolution occurs on a porous gas diffusion electrode at reasonably high current densities, its structure is destroyed. The solid polymer electrolyte system has better prospects than the alkaline system for functioning in a regenerative mode because the electrolyte in the former is a proton-conducting membrane and in the latter is a liquid electrolyte held in a matrix. The other system which shows promise is the regenerative solid oxide fuel cell[60] (Section 3.5). The alternative is to have separate units for water electrolysis and fuel cell power generation.

Regenerative solid polymer electrolyte fuel cells operating at the relatively low current densities which are acceptable for NASA's Pathfinder Mission can attain high efficiencies. Efficiency is more important than power density for NASA's application because about 90% of the weight of the energy storage system will be the weight of the reactants for the required duty cycle of the mission. Two approaches may be taken to develop electrocatalysts for this regenerative fuel cell. One is based on the expectation that an alloy electrocatalyst of the type Pt-Ir will exhibit reasonably high activities because Pt is a good electrocatalyst for oxygen reduction, while Ir exhibits an excellent electrocatalytic activity for oxygen evolution. The other is to develop an electrode with two electroactive layers. The layer which will be closer to the electrolyte will have the electrocatalyst for oxygen evolution (e.g., the PtRuTaO catalyst referred to above, embedded on a metallic gauze), and the layer on the backside for oxygen reduction will have a Pt electrocatalyst. Success in developing regenerative fuel cells will be a major advance in the 21st century.

3.4. Molten Carbonate Fuel Cells

3.4.1. Perovskites for Oxygen Cathodes

Lithiated nickel oxide is widely used at the present time as the electrocatalyst for the oxygen electrode. It functions quite satisfactorily in terms of a low overpotential for oxygen reduction. The main problem is its stability in the highly corrosive electrolyte (62% Li_2CO_3, 38% K_2CO_3) that is currently used. Recently, it has been shown at Energy Research Corporation that this electrode material is more stable in Na_2CO_3-K_2CO_3 than in the currently used electrolyte. The other approach has been to select and evaluate mixed metal oxides, mainly of the perovskite (ABO_3) family. Over 50 such materials have been tested, and the most promising are manganese-doped lithium ferrite and magnesium-doped lithium manganate.

3.4.2. Alternate Electrolytes

There have been hardly any efforts to find alternatives to the 62% Li_2CO_3, 38% K_2CO_3 (mol %) electrolyte which is held in a lithium aluminate tile (matrix). In the preceding subsection, it was mentioned that there has been some work on testing Na_2CO_3-K_2CO_3 as the electrolyte and that in this electrolyte, the cathode material is more stable. One interesting fact is that in lithium-rich melts the peroxide ion is the stable species, which is formed according to

$$CO_3^{2-} + \tfrac{1}{2}O_2 \Leftrightarrow CO_2 + O_2^{2-} \qquad (8)$$

In K_2CO_3-rich melts, the superoxide ion is formed:

$$2CO_3^{2-} + 3O_2 \Leftrightarrow 4O_2^- + 2CO_2 \qquad (9)$$

Though a considerable amount of work has been done on the electrode kinetics of oxygen reduction in molten carbonate, more work is necessary for a detailed understanding of the intermediate steps in electrolytes where these two types of stable species are formed. Spectroscopic methods will be useful to identify the roles of the peroxide and superoxide species.

3.4.3. Stable and Low-Cost Anode Materials

The Ni–Cr (1–10% Cr) anodes perform quite satisfactorily. The main problems are high cost and susceptibility to sintering and mechanical deformation. The chromium in the alloy is oxidized to $LiCrO_2$ and to a great extent prevents sintering, but the mechanical creep problem still remains. Metal oxides such as Al_2O_3, $LiAlO_2$, and ZrO_2 have been added to inhibit sintering and creeping. Replacing Ni with low-cost Cu provides satisfactory electrochemical performance, but the sintering and mechanical problems are more severe. Further, copper is not as good as nickel as an internal-reforming catalyst. Metal-ceramic composites need to be investigated more so than metal-plated ceramics. Oxide additions to metals is also another promising approach.

3.5. Solid Oxide Fuel Cells

3.5.1. New Materials and Monolithic Structures

Progress in developing solid oxide fuel cell technology has been rapid, particularly due to the pioneering efforts at Westinghouse Electric Corporation. The attractive feature of this type of fuel cell is that it is a two-phase (solid and gas) system, and problems encountered in dealing with three-phase systems, such as electrolyte flooding, are minimized. The promising areas of work on materials for solid oxide fuel cells are:

(i) Anodes: Materials with mixed electronic and ionic conductivity are needed to enhance the three-dimensional reaction zone, as in the case of the solid polymer electrolyte fuel cell. Liou and Worrell have demonstrated that reduced titania in a ZrO_2–Y_2O_3 environment exhibits such behavior.[61] More work in this direction is essential.

(ii) Cathodes: Strontium-doped lanthanum manganate is a satisfactory material. The remaining problem is the slow diffusion of manganese into the electrolyte. About half the cell resistance is in the cathode. Alternate perovskite materials with better electronic conductivity need to be investigated.

(iii) Electrolytes: One of the goals has always been to find electrolytes which can be used at a lower operating temperature (say, 700°C). Lower operating temperatures are favorable from the points of view of better thermal compatibility and stability of cell components. Some electrolytes which initially showed promise are doped CeO_2 and doped Bi_2O_3. These materials have higher ionic conductivities than Y_2O_3-doped ZrO_2, but are susceptible to reduction under reducing environments near the anode. There is a considerable effort in several directions to develop solid-state ionic conductors. The search for such materials as electrolytes for fuel cells must continue.

(iv) Interconnection: Magnesium-doped lanthanum chormite has proven to perform well as the interconnection material. Some minor problems which still exist are (a) the slight decrease in electronic conductivity at low partial pressures of oxygen; and (b) the lower thermal expansion coefficient of this material as compared to that of the calcia-supported zirconia (support tube material). Efforts are under way to match these thermal expansion coefficients by increasing the magnesium content of the interconnection material. Alternate perovskites for interconnection materials are worthy of investigation.

The Westinghouse design of solid oxide fuel cells is based on a tubular design. The cathode, electrolyte, anode, and interconnection layers are deposited using electrochemical vapor-deposition or chemical vapor-deposition techniques. The fabrication technique is the major contributor to the capital cost of solid oxide fuel cells. Argonne National Laboratory, in cooperation with Allied Signal, is developing a solid oxide fuel cell system with a novel monolithic structure (honeycomb design). There is no support tube in this structure; thus, the weight and volume of this system can be greatly reduced. Further, the fabrication techniques for this system only involve tape-casting of the separate layers and bonding of these layers during the sintering operation. Until the present time, good performances have been obtained in single cells but not with multicell stacks. The monolithic solid oxide fuel cell system shows promise of being developed in the 21st century.

Another design which has always been appealing is a flat-plate structure because the assembly of a multicell stack (bipolar) should be relatively simple. The main difficulties have been with sealing of the edges. It is difficult to find gasket materials or sealants which will be able to withstand the high operating temperature. With such a design, which Japanese research workers are investigating, plasma deposition of thin layers of electrodes, electrolyte, and interconnection becomes feasible.

3.5.2. Regenerative Fuel Cells

An attractive feature of the solid oxide fuel cell is that it is the only system which functions well in both the fuel cell and the electrolysis mode. The dual-function mode has been demonstrated by Brown, Bovin and Cie and by Westinghouse Electric Corporation in the 1960s and 1970s. With the increasing interest in hydrogen energy technologies, particularly due to environmental problems, the photovoltaic systems coupled with solid oxide fuel cell systems could provide the means for solar to electric energy conversion for terrestrial applications in the most efficient manner in the 21st century. The direction of work, which has been outlined for primary fuel cells, will be equally applicable for regenerative fuel cells. Regenerative fuel cells could also play an important role in space and defense applications in the 21st century.

3.6. Direct Methanol Fuel Cells

3.6.1. Rationale for Developing Direct Methanol Fuel Cells for Electric Vehicles

The anodic oxidation of methanol in fuel cell power plants always has been and will be the fuel cell researchers' dream until it is accomplished, demonstrated, and commercialized. There are several incentives for researching and developing this type of fuel cell power plant. Firstly, it would enable efficient utilization of the abundant primary energy resources of natural gas and coal in the USA and a reduction in oil imports, which comprise over 50% of the present oil consumption in the USA. Technologically and economically, methanol is the idea liquid fuel which can be synthesized on a large scale from these primary fuels. Second, the alcohols (methanol and ethanol) can readily replace gasoline as the fuel for internal combustion and diesel engine-powered vehicles, with minor modifications to these engines. The present fuel distribution network will hardly need any alterations. During the transition period from the present until methanol fuel cell power plants are developed and commericalized, methanol can easily replace gasoline (the reason for excluding ethanol is that ethanol is better produced from biomass, which is not as abundant a resource as natural gas or coal). Third, the rapidly growing environmental problems (global warming due to increasing CO_2 levels, NO_x emissions, and ozone formation) make it imperative to use alternate fuels (compressed natural gas, methanol, ethanol, and hydrogen) in internal combustion engine-powered vehicles in the short term and in electric vehicles (hybrid fuel cell/battery powered) in the long term. Use of the above-mentioned alternate fuels in the conventional manner will have only relatively small effects on the environmental problems, unlike those in the case of electric vehicles. Battery-powered vehicles will have a smaller range (driving distance between charging) than the internal combustion engine-powered vehicles because even the most advanced batteries (sodium sulfur and zinc/bromine) are limited in terms of energy density. An additional problem with batter-powered vehicles is that the lifetime of the power source is considerably less than that of internal combustion engine-powered vehicles—a minimum of 1000 deep charge/discharge cycles will be required over a five-year period, and even the most advanced batteries for electric vehicles have not reached this goal. With the fuel cell power source, the lifetime should not be a problem.

However, due to (i) the lower power density but higher energy density of a fuel cell as compared to batteries and (ii) the higher cost of fuel cell cells, a fuel cell/battery hybrid power source will be the ideal power source for electric vehicles. In such vehicles, the battery has to provide the power only for startup and acceleration and is charged by the fuel cell during cruising; thus, the lifetime of the battery is greatly extended. The CO_2 problem will still exist with methanol-powered vehicles. However, because of the higher efficiency (by at least a factor of two) for methanol utilization in fuel cells as compared to that in internal combustion engines, the CO_2 problem will be reduced by at least a factor of two. However, the greater impact is that the NO_x and ozone problems will be practically eliminated with methanol fuel cell-powered vehicles.

3.6.2. Electrocatalysts for Methanol Oxidation in Phosphoric Acid and Solid Polymer Electrolyte Fuel Cells

The most attractive fuel cell systems for transportation applications are the ones with phosphoric acid and proton-conducting membranes (solid polymer) as the electrolytes.[37] The phosphoric acid fuel cell is limited in terms of power density (power/weight and power/volume) as compared with the solid polymer electrolyte fuel cell. Thus, the former is more applicable as a power source for bigger vehicles such as buses, trucks, and fleet vehicles while the latter is for passenger vehicles. An additional advantage of the former for commercial vehicles is that the average operating time for these vehicles is about 12–16 h/day while for passenger vehicles it is about 1 h/day. The higher-temperature operation of the phosphoric acid fuel cell is, thus, more suited for the commercial vehicles, and the converse (solid polymer electrolyte fuel cell) for the passenger vehicles.

The current status of and directions of research on electrocatalysis of methanol oxidation are adequately presented in Section 2.6. Thus, in this section, it is only stressed that these areas of research should and will continue until we can unravel the mysteries of the low electrocatalytic activities and poisoning problems. The bifunctional metal/metal oxide electrocatalysts, where the rates of the reaction of adsorbed CO or COH species (on platinum) and OH species (on the metal oxide such as RuOx) can be very fast, appear to be the most promising. One advantage of phosphoric acid over the solid polymer electrolyte may be the considerably higher operating temperature of the former, which may greatly accelerate the kinetics of the desorption of the intermediates. Spectroscopic methods (FTIR) have been most useful in examining the intermediate species. The other major challenge in the 21st century will be to inhibit the diffusion of the methanol from the anode to the cathode and its depolarizing effect on the oxygen reduction reaction: the answer will be to find an oxygen reduction electrocatalyst inert toward methanol oxidation.

REFERENCES

1. G. J. Young, *Fuel Cells*, Vols. 1 and 2, Reinhold, New York (1960 and 1963).
2. E. W. Justi and A. W. Winsel, *Kalte Verbrennung*, Franz Steiner Verlag, Weisbaden (1962).

3. B. S. Baker (ed.), *Hydrocarbon Fuel Cell Technology*, Academic Press, New York (1965).

4. K. R. Williams (ed.), *An Introduction to Fuel Cells*, Elsevier, New York (1966).

5. H. A. Liebhafsky and E. J. Cairns, *Fuel Cells and Fuel Cell Batteries*, John Wiley & Sons, New York (1968)

6. C. Berger (ed.), *Handbook of Fuel Cell Technology*, Prentice-Hall, Englewood Cliffs, New Jersey (1968).

7. J. O'M. Bockris and S. Srinivasan, *Fuel Cells: Their Electrochemistry*, McGraw-Hill, New York (1969).

8. S. S. Penner (ed.), *Assessment of Research Needs for Advanced Fuel Cells*, *Energy* 11(1/2) (1986).

9. K. Kinoshita, F. R. McLarnon, and E. J. Cairns, *Fuel Cells—A Handbook*, Office of Scientific and Technical Information, United States Department of Energy, METC-88/6096 (1988).

10. A. J. Appleby and F. R. Foulkes, *Fuel Cell Handbook*, Van Nostrand Reinhold, New York (1989).

11. D. G. Lovering (ed.), *Fuel Cells—Grove Anniversary Symposium '89*, Elsevier, London (1990).

12. S. Srinivasan, E. A. Ticianelli, C. R. Derouin, and A. Redondo, *J. Power Sources* 22, 359 (1988).

13. S. Srinivasan, S. Somasundaram, D. H. Swan, H. Koch, D. J. Manko, M. A. Enayetullah, and A. J. Appleby, *Proceedings of Symposium on Fuel Cells* (R. E. White and A. J. Appleby, eds.), A.I.Ch.E Meeting, San Francisco, California, Nov. 6-7, 1989.

14. K. Prater, in *Fuel Cells—Grove Anniversary Symposium '89* (D. G. Lovering ed.), Elsevier, London (1990).

15. K. Kordesch, in: *Handbook of Fuel Cell Technology* (C. Berger, ed.), pp. 361-424, Prentice-Hall, Englewood Cliffs, New Jersey (1968).

16. W. Juda, 134th Meeting of the Electrochemical Society, Montreal, Canada, October, 1968.

17. S. Srinivasan, *J. Electrochem. Soc.* 136, 41C (1989).

18. P. N. Ross, Oxygen Reduction on Supported Pt Alloys and Intermetallic Compounds in Phosphoric Acid, EPRI-EM-1553, Lawrence Berkeley Laboratory, Palo Alto, California (1980).

19. V. M. Jalan and D. A. Landsman, U.S. Patent 4,186,110 (1980); V. M. Jalan, D. A. Landsman, and J. M. Lee, U.S. Patent 4,192,907 (1980); V. M. Jalan, U.S. Patent 4,202,934 (1980).

20. V. Jalan, Extended Abstracts, 161st Meeting of the Electrochemical Society, Montreal, Canada, May 9-14, 1982, p. 581.

21. V. Jalan and E. J. Taylor, *J. Electrochem. Soc.* 130, 2299 (1983).

22. C. J. Maggiore, P. J. Hyde and S. Srinivasan, Extended Abstracts, 161st Meeting of the Electrochemical Society, Montreal, Canada, May 9-14, 1982.

23. M. T. Paffett, J. G. Berry, and S. Gottesfeld, *J. Electrochem. Soc.* 135, 1431 (1988).

24. D. A. Landsman and F. J. Luczak, U.S. Patent 4,316,944 (1982); U.S. Patent 4,373,014 (1983).

25. F. J. Luczak and D. A. Landsman, U.S. Patent 4,447,586 (1984).

26. P. Stonehart and J. P. McDonald, Stability of Acid Fuel Cell Cathode Materials, Final Report, EPRI RP-1200-2, Lawrence Berkeley Laboratory, Palo Alto, California (1981).

27. M. L. Warshay and P. R. Prokopious, in: *Fuel Cells—Grove Anniversary Symposium '89* (D. G. Lovering, ed.), Elsevier, London (1990).

28. E. R. Gonzalez, K.-L. Hsueh, and S. Srinivasan, *J. Electrochem. Soc.* 130, 1 (1983).

29. A. J. Appleby and B. S. Baker, *J. Electrochem. Soc.* 125, 404 (1978).

30. B. S. Baker and H. C. Maru, in: *Progress in Battery Solar Cells*, Vol. 5, JEC Press (1984).

31. A. J. Appleby, *Progress in Batteries and Solar Cells*, (H. Shimotake *et al.*, eds.), p. 246, JEC Press, Cleveland, Ohio (1984).

32. E. A. Ticianelli, C. R. Derouin, A. Redondo, and S. Srinivasan, *J. Electrochem. Soc.* 135, 2209 (1988).

33. E. A. Ticianelli, C. R. Derouin, and S. Srinivasan, *J. Electroanal. Chem.* 251, 275 (1988).

34. S. Srinivasan, M. A. Enayetullah, S. Somasundaram, D. H. Swan, D. J. Manko, H. Koch, and A. J. Appleby, *Proc. I.E.C.E.C.* 24, 1623 (1989).

35. S. Srinivasan, D. J. Manko, H. Koch, M. A. Enayetullah, and A. J. Appleby, *J. Power Sources* 29, 367 (1990).

36. A. Parthasarathy, C. R. Martin and S. Srinivasan, *J. Electrochem. Soc.* 138, 916 (1991).

37. H. S. Murray and J. R. Huff, Fuel Cell/Battery Hybrid Vehicle Assessment, Los Alamos National Laboratory Report, #LA-10948-MS, UC-96 (1987).

38. I. D. Raistrick, U.S. Patent 4,876,115 (1989).

39. L. A. Knerr, E. J. Reid, and F. Solomon, Extended Abstracts, 173rd Meeting of the Electrochemical Society, Atlanta, Georgia, May 15-20, 1988, p. 19.

40. J. Zagal, P. Bindra, and E. Yeager, *J. Electrochem. Soc.* **127,** 1506 (1980).
41. J. A. S. Bett, H. R. Kunz, S. W. Smith, and L. L. Vane Dine, Investigations of Alloy Catalysts and Redox Catalysts for Phosphoric Acid. Electrochemical Systems, FCR-7157f, prepared by International Fuel Cells for Los Alamos National laboratory under Contract #9-X13-D6271-1(1985).
42. J. A. Shropshire, *J. Electroanal. Chem.* **9,** 90 (1965).
43. B. D. McNicol, *Proceedings of the Workshop on the Electrocatalysis of Fuel Cells* (W. E. O'Grady, S. Srinivasan, and R. F. Dudley, eds.), Brookhaven National Laboratory, Upton, New York, May 15–16, 1973, Proceedings Vol. 79-2, pp. 93–113, The Electrochemical Society, Princeton, New Jersey (1978).
44. M. Watanabe and S. Motoo, *J. Electroanal. Chem.* **69,** 429 (1975).
45. M. Watanabe, N. Furuya, and S. Motoo, *J. Electroanal. Chem.* **191,** 367 (1985).
46. R. Adzic, W. E. O'Grady, and S. Srinivasan, *J. Electrochem. Soc.* **128,** 913 (1981).
47. J. Wang, H. Nakajima, and H. Kita, *J. Electroanal. Chem.* **250,** 213 (1988).
48. A. J. Appleby, U.S. Patent 4,610,938 (1986).
49. E. Yeager, M. Razak, A. Razak, and E. Yeager, *J. Electrochem. Soc.* **136,** 385 (1989).
50. S. Srinivasan, C. R. Derouin and I. D. Raistrick, Extended abstracts, 169th Meeting of the Electrochemical Society, Boston, Massachusetts, May 4–9, 1986, Abstract 537.
51. A. J. Appleby, personal communication.
52. A. C. C. Tseung and K. L. K. Yeung, *J. Electrochem. Soc.* **125,** 1003 (1978).
53. M. Razaq, A. Razaq, E. Yeager, D. D. DesMarteau, and S. Singh, *J. Appl. Electrochem.* **17,** 1057 (1987).
54. F. Weigel and G. Hoffmann, *J. Less-Common Met.* **44,** 99 (1976).
55. O. Nakamura, T. Kokama, I. Ogino, and Y. Miyake, *Chem. Lett.* **1979,** 17.
56. J. L. Briant and G. C. Farrington, Extended Abstracts, Oct. 15–20, 1978.
57. G. C. Farrington, J. L. Briant, M. W. Breiter, W. L. Roth, *J. Solid State Chem.* **24,** 311 (1978).
58. A. J. Appleby, *J. Power Sources* **22,** 377 (1988).
59. R. S. Yeo, J. Orehotsky, W. Visscher, and S. Srinivasan, *J. Electrochem. Soc.* **128,** 1900 (1981).
60. A. O. Isenberg, *Solid State Ionics* **3–4,** 431 (1981).
61. S. S. Liou and W. L. Worrell, *Proceedings of the First International Symposium on Solid Oxide Fuel Cells* (S. C. Singhal, ed.), Proceedings Vol. 89–11, pp. 81–89, The Electrochemical Society, Pennington, New Jersey (1989).

37

XAS Techniques for Investigation of Materials for Energy Conversion and Storage

James McBreen

1. INTRODUCTION

As in other areas of materials science, advances in batteries and fuel cells depend on our understanding and control of processes at the atomic and molecular level. In the past, a major impediment to achieving such an understanding and control for electrochemical systems was that the structure of the electrode/electrolyte interface had eluded experimental verification. This has been a particular hindrance to investigators because other aspects of electrochemical processes such as the overall reaction rate (the electrical current) and the driving force for the reaction (the electrode potential) can be controlled with great accuracy. Fortunately, techniques capable of probing the liquid/solid interface to aid in elucidating electrochemical phenomena are emerging rapidly. There are excellent recent reviews of these new optical,[1,2] X-ray,[3,4] and scanning tunneling microscopy techniques.[5] The opportunity now exists to utilize advanced instrumentation to define detailed features, participating chemical species, and interfacial structure with a precision heretofore not possible. The field stands at much the same position as investigation of the gas-solid interface stood 25 years ago. Progress in understanding the gas/solid interface has been coupled with the availability of surface analytical techniques such as X-ray photoelectron spectroscopy (XPS) and low-energy electron diffraction (LEED). Analogous techniques are now available for the solid/liquid interface.

Investigations of batteries and fuel cells are even more complicated. In addition to the structure of the electrode/electrolyte interface, the structures of reactants, products, catalysts, and electrolytes are often not known. Structural determinations are difficult because often these materials are amorphous. Many battery materials are hydrated oxides so meaningful structure determinations can only be made in the electrolyte. Recent work has shown that X-ray absorption spectroscopy (XAS) is a very powerful technique for *in situ* studies of these materials. This chapter describes the XAS technique and its application to problems related to batteries and fuel cells.

James McBreen • Department of Applied Science, Brookhaven National Laboratory, Upton, New York 11973.

Electrochemistry in Transition, edited by Oliver J. Murphy *et al.* Plenum Press, New York, 1992.

2. X-RAY ABSORPTION SPECTROSCOPY

The principles of XAS are best described by using concrete examples. X-ray absorption spectroscopy is simply the accurate determination of the X-ray absorption coefficient of a material as a function of photon energy, in an energy range that is below and above the absorption edge of one of the elements in the material. The simplest method for doing this is to perform a transmission XAS experiment.

2.1. Transmission XAS

The experimental aspects of a transmission XAS experiment are relatively simple. Figure 1 is a schematic representation of the experimental configuration. It consists of an X-ray source, a double crystal monochromator, a thin sample of the material, detectors for monitoring the beam intensity before and after it passes through the material, and a data acquisition system.

The data acquisition system is used for several purposes. These include stepping the monochromator to pass the desired photon energies, alignment of the sample in the beam, and monitoring the signals from the detectors. This is the scheme used at the National Synchrotron Light Source (NSLS) at Brookhaven National Laboratory (BNL). The time for obtaining a full spectrum is typically 20 minutes and is mainly limited by the dead time needed for rotation of the crystals in the monochromator.

2.2. Time-Resolved XAS

The technique of dispersive extended X-ray absorption fine structure (EXAFS) permits time-resolved spectral acquisition with time scales as low as 5 ms.[6] This is the scheme used on the dispersive beam line at LURE-CNRS in Orsay, France. The X-ray optics consists of a bent triangular-shaped Si(311) monochromator crystal to focus and disperse the quasi-parallel polychromatic X-ray beam from a positron storage ring. The bent crystal yields a correlation between the photon energy and the direction of the photon beam. This is shown schematically in Fig. 2. As a result, the beam converges to a focal spot, where the sample is located. On passing through the sample, the beam diverges toward a position-sensitive detector consisting of 1024 sensing elements spaced over 2.56 cm. The photon energy–reflecting angle correlation yields a pixel number–energy correlation in the detector array. Since there is no movement in the monochromator, data acquisition times are limited only by the response of the detector. At present, the facility at LURE has a band pass of about 700 eV. Thus, XAS can only be recorded to about 600 eV above the absorption edge. However, the technique has the great advantage of time-resolved measurements and a resolution of 10 meV for energy shifts at the absorption edge.[7] This is due to the absence of mechanical movement in the monochromator.

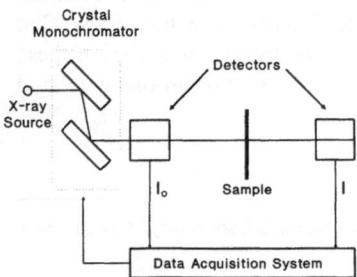

FIGURE 1. Experimental setup for transmission EXAFS measurements.

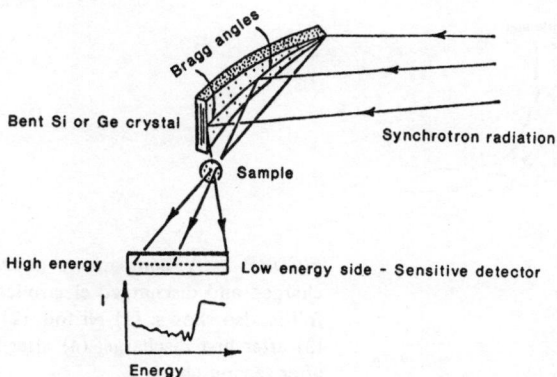

FIGURE 2. Experimental setup for dispersive EXAFS measurements.

2.3. XAS Spectra

Figure 3 shows a typical X-ray absorption spectrum for β-Ni(OH)$_2$ taken at the K-absorption edge for Ni. The K-edge is due to the ejection of a $1s$ electron from the Ni atom core. At low energies (<50 eV beyond the edge), the electron undergoes multiple scattering and transitions to empty states in the vicinity of the Fermi level. This region of the spectrum is called the X-ray absorption near-edge structure (XANES) region. XANES gives information on the oxidation state of the excited atom and on the type of coordination around the atom. The oscillations in the absorption coefficient at higher energies constitute the extended X-ray absorption fine structure (EXAFS). The EXAFS is a final-state interference effect between the photoelectron wave of the excited atom and a small fraction of that wave that is backscattered by surrounding atoms. Fortunately, the theory of EXAFS has been worked out in detail.[8-10] Analysis of the data gives information on the local bonding around the absorbing atom.

XANES

The fine structure in the XANES region can be explained in terms of (i) transition of the ejected photoelectron to unoccupied states in the vicinity of the Fermi level, and (ii) the

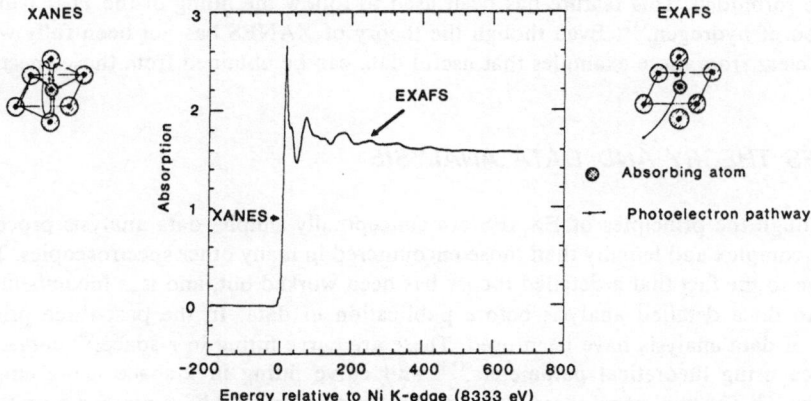

FIGURE 3. XAS for β-Ni(OH)$_2$; XANES and EXAFS regions are indicated; the multiple scattering (left) and backscattering (right) processes are shown schematically.

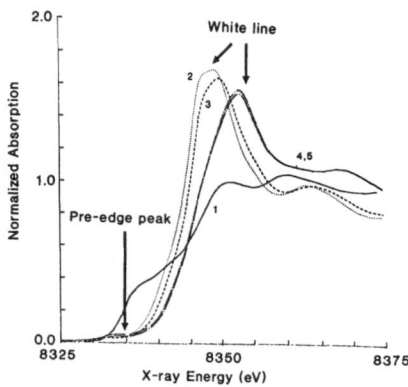

FIGURE 4. A comparison of XANES spectra for charged and discharged electrodes; a spectrum for Ni foil is also shown. (1) Ni foil, (2) fresh dry electrode, (3) after first discharge, (4) after first charge, and (5) after second charge.

long mean free path of the low-energy photoelectron, which results in multiple scattering around the excited atom. Because of the dipole selection rule, the shape of the edge yields information on both the type and symmetry of the ligands around the excited atom. Edge shifts due to core–hole interactions are indicative of changes in oxidation state. Thus, XANES yields important chemical information about the absorbing atom.

Figure 4 shows XANES spectra for nickel foil and various charged and discharged nickel oxide electrodes.[11] The edge shift to higher energies for the charged material can be clearly seen. The other features of the XANES are a small pre-edge peak and the white line. The pre-edge peak is due to the $1s \to 3d$ transition, and the white line to multiple scattering of the photoelectron in the NiO_6 octahedra. Normally, in the oxidized material the white line peak should be larger since there are more empty states at the Fermi level. However, in this case it is smaller. For transition metals the $1s \to 3d$ transition is normally forbidden by the dipole selection rule. The low-intensity peak is due to the weaker quadrupole transitions. The increase in the pre-edge peak and the decrease in white line intensity in the oxidized material are consistent with a distorted octahedral coordination. With the distorted coordination there is hybridization of p and d states, which makes the $1s \to 3d$ transition partially allowed. The same distortion reduces multiple scattering and decreases the intensity of the white line.

In the case of XANES for platinum catalysts at the L_3 edge, the white line is due to transitions between $2p_{3/2}$ and $5d_{5/2}$ states. At the L_2 edge, transitions into the empty $5d_{5/2}$ states are forbidden. This feature has been used to follow the filling of the Pt d bands on adsorption of hydrogen.[12] Even though the theory of XANES has not been fully worked out, it is clear from these examples that useful data can be obtained from these spectra.

3. EXAFS THEORY AND DATA ANALYSIS

Although the principles of EXAFS are conceptually simple, data analysis procedures are more complex and lengthy than those encountered in many other spectroscopies. This is partly due to the fact that a detailed theory has been worked out, and it is incumbent upon authors to do a detailed analysis before publication of data. In the past three principal methods of data analysis have been used. These are curve fitting in r-space,[9] curve fitting in k-space using theoretical parameters,[13] and curve fitting in k-space using empirical parameters.[14] The goal of all these methods is to extract the EXAFS functions from the raw data and to determine the structural parameters of interest from the extracted functions. The method of fitting in k-space using empirical parameters is discussed below.

Since meaningful data analysis cannot be done without understanding the theory, both will be discussed together. This will be done using results obtained on a β-$Ni(OH)_2$ sample. In the pre-edge region, X-ray absorption can be described by the relationship

$$I = I_0 \, e^{-\mu x} \tag{1}$$

where μ is the X-ray absorption coefficient of the sample, and x is the sample thickness. When the X-ray energy (8333 eV) is sufficient to liberate an inner nickel K-shell electron, there is an abrupt increase in the absorption (see Fig. 3). At low energies (<50 eV), the electron undergoes multiple scattering in the vicinity of the excited atom. This process contributes to the XANES. At higher energies, the electrons are excited to the continuum. The ejected photoelectron travels as an outgoing spherical wave with a wavelength (λ):

$$\lambda = \frac{2\pi}{k} \tag{2}$$

where k, the photoelectron wave vector, is given as

$$k = \left[\frac{2m}{\hbar^2} (h\nu - E_b - E_0) \right]^{1/2} \tag{3}$$

where m is the electron mass, $\hbar$ is Planck's constant divided by 2π, ν is the photon frequency, E_b is the binding energy of the electron, and E_0 is a correction to E_b caused by atomic potentials. If the nickel atoms had no neighbors, such as in an inert gas, then μ would continue to decrease smoothly beyond the absorption edge. However, neighboring oxygen atoms in the $Ni(OH)_2$ backscatter a small fraction of the photoelectron wave. The result is the EXAFS, which appears as oscillations in μ versus E. The EXAFS is a final-state interference effect involving backscattering of the photoelectron wave by neighboring atoms. The EXAFS function $\chi(k)$ is

$$\chi = \frac{\mu(E) - \mu_0(E)}{\mu_0(E)} \tag{4}$$

where μ and μ_0 are the X-ray absorption coefficients of the absorbing atom in the material of interest and in the free state, respectively. The difference $\mu - \mu_0$ depends on the local structure of the absorbing atom and represents the EXAFS. The division by μ_0 normalizes the EXAFS to a per atom basis. The first part of the data analysis is removal of the background, abstraction of the EXAFS, and normalization of the EXAFS.

3.1. Abstraction of the EXAFS

3.1.1. Pre-Edge Removal and Inner Potential Determination

The pre-edge absorption is removed by fitting the pre-edge data from -200 to -60 eV to a quadratic function, extrapolating it above the edge, and then subtracting it from the raw X-ray absorption data. The fit and extrapolation are shown in Fig. 5. To do EXAFS calculations in k-space, it is necessary to estimate the inner potential. This is determined from the inflection point at the absorption edge.

3.1.2. Background Removal

Background removal is the separation of the low-frequency components of the background from the higher-frequency oscillations due to the EXAFS. This involves the use of

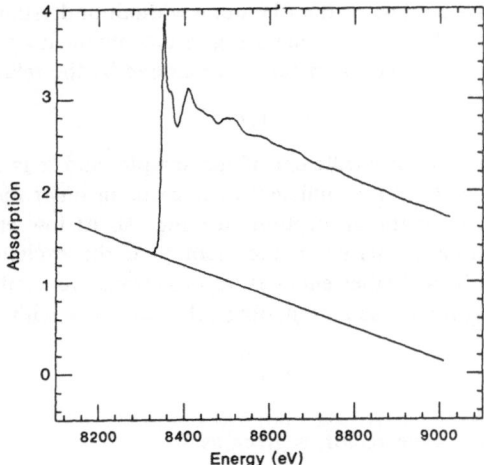

FIGURE 5. Pre-edge subtraction.

a cubic spline function $f(E)$ technique. The first derivative of the calculated background is used to determine the quality of the separation. The generation of the function $f(E)$ is controlled by a user-specified parameter, S, known as a smoothing parameter that determines the "tightness" of fit. An initial guess is made for the smoothing parameter, and the first derivative of the calculated background is checked to see if it has frequency components belonging to the EXAFS data. The optimum value of S is chosen such that in the first derivative of the calculated background, the EXAFS oscillations begin to disappear. Another method to check $f(E)$ is to examine the Fourier transform of the EXAFS. Values of S that are too low leave some of the low-frequency oscillations of $f(E)$ in the data. These appear as peaks at low r values in the Fourier transform. Values of S that are too large remove part of the EXAFS and decrease the magnitude of the peaks. There should be no oscillations in the background that correspond to the EXAFS. The final step is normalization of the EXAFS. This is done by dividing the EXAFS by the step height at 50 eV above the edge. This yields the final EXAFS plot, shown in Fig. 6.

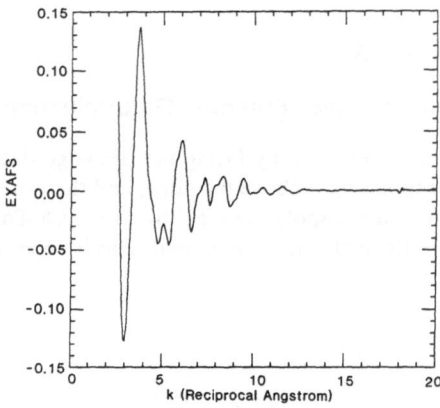

FIGURE 6. Final EXAFS or chi plot.

3.2. The EXAFS Spectrum

The EXAFS spectrum is the superimposition of contributions from different coordination shells to the backscattering process. The theoretical expression that relates the measured EXAFS to the structural parameters is given by

$$\chi(k) = \sum_j A_j(k) \sin[2kR_j + \phi_j(k)] \tag{5}$$

where j refers to the jth coordination shell, R_j is the average distance between the absorbing atom and the neighboring atoms in the jth shell, $\phi(k)$ is the phase shift suffered by the photoelectron in the scattering process, and $A_j(k)$ is the amplitude function, which is expressed as

$$A_j(k) = \left(\frac{N_j}{kR_j^2}\right) S_0^2(k) F_j(\pi, k) e^{-2(R_j-\Delta)/\lambda_j - 2\sigma_j^2 k^2} \tag{6}$$

Here N_j is the average coordination number, $F_j(k)$ is the backscattering amplitude of the atoms in the jth shell, σ_j^2 is a Debye–Waller term which accounts for the static and thermal disorder present in the materials, $S_0^2(k)$ is an amplitude reduction term which accounts for relaxation of the absorbing atom and multielectron excitations (shake-up/off) at the absorbing atoms, λ is the mean free path of the photoelectron, and Δ is a correction for the fact that S_0^2 and $F_j(k)$ already account for the photoelectron losses in the first shell.

3.2.1. The Fourier Transformation

The Fourier transform of the EXAFS data leads to the extraction of the information about the individual coordination shells and is given by

$$\Theta_n(r) = \frac{1}{\sqrt{2\pi}} \int_{k_{min}}^{k_{max}} k^n \chi(k) e^{2ikr} \, dk \tag{7}$$

The function $\chi(k)$ is multiplied by a factor k^n to equalize the envelope of $\chi(k)$ over the transformation range. The value of n is normally chosen in the range from 1 to 3 depending upon the amplitude variation and the signal-to-noise ratio of the measured data. The function $\Theta_n(r)$ is called the radial structure function and contains a series of peaks which reflect the local structure. To reduce termination errors which accrue from the finite range of the Fourier transformation, the values of k_{min} and k_{max} are chosen to coincide with nodes in the $\chi(k)$ function.

3.2.2. The Radial Structure Function

Figure 7 is a radial structure function from the data in Fig. 6, where the Fourier transformation was done over the range 3.6–15.25 Å^{-1}. This can be best understood by considering the structure of β-Ni(OH)$_2$, which is shown in Fig. 8.[15,16] β-Ni(OH)$_2$ has a layered hexagonal brucite structure. Within the plane each Ni atom is octahedrally coordinated with six oxygen atoms, three above the plane and three below the plane. Within the plane the Ni atoms are coordinated with three nearest shells of Ni atoms, each containing six atoms. These are indicated by 1, 2, and 3 in Fig. 8A. The three main peaks in Fig. 7 are due to the Ni—O coordination and to the first and third Ni coordination shells. The second Ni coordination shell and the Ni atoms in the neighboring planes cannt be seen because of the $1/R^2$ dependence of the EXAFS amplitude. The third Ni shell can be seen because of a special "focusing effect" caused by forward scattering by the first-shell Ni atoms that shadow the third shell. The r values for the peaks are slightly lower than the actual coordination distances because of the phase shift term in Eq. (5).

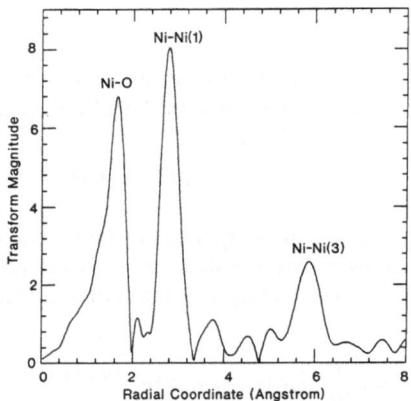

FIGURE 7. A Fourier transform of the data in Fig. 6.

3.2.3. The Phase Shift

The phase shift in Eq. (5) contains contributions from both the absorbing and backscattering atoms. The backscattered photoelectron experiences the central atom phase shift twice, once going out and once coming back. It encounters the phase shift of the backscatterer only once. Figure 9 shows theoretical scattering and absorbing atom phase shifts for various atoms.[8] The behavior of O is similar to that of C, and first-row transition metals such as Ni behave similarly to Ge. In the case of low-Z elements, we can distinguish between backscattering by O and other elements such as P, S, and Cl since the phase difference is about 1.57 rad.

3.2.4. Backscattering Amplitude

The amplitude function F_j in Eq. (6) depends solely on the backscattering atom. Figure 10 shows backscattering amplitudes for representative elements.[8] In the case of low-Z elements, F_j decreases rapidly with k. First-row transition elements behave similarly to Ge, with a peak at intermediate k values. Heavier elements such as third-row transition metals display more complicated behavior.

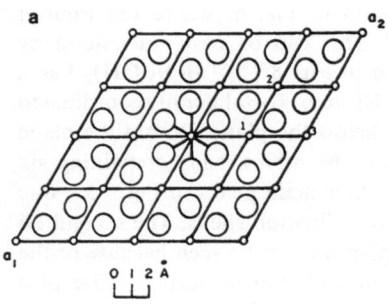

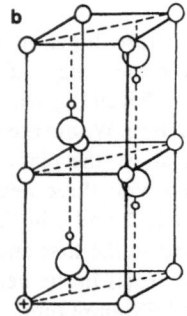

FIGURE 8. The structure of β-Ni(OH)$_2$: (A) within the plane, (B) stacking along c-axis.

O Ni   O-o OH

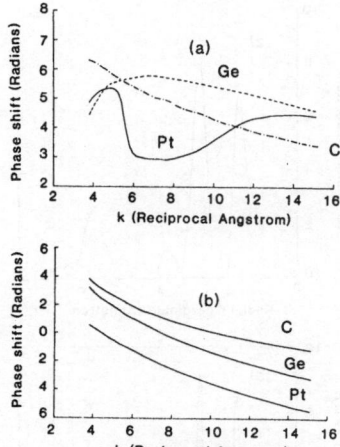

FIGURE 9. (a) Theoretical scattering phase shifts for C, Ge, and Pt. (b) Theoretical central atom phase shifts for C, Ge, and Pt.

3.2.5. Transferability of Phase and Amplitude

The present data analysis is based on the transferability of phase and amplitude parameters.The parameters are taken from known structures of the absorber–backscatterer pair and used in the calculations. In this analysis, NiO and Ni foil were used. Both are excellent standards since they have regular coordination with unique bond lengths. The great advantage of using empirical standards is that the data from the unknown and the standard can be analyzed in the same way. This eliminates errors due to background subtraction, normalization, etc.

3.2.6. The Imaginary Part of the Fourier Transform

In addition to the real part shown in Fig. 7, the Fourier transform has an imaginary part. Both parts are shown in Fig. 11. The real part contains amplitude information, and the imaginary part contains both phase and amplitude information.

3.2.7. The Inverse Fourier Transform

The information on a single shell can be obtained by applying a window function in the radial structure function and performing an inverse Fourier transformation from r-space to k-space:

$$k^n \chi_j(k) = \frac{1}{\sqrt{2\pi}} \int_{R_j - \frac{1}{2}\Delta R_j}^{R_j + \frac{1}{2}\Delta R_j} \Theta_n(r) e^{-2ikr} dr \qquad (8)$$

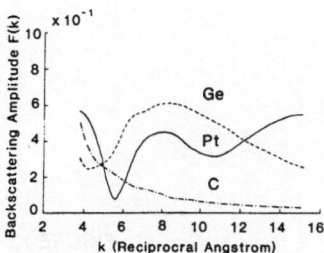

FIGURE 10. Theoretical backscattering amplitudes for C, Ge, and Pt.

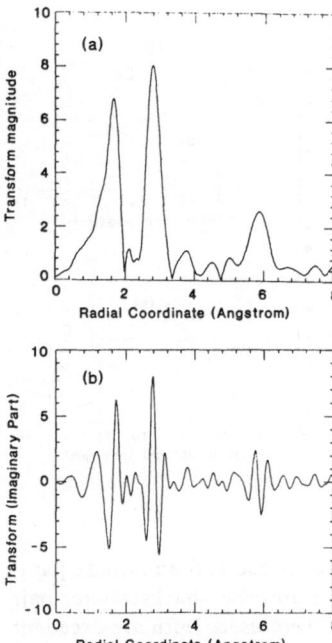

FIGURE 11. The Fourier transform: (a) real part, (b) imaginary part.

where R'_j is the position of the peak, and R_j is the width of the window function. Usually, R_j is chosen between nodes in the imaginary part of the Fourier transform on either side of the peak. A back transform for the second peak in Fig. 7 is shown in Fig. 12. The result is a pure sinusoidal EXAFS from a single shell. The amplitude has a maximum at $k = 8$ Å^{-1}, which is typical for a backscatterer such as Ni.

3.2.8. Calculation of Structural Parameters

The remainder of the data analysis involves least-squares fits to these back-transformed peaks in k-space. Phase and amplitude parameters from the standard compounds are used. The parameters calculated are N, R, $\Delta\sigma^2$, and ΔE_0; $\Delta\sigma^2$ is determined relative to the standard.

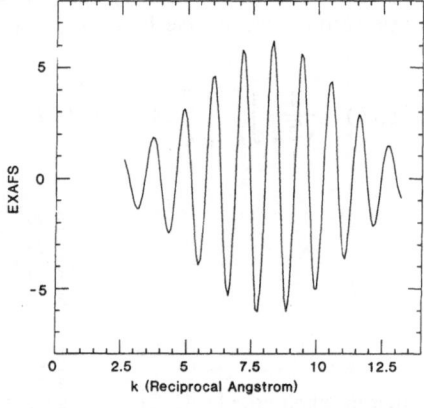

FIGURE 12. Back transform of second peak in Fig. 7.

In Eq. (5) $\chi(k)$ is a function of k and not E. Equation (3) shows that k depends on E_0, and E_0 is not related to any simple feature of the spectrum such as the inflection point. Accordingly, E_0 is treated as an adjustable parameter in the data analysis. This can compensate for differences in bonding between the unknown and the standard. The fitting procedure is a highly iterative process. The following is a typical procedure. Initially, the data are fitted in k-space, and all four parameters, N, R, $\Delta\sigma^2$, and E_0, are allowed to vary. When a good fit is found, R and E_0 are kept constant, and N and $\Delta\sigma^2$ are allowed to vary. After N and $\Delta\sigma^2$ are optimized, they are fixed, and R and E_0 are allowed to vary. Fits in k-space are done with various k weighting factors, typically k^1 and k^3. Fits are also checked in r-space using both the real and imaginary parts of the Fourier transform. The latter is an important check for the presence of overlapping coordination shells and the correctness of the assumptions about the nature of the coordinating atom. Analysis of data with overlapping coordination shells is much more complicated.

3.2.9. The Phase-Corrected Fourier Transform

High-Z scatterers can introduce side lobes on peaks in k^n-weighted Fourier transforms. Use of a phase- and amplitude-corrected Fourier transformation minimizes this. Such a corrected Fourier transformation is given by

$$\int_{k_1}^{k_2} \frac{\chi(k)e^{-i\phi_j(k)}e^{2ikr}}{F_j(\pi, k)} \, dk \tag{9}$$

Once again, $\phi_j(k)$ and $F_j(\pi, k)$ are obtained from suitable reference compounds.

4. EXAFS STUDIES OF BATTERY AND FUEL CELL MATERIALS

Since 1985 several XAS investigations of battery and fuel cell materials have been made at Brookhaven National Laboratory. *In situ* studies have included investigations of carbon-supported platinum fuel cell electrodes in several acids,[17] nickel oxide battery electrodes in strong alkali,[11] and the formation of supersaturated zincate solutions at discharging zinc electrodes in $12M$ KOH.[18] Preliminary studies have been done on underpotential deposited (upd) Cu and Pb monolayers,[19] the products of methanol oxidation on carbon-supported platinum in $1N$ H_2SO_4, and mossy zinc deposited in alkaline electrolyte. Time-resolved EXAFS has been coupled with cyclic voltammetry on nickel oxide[20] and carbon-supported platinum electrodes.[17] In addition, extensive *ex situ* EXAFS studies have been done on complexes in aqueous zinc bromide electrolytes and on the pyrolysis products of Fe and Co macrocycles.[21] More recently, work has been done on RbBr and $ZnBr_2$ in poly(ethylene oxide)-based electrolytes and on various forms of PbO_2. Some of this work will be reviewed here.

4.1. In situ EXAFS Studies of Nickel Oxide Electrodes

In situ transmission X-ray absorption spectroscopy was used to investigate changes in the structure of $Ni(OH)_2$ as it was cycled as a nonsintered nickel oxide electrode in concentrated alkali.[11] Since the charged products have very diffuse X-ray diffraction patterns, little was known about their structure. The most intense lines in the diffraction pattern are the first- and second-order reflections from the basal plane. These indicate that the structure shown in Fig. 8 expands along the c-axis on charge. The other diffuse lines indicate that there is a concomitant contraction in the a-axis. However, no details of the structure within the a-axis were known. In this investigation the changes occurring within the basal plane, during charge, were elucidated for the first time.

4.1.1. The Electrochemical Cell

The electrochemical cell for the *in situ* studies is shown in Fig. 13.[22] The cell consisted of a thin nonsintered nickel oxide electrode 25.4 mm in diameter and 0.25 mm thick. The electrode composition was 69.1 mg β-Ni(OH)$_2$, 5.4 mg Co(OH)$_2$, 32.4 mg graphite, 11.6 mg carbon fibers, and 12.4 mg plastic binder. The electrolyte was $8.4M$ KOH $+ 0.5M$ LiOH, which was absorbed in filter paper separators. The counter electrode was of thin Grafoil, and the reference electrode was a zinc wire. The cells were charged at the 10-h rate (2 mA) for 16 h. This was followed by discharge at the 3.3-h rate (6 mA) to 1.0 V versus the zinc reference. On the second charge the cells were recharged at 3 mA for 7 h. Cells were removed at the end of both charges and at the end of discharge and X-ray absorption measurements were made.

4.1.2. EXAFS of Charged Electrodes

Figure 14 shows the radial structure functions (RSF) for a charged and a discharged nickel oxide electrode. The RSF for the charged electrode has all the features of a discharged electrode. However, all peaks are shifted to lower r values, indicating contractions in interatomic distances within the basal plane. The RSF for a recharged electrode was essentially identical to that found after the first charge. Table 1 gives the structural parameters determined from the first and second peaks of the RSF for the charged and discharged electrodes. Every attempt to fit the first and second peaks in the RSF for the charged material with a single shell failed to produce a satisfactory fit with meaningful structural parameters. With further analysis the best fit was found with two shells. In the case of the discharged material, a two-shell fit was also needed for the second peak in the RSF. These results and the XANES features (see Section 2.3) show the complementary nature of the XANES and EXAFS results. They are self-consistent in that both indicate a distorted coordination for the charged material.

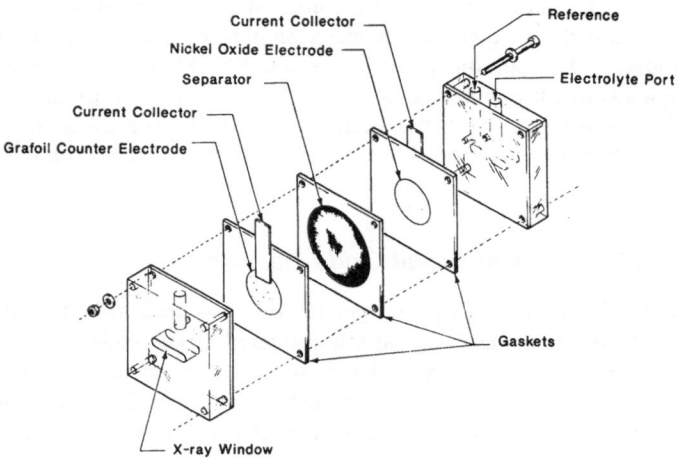

FIGURE 13. Cell for *in situ* X-ray absorption measurements on nickel oxide electrodes.

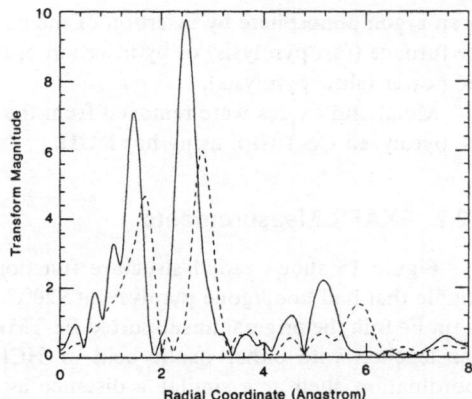

FIGURE 14. A comparison of radial structure functions for charged (——) and discharged (- - -) nickel oxide electrodes.

4.2. Pyrolysis Products of Metal Macrocycles

Absorbed N_4-chelates of transition metals such as Fe and Co have a high activity for electrocatalysis of oxygen reduction. However, they are unstable. It is well known that pyrolysis of these materials at temperatures between 300 and 1200°C greatly increases the stability and activity of these electrocatalysts. Activities comparable to that of platinum have been reported.[23] These catalysts have already found use in commercially available zinc/air hearing aid batteries and have been demonstrated in air depolarized chloralkali cells. Furthermore, these electrocatalysts are specific for oxygen and would be ideal for use in direct methanol fuel cells. Because of the great promise of these catalysts, it is vital to know their structure. This would permit tailoring of new electrocatalyst structures.

In this study the structure of pyrolyzed Fe and Co tetramethoxyphenylporphyrins (TMPP) on Vulcan XC-72 carbon was investigated using X-ray absorption spectroscopy.[21] Early in the investigation, it became apparent that pyrolysis yielded a mixture of products which included metals and oxides. Chemical leaching of these did not reduce the electrocatalytic activity. However, it revealed another metal compound which is the electrocatalyst. This compound is amorphous so EXAFS was an ideal technique for investigating its structure.

4.2.1. Catalyst Preparation

H_2-TMPP was synthesized, and the Fe and Co compounds were prepared in an insertion reaction. These were absorbed on carbon from acetone solutions, and the pyrolysis was done

TABLE 1
Structural Parameters for Nickel Oxide Electrodes

State of charge	Shell	R (Å)	N	$\Delta\sigma^2$ (Å^2)	ΔE_0 (eV)
Charged	Ni—O(1)	1.88	4.1	−0.0039	7.31
	Ni—O(2)	2.07	2.2	0.010	0.32
	Ni—Ni(1)	2.82	4.7	−0.0023	6.55
	Ni—Ni(2)	3.13	1.0	0.0072	−0.09
Discharged	Ni—O	2.05	5.7	0.0016	1.92
	Ni—Ni(1)	2.82	1.0	0.0009	−7.92
	Ni—Ni(2)	3.13	5.0	0.0007	0.08

in an argon atmosphere by insertion of the boat containing the catalyst into the hot zone of the furnace (fast pyrolysis) or by insertion of the boat in a cold furnace and then turning on the power (slow pyrolysis).

Metal and oxides were removed from the pyrolyzed Fe-TMPP using hot HCl and from the pyrolyzed Co-TMPP using hot KOH.

4.2.2. EXAFS Measurements

Figure 15 shows radial structure functions for a Fe-TMPP on Vulcan XC-72 carbon sample that had undergone pyrolysis at 920°C. Also shown are radial structure functions for 7-μm Fe foil, the original unsupported Fe-TMPP, and samples of the pyrolyzed material that was leached with either oxalic acid or HCl. Leaching reveals a material with a single coordination shell at a similar r distance as that for the original unsupported Fe-TMPP material. This is the catalyst.

Figure 16 shows a radial structure function for a Co-TMPP on Vulcan XC-72 sample that had undergone slow pyrolysis at 800°C. Also shown are radial structure functions for a 7-μm Co foil, unpyrolyzed Co-TMPP, and pyrolyzed material that was leached with KOH. Pyrolysis evidently produces Co metal. Removal of this in hot KOH reveals the catalyst, which has a similar structure to the core of the original macrocycle.

The experimental data for both the pyrolyzed Fe-TMPP and Co-TMPP materials could be fitted to a metal atom coordinated with four nitrogens in much the same way as the metal core of the original macrocycle. However, no nitrogen could be detected by either surface EXAFS or XPS. This does not agree with more recent XPS results on pyrolyzed iron porphyrins,[24] where evidence for nitrogen was found. Thus, at present there is some uncertainty about the identity of the coordinating atoms around the transition metal. What can be said with a reasonable degree of certainty is that monodispersed Fe and Co atoms are the sites for electrocatalysis. No evidence of metal–metal interactions could be found.

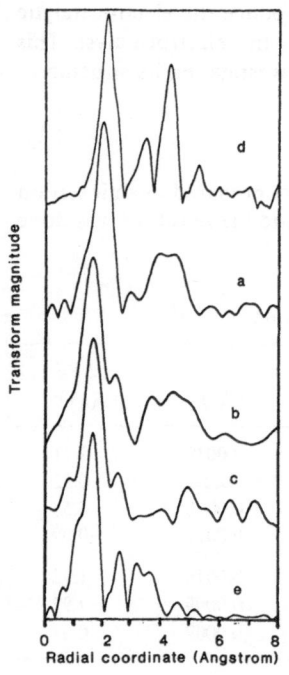

FIGURE 15. Radial structure functions for Fe-TMPP on Vulcan XC-72 after slow pyrolysis at 920°C (a) and for the same material leached with oxalic acid (b) and after a leach with hydrochloric acid (c). Radial structure functions for 7-μm Fe foil (d) and unsupported Fe-TMPP (e) are shown for comparison.

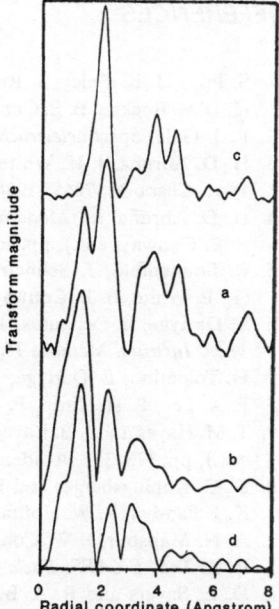

FIGURE 16. Radial structure functions for Co-TMPP on Vulcan XC-72 after slow pyrolysis at 800°C (a) and for the same material after a leach with KOH (b). Radial structure functions for 7-μm Co foil (c) and untreated Co-TMPP (d) are shown for comparison.

This would indicate that bridged bimetallic sites with critical geometry, as have been proposed,[25] are not necessary for oxygen reduction electrocatalysis. Thus, the prospects for substitutes for platinum in fuel cells are bright.

5. CONCLUSIONS AND FUTURE WORK

The application of XAS to materials problems related to batteries and fuel cells has only begun. The work discussed here shows that XAS can be applied to practically all aspects of these devices. The great advantages are that measurements can be done *in situ* and that the technique is element specific and gives chemical information. Thus, the chemical environment of a component in an electrode, such as an additive, can be probed. Since EXAFS probes only short-range order, it is a powerful tool for the study of amorphous materials, electrolytes, and adsorbed species. More widespread use will be made of these techniques in the future. This will put the field of energy conversion and storage on a more firm scientific basis and attract scientists from other disciplines to electrochemistry.

ACKNOWLEDGMENTS

The author is grateful to Dr. D. C. Koningsberger (Eindhoven University, The Netherlands) for providing many of the programs for data analysis and for giving guidance on their use.

The author gratefully acknowledges support of the U.S. Department of Energy, Division of Materials Sciences, under contract number DE-FG05-89ER45384 for its role in development and operation of Beam Line X-11 at the National Synchrotron Light Source (NSLS). The NSLS is supported by the Department of Energy, Division of Materials Sciences, under contract number DE-AC02-76CH00016. The author also acknowledges funds from a grant from the NATO Scientific Affairs Division (grant no. 86/532).

REFERENCES

1. S. Pons, J. K. Foley, J. Russell, and M. Seversen, in: *Modern Aspects of Electrochemistry*, No. 17, (J. O'M. Bockris, B. E. Conway, and R. E. White, eds.), pp. 223–302, Plenum Press, New York (1985).
2. R. J. Gale, *Spectroelectrochemistry*, Plenum Press, New York (1988).
3. H. D. Abruña, J. H. White, M. J. Albarelli, G. M. Bommarito, M. J. Bedzyk, and M. McMillan, *J. Phys. Chem.* **92**, 7045 (1988).
4. H. D. Abruña, in: *Modern Aspects of Electrochemistry*, No. 20 (J. O'M. Bockris, R. E. White, and B. E. Conway, eds.), pp. 265–326, Plenum Press, New York (1989).
5. R. Sonnenfeld, J. Schneir, and P. K. Hansma, in: *Modern Aspects of Electrochemistry*, No. 21 (R. E. White, B. E. Conway, and J. O'M. Bockris, eds.), pp. 1–28, Plenum Press, New York (1990).
6. E. Dartyge, L. Depautex, J. M. Dubuisson, A. Fontaine, A. Jucha, P. Leboucher, and G. Tourillon, *Nucl. Instrum. Methods Phys. Res., Sect. A* **246**, 452 (1986).
7. H. Tolentino, E. Dartyge, A. Fontaine, and G. Tourillon, *J. Appl. Crystallogr.* **21**, 15 (1988).
8. P. A. Lee, P. H. Citrin, P. Eisenberger, and B. M. Kincaid, *Rev. Mod. Phys.* **53**, 769 (1981).
9. T. M. Hayes and J. B. Boyce, in: *Solid State Physics*, Vol. 37 (H. Ehrenreich, F. Seitz, and D. Turnbull, eds.), pp. 173–351, Academic Press, New York (1982).
10. D. C. Koningsberger and D. C. Prins, *X-Ray Absorption*, Wiley, New York (1988).
11. K. I. Pandya, R. W. Hoffman, J. McBreen, and W. E. O'Grady, *J. Electrochem. Soc.* **137**, 383 (1990).
12. A. N. Mansour, J. W. Cook, Jr., and D. E. Sayers, *J. Phys. Chem.* **88**, 2330 (1984).
13. B. K. Teo, *EXAFS: Basic Principles and Data Analysis*, Springer-Verlag, Berlin (1985).
14. D. E. Sayers and B. A. Bunker, in: *X-Ray Absorption* (D. C. Koningsberger and R. Prins, eds.), pp. 211–253, John Wiley & Sons, New York (1988).
15. R. S. McEwen, *J. Phys. Chem.* **75**, 1782 (1971).
16. A. Szytula, A. Murasik, and M. Balanda, *Phys. Status Solidi* **43**, 125 (1971).
17. W. E. O'Grady, J. McBreen, G. Tourillon, E. Dartyge, and A. Fontaine, Extended Abstracts, Vol. 88-1, p. 621, The Electrochemical Society, Pennington, New Jersey (1988).
18. J. McBreen, W. E. O'Grady, G. Tourillon, E. Dartyge, A. Fontaine, and K. I. Pandya, Extended Abstracts, Vol. 89-2, The Electrochemical Society, Pennington, New Jersey (1989).
19. J. McBreen, W. E. O'Grady, G. Tourillon, E. Dartyge, and A. Fontaine, Extended Abstracts, Vol. 90-1, The Electrochemical Society, Pennington, New Jersey (1990).
20. J. McBreen, W. E. O'Grady, G. Tourillon, E. Dartyge, A. Fontaine, and K. I. Pandya, *J. Phys. Chem.* **93**, 6308 (1989).
21. J. McBreen, W. E. O'Grady, D. E. Sayers, C. Y. Yang, and K. I. Pandya, in: *Proceedings of Symposium on Electrode Materials and Processes for Energy Conversion and Storage* (S. Srinivasan, S. Wagner, and H. Wroblowa, eds.), Proceedings Vol. 87-12, pp. 182–197, The Electrochemical Society, Pennington, New Jersey (1987).
22. J. McBreen, W. E. O'Grady, K. I. Pandya, R. W. Hoffman, and D. E. Sayers, *Langmuir* **3**, 428 (1987).
23. J. A. S. Bett, H. R. Kunz, S. W. Smith, and L. L. Van Dine, Investigation of Alloy Catalysts and Redox Catalysts for Phosphoric Acid Electrochemical Systems, Final Report to Los Alamos National Laboratory, Contract No. 9-X13-D6271-1 (1985).
24. T. Sawaguchi, T. Itabashi, T. Matsue, and I. Uchida, *J. Electroanal. Chem.* **279**, 219 (1990).
25. J. P. Collman, P. Denisevich, Y. Konai, M. Marrocco, C. Koval, and F. C. Anson, *J. Am. Chem. Soc.* **102**, 6027 (1980).

38

Electrocatalysis on SPE Membrane Electrodes

Hideaki Kita, Hiroshi Nakajima, and Katsuaki Shimazu

1. INTRODUCTION

Solid polymer electrolytes (SPE) are very attractive materials in the field of electrolysis. Nafion (Du Pont Company) is a typical cation-exchange membrane, and its electrochemical properties, especially its ionic conductance, have been extensively studied.[1-4] Furthermore, an attempt has been made to use the membrane as an electrode by bonding a metal layer on its surface.[5-7] We have reported that the platinum bonded to Nafion shows a similar voltammogram to that of platinum metal, independent of the presence or absence of electrolytic solution on the platinum side of the membrane.[8] The absence of electrolytic solution on the platinum side provides a great advantage in that hydrogen or oxygen can adsorb directly from the gas phase and then be oxidized or reduced electrochemically, with no involvement of transfer processes in the solution. In fact, polarization of the electrode in this condition causes a large current exceeding the limiting value expected in the presence of electrolytic solution.[8]

In a series of papers, we have reported the ionization of H_2 and reduction of O_2 on Pt-SPE,[8,9] the oxidation of CO on Au-SPE,[10] the morphological study of both electrodes,[11] and the oxidation of CH_3OH vapor on Pt-SPE electrode.[12]

In this chapter, we discuss the electrocatalysis of the metal-bonded SPE electrodes based upon the results that we have obtained and report the remarkable effect that introduction of molybdenum species onto the Pt-SPE electrode has on CH_3OH oxidation.

2. EXPERIMENTAL

Pt-SPE was prepared by the method of Takenaka and Torikai.[13] A sheet of SPE (Nafion 120, 315, 324, or 425 or Asahi Kasei A201) was first immersed in boiling water for 0.2–2 h and then mounted at the bottom of a cylinder so that the upper side faced a 0.01–0.05M

Hideaki Kita, Hiroshi Nakajima, and Katsuaki Shimazu ● Department of Chemistry, Faculty of Science, Hokkaido University, Sapporo 060, Japan.

Electrochemistry in Transition, edited by Oliver J. Murphy *et al.* Plenum Press, New York, 1992.

H_2PtCl_6 or $HAuCl_4$ aqueous solution in the cylinder and the lower side faced a 0.1–$1.3M$ sodium borohydride or $0.1M$ hydrazine alkaline aqueous solution (ca. 50 ml) in a beaker. Metal ions were reduced at the upper side of the membrane. The apparent density of the platinum or gold bonded onto the SPE was 4–12 mg cm^2.

The cell used is of the three-compartment type, consisting of reference, counter, and test electrode compartments. The test electrode compartment is a gas chamber, divided by the metal-SPE membrane electrode. The metal side of the SPE electrode directly faces the gas phase (H_2, O_2, CO, CH_3OH, or N_2) and serves as a test electrode. The exposed area of the test electrode was 3.14 cm^2 in most cases. The reference electrode was the hydrogen electrode, and potentials in the text are referred to the reversible hydrogen electrode (RHE).

Electrolyte solutions were prepared from NaOH, H_2SO_4, $HClO_4$ (Wako Pure Chemicals Co., suprapure reagents), and purified water (Millipore, Milli Q system). Carbon monoxide from Seitetsu Kagaku Co. ($3N$ purity) was passed through a trap cooled to near the boiling point of liquid N_2. Hydrogen and N_2 were purified with respective commercial purifiers. Methanol vapor was supplied by passing N_2 through liquid CH_3OH at various temperatures.

Exhaust gas was analyzed, when required, using a mass spectrometer (Hitachi, GM-5) and gas chromatograph (TCD/SiO$_2$ for CO$_2$ and FID/APS201 for HCHO and HCOOCH$_3$, respectively).

Electrochemical measurements were carried out by a potential sweep or potentiostatic method at room temperature.

3. RESULTS AND DISCUSSION

3.1. Characterization of SPE Electrodes

3.1.1. Voltammogram

We first examined whether the metal-SPE electrode works when its metal side faces the gas phase. The cyclic voltammogram observed at the Pt-SPE electrode was exactly the same as those observed at a platinum electrode immersed in solution. This demonstrates that the metal-SPE electrode works properly. This was also the case for the Au-SPE electrode.

3.1.2. Roughness Factor

The roughness factor (r.f.) is also one of the important factors for the understanding of the characteristics of the electrode. From the quantity of electricity required for the oxidation of the adsorbed hydrogen (Pt-SPE) or the reduction of the surface oxide (Au-SPE), the roughness factor was calculated as listed in Tables 1 and 2 by assuming 0.21 or 0.42 mC, respectively, for a true area of 1 cm^2.[10,11] We can conclude from the tables that the r.f. of the cation-exchange SPE (Nafion) electrode is several times larger when the bare SPE side is in contact with acid solution than when it is in contact with alkaline solution, and the r.f. of the anion-exchange SPE (A201) electrode exhibits the opposite trend. Such a systematic difference suggests that the electrochemically active zone is located around the Pt/SPE interface just inside the membrane. This is also the case with Au-SPE electrodes as shown in Table 2. Since the r.f. of a Pt or Au metal electrode is not affected by the solution acidity or basicity, the differences observed at the metal-SPE electrodes lead us to conclude the existence of reaction zones as stated above. In comparison with the r.f.'s of Pt-SPE electrodes, those of the Au-SPE electrodes are always smaller, by up to ten times, independent of the kind of SPE and the solution composition.

TABLE 1

Roughness Factors (r.f.) of Pt-SPE and
Pt Electrodes[a]

Electrode	Solution	r.f.
Pt-Nafion 315	0.5M H_2SO_4	381 ± 122
	1M NaOH	64 ± 12
Pt-A201	0.5M H_2SO_4	150
	1M $HClO_4$	108 ± 8
	1M NaOH	710 ± 91
Pt Metal	0.1M $HClO_4$	1.3
	0.1M NaOH	1.2

[a] Sweep rates: 50–100 mV/s.

3.1.3. Morphological Measurements

Nafion 315 and 324 have a roughened surface on one side, at which metal was deposited. The Pt layer is densely packed near the surface while the Au layer is loosely packed and reaches deeper into the SPE membrane. To elucidate the details, transmission electron microscopy (TEM) observations were carried out.[11] Figure 1 shows typical morphological features of Pt embedded just below the surface. Small Pt crystallites (<10 nm in size) are connected with each other. On the other hand, Au reveals an entirely different crystal habit. Au crystallites grow much larger, some of them reaching a few hundred nanometers in size. These crystallites have an angular shape and are sometimes definitely hexagonal. This indicates that Au crystallites grow as single crystals.

It is interesting to consider the ion cluster network model for Nafion presented by Hsu and co-workers.[14,15] In this model, the ionic conduction takes place through spherical ion clusters of ca. 4-nm diameter which are connected by channels of ca. 1-nm diameter. The morphological features of Pt resemble the ion cluster network model, so that Pt seems to deposit in the ion clusters and channels. On the other hand, Au deposits in large crystals, exceeding by far the ion cluster size, with an angular shape characteristic of single crystals. Both metals were plated under the same conditions, and there was not a noticeable difference in the plating rates. Hence, the morphological difference must be attributed to a property of the metal itself.

TABLE 2

Roughness Factors (r.f.) of Au-SPE and Au Electrodes

Electrode	Solution	r.f.	ϕ_+[a] (V)	Sweep rate (mV/s)
Au-Nafion 315 (cation type)	1M $HClO_4$	61	1.8	60
	1M NaOH	22	1.5	50
Au-Nafion 425 (cation type)	1M $HClO_4$	67	1.8	50
	1M NaOH	19	1.5	60
Au-A201 (anion typoe)	0.5M H_2SO_4	18	1.8	60
	1M NaOH	72	1.5	60
Au metal	1M NaOH	1.65	1.6	120

[a] ϕ_+ denotes the positive potential limit at which the potential sweep is reversed. The O_2 evolution starts taking place at potentials larger than ϕ_+.

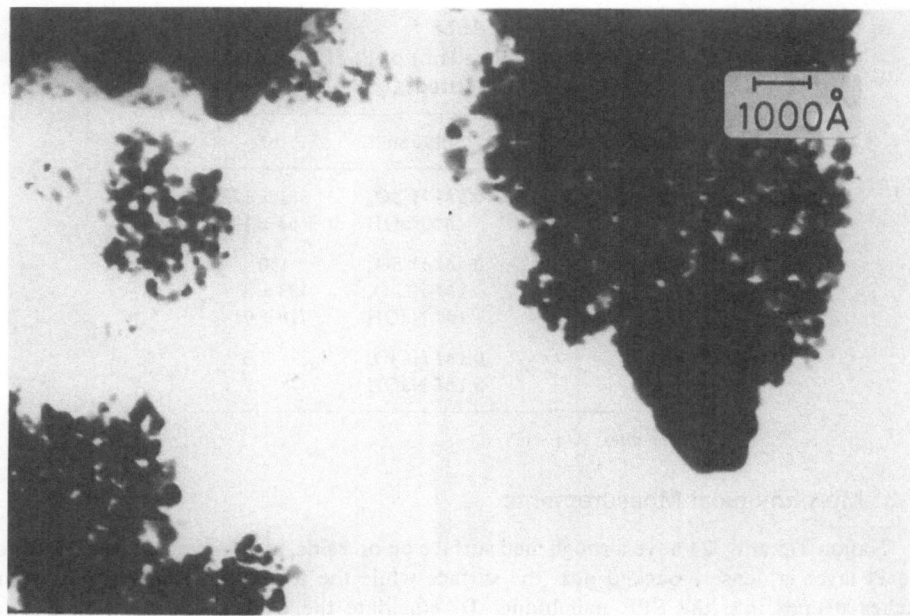

FIGURE 1. TEM photograph of cross section of Pt-SPE (Nafion 315) electrode (just below the surface). Each Pt particle is less than 10 nm in diameter.

3.2. Electrode Reactions at Metal-SPE Electrodes

3.2.1. H$_2$ and O$_2$ Electrode Reactions

When the gas chamber of the cell was filled with H$_2$ or O$_2$, the H$_2$ ionization or O$_2$ reduction reaction took place at potentials between 0 and 1.0 V. The polarization curves thus obtained on Pt-SPE electrodes are discussed below.

In Fig. 2, curve Pt-SPE I (for an SPE electrode with Pt deposited on one side) shows the i–ϕ relation for the hydrogen ionization reaction when the gas chamber was filled with

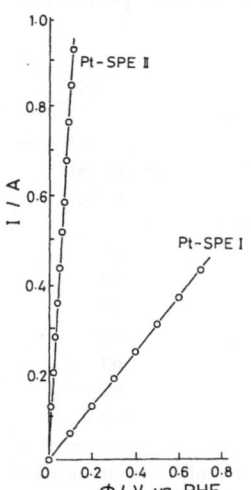

FIGURE 2. Anodic i–ϕ curves (H$_2 \rightarrow$ 2H$^+$ + 2e^-) on Pt-SPE I and II. Potentiostatic steady polarization, 1M HClO$_4$ in the counter electrode compartment, H$_2$ flow (1 atm) in the gas chamber. Apparent surface area, 14.5 cm^2.

H_2 gas alone. The i–ϕ relation was independent of sweep rate and showed an ohmic relation between i and ϕ, with a slope of 1.68 Ω. The slope was surprisingly independent of cathodic and anodic reactions (evolution and ionization of hydrogen).

Another example is the reduction of oxygen which took place when the gas chamber was filled with oxygen alone. The i–ϕ relation was also ohmic at $\phi < 0.8$ V, with a slope of 2.1 Ω. The value of the slope is almost the same as observed in the H_2 ionization and seems to be determined by the amount of the bonded platinum, the resistance of the membrane, the electrolyte concentration, and the geometry of the reference electrode in the cell.

Next we used another SPE electrode with platinum on both sides, Pt-SPE II (apparent surface area, 14.5 cm^2), where the platinum layer facing the electrolytic solution worked as a reference electrode. In this case, the solution resistance is completely excluded. Results for the hydrogen ionization reaction when the gas chamber was fed a flow of atmospheric hydrogen gas are shown in Fig. 2 in comparison with those at Pt-SPE I. We have again an ohmic relation for the hydrogen ionization reaction but with a much steeper slope than that for the reaction on Pt-SPE I. It is seen that the current far exceeds the limiting value (ca. 1 mA/cm$^{2(16)}$) for a usual gas electrode in solution. The reciprocal of the slope on Pt-SPE II gives a resistance of 0.127 Ω, which is much lower than the value of 1.68 Ω for Pt-SPE I. The decrease of 1.55 Ω must be attributed to the solution resistance, which is calculated as 1.31 Ω from a specific conductivity of 0.343 Ω^{-1} cm^{-1} ($1M$ HClO$_4$), a surface area of 14.5 cm^2, and a distance of 6.5 cm between the counter electrode of platinized Pt foil and the membrane surface (in this case, the Luggin capillary was absent). Hence, the value of 0.127 Ω of Pt-SPE II will be taken as mainly the resistance of the membrane.

3.2.2. Electrochemical Oxidation of CO

The electrochemical behavior of CO has been receiving considerable interest because of its potential application to fuel cells and the self-poisoning phenomena widely found in the electrochemical oxidation of organic substances. Our recent study[17] of the oxidation of CO at an Au electrode in alkaline solution shows that (i) the reaction proceeds by the reactant pair mechanism, where an adsorbed CO—OH$^-$ pair undergoes one-electron oxidation in the rate-determining step, (ii) a poisoning species, CO(a), forms at pH > 11 and $\phi < 0.25$ V (versus RHE), and (iii) the catalytic activity is higher than that of Cu, Ag, and Pt, being ca. 10^3 times larger in the absence of CO(a) than that of Pt. Thus, the Au-SPE electrode is expected to show enhanced activity for the electrochemical oxidation of CO.

Curve 1 in Fig. 3 represents the voltammogram on an Au-SPE (A201, anion exchange) electrode where the Au side is in contact with a flow of 1 atm CO and the other side (bare

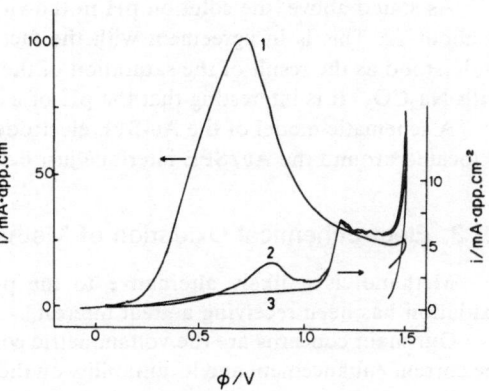

FIGURE 3. Voltammograms at Au-A201 in contact with $1M$ NaOH. (1) After aging at −0.1 V for 10 min under 1 atm CO flow; (2) after a further 30 min under 1 atm N$_2$ flow; (3) 2nd sweep following (2); sweep rate, 60 mV/s.

SPE) is in contact with $1M$ NaOH solution. A maximum current of 60 mA is reached for the unit apparent surface area, which is ca. 10^2 times larger than the corresponding value of 0.65 mA/cm^2 observed at an Au electrode immersed in the same solution saturated with 1 atm CO. The latter current is limited by the diffusion of the dissolved CO.[17] Thus, the Au-SPE electrode proved to be very effective in the oxidation of CO.

We have reported that a poisoning species forms on an Au electrode in alkaline solution during CO oxidation[11] and that, from the *in situ* IR spectra, the species is linearly bonded CO[18,19] [denoted simply as CO(a) in the following].

In the present system, the formation of CO(a) is also confirmed. After the cyclic potential sweeps between -0.1 and 1.5 V, the electrode was kept at -0.1 V for 10 min (t_{ad}) under the flow of CO at 1 atm in the gas chamber and then for another 30 min under N$_2$ flow at 1 atm. The first voltammogram thus obtained shows no oxidation of bulk CO, but a small peak at ca. 0.8 V is observed, as illustrated in Fig. 3 (curve 2). This small peak suggests the presence of a strongly adsorbed species remaining on the surface. The second voltammogram (Fig. 3, curve 3) is the same as that observed in the absence of CO, indicating that the CO(a) is removed by the first potential sweep. The quantity of electricity required for the oxidation of CO(a), $Q_{CO(a)}$, was examined as a function of the adsorption potential. In these experiments the electrode was pretreated by cyclic potential sweeps between the adsorption potential and 1.5 V. The formation of CO(a) at the Au-A201 electrode was found to take place at potentials more negative than 0.1 V, but the amount formed for the true unit area was several times smaller than that observed[17] at an Au electrode over the whole potential range. Comparison of the present result at -0.2 V with the previously reported pH dependence of $Q_{CO(a)}$ at an Au electrode[17] shows that $Q_{CO(a)}$ at the Au-A201 electrode corresponds to the value for the metal electrode at a pH of ca. 12 though the membrane is in contact with $1M$ NaOH (pH = 13.6).

The amount of the product, that is, CO$_2$, in the flowing gas was examined by a volumetric method in which a liquid N$_2$-cooled trap and vacuum system were used. The amount of CO$_2$ produced in successive time intervals and the corresponding electricity passed gives the current efficiency as a function of the reaction time. The efficiency at Au-A201 in contract with $1M$ NaOH gradually increases from ca. 45% and reaches ca. 60% after 2 h. The remaining 40% reacts with OH$^-$, and the resulting CO$_3^{2-}$ diffuses through the membrane into the solution. However, addition of $0.1M$ Na$_2$CO$_3$ into $1M$ NaOH raises the efficiency to a value more than 90%. Hence, the main process will be

$$CO + 2OH^- \rightarrow CO_2 + H_2O + 2e^-$$

It is worthwhile mentioning that the elimination of the product as CO$_2$ prevents the change of the pH of alkaline media. In a usual electrolysis, the CO$_2$ produced neutralizes an alkaline solution and the solution pH decreases.

As stated above, the solution pH in the vicinity of the electrochemically active surface is about 12. This is in agreement with the fact that the product evolves as CO$_2$, which is understood as the result of the saturation of the solution in the vicinity of the active surface with Na$_2$CO$_3$. It is interesting that the pH of a saturated Na$_2$CO$_3$ solution is ca. 12 at 25°C.

A schematic model of the Au-SPE electrode is shown in Fig. 4, where the reaction zone is located around the Au/SPE interface just below the surface.

3.2.3. Electrochemical Oxidation of Methanol

Methanol is a likely alternative to the petroleum-derived fuels. Its electrochemical oxidation has been receiving a great interest.

Our main concerns are the voltammetric comparison of a Pt and the Pt-SPE electrodes, the current enhancement and its durability on the SPE electrode, and the product distribution.

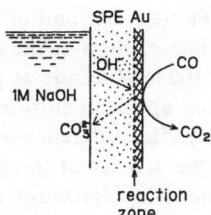

FIGURE 4. A model of the Au-SPE electrode.

The product distribution at a Pt electrode reported by different authors varies depending on the experimental conditions. For example, the CO_2 fraction in the products, which is taken as a measure of the methanol combustion, has been reported to be 60–100% in acidic solution[20-22] and 30% in alkaline solution.[20] Takahashi and co-workers[23] showed that the current fraction of CO_2 increases with the reaction time and with the r.f. of a platinized Pt electrode in $1M$ H_2SO_4.

Methanol oxidation on a metal-SPE electrode has been reported by Aramata and Ohnishi.[24] The initial catalytic activity of platinum on either a cation- or anion-exchange SPE membrane is comparable to that of a platinized Pt. After an initial deactivation during a short time, the Pt-SPE electrode retains a high activity for a long time, whereas platinized Pt loses its catalytic activity to a great extent during polarization. Aramata and Ohnishi[24] fed the methanol from a $1M$ $CH_3OH + 1M$ $HClO_4$ solution in contact with the platinum side of the Pt-SPE electrode. In the present study, we used methanol vapor as in the case of CO oxidation.

Figure 5 shows the voltammogram for CH_3OH_{vap}/Pt-SPE$/0.1M$ $HClO_4$ (curve 2). The oxidation starts at ca. 0.5 V and is accelerated sharply to a large current even at a low partial pressure of methanol, P_M, of several torrs. The oxidation current shows a maximum at 0.76 V and then increases again with further positive polarization. The figure also compares the voltammogram for CH_3OH (3 torr)$/Pt$-SPE$/0.1M$ $HClO_4$ with that obtained at a Pt electrode in $2mM$ $CH_3OH + 0.1M$ $HClO_4$ (curve 1). The main peaks on the anodic sweep at the two electrodes are almost the same. Thus, both electrodes are taken to function fundamentally in the same way for the methanol oxidation. The current appears 10^3 times larger at the Pt-SPE electrode than at a Pt electrode. The Pt-SPE electrode keeps its catalytic activity high even after a long polarization of 20 h. The current at a Pt electrode, however, decreases very rapidly, dropping by a factor of 50 in an hour.

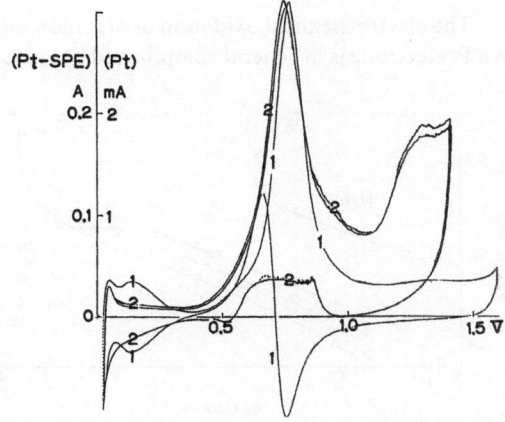

FIGURE 5. Cyclic voltammograms for methanol oxidation at Pt-SPE and Pt electrodes.

Figure 6 shows Tafel plots at a Pt electrode and at the Pt-SPE electrode. The currents are those observed at 2 min after the interruption of the anodic sweep at various potentials. The Tafel line holds with a slope of 100 mV/decade at potentials below ca. 0.65 V at both electrodes. The current density per unit apparent surface area is 2×10^2 times larger at the Pt-SPE electrode than at a Pt electrode. The difference will become much larger at a longer polarization because of the decay of the activity of the latter electrode. The solid circles in Fig. 6 represent the results obtained after 1-h polarization at the Pt-SPE electrode. The current decay is very small.

In Fig. 6, the oxidation current of the Tafel relation is not affected by the methanol partial pressure, P_M. At polarizations higher than $0.6 \sim 0.65$ V, the current tends to approach a limiting value which depends sensitively on P_M. The limiting current, i_L, increases proportionally to P_M as expected from the diffusion control of methanol through the channels but becomes constant at $P_M >$ ca. 50 torr. In the latter region it is likely that the water permeation through the SPE membrane controls the methanol oxidation since the feed gas does not contain water vapor.

Product distribution was examined under various conditions at the SPE electrodes by collecting the exhaust gas for several minutes to one hour, depending on the polarization potential. The yield of CO_2 at 1.0 V increases with decreasing P_M and becomes almost 100% at $P_M < 10$ torr. The sum of the yields for CO_2 and $HCOOCH_3$ reaches almost 100% over the whole range of P_M studied, and therefore the amounts of other products are expected to be negligibly small. In fact, formaldehyde is detected only in a trace amount. The water vapor pressure, P_w, has no effect on the product distribution at low P_M. The CO_2 yield is ca. 95% up to $P_w = 19$ torr, and that of $HCOOCH_3$ is almost zero. This is attributed to a sufficient supply of water by the permeation through the SPE membrane.

These facts can be understood in terms of the following scheme:

$$CH_3OH + H_2O \rightarrow CO_2 + 6H^+ + 6e^- \tag{1}$$

$$CH_3OH + H_2O \rightarrow HCOOH + 4H^+ + 4e^- \tag{2a}$$

$$CH_3OH + HCOOH \rightarrow HCOOCH_3 + H_2O \tag{2b}$$

At low P_M, the amount of water supplied through the SPE membrane is enough for the methanol oxidation (Eq. 1). On the other hand, at high P_M, methanol does not receive a sufficient amount of water for reaction (1) and is partially oxidized to HCOOH (Eq. 2a), which is esterified in reaction (2b), where the water liberated is consumed in reactions (1) and (2a). The esterification will take place rapidly in the strongly acidic media of the SPE used.

3.3. Methanol Oxidation at a Molybdenum-Modified Pt-SPE Electrode

The electrochemical oxidation of organic substances, even such a simple one as methanol, on a Pt electrode is in general complicated because of by-product formation and self-poisoning

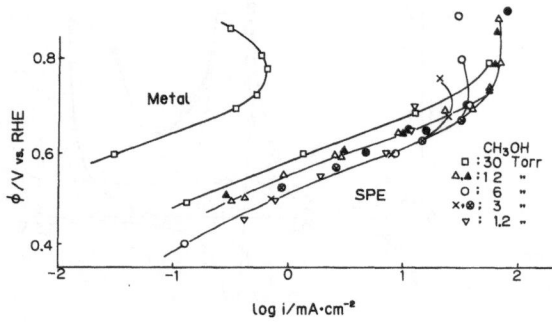

FIGURE 6. Tafel plots at Pt and Pt-SPE electrodes (2 min, 0.1M HClO$_4$). Current density is current/apparent surface area.

phenomena. A number of reports have described attempts to reduce the poisoning effect or to enhance the catalytic activity of Pt by introducing a second element on the surface. The number of second elements studied so far by various authors reaches more than 30 as reviewed by us.[25] Among these, molybdenum roused our interest because of its remarkable effect reported by Shropshire.[26] The presence of molybdate reduces the polarization by ca. 0.3 V at 82°C. Shropshire's results were explained by a cycle involving the chemical reaction of CH_3OH with molybdate and subsequent electrooxidation of the reduced molybdate. This mechanism is known as a "surface redox mechanism."

The effect, however, does not seem to have attracted much attention. Recently, we[27] reconfirmed the catalysis by molybdenum species adsorbed on a platinized Pt electrode, which leads to an acceleration in the methanol oxidation by 2×10^3 to 10^4 times in $0.5 M$ H_2SO_4 at 0.1 to 0.3 V (versus RHE) and 82°C. From a detailed analysis of the voltammograms, it is concluded that the methanol oxidation is catalyzed by the adsorbed redox couple Mo(IV)/(III). Mo(VI) is catalytically inactive, in contradiction to the conclusion of Shropshire.[26] The presence of Mo(IV) and Mo(III) was discussed in thermodynamic terms and supported by XPS measurements.

In the present study, a molybdenum-modified Pt-SPE electrode was used. The Pt-SPE electrode was first polarized at 0.05 V for about 1 h under a N_2 flow in the gas phase and then was modified with molybdenum species at the same potential by introducing $8 mM$ $Na_2MoO_4 + 0.5 M$ H_2SO_4 solution into the gas phase. After 1 min, the electrode at the same potential was washed with N_2-saturated H_2SO_4 and finally exposed to a N_2 flow for more than 20 min. Then methanol vapor at a pressure of 24 torr was fed, and the potential was shifted in a stepwise manner. The current after 5 min at each potential was recorded.

The Tafel plot is shown in Fig. 7. The numbers show the order of the measurements. The Tafel plot shows a complicated behavior and is divided into three regions, I (nos. 1 to 8 at low potentials), II (nos. 9 to 14 at middle potentials), and III (nos. 15 to 19 at high potentials). Region I shows that the methanol oxidation proceeds even near the reversible potential of CH_3OH/CO_2 ($\phi_0 = 0.04$ V[28]), which is a new finding at the molybdenum-modified Pt-SPE electrode. The decrease in the polarization potential amounts to ca. 0.3 V compared with that in region III. It is interesting to note that the current–potential relation below 0.1 V is rather stable and reversible. The current above 0.1 V, however, tends to decrease rapidly with time as indicated by the arrows (nos. 3 to 8). Once the potential reaches ca. 0.3 V, a following Tafel plot leads to region II, which still shows a considerable decrease in the polarization potential compared with that in region III, ca. 0.15 V. Again, the current–potential relation between 0.2 and 0.3 V (nos. 9 to 12) is stable and reversible. The current above 0.3 V becomes time dependent and finally gives region III at potentials more positive

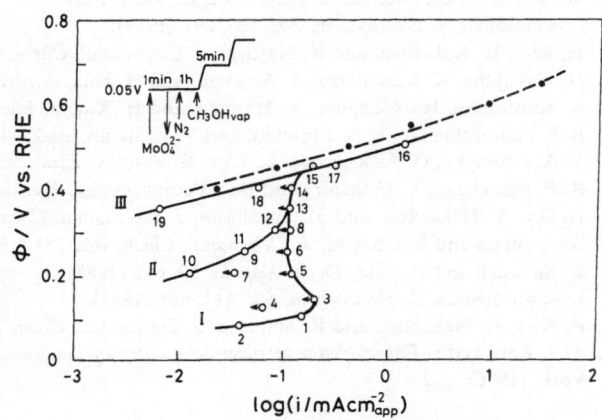

FIGURE 7. Tafel plot for the methanol oxidation at 25°C on Mo-modified (——) and unmodified (- - -) Pt-SPE electrodes. Gas phase: CH_3OH (24 Torr) + H_2O (15 Torr); r.f. = 830.

than 0.4 V. The Tafel plot in this region is the same as that obtained at the Pt-SPE electrode without the molybdenum species (Fig. 6).

Similar results are obtained at 80°C. The catalytic effect of the molybdenum species is further enhanced. The current in region II increases by a factor of ca. 10^2 as compared to that at 25°C.

4. CONCLUDING REMARKS

The present study shows many advantages of the SPE electrode. It has a high roughness factor of several hundreds for Pt, a constant catalytic activity without a decay as observed at a Pt electrode, a high current density for electrode reactions of sparingly soluble species, and many features characteristic of the respective electrode reactions. For example, the CO oxidation at an Au-SPE electrode in alkaline media proceeds without a pH change because the product leaves the electrode as CO_2 in the gas phase, not penetrating the membrane as CO_3^{2-}. The molybdenum-modified Pt-SPE electrode reveals an excellent catalytic activity, which is much higher than that of the molybdenum-modified Pt electrode.

Thus, the SPE membrane electrode has promising features for its use in various electrode reactions, especially in fuel cells.

REFERENCES

1. Z. Twardowski, H. L. Yeager, and B. O'Dell, *J. Electrochem. Soc.* **129**, 328 (1982).
2. R. S. Yeo, *J. Electrochem. Soc.* **130**, 535 (1983).
3. W. Y. Hsu and T. D. Gierke, *J. Membr. Sci.* **13**, 307 (1983).
4. A. Steck and H. L. Yeager, *J. Electrochem. Soc.* **130**, 1297 (1983).
5. Z. Ogumi, K. Nishio, and S. Yoshizawa, *Electrochim. Acta* **26**, 1779 (1981).
6. I. Bergman, *J. Electroanal. Chem.* **157**, 59 (1983).
7. A. Katayama-Aramata and R. Ohnishi, *J. Am. Chem. Soc.* **105**, 658 (1983).
8. A. Katayama-Aramata, H. Nakajima, K. Fujikawa, and H. Kita, *Electrochim. Acta* **28**, 777 (1983).
9. H. Kita, K. Fujikawa, and H. Nakajima, *Electrochim. Acta* **29**, 1721 (1984).
10. H. Kita and H. Nakajima, *Electrochim. Acta* **31**, 193 (1986).
11. H. Nakajima, Y. Takakuwa, H. Kikuchi, K. Fujikawa, and H. Kita, *Electrochim. Acta* **32**, 791 (1987).
12. H. Nakajima and H. Kita, *Electrochim. Acta* **33**, 521 (1988).
13. H. Takenaka and E. Torikai, Kokai Tokkyo Koho (Japan Patent) **55**, 38934 (1980).
14. W. Y. Hsu and T. D. Gierke, *J. Membr. Sci.* **13**, 307 (1983).
15. W. Y. Hsu and T. Berzins, *J. Polym. Sci.* **23**, 933 (1985).
16. S. Schuldiner, *J. Electrochem. Soc.* **106**, 891 (1959).
17. H. Kita, H. Nakajima, and K. Hayashi, *J. Electroanal. Chem.* **190**, 141 (1985).
18. H. Nakajima, K. Kunimatsu, A. Aramata, and H. Kita, *J. Electroanal. Chem.* **201**, 175 (1986).
19. K. Kunimatsu, H. Nakajima, A. Aramata, and H. Kita, *J. Electroanal. Chem.* **207**, 293 (1986).
20. B. I. Podlovchenko, A. N. Frumkin, and V. F. Stenin, *Elektrokhimiya* **4**, 339 (1968).
21. V. A. Gromyko, O. A. Khazova, and Yu. B. Vasiliev, *Elektrokhimiya* **12**, 1352 (1976).
22. R. P. Petukhova, V. F. Stenin, and B. I. Podlovchenko, *Elecktrokhimiya* **14**, 755 (1978).
23. K. Ota, Y. Nakagawa, and M. Takahashi, *J. Electroanal. Chem.* **179**, 179 (1984).
24. A. Aramata and R. Ohnishi, *J. Electroanal. Chem.* **162**, 153 (1984).
25. K. Shimazu and H. Kita, *Denki Kagaku* **53**, 652 (1985).
26. J. A. Shropshire, *J. Electrochem. Soc.* **112**, 465 (1965).
27. H. Kita, H. Nakajima, and K. Shimazu, *J. Electroanal. Chem.* **248**, 181 (1988).
28. A. J. Bard (ed.), *Encyclopedia of Electrochemistry of the Elements*, Vol. VII, Marcel Dekker, New York, (1976), p. 2.

39

Advances in Aluminum–Air Salt Water Batteries

B. M. L. Rao, W. Kobasz, W. H. Hoge, R. P. Hamlen, W. Halliop, and N. P. Fitzpatrick

1. INTRODUCTION

The aluminum–air cell belongs to the category of metal–air cells.[1-4] Of these, zinc–air batteries operating in alkaline electrolytes have been developed for industrial (railway signaling, telephone branch exchange) and commercial (hearing aid, powering light buoys) applications. The energy densities of the industrial batteries are 150–300 W h/kg and 175–400 W h/liter. The aluminum–air cell potentially has a higher energy density than the zinc–air cell. Therefore, successful development of the aluminum–air cell would provide a significant new battery product opportunity. The aluminum–air products range from salt water batteries for consumer use to alkaline electrolyte batteries for industrial applications. Also, the aluminum–air battery is under development as a power source for electric vehicles. This is because aluminum contains approximately one-half the energy content of gasoline per unit weight and three times the energy per unit volume. In this chapter, we discuss advances in salt water-based aluminum–air batteries.

Aluminum–air batteries have been under development in the Alcan laboratories since 1981. The results of the progress have been published[5-9] by the Kingston Research and Development Center, Kingston, Ontario, Canada, and by the Banbury laboratories, Banbury, United Kingdom. Alupower, a wholly owned subsidiary of Alcan, was formed in 1985 to commercialize aluminum-based energy-related products.

2. MERITS OF ALUMINUM AS AN ANODE METAL IN BATTERIES

The theoretical energy density of aluminum is higher than that of magnesium and zinc anodes on a weight basis, and its volumetric energy density is higher than even that of the lithium anode (Tables 1 and 2). The steady-state operating voltage of aluminum is slightly

B. M. L. Rao, W. Kobasz, W. H. Hoge, and R. P. Hamlen • Alupower Incorporated, Warren, New Jersey 07059.   *W. Halliop and N. P. Fitzpatrick* • Alcan International Ltd., Kingston, Ontario, Canada K7L 4Z4.

Electrochemistry in Transition, edited by Oliver J. Murphy *et al.* Plenum Press, New York, 1992.

TABLE 1
Comparison of Capacity Densities of Anode Metals

Anode metal	Valence change	Capacity density	
		A h/g	A h/cm^3
Aluminum	3	2.98	8.05
Magnesium	2	2.2	2.83
Zinc	2	0.82	5.82
Lithium	1	3.86	2.06

TABLE 2
Comparison of Energy Densities of Anode Metals[a]

Anode metal	Half-cell potential[b] (V)	Energy density	
		W h/g	W h/cm^3
Aluminum			
In salt water[c]	−1.5	4.47	12.07
In alkaline medium[d]	−1.65–2.1	4.9–6.3	13.2–16.9
Magnesium			
In salt water[c,e]	−1.65	3.63	4.7
Zinc			
In salt water[c]	−0.76	0.62	4.44
In alkaline medium[d]	−1.2	0.98	6.98
Lithium			
In nonaqueous electrolyte	−3.0	11.58	6.18

[a] Based on 100% material utilization.
[b] Potential vs. Hg/HgCl.
[c] Ph = 7.
[d] pH > 14.
[e] Passivates in alkali.

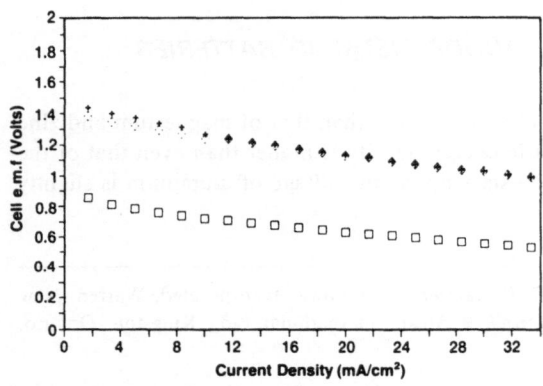

FIGURE 1. Metal–air cell polarization curves, □, Zn–air cell; +, Mg–air cell; ◇, Al–air cell. 12% NaCl, 0.32-cm gap, 20°C.

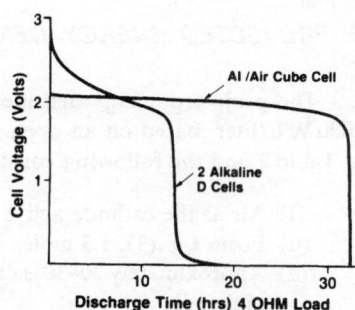

FIGURE 2. Comparison of discharge curves of aluminum–air cube cell (221 g/90 g empty) and two alkaline D cells (261 g).

lower than that of magnesium but is higher than that of zinc in salt water (Fig. 1). These features, and the higher kinetic stability (lower corrosion rate) of aluminum as compared to magnesium under open-circuit condition, render aluminum an attractive candidate for metal-air batteries.

Figure 2 shows that aluminum–air batteries (221 g) have a longer life than alkaline zinc–manganese dioxide D cells (261 g) under comparable discharge conditions (4-Ω load). This is a simple demonstration of the high energy density available from an aluminum–air battery.

3. ALUMINUM–AIR CELL REACTIONS

Figure 3 presents a schematic of an aluminum–air cell. The cell consists of an aluminum alloy anode and a gas diffusion-type air cathode separated by an electrolyte gap. The anode and the cathode half-cell reactions are presented in Eqs. (1) and (2), respectively:

$$Al \rightarrow Al^{3+} + 3e^- \tag{1}$$

$$O_2 + 2H_2O + 4e^- \rightarrow 4(OH)^- \tag{2}$$

The overall cell reaction is

$$4Al + 3O_2 + 6H_2O \rightarrow 4Al(OH)_3 \tag{3}$$

The observed open-circuit voltage of the aluminum–air cell in salt water (pH 7) is 1.5 V. Aluminum hydroxide is precipitated in the cell as the reaction product.

The air cathode of the cell is designed to operate with atmospheric oxygen. Therefore, the cathode active material is abundantly available at no extra cost. The life of the cell is determined by the quantity of aluminum and water that is consumed during the cell reaction.

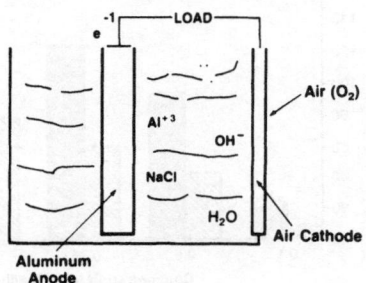

FIGURE 3. The aluminum–air battery chemical reactions. Reaction: anode $Al = Al^{+3} + 3e^-$, cathode $O_2 + 2 H_2O + 4e^- = 4 OH^-$, overall $4 Al + 6 H_2O + 3 O_2 = 4 Al(OH)_3$.

4. PROJECTED ENERGY DENSITY

The projected energy densities of an aluminum–air cell are 450–700 W h/kg and 650–1000 W h/liter, based on an open-circuit voltage of 1.5 V. This is based on the information in Table 2 and the following considerations:

(i) Air as the cathode active material does not contribute to cell weight.
(ii) From Eq. (3), 1.5 moles of water are consumed per mole of aluminum reacted.
(iii) Approximately 20–30% of the theoretical energy density may be realized in commercial cells.

5. ADVANTAGES OF ALUMINUM–AIR SALT WATER BATTERIES

Our interest in aluminum–air salt water batteries stems from the following potential advantages over the zinc–air alkaline electrolyte system.

5.1. Infinite Shelf Life on Dry Storage

The aluminum–air system can be configured as a "reserve" or "standby" battery. In this case, the battery would be stored dry, and the electrolyte would be added by the consumer prior to the use of the battery—known as "activation." In the absence of parasitic reactions, the battery would have an infinite shelf life.

5.2. Use of Innocuous Electrolyte

From the consumer safety point of view, it is desirable to develop reserve batteries capable of being activated by the addition of a noncorrosive, innocuous, commonly available solution. This is important since the consumer has to do the filling, handling, and disposing of the used cells. An aqueous solution of common salt or sea water fits this requirement.

Therefore, we are developing aluminum–air batteries based on the use of a sodium chloride solution as the electrolyte. We recognize that the conductivity of a sodium chloride solution is five to ten times lower than that of an alkaline electrolyte (Fig. 4) and that the cell would have a higher internal resistance. We consider that the advantage of an innocuous

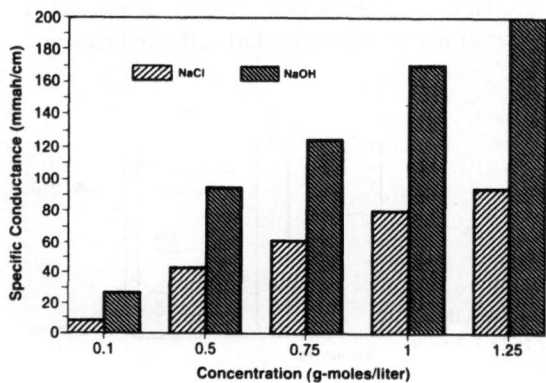

FIGURE 4. Conductivity of aqueous electrolytes (20°C).

electrolyte is more important than the penalty of high electrolyte resistance in some consumer applications.

5.3. Safety

The product of the aluminum–air cell reaction is aluminum hydroxide (Eq. 3). Aluminum hydroxide is nontoxic and is used in pharmaceuticals, cosmetics, and other household products. The pH of the saline solution next to the cathode is expected to rise because of the formation of hydroxyl ions from the reduction of oxygen at the air cathode (Eq. 2). However, the solution does not rise above 12 (mildly alkaline) due to the buffering action of aluminum hydroxide. Thus, the safety of the electrolyte is automatically maintained.

5.4. Environmental Compatibility

Zinc anodes used in alkaline cells (i.e., zinc–air, zinc–manganese dioxide) normally contain 1.5 to 4 wt. % mercury. This toxic mercury is added to reduce the shelf-storage at the zinc anode in the highly corrosive alkaline electrolyte. There is an increasing environmental concern over the indiscriminate disposal of used zinc-based batteries because of their mercury content. Aluminum–air salt water cells do not contain mercury or alkali. As indicated in the previous section, the cell reaction product is nontoxic. Nonetheless, we recommend that aluminum–air batteries be used in open areas with sufficient ventilation, since some hydrogen is produced during use. We are examining approaches to reduce the amount of hydrogen generated and to eliminate its release to the environment.

5.5. Reusability

The aluminum–air reserve battery is activated by adding salt water prior to use. However, if the aluminum is not fully consumed during initial use of the battery, the battery may be emptied, rinsed with fresh water, and stored for reuse a number of times before the full capacity is realized. As a result of the simplicity, safety, and environmental compatibility of aluminum–air salt water batteries, we have produced a number of "Aluminum–Air Power Cell Kits." These educational kits are for hands-on teaching of the electrochemical principles of batteries.

6. PREVIOUS WORK ON ALUMINUM–AIR SALT WATER BATTERIES

Published literature on aluminum–air cells has been recently reviewed by Littauer and Cooper.[1] Ritschel and Vielstich[2] have reported on aluminum–air cells in neutral and near-neutral electrolytes.[2] Despić and Milanovich[3] and Cooper and Homsy[4] have operated a pumped electrolyte system with a mechanically rechargeable anode. Other important references in this area are found in a number of publications from the Alcan laboratories.[5-13]

Aluminum–air cells with specific energies of 150–200 W h/kg are under development for stationary power sources, for marine applications, and for electric vehicles. Alcan has participated in the U.S. Department of Energy program as a subcontractor to Eltech Systems Corporation in the area of electrolyte management for alkaline battery systems. Alcan has also developed a mechanically rechargeable saline aluminum–air power pack for use as a hybrid electric vehicle power source in combination with a conventional lead–acid battery. In this application, the aluminum–air battery serves as a range extender/recharging unit.

7. CELL COMPONENT AND SYSTEM DEVELOPMENT

In the following sections, the key areas that impact on energy density, performance, operating life, and cost are discussed.

7.1. Aluminum Anodes

Preferably, aluminum anodes would operate near the thermodynamic potential (-1.66 V in saline and 2.7 V in alkali),[14] with low polarization and corrosion both at open circuit and under load. However, practical aluminum electrodes operate at a significantly lower potential. This is because:

(a) Aluminum is normally covered by an oxide film. This oxide film causes a delay in reaching a steady-state voltage due to internal resistance.
(b) Aluminum undergoes a parasitic corrosion reaction, resulting in less than 100% utilization of the metal and the evolution of hydrogen.

The above factors reduce both voltage and the utilization efficiency of the aluminum anode. In the research for solutions to overcome these inefficiencies, the electrochemical behavior of a number of aluminum alloys was studied.[15,16] The investigation showed that alloying with certain metals (gallium, mercury) improved the voltage.[8,9,11-13] The electrochemical behavior of super-purity aluminum (99.999% pure) was first studied in saline and alkaline electrolytes. Then a number of alloys were prepared by the controlled, low-level (50–1000 ppm) addition of alloying elements to the super-purity aluminum. These alloys were then characterized both metallurgically and electrochemically.

This study led to a reduction in the passivation and corrosion problems. Additives tested were gallium, mercury, thallium, and indium to aid the aluminum dissolution reaction. However, mercury and thallium additions are not used because of their toxicity. This route of controlling the behavior of the aluminum anode by the addition of alloying agents is known as "activation," and new compositions have been produced for use in alkaline and saline electrolyte batteries. Figure 5 illustrates the observed trend in the improvement in the aluminum anode behavior in an alkaline electrolyte. Similar trends are observed in electrolyte solutions. Addition of the alloying elements to the solution[17,18] was less effective than the alloying techniques because of the difficulty in controlling the concentration.

The mechanism of activation of aluminum is related to:

(i) moderating the thickness of the oxide film,
(ii) reducing the rate of the direct reduction of water by aluminum, and
(iii) controlling the dissolution morphology.

For example, the surface of an aluminum–gallium alloy becomes enriched with gallium during anodic polarization. This reduces the oxide film barrier and increases the voltage efficiency.

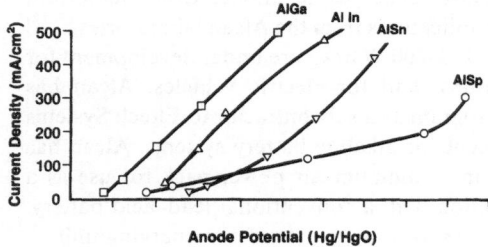

FIGURE 5. Microcell polarization curves comparing the performance of super-purity Al (AlSp) with 200 ppm Ga, In, and Sn binary alloys in $4M$ NaOH at 50°C.[8]

TABLE 3
Aluminum Alloy Polarization Data in 12% Sodium Chloride Solution at 20°C

	E vs. Hg/HgCl (V) at polarization current (mA/cm^2):				
	0	10	20	50	100
Target	−1.66	−1.52	−1.51	−1.50	−1.45
Sample	−1.52	−1.50	−1.49	−1.48	−1.45

The approach of the controlled, low-level addition of selected alloying elements has resulted in a number of new proprietary aluminum alloys with high voltages and high utilization efficiencies in salt water. Table 3 gives the steady-state voltage performance of a proprietary aluminum alloy in 12% sodium chloride solution with an 80% utilization at 10 mA/cm^2.

Iron and copper increase the parasitic corrosion of aluminum. From an economic point of view, it is desirable to develop aluminum anodes based on high-purity smelter-grade metal (99.8%) instead of the higher-cost super-purity metal (99.999%). If iron tolerance could be increased to 0.05%, then smelter-grade aluminum could be used. This would reduce the cost of the anode metal.

7.2. Electrolyte and Reaction Product Management

The conductivity of salt water is lower than that of an alkaline electrolyte. To minimize the power loss due to internal resistance, a narrow anode–cathode gap is required.

The anodic dissolution of aluminum in saline solutions results initially in soluble complexes with either hydroxyl or chloride ions, and subsequently in an aluminum hydroxide gelatinous precipitate. The gel product (hydrated aluminum hydroxide) has some negative effects:

(i) It can cause anode passivation due to the accumulation of aluminum hydroxide on the electrode.

(ii) It binds water and increases the water requirement. For example, gelation increases the water requirement from 0.33 ml/A h (Eq. 3) to approximately 5 ml/A h. This increased water requirement reduces both the gravimetric (W h/kg) and the volumetric (W h/liter) energy density.

(iii) Gelation makes cleaning the cell for reuse more difficult.

In attempting to minimize the problems, the following approaches have been tried:

(i) Use of mixed-salt electrolytes. Addition of sodium phosphate, sodium sulfate, sodium fluoride, and sodium bicarbonate to sodium chloride solutions increases the compactness of aluminum hydroxide and therefore reduces the water require-ments.[20]

(ii) Use of a crystallizer/separator, involving seeding and then the precipitation of large particles from the electrolyte. The separator retains the precipitated solids, and the remaining liquid is circulated back to cells. This approach has been successfully used in aluminum–air alkaline electrolyte batteries.[8,9,19]

(iii) Generation of turbulence by various means. The effect of inducing turbulence by reciprocal[8] pumping of the electrolyte through the cell is shown in Fig. 6.

We are also examining the addition of flocculating agents to precipitate the aluminum hydroxide.

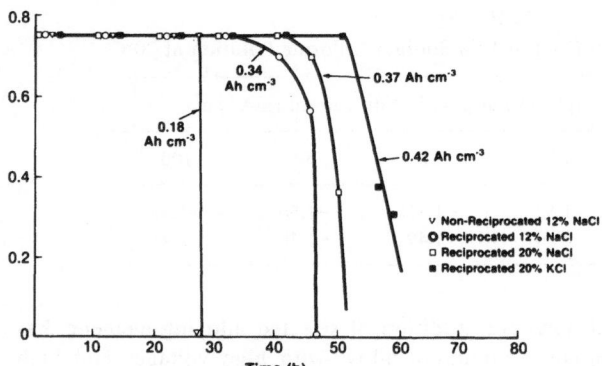

FIGURE 6. The effect of reciproca-
tion on electrolyte capacity for a
1-mm-gap cell as derived from the
current–time curves.[8]

7.3. Air Cathode Development

Gas diffusion-type air cathodes of suitable performance characteristics are available on the market. Further advances are anticipated in volume production and lower cost.

Alupower has a new, continuous process for making multilayered, laminated gas diffusion-type cathodes. A cross section of an air cathode is presented in Fig. 7. The new air cathodes can be produced in various forms, including;

(i) a reactive carbon layer sandwiched between a nickel grid current collector and a porous polyethylene film (AC-44), and

(ii) a four-ply construction consisting of a nickel grid with a carbon layer on each side, plus a porous fluoropolymer film attached to one surface. Production of 8- to 11-inch wide cathodes can be routinely handled in lengths of 100 feet or more.

A comparison of the performance of Alupower air cathodes with that of some other commercially available air cathode samples is shown in Figs. 8 and 9. The polarization curves show that Alupower cathodes are comparable in performance in saline solutions (0.002–0.04 W/cm^2) and in alkaline electrolytes (0.08–0.12 W/cm^2) over the current density ranges of interest for Alupower products.

The following is a list of aluminum air batteries and related products that are the results of developments at Alupower and the Alcan laboratories:

1. Aluminum alloy anodes for saline and alkaline battery use
2. Air cathodes
3. Aluminum–air salt water batteries:
   - *NightStar Reserve Light.* A portable emergency light source with a life of 240 hours.
   - *Barge Mooring Light.* A lighting system specifically designed for marking moored barges and barge fleets. The battery has a special bulb/lens combination meeting U.S. Coast Guard requirements and a solar switch to turn the light off during daylight hours. The performance data for the NightStar series of batteries are presented in Fig. 10. The energy density from NightStar batteries ranges from 140 to 200 W h/kg and 110 to 160 W h/liter in the operating range of 2.8 to 2.0.
   - *Reserve Light One—Strobe Light.* A warning or safety beacon for motorists, campers, and boaters. The light flashes brightly for 48 hours.

FIGURE 7. Cross section of Alupower AC45 air cathode. Total thickness is about 0.020 in. 1. Reaction layer made by impregnating a porous fibrous web with a proprietary mixture of carbon, binder catalyst and additives. There is considerable latitude in formulating this reaction layer. 2. Current collection of nickel expanded mesh. 3. Second reaction layer. 4. A microporous film, usually of fluoropolymer or polyolefin. For some applications this film is omitted. Any of the layers can be omitted or replaced by substitutes, with some limitation.

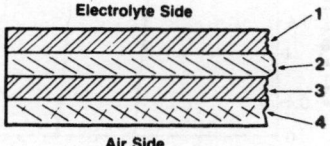

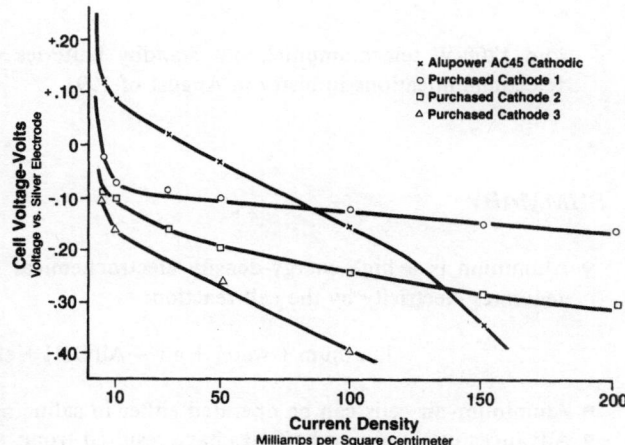

FIGURE 8. Polarization curves for Alupower air cathode and other commercially available air cathodes in 4M KOH electrolyte.

- *Flexilight.* A "standby" light source that operates for 12 hours of constant light and provides 48 hours of service as a resuable light.
- *Landwatt.* A high-rate (Fig. 11) modular cell capable of being assembled in series and parallel circuits.

4. Aluminum–air alkaline batteries. Alcan research and development efforts have resulted in a number of aluminum–air alkaline batteries that have higher power and energy density capabilities than salt water batteries. The most advanced of these are the 600-

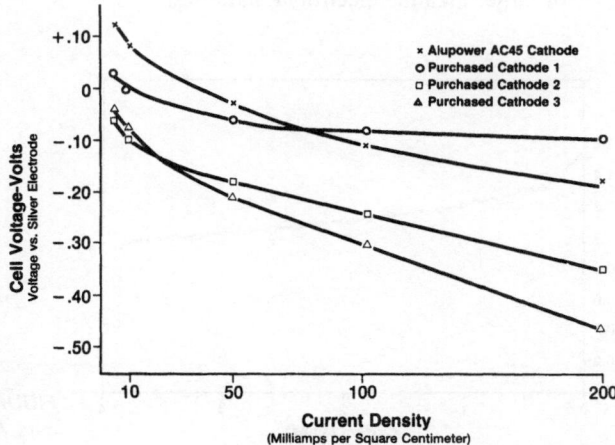

FIGURE 9. Polarization curves for Alupower air cathode and other commercially available air cathodes in 12% NaCl electrolyte.

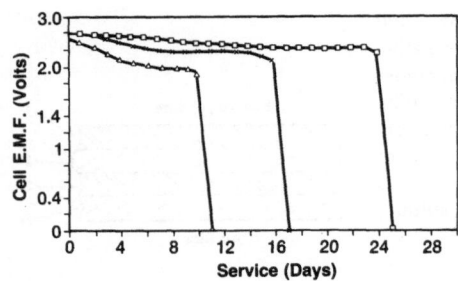

FIGURE 10. Barge cell performance at 70°F. □ 450 mA; ×, 800 mA; △, 1.15 A.

and 1200-W telecommunications standby batteries which were introduced to the telecommunications industry in August of 1991.

9. SUMMARY

- Aluminum is a high-energy-density electrochemical fuel for metal–air batteries. It generates electricity by the cell reaction:

$$aluminum + water + air \rightarrow Al(OH) + electricity$$

- Aluminum–air cells can be operated either in saline or alkaline electrolytes.
- Advances in aluminum batteries have resulted from:
    (i) development of high-efficiency aluminum alloy electrodes, and
    (ii) efficient electrolyte management to maximize cell life.
- Batteries based on the aluminum–air system have the following advantages:
    (i) infinite shelf life prior to activation
    (ii) Use of an innocuous electrolyte
    (iii) Safety
    (iv) Environmental compatibility
    (v) Reusability.
- Alupower has combined the advances and advantages with new air cathodes to produce a number of new aluminum–air salt water batteries and products, along with a range of larger alkaline electrolyte batteries.

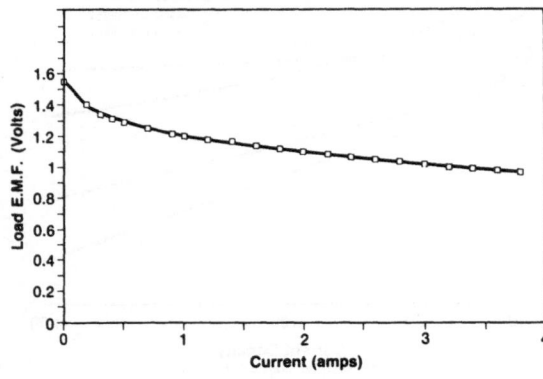

FIGURE 11. Landwatt cell polarization data. Al–air cell, 180 cm^2, 12% NaCl, 20°C.

REFERENCES

1. E. L. Litlauer and J. F. Cooper, in: *Handbook of Batteries and Fuel Cells* (D. Linden, ed.), Chapter 30, McGraw-Hill, New York (1984).
2. M. Ritschel and W. Vielstich, *Electrochim. Acta* **24**, 885 (1979).
3. A. R. Despić and P. D. Milanovich, *Recl. Trav. Inst. Sci. Tech. Acad. Serbe Sci. Arts* **12**(1), (1979).
4. J. F. Cooper and R. V. Homsy, UCRL-86560, August 1981; J. F. Cooper, International Reactive Metal–Air Battery Workshop, Palo Alto, California, May 18, 1982.
5. N. P. Fitzpatrick, F. N. Smith, and P. W. Jefferey, SAE, Proceedings of the International Congress and Exposition, Detroit, Michigan, February 28, 1983, Paper No. 830290.
6. B. Achaqrya and N. P. Fitzpatrick, International Conference on Aluminum, New Delhi, India, October 30, 1985.
7. G. M. Scamans, *Chem. Ind.* **6**, 192 (1986).
8. G. M. Scamans, W. B. O'Callaghan, N. P. Fitzpatrick, and R. P. Hamlen, Proceedings of the 21st IECEC Conference, San Diego, California, August 1986, p. 1057.
9. W. B. O'Callaghan and S. R. Knowles, Extended Abstracts of the Electrochemical Society Meeting, Las Vegas, Nevada, October, 1985, Abstract 8, pp. 15–16.
10. N. P. Fitzpatrick and G. M. Scamans, *New Scientist* **111**, 34 (1986).
11. P. W. Jeffery and W. Halliop, Extended Abstracts of the Electrochemical Society Meeting, Honolulu, Hawaii, October, 1987, Abstract 134, p. 193.
12. C. D. S. Tuck, J. A. Hunter, and G. M. Scamans, *J. Electrochem Soc.* **134**, 2970, 1987.
13. W. Halliop, Extended Abstracts of the 38th Meeting of the International Society of Electrochemistry, Maastricht, The Netherlands, September, 1987, Abstract 7.10, pp. 766–767.
14. M. Pourbaix, *Atlas d'Equilibres Electrochimiques*, Pergamon Press, Oxford, England, 1966, pp. 168–176.
15. J. T. Reding and J. J. Newport, *Materials Protection* **5**, 15 (1966).
16. M. J. Prior *et al.*, U.S. Patents (1965–1968).
17. W. Boehenstedt, *J. Power Sources* **5**, 245 (1980).
18. A. R. Despić, R. M. Stevanović, and A. M. Vorkapić, ISE Meeting, Berkeley, California, Abstract A2-19.
19. A. Mimoni, LLNL Preprint UCRL 92281, Lawrence Livermore National Laboratory, Livermore, California, March 1985.
20. D. M. Dražić, A. R. Despić, S. Zečević, M. Atanacković, and I. Illiev, in: *Power Sources 7* (D. H. Collins, ed.), p. 353, Academic Press, London (1978).

40

Organic Electrolytes of Rechargeable Lithium Batteries

Yoshiharu Matsuda

1. INTRODUCTION

Research on lithium (Li) batteries was first begun on primary cells, and commercial primary batteries appeared in the early 1970s. These were $Li/(CF)_n$, Li/MnO_2, and $Li/SOCl_2$ batteries. However, the technology of rechargeable Li batteries is more difficult than that of primary Li batteries. At present, many prototype rechargeable Li batteries have been developed, and some of them have been used in limited special fields. Further work on a stable and conductive electrolyte, positive and negative electrodes, and cell construction is still needed in order to develop rechargeable Li batteries for commercial use. In this chapter, the problems involved in the development of organic electrolytes for rechargeable Li batteries are described. Many subjects are presented and discussed since numerous kinds of rechargeable Li batteries have been considered.

2. PRIMARY LITHIUM BATTERIES AND PROTOTYPE RECHARGEABLE LITHIUM BATTERIES

2.1. Primary Lithium Batteries

Commercial primary Li batteries, which are now well known, are listed in Table 1. The most popular battery on the market is Li/MnO_2. This cell is coin-shaped or of the cylinder type. The latter has a wide electrode area and a high power density and is thus applicable as an electrical power source for automatic cameras.[5] The anodic reaction of Li/MnO_2 (3.0 V), $Li/(CF)_n$ (3.0 V), and Li/CuO (1.5 V) is shown in Eq. (1), and the cathodic reactions are represented in Eqs. (2)–(4), respectively.

$$Li \rightarrow Li^+ + e^- \tag{1}$$

$$Li + Mn(IV)O_2 + e^- \rightarrow LiMn(III)O_2 \tag{2}$$

$$Li^+ + (CF)_n + e^- \rightarrow Li(CF)_n \tag{3}$$

$$2Li^+ + CuO + 2e^- \rightarrow Cu + Li_2O \tag{4}$$

Yoshiharu Matsuda • Department of Applied Chemistry, Faculty of Engineering, Yamaguchi University, Ube 755, Japan.

Electrochemistry in Transition, edited by Oliver J. Murphy *et al.* Plenum Press, New York, 1992.

TABLE 1
Primary Li Batteries

Negative electrode	Electrolyte solution	Positive electrode	Reference(s)
Li	$LiClO_4/PC-DME^a$	MnO_2	1
Li	$LiBF_4/BL^b$	$(CF)_n$	2
Li	Organic electrolyte	CuO or CuS	3, 4

^{a} PC, Propylene carbonate; DME, 1,2-dimethoxyethane.
^{b} BL, γ-Butyrolactone.

2.2. Prototype Rechargeable Lithium Batteries

Some prototype rechargeable Li batteries are listed in Table 2. Many creative attempts have been made in order to achieve practical rechargeability of these batteries. As an example, the Li/MoS_2 cell by Moli Energy employs MoS_2, an intercalation compound, and $LiAsF_6$ as the solute, which forms an Li^+-permeable stable film on the Li electrode. The Li–Al/TiS_2 cell by Hitachi Maxell employs $LiPF_6$ as the solute, 4-methyldioxolane (4-MeDOL) instead of dioxolane (DOL) as the solvent, and hexamethylphosphoric triamide (HMPA) to inhibit the decomposition and polymerization of the solvent. In the other cells in Table 2, new materials are employed. These are Li–Wood's metal, linear-graphite hybrid (LGH), polyaniline, and poly(ethylene oxide) (PEO).

The reaction of the Li electrode is the process shown in Eq. (5):

$$Li \underset{\text{charge}}{\overset{\text{discharge}}{\rightleftharpoons}} Li^+ + e^- \qquad (5)$$

The reaction of the positive electrode when the latter consists of an intercalation compound is[14]

$$xLi^+ + TiS_2 + xe^- \underset{\text{discharge}}{\overset{\text{charge}}{\rightleftharpoons}} Li_x TiS_2 \qquad (0 < x < 1) \qquad (6)$$

TABLE 2
Prototype Rechargeable Li Batteries

Negative electrode	Electrolyte solution	Positive electrode	Reference(s)
Li	$LiAsF_6/PC^a$	MoS_2	6, 7
Li–Al	$LiPF_6$/4-MeDOL–DME + HMPA^b	TiS_2	8
Li–Al	Organic electrolyte	MnO_2	9
Li–Wood's metal (Bi–Pb–Sn–Cd)	$LiClO_4/PC$	C	10
Li–LGH^c	Organic electrolyte	VO_x	11
Li–Al	$LiClO_4/PC$	Polyaniline	12
Li	$LiClO_4/PEO^d$	VO_x or TiS_2	13

^{a} PC, Propylene carbonate.
^{b} 4-MeDOL, 4-Methyldioxolane; HMPA, hexamethylphosphoric triamide.
^{c} LGH, Linear-graphite hybrid.

The charge–discharge reaction of the polyaniline positive electrode is[12]

$$-\bigcirc-\ddot{N}- \ + \ BF_4^- \ \underset{discharge}{\overset{charge}{\rightleftharpoons}} \ =\bigcirc=\ddot{N}= \tag{7}$$
$$\qquad\qquad H \qquad\qquad\qquad\qquad\qquad H \ \ BF_4^-$$

In general, expected characteristics of practical rechargeable Li batteries are as follows:

(a) Deep charge–discharge cycles and a long cycle life
(b) High power and high energy densities
(c) Thin electrodes
(d) Excellent charge–discharge capability of the Li electrode
(e) High conductivity and safety of the electrolyte
(f) Safety, nontoxicity, reliability, and low cost of the system.

3. ORGANIC ELECTROLYTE SOLUTIONS

Many kinds of organic solvents have been examined as electrolytes for lithium batteries.[15] The physical properties of common solvents for Li batteries are given in Table 3.[16] Molecular weight is a fundamental property that is associated with the concentration of the molecule and is related to vapor pressure and viscosity (η). Usually, low viscosity of the solution contributes to high mobilities of ions in the solution. A solvent with high dielectric constant (ε_r) is preferred to attain a high degree of dissociation of the solute.[17,18] Low melting point and high boiling point are desirable. The solubility of the electrolytic salt in the solvent is of special importance and is closely related to the dipole moment of the solvent. In addition, the solvent should be safe and nontoxic.

In many cases, the solvents employed in rechargeable Li batteries are mixed systems of two or three solvents. In such a mixed-solvent system, proper donicity (D.N.) of the solvent forming suitable radii of the solvated ions is important for high dissociation[19] and mobility of ions and for high cycling efficiency. However, chemical and electrochemical stability of the solvents is very important for the mixed-solvent systems. In Fig. 1, the molar conductivity of 1 mol/dm^3 LiClO$_4$ in a propylene carbonate (PC)–tetrahydrofuran (THF) mixture is compared with that in a PC–1,2-dimethoxyethane (DME) mixture. As the values of ε_r and η for pure THF are similar to those for DME, the variations of ε_r and η in PC–THF are also similar to those in PC–DME.[19] However, the molar conductivity in PC–THF was relatively low. In both systems Li$^+$ is probably coordinated by the ethers in preference to PC. It is considered that the primary solvation number of DME per Li$^+$ is two in PC–DME,[20] and that of THF is four in PC–THF.[21] That is, the ionic radius of the solvated Li$^+$ would be larger in PC–THF than in PC–DME since there is not a large difference in molecular volume between THF and DME. It is considered that there is little difference in the degree of association between the two systems containing the same concentrations of the ethers. Thus, the relatively large size of the Li$^+$–THF solvate would be responsible for the lower conductance in PC–THF than in PC–DME. An analogous explanation applies in the case of sulfolane (SL)–DME and SL–THF shown in Fig. 2. The SL–THF/LiClO$_4$ (1 mol/dm^3) system shows a conductivity maximum at a THF concentration of about 60 mol % (m/o). However, the maximum value of the conductivity in SL–THF is only 70% of that in SL–DME. On the other hand, the conductance in dimethyl sulfoxide (DMSO)–THF was quite high, unlike the cases of PC–THF and SL–THF (Fig. 3). The conductivity of LiClO$_4$ in the DMSO–THF system varied with the solvent composition in a similar manner and to the same extent as in the DMSO–DME system, except for the composition region of fairly high ether concentration. This means that the ion–solvent interaction, or the solvation, in the two systems

TABLE 3
Physical Properties of Organic Solvents[a]

Solvent	Molecular weight	mp (°C, 1 atm)	bp (°C, 1 atm)	Density (g/cm³)	Relative dielectric constant	Viscosity (cP)	Dipole moment (debye)	D.N.[b]	A.N.[c]
Acetonitrile (AN)	41.1	−45.72	81.77	0.783	38	0.345	3.94	14.1	18.9
γ-Butyrolactone (BL)	86.1	−42	206	1.125	39.1	1.751	4.12	—	—
1,2-Dimethoxyethane (DME)	90.1	−58	84.7	0.86	7.20	0.455	1.07	24	19.3
Dimethyl sulfoxide (DMSO)	78.1	18.42	189	1.0955	46.45	1.991	3.96	29.8	—
1,3-Dioxolane (DOL)	74.1	−95	78	1.0600[d]	6.79[e]	0.58	—	—	—
Ethylene carbonate (EC)	88.1	39–40	248	1.3218[f]	89.6[g]	1.86[g]	4.80	16.4	—
2-Methyltetrahydrofuran (2-MeTHF)	86.1	—	78–80	0.848	6.24	0.457	—	—	—
Propylene carbonate (PC)	102.1	−49.2	241.7	1.198	64.4	2.530	5.21	15.1	18.3
Sulfolane (SL)	120.2	28.86	287.3	1.2619[e]	42.5[e]	9.87[e]	4.7	14.8	19.3
Tetrahydrofuran (THF)	72.1	−108.5	66	0.888	7.25[e]	0.46[e]	1.71	20.0	8.0

[a] At 25°C, except where otherwise noted.
[b] Donor number.
[c] Acceptor number.
[d] At 20°C.
[e] At 30°C.
[f] At 39°C.
[g] At 40°C.

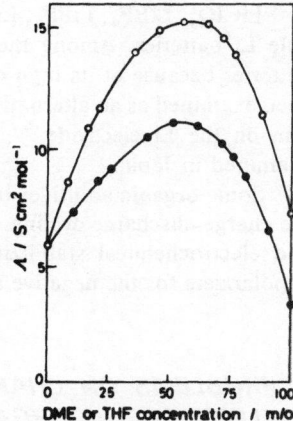

FIGURE 1. Variation of molar conductivity of 1 mol/dm³ LiClO₄ with solvent composition, 30°C: ○, PC-DME; ●, PC-THF. [Y. Matsuda, M. Morita, and F. Tachihara, *Bull. Chem. Soc. Jpn.* **59**, 1967–1973 (1986).]

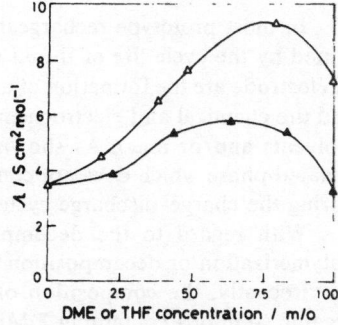

FIGURE 2. Variation of molar conductivity of 1 mol/dm³ LiClO₄ with solvent composition, 30°C: ○, SL–DME; ▲, SL-THF. [Y. Matsuda, M. Morita, and F. Tachihara, *Bull. Chem. Soc. Jpn.* **59**, 1967–1973 (1986).]

is almost equivalent. As DMSO has higher D.N. than DME or THF, it is reasonable to assume that DMSO coordinates preferentially to Li⁺ in both systems. This description is consistent with the experimental results shown in Fig. 3. Besides the relatively low viscosity, the strong affinity of DMSO for Li⁺, which leads to a smaller size of the solvated Li⁺, might be related to the high conductance of LiClO₄ in neat DMSO compared with that in neat PC or SL.

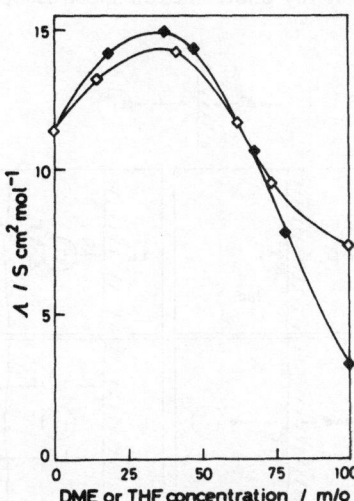

FIGURE 3. Variation of molar conductivity of 1 mol/dm³ LiClO₄ with solvent composition, 30°C: ◇, DMSO-DME; ◆, DMSO-THF. [Y. Matsuda, M. Morita, and F. Tachihara, *Bull. Chem. Soc. Jpn.* **59**, 1967–1973 (1986).]

$LiClO_4$, $LiBF_4$, $LiPF_6$, $LiAsF_6$, and $LiCF_3SO_3$ have been used as the solute in rechargeable Li batteries. Among these, it seems that $LiClO_4$ is not suitable for rechargeable Li batteries because of its high oxygen content. In the United States and Canada, $LiAsF_6$ has been examined as an alternative electrolytic salt due to the resulting formation of a protective film on the Li electrode.[6,7,15] By contrast, $LiBF_4$[2,14,17] and $LiPF_6$[7] have been mainly examined in Japan.

Some organic additives in the electrolytic solution sometimes have beneficial effects on the charge–discharge cycling of the Li electrode. These additives would behave as chemical and electrochemical stabilizing reagents, inhibitors of polymerization of the solvent, or depolarizers for the negative and/or positive electrodes.

4. PROBLEMS ON CHARGE–DISCHARGE CYCLING OF THE LITHIUM ELECTRODE IN ORGANIC ELECTROLYTE SOLUTIONS

In most prototype rechargeable Li batteries, the charge–discharge cycling life is determined by the cycle life of the Li electrode. The main reasons for the short cycle life of the Li electrode are the formation of an insulating film or an isolated Li phase on the Li electrode and the chemical and electrochemical reactions between the Li electrode and the electrolyte (solvents and/or ions). As shown in Fig. 4, the deposition of an insulating film (a) or an isolated phase which does not contribute to the discharge (b) forms on the electrode surface during the charge–discharge cycle.[22,23]

With regard to the decomposition of the electrolytic solutions, an example is the polymerization or decomposition of DOL as shown in Scheme 1.[24]

Recently, the composition of the films formed on the surface of the Li electrode in organic solutions containing DME and THF has been reported.[25] The mechanisms of the film formation are shown in Schemes 2 and 3.[25] These problems should be solved in order to realize a rechargeable Li battery. Furthermore, lithium hydroxide (LiOH) forms on the surface of the Li electrode in the presence of $0.005M$ H_2O:

$$Li_2O \xrightarrow{H_2O} 2LiOH \xrightarrow{2H_2O} 2LiOH \cdot H_2O \tag{8}$$

Also, the reactions of $LiClO_4$, $LiAsF_6$, and $LiCF_3SO_3$ on the Li surface have been studied by X-ray photoelectron spectroscopy (XPS).[25]

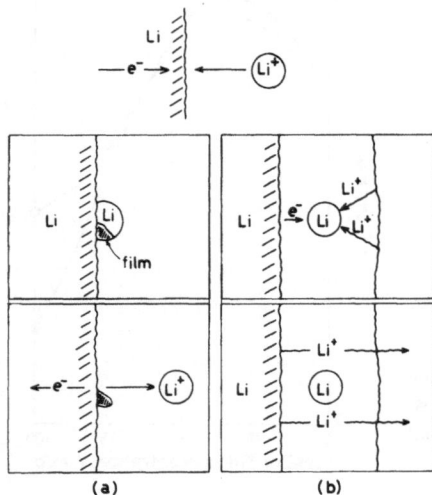

FIGURE 4. Schematic illustration of the processes involved in efficiency loss in the Li cycling: (a) deposition of an insulating film on the Li electrode; (b) formation of an isolated Li phase on the Li electrode. [Y. Matsuda, *Denki Kagaku* **55**, 111–119 (1987).]

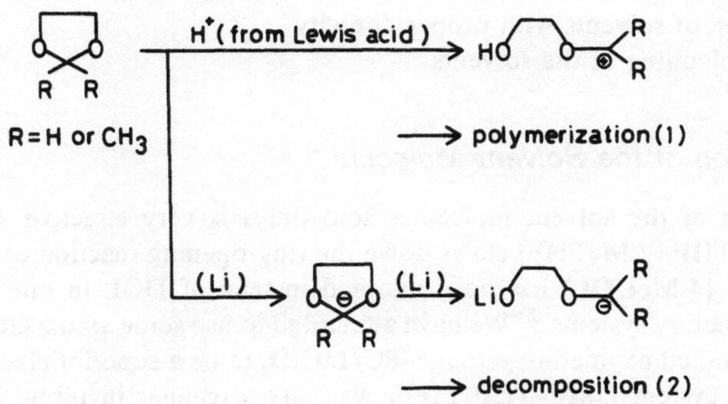

H⁺ (from Lewis acid) →

R = H or CH₃

→ polymerization (1)

(Li) → (Li) → LiO

→ decomposition (2)

SCHEME 1

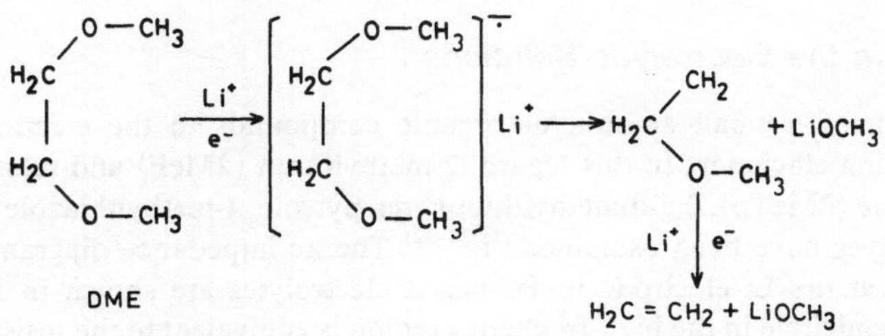

DME

SCHEME 2

THF

hydrogen abstraction

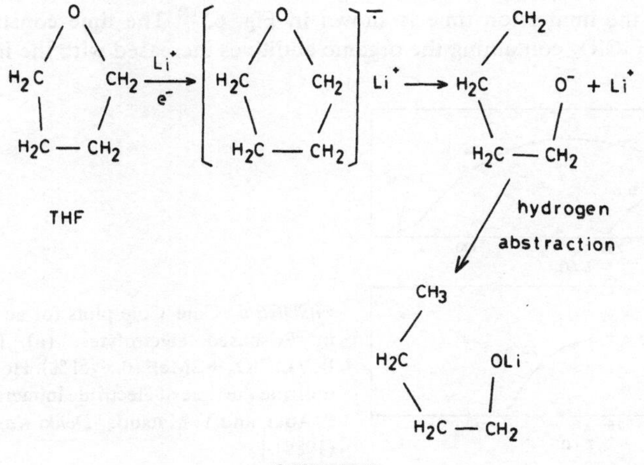

SCHEME 3

5. IMPROVED TECHNOLOGIES FOR RECHARGEABLE LITHIUM BATTERIES

With further work on rechargeable Li batteries, some improvements in the technology are expected. These would involve the following aspects:

(a) Modification of the solvent molecule
(b) Additives in the electrolytic solutions
(c) Selection of chemically and electrochemically stable solvents
(d) Selection of electrolytic salts
(e) Adoption of Li-alloy or Li-carbon electrodes

(f) Adoption of solvents with proper donicity

(g) Proper blending of the solvents

5.1. Modification of the Solvent Molecule

Modification of the solvent molecules sometimes is very effective. Substitution of a methyl group in THF (2-MeTHF) slows down the ring-opening reaction of THF.[15] Methyl-substituted DOL (4-MeDOL) has been adopted instead of DOL in one of the prototype rechargeable Li battery systems.[8] We have attempted to use some alkoxyethanes in the place of DME and found ethoxymethoxyethane–PC/LiClO₄ to be a superior electrolyte system.[26] The mixed-ether system DME-THF/LiPF₆ was also excellent in terms of the Li cycling efficiency.[27]

5.2. Additives in the Electrolytic Solutions

The addition of a small amount of organic compounds to the electrolytic solution improves Li cycling efficiency. In this regard, 2-methylfuran (2MeF) and related chemicals, 2-methylthiophene (2MeTp), 2,5-dimethylthiophene, pyrrole, 4-methylthiazole (4MeTz), and quinone-imine dyes have been examined.[15,28-32] The ac impedance diagrams (Cole–Cole plots) measured at the Li electrode in PC-based electrolytes are shown in Fig. 5.[33] The diameter of the semicircle in the high-frequency region is equivalent to the resistance between the Li electrode and the electrolyte solution. This resistance increased with immersion time, and it decreased with addition of 2MeF. The variation of the time constant ($R \cdot Cd$) of the impedance with the immersion time is shown in Fig. 6.[33] The time constants for the Li electrode in PC/LiClO₄ containing the organic additives increased with the immersion time.

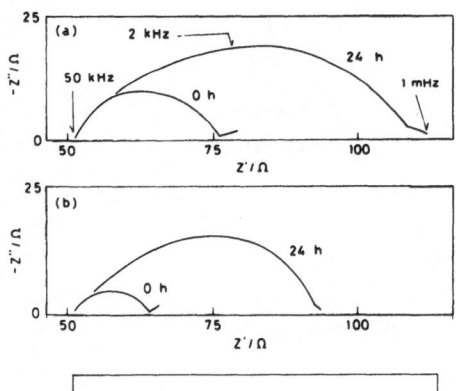

FIGURE 5. Cole–Cole plots for ac impedance at Li in PC-based electrolytes: (a) PC/LiClO₄; (b) PC/LiClO₄ + 2MeF (0.5 vol %). Hours on the curves indicate the time of electrode immersion. [M. Morita, S. Aoki, and Y. Matsuda, *Denki Kagaku* **57**, 523–526 (1989).]

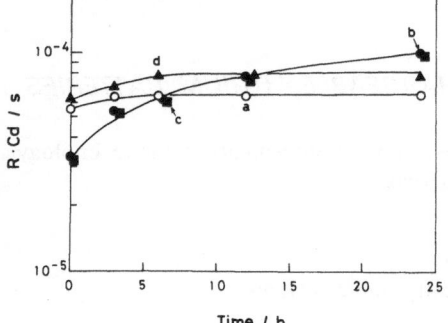

FIGURE 6. Variation of the time constant ($R \cdot Cd$) with the immersion time: (a) PC/LiClO₄; (b) PC/LiClO₄ + 2MeF(0.5 vol %); (c) PC/LiClO₄ + 2MeTp(0.5 vol %); (d) PC/LiClO₄ + 4MeTz(0.5 vol %). [M. Morita, S. Aoki, and Y. Matsuda, *Denki Kagaku* **57**, 523–526 (1989).]

Moreover, these variations predominated for the electrode in contact with the solution containing 2MeF or 2MeTp. Since the variation of the time constant reflects the changes in the composition of the interface between the Li electrode and the electrolyte solution, the composition of the film on the Li electrode in PC/LiClO₄ containing 2MeF or 2MeTp evidently changed with the immersion time.[33]

In some cases, the organic additives affect the performance of the positive electrode. The addition of the crown ether 12-crown-4 or 15-crown-5 to PC/LiClO₄ increases the molar conductivity and improves the charge–discharge characteristics of the TiS₂ electrode.[34]

5.3. Selection of Chemically and Electrochemically Stable Solvents

One approach to developing a stable electrolyte which is chemically unreactive toward Li is to blend stable solvents. Recently, based on this theory, ethylene carbonate (EC) has been blended into the PC-2MeTHF system. SL[35,36] and DMSO[37] are chemically and electrochemically stable, having dielectric constants of 40 or higher. Therefore, these are promising solvents for such electrolytes. However, both SL and DMSO have high melting points and high viscosities at ambient temperature; both these properties are unsuitable for battery electrolytes. Thus, some ethers that have low melting points and low viscosities would have to be blended into SL or DMSO in order to improve the physicochemical properties of the electrolyte solution. Recently, the use of 1,3-dimethoxypropane as the blending co-solvent for a rechargeable Li battery has been reported.[38]

5.4. Selection of Electrolytic Salts

The selection of electrolytic salts is very important. LiBF₄ is stable, but its conductivity is relatively low in organic systems. In many cases, the charge–discharge cycle life of the Li electrode is rather long in LiAsF₆ solutions, but in some cases LiAsF₆ is not suitable owing to the toxicity.[39] From the point of view of obtaining a highly conductive solution, LiPF₆ is an excellent salt, but it is rather reactive, and the addition of inhibitors is necessary with the use of this salt. The problems on the selection of electrolytic salts still remain to be solved.

5.5. Adoption of Li-Alloy or Li–Carbon Electrodes

One of the methods to improve the charge–discharge cycle efficiency of the Li negative electrode is adoption of Li alloys or Li-linear-graphite hybrid (Li-LGH).[11] The coulombic efficiencies of Li cycling in DMSO-DME (1:1 by volume) containing $1M$ LiClO₄, LiBF₄, and LiPF₆ are shown in Fig. 7.[40] The cycling efficiencies were affected by the coexisting

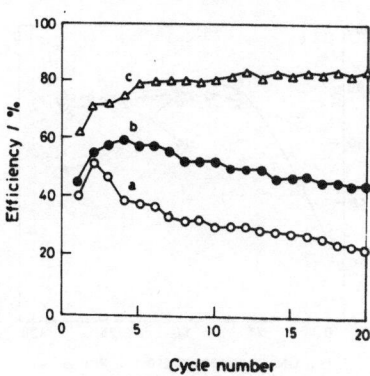

FIGURE 7. Lithium cycling efficiency on Ni substrate in DMSO-DME (1:1 by volume) containing 1 mol/dm³ LiClO₄ (a), LiBF₄ (b), and LiPF₆ (c); $i_p = i_s = 2$ mA/cm², $Q = 0.2$ C/cm². [M. Morita, F. Tachihara, and Y. Matsuda, in: *Proceedings of the 32nd International Power Sources Symposium*, pp. 124–134, The Electrochemical Society, Pennington, New Jersey (1986).]

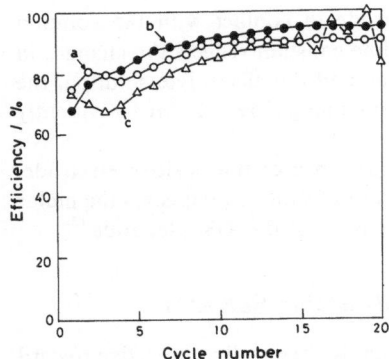

FIGURE 8. Lithium cycling efficiency on Al substrate in DMSO-DME (1:1 by volume) containing 1 mol/dm^3 LiClO$_4$ (a), LiBF$_4$ (b), and LiPF$_6$ (c); $i_p = i_s = 2$ mA/cm^2, $Q = 0.2$ C/cm^2. [M. Morita, F. Tachihara, and Y. Matsuda, in: *Proceedings of the 32nd International Power Sources Symposium*, pp. 124–134, The Electrochemical Society, Pennington, New Jersey (1986).]

anions, and the efficiency in LiPF$_6$ was much higher than that in LiClO$_4$ or LiBF$_4$. In contrast, high efficiencies have been obtained in cells with Li-alloy negative electrodes. The results for the Li–Al electrode in DMSO-DME containing Li salts are shown in Fig. 8.[40] The Li–Al electrode suppressed the influence of the anions, and its charge–discharge efficiency is elevated as compared to that in Fig. 7.

5.6. Adoption of Solvents with Proper Donicity

The donicity of the solvents is very important because the size of the Li$^+$ ion is very small, and Li$^+$ is easily solvated preferentially by the solvent whose donicity is highest in a mixed-solvent system. The coulombic efficiency of the charge–discharge cycle is affected by the reactivity of the solvent which solvates Li$^+$.[41]

Figure 9[41] shows the coulombic efficiency of Li (average of 20 cycles) on a Ni substrate in mixed-solvent systems as a function of the solvent composition. In PC-DME and SL-DME, maximum efficiencies were observed in the solutions containing 50 vol % DME, but the dependences of the efficiencies upon the DME concentration were rather small in these systems. On the other hand, the average efficiencies in DMSO-DME and 3-methyl-1,3-oxazolidine-2-one (MO)-DME varied markedly with the solvent composition. This is attributed to changes in the states of the Li$^+$ solvates. In the PC-DME and SL-DME systems, Li$^+$ tends to solvate with DME, whose DN is relatively high. DMSO and MO (DN 27) have higher DNs than DME, so Li$^+$-DMSO and Li$^+$-MO solvates will be mainly present in DMSO-DME and MO-DME, respectively. These effects of solvent composition on coulombic efficiency are related to the fact that the coulombic efficiency is much influenced by the chemical or electrochemical stability of the solvent that coordinates to Li$^+$.

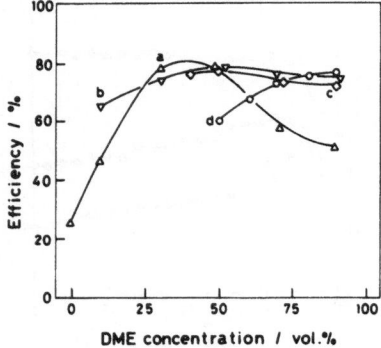

FIGURE 9. Correlation between average coulombic efficiency and DME concentration in the solvent: (a) DMSO-DME/LiPF$_6$; (b) PC-DME/LiPF$_6$; (c) SL-DME/LiPF$_6$; (d) MO-DME/LiPF$_6$ (MO = 3-methyl-1,3-oxazolidine-2-one). [Y. Matsuda, S. Aoki, and M. Morita, *Proceedings of the 33rd International Power Sources Symposium*, pp. 26–31, The Electrochemical Society, Pennington, New Jersey (1988).]

6. SOLID POLYMER ELECTROLYTES

It has been reported that poly(ethylene oxide) (PEO) forms complexes with various alkali-metal salts.[42] Appreciable conductivities for the alkali cations of these materials were demonstrated, and the prospect of using solid polymer electrolytes in Li batteries was presented in 1979.[43] A new Li battery design based on such thin-film polymer electrolytes has appeared and a 10-W h all-solid-state battery using a PEO-based electrolyte working at 80–100°C has been recently reported.[13]

Polymer-based macromolecules, such as polysiloxanes, polyphosphazenes, and polyaziridines, are alternatives to PEO. The electrolytic salts typically combined with the polymer-based macromolecules include $LiClO_4$ and $LiCF_3SO_3$. However, many compounds are useful, including fluoride compounds such as $LiAsF_6$ and $LiPF_6$, boron compounds, $LiB(C_6H_5)_4$, and $LiB(C\equiv CR)_4$. These solid electrolytes are referred to as complexes. The conductivity of $(PEO)_9LiCF_3SO_3$ is about 10^{-5} S/cm at 80°C.[44] Recently, solid electrolytes consisting of a PEO-grafted poly(methyl methacrylate) and Li salts containing poly(ethylene glycol) have shown high conductivity of about 5×10^{-5} S/cm at 30°C.[45,46]

The development of new, stable, and highly conductive electrolytes will contribute to efforts to produce rechargeable Li batteries with high power and energy densities.

7. FUTURE PROSPECTS

Many kinds of primary Li batteries have been used on a worldwide level, and some prototype rechargeable Li batteries have been applied in limited fields. However, further improvements are desirable to produce practical rechargeable Li batteries with high power and energy densities in a wide variety of applications.

One important approach is the selection of a suitable combination of the negative and positive electrodes, solvents, and electrolytic salt. The resulting combination should yield charge–discharge cycling of the Li electrode with high efficiency and uniform surface morphology.

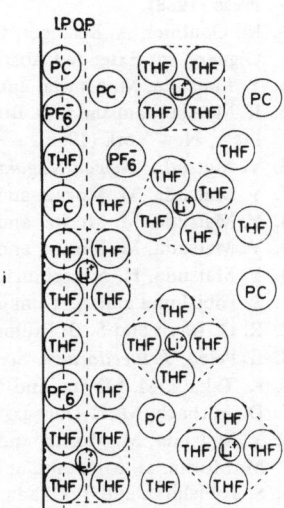

FIGURE 10. Electrical double-layer structure in the Li electrode/electrolyte solution interphase. [Y. Matsuda, *Nippon Kagaku Kaishi* (*J. Chem. Soc. Jpn., Chem. Ind. Chem.*) **1989**, 1–14.]

On a basic level, the study of the electrical double-layer structure between the electrode and the electrolyte in Li batteries should be pursued further. For example, the electrical double-layer structure between the Li electrode and a PC–THF/LiPF$_6$ solution would be as shown in Fig. 10.[16] Since Li$^+$ is solvated with four molecules of THF, THF preferentially adsorbs on or comes into contact with the Li electrode and reacts with Li more easily than PC. At the same time, PF$_6^-$ adsorbs specifically on the Li electrode and affects the kinetics of charge–discharge reactions of the Li electrode. The structure of the electrical double layer in Li batteries is an almost unexplored field.

Finally, safety, including nontoxicity, is an important consideration. The balance between practical capabilities (energy and power densities, cycling life, etc.) and safety will be more important in the future.

REFERENCES

1. H. Ikeda, S. Ueno, T. Saito, S. Nakaido, and H. Tamura, *Denki Kagaku* **45**, 391 (1977).
2. R. Okazaki, K. Aoki, K. Tsubaki, T. Iijima, and A. Morita, *National Technical Report* **24**(2), 281 (1978).
3. R. Bates and Y. Jumel, in: *Lithium Batteries* (J. P. Gabano, ed.), pp. 73–95, Academic Press, New York (1983).
4. N. Magalit, in: *Lithium Batteries* (J. P. Gabano, ed.), pp. 137–168, Academic Press, New York (1983).
5. H. Ikeda and N. Furukawa, in: *Practical Lithium Batteries* (Y. Matsuda and C. R. Schlaikjer, eds.), pp. 57–64, JEC Press (1988).
6. J. A. Stiles, Extended Abstracts, 3rd International Meeting on Li Batteries, 1986, pp. 189–194.
7. J. A. R. Stiles, *New Mater. New Process* **3**, 89 (1985).
8. I. Yoshimitsu, S. Kitagawa, K. Kajita, and T. Manabe, Extended Abstracts, 26th Battery Symposium Japan, 1985, pp. 49–52.
9. N. Furukawa, T. Saito, K. Teraji, I. Nakane, and T. Nohma, in: *Proceedings of the Symposium on Primary and Secondary Ambient Temperature Lithium Batteries* (J. P. Gabano, Z. Takehara, and P. Bro, eds.), pp. 557–564, The Electrochemical Society, Pennington, New Jersey (1988).
10. N. Koshiba, H. Hayakawa, and K. Momose, Extended Abstracts, 26th Battery Symposium Japan, 1985, pp. 145–148.
11. K. Inada, K. Ikeda, S. Inomata, T. Nishii, M. Miyabayashi, and H. Yui, in: *Practical Lithium Batteries* (Y. Matsuda and C. R. Schlaikjer, eds.), pp. 96–99, JEC Press (1988).
12. T. Kita, in: *Practical Lithium Batteries* (Y. Matsuda and C. R. Schlaikjer, eds.), pp. 124–128, JEC Press (1988).
13. M. Gauthier, A. Belanger, O. Fauteux, M. Diuval, B. Kapfer, M. Robitaille, R. Bellemare, and Y. Giguere, in: Extended Abstracts, 3rd International Meeting on Li Batteries, 1986, pp. 238–241.
14. Y. Matsuda, M. Morita, and K. Takata, *J. Electrochem. Soc.* **131**, 1991 (1984).
15. K. M. Abraham and S. B. Brummer, in: *Lithium Batteries* (J. P. Gabano, ed.), pp. 371–406, Academic Press, New York (1983).
16. Y. Matsuda, *Nippon Kagaku Kaishi* **1989**, 1.
17. Y. Matsuda, M. Morita, and T. Yamashita, *J. Electrochem. Soc.* **131**, 2821 (1984).
18. Y. Matsuda, M. Morita, and K. Kosaka, *J. Electrochem. Soc.* **130**, 101 (1983).
19. Y. Matsuda, M. Morita, and F. Tachihara, *Bull. Chem. Soc. Jpn.* **59**, 1967 (1986).
20. Y. Matsuda, H. Nakashima, M. Morita, and Y. Takasu, *J. Electrochem. Soc.* **128**, 2552 (1981).
21. S. Tobishima and A. Yamaji, *Electrochim. Acta* **28**, 1067 (1983).
22. R. D. Rauh and S. B. Brummer, *Electrochim. Acta* **22**, 75 (1977).
23. E. Peled, *J. Electrochem. Soc.* **126**, 2047 (1979).
24. K. Takata, M. Morita, and Y. Matsuda, *J. Electrochem. Soc.* **131**, 126 (1985).
25. D. Aurbach, M. L. Daroux, P. W. Fagny, and E. Yeager, *J. Electrochem. Soc.* **135**, 1863 (1988).
26. Y. Matsuda, M. Morita, and S. Kanameda, *Denki Kagaku* **52**, 702 (1984).
27. M. Morita, H. Miyazaki, and Y. Matsuda, *Electrochim. Acta* **31**, 573 (1986).
28. S. Tobishima and T. Okada, *Denki Kagaku* **53**, 742 (1985).
29. K. M. Abraham, *J. Power Sources* **14**, 179 (1985).

30. Y. Matsuda, H. Hayashida, and M. Morita, in: *Proceedings of the Symposium on Primary and Secondary Ambient Temperature Lithium Batteries* (J. P. Gabano, Z. Takehara, and P. Bro, eds.), pp. 610-617, The Electrochemical Society, Pennington, New Jersey (1988).

31. Y. Matsuda and M. Morita, *Prog. Batteries Sol. Cells* **7**, 266 (1988).

32. Y. Matsuda and M. Morita, *J. Power Sources* **26**, 579 (1989).

33. M. Morita, S. Aoki, and Y. Matsuda, *Denki Kagaku* **57**, 523 (1989).

34. Y. Matsuda, H. Hayashida, and M. Morita, *Denki Kagaku* **53**, 628 (1985).

35. Y. Matsuda, M. Morita, K. Yamada, and K. Hirai, *J. Electrochem. Soc.* **132**, 2538 (1985).

36. M. Morita, Y. Okada, and Y. Matsuda, *J. Electrochem. Soc.* **134**, 2665 (1987).

37. M. Morita, F. Tachihara, and Y. Matsuda, *Electrochim. Acta* **32**, 299 (1978).

38. J. S. Foos and T. J. Stolki, *J. Electrochem. Soc.* **135**, 2769 (1988).

39. Y. Matsuda, *Prog. Batteries Sol. Cells* **6**, 17 (1987).

40. Y. Matsuda, *Proceedings of the 32nd International Power Sources Symposium*, pp. 124-135, Electrochemistry Society, New Jersey (1986).

41. Y. Matsuda and M. Morita, *Proceedings of the 33rd International Power Sources Symposium*, pp. 26-31, Electrochemistry Society, New Jersey (1988).

42. D. E. Fenton, J. M. Parker, and P. V. Wright, *Polymer* **14**, 589 (1973).

43. M. Armand, J. M. Chabagno, and M. Duclot, in: *Fast Ion Transport in Solids* (P. Vashista, J. N. Mundy, and G. K. Shenoy, eds.), pp. 131-136, Elsevier, New York (1979).

44. L. L. Yang, R. Hug, and G. C. Farrington, Extended Abstracts, 3rd International Meeting on Li Batteries, 1986, pp. 223-224.

45. M. Morita, M. Motoda, Y. Matsuda, T. Takahashi, and H. Ashitaka, Extended Abstracts, 29th Battery Symposium Japan, 1988, pp. 199-200.

46. M. Morita, T. Fukumasa, M. Motoda, H. Tsutsumi, Y. Matsuda, T. Takahashi, and H. Ashitaka, *J. Electrochem. Soc.* **137**, 3401 (1990).

41

Mechanism of the Cathodic Discharge of MnO₂ in Alkaline Electrolyte

Terrell N. Andersen and Joseph M. Derby

1. INTRODUCTION

1.1. Background

Development of portable electric and electronic devices over the past several decades has led to the need for increasing drain rates from consumer batteries. To meet this need, $Zn-MnO_2$ batteries with $ZnCl_2$ electrolyte (heavy-duty Leclanché cells) and KOH electrolyte (alkaline cells) have superseded the common zinc–carbon or Leclanché cell, which utilizes an NH_4Cl electrolyte. The alkaline cell has the superior heavy-drain discharge characteristics of these batteries.[1]

Alkaline-battery discharge parallels that of MnO_2, and, correspondingly, the discharge depends strongly on the nature of the MnO_2 used in mixture with graphite as the cathode. Hence, the MnO_2 market for alkaline cells is driven by the so-called battery activity of the MnO_2.†

1.2. Battery Activity

As background for defining battery activity and discussing discharge mechanism, consider the alkaline-cell discharge reactions:

$$Zn + 2OH^- \rightarrow ZnO + H_2O + 2e^- \quad \text{(anode)} \tag{1}$$

$$MnO_2 + H_2O + e^- \rightarrow MnOOH + OH^- \quad \text{(cathode, 1st } e^-) \tag{2a}$$

$$MnOOH + H_2O + e^- \rightarrow Mn(OH)_2 + OH^- \quad \text{(cathode, 2nd } e^-) \tag{2b}$$

The two-stage cathodic reaction is manifest in the two sequential segments of the MnO_2 discharge curve (Fig. 1). Above ~ -0.34 V versus Hg/HgO in $9M$ KOH (or between the initial battery voltage of approximately 1.5 V and 1.0 V), cathodic reduction of MnO_2 is

†Other characteristics of MnO_2 are also important, such as purity; for example, trace levels of various heavy metals cause H_2 gassing at the anode, which is destructive to the battery (Ref. 2).

Terrell N. Andersen and Joseph M. Derby ● Kerr-McGee Technology Division, Oklahoma City, Oklahoma 73125.

Electrochemistry in Transition, edited by Oliver J. Murphy *et al.* Plenum Press, New York, 1992.

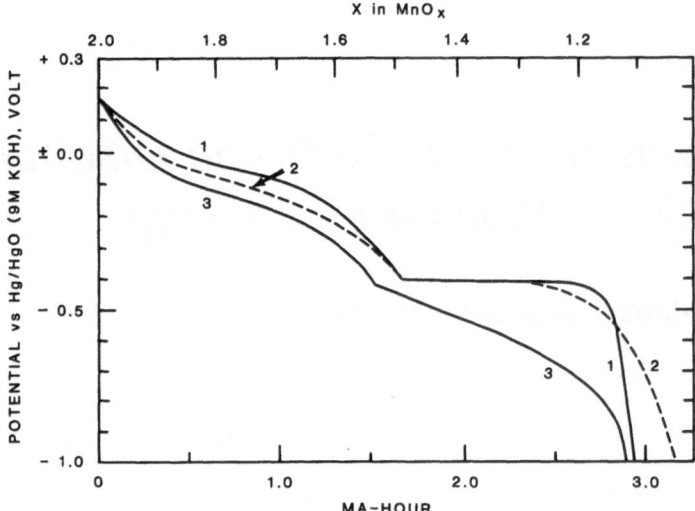

FIGURE 1. Discharge curves for thin films of EMD on graphite substrate in $9M$ KOH at 23°C. Current density: (1) 0.01 mA/cm²; (2) 0.03 mA/cm²; (3) 0.28 mA/cm². (From Kozawa and Yeager,[3] reprinted by permission of the publisher, The Electrochemical Society, Inc.)

constrained thermodynamically to Eq. (2a). Below about -0.34 V versus Hg/HgO or 1.0 V versus Zn, which is the reversible potential for Eq. (2b), both reactions (2a) and (2b) can occur at the cathode. At very low current densities, most of the charge from the first-electron reduction of MnO_2 is realized before the second-electron reduction begins (curves 1 and 2 of Fig. 1), but at high current densities the overvoltage is great enough to take the cell voltage into the second-electron regime before the first-electron reduction is complete (curve 3).

The useful cell capacity lies within the first-electron reduction, because of the voltage demand of most portable devices. Therefore, battery activity is defined as the discharge capacity or the charge that may be extracted per unit mass of MnO_2 during reduction to a given potential end point, which usually is taken to be at or near 1.0 V versus Zn (at or near the end of the first electron discharge). Therefore, this chapter will focus mainly on the first electron discharge of MnO_2.

MnO_2 battery activity naturally depends on the electrode configuration and on various discharge parameters, such as the current density and temperature, so a certain degree of uniformity in experimental setup and procedure is needed to compare results among different laboratories. Reference 4 elaborates on the details of MnO_2 discharge testing. Laboratory tests usually employ thin pressed pellets of powdered MnO_2–graphite mixtures which contain a predominance of graphite. These cathode pellets are discharged in cells flooded with $\sim 9M$ KOH at current densities of 10–40 mA/g of MnO_2. This procedure results in discharge curves that are similar to actual battery discharge curves during standard tests,[1,4] but prevents kinetic limitations by electronic conductivity in the solid or ionic conductivity in the electrolyte. Thereby, the intrinsic discharge properties of the MnO_2 are manifest. Discharge of a dry cell often is a mixed process.

Manganese(IV) oxide exists in many crystallographic forms, some of which also contain low concentrations of other cations.[5,6] Of these natural and synthetic substances, the γ-MnO_2 forms and the structurally similar ε-MnO_2 and ρ-MnO_2 are battery active and, consequently, are those used in the battery industry.[5-7] Battery-active MnO_2 has three origins, natural manganese dioxide (NMD), chemical manganese dioxide (CMD), and electrolytic manganese dioxide (EMD). The battery activities of these so-called battery materials can differ greatly

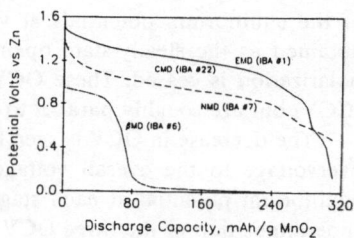

FIGURE 2. Discharge curves for four types of MnO$_2$ in 9M KOH at 22°C. Cathodes consisted of MnO$_2$-graphite pellets in an MnO$_2$:C ratio of 1:3. Discharge rate = 20 mA/g of MnO$_2$. Procedure given in Ref. 4. (S. F. Burkhardt, unpublished results.)

from one another in a given application.[1,4,8,9] EMD generally exhibits the greatest alkaline discharge capacity, as shown by Fig. 2 for four typical materials, and, consequently, EMD is used exclusively in alkaline cells. Even EMD can exhibit a varied alkaline discharge capacity, depending on the conditions under which it is deposited, such as electrolyte temperature, current density, anode composition, and electrolyte composition (see Fig. 3 and Ref. 10). Detrimental effects of low temperature, high current density, certain impurities, and carbon or lead anodes (as opposed to titanium anodes) in EMD deposition cells are well known.[10-14]

The question of discharge mechanism is key in understanding the reasons for such varied discharge capacity.

2. MECHANISM OF DISCHARGE

2.1. Roles of Thermodynamics and Kinetics

The various manganese dioxides differ in battery activity, both for thermodynamic or equilibrium reasons and for kinetic reasons. Thermodynamic differences are studied by means

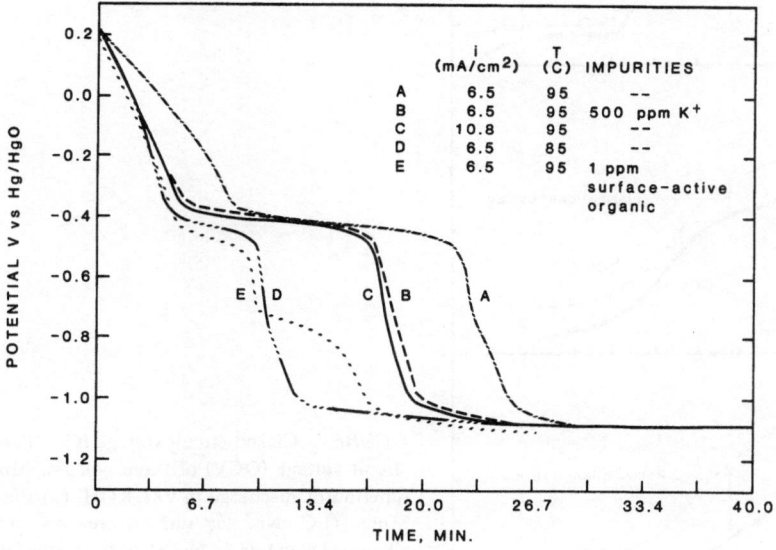

FIGURE 3. Discharge curves for thin films of EMD deposited on Pt from 0.3M H$_2$SO$_4$ + 0.3M MnSO$_4$ and under the various conditions stated in the upper right. Amount of deposit = 0.8 C/cm^2. Discharge rate = 0.22 mA/cm^2.

of the equilibrium potentials at various stages of discharge. The equilibrium potential is obtained as the steady-state open-circuit voltage (OCV) or total recovery potential when polarization is ceased. These OCVs are positive of the discharge or closed-circuit voltage (CCV) but are roughly parallel to the latter (Fig. 4).

The decrease in OCV is seen to contribute as much as or more than the polarization or overvoltage to the overall voltage decrease during discharge. Furthermore, the OCV or equilibrium potential at each stage of discharge differs with type of MnO_2, as seen from superimposition of the three OCV curves of Fig. 4, which is shown in Fig. 5.

2.2. Equilibrium Potential as a Diagnostic Mechanistic Tool

The equilibrium potential versus amount of discharge has proven valuable for studying the first-electron discharge mechanism, because of several diagnostic features:

1. a continuous decrease in voltage with discharge,
2. a variation with type of MnO_2,
3. a variation with pH of the electrolyte, and
4. failure to follow a simple Nernstian behavior.

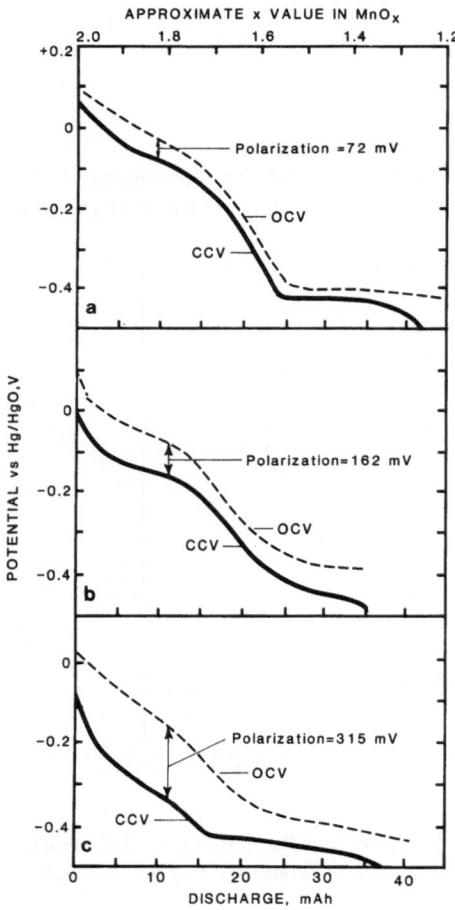

FIGURE 4. Closed-circuit voltage (CCV) and open-circuit voltage (OCV) of three types of MnO_2 as a function of discharge in $9M$ KOH. (a) Electrolytic MnO_2 (I.C. No. 1); surface area $58.5 \, m^2/g$. (b) Chemical MnO_2 (I.C. No. 8); surface area $70.5 \, m^2/g$. (c) Natural Ore (I.C. No. 7); surface area $22.0 \, m^2/g$. (From Kozawa,[15] reprinted by permission of the publisher, the I.C. MnO_2 Sample Office.)

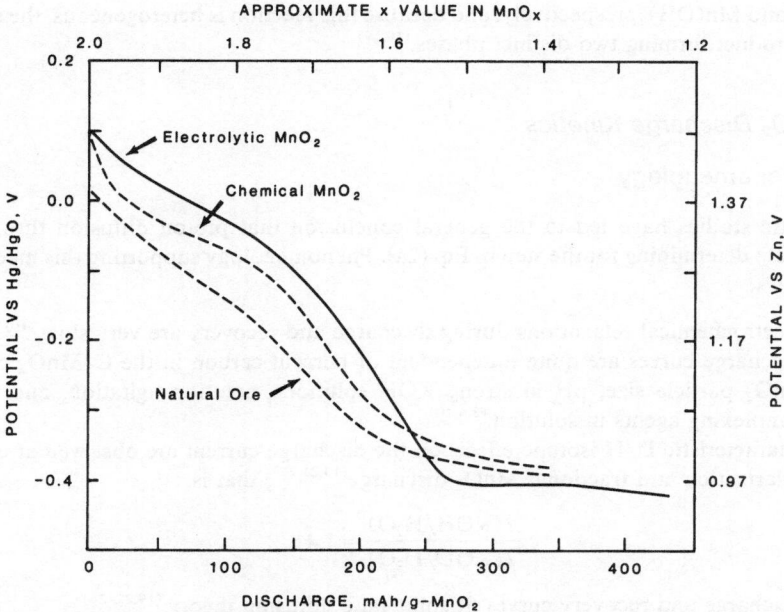

FIGURE 5. Open-circuit voltage curves from Fig. 4 superimposed. (From Kozawa,[15] reprinted by permission of the publisher, the I.C. MnO₂ Sample Office.)

These features can best be discussed by reference to an applicable expression for the equilibrium electrode potential during the first-electron transfer, $E_{e,1}$[16]:

$$E_{e,1} = \text{OCV} = E_1^\circ - \frac{2.3RT}{F} \cdot \text{pH} + \frac{RT}{F} \ln \frac{1-r}{r} - \frac{\phi(r-1)}{F} \qquad (3)$$

Here r is the fraction of the Mn atoms in the reduced ($+3$) state, ϕ is a molar lattice interaction energy, F is Faraday's constant, and the other terms have their usual significance.

The sloping potential-versus-discharge function is indicative of a homogeneous solid phase[17,18]; that is, MnO₂ and MnOOH do not form two distinct phases (otherwise, the activities of those species would be "one" and the equilibrium curve would be flat). Rather, the solid phase is a homogeneous mixture of Mn^{4+}, Mn^{3+}, O^{2-}, and OH^- ions. It is on this basis that the logarithmic term in r arises.

The Mn^{3+} ions formed upon reduction, being larger than the original Mn^{4+} ions, induce strain into the lattice, which results in gradual dilation of the original lattice and, finally, in conversion to the α-Mn₂O₃ structure (details are referenced in Liang's review[19]). This effect is identified as the source of misfit with a classical Nernst equation and is accounted for in the term linear in r, as introduced by Ruetschi.[16] Different ways of introducing the changing lattice have been published by others, as reviewed by Desai et al.[20]

The pH dependence of Eq. (3) follows naturally from Eq. (2a) and the fact that hydroxide ions in the electrolyte are in great excess.

E_1° varies with the MnO₂ structure, as Fig. 5 exemplifies. This variation will be discussed further after the kinetics of discharge are addressed.

The equilibrium potential for manganese dioxide during the second-electron transfer lacks all of the exciting differences shown by the first-electron transfer, except the pH dependence (Fig. 5). That is, the curve is flat (not sloping) and does not depend sensibly on the type of MnO₂. This is because the reactant and product are the same in each case, namely,

MnOOH and $Mn(OH)_2$, respectively, and because this reaction is heterogeneous, the reactant and the product forming two distinct phases.[10,19]

2.3. MnO₂ Discharge Kinetics

2.3.1. Phenomenology

Kinetic studies have led to the general conclusion that proton diffusion through the lattice is rate determining for the step in Eq. (2a). Phenomenology supporting this mechanism is as follows:

1. Electrochemical relaxations during discharge and recovery are very slow.[10]
2. Discharge curves are quite independent of percent carbon in the C–MnO₂ cathode, MnO₂ particle size, pH in strong KOH solutions, solution agitation, and Mn-ion complexing agents in solution.[3,4,10]
3. Characteristic D/H isotope effects on the discharge current are observed at constant polarization and fractional MnO₂ discharge[10,21,22]; that is,

$$\frac{I(KOH/H_2O)}{I(KOD/D_2O)} \simeq 1.41$$

4. Discharge and recovery curves fit solid-state diffusion theory.[19,22-26]

Conceptually, the diffusion involves solid-state proton transfer between O^{2-} ions and simultaneous electron transfer between Mn^{4+} ions, as shown schematically in Fig. 6. As the electrons are transferred, the Mn^{4+} ions on which they reside become Mn^{3+} ions. In order to maintain microscopic charge neutrality, the protons and Mn^{3+} ions are in the same general proximity, although the electronic structure allows the electrons and protons to move somewhat independently.[27] Nevertheless, the concentration gradients of both are equivalent, as represented by the gradient in the lower half of Fig. 6. Mathematical treatment is based on solution of Fick's diffusion equation (see Ref. 28) for semi-infinite linear diffusion with initial and boundary conditions appropriate to the experiment. The potential is related to the surface-concentration ratio of Mn^{4+} ions to Mn^{3+} ions through a Nernst type equation such as Eq. (3); that is, the entire potential change during charge and discharge is associated with

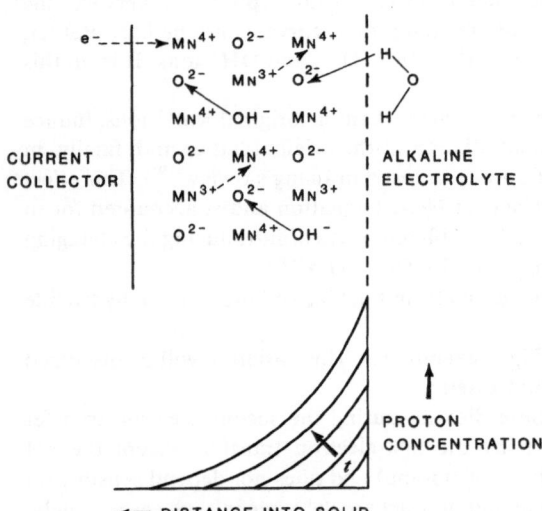

FIGURE 6. Schematic representation of MnO₂ discharge in alkaline solution (top) and corresponding proton gradient in the oxide (bottom). Movement of electrons (→) and protons (←) in oxide is shown.

concentration overvoltage, and experiments are performed such as to remove ohmic potential drop, if significant. All the treatments thus far have restricted concentration effects to the logarithmic term in Eq. (3), and none have considered the linear term in r.

Correspondingly, a single solution of the diffusion equation would not be expected to fit discharge over a broad range of reduction, r, and experimental application has generally been limited to shallow discharge.

2.3.2. Quantitative Treatments of MnO₂ Polarization

Diffusion of protons in the MnO₂ lattice has been treated for galvanostatic,[22-26] potentiostatic,[22,29] and impedance[30,31] studies as well as for chemical leaching studies.[32,33] Here we will focus on the work that is directly related to electrochemical discharge.

Scott[24] and Kornfeil[25] solved the diffusion equation for galvanostatic discharge and open-circuit recovery after such discharge. Assuming that the initial state of reduction (i.e., the Mn^{3+} concentration) is zero, these investigators obtained for the potential–time function:

$$E_t = E_0 + K \log \frac{[Mn^{4+}]}{[Mn^{3+}]}$$

$$= E_0 + K \log \left[\frac{SF\sqrt{\pi D}}{2IV(\sqrt{t} - \sqrt{t - \tau})} - 1 \right] \qquad (4)$$

Here E_t is the oxide potential at time t, and E_0 is the potential at time zero; D is the diffusion coefficient of protons in the oxide phase; I is the discharge current; V is the molar volume of MnO₂ (assumed to be constant as MnO₂ is converted to MnOOH); F is Faraday's constant, S is the electrochemically active surface area of the solid; t is the time measured from beginning of discharge; and τ is the total discharge time. Thus, $t - \tau$ is the recovery time, which is taken as zero for $t < \tau$.

Kornfeil[25] allowed E_0, K, and $S\sqrt{D}$ to be parameters and determined them from the data. He then showed that, with these parameters, Eq. (4) fit both the discharge and recovery over a fairly substantial time range. He also showed that these same constants fit discharge and recovery data for different discharge times and currents. Interestingly, K was determined to be 0.068 V at 25°C rather than the 0.059 V derived from first principles.

As written, Eq. (4) did not fit discharge curves in the very early stages or recovery curves in the late stages. The early misfit was due to the fact that the Mn oxidation state at the beginning of the test was slightly less than 4 (as is the case with all battery-grade manganese dioxides). The misfit after 20 minutes of open-circuit recovery was due to the fact that the diffusion lengths through the solid particles or thin films of MnO₂ are so small that infinite diffusion does not apply. The latter problem was addressed by Era et al.,[23] who modified Eq. (4) accordingly. Gabano et al.[22] corrected the initial conditions to allow for part of the Mn already being in the +3 oxidation state and thereby developed the following expression for galvanostatic discharge:

$$E_t = E_0 + \frac{2.3RT}{F} \log \frac{x_0 - \frac{3}{2} - KI\sqrt{t}}{2 - x_0 + KI\sqrt{t}} \qquad (5)$$

Here x_0 is the mole ratio of oxygen to manganese in the initial oxide, i.e., $(n_O/n_{Mn})_i$. For stoichiometric MnO₂, x_0 would be 2, but for actual manganese dioxides x_0 varies from about 1.89 to 1.97. K contains the diffusion parameters as well as conversion constants; that is,

$$K = \frac{V}{FS\sqrt{\pi D}} \qquad (6)$$

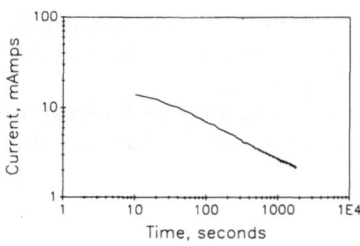

With the *correct* $S\sqrt{D}$, which is the only adjustable parameter in Eq. (5), a plot of the logarithmic term against E_t gives a slope of $2.3RT/F$, which Gabano *et al.* found.[22]

Diffusion treatment applied to cathodic potential-step experiments explains the current decay with time according to Eq. (7)[22]:

$$\log I = -0.5 \log t + \log(x_0 - x_s)/K \tag{7}$$

K was defined above, and x_s is the surface ratio of O to Mn, $(n_O/n_{Mn})_s$. The value of $x_0 - x_s$ in Eq. (7) is defined by the potential step through the equilibrium expressions for E at both values of x; that is,

$$E(x_i) = E_0 + \frac{2.3RT}{F} \log \left[\frac{x_i - \frac{3}{2}}{2 - x_i} \right] \tag{8}$$

Substituting x_0 and x_s, respectively, into Eq. (8) and subtracting the two resulting equations yields

$$\Delta E = E(x_0) - E(x_s) = \frac{2.3RT}{F} \log \frac{(x_0 - \frac{3}{2})(2 - x_s)}{(x_s - \frac{3}{2})(2 - x_0)} \tag{9}$$

By means of Eq. (9), x_s can be determined from ΔE, the potential step, and x_0, which is determined from assays of the initial material for total manganese and total oxidizing power—i.e., the amount of reducing agent such as ferrous or oxalate ions needed to dissolve the sample.[10,34]

With a knowledge of x_s, the potentiostatic method is applied to yield diffusion data through Eq. (7), the log I–log t curve yielding a slope of $-\frac{1}{2}$ (see Fig. 7) and the value of I at 1 s yielding K, which in turn yields $S\sqrt{D}$.

Gabano *et al.*[22] and Brouillet *et al.*[29] verified the application of the potentiostatic method by developing log I–log t slopes of $-\frac{1}{2}$. Gabano *et al.*[22] showed that different ΔE values yield the same K and demonstrated that K varies directly with the weight of manganese dioxide sample (through S). They also showed the self-consistency of the potentiostatic and galvanostatic methods on a given sample and demonstrated the expected isotope effect in $S\sqrt{D}$ for protonated versus deuterated solutions.

Wruck[26] developed a more general model than his predecessors. He considered a more realistic $E_{e,1}$, versus r relationship, ohmic losses, charge transfer kinetics, proton diffusion in the oxide phase, and water diffusion in the pores. Wruck's model allowed a satisfactory fit of constant-current discharge curves for cylindrical composite cathodes in alkaline solutions.

3. STRUCTURAL FEATURES RESPONSIBLE FOR BATTERY ACTIVITY

The origin of battery activity has been of vital interest for years. The molecular structure, solid-state structure, lattice water, and pore structure have been identified as important factors in determining discharge capacity, although the relative role of each factor and the mechanism involved for all types of MnO$_2$ is far from resolved. In this section we will briefly review some of the battery activity factors and corresponding models and illustrate some of the successes and questions in applying them.

3.1. Application of S√D

The diffusion treatment, summarized in the last section, has been found sufficiently applicable that $S\sqrt{D}$ is considered to be a battery-activity parameter. The two factors clearly represent two different structural/dynamic aspects of manganese dioxide—that is, the pores/interstices that transport water and hydroxide ions between the bulk solution and the individual solid particles as represented by S and the solid-state structure itself, which determines D.

Methods for separating D and S usually have addressed independent means of estimating surface area. The most common assumption has been that the electrochemically active surface area, S, is equivalent to the N$_2$ BET surface area,[22,26,30,35,36] which also is equivalent to the area determined by adsorption of several other gases.

Such surface areas range from 10 to 100 m^2/g of MnO$_2$ and usually result in solid-state proton diffusion coefficients of $\sim 10^{-16}$-10^{-20} cm^2/s.

Several investigators[31-33,37] have argued that the predominating pores in EMD, which are smaller than 10 nm in diameter,[38] cannot contribute to electrochemical discharge. These investigators usually assume that the geometric surface of the MnO$_2$ particles (~ 30 μm in diameter) is the appropriate surface. Since the geometric surface area is only about 10^{-3} of the gas adsorption area, the diffusion coefficients corresponding to the smaller surface area are $\sim 10^{-7}$-10^{-12} cm^2/s. Kahil et al.[37] have supported their argument for the smaller surface area and larger D value by independently determining a proton diffusion coefficient of 10^{-10} cm^2/s directly from NMR measurement, which they suggest directly determines the transition time for proton hopping between adjacent oxide ions in the MnO$_2$. Malati et al.[32] supported their assumption with MnO$_2$ leaching, which accelerated with a decrease in particle size.

Some other studies appear to support the gas adsorption area. Kozawa[35] found the same MnO$_2$ surface area available to Zn^{2+}-ion adsorption from solution as is available to gas adsorption. The ion adsorption time (~ 12 h) was of the same order as many discharge tests. Kozawa[39] also found that the MnO$_2$ absorbs essentially the same volume of water as gas in a relatively short time (~ 1 h). If the small pores are not electrochemically active, it is difficult to explain the large body of data that correlates battery activity with porosity, as will be discussed in Section 3.3.

Little work has been published that quantitatively analyzes the relationship between battery activity and values of S or D. Kozawa and Powers[36] studied differences in discharge behavior between an EMD and the β-like MnO$_2$ obtained after heat-treating the EMD. In alkaline solution they found that surface area (from gas adsorption), equilibrium potential, and polarization (manifest in the proton diffusion coefficient) contribute to the battery-activated differences.

Most investigators agree that identification of the electrochemically effective surface area is not settled and would be a difficult task at best.

Using the N$_2$ BET surface area for S, Gabano et al.[22] determined the proton diffusion coefficient in $1M$ KOH at ambient temperature for one sample of each of three types of MnO$_2$—an electrochemical, chemical, and natural-ore sample. We made similar determinations in our laboratory in $5M$ NH$_4$Cl + $1M$NH$_4$OH solution. Both sets of results are shown in Table 1. Both surface area and the diffusion coefficient appear to play a role. Data on more samples would be desirable, but the question of the appropriate surface area also needs resolution.

TABLE 1
Parameters for Diffusion of Protons in MnO_2

Study	Type of MnO_2	$S\sqrt{D}$ $(cm^3/s^{1/2}\ g)$	S (m^2/g)	D (cm^2/s)
Ref. 22	Electrolytic	5.07×10^{-4}	55	8×10^{-19}
	Chemical	2.45×10^{-4}	30	6.7×10^{-19}
	Natural	1.43×10^{-4}	18	7×10^{-19}
Present study	Electrolytic (IBA 1)	8.38×10^{-5}	42	4.0×10^{-20}
	Chemical (IBA 22)	2.47×10^{-5}	54	2.12×10^{-21}
	Natural (IBA 7)	6.14×10^{-5}	8.2	4.47×10^{-19}

3.2. Cation Vacancy Theory

3.2.1. Definition

The cation vacancy theory of Ruetschi[40] is probably the most fully developed theory published to account for battery activity. Quantitative application has successfully rationalized the vast difference in battery activity between γ- or ε-MnO_2 and nonactive β-MnO_2 (demonstrated in Fig. 2) and the continuous decrease in battery activity imparted to γ-MnO_2 and ε-MnO_2 as they are heat-treated to change them successively into an inactive β-like MnO_2.

After carefully examining MnO_2 compositional and structural data, Ruetschi focused on the structural or combined water and the associated oxide structure as the basic factor responsible for battery activity. Structural water is all that water progressively removed from predried manganese dioxide by heating from about 100°C (the predrying temperature) to 400°C or more; no more water is available after such water is removed.

All MnO_2 structures are comprised of MnO_6 octahedra, arranged so as to share edges and corners. Since O^{2-} ions are much larger than Mn^{4+} ions, the lattice resembles a dense packing of O^{2-} ions, with the small Mn^{4+} ions arranged in a more or less orderly fashion in edge-shared MnO_6 octahedral chains between O^{2-} layers.

According to the cation vacancy theory, some of the Mn^{4+} ions are missing, leaving a cation vacancy (see Fig. 8). For each vacancy four protons are present on adjacent O^{2-} ions to balance the charge, the protons being present as OH^- ions. Being of practically the same

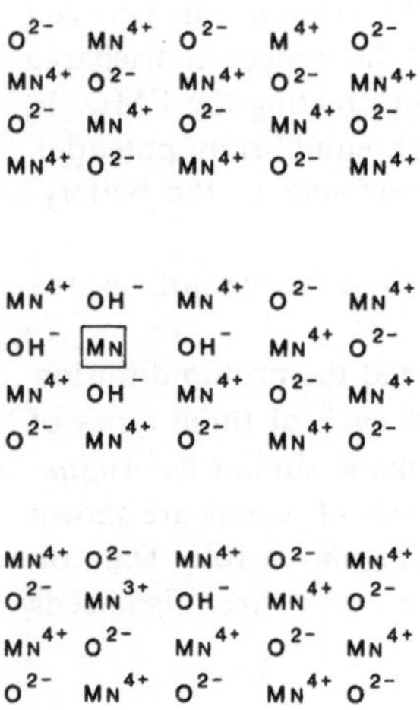

FIGURE 8. Two-dimensional schematic diagram of MnO_2 lattice, showing cation vacancy (*center*), reduced Mn as Mn^{3+} ion (*bottom*), and charge-compensating OH^- ions as compositional water.

size as O^{2-} ions, the OH^- ions do not significantly alter the lattice. Some Mn^{4+} ions are present in the reduced form, Mn^{3+}, and each such Mn^{3+} ion requires an adjacent H^+ (as an OH^- ion) for charge compensation. During reduction all the Mn^{4+} ions are reduced to Mn^{3+} ions. No other form of Mn is considered besides Mn^{4+} and Mn^{3+} ions. In particular, Mn^{2+} ions are considered unlikely based on thermodynamic and steric grounds, and chemical analyses bear this out.†

Thus, the MnO_2 lattice is comprised entirely of Mn^{4+}, Mn^{3+}, O^{2-}, and OH^- ions and manganese-ion vacancies, $\boxed{Mn}$. All the structural water is present in the form of OH^- ions associated with either Mn^{4+} vacancies or Mn^{3+} ions.

The formula of manganese dioxide may be formalized on the basis of the fraction of Mn^{4+}-ion vacancies, $\boxed{Mn}$ (which shall be denoted here by z), and the fraction of Mn^{4+} ions in the reduced state, $+3$ (denoted by y).‡ On the basis of a total of two oxygen atoms per molecule of manganese dioxide, the molecular formula is:

$$\text{Molecular formula} = Mn^{4+}_{1-z-y} \cdot Mn^{3+}_{y} \cdot O^{2-}_{2-4z-y} \cdot OH^-_{4z+y} \tag{10}$$

Values of z and y may be determined from any two of the following three assays: average oxidation state of Mn, total Mn, and lattice water. The average oxidation state is determined by means of dissolution–titration, such as with ferrous, oxalate, or arsenite ions, which reduce all Mn^{4+} and Mn^{3+} ions to the $+2$ state as they dissolve the oxide. Account must be taken of the impurities in manganese dioxide before raw assays are applied to the calculation of z and y. In the case of EMDs, these impurities exist at the ~1% level and consist mainly of sulfate salts of alkali and alkaline-earth cations adsorbed by the EMD, but also include tiny amounts of heavy-metal salts including $Mn(OH)_2$. CMDs also contain only about 1% impurities, but NMDs may contain ~10% impurities, most of which are gangue materials such as silica, alumina, and iron oxide and other acid-insoluble substances.

Chemical analyses and resultant z and y values for a good-grade commercial EMD sample and a battery-inactive β-MnO_2 are shown in Table 2.

TABLE 2
Cation Vacancy Model Parameters for a
Typical High-Quality EMD and a
Highly Crystalline β-MnO_2

Parameter	EMD	β-MnO_2 (IBA 6)
% MnO_2	92.8[a]	>99[b]
% Mn	61.8[a]	>63[b]
% Lattice water	3.65[a]	≤0.15[a]
z (mole fraction)	0.06	<0.01
y (mole fraction)	0.08	<0.01

[a] Determinations made in this laboratory; values are on an impurity-free basis.
[b] Reference 34.

†For example, in this laboratory: (1) Mn^{2+} ions leached from numerous EMD samples with $1N$ H_2SO_4 accounted for less than 0.5% of the EMD weight; (2) leaching several EMD samples with $2M$ NaCN solution (which strongly complexes Mn^{2+} ions but not Mn^{3+} ions) resulted in less than 0.01% Mn^{2+}; (3) leaching an EMD sample with NH_4Cl or $NH_4Cl + ZnCl_2$ solution at pH 5 resulted in less than 0.01% Mn^{2+}. The small amounts of Mn^{2+} present probably arise from residual adsorbed Mn^{2+} ions from the EMD deposition bath ($MnSO_4$ in dilute H_2SO_4) that are converted to surface $Mn(OH)_2$ during the neutralization.[41]

‡In Ruetschi's papers the cation vacancy fraction is termed x, but we call it z in this chapter to distinguish it from the mole ratio of oxygen to manganese, which we previously defined as x.

The fraction of Mn^{4+} ions missing (z) and reduced (y) are the averages determined from the three given assays, permuted two at a time. Consistency in the z and y values from the three pairs of assays is, in itself, a test of the applicability of the model.

3.2.2. Effect of $\boxed{Mn}$/Compositional Water on Battery Activity

Manganese vacancies with their associated OH groups assist battery activity in two ways:
1. They impart a greater free energy to the MnO_2 lattice, which results in a more negative free energy of reaction for Eq. (2a) and, correspondingly, a more positive open-circuit potential at any given stage of reduction
2. They increase the diffusion rate of protons through the lattice, which decreases the concentration overvoltage or polarization.

In case 1, the cation vacancies affect both the logarithmic and linear terms in r of Eq. (3). Statistical-mechanical derivations of the electrode potential have been given by several investigators.[16,42,43] The cation vacancy population [z in Eq. (10)] determines the number of sites available for protons to occupy during reduction and thereby influences both r, at a given stage of reduction, and the statistical distribution of protons on such sites. The statistical contribution to the partition function is thus determined, and the logarithmic term results.

The potential energy term also varies with r and, thereby, with the cation vacancy population, through variation of the average proton potential energy with $\boxed{Mn}$ fraction.[16]

A recent treatment by Ruetschi and Giovanoli[44] explains how the cation vacancies assist proton diffusion. Vacancies allow proton movement in the a and b directions of the orthorhombic lattice whereas movement in a vacancy-free lattice is restricted to the c direction by virtue of the symmetry at the oxygen atoms—that is, protons can move between only the half of the oxygen atoms that occupy pyramidal positions; the oxygen atoms at planar sites cannot participate.

3.2.3. Experimental Application/Verification of Cation Vacancy Theory

In previous sections, the interrelationship of Mn content, average Mn oxidation state (or x in MnO_x), and compositional water were discussed, as well as the electrode potential and proton diffusion. Some other properties related to compositional water include absolute density, theoretical discharge capacity, and electronic conductivity. The electronic conductivity decreases with an increase in compositional water, either from lattice vacancies or partial MnO_2 reduction.[45] This inverse correlation is attributed to the fact that electronic conduction occurs by electrons tunneling or hopping from Mn^{4+} to Mn^{4+} ion, and, as the cation vacancy population increases, the average distance between Mn^{4+} ions increases, thus decreasing the rate of electron transfer.[40,45,46]

In this chapter, we are primarily interested in rationalizing differences in battery activity. The cation vacancy theory is highly successful in explaining the changes that accompany heat treatment of electrolytic manganese dioxide at successively higher temperatures.[40] With increasing temperature, compositional water is increasingly expelled. At about 375°C, the crystal structure also changes, as noted in the X-ray diffraction patterns by the appearance of the 3-Å β-MnO_2 peak and the disappearance of the 4.1-Å γ-MnO_2 peak (Fig. 9). Various MnO_2 samples that have been heat-treated exhibit electronic conductivities, equilibrium potentials, and densities characteristic of the amount of remaining combined water and, thus, of the temperature at which they were heated.[40,45,47,48,50,51] The discharge capacity of EMD generally decreases after heat treatment to successively higher temperatures with its progressive removal of structural water.[51,52] However, there have been a few cases in which the discharge capacity in $ZnCl_2$–NH_4Cl electrolytes increases after heat treatment (e.g., see Ref. 49). This

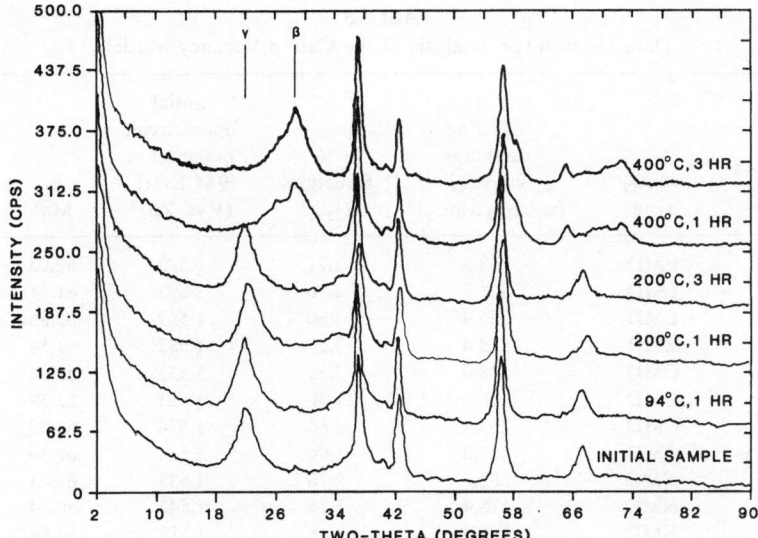

FIGURE 9. X-ray diffraction scan of powdered EMD sample as-produced and after heat treatment in air under various conditions. Cu X-ray tube. Scan rate $=1° 2\theta$ per minute. Characteristic peaks are shown for γ-MnO₂ and β-MnO₂.

is because the pore size increased with heating, and the pore size is critical in neutral discharge whereas solid-state proton transport is critical in alkaline discharge (see Section 3.3).

The success of the cation vacancy theory in explaining the changes accompanying heat-treatment of EMD led us to the more general question of model applicability to the various generic types of MnO₂. We here consider the differences in alkaline discharge capacity of the three major MnO₂ forms—EMD, CMD, and NMD—and battery-inactive β-MnO₂ with measured solid-state parameters. Table 3 is a compilation of data that we used to determine the efficacy of the model.

Only partial correlation is found between alkaline discharge capacity and compositional water (Fig. 10), which, considering the approximately equivalent Mn oxidation state in each case, is a measure of the cation vacancy population. The principal discrepancy appears to be the relative position of the NMDs compared to the EMDs and CMDs. Impurities may play an inhibiting role, as NMDs contain about 10% impurity compared to only 1% impurities in the cases of EMD and CMD and less than 1% impurity for the β-MnO₂. Such an impurity effect would have to be much larger than the error from not normalizing the current or the discharge capacity for the amount of impurity-free MnO₂ present. Such normalization would only increase the NMD discharge capacities by 15–25 mA h/g, whereas the apparent dis-

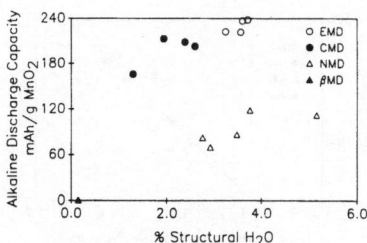

FIGURE 10. Discharge capacities to 1.0 V versus Zn of various types of MnO₂ samples in 9M KOH as a function of structural water.

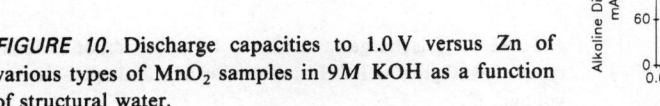

TABLE 3
Data Used in the Analysis of the Cation Vacancy Model

Sample	MnO$_2$ type	Alkaline discharge capacity (mA h/g MnO$_2$)[a]	% Structural H$_2$O[b]	Initial open-circuit potential in 9M KOH (V vs. Zn)[a]	% Mn[c]	% MnO$_2$[c]
IBA 1	EMD	238.3	3.71	1.592	62.05	91.62
IBA 10	EMD	221.9	3.56	1.593	61.61	92.32
IBA 23	EMD	236.9	3.60	1.599	60.86	92.09
IBA 24	EMD	222.4	3.24	1.582	61.54	91.89
IBA 12	CMD	208.8	2.39	1.533	61.98	91.60
IBA 22	CMD	166.0	1.30	1.493	62.38	92.46
Commercial A	CMD	203.0	2.60	1.574	61.00	92.10
Commercial B	CMD	213.0	1.94	1.534	61.30	92.50
IBA 7	NMD	82.0	2.76	1.533	61.43	92.65
IBA 13	NMD	118.4	3.76	1.542	60.84	94.09
IBA 27	NMD	86.0	3.48	1.512	54.68	81.42
IBA 28	NMD	69.4	2.92	1.500	57.00	85.14
IBA 29	NMD	112.1	5.16	1.536	52.59	80.05
IBA 6	β	0.5	0.15	1.374	63.34	100.00

[a] Data taken from Ref. 4 or measured in this laboratory by the method of Ref. 4.
[b] Data averaged from several measurements made in this laboratory and/or from other publications. Structural H$_2$O = H$_2$O removed at 750°C after sample was dried at 110°C for several hours in forced air oven.
[c] Data taken from Ref. 34 or measured in this laboratory by method of Ref. 34.

crepancy between the various MnO$_2$ types in Fig. 10 is about 120 mA h/g of MnO$_2$. Interestingly, the equilibrium potentials appear to correlate better with the discharge capacities than do the compositional waters (Fig. 11), although there still is discrepancy in the capacities of the NMDs relative to those of the CMDs. Figure 11 shows data for numerous additional EMD samples reported in Ref. 4 but not in Table 3. All the EMD samples lie in the same range in Fig. 11.

Figure 12 shows no trend in percent MnO$_2$ (Mn oxidation state) or Mn with type of manganese dioxide, except for several NMDs, which are very impure. In this figure, the compositional parameter was normalized for the impurities. Still, the capacities of the NMDs appear displaced negatively from those of the other types of MnO$_2$. Figure 12, like Fig. 11, contains data for the many additional EMD samples reported in Refs 4 and 23 but not in Table 3.

The differences in alkaline discharge capacity among the various EMDs are small compared to those differences between the different types of MnO$_2$ (Figs 10–12) or between EMDs before and after significant heat treatment. Also, there is no apparent correlation between the discharge capacities of these EMDs and the molecular structural data.

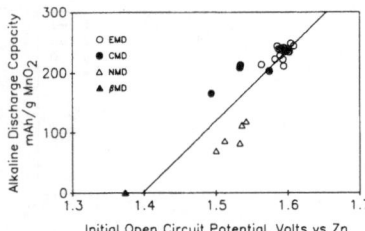

FIGURE 11. Discharge capacities to 1.0 versus Zn of various types of MnO$_2$ in 9M KOH as a function of initial open-circuit voltage, E_i.

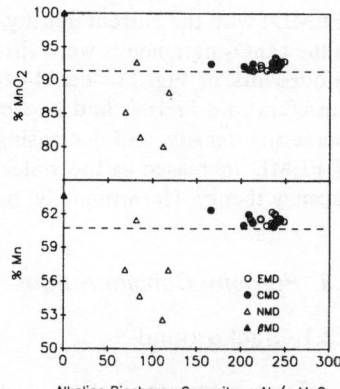

FIGURE 12. Discharge capacities to 1.0 V versus Zn of various types of MnO₂ in 9M KOH as a function of MnO₂ and Mn content.

To study such noncorrelation further, we deposited a suite of EMD samples under different cell conditions and finished them in the same manner with respect to grind and neutralization medium. (The IBA samples are not equivalent in these latter respects.) These tests were prompted by the well-known fact that discharge capacities can be lowered by 5–10% and more by lowering the electrolyte temperature or raising the current density of deposition.[10]

Figure 13 shows compositional data for the samples relevant to cation vacancies and initial oxidation state of the Mn. Results appear to show that factors other than compositional water or cation vacancies are responsible for the differences in alkaline discharge capacity.

In seeming contrast to our results of Fig. 13, Preisler[65] found variation in the composition (MnO₂, Mn, and structural water) and absolute density of an EMD and fibrous EMD

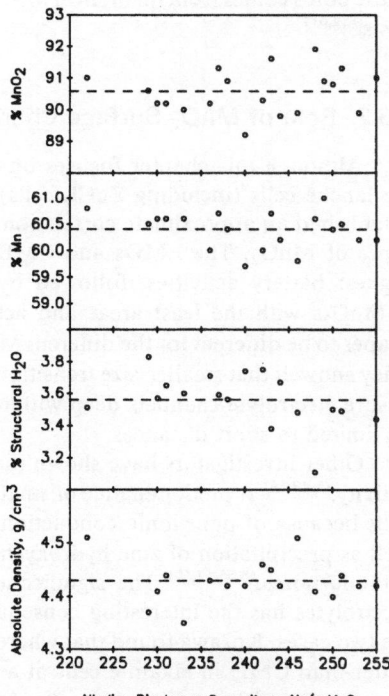

FIGURE 13. Discharge capacities to 1.0 V versus Zn of various EMD samples in 9M KOH as functions of chemical composition and absolute density.

(FEMD) with the current density at which the EMDs were deposited. However, the changes in the MnO_2 parameters were virtually imperceptible over the current density range included in our work of Fig. 13—i.e., 4–10 mA/cm^2. Only as the current density was lowered below 4 mA/cm^2 did Preisler find a readily measurable trend of increasing MnO_2 and Mn content, increasing density, and decreasing structural water content. He also found that the potential of FEMD increased as the water content increased, which is in agreement with the cation vacancy theory. Unfortunately, battery activities were not determined in this work.

3.3. Porosity Considerations

3.3.1. Background

Even before the cation vacancy theory was born, researchers noted that porosity or surface features of MnO_2 varied significantly with MnO_2 origin, deposition conditions (in the case of EMD), and heat treating of γ-MnO_2 to produce a β-like MnO_2. Correspondingly the impact of porosity on battery activity was noted.[10,36,52,54] Porosity features include gas sorption, water desorption, mercury intrusion (porosimetry), and ion exchange. The significance of surface involvement is comprehendible from the fact that powdered MnO_2 has a surface area of ~10–100 m^2/g, and 10–20% of the atoms occupy a surface position.[17] Diagnostic surface methods involve both capacity of the MnO_2 for liquids or gases and the shapes of the isotherms and porograms. Analyses of these curves with surface tension theory reveal voids with effective radii from <2 nm to more than 10^5 nm. These voids are usually classified according to their radius as micropores ($r \leq 2$ nm), transitional pores or mesopores ($r \simeq 2$–20 nm), macropores ($r \simeq 20$–100 nm), and interstices between grains ($r \geq$ 100 nm).[56,57] The pores smaller than macropores contribute most heavily to the surface area and also are those of most relevance regarding battery activity.[38,57–59] Most definitions of these pores comes from theoretical interpretation of isotherms[36,52,53,56,57] and other diagnostic tests.[59,61]

3.3.2. Role of MnO_2 Surface Characteristics in $ZnCl_2$ and NH_4Cl–$ZnCl_2$ Cells

Although this chapter focuses on alkaline cells, the historically significant findings for Leclanché cells (including $ZnCl_2$ cells) must be considered as background. Earlier work[52] established an approximate correlation between battery activity and surface area for various types of MnO_2. The EMDs and CMDs possessed both the greatest surface areas and the highest battery activities, followed by activated ores, active ores (NMDs), and, finally, β-MnO_2s with the least areas and activities. These workers[52] also showed the isotherm shapes to be different for the different MnO_2 forms and discussed differences in pore structure. They showed that smaller-size transitional pores are important for battery activity. Such pores ensure electrolyte channels deep within MnO_2 particles so that slow solid-state diffusion can be limited to short distances.

Other investigators have shown that an optimum pore size is important to good battery activity.[8,49,56] A predominance of small pores leads to impeded battery activity in Leclanché cells because of poor ionic conduction and concentration (pH) overvoltage in the pores as well as precipitation of zinc hydroxychloride or zinc ammonium chloride at the entrance to cathode pores.[8,55,57,60] The significant difference in properties of Leclanché and alkaline electrolytes has the interesting consequence that different MnO_2 forms can be optimum in the two cases. Kozawa found that whereas EMD with mean pore diameter of 4 nm performed better than CMD in alkaline cells at a current density of 10 mA/g of MnO_2, CMD with its larger pores of 8-nm diameter performed better in $ZnCl_2$ cells.[8]

3.3.3. Dependence of Alkaline Discharge on MnO$_2$ Surface Properties

Alkaline-cell electrolyte, $9M$ KOH, largely overcomes the ionic resistance, pH-polariz-ation, and salt-precipitation problems of the near-neutral ZnCl$_2$–NH$_4$Cl electrolytes.[8,60] This allows the solution-based processes in the cathode to accelerate to the point that the solid-state proton diffusion becomes rate determining in many cases. Hence, many investigators have used polarization curves directly to determine the diffusion coefficient of protons in MnO$_2$.

Nevertheless, porosity is manifest in various observations, the best example perhaps being the approximately inverse correlation between surface area and alkaline discharge capacity of *EMDs*.[10] Pore volumes measured directly and by moisture (bulk water) release on drying have been found to parallel the surface areas qualitatively.[10] Alkaline discharge capacity versus surface area is plotted in Fig. 14 for the EMD samples deposited in this laboratory that were also evaluated for compositional differences in Fig. 13. Whereas the compositional differences are, for the most part, within experimental imprecision ($\sim$1%), the surface areas vary by more than a factor of 3. The spread in data is not unexpected in light of the several different depositional parameters varied to change the battery activity and in light of the fact that MnO$_2$ samples with different pore characteristics can produce equivalent surface areas.[36,56]

Results such as those of Fig. 14 are attributed by some to rate control by OH$^-$-ion conduction in transitional pores and micropores. It has been suggested that the ionic diffusion coefficient is much smaller in pores less than approximately 10 nm in diameter than in bulk solution,[60] the reason being that water on the walls of the pores is adsorbed on the MnO$_2$ and diffusion is, correspondingly, constrained (see also Ref. 61). EMD samples having a predominance of micropores have been found to yield poorer alkaline-battery activities than ones with transitional pores (or mesopores).[56] The issue of micropores is a bit cloudy, as one study[62] indicates that such pores are created by drying the sample above ambient temperature in preparation for sorption tests. However, EMD manufacturers usually dry the final product at temperatures greater than ambient, so it appears that EMD is supplied with some micropores.

Scanning electron microscopy (SEM) photographs of a great many EMD samples at 50,000 $\times$ suggest that low-surface-area, high-capacity material contains a network of wormlike pores that are about 10 nm in diameter (in the transitional pore range) and 400 nm long, whereas high-surface-area, lower-capacity EMD lacks this feature.[60]

If one adheres to solid-state proton diffusion as the rate-determining step in EMD alkaline discharge, one may still rationalize the inverse surface area–capacity relationship of Fig. 14 in terms of crystallite boundaries, which both increase surface area and impede solid-state proton diffusion.

As EMD is deposited, more or less ordered microdomains are formed,[63] which are much smaller than the grains formed upon grinding the material. By sonicating a suspension of EMD, we have subdivided powdered EMD further to produce tiny prismatic crystallites as well as bundles of such crystallites, which are visible in transmission electron microphoto-graphs (see Fig. 15).

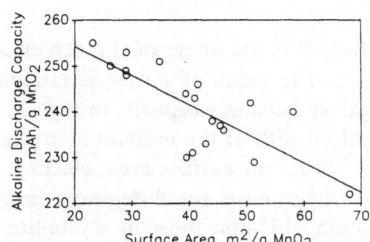

FIGURE 14. Discharge capacities to 1.0 V versus Zn of various EMD samples in $9M$ KOH as a function of N$_2$ BET surface area.

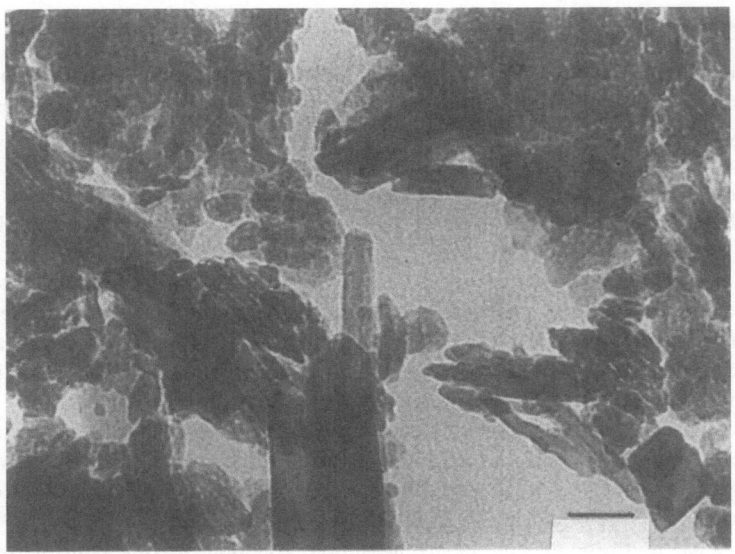

FIGURE 15. Transmission electron photomicrograph of sonicated EMD sample. Sample = IBA No. 1. The bar length is 0.025 μm.

The width of the individual crystallites is about 5–10 nm (50–100 Å). The surface area of EMD calculated from the surface of such individual crystallites is given approximately by

$$S(\text{m}^2/\text{g}^-) = 4 \times 10^4/w\rho \tag{11}$$

Here, w is the width of the crystallites in angstroms, and ρ is the density, assumed to be the theoretical crystallographic density of about 4.8 g/cm^3. Taking w as 50–100 Å, one determines S to be 85–170 m^2/g. The N$_2$ BET surface areas of the EMDs without sonication are 25–70 m^2/g. If one supposes that the measured surface area is due to the observed surfaces in Fig. 15, then one could account for the measured surface area on the basis of bundles of crystallites.

X-ray diffraction (XRD) patterns of EMDs deposited under different conditions differ mainly in the breadth of the 4.1-Å peak, which corresponds to the (110) plane of the orthorhombic lattice. Furthermore, this peak is broader for EMD of low discharge capacity than of high discharge capacity—for example, low temperature of deposition gives a broader peak than high temperature.[10] Line broadening is quantified in terms of peak width at the half-height intensity, which we shall designate as $\Delta\theta$ (radians on 2θ scale). Broadening may be caused by narrower crystallites in the direction perpendicular to the plane in question. The thickness of the crystallites, t, is given by the Scherrer formula,[64] i.e.,

$$t = \frac{0.9\lambda}{\Delta\theta \cos\theta} \tag{12}$$

where θ is the angle of the center of the peak, and λ is the wavelength of the X-radiation used. The value of t was determined for several of the samples of Fig. 14 and is plotted against discharge capacity in Fig. 16. Values of t are seen to be of the same order of magnitude as the width of the individual crystallites observed in Fig. 15.

From the surface area, electron microscope, and XRD considerations, one could be led to a hypothesis for differences in alkaline discharge capacity of various EMDs based on proton diffusion through crystallite–crystallite interfaces as the rate-determining step. The

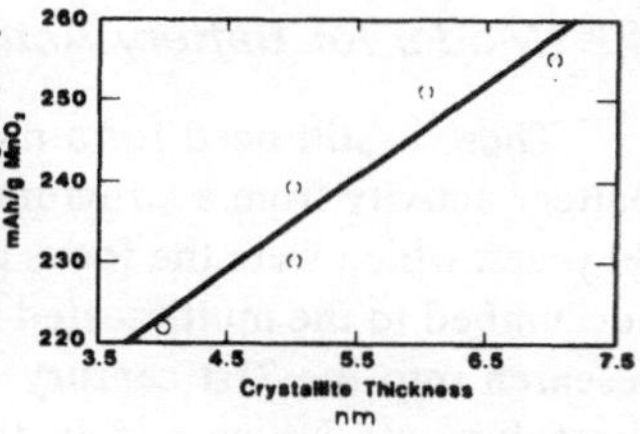

FIGURE 16. Discharge capacities to 1.0 V versus Zn of various EMD samples in 9 M KOH as a function of crystallite thickness, t, normal to (110) plane.

XRD linewidth determinations and TEM photographs indicate that less battery-active EMD (e.g., that deposited at low temperature or at elevated current density) has smaller and narrower crystallites and crystallite bundles and, hence, more interfaces per unit length of diffusion or per gram of EMD than more battery-active EMD.

The cause-and-effect relationship between deposition parameters and crystallite-bundle size has been attributed to a competition between crystallite nucleation and growth,[57] rapid nucleation leading to the smaller and more porous crystallite bundles. Preisler[65] has provided a considerably more detailed explanation, taking into account the probable rate-determining steps during MnO₂ deposition of deprotonation of hydrated Mn⁴⁺ ions and diffusion through solid manganese oxide phases formed by precipitation.

4. NEW HORIZONS IN ELECTROCHEMICAL SCIENCE AND TECHNOLOGY

4.1. Overview

The annual small-battery market is $11 billion worldwide and growing at a rate of 9% per year.[66] More than three-fourths of this market involves MnO₂-containing cells. This trend likely will continue into the next century, fueled by people's desire for portable electronic and motorized devices. Fundamental understanding of MnO₂ electrochemistry is still quite limited, so research is expected to continue at the rapid pace set and maintained over the past several decades. Although new battery systems are being introduced continuously, MnO₂ is remaining the cathode of choice in many of them, including the systems that seem destined for large-volume consumer use. These new systems will require much research, and this will offset any slowdown in activity on the Zn–MnO₂ systems, which are perhaps nearing their fundamental limits at low to moderate drain rates—that is, up to 85–90% of theoretical first-electron capacity.[55] Rapid drain rates, as needed for photoflash applications, will provide the opportunity to pursue kinetics for profit. Basic understanding in other areas, such as abrasity to tools and dies during cathode formation, and density and flow under pressure, will offer another dimension for structural studies.

Research into rechargeability has started intensively and will continue into the next century. At this time, the convenience of primary cells appears to overpower the conservation aspects of rechargeable cells, but environmental concerns and depletion of resources should shift the emphasis in the 21st century. Rechargeable small cells have been the focus primarily in high-power-density applications such as power tools. Although nickel–cadmium and lead–acid cells have dominated this market, the development of spiral (jellyroll) designs in rechargeable alkaline[67] and Li–MnO₂ cells[68] promises to provide the impetus for research in the fast end of the discharge regime.

Battery developers are set to initiate flat-plate secondary alkaline cells. Such cells would offer great weight advantage over the lead–acid battery in the $12 billion "large" secondary-battery market. This move will further provide impetus for 21st century research on fast MnO₂ discharge.

4.2. Model for Battery Activity

There is still need for a more complete yet understandable model that describes MnO_2 battery activity from a structural basis, despite the intensive and fruitful efforts over the past 30 years, which were the focus of this chapter. The great complexities of MnO_2 have partially succumbed to the multifaceted structural methods and ingenious theoretical approaches, but research into the 21st century will be required to unravel the relative roles of processes in the solid-state lattice and in the pores (or mesostructure). A self-consistent model is still needed to explain effects of the various deposition parameters, such as current density, temperature, and electrolyte composition, on discharge. Particulates suspended in the deposition bath appear to have their own interesting influence on the deposition process and deposit,[2,69] and this area has been studied very little.

Further development of each existent working model and more experimental studies to develop the needed supportive data such as Ref. 65 should aid the modeling effort. However, conceptual innovations may be needed to satisfactorily define and understand such enigmas as the dynamic role of small pores, the electrochemically active surface, and the solid–liquid interfacial electrochemistry in general.

Future modeling should also address the cathode macrostructure, which is MnO_2 particles, graphite particles, and electrolyte compressed together. Although influence of the macrostructure is conveniently (experimentally) excluded in order to focus on intrinsic MnO_2 effects, there is need to better understand the behavior of cathode discharge as a function of electrode dimension, particle size, and electronic conductivity. Works such as those of Coleman,[70] Wruck,[26] and Atlung and Jacobsen[30] are relevant in this regard.

After the MnO_2 structure is adequately related to the discharge kinetics, the MnO_2 synthesis methods and parameters still need to be related back to the structure. The kinetics and mechanism of EMD deposition have been addressed,[65,71,72,81] but to a far less extent than discharge. Preisler[65] has considered all aspects, and further study along those lines should prove fruitful—particularly if more facts are developed with which to equip such theories.

4.3. New MnO₂-Containing Battery Systems

MnO_2 has stood the test of time as a successful cathode. Its low cost coupled with favorable performance features such as flexibility with regard to electrolyte, acceptable energy density, long shelf life, and nontoxicity has identified it as a competitive cathode not only in the various aqueous zinc batteries but also in other aqueous and nonaqueous systems as well. Table 4 shows the commercialization history of MnO_2 batteries over the past century. The indication is that an influx of new MnO_2-containing batteries will be commercialized near the turn of the century. Several other systems, not listed, are in the experimental stage.

Each battery system performs best for its intended use with an MnO_2 structure optimized for that system and use. This chapter focused on the advantage of low-surface-area EMDs over other EMDs as well as CMDs, NMDs, and battery-inactive MnO_2 forms in alkaline-cell applications. In this case, high equilibrium potential and solid-state structure amenable to fast proton diffusion are important. As also discussed, $ZnCl_2$ and NH_4Cl (Leclanché) cells often call for MnO_2 with larger pore size than alkaline cells, and this need is better met by some CMDs than EMDs.[8]

In the case of lithium–MnO_2 cells, research is starting to show advantages of some MnO_2 structures over others, as demonstrated in several of the papers at a recent symposium on rechargeable lithium batteries.[68] As these and other MnO_2 cells develop further, research into discharge mechanisms will intensify to optimize MnO_2 structure for the particular system.

TABLE 4
Commercialization History of MnO$_2$ Batteries

MnO$_2$ battery type	Commercialization date
Leclanché	1890
ZnCl$_2$	1950
Alkaline	1960
Lithium (Nonaqueous)	1976
Lithium rechargeable (Nonaqueous)	1988
Rechargeable alkaline	1990s[a]
"RAM" alkaline	
"Modified MnO$_2$" cell	
Hydrogen rechargeable	1990s[a]
Aluminum seawater	1990s[a]

[a] Estimated dates.

4.4. Rechargeable MnO₂-Containing Cells

Rechargeable battery research should become more prominent in the near future, because of the three new secondary systems being commercialized (see Table 4). Of these new systems, the rechargeable alkaline MnO$_2$ cell has many features that parallel those of primary alkaline batteries.

Although the first-electron discharge of MnO$_2$ is reversible, the electrode cannot be reversed at will. This is because strain is induced into the lattice by the Mn^{3+} ions formed upon reduction (see Section 2.2), and conversely, during recharge, the lattice shrinks as the Mn^{3+} ions are oxidized to the smaller Mn^{4+} ions. These physical oscillations tend to disintegrate the cathode and destroy its rechargeability, unless special precautions are taken. The swelling/shrinking increases with depth of discharge. Second-electron reduction of the MnO$_2$ cathode is irreversible on chemical grounds, and the Zn–MnO$_2$ cells cannot be recharged from such a state. McBreen[73] has studied the reduction–reoxidation processes in detail.

Amano et al.[74] and Kordesch and co-workers[75-77] solved the first-electron rechargeability problems by limiting the depth of MnO$_2$ discharge to half the first electron, by adding fibers and other special binders to minimize cathode swelling, by constraining the cathode swelling with a supportive metal cage, and by controlling the recharge potential to prevent formation of deleterious manganates. They controlled the depth of discharge by limiting the amount of zinc anode present.[74] The overall cell recyclability thereby becomes limited by zinc anode recyclability. Recyclability of EMD in the Kordesch sense is strongly influenced by foreign ions (such as titanium) incorporated into the MnO$_2$ lattice via doping of the deposition bath with related solids.[78]

Wroblowa and co-workers[79,80] induced rechargeability through modification of the MnO$_2$ lattice. This was done by doping it in various ways with a small percentage of bismuth oxide or lead oxide. These substances convert the MnO$_2$ into an open, layered structure which is stabilized by the inserted ions during the discharge–recharge cycle.

The latter chemical-doping method leads to great rechargeability—i.e., to more than 2500 cycles at 1.6 electron change per MnO$_2$ molecule. However, zinc ions, such as would be produced by a zinc anode, lower the depth of discharge by more than 50%. The MnO$_2$ potential remains positive of 1.0 V versus Zn during the entire cycle, which is greater than the open-circuit potential for the second-electron transfer. Thus, the modifications of the layered structure change the electrochemistry from that of the conventional MnO$_2$–alkaline system.

Understanding the remarkable effects of doping MnO_2 on its electrochemistry will undoubtedly be one of the research directions during the next century.

4.5. Second-Electron Discharge of MnO₂

Another future area of research may be aimed at the second-electron discharge in primary cells. Electronics manufacturers are designing equipment that will operate at lower voltages, thus making the second-electron reduction reaction significant. The basic phenomenology of the second-electron discharge has been identified by Kozawa,[10] who concluded that the mechanism involved dissolution of Mn(III) oxide followed by electrochemical reduction of the soluble Mn(III) at the cathodic current collector to form $Mn(OH)_2$. However, MnO_2 structure has not yet been related to the second-electron reduction, since battery-powered electronic equipment has been traditionally designed to function at multiples of 1.0–1.5 V.

ACKNOWLEDGMENT

We wish to thank J. M. Berry, who performed much of the laboratory test work and prepared many of the figures. We also thank the Analytical Chemistry and the Product Development groups at Kerr–McGee Technical Center for their technical and measurement support.

REFERENCES

1. K. V. Kordesch, in: *Batteries, Volume 1, Manganese Dioxide* (K. V. Kordesch, ed.), Chapter 2, Marcel Dekker, New York (1974).
2. K. Takahashi and Y. Nakayama, in: *Proceedings of the Symposium on Manganese Dioxide Electrode Theory and Practice for Electrochemical Applications* (B. Schumm, Jr., M. P. Grotheer, M. L. Middaugh, and J. C. Hunter, eds.), pp. 208–222, The Electrochemical Society, Pennington, New Jersey (1985).
3. A. Kozawa and J. F. Yeager, *J. Electrochem. Soc.* **112,** 959 (1965).
4. S. F. Burkhardt, in: *Handbook of Manganese Dioxides, Battery Grade,* (D. Glover, B. Schumm, Jr., and A. Kozawa, eds.), pp. 217–236, The International Battery Materials Association, Cleveland, Ohio (1989).
5. R. Carbonnier, in: *Handbook of Manganese Dioxides, Battery Grade* (D. Glover, B. Schumm, Jr., and A. Kozawa, eds.), pp. 1–13, The International Battery Materials Association, Cleveland, Ohio (1989).
6. R. G. Burns and V. M. Burns, in: *Manganese Dioxide Symposium, Vol. 2, Tokyo, 1980* (B. Schumm, Jr., H. M. Joseph, and A. Kozawa, eds.), pp. 97–112, I.C. MnO_2 Sample Office, Cleveland, Ohio (1981).
7. V. M. Burns, R. G. Burns, and W. K. Zwicker, in: *Manganese Dioxide Symposium, Vol. 1, Cleveland, 1975* (compiled by A. Kozawa and R. J. Brodd), pp. 288–305, I.C. Sample Office, c/o Union Carbide Corporation, Cleveland, Ohio (1975).
8. A. Kozawa, in: *Progress in Batteries and Solar Cells*, Vol. 7 (IBA Hawaii Meeting), (M. Nagayama and A. Kozawa, eds.), pp. 2–19, JEC Press and IBA Inc., Cleveland, Ohio (1988).
9. D. S. Freeman, P. F. Pelter, F. L. Tye, and L. L. Wood, *J. Appl. Electrochem.* **1,** 127–136 (1971).
10. A. Kozawa, in: *Batteries, Volume 1, Manganese Dioxide* (K. V. Kordesch, ed.), Chapter 3, Marcel Dekker, New York (1974).
11. A. Kozawa, in: *Electrochemistry of Manganese Dioxide and Manganese Dioxide Batteries in Japan* (S. Yoshizawa, K. Takahashi, and A. Kozawa, eds.), pp. 57–114, U.S. Branch Office of the Electrochemistry Society of Japan, Cleveland, Ohio (1973).
12. G. Bewer, H. Debrodt, and H. Herbst, *J. Metals* **1982** (January), 37.

13. C. F. B. Coetzee and W. A. M. Te Riele, The Production of Electrolytic Manganese Dioxide from Ferromanganese Furnace Sludge, Report M60D, Council for Mineral Technology, Randberg, South Africa (1984).

14. E. A. Kalinovskii, Yu. K. Rossinskii, and Zh. M. Kebadze, *J. Appl. Chem. USSR* **61**, 258 (1988).

15. A. Kozawa, in: *Manganese Dioxide Symposium, Vol. 2, Tokyo, 1980* (B. Schumm, Jr., H. M. Joseph, and A. Kozawa, eds.), pp. 559–574, I.C. MnO₂ Sample Office, Cleveland, Ohio (1981).

16. P. Ruetschi, *J. Electrochem. Soc.* **135**, 2657 (1988).

17. A. Kozawa, in: *Manganese Dioxide Symposium, Vol. 2, Tokyo, 1980* (B. Schumm, Jr., H. M. Joseph, and A. Kozawa, eds.), pp. 321–334, I.C. MnO₂ Sample Office, Cleveland, Ohio (1981).

18. K. J. Vetter and N. Jaeger, *Electrochim. Acta* **11**, 401 (1966).

19. C. C. Liang, in: *Encyclopedia of Electrochemistry of the Elements*, Vol. 1 (A. J. Bard, ed.) Chapter I-6, Marcel Dekker, New York (1973).

20. B. D. Desai, R. A. S. Dhume, and V. N. Kamat Dalal, *J. Appl. Electrochem.* **18**, 62 (1988).

21. A. Kozawa and R. A. Powers, *J. Electrochem. Soc.* **115**, 122 (1968).

22. J. P. Gabano, J. Seguret, and J. F. Laurent, *J. Electrochem. Soc.* **117**, 147 (1970).

23. A. Era, Z. Takehara, and S. Yoshizawa, *Electrochim. Acta* **12**, 1199 (1967).

24. A. B. Scott, *J. Electrochem. Soc.* **107**, 941 (1960).

25. F. Kornfeil, *J. Electrochem. Soc.* **109**, 349 (1962).

26. W. J. Wruck, The Characterization and Modeling of the Alkaline MnO₂ Cathode, Ph.D thesis, The University of Wisconsin—Madison (1984).

27. J. B. Goodenough, in: *Proceedings of the Symposium on Manganese Dioxide Electrode Theory and Practice for Electrochemical Applications* (B. Schumm, Jr., M. P. Grotheer, R. L. Middaugh, and J. C. Hunter, eds.), pp. 77–96, The Electrochemical Society, Pennington, New Jersey (1985).

28. J. Goodisman, *Electrochemistry: Theoretical Foundations*, pp. 252–265, John Wiley & Sons, New York (1987).

29. P. Brouillet, A. Grund, F. Jolas, and R. Mellet, in: *Batteries 2* (D. H. Collins, ed.), pp. 189–199, Pergamon Press, Oxford (1965).

30. S. Atlung and T. Jacobsen, *Electrochim. Acta* **21**, 575 (1976).

31. J. Euler, *Electrochim. Acta* **4**, 27 (1961).

32. M. A. Malati, M. W. Rophael, and I. I. Bhayat, *Electrochim. Acta* **26**, 239 (1981).

33. J. J. Laragne and J. Brenet, *Bull. Soc. Chim* **9**, 2455 (1968).

34. D. Glover, B. Schumm, Jr., and A. Kozawa (eds.), *Handbook of Manganese Dioxides, Battery Grade*, pp. 21–46, The International Battery Materials Association, Cleveland, Ohio (1989).

35. A. Kozawa, *J. Electrochem. Soc.* **106**, 552 (1959).

36. A. Kozawa and R. A. Powers, *J. Electrochem. Soc.* **5**, 535 (1967).

37. H. Kahil, F. Dalard, and J. Guitton, *Surface. Technol.* **16**, 331 (1982).

38. M. A. Dakri, F. L. Tye, and J. L. Whiteman, in: *Power Sources 1966* (D. H. Collins, ed.), pp. 65–81, Pergamon Press, New York (1967).

39. A. Kozawa, in: *Measurements on Battery Materials, Atlanta, Georgia, 1988* (D. Glover, B. Schumm, Jr., and A. Kozawa, eds.), pp. 290–304, The International Battery Materials Association, Cleveland, Ohio (1988)

40. P. Ruetschi, *J. Electrochem. Soc.* **131**, 2737 (1984).

41. G. E. Van De Steeg, Private communication.

42. W. C. Maskell, J. E. A. Shaw, and F. L. Tye, *Electrochim. Acta.* **28**, 225 (1983).

43. W. C. Maskel, J. E. A. Shaw, and F. L. Tye, *Electrochim. Acta* **28**, 231 (1983).

44. P. Ruetschi and R. Giovanoli, *J. Electrochem. Soc.* **135**, 2663 (1988).

45. E. Preisler, *J. Appl. Electrochem.* **6**, 311 (1976).

46. J. Brenet and P. Faber, *J. Power Sources* **4**, 203 (1979).

47. E. Preisler, in: *Manganese Dioxide Symposium Vol. 2, Tokyo, 1980* (B. Schumm, Jr., H. M. Joseph, and A. Kozawa, eds.), pp. 184–206, I.C. MnO₂ Sample Office, Cleveland, Ohio (1982).

48. A. J. Brown, F. L. Tye, and L. L. Wood, *J. Electroanal. Chem. Interfacial Electrochem.* **122**, 337 (1981).

49. J. Koshiba, in: *Progress in Batteries and Solar Cells*, Vol. 7 (IBA, Hawaii Meeting, 1988) (M. Nagayama and A. Kozawa, eds.), pp. 38–43, JEC Press and IBA Inc., Cleveland, Ohio (1988).

50. A. Kozawa, in: *Proceedings of the 2nd Battery Material Symposium, Graz, 1985* (K. V. Kordesch and A. Kozawa, eds.), pp. 545–563, International Battery Association, Cleveland, Ohio (1985).

51. E. Preisler, in: *Battery Material Symposium, Vol. 1, Brussels 1983* (A. Kozawa and N. Nagayama, eds.), pp. 143-160, International Battery Materials Association, Cleveland, Ohio (1984).

52. C. St. Claire-Smith, J. A. Lee, and F. L. Tye, in: *Manganese Dioxide Symposium, Vol. 1, Cleveland, 1975* (compiled by A. Kozawa and R. J. Brodd), I.C. Sample Office, Cleveland, Ohio (1975).

53. J. A. Lee and C. E. Newnham, *J. Colloid Interface Sci.* **56,** 391 (1976).

54. J. P. Brenet, M. Cyrankowska, G. Ritzler, R. Sada, and K. Traore, in: *Manganese Dioxide Symposium, Vol. 1, Cleveland, 1975* (compiled by A. Kozawa and R. J. Brodd), pp. 1-24, I.C. Sample Office, Cleveland, Ohio (1975).

55. A. Kozawa, in: *Comprehensive Treatise of Electrochemistry,* Vol. 3 (J. O'M. Bockris, B. E. Conway, E. Yeager, and R. E. White, eds.), pp. 207-218, Plenum Press, New York (1981).

56. M. R. Tarasevich, E. J. Shkol'nikov, V. E. Kazarinov, L. P. Esayan, Z. M. Buzova, B. D. Gurvitz, and N. I. Hoteeva, in: *Progress in Batteries and Solar Cells,* Vol. 7 (IBA Hawaii Meeting, 1988) (M. Nagayama and A. Kozawa, eds), pp. 49-57, JEC Press and IBA Inc., Cleveland, Ohio (1988).

57. A. Kozawa, from: The IBA 1989 MnO_2 Lecture, May 1989.

58. A. Kozawa, in: *Manganese Dioxide Symposium, Vol. 1, Cleveland, 1975* (compiled by A. Kozawa and R. J. Brodd), pp. 470-518, I.C. Sample Office, Cleveland, Ohio (1975).

59. A. Kozawa, in: *Handbook of Manganese Dioxides, Battery Grade* (D. Glover, B. Schumm, Jr., and A. Kozawa, eds.), pp. 267-273, International Battery Materials Association, Cleveland, Ohio (1989).

60. B. Schumm, Jr., personal communication (1989).

61. A. Kozawa, in: *Progress in Batteries and Solar Cells,* Vol. 7 (IBA Hawaii Meeting, 1988) (M. Nagayama and A. Kozawa, eds.), pp. 59-65, JEC Press and IBA Inc., Cleveland, Ohio (1988).

62. A. J. Brown, C. R. St. Claire-Smith, F. L. Tye, and J. L. Whiteman, *J. Colloid Interface Sci.* **51,** 516 (1975).

63. T. Ohzuku and T. Hirai, in: *Progress in Batteries and Solar Cells,* Vol. 7 (M. Nagayama and A. Kozawe, eds.), pp. 31-36, JEC Press and IBA Inc., Cleveland, Ohio (1988).

64. B. D. Cullity, *Elements of X-ray Diffraction,* Addison-Wesley Publishing Co., Reading, Massachusetts (1956), pp. 96-99.

65. E. Preisler, in: *Proceedings of the 2nd Battery Material Symposium, Graz, 1985* (K. V. Kordesch and A. Kozawa, eds.), pp. 247-266, International Battery Association, Cleveland, Ohio (1985).

66. Data compiled by Battery Technologies, Inc., Mississauga, Ontario, Canada (1988).

67. Kordesch, J. Gsellmann, W. Harer, W. Taucher, and K. Tomantschger, in: *Progress in Batteries and Solar Cells,* Vol. 7 (IBA Hawaii Meeting, 1988) (M. Nagayama and A. Kozawa, eds.), pp. 111-124, JEC Press and IBA Inc., Cleveland, Ohio (1988).

68. IBA Los Angeles Meeting, May 12-13, 1989.

69. M. Yoshio and H. Noguchi, in: *Proceedings of the 2nd Battery Material Symposium, Vol. 2, Graz, 1985* (K. V. Kordesch and A. Kozawa, eds.), pp. 267-277, International Battery Association, Cleveland, Ohio (1985).

70. J. J. Coleman, *Trans. Electrochem. Soc.* **90,** 545 (1946).

71. M. Fleischmann, H. R. Thirsk, and I. M. Tordesillas, *Trans. Faraday Soc.* **58,** 1865 (1962).

72. R. L. Paul and A. Cartwright, *J. Electroanal. Chem.* **201,** 113 (1986).

73. J. McBreen, *Power Sources,* Vol. 5 (*Proceedings of the 9th International Symposium, 1974*) (D. H. Collins, ed.), pp. 525-538, Academic Press, London (1975).

74. Y. Amano, H. Ogawa, Y. Umeo, T. Ohira, and K. Muvakami, U.S. Patent 3,530,496, September 22, 1970.

75. K. Kordesch, in: *Comprehensive Treatise of Electrochemistry,* Vol. 3 (J. O'M. Bockris, B. E. Conway, E. Yeager, and R. E. White, eds.), pp. 219-232, Plenum Press, New York (1981).

76. K. Kordesch, U.S. Patent 2,962,540 (1960).

77. K. Kordesch and J. Gsellmann, U.S. Patent 4,384,029 (1983).

78. S. Fletcher, J. Galea, J. A. Hamilton, T. Tran, and R. Woods, *J. Electrochem. Soc.* **133,** 1277 (1986).

79. H. S. Wroblowa, N. Gupta, and Y-F. Yao, in: *Proceedings of the 2nd Battery Material Symposium, Graz, 1985* (K. V. Kordesch and A. Kozawa, eds.), pp. 203-220, International Battery Association, Cleveland, Ohio (1985).

80. H. S. Wroblowa, Extended Abstracts, Meeting of the Electrochemical Society, San Diego, California, October 1986, Abstract No. 31.

81. A. Grzegorzewski and K. E. Heusler, *J. Electroanal. Chem.* **228,** 455 (1987).

Index